THE EPIGENOME AND DEVELOPMENTAL ORIGINS OF HEALTH AND DISEASE

THE EPIGENOME AND DEVELOPMENTAL ORIGINS OF HEALTH AND DISEASE

Edited by

CHERYL S. ROSENFELD

AMSTERDAM • BOSTON • HEIDELBERG • LONDON
NEW YORK • OXFORD • PARIS • SAN DIEGO
SAN FRANCISCO • SINGAPORE • SYDNEY • TOKYO
Academic Press is an imprint of Elsevier

Academic Press is an imprint of Elsevier
125 London Wall, London EC2Y 5AS, UK
525 B Street, Suite 1800, San Diego, CA 92101-4495, USA
225 Wyman Street, Waltham, MA 02451, USA
The Boulevard, Langford Lane, Kidlington, Oxford OX5 1GB, UK

Copyright © 2016 Elsevier Inc. All rights reserved.

No part of this publication may be reproduced or transmitted in any form or by any means, electronic or mechanical, including photocopying, recording, or any information storage and retrieval system, without permission in writing from the publisher. Details on how to seek permission, further information about the Publisher's permissions policies and our arrangements with organizations such as the Copyright Clearance Center and the Copyright Licensing Agency, can be found at our website: www.elsevier.com/permissions.

This book and the individual contributions contained in it are protected under copyright by the Publisher (other than as may be noted herein).

Notices
Knowledge and best practice in this field are constantly changing. As new research and experience broaden our understanding, changes in research methods, professional practices, or medical treatment may become necessary.

Practitioners and researchers must always rely on their own experience and knowledge in evaluating and using any information, methods, compounds, or experiments described herein. In using such information or methods they should be mindful of their own safety and the safety of others, including parties for whom they have a professional responsibility.

To the fullest extent of the law, neither the Publisher nor the authors, contributors, or editors, assume any liability for any injury and/or damage to persons or property as a matter of products liability, negligence or otherwise, or from any use or operation of any methods, products, instructions, or ideas contained in the material herein.

ISBN: 978-0-12-801383-0

British Library Cataloguing-in-Publication Data
A catalogue record for this book is available from the British Library

Library of Congress Cataloging-in-Publication Data
A catalog record for this book is available from the Library of Congress

For information on all Academic Press publications
visit our website at http://store.elsevier.com/

Typeset by TNQ Books and Journals
www.tnq.co.in

Dedication

I dedicate this book to my father, Robert L. Rosenfeld, who passed away on February 5, 2005. From early childhood onwards, he encouraged me to pursue my interests in science and medicine and taught me that a good start and supportive environment can last a lifetime.

Contents

5. Perinatal Neurohormonal Programming and Endocrine Disruption

CHERYL S. ROSENFELD

6. Parental Nutrition and Developmental Origins of Health and Disease

MIGUEL A. VELAZQUEZ, CONGSHAN SUN AND TOM P. FLEMING

7. Maternal Prenatal Stress and the Developmental Origins of Mental Health: The Role of Epigenetics

KIERAN J. O'DONNELL AND VIVETTE GLOVER

8. Epigenetics in the Developmental Origin of Cardiovascular Disorders

JEFFREY S. GILBERT

9. Developmental Effects of Endocrine-Disrupting Chemicals in the Ovary and on Female Fertility

MEHMET UZUMCU AND APARNA MAHAKALI ZAMA

10. Developmental and Epigenetic Origins of Male Reproductive Pathologies

CARLOS GUERRERO-BOSAGNA

11. Developmental Origins of Childhood Asthma and Allergic Conditions—Is There Evidence of Epigenetic Regulation?

STEVE TURNER

12. Immune Disorders, Epigenetics, and the Developmental Origins of Health and Disease

RODNEY R. DIETERT AND AMRIE C. GRAMMER

13. Neurobehavioral Disorders and Developmental Origins of Health and Disease

CURT A. SANDMAN, QUETZAL A. CLASS, LAURA M. GLYNN AND ELYSIA POGGI DAVIS

14. Metabolic Disorders and Developmental Origins of Health and Disease

SARA E. PINNEY

15. The Developmental Origins of Renal Dysfunction

JAMES S.M. CUFFE, SARAH L. WALTON AND KAREN M. MORITZ

16. Cancer and Developmental Origins of Health and Disease—Epigenetic Reprogramming as a Mediator

SHUK-MEI HO, ANA CHEONG, SARAH TO, VINOTHINI JANAKIRAM, PHERUZA TARAPORE AND YUET-KIN LEUNG

17. Epigenetic Regulation of Gastrointestinal Epithelial Barrier and Developmental Origins of Health and Disease

J.P. LALLÈS, C. MICHEL, V. THEODOROU AND J.P. SEGAIN

18. How the Father Might Epigenetically Program the Risk for Developmental Origins of Health and Disease Effects in His Offspring

KRISTIN E. MURPHY, TIMOTHY G. JENKINS AND DOUGLAS T. CARRELL

19. Linkage between In Utero Environmental Changes and Preterm Birth

MARKUS VELTEN AND LYNETTE K. ROGERS

20. Sexual Dimorphism and DOHaD through the Lens of Epigenetics: Genetic, Ancestral, Developmental, and Environmental Origins from Previous to the Next Generation(s)

CLAUDINE JUNIEN, SARA FNEICH, POLINA PANCHENKO, SARAH VOISIN AND ANNE GABORY

21. Transgenerational Epigenetic Inheritance: Past Exposures, Future Diseases

CARLOS GUERRERO-BOSAGNA

22. The Placenta and Developmental Origins of Health and Disease

JANE K. CLEAL AND ROHAN M. LEWIS

23. The Moral and Legal Relevance of DOHaD Effects for Pregnant Mothers

MICHELE LOI AND MARIANNA NOBILE

24. Introduction to *Moms in Motion* (*MIM*)

CHERYL S. ROSENFELD

25. Reversing Harmful Developmental Origins of Health and Disease Effects

MEI-WEI CHANG, SUSAN NITZKE, ROGER BROWN, M JEAN BRANCHEAU EGAN, KOBRA EGHTEDARY AND CHERYL S. ROSENFELD

26. Informational Resources for Developmental Origins of Health and Disease Research

CHERYL S. ROSENFELD

Index 523

List of Contributors

Roger Brown School of Nursing, University of Wisconsin-Madison, Madison, WI, USA

Tatjana Buklijas Liggins Institute, The University of Auckland, Auckland, New Zealand

Douglas T. Carrell Department of Surgery (Urology), University of Utah School of Medicine, Salt Lake City, UT, USA

Mei-Wei Chang Michigan State University College of Nursing, East Lansing, MI, USA

Ana Cheong Department of Environmental Health, University of Cincinnati College of Medicine, Cincinnati, OH, USA; Center for Environmental Genetics, University of Cincinnati Medical Center, Cincinnati, OH, USA

Quetzal A. Class Department of Psychological and Brain Sciences, Indiana University, Bloomington, IN, USA

Jane K. Cleal Institute of Developmental Sciences, University of Southampton, Southampton, UK

James S.M. Cuffe School of Biomedical Science, The University of Queensland, St Lucia, QLD, Australia

Elysia Poggi Davis Department of Psychiatry and Human Behavior, University of California, Irvine, CA, USA; Department of Psychology, University of Denver, Denver, CO, USA

Rodney R. Dietert Department of Microbiology and Immunology, College of Veterinary Medicine, Cornell University, Ithaca, NY, USA

M Jean Brancheau Egan Michigan Department of Community Health, WIC Division, Lansing, MI, USA

Kobra Eghtedary Michigan Department of Community Health, WIC Division, Lansing, MI, USA

Tom P. Fleming Centre for Biological Sciences, University of Southampton, Southampton General Hospital, Southampton, UK

Sara Fneich INRA, UMR1198 Biologie du Développement et Reproduction, Jouy-en-Josas, France

Anne Gabory INRA, UMR1198 Biologie du Développement et Reproduction, Jouy-en-Josas, France

Jeffrey S. Gilbert Department of Biomedical Sciences, University of Minnesota Medical School, Duluth, MN, USA

Vivette Glover Institute of Reproductive and Development Biology, Imperial College London, London, UK

Peter D. Gluckman Liggins Institute, The University of Auckland, Auckland, New Zealand

Laura M. Glynn Department of Psychiatry and Human Behavior, University of California, Irvine, CA, USA; Department of Psychology, Chapman University, Orange, CA, USA

Amrie C. Grammer University of Virginia Research Park, VA, USA

Carlos Guerrero-Bosagna Avian Behavioral Genomics and Physiology Group, IFM Biology, Linköping University, Linköping, Sweden

Mark A. Hanson Institute of Developmental Sciences, University of Southampton and NIHR Nutrition Biomedical Research Centre, University Hospital Southampton, Southampton, UK

Shuk-Mei Ho Department of Environmental Health, University of Cincinnati College of Medicine, Cincinnati, OH, USA; Center for Environmental Genetics, University of Cincinnati Medical Center, Cincinnati, OH, USA; Cincinnati Cancer Center, Cincinnati, OH, USA; Cincinnati Veteran Affairs Medical Center, Cincinnati, OH, USA

Vinothini Janakiram Department of Environmental Health, University of Cincinnati College of Medicine, Cincinnati, OH, USA; Center for Environmental Genetics, University of Cincinnati Medical Center, Cincinnati, OH, USA

Timothy G. Jenkins Department of Surgery (Urology), University of Utah School of Medicine, Salt Lake City, UT, USA

Claudine Junien INRA, UMR1198 Biologie du Développement et Reproduction, Jouy-en-Josas, France

J.P. Lallès Institut National de la Recherche Agronomique, UR1341 ADNC, Saint Gilles, France; Centre de Recherche en Nutrition Humaine-Ouest, Nantes, France

Yuet-Kin Leung Department of Environmental Health, University of Cincinnati College of Medicine, Cincinnati, OH, USA; Center for Environmental Genetics, University of Cincinnati Medical Center, Cincinnati, OH, USA; Cincinnati Cancer Center, Cincinnati, OH, USA

Rohan M. Lewis Institute of Developmental Sciences, University of Southampton, Southampton, UK

Michele Loi Centro de Estudos Humanísticos, Universidade do Minho, Campus de Gualtar, Braga, Portugal

C. Michel Centre de Recherche en Nutrition Humaine-Ouest, Nantes, France; Institut National de la Recherche Agronomique/Université de Nantes, UMR1280, Nantes, France; Institut des Maladies de l'Appareil Digestif, Nantes, France

Karen M. Moritz School of Biomedical Science, The University of Queensland, St Lucia, QLD, Australia

Kristin E. Murphy Department of Surgery (Urology), University of Utah School of Medicine, Salt Lake City, UT, USA

Susan Nitzke Department of Nutritional Sciences, University of Wisconsin-Madison, Madison, WI, USA

Marianna Nobile Universita' degli Studi di Milano-Bicocca, Dipartimento dei Sistemi Giuridici, Milano, Italy

Kieran J. O'Donnell The Ludmer Centre for Neuroinformatics and Mental Health, Douglas Mental University Institute, McGill University, Montreal, QC, Canada

Polina Panchenko INRA, UMR1198 Biologie du Développement et Reproduction, Jouy-en-Josas, France

Sara E. Pinney Department of Pediatrics, Perelman School of Medicine at the University of Pennsylvania, Division of Endocrinology and Diabetes, The Children's Hospital of Philadelphia, Philadelphia, PA, USA

Ken Resnicow University of Michigan School of Public Health, University of Michigan, Ann Arbor, MI, USA

Lynette K. Rogers Center for Perinatal Research, The Research Institute at Nationwide Children's Hospital, Columbus, Ohio

Cheryl S. Rosenfeld Department of Bond Life Sciences Center, Department of Biomedical Sciences, Genetics Area Program, Thompson Center for Autism and Neurobehavioral Disorders, University of Missouri, Columbia, MO, USA

Lewis P. Rubin Department of Pediatrics, Texas Tech University Health Sciences Center El Paso, Paul L. Foster School of Medicine, El Paso, TX, USA

Curt A. Sandman Department of Psychiatry and Human Behavior, University of California, Irvine, CA, USA

J.P. Segain Centre de Recherche en Nutrition Humaine-Ouest, Nantes, France; Institut National de la Recherche Agronomique/Université de Nantes, UMR1280, Nantes, France; Institut des Maladies de l'Appareil Digestif, Nantes, France

Congshan Sun Centre for Biological Sciences, University of Southampton, Southampton General Hospital, Southampton, UK

Martha Susiarjo Department of Cell and Developmental Biology, Perelman School of Medicine, University of Pennsylvania, Philadelphia, PA, USA

Pheruza Tarapore Department of Environmental Health, University of Cincinnati College of Medicine, Cincinnati, OH, USA; Center for Environmental Genetics, University of Cincinnati Medical Center, Cincinnati, OH, USA; Cincinnati Cancer Center, Cincinnati, OH, USA

V. Theodorou Institut National de la Recherche Agronomique, UMR Toxalim, Toulouse, France

Sarah To Department of Environmental Health, University of Cincinnati College of Medicine, Cincinnati, OH, USA

Steve Turner Child Health, Royal Aberdeen Children's Hospital, Aberdeen, UK

Mehmet Uzumcu Department of Animal Sciences, Rutgers, The State University of New Jersey, New Brunswick, NJ, USA

Miguel A. Velazquez Centre for Biological Sciences, University of Southampton, Southampton General Hospital, Southampton, UK

Markus Velten Department of Anesthesiology and Intensive Care Medicine, Rheinische Friedrich-Wilhelms-University Bonn, Germany

Sarah Voisin INRA, UMR1198 Biologie du Développement et Reproduction, Jouy-en-Josas, France

Sarah L. Walton School of Biomedical Science, The University of Queensland, St Lucia, QLD, Australia

Aparna Mahakali Zama Department of Animal Sciences, Rutgers, The State University of New Jersey, New Brunswick, NJ, USA

Acknowledgments

This book would not have been possible without all of the coauthors who kindly shared their knowledge and passion for the various areas. Ms Lisa Eppich and Ms Catherine A. Van Der Laan at Elsevier were integral in making this book come to life.

The concept of developmental origins of health and disease (DOHaD) was first clearly articulated by the late Sir David Barker. It was thus appropriate that it was originally termed the "Barker hypothesis" but subsequently changed to "fetal origin of adult disease (FOAD)," and most recently to "developmental origins of adult health and disease." Since its conception, the DOHaD concept has gained currency and led to paradigm shifts in how scientists and clinicians view a variety of noncommunicable diseases. Correspondingly, it has paved the way for new avenues of diagnosis, prevention, and treatment strategies.

I am grateful to the many teachers, mentors, colleagues, and friends who along the way fostered my interests and curiosity in science and medicine. In particular, I am grateful to the late Mrs Patricia Murphy whose enthusiasm and wonderment for biology were contagious. My PhD mentor, Dr Dennis Lubahn, was incredibly supportive of me and my research ideas. The lessons I learned in his laboratory have stayed with me all of these years. Most of all, I am thankful to Dr R. Michael Roberts. For over 20 years, he has been a wonderful mentor, colleague, and friend.

I am thankful to Dr Deborah Wagner, my friend and former classmate, for allowing me over various holidays to serve as a relief veterinarian at her animal hospital. As veterinary students, we made a "Forrest Gump" pact that if she were to ever open her animal hospital, I would be her first mate. It has been rewarding to be able to indulge my veterinary interests and keep in touch with advances in clinical medicine. These experiences have helped shape my thinking and research directions.

Finally, I am grateful to my mother, sister, brother, nieces and nephews, and other family members who walk on two and four legs.

Cheryl S. Rosenfeld

CHAPTER

1

The Developmental Origins of Health and Disease (DOHaD) Concept: Past, Present, and Future

Peter D. Gluckman[1], *Tatjana Buklijas*[1], *Mark A. Hanson*[2]

[1]Liggins Institute, The University of Auckland, Auckland, New Zealand; [2]Institute of Developmental Sciences, University of Southampton and NIHR Nutrition Biomedical Research Centre, University Hospital Southampton, Southampton, UK

OUTLINE

INTRODUCTION

The overarching argument of the conceptual paradigm and the research field of developmental origins of health and disease (DOHaD) is that the state of health and risk from disease in later childhood and adult life is significantly affected by environmental factors acting during the preconceptional, prenatal, and/or early postnatal periods. The emphasis has been on obesity, type 2 diabetes mellitus, and cardiovascular disease, but a significant body of work has also been focused on endocrine cancers, osteoporosis and frailty in the elderly, mental health, cognitive function, respiratory disease, immune function, and allergy. While the field as currently

The Epigenome and Developmental Origins of Health and Disease
http://dx.doi.org/10.1016/B978-0-12-801383-0.00001-3

Copyright © 2016 Elsevier Inc. All rights reserved.

constructed is just over two decades old, it is based on research that goes back to the 1930s. In this chapter, we bring together research traditions, concepts, and approaches that, over the last 80 years, have explored the question of prenatal and early postnatal environmental influences that impact health and disease in later life and look forward to emergent areas of attention and application of the concept.

THE ORIGINS OF THE FIELD

The idea that experiences in early life influence health in later life may be found throughout the history of Western medicine: well into the 1800s, it was believed that anything that a mother saw, touched, ate, or even imagined—collectively known as "maternal impressions"—had a capacity to permanently influence the developing organism [1,2]. In the early 1800s, the common view was that a new organism was created in a process called "generation," out of maternal and paternal contributions as well as various experiences that the mother had during (and even before) pregnancy [3]. But in the nineteenth century, "generation" was replaced with "reproduction," built on the new idea of "heredity." The noun "heredity" was first used in the 1830s, to describe the transmission of parental qualities during conception, and at the same time to make a distinction between those inherited qualities and the properties that emerged during development [4]. Scientists studied where heredity resided within the cell; how hereditary particles were distributed among cells and their quantity was prevented from doubling in each successive generation; and to what extent were hereditary elements in the cells sensitive to developmental influences [5]. August Weismann's work provided the conceptual basis for new thinking about heredity and development: while germ cells produced somatic cells, he argued, they were not affected by anything that somatic cells acquired or learned [6]. It followed that each individual was born with a certain predisposition toward disease; no environmental modifications could improve one's outlook. The best one could do to improve one's offspring's chances was to reproduce with a person of "better heredity."

The early 1900s were the heyday of "hard heredity," exemplified by the emergence of genetics, an experimental discipline concerned with mechanisms of heredity, and the dominance of the social program of eugenics, seeking to reform society through rationalizing human reproduction [7]. But the deep economic crisis of the 1930s made it obvious that environmental conditions had a strong influence on the emergence and prevalence of disease [8]. New epidemiological work suggested that the conditions of early life played a role at least as important as heredity. A landmark paper by the Scottish epidemiologist William Ogilvy Kermack and colleagues in 1934 argued that "the data behaved as if the expectation of life was determined by the conditions existing during the years 0–15 (…) the health of the man is determined preponderantly by the physical constitution which the child has built up" [9]. The recognition of the importance of environmental conditions and the apparent demise of eugenics did not, however, mean the fall of the genetic model, which remained dominant through the twentieth century [8].

The Second World War was a pivotal event in the making of the developmental approach to the study of health and disease, conceptualized in the early twenty-first century as DOHaD. Even before the war, physiologists, teratologists, and agricultural scientists collected experimental evidence showing that manipulating the life conditions of pregnant animals permanently affected the patterns of growth and the phenotype of their offspring [10–12]. Interventions in humans were, for obvious reasons, too subtle to produce substantial differences, and they also focused on maternal mortality and morbidity rather than longer-term outcomes in the offspring [13]. But wartime famines provided rare "natural experiments" by exposing thousands of women to periods of severe undernutrition, in some cases sharply delineated [14,15]. The longest and the most severe famine took place in Leningrad under German siege (between September 1941 and January 1944) [14], but the clearest data came from Rotterdam and The Hague, two cities in northwestern Holland that had suffered food shortages during German reprisals from September 1944 to May 1945, in what became known as the Dutch Winter Famine [15,16]. Data collected showed that starvation in the last trimester of pregnancy caused a reduction in the birth weight of the offspring, while famine around conception increased the chance of miscarriage and malformation. Postwar Germany provided an opportunity for the British team working on the intersection of physiology, nutritional science, and pediatrics (Robert McCance, Elsie Widdowson, and Rex Dean) to study how low food rations and lack of food variety influenced lactation in new mothers, infant birth weight, and childhood growth [17]. Back in their Cambridge laboratories, Widdowson and McCance tested their clinical findings in animal experiments and demonstrated that the size of the litter and the rate of offspring growth depended on maternal nutrition. Interestingly, prenatal and early postnatal (preweaning) nutrition did not affect just the weight that the pups attained by adulthood: it also influenced susceptibility to infections, body proportions, and timing of reproductive maturation, as well as behavior [18]. Their results supported the theory of "critical" or "sensitive periods," popular across disciplines as diverse as ethology (behavioral studies), linguistics, child psychology, and physiology, according to which each organ or tissue has a distinctive period of critical differentiation as well as a period of maximum growth, during which these organs and tissues are highly sensitive to injury [11].

But, as the world recovered from the wartime trauma, in the 1960s and 1970s interest in the relationship between prenatal and perinatal influences and later health and disease waned. In this period, fetal physiologists largely focused on questions that emphasized fetal autonomy rather than interplay between the environment and the developing organism. They studied, for example, fetal respiratory movements, fetal endocrine growth mechanisms, or fetal control of the onset of labor [19]. It was mostly researchers with a strong interest in socioeconomic determinants of health inequalities who pursued questions of the interaction between environment and development. At the University of Birmingham's Department of Social Medicine, under Professor Thomas McKeown, a young David Barker completed his PhD thesis on "Prenatal influences and subnormal intelligence" (1966) [20]. He found that children of all levels of "subnormal" intelligence (classified as IQ under 75) had a birth weight lower than expected. Interestingly, the "normal" siblings of all but the most severely "subnormal" children (IQ less than 50) were born at low birth weight too. These low birth weights of both "normal" and "subnormal" children, Barker suggested, reflected "influences which affect the intra-uterine lives of all children in their families."

At the same time, a pair of South African political immigrants to the United States (via the University of Manchester's Department of Social Medicine), the Columbia University epidemiologists Zena Stein and Mervyn Susser, undertook a large program of study of the influence of maternal nutrition on "mental competence" [21,22]. One part of their research was an intervention study of providing food supplements to pregnant women drawn from a population with a high frequency of low birth weight; the other, an observational study based on the Dutch Winter Famine cohort. While both of these studies produced negative results, they rekindled interest in the Dutch Winter Famine cohort, which from then onwards would play a key role in the study of developmental, as well as transgenerational, influences upon adult health.

The first study (by Stein, Susser, and the Amsterdam researcher Anita Ravelli) to associate undernutrition during gestation in the Dutch Winter Famine cohort with manifestations of metabolic disease was published in 1976, showing that young men exposed in early pregnancy had significantly higher rates of obesity, while men exposed in late pregnancy (and first months of infancy) had lower obesity rates [23]. Women who suffered famine during first and second trimester in utero had offspring whose birth weight was lower than the birth weight of offspring born to women who had not been exposed to famine as fetuses; but the offspring of women exposed during the third trimester showed no reduction in birth weight [24]. Overall, from the mid-1970s onwards multiple historical and prospective epidemiological studies studied the relationship between maternal morbidity, infant mortality and birth weight, and morbidity and mortality from cardiovascular disease [25–28]. In North America, Millicent Higgins was the first to employ a long-term formal birth cohort to examine later-life outcomes, reporting in 1980 that the male offspring of women with toxemia developed significantly higher blood pressures by 20 years of age. By the mid-1980s, reports linking low birth weight and later hypertension started to merge [27,28].

Innovative conceptual developments took place in Eastern Germany, where the Berlin endocrinologist Günther Dörner studied differences in phenotypes between men born before, during, and after the Second World War and showed association between nutrition and later prevalence of metabolic and cardiovascular disease [29–31]. He became particularly interested in the effects of maternal stress and gestational diabetes on the offspring [32,33]. Dörner argued that the concentrations of hormones, neurotransmitters, and metabolites during the early development "preprogrammed" feedback loops control reproduction and development [34,35]. Inappropriately set feedback loops could then trigger disease. Dörner's "*functional teratology*," as he termed it, may be compared to the proposal by another endocrinologist in this period, Norbert Freinkel, who spoke of *metabolic teratogenesis* when describing the effect of gestational diabetes on the next generation [36]. Around the same time, Frans Van Assche and colleagues used experimental models of gestational diabetes (in which part of the pancreatic β cells were chemically destroyed) to show that prenatal diabetogenic conditions permanently affected the adult offspring's ability to handle situations stressing their glucose metabolism, such as pregnancy, and also had similar transgenerational effects [37].

It was, however, the work of David Barker in the 1980s, by then at the University of Southampton's MRC Environmental Epidemiology Unit, that prompted this diverse collection of observations to coalesce into a field. In 1985, Barker's unit produced a compendium of maps that used color gradation to show differences in mortality from selected diseases across the counties of England and Wales [38]. The mortality from cardiovascular disease appeared highest in poor and lowest in rich areas, a finding that opposed the view established by the Framingham Heart Study, according to which affluence caused the modern "epidemic" of heart diseases [39]. Barker, who since his Birmingham days with McKeown had been interested in prenatal influences especially nutrition, observed a strong geographical relationship between the areas of high infant mortality in the early twentieth century and the areas of high cardiovascular mortality between 1968 and 1978 [40]. Barker's best-known study used the largest and most detailed set of records on infant welfare that could be found in the United Kingdom, those from the county of Hertfordshire from 1911 onwards. These records contained information about birth weight and early growth and development, and became the basis of a large study cohort. The follow-up of the Hertfordshire infants showed that death rates for cardiovascular disease fell progressively from those weighing 2.5 kg or less at birth to those weighing 4.3 kg, with a slight increase in the heaviest group (above 4.3 kg) [41]. Blood pressure and cholesterol levels showed equivalent trends.

Barker's work attracted wide interest. The Southampton team began collaboration with the Amsterdam group studying the Dutch Winter Famine cohort. Other studies, both retrospective and prospective, followed worldwide. Early on, Barker's work came to the attention of the doyen of fetal physiology Geoffrey Dawes, who was near retirement [42]. In Dawes's view, Barker's findings opened up new research opportunities. Dawes introduced Barker to his colleagues working in fetal physiology at a workshop meeting near La Spezia, Italy, in 1989. The proceedings, published under the title *Fetal Autonomy and Adaptation*, make interesting reading: the physiologists were clearly skeptical about Barker's observations, but resolved to test them experimentally. The first workshop on "fetal origins of adult disease" took place in Sydney in 1994, and it brought a broader group of perinatal scientists in contact with Barker's group [43]. This meeting launched a fast-growing series of annual workshops. Under David Barker's leadership, foremost investigators of that period created the Council for the Fetal Origins of Adult Disease (FOAD) with John Challis as its first chair. The first global congress on FOAD was held in Mumbai, India, in 2001. At the second congress, in Brighton, UK, in 2003, it was decided the council would be reformed into an academic society, Developmental Origins of Health and Disease (DOHaD), with Peter Gluckman as the founding president and Mark Hanson as secretary.

CONCEPTUAL DEVELOPMENTS AND EXPERIMENTAL OBSERVATIONS

A major conceptual development in the DOHaD field took place in 1992 when David Barker and Nicholas Hales proposed their "thrifty phenotype" hypothesis [44]. "Thrifty phenotype" was a developmental alternative to the "thrifty genotype" explanation of the modern epidemic of noncommunicable diseases (NCDs), proposed in the 1960s by a medical geneticist James Neel [45,46] interested in the evolution of contemporary human populations. Neel argued that in the past "thrifty genes" had been selected because they provided advantage in the time of famine; but in the affluent contemporary world they only increased disease risk. The "thrifty phenotype," by contrast, placed an emphasis on development, arguing that nutritionally inadequate conditions in pregnancy not only affected fetal growth but also induced permanent changes in insulin secretory capacity and in glucose metabolism. While well adapted for famine, in a nutritionally rich postnatal environment the individual would have a higher risk of metabolic disease.

The initial focus of the field—in David Barker's work and in studies of the Dutch Winter Famine cohort, as well as, earlier, in postwar undernutrition studies—was on the consequences of low birth weight. For this reason, experimental studies induced intrauterine growth restriction (IUGR) in animal models, while clinical studies compared the outcomes of normal-weight and IUGR offspring [47–50]. The centrality of low birth weight to Barker's hypothesis, however, became a problem. First, the supposed link between low birth weight and adult NCD contradicted the observed ecological trends: post–Second World War, the incidence of NCDs increased precisely in those countries that also had high average birth weights, such as Norway, Finland, and Scotland [51]. Second, disciplinary divisions hindered recognition of this problem. Most early epidemiological work studied processes associated with suboptimal nutritional conditions or excessive maternal stress and later disease risk; at the same time, the class of phenomena highly relevant for the modern, developed world, associated with gestational diabetes, maternal obesity, and infant overfeeding, was being studied by endocrinologists and obstetricians. The relative importance of these pathways clearly differed in different populations and at different stages of the nutritional transition. Yet the two disciplinary groups, epidemiologists and clinicians, presented their results at different venues and rarely talked to each other. Third, many epidemiological studies produced data inconsistent with or even contrary to Barker's hypothesis [52]. Some even argued that the observed differences did not exist, i.e., that they were results of errors, confounding, and inappropriate study design [53]. Even the studies of historical famines did not at all support the "thrifty phenotype": for instance, studies on survivors of the Leningrad siege found no difference in glucose tolerance, lipid concentrations, hypertension, or cardiovascular disease rates between those groups exposed and those who were not exposed to famine during development [54]. A Finnish historical study found no effect of famine exposure upon survival in adulthood [55]. Even studies based on the Dutch Winter Famine cohorts produced results that contradicted previous findings and the main hypothesis: for example, that glucose tolerance was lower in participants exposed in mid- to late gestation rather than early to mid-gestation [56,57]. Critics maintained that Barker's hypothesis was neither precisely formulated nor consistent, with regard to the timing of critical events at the stage of gestation when undernutrition might have been expected to produce the specific impact [58]. They emphasized the need to move from epidemiological to mechanistic studies, because retrospective studies rarely provided sufficiently detailed data, prospective studies took too long, and neither could explain how, for example, a modification in nutrition in mid-gestation may lead to a change in blood pressure decades later [57].

Rather than undermining the field, these challenges to the DOHaD concept encouraged further research and inspired conceptual thinking. Researchers recognized that there was indeed an excessive emphasis on birth weight, while in reality early life events informed later disease outcomes in multiple ways, depending on the type and timing of the insult [59]. Birth weight was only one of many possible proxies for variable intrauterine effects [60]. "Programming" (a term introduced by Alan Lucas [61], a decade and a half after Dörner's "preprogramming" [34,35], which, while popular, is problematic for its inference of a predetermined developmental course rather than a plastic one) [62] could, and did, operate in the absence of effects on birth weight. For example, one study showed that variation in maternal nutrition influenced childhood carotid intima media thickness independently of birth weight [63]. Insulin resistance, predicted by the "thrifty phenotype" hypothesis to be a major adaptation to adverse prenatal nutritional conditions, only appeared some years after birth. In fact, growth-retarded babies tended to have greater insulin sensitivity [64]. A more sophisticated understanding of the developmental events was therefore required.

In the early 2000s, the Cambridge ethologist Patrick Bateson, first alone [65] and then together with the authors and others [66], proposed a comprehensive hypothesis that placed the DOHaD phenomenon firmly within the evolutionary framework of developmental plasticity and allowed for multiple pathways of induction. We argued that, within the normal range, environmental cues received during development influenced the developmental path taken by the organism. The path was adaptive if the later postnatal environment matched the prenatal one, but if the environment changed or the prediction proved to be wrong, it would turn out to be pathological. The model was gradually refined to take into account the observations that the environments that mattered were those of later childhood, after weaning and up to the age of peak reproductive fitness at the end of the second decade of life [62,67]. This concept, termed the *predictive adaptive response*, could account for findings previously deemed paradoxical or inexplicable [68,69]. For instance, infants born during the Leningrad famine grew up in a nutritional environment much poorer than the Dutch Winter Famine babies; because their "nutritional forecast" was correct, the lack of increase in the incidence of cardiovascular or metabolic disease was not surprising. In Jamaica, children born small were more likely to respond to malnutrition with marasmus/wasting, while those born larger responded with kwashiorkor (syndrome characterized by altered protein, amino acid, and lipid metabolism in addition to wasting). Kwashiorkor is a more severe condition, associated with higher mortality than marasmus, so it may be argued that larger babies forecast a more plentiful nutritional environment during their development and were maladapted to low nutritional planes [70]. Although this heuristic model has received criticisms, these have largely been addressed [67].

We now argue that there are two categories of pathways by which developmental factors induce later disease risk [62,71]. The induction by normative exposures involving maternal stress or reduced fetal nutrition could be interpreted using the predictive adaptive model and is thus best framed in terms of an evolved adaptive response, ultimately disadvantageous in the mismatched modern obesogenic environment. However, maternal obesity, gestational diabetes, and infant overfeeding could be interpreted as evolutionary novel exposures, likely inducing long-term effects through alternative mechanisms including the adipogenic effect of fetal hyperinsulinemia [72].

Theoretical work has both fed off and inspired further expansion and refinement of molecular human and experimental animal research. Experimental studies demonstrated how intrauterine challenges could lead to metabolic disease in adult animals [73,74]. The early models used a variety of maternal nutritional challenges in rodents [75,76]. In general, these models showed that maternal undernutrition led to offspring obesity, hypertension, and insulin resistance. A feature of these studies is that they induced an integrated, relatively stereotypic phenotype in the offspring, with common features including hyperphagia, altered energy expenditure [77], fat preference in the diet [78,79], and altered timing of puberty [80], as well as the endothelial dysfunction, hypertension, insulin resistance, and obesity [73]. Importantly, Vickers, then others, showed that long-term effects of maternal undernutrition could be reversed by neonatal leptin treatment [81]. This finding was interpreted as experimental support of the predictive adaptive response model, a conclusion reinforced when other approaches to reversal were also demonstrated to be effective [82]. As it became clear clinically that there were at least two major pathways to long-term developmental effects, researchers started to study offspring of experimental animals exposed to high fat [83,84] or high fructose concentrations [85,86]. These experiments confirmed that the offspring of mothers under such exposures developed obesity and insulin resistance as adults. A limited amount of work was also done in large-animal models [87,88].

At the same time, a considerable number of experimental studies examined the impact of maternal stress and care on the offspring phenotype. Pregnant animals were exposed to dexamethasone, to mimic maternal stress; their offspring exhibited phenotypic outcomes similar to those seen with maternal nutritional manipulation [89]. Jonathan Seckl, Michael Meaney, Frances Champagne, and colleagues examined the impact of maternal care upon their offspring's neuroendocrine and behavioral development using the highly influential model of rat dams that exhibited either high licking/grooming (LG) or low LG behavior [90,91]. Importantly, they showed that the maternal behavior changed the expression of a glucocorticoid receptor gene (through epigenetic modification), and thus the offspring's response to stress, and that the difference persisted into adulthood, yet could be eliminated by cross-fostering, to high from a low-LG or to low from a high-LG mother [91]. The phenotype produced by different levels of stress exposure could also be reversed by manipulation of the epigenetic change (in the glucocorticoid receptor gene) using a histone deacetylase inhibitor. Together, these studies pointed to the long-term effects of maternal nutritional and hormonal signals upon the offspring phenotype.

DOHaD AND EPIGENETICS

The success of DOHaD over the past decade was in no small part related to the application of epigenetics to explain the relationship between developmental exposure and later risk from disease in molecular terms. Epigenetics is today usually defined as "the study of mitotically and/or meiotically heritable changes in gene function that cannot be explained by changes in DNA section" [92]. Although the name epigenetics goes back to the mid–twentieth century and the work of Conrad Waddington (who used it to describe the study of causal mechanisms by which the genes bring about phenotypic effects), the discipline started growing from the mid-1990s onwards [93]. Today the discipline is well established though somewhat controversial: while mitotic heritability of epigenetic marks is widely accepted, the existence and role of transgenerational epigenetic inheritance, well documented in some species and especially in plants [94], remains contentious, with some experts doubting its significance.

Paradoxically, it was the publication of the human genome and the rise of genomics in the 1990s and 2000s that gave epigenetics a boost. Throughout "the century of the gene" [95], it was believed that once the DNA sequence was revealed, the variation in disease risk at the population level would be explained. But a decade later and following many genome-wide association studies as well other genomic and genetic studies, it is clear that, while some single nucleotide polymorphisms are indeed associated with the risk of NCDs, those linked to large effects in individuals are rare in the population [96,97]. The quest for common variants that contribute smaller effects continues, but the interest in other explanations of disease causation—in particular, developmental ones—has increased.

Epigenetics provided DOHaD with tools to show precisely how the developmental environment modulates gene transcription, producing a long-term effect on gene expression and on phenotypic outcome. For epigenetics, DOHaD offered a repository of clinically important research problems. While DOHaD studies in the 1990s largely focused on explaining the effects of modifications in developmental environment in functional terms (e.g., reduced renal nephron number [98], altered hormonal sensitivity [99], altered endothelial function [100], or altered hepatic metabolic activity [101]), by the mid-2000s, and particularly after the seminal work on the LG/HG mice [91], attention shifted to molecular epigenetics.

While there is a large range of epigenetic modifications, DNA methylation and histone modification are the best studied. The exposures that received the most research attention have been maternal stress and maternal nutrition. Studies in rats have shown, for example, that a change in the maternal diet altered DNA methylation and histone modification in the 5′ regulatory regions of specific nonimprinted genes [102–105]. Induced changes could be prevented by nutritional interventions in pregnancy [102] or changed by hormonal modifications in the juvenile period [106].

In contrast to good experimental evidence from animal models, data supporting this argument for humans have been scarcer. Perhaps the most definitive and influential study was that of Godfrey and colleagues [107] on two independent cohorts of children aged 6 or 9 years. Using DNA obtained from the umbilical cord at birth and taking a relatively unbiased discovery approach, the authors found a strong association, consistent across cohorts, between the methylation of a particular site in the RXRα promoter region and the degree of adiposity 6–9 years later. They also demonstrated that this epigenetic change was associated with maternal carbohydrate intake in early pregnancy: mothers with lower carbohydrate intake had higher methylation in the RXRα promoter region. Later, the group showed in vitro that epigenetic changes induced nutritionally at this region of the promoter did not affect adipocyte differentiation, but they did alter insulin sensitivity and glucose metabolism in these same adipocytes when they were fully differentiated [108]. Since then, others, e.g., Harvey et al. [109], have linked other epigenetic changes at birth to other later phenotypic effects. The discussed two studies have suggested that, independent of birth weight, a sizable portion of phenotypic variance in body composition in late childhood is determined by normative in utero exposures, especially those operating in early pregnancy. Such observations, together with epidemiological analysis of periconceptional nutrition [110] and experimental observations [87,111] of periconceptional undernutrition, are shifting attention to the periconceptional and preconceptional period.

THE WIDER DOHaD RESEARCH AGENDA

When thinking about the present and future DOHaD research, three major research avenues come to mind: (1) broadening and refining the epigenetic approach: studying a range of epigenetic marks; studying epigenetic inheritance through the male line as well as inheritance across multiple generations; (2) considering preventative interventions, targeted at epigenetic modifications; (3) expanding the range of studied exposures from nutrition and stress to include a variety of environmental variables. These avenues are based on recent research. For example, nongenetic inheritance through the male line has become an intense area of study. A limited number of experimental [112,113] and epidemiological [114] studies suggest that paternal factors may influence offspring health too [115]. Such studies are in their infancy, but there is growing evidence showing that epigenetic marks are transmitted across generations via the sperm. Intergenerational maternal effects have been more clearly documented and there are multiple mechanisms by which such transduction might occur [116]. Epigenetic marks can be induced de novo (under persisting environmental influences) in each generation [117] or can be transmitted directly to the offspring [118].

The experimental data and, particularly, recent human data linking maternal state to the offspring's epigenetic state [119] have allowed investigators to consider preventative interventions before and during pregnancy. An increasing number of clinical studies have commenced in which maternal nutrition is manipulated to explore the impact of nutritional variations on the offspring. Animal studies continue to explore potentially applicable techniques to reverse conditioning in human infants. The applicability of such interventions will rely heavily on validation of epigenetic biomarkers to monitor effects, given the time it will take for the phenotype to emerge.

Such studies are all highly experimental and preliminary, but they point to the direction in which DOHaD research will progress. There will be ongoing expansion of epigenetic methodologies to inform both experimental studies and clinical human studies. The skill set of investigators engaged in DOHaD research is thus likely to shift significantly in the next decade.

Regarding the range of the studied exposures, DOHaD had its early origins in the study of teratogenesis; more recently, environmental toxicologists have embraced the DOHaD paradigm [120]. They, too, have seen that subtle levels of toxins such as air pollutants, or endocrine disruptors such as bisphenol A, can affect fetal epigenetic states and, at least experimentally, induce longer-term phenotypic change not dissimilar to that found in the more traditional DOHaD domain [121]. The interaction between epigenetic processes and the environment is of much current interest [122].

Globally, concern has been increasing over the effects of air pollution, as well as many industrial and agricultural chemicals, on the fetus. While their effects on the fetal epigenetic state are undisputed, the clinical significance of such observations is less clear. This lack of clarity stresses the importance of new prospective clinical cohorts, preferably starting before conception, providing biological samples to measure the exposome of mother and infant, measuring the epigenetic state of the infant as well as its emergent phenotype over time. Such studies are complex and expensive but important [123,124].

DOHaD AND PUBLIC POLICY

From its origins, DOHaD was focused on the causes of NCDs, especially cardiovascular and metabolic disease. Today NCDs—especially diabetes, cardiovascular disease, chronic lung disease, and common forms of cancer—account for around 60% of all deaths worldwide [125]. A high proportion of these diseases occurs in Asia, especially China and India, but a rise in the near future in parts of the world poorly equipped to deal with them—e.g., sub-Saharan Africa—has been predicted [126]. While the rise in prevalence has to do with the Western lifestyle, urbanization, and greater economic prosperity, the newly emerging epidemic is caused not by these factors as such, but by the magnitude of the recent change in behavior and environment. Scholars in DOHaD from Barker [127] onward [128] have long argued that a life course approach is important in understanding both the origins and prevention of the obesity and NCD epidemics.

However, despite the wealth of experimental, epidemiological, and clinical data supporting the DOHaD concept, the latter has gained essentially no traction within the public health community. The reasons for this failure have been discussed above and in other articles [62], and they include the overemphasis on low birth weight, lack of a conceptual framework, and a lack of underlying mechanisms. But the evidence of multiple pathways, operating independently of birth weight; the construction of a conceptual framework; and the elaboration of the epigenetic mechanisms have all addressed these concerns. Perhaps the most difficult problem was obtaining an estimate of the weight of the developmental effect. But the calculations of Barker and colleagues suggest that, using the proxy of low birth weight as a marker of a poor developmental environment, the risk of cardiovascular disease is increased by a factor of about sevenfold in adults with birth weight at the low end of the range. However, studies such as Godfrey's showed that this pathway was not pathological and exceptional but rather normative and ubiquitous, operating in uncomplicated pregnancies across the normal developmental range. The rising rates of maternal obesity and gestational diabetes point to the prevalence of nonadaptive responses, no longer restricted to the developed world, but of high importance in the developing world, where the double burden of unbalanced nutrition is a growing concern. The normative nature of the exposures pointed to the overlap with the reproductive–maternal–neonatal child health agenda, which the World Health Organisation (WHO) and other organizations recognize as linked to the millennium development goals [129].

In 2011, a landmark event occurred when the United Nations General Assembly adopted the resolution titled "Political declaration of the high-level meeting of the General Assembly on the prevention and control of non-communicable diseases (document A/66/L.1)." For the first time, the influence of early life course events was recognized by the international community and a specific clause explained the DOHaD concept (see Box 1). This declaration led the WHO and its regional divisions to start considering life course biology and its import in greater depth. With an increased attention, nongovernmental organizations interested in NCDs have started taking notice. Private- as well as public-sector research is increasingly engaged.

In 2014, the Director General of the WHO announced the establishment of a Commission to End Childhood Obesity. The background paper made it clear that the life course approach, of which DOHaD forms a part, made an important basis of this initiative. The announcement recognized the longer-term benefits of the primary prevention of childhood obesity. The commission and its working groups include many members with a depth of experience in DOHaD-related research. The DOHaD research is at last poised to influence public health.

BOX 1

Clause 26 of the Political Declaration of the High-Level Meeting of the General Assembly on the Prevention and Control of Noncommunicable Diseases (Document A/66/L.1). UN General Assembly, 66th Session, Follow-Up to the Outcome of the Millennium Summit, September 16, 2011 (http://www.un.org/ga/search/view_doc.asp?symbol=A/66/L.1)

(We) note also with concern that maternal and child health is inextricably linked with NCDs and their risk factors, specifically as prenatal malnutrition and low birth weight create a predisposition to obesity, high blood pressure, heart disease and diabetes later in life; and that pregnancy conditions, such as maternal obesity and gestational diabetes, are associated with similar risks in both the mother and her offspring.

References

[1] Shildrick M. Maternal imagination: reconceiving first impressions. Rethink Hist 2000;4(3):243–60.

[2] Hanson C. A cultural history of pregnancy. Basingstoke (UK): Palgrave; 2004.

[3] Hopwood N. Embryology. In: Bowler PJ, Pickstone JV, editors. The Cambridge history of science. The modern biological and earth sciences, vol. 6. Cambridge: Cambridge University Press; 2009. p. 285–315.

[4] López-Beltrán C. The medical origins of heredity. In: Müller-Wille S, Rheinberger H-J, editors. Heredity produced: at the crossroads of biology, politics and culture, 1500–1870. Cambridge (MA): Massachusetts Institute of Technology; 2007. p. 105–32.

[5] Farley J. Gametes and spores: ideas about sexual reproduction, 1750–1914. Baltimore (MD): Johns Hopkins University Press; 1982.

[6] Müller-Wille S, Rheinberger H-J. A cultural history of heredity. Chicago: University of Chicago Press; 2012.

[7] Paul DB. Controlling human heredity: 1865 to the present. Atlantic Highlands (NJ): Humanities Press; 1995.

[8] Kevles D. In the name of eugenics: genetics and the uses of human heredity. New York: Alfred Knopf; 1985.

[9] Smith GD, Kuh D, Kermack WO. Commentary: William Ogilvy Kermack and the childhood origins of adult health and disease. Int J Epidemiol 2001;30(4):696–703.

[10] Walton AJ, Hammond J. The maternal effects on growth and conformation in Shire horses-Shetland pony crosses. Proc R Soc B Biol Sci 1938;125:311–35.

[11] Kalter H. Teratology in the 20th century: environmental causes of congenital malformations in humans and how they were established. Neurotoxicol Teratol 2003;25:131–282.

[12] Barcroft J. Researches on pre-natal life. Oxford: Blackwell; 1946.

[13] Williams S. Relief and research: the nutrition work of the National Birthday Trust Fund, 1935–9. In: Smith DF, editor. Nutrition in Britain: science, scientists and politics in the twentieth century. London: Routledge; 1997. p. 99–122.

[14] Antonov AN. Children born during the siege of Leningrad in 1942. J Pediatr 1947;30(3):250–9.

[15] Smith CA. Effects of maternal undernutrition upon the newborn infant in Holland (1944–45). J Pediatr 1947;30(3):229–43.

[16] Burger GCE, Drummond JC, Sandstead HR. Nutrition and starvation in the Western Netherlands. The Hague: General State Printing Office; 1948.

[17] Dean RFA. Studies of undernutrition Wuppertal 1946-9. XXVIII. The size of the baby at birth and the yield of breast milk. Spec Rep Ser Med Res Counc (G B) 1951;275:346–78.

[18] McCance RA, Widdowson EM. The determinants of growth and form. Proc R Soc Lond Ser B Containing Papers Biol Character 1974;185(78):1–17.

[19] Gluckman PD, Buklijas T. Sir Graham Collingwood (Mont) Liggins. 24 June 1926–24 August 2010. Biogr Mem Fellows R Soc 2013;59:195–214.

[20] Barker DJ. Low intelligence. Its relation to length of gestation and rate of foetal growth. Br J Prev Soc Med 1966;20(2):58–66.

[21] Neugebauer R, Paneth N. Epidemiology and the wider world: celebrating Zena Stein and Mervyn Susser. Paediatr Perinat Epidemiol 1992;6(2):122–32.

[22] Stein Z, Susser M, Saenger G, Marolla F. Famine and human development: the Dutch Winter Famine of 1944–1945. New York: Oxford University Press; 1975.

[23] Ravelli GP, Stein ZA, Susser MW. Obesity in young men after famine exposure in utero and early infancy. N Engl J Med 1976;295(7):349–53.

[24] Lumey LH. Decreased birthweights in infants after maternal in utero exposure to the Dutch famine of 1944–1945. Paediatr Perinat Epidemiol 1992;6(2):240–53.

[25] Forsdahl A. Are poor living conditions in childhood and adolescence an important risk factor for arteriosclerotic heart disease? Br J Prev Soc Med 1977;31(2):91–5.

[26] Higgins M, Keller J, Moore F, Ostrander L, Metzner H, Stock L. Studies of blood pressure in Tecumseh, Michigan. I. Blood pressure in young people and its relationship to personal and familial characteristics and complications of pregnancy in mothers. Am J Epidemiol 1980;111:142–55.

[27] Gennser G, Rymark P, Isberg PE. Low birth weight and risk of high blood pressure in adulthood. Br Med J Clin Res Ed 1988;296(6635):1498–500.

[28] Wadsworth ME, Cripps HA, Midwinter RE, Colley JR. Blood pressure in a national birth cohort at the age of 36 related to social and familial factors, smoking, and body mass. Br Med J Clin Res Ed 1985;291(6508):1534–8.

[29] Dörner G. Die mögliche Bedeutung der prä- und/oder perinatalen Ernährung für die Pathogenese der Obesitas. Acta biologica medica Ger 1973;30:19–22.

[30] Dörner G, Haller K, Leonhardt M. Zur möglichen Bedeutung der prä- und/oder früh postnatalen Ernährung für die Pathogenese der Arterioskleroze. Acta Biol Med Ger 1973;31:31–5.

[31] Dörner G, Mohnike A. Zur möglichen Bedeutung der prä- und/oder frühpostnatalen Ernährung für die Pathogenese der Diabetes Mellitus. Acta Biol Med Ger 1973;31:7–10.

[32] Rohde W, Ohkawa T, Dobashi K, Arai K, Okinaga S, Dörner G. Acute effects of maternal stress on fetal blood catecholamines and hypothalamic LH-RH content. Exp Clin Endocrinol 1983;82(3):268–74.

[33] Dörner G, Plagemann A, Rückert J, et al. Teratogenetic maternofoetal transmission and prevention of diabetes susceptibility. Exp Clin Endocrinol 1988;91(3):247–58.

[34] Koletzko B. Developmental origins of adult disease: Barker's or Dörner's hypothesis? Am J Hum Biol 2005;17(3):381–2.

[35] Dörner G. Perinatal hormone levels and brain organization. In: Stumpf W, Grant LD, editors. Anatomical neuroendocrinology. Basel: Karger; 1975. p. 245–52.

[36] Freinkel N. Banting Lecture 1980. Of pregnancy and progeny. Diabetes 1980;29(12):1023–35.

[37] Van Assche FA, Holemans K, Aerts L. Long-term consequences for offspring of diabetes during pregnancy. Br Med Bull 2001;60:173–82.

[38] Gardner M, Winter P, Barker D. Atlas of mortality from selected diseases in England and Wales, 1968–1978. Chichester: Wiley; 1984.

[39] Levy D, Brink S. A change of heart: how the Framingham heart study helped unravel the mysteries of cardiovascular disease. 1st ed. New York: Knopf; 2005.

[40] Barker DJ, Osmond C. Infant mortality, childhood nutrition, and ischaemic heart disease in England and Wales. Lancet 1986;1(8489):1077–81.

[41] Barker DJP. Mothers, babies and disease in later life. London: BMJ Publishing Group; 1994.

[42] Dawes GS, Borruto F, Zacutti A, editors. Fetal autonomy and adaptation. Chichester: Wiley; 1990.

[43] Barker DJ, Gluckman PD, Robinson JS. Conference report: fetal origins of adult disease–report of the first international study group, Sydney, 29–30 October 1994. Placenta 1995;16(3):317–20.

[44] Hales CN, Barker DJ. Type 2 (non-insulin-dependent) diabetes mellitus: the thrifty phenotype hypothesis. Diabetologia 1992;35(7):595–601.

[45] Neel JV. Diabetes mellitus: a "thrifty" genotype rendered detrimental by "progress"? Am J Hum Genet 1962;14(4):353–62.
[46] Neel JV. The study of natural selection in primitive and civilized populations. Hum Biol 1958; 30(1):43–72.
[47] De Prins FA, Van Assche FA. Intrauterine growth retardation and development of endocrine pancreas in the experimental rat. Biol Neonat 1982;41:16–21.
[48] Cogswell ME, Yip R. The influence of fetal and maternal factors on the distribution of birthweight. Semin Perinatol 1995;19(3):222–40.
[49] Curhan GC, Chertow GM, Willett WC, et al. Birth weight and adult hypertension and obesity in women. Circulation 1996;94(6):1310–5.
[50] Curhan GC, Willett WC, Rimm EB, Spiegelman D, Ascherio AL, Stampfer MJ. Birth weight and adult hypertension, diabetes mellitus, and obesity in US men. Circulation 1996;94(12):3246–50.
[51] Kramer MS, Joseph KS. Enigma of fetal/infant-origins hypothesis. Lancet 1996;348(9037):1254–5.
[52] Paneth N, Susser M. Early origin of coronary heart disease (the "Barker hypothesis"). Br Med J 1995;310:411–2.
[53] Huxley R, Neil A, Collins R. Unravelling the fetal origins hypothesis: is there really an inverse association between birthweight and subsequent blood pressure? Lancet 2002;360(9334):659–65.
[54] Stanner SA, Bulmer K, Andres C, et al. Does malnutrition in utero determine diabetes and coronary heart disease in adulthood? Results from the Leningrad siege study, a cross sectional study. BMJ 1997;315(7119):1342–8.
[55] Kannisto V, Christensen K, Vaupel JW. No increased mortality in later life for cohorts born during famine. Am J Epidemiol 1997;145(11):987–94.
[56] Ravelli AC, Van der Meulen JHP, Michels RPJ, et al. Glucose tolerance in adults after prenatal exposure to famine. Lancet 1998;351:173–7.
[57] Lumey LH. Glucose tolerance in adults after prenatal exposure to famine. Lancet 2001;357(9254):472–3.
[58] Editorial. An overstretched hypothesis? Lancet 2001;357(9254):405.
[59] Gluckman PD, Pinal C. Comment on "An overstretched hypothesis?". [Lancet 2001] Lancet 2001;357(9270). 1798.
[60] Kuzawa CW, Gluckman PD, Hanson MA. Developmental perspectives on the origin of obesity. In: Fantuzzi G, Mazzone T, editors. Adipose tissue and adipokines in health and disease. Totowa (NJ): Humana Press; 2007. p. 207–19.
[61] Lucas A. Programming by early nutrition in man. Ciba Found Symp 1991;156:38–50.
[62] Hanson MA, Gluckman PD. Early developmental conditioning of later health and disease: physiology or pathophysiology? Physiol Rev 2014;94:1027–76.
[63] Gale CR, Jiang B, Robinson SM, Godfrey KM, Law CM, Martyn CN. Maternal diet during pregnancy and carotid intima-media thickness in children. Arterioscler Thromb Vasc Biol 2006;26:1877–82.
[64] Mericq V, Ong KK, Bazaes RA, et al. Longitudinal changes in insulin sensitivity and secretion from birth to age three years in small- and appropriate-for-gestational-age children. Diabetologia 2005;48:2609–14.
[65] Bateson P. Fetal experience and good adult design. Int J Epidemiol 2001;30(5):928–34.
[66] Bateson P, Barker D, Clutton-Brock T, et al. Developmental plasticity and human health. Nature 2004;430:419–21.
[67] Bateson P, Gluckman P, Hanson M. The biology of developmental plasticity and the Predictive Adaptive Response hypothesis. J Physiol 2014;592(11):2357–68.
[68] Hanson M, Gluckman P. The human camel: the concept of predictive adaptive responses and the obesity epidemic. Pract Diabetes Int 2003;20(8):267–8.
[69] Gluckman PD, Hanson MA, Spencer HG. Predictive adaptive responses and human evolution. Trends Ecol Evol 2005;20(10):527–33.
[70] Forrester TE, Badaloo AV, Boyne MS, et al. Prenatal factors contribute to emergence of kwashiorkor or marasmus in response to severe undernutrition: evidence for the predictive adaptation model. PLoS One 2012;7(4):e35907.
[71] Kuzawa CW, Gluckman PD, Hanson MA, Beedle AS, Stearns SC, Koella JC. Evolution, developmental plasticity, and metabolic disease. In: Stearns SC, Koella JC, editors. Evolution in health and disease. 2nd ed. Oxford: Oxford University Press; 2007. p. 253–64.
[72] Hanson MA, Gluckman PD, Ma RCW, Matzen P, Biesma RG. Early life opportunities for prevention of diabetes in low and middle income countries. BMC Public Health 2012;12:1025.
[73] Vickers MH, Breier BH, Cutfield WS, Hofman PL, Gluckman PD. Fetal origins of hyperphagia, obesity, and hypertension and postnatal amplification by hypercaloric nutrition. Am J Physiol 2000;279:E83–7.
[74] McMillen IC, Robinson JS. Developmental origins of the metabolic syndrome: prediction, plasticity, and programming. Physiol Rev 2005;85:571–633.
[75] Langley SC, Jackson AA. Increased systolic blood pressure in adult rats induced by fetal exposure to maternal low protein diets. Clin Sci 1994;86:217–22.
[76] Woodall SM, Johnston BM, Breier BH, Gluckman PD. Chronic maternal undernutrition in the rat leads to delayed postnatal growth and elevated blood pressure of offspring. Pediatr Res 1996;40:438–43.

[77] Vickers MH, Breier BH, McCarthy D, Gluckman PD. Sedentary behavior during postnatal life is determined by the prenatal environment and exacerbated by postnatal hypercaloric nutrition. Am J Physiol 2003;285(1):R271–3.

[78] Bellinger L, Lilley C, Langley-Evans SC. Prenatal exposure to a low protein diet programmes a preference for high-fat foods in the young adult rat. Br J Nutr 2004;92:513–20.

[79] Bayol SA, Farrington SJ, Stickland NC. A maternal 'junk food' diet in pregnancy and lactation promotes an exacerbated taste for 'junk food' and a greater propensity for obesity in rat offspring. Br J Nutr 2007;98(4):843–51.

[80] Sloboda DM, Howie GJ, Pleasants A, Gluckman PD, Vickers MH. Pre- and postnatal nutritional histories influence reproductive maturation and ovarian function in the rat. PLoS One 2009;4(8):e6744.

[81] Vickers MH, Gluckman PD, Coveny AH, et al. Neonatal leptin treatment reverses developmental programming. Endocrinology 2005;146:4211–6.

[82] Gray C, Li M, Reynolds CM, Vickers MH. Pre-weaning growth hormone treatment reverses hypertension and endothelial dysfunction in adult male offspring of mothers undernourished during pregnancy. PLoS One 2013;8(1):e53505.

[83] Howie GJ, Sloboda DM, Kamal T, Vickers MH. Maternal nutritional history predicts obesity in adult offspring independent of postnatal diet. J Physiol 2009;587(4):905–15.

[84] Elahi MM, Cagampang FR, Mukhter D, Anthony FW, Ohri SK, Hanson MA. Long-term maternal high-fat feeding from weaning through pregnancy and lactation predisposes offspring to hypertension, raised plasma lipids and fatty liver in mice. Br J Nutr 2009;102:514–9.

[85] Vickers MH, Clayton ZE, Yap C, Sloboda DM. Maternal fructose intake during pregnancy and lactation alters placental growth and leads to sex-specific changes in fetal and neonatal endocrine function. Endocrinology 2011;152(4):1378–87.

[86] Flynn ER, Alexander BT, Lee J, Hutchens ZM, Maric-Bilkan C. High-fat/fructose feeding during prenatal and postnatal development in female rats increases susceptibility to renal and metabolic injury later in life. Am J Physiol. Regul Integr Comp Physiol 2013;304(4):R278–85.

[87] Todd SE, Oliver M, Jaquiery AL, Bloomfield FH, Harding JE. Periconceptional undernutrition of ewes impairs glucose tolerance in their adult offspring. Pediatr Res 2009;65:409–13.

[88] Cleal JK, Poore KR, Boullin JP, et al. Mismatched pre- and postnatal nutrition leads to cardiovascular dysfunction and altered renal function in adulthood. Proc Natl Acad Sci USA 2007;104:9529–33.

[89] Seckl JR, Cleasby M, Nyirenda MJ. Glucocorticoids, 11β-hydroxysteroid dehydrogenase, and fetal programming. Kidney Int 2000;57:1412–7.

[90] Meaney MJ. Maternal care, gene expression, and the transmission of individual differences in stress reactivity across generations. Annu Rev Neurosci 2001;24:1161–92.

[91] Weaver ICG, Cervoni N, Champagne FA, et al. Epigenetic programming by maternal behavior. Nat Neurosci 2004;7(8):847–54.

[92] Riggs AD, Martienssen RA, Russo VEA. Introduction. In: Russo VEA, Martienssen RA, Riggs AD, editors. Epigenetic mechanisms of gene regulation. Cold Spring Harbor: Cold Spring Harbor Laboratory Press; 1996. p. 1–4.

[93] Haig D. The (dual) origin of epigenetics. Cold Spring Harb Symp Quant Biol 2004;69:67–70.

[94] Herman JJ, Sultan SE. Adaptive transgenerational plasticity in plants: case studies, mechanisms, and implications for natural populations. Front Plant Sci 2011;2.

[95] Fox Keller E. Century of the gene. Cambridge (MA): Harvard University Press; 2000.

[96] Maher B. Personal genomes: the case of the missing heritability. Nature 2008;456(7218):18–21.

[97] Manolio TA, Collins FS, Cox NJ, et al. Finding the missing heritability of complex diseases. Nature 2009;461(7265):747–53.

[98] Brenner BM, Chertow GM. Congenital oligonephropathy and the etiology of adult hypertension and progressive renal injury. Am J Kidney Dis 1994;23(2):171–5.

[99] Bauer MK, Breier BH, Harding JE, Veldhuis JD, Gluckman PD. The fetal somatotropic axis during long term maternal undernutrition in sheep: evidence for nutritional regulation in utero. Endocrinology 1995;136(3):1250–7.

[100] Ozaki T, Hawkins P, Nishina H, Steyn C, Poston L, Hanson MA. Effects of undernutrition in early pregnancy on systemic small artery function in late-gestation fetal sheep. Am J Obstet Gynecol 2000;183(5):1301–7.

[101] Ozanne SE, Smith GD, Tikerpae J, Hales CN. Altered regulation of hepatic glucose output in the male offspring of protein-malnourished rat dams. Am J Physiol Endocrinol Metab Gastrointest Physiol 1996;270:E559–664.

[102] Lillycrop KA, Phillips ES, Jackson AA, Hanson MA, Burdge GC. Dietary protein restriction of pregnant rats induces and folic acid supplementation prevents epigenetic modification of hepatic gene expression in the offspring. J Nutr 2005;135:1382–6.

[103] Bogdarina I, Welham S, King PJ, Burns SP, Clark AJ. Epigenetic modification of the renin-angiotensin system in the fetal programming of hypertension. Circ Res 2007;100:520–6.

[104] Lillycrop KA, Slater-Jefferies JL, Hanson MA, Godfrey KM, Jackson AA, Burdge GC. Induction of altered epigenetic regulation of the hepatic glucocorticoid receptor in the offspring of rats fed a protein-restricted diet during pregnancy suggests that reduced DNA methyltransferase-1 expression is involved in impaired DNA methylation and changes in histone modifications. Br J Nutr 2007;97(6):1064–73.

[105] Lillycrop KA, Phillips ES, Torrens C, Hanson MA, Jackson AA, Burdge GC. Feeding pregnant rats a protein-restricted diet persistently alters the methylation of specific cytosines in the hepatic PPARα promoter of the offspring. Br J Nutr 2008;100(2):278–82.

[106] Gluckman PD, Lillycrop KA, Vickers MH, et al. Metabolic plasticity during mammalian development is directionally dependent on early nutritional status. Proc Natl Acad Sci USA 2007;104(31):12796–800.

[107] Godfrey KM, Sheppard A, Gluckman PD, et al. Epigenetic gene promoter methylation at birth is associated with child's later adiposity. Diabetes 2011;60: 1528–34.

[108] Ngo S, Li X, O'Neill R, Bhoothpur C, Gluckman P, Sheppard A. Elevated S-adenosylhomocysteine alters adipocyte functionality with corresponding changes in gene expression and associated epigenetic marks. Diabetes 2014;63:2273–83.

[109] Harvey N, Lillycrop K, Garratt E, et al. Evaluation of methylation status of the eNOS promoter at birth in relation to childhood bone mineral content. Calcif Tissue Int 2012;90(2):120–7.

[110] Morton SMB, Gluckman PD, Hanson MA. Maternal nutrition and fetal growth and development. In: Gluckman PD, Hanson MA, editors. Developmental origins of health and disease. Cambridge: Cambridge University Press; 2006. p. 98–129.

[111] Watkins AJ, Wilkins A, Cunningham C, et al. Low protein diet fed exclusively during mouse oocyte maturation leads to behavioural and cardiovascular abnormalities in offspring. J Physiol 2008;586(8):2231–44.

[112] Ng S-F, Lin RCY, Laybutt DR, Barres R, Owens JA, Morris MJ. Chronic high-fat diet in fathers programs β-cell dysfunction in female rat offspring. Nature 2010;467(7318):963–6.

[113] Guerrero-Bosagna C, Settles M, Lucker B, Skinner MK. Epigenetic transgenerational actions of vinclozolin on promoter regions of the sperm epigenome. PLoS One 2010;5(9):e13100.

[114] Pembrey M, Saffery R, Bygren LO, Network in Epigenetic Epidemiology. Human transgenerational responses to early-life experience: potential impact on development, health and biomedical research. J Med Genet 2014;51(9):563–72.

[115] Hughes V. Epigenetics: the sins of the father. Nature 2014;507:22–4.

[116] Gluckman PD, Hanson MA, Beedle AS. Non-genomic transgenerational inheritance of disease risk. BioEssays 2007;29(2):145–54.

[117] Burdge G, Hoile S, Uller T, Gluckman PD, Hanson MA. Progressive, transgenerational changes in offspring phenotype and epigenotype following nutritional transition. PLoS One 2011;6(11):e28282.

[118] Danchin É, Charmantier A, Champagne FA, Mesoudi A, Pujol B, Blanchet S. Beyond DNA: integrating inclusive inheritance into an extended theory of evolution. Nat Rev Genet 2011;12(7):475–86.

[119] Hanson MA, Gluckman PD, Godfrey KM. Developmental epigenetics and risks of later non-communicable disease. In: Seckl JR, Christen Y, editors. Hormones, intrauterine health and programming. Springer; 2014. p. 175–83.

[120] Schug TT, Barouki R, Gluckman PD, Grandjean P, Hanson M, Heindel JJ. PPTOX III: environmental stressors in the developmental origins of disease—evidence and mechanisms. Toxicol Sci 2013;131(2):343–50.

[121] Cagampang FR, Torrens C, Anthony F, Hanson M. Developmental exposure to bisphenol A leads to cardiometabolic dysfunction in adult mouse offspring. J Dev Origins Health Dis 2012;3(4):287–92.

[122] Feil R, Fraga MF. Epigenetics and the environment: emerging patterns and implications. Nat Rev Genet 2012;13:97–109.

[123] Barouki R, Gluckman PD, Grandjean P, Hanson M, Heindel JJ. Developmental origins of non-communicable disease: Implications for research and public health. Environ Health 2012;11:42.

[124] Dietert RR. Developmental immunotoxicity, perinatal programming and noncommunicable diseases: focus on human studies. Adv Med 2014. Article ID 867805. http://dx.doi.org/10.1155/2014/867805.

[125] Narayan KM, Ali MK, Koplan JP. Global noncommunicable disease: where worlds meet. N Engl J Med 2010;363(13):1196–8.

[126] Alwan AD, Galea G, Stuckler D. Development at risk: addressing noncommunicable diseases at the United Nations high-level meeting. Bull World Health Organ 2011;89. 546–546a.

[127] Barker DJP. Past obstacles and future promise. In: Gluckman P, Hanson M, editors. Developmental origins of health and disease. Cambridge: Cambridge University Press; 2006. p. 481–95.

[128] Gluckman PD, Hanson M, Zimmet P, Forrester T. Losing the war against obesity: the need for a developmental perspective. Sci Transl Med 2011; 3(93):93cm19.

[129] Requejo JH, Bryce J, Barros AJ, et al. Countdown to 2015 and beyond: fulfilling the health agenda for women and children. Lancet 2014;385(9966):466–76.

CHAPTER

2

Historical Perspective of Developmental Origins of Health and Disease in Humans

Lewis P. Rubin

Department of Pediatrics, Texas Tech University Health Sciences Center El Paso, Paul L. Foster School of Medicine, El Paso, TX, USA

OUTLINE

INTRODUCTION

Adverse social and psychosocial circumstances, including exposure to social, economic, and psychological stressors, have been associated with a variety of poor health outcomes in different parts of the world and in a range of ethnic and age groups. During the last two decades, epidemiological evidence that early life stressors increase susceptibility to disease later in life has become supported by a wealth of prospective clinical data, experimental models, and an emerging appreciation of the underlying molecular and developmental mechanisms, most recently epigenetic regulation of gene expression. This paradigm is now known as the developmental origins of health and disease (DOHaD) [1]. A central tenet of this framework is that early life psychosocial stress has an indirect lasting impact on physiological wear and tear via health behaviors, adiposity, and socioeconomic factors in adulthood [2,3].

The Epigenome and Developmental Origins of Health and Disease
http://dx.doi.org/10.1016/B978-0-12-801383-0.00002-5

Copyright © 2016 Elsevier Inc. All rights reserved.

Psychosocial stress may also lead to preterm birth by altering immune system function either independently or in interaction with neuroendocrine dysfunction.

The association between birth weight and development of traditionally adult-onset diseases, such as type 2 diabetes mellitus and cardiovascular diseases, was first demonstrated 25 years ago by Hales [4]. In the decades since, the hypothesis that environmental factors act on the genome to create differences in vulnerability or resilience has become a central concept in health-related research, especially maternal child health [5,6]. One underlying biological mechanism for effects of fetal deprivation and stress is environmentally inducible epigenetic change.

Correspondingly, investigation of biological mechanisms of health and disease has increasingly emphasized the environmental context of the individual. A "critical period" refers to a time window when developmental changes in the organism (or subsystems) toward increasing complexity, greater plasticity, and more efficient functioning occur rapidly and may be most easily modified either in favorable or unfavorable directions. Also known as "biological programming" or as a "latency model," these critical, environmentally sensitive periods underlie the developmental origins of adult disease. The critical developmental period model also includes the possibility that effects of an exposure in development may be dramatically changed by later physiological or psychological stressors. This expansion of critical period (fetal) effects with later-life effect modifiers provides a plausible framework to approach the interactions between early and later-life risk factors [7].

A related concept, the "thrifty phenotype" hypothesis [8], states that the thrifty aspects of adaptation to a nutrient-limited environment can induce later unhealthy permanent changes. These include reduced capacity for insulin secretion and insulin resistance that, combined with obesity, aging, and physical inactivity, are important factors in determining type 2 diabetes in a nutritionally rich environment. Indeed, a key tenet of the developmental/fetal origins paradigm is that fetal undernutrition in mid-to-late gestation impairs fetal growth and can program later-life deleterious health outcomes including disturbed somatic growth, metabolic stress, aberrant glucose tolerance and type 2 diabetes mellitus, and cardiovascular disease. In the 1960s, Widdowson and McCance [9] were among the first to demonstrate that brief periods of undernutrition during critical developmental times are not necessarily followed by "catch-up growth." They showed that the earlier in life rats were exposed to malnutrition, the more serious and permanent were later effects [9].

Although different models invoked to describe and explain biopsychosocial stress significantly overlap, some have been developed to understand conditions or effects in specific populations and communities. The underlying biological mechanisms involve environmentally induced epigenetic changes in gene expression, which may be passed transgenerationally. This chapter addresses some complexities inherent in stress assessment and reviews the historical evolution of molecular, especially epigenetic, mechanisms that mediate effects of physical, nutritional, psychological, and social stress.

Cotemporal with the recognition that epigenetics is the fundamental mechanism of an organism's adaptation, several biopsychosocial frameworks have advanced understanding of developmental origins of health and disease. Allostasis, which incorporates hormonal responses to predictable environmental changes, provides an integrative framework. Geronimus's concept of "weathering" [10] aims to explain how socially structured, repeated stress can accumulate and increase disease vulnerability in African Americans. Weathering emphasizes the importance of internalized/interpersonal racism as a driver of racial outcome disparities. Similarly, for Mexican immigrants and Mexican Americans, an "acculturation" framework has proven to be especially useful for exploring health disparities, including preterm birth and neuropsychiatric risk in childhood [11].

Although traditional theory and methods have focused separately on how social and physical environmental factors affect children's health, evolving research has underscored the importance of integrated approaches [12]. Fundamentally, socioeconomic and physical/chemical environmental dimensions interact, stress being a permissive factor for susceptibility to certain immunological and toxic insults. In addition, unhealthy residential environments and psychosocial stress can covary [13], potentiating adverse effects on health and development. As an example, prenatal stress and environmental metal exposures synergistically alter maternal diurnal cortisol during pregnancy and stress responses in the offspring [14], although these interactions remain poorly understood. Environmental pollutants, such as pesticides and endocrine disruptors, can produce birth defects, impaired fecundity, infertility, and altered somatic growth and neurodevelopment [15].

This chapter considers the development of contemporary concepts of the biopsychosocial underpinnings of later health and disease. It then reviews the development of the current understanding of epigenetics as one of the fundamental biological mechanisms of plasticity and adaptation.

SOCIOECONOMIC STRESSORS AND INFANT AND CHILD HEALTH

The American psychologist James Garbarino coined the term *socially toxic environments* to describe conditions of poverty and violence [16]. This heuristic is particularly germane for examining adverse effects on maternal and child health. Despite improvements in maternal health care, socioeconomic differences in birth outcomes remain pervasive, show substantial variation by racial or ethnic subgroup, and are associated with disadvantage measured at multiple levels (individual/family, neighborhood) and time points (childhood, adulthood), and with adverse health behaviors that are themselves socially patterned [17]. The ongoing longitudinal Adverse Childhood Experiences Study of adults has uncovered significant associations among chronic conditions, quality of life and life expectancy in adulthood, and trauma and stress associated with adverse childhood experiences. The latter include physical or emotional abuse or neglect, deprivation, or exposure to violence [18]. A growing body of evidence indicates that there exists a stepwise gradient in health according to socioeconomic status (SES); that is, people in each class commonly have poorer health outcomes than those just above them and have better outcomes than those below [19,20]. The shape of this relationship between SES and health actually may be curvilinear, decreasing health outcomes becoming more common at even higher SES levels [21]. Socioeconomic gradients, and their implications for *relative* social status, therefore, inform explanations of how economics interacts with psychosocial stress and, thereby, epigenetic pathways.

Wilkinson and Picket have argued that relative deprivation is the core mechanism of how income impacts health in societies, i.e., why many problems associated with relative deprivation are more prevalent in more unequal societies [22,23]. Specific associated deleterious health outcomes include preterm birth, low birth weight (LBW), and child mortality. In economics, social science, and policy, a commonly utilized measure of statistical dispersion representing the income distribution of a country's population is the Gini Index [24] developed by the statistician and sociologist Corrado Gini in 1912. Modifications of the Gini Index recently have begun to be utilized as an indicator of income or wealth inequality in health research.

Theoretically, health effects of income inequality, measured by the Gini Index, could result from inequalities in access to opportunities and material goods or lack of systemic investment in social and physical infrastructure, or from psychosocial stress. For nearly a century, socioeconomic deprivation in Europe and South Asia has been linked to fetal growth. Conversely, improved socioeconomic conditions and nutrition can be associated with increasing birth weight, as occurred in post-World War II Japan. Observations on extreme deprivation during World War II and longitudinal analysis of these historical cohorts have profoundly informed the development of developmental origins theory later in the twentieth century. During the Dutch Hunger Winter of 1943–1944, the average birth weight declined 200 g [25] and during the Nazi siege of Leningrad by 500 g [26]. Moreover, follow-up of the Dutch Hunger Winter birth cohort of individuals prenatally exposed to the famine shows deleterious transgenerational health effects [27].

The second, not mutually exclusive, explanation for effects of scarcity and poverty on health outcomes is psychosocial (neurobiological) stress. In fact, the apparently deep-seated social problems associated with poverty, relative deprivation, or low social status appear to be more common in more stratified societies [22]. Income inequality theory emphasizes that *relative deprivation* is accompanied by psychosocial deprivation and leads to stress, poor health behaviors, and, consequently, poor health. In this theory, relative deprivation and social stratification, more than direct effects of deprivation of physical living conditions, are the causal link with poor health outcomes [28]. A study of associations between longitudinal neighborhood poverty trajectories and preterm birth compared neighborhoods with long-term low poverty, those with long-term high poverty, and those that experienced increasing poverty early in the study period; the latter two neighborhoods had 41% and 37% increased odds of preterm birth (95% confidence interval [CI] = 1.18, 1.69 and 1.09, 1.72, respectively) [29]. The finding that high (compared with low) cross-sectional neighborhood poverty was not associated with preterm birth (OR = 1.08; 95% CI = 0.91, 1.28) [29] supports a relative deprivation thesis.

ALLOSTASIS AND ALLOSTATIC LOAD

Adaptation to stress comes at a price. Developmental plasticity has evolved to match an organism to its early environment. But a mismatch between phenotypic outcomes of adaptive plasticity and environmental conditions later in life increases the risk of metabolic and cardiovascular diseases. The concept of allostasis (stability through adaptation) and accumulated life stress, McEwen's allostatic load [30], provides a versatile paradigm to understand childhood and adult outcomes. The benefits of adaptation (allostasis) and the costs of adaptation (allostatic load) lead to different trade-offs in health and disease, reinforcing a Darwinian concept of stress [30].

The allostatic load is essentially a composite indicator of accumulated stress-induced biological risk whereby chronic or recurrent stress leads to cascading, potentially irreversible changes in biological stress-regulatory systems. Over time, the effects of early life stressors may lead to individual differences in cumulative effects of stress and in stress-related physical and psychological disorders [31]. Allostatic load informs several overlapping conceptual frameworks including weathering, acculturation, cumulative physiological dysregulation, and the biological risk profile.

Chronic stress exposure with continuous activation of stress response systems and reduced access to social coping mechanisms can lead to increased allostatic load and premature cellular and body aging, the latter via cumulative oxidant stress and telomere shortening. Early life stress alters brain structure and neuronal connectivity [32,33]. Based on experimental animal models of stress vulnerability and resilience [34], physical and neurogenic stressors can activate different brain circuits but they share the adaptive value of harnessing the stress response; stress response activation is high for moderate and temporally confined stress exposures but becomes uncertain if stressors are very severe or chronic. Moreover, unpredictable stressors can be more maladaptive for later health than predictable stressors.

McEwen and colleagues [31] have argued that allostatic load can be quantified by cataloging specific biomarkers across major biological regulatory systems, including the cardiovascular, metabolic, endocrine, and immune systems. This quantification of allostatic load has explanatory power in understanding the relationship between specific stressors and physical health, initially using data from the MacArthur Studies of Successful Aging [35] and National Health and Nutrition Examination Survey (NHANES) data for adults [10]. More recently, allostatic load has been employed to predict risk of adverse reproductive [36] and pediatric health outcomes [2], although empirical evidence on its utility in pregnancy has been conflicting [37,38]. Incorporation of biomarkers for stress, aging, and immune responses may permit a more comprehensive and predictive allostatic load assessment.

WEATHERING AND AFRICAN AMERICAN HEALTH DISPARITIES

In the United States (US), epidemiological research in the 1990s began to devote attention to the relationships that exist among racism, social class, and health, initially focusing on racial differences. Especially in terms of the American black experience of race, Geronimus's concept of "weathering" [10,39] was devised to explain how socially structured, repeated stress can accumulate and increase disease vulnerability. Weathering builds on earlier research on the effects of insult accumulation; it is founded in cumulative disadvantage theory as a framework to explain physical consequences of social inequality. Indeed, chronic stressors have been recognized as particularly salient for poor and minority women, who also experience the highest rates of adverse birth outcomes [40]. In the US, despite steady decrease in infant mortality over recent decades, significant racial disparities persist, which suggests that factors that drive rates down may be different from the factors that drive them apart [41].

An important emphasis of weathering is the focus on internalized/interpersonal racism and cumulative socioeconomic disadvantage to explain racial disparities in outcomes, including the dramatic black/white differences in birth outcomes in the US. The framework facilitates a reorientation of perinatal psychosocial research from the measurement of acute stress, e.g., counts of life events occurring during pregnancy, to measures of chronic stressors [40,42]. The biological underpinning is that increased "wear and tear" on an individual subjected to repeated challenge or stress chronically taxes the hypothalamic–pituitary–adrenal (HPA) axis leading ultimately to physical and psychological disease vulnerability.

Although race is very much a social construct [43], several possible linkages might explain associations between phenotypic race and subsequent health or disease—latent variable institutionalized racism, which may explain the racial differences in a wide variety of socioeconomic exposures; periconceptional maternal health and support; interpersonal racism; possible correlations between phenotypic race and genotype; and epigenetic gene regulation, i.e., genetic predisposition or gene-environment/epigenetics can modify risk.

Social disparities certainly explain some racial health disparities. In a series of studies in Chicago, between-neighborhood variation [44] in concentrated poverty and low social support predicted the relationship between LBW [45] and maternal age [46]. For urban residents, neighborhood also exerted a powerful influence on diet.

A weathering pattern of age-related birth outcomes is particularly observed in African American women. Maternal age is an important determinant of birth outcomes, in part, presumably because it is a measure of a mother's level of biological and psychosocial preparation for childbearing. However, in contrast to the J- or U-shaped relationship that exists between maternal age and LBW rates among non-Hispanic white, Mexican American, and non-US-born black women [47], LBW rates among US-born African Americans are lowest in their teens and rise with increasing age [48]. A population-based test of the weathering hypothesis conducted in first births in Michigan showed that African American women, on average, and those who reside in low-income areas, in particular, experience worsening health profiles between their teens and young adulthood [39]. That profile contributes to an increased three- and fourfold risk of LBW and very LBW (VLBW) births, respectively, with advancing maternal age as well as to the black-white prematurity gap [39,49].

Further evidence of weathering is the persistence of a black disadvantage in infant mortality for higher SES individuals. This finding is attributable entirely to the much higher incidence of LBW (primarily VLBW) among black infants [50,51]. The persistence of increased risk of LBW and VLBW among African American infants, even among wealthier and educated individuals, indicates a current imperfect understanding of the biopsychosocial determinants of preterm birth and current limitations on preventing prematurity.

ACCULTURATION AND THE HISPANIC PARADOX

Whereas weathering can provide a coherent, multidimensional approach to persistent outcome disparities for African American women and children, consideration of health outcomes in ethnic immigration presents unique challenges. Immigrant health research employs theoretical frameworks (e.g., acculturation) that emphasize cultural explanations but, until more recently, a "social determinants of health" framework, which emphasizes social and structural explanations, has been less commonly used [52]. Ethnicity is a broad category that may incorporate race, cultural tradition, common population history, religion, language, and, often, a shared genetic heritage. Largely driven by births in the Hispanic population, children in immigrant families will come to represent one-third of all US children and account for almost all the growth in the US workforce over the next 40 years [53]. Nevertheless, in the Americas, Hispanic or Latino ethnicity is an imprecise concept. Similarly, acculturation models have methodological limitations, and investigation into possible epigenetic mechanisms is only beginning.

A notable observation is that health outcomes for Hispanics/Latinos in the US, despite frequent socioeconomic disadvantage, are generally more favorable than for other ethnic/racial groups such as African and Native Americans. This epidemiological paradox, originally proposed by Markides and Coreil [54], has been framed as a Hispanic paradox or healthy immigrant paradox. More specifically, for maternal child health a "Latina epidemiologic paradox" refers to the finding that Hispanic mothers in the US can have a similar or lower risk for delivering an LBW infant than non-Hispanic white mothers. An important context is that non-Hispanic white and overall LBW and preterm birth risk in the US remains relatively high compared to economically similar countries. In this sense, an acculturation context is valuable for examining the generally observed intergenerational decline in health among Hispanic Americans. An important caveat is the examination of more granular health outcome data for US. Hispanics show substantial heterogeneity among countries of origin and number of generations living in the US. Longer residence for

Mexicans in the US, individually and across generations, appears to erase any immigrant health advantage.

The US Hispanic paradox in infant and child outcomes is, in fact, a relatively new epidemiological phenomenon. Infant mortality rates for the largely Mexican Southwest US Spanish-surname population were not always similar to, or lower than, those for non-Hispanic whites. Throughout the first half of the twentieth century, they were much higher than those of whites, due primarily to shifts in postneonatal mortality [55,56]. A change apparently occurred by 1980, when demographic data sets showed parity or near-parity between infant mortality rates of Hispanic and non-Hispanic white populations in Texas [55] and California [57].

Acculturation is a transitional process that occurs as immigrant groups are exposed to beliefs, traits, and lifestyles of the mainstream culture. In the mid-twentieth century, Warner and Strole [58] described acculturation as a unidirectional process that occurs as ethnic groups unlearn presumably inferior culturally based behaviors. Since that time, the nativity composition of the US population has changed substantially, largely as a result of immigration from Latin America and Asia following adoption of the Immigration Act of 1965. Currently, foreign-born African, Cuban, Mexican, and Chinese women appear to have significantly lower risks of infant mortality, LBW, and preterm birth [59]. In addition, subsequent empirical findings have uncovered an immigrant advantage phenomenon.

In some instances, acculturation to the "core culture" may subject individuals to unfavorable social conditions including societal and financial stressors, i.e., allostatic load [60] and attendant poor outcomes. Mexican acculturation may also be accompanied by loss of social supports, strong family ties, and group identity. Accordingly, health outcomes for immigrants might be bimodal, that is, a combination of the "healthy immigrant" effect (i.e., newer immigrants are healthier) and the effects of acculturation—the longer Mexican immigrants reside in the US, the greater their likelihood of losing culture-related protective health features [61–65]. For example, recent immigrants have better mental health than natives ("immigration advantage") and Mexican immigrants have lower reported rates of psychopathology than the overall US population. This observation is less true for Cubans and not true for Puerto Ricans living in the US mainland [66]. Data are also inconsistent whether latent sociocultural advantages conferred on Mexican Americans who live in high-density Mexican American neighborhoods outweigh the disadvantages conferred by high poverty in many of those neighborhoods.

Consequently, the acculturation model has been particularly applicable for understanding health status and intergenerational effects for Mexican women who have migrated to the US or live on the US/Mexico border and their offspring. The greater part of acculturation literature concerns these populations. Indeed, the few studies comparing the effect of acculturation across Latino/Hispanic subgroups suggest the experience of acculturation and its effects on health outcomes may be different for Mexicans compared with different Caribbean, Central, or South American Latino groups [67–69]. For Mexican Americans, a graded effect of residence in the US is seen particularly in pregnancy complications such as excessive gestational weight gain [70] and pediatric morbidities. More generally, Mexican American women are observed to have different health profiles stratified by generation in the US, which serves as a surrogate measure for acculturation. In particular, LBW increases from first- to third-generation residence.

Acculturation measures have variously incorporated English-language use, nativity, social integration, and social assimilation [63,65,71]. Explanations focusing on immigration as positively selective on good health [72,73], i.e., selectively healthy immigrant women of childbearing age are more likely to give birth to healthy infants in the US compared with their native-born counterparts who are not selectively healthy in the same way, do not appear to explain the intergenerational rise in deleterious pregnancy outcomes in Mexican Americans.

Research challenges in acculturation explanations remain. Many acculturation measures such as language, generation, or self-reported ethnic identity are, at best, proxies that do not fully capture acculturation. Acculturation instruments for children and adolescents need more sensitive items to discriminate linguistic differences, measure other factors, and describe allosteric load and epigenetic effects. The Hispanic Community Health Study/Study of Latino Youth is a multisite epidemiologic study of obesity and cardiometabolic risk among US Hispanic/Latino children. The SOL Youth Study incorporates a parent/child socioecological framework, social cognitive theory, family systems theory, and acculturation research to refine a predictive conceptual model for Hispanic children and adolescent health outcomes [74]. Finally, the extent to which intergenerational differences in reproductive and childhood outcomes are attributable to the intrauterine environment and fetal programming is poorly understood [75].

A LIFE COURSE APPROACH AND THE DEVELOPMENTAL ORIGINS OF HEALTH AND DISEASE

Longitudinal models of health disparities have developed: (1) from a programming perspective in which early life (including fetal) exposures influence health over the lifespan and (2) emphasis on cumulative pathways that conceptualize health decline as a result of cumulative wear and tear to allostatic systems. Both perspectives build on a stress paradigm and both can be synthesized in a life course perspective [7] or "life history theory." In effect, disparities in birth outcomes are the consequences of differential developmental trajectories set forth by early life experiences *and* cumulative allostatic load over the life course. Therefore, life course epidemiology is concerned with risk from physical or social exposures during gestation, childhood, adolescence, young adulthood, and later adult life for long-term health [7,76,77]. The aim is to elucidate biological, behavioral, and psychosocial processes that operate across an individual's lifespan and across generations. The proposition is that various biological and social factors throughout life independently, cumulatively, and interactively influence health and disease in adult life.

The catalyst for a life course approach applied to health outcomes stemmed from a revival of interest in the role of early life factors in cardiovascular and other chronic diseases. Foundational ecological and historical cohort studies followed by investigations in experimental animal models led to the "fetal origins hypothesis" now known as DOHaD and associated with David Barker [78,79], Peter Gluckman [1], and others. The original observations pointed to antecedents of cardiovascular disease in England [80] and have been followed by comprehensive cohort studies from other localities.

According to this hypothesis, which has evolved into a scientific paradigm, environmental exposures during early development such as undernutrition affect later pathophysiological processes associated with chronic, especially noncommunicable, disease [81] and "programming" the structure or function of organs, tissues, or body systems. This idea of "biological programming" emerged as an alternative, complementary concept to an adult lifestyle approach to adult chronic disease that focuses on how adult behaviors (such as smoking, diet, exercise, alcohol consumption) affect the onset and progression of diseases in adulthood—that adaptive pathways leading to later risk accords with current concepts of evolutionary developmental biology, especially those concerning parental effects [81]. Accumulation of risk factors over the life course leads to cumulative health disadvantage, including obesity, diabetes, and cardiovascular disease.

Compared with excellent, very good, or good childhood health, poor childhood health has been associated with more than threefold greater odds of having poor adult self-rated health and twice the risk of a work-limiting disability or a chronic health condition; importantly, these associations are independent of childhood and current socioeconomic position or health-related risk behaviors [82]. Of particular importance in child health, unfavorable birth outcomes and psychosocial deprivation early in life have been associated with adverse health and developmental outcomes throughout the life course and changes in CNS architecture [83,84].

EPIGENETIC REGULATION AND EFFECTS OF PSYCHOSOCIAL STRESS

In modern biology, adaptive evolutionary change is based on neo-Darwinian evolutionary theory guided exclusively by natural selection. Throughout most of twentieth century, neo-Darwinian biology, inheritance of outcomes of environmentally instructive processes does not occur. Although the late eighteenth and early nineteenth century French academician Jean-Baptiste Lamarck has become associated with the proposition of heritability of acquired characteristics, the original Darwinian and Lamarckian theories have been revised in recent decades. In the early twentieth century, biologists accepted the chromosome theory of heredity and the theory of mutation, founded on the experiments of Gregor Mendel, Hugo de Vries, Thomas Hunt Morgan, August Weismann, and others. Following the synthesis of genetics and Darwinian ideas in the 1930s and 1940s, Darwinism has come to mean the theory that evolutionary change results from natural selection of the random genetic variations generated by mutation and recombination [85].

A death knell for Lamarckian influence in biology came in the 1920s with discrediting of Paul Kammerer's work on inheritance of acquired traits in the amphibian midwife toad [86], although that work's initial refutation may have been too final. Nevertheless, in the Soviet Union (USSR), the agronomist T. D. Lysenko and his supporters exclusively promoted inheritance of acquired traits for several decades [87]. But, for much of the twentieth century, the historical verdict has been that Lamarck got it wrong. The discovery of epigenetic mechanisms in recent decades has led to new complexities and some synthesis of the interactive roles of nature and nurture in heredity.

Epigenetics studies changes in gene expression that occur without changes in DNA sequence. The term *epigenetics* was first used in 1940 by Conrad Waddington to define the "interaction of genes with their environment which bring the phenotype into being" [88]. In effect, the functional history of a gene in one generation can influence its expression in the next. Inherited epigenetic changes in the structure of chromatin can influence neo-Darwinian evolution as well as cause a type of "Lamarckian" inheritance [89]. Partly prompted by the observation of X chromosome inactivation in mammals, in the 1970s and advancing in the subsequent decade, Robin Holliday proposed that specific sequence DNA methylation has an important role in developmental gene regulation and these modifications may be stably inherited [90]. Environmental regulation of DNA methylation as a mechanism of gene expression has since assumed a central role in investigation of carcinogenesis, development, and aging. Since approximately 2000, research on epigenetics has accelerated (Figure 1) and prompted publication of several new scientific journals including *Epigenetics, Epigenetics Chromatin, Epigenomics,* and *Clinical Epigenetics.*

Epigenetic change is developmental and environmental. Mechanisms underlying the developmental origins of disease and psychosocial models

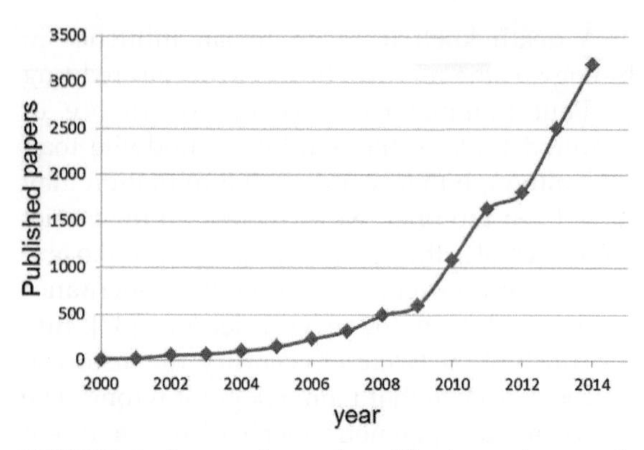

FIGURE 1 Increase in yearly publications using search words "Epigenetics OR Epigenomics" and year 2000–2014; PubMed (http://www.ncbi.nlm.nih.gov). Prior to the year 2000, fewer than 10 scientific reports were published annually.

of disease risk involve environmental rather than developmental epigenetics. A mechanism is the developing HPA stress axis is exquisitely sensitive to regulation by social forces represented at the level of the epigenome [20]. Evidence is rapidly mounting in support of the influence of environment and "lifestyle" (including nutrition, behavior, stress, physical activity, working habits, smoking, and alcohol consumption) on epigenetic change. Specifically, DNA methylation marks have been postulated as a mechanism for the enduring effects of the prenatal environment. In essence, a neonate's methylome contains a molecular memory of the in utero experience. However, interindividual variation in methylation also arises from DNA sequence polymorphisms that result in methylation quantitative trait loci (methQTLs). Teh et al. [91] surveyed the genotypes and DNA methylomes of 237 neonates and found 1423 punctuate regions of the methylome that were highly variable across individuals, termed variably methylated regions (VMRs), against a backdrop of homogeneity. MethQTLs were readily detected in neonatal methylomes, and genotype alone best explained ~25% of the VMRs. The best explanation for 75% of VMRs was the interaction of genotype with different in utero environments, including maternal smoking, maternal depression, maternal body mass index (BMI), infant birth weight, gestational age, and birth order [91].

Epigenetic modifications include DNA methylation (to date, the most extensively investigated), covalent modifications of histone tails, and effects mediated by noncoding RNAs such as microRNAs (miRNAs) and long noncoding RNAs (lncRNAs). DNA methylation functions in regulating chromatin structure and remodeling, X-chromosome inactivation, genomic imprinting, chromosome stability, and gene transcription. Addition of a methyl group to cytosine within the DNA sequence as 5-methylcytosine (5MeC) represents 2–5% of all cytosines in mammalian genomes and is found principally at CpG dinucleotides. In a sense, 5MeC may be considered as a fifth (developmentally and environmentally regulated) ribonucleotide. Generally, gene promoter hypermethylation decreases gene expression. On the other hand, hypomethylation of a noncoding region has been linked to chromosome instability. One category of developmental epigenetics is "genomic imprinting," a genetic phenomenon by which certain genes are expressed in a parent-of-origin-specific manner via methylation of the unexpressed allele [92].

Epigenetic marks may be transmitted across generations, either directly by persisting through meiosis or indirectly through replication in the next generation of the conditions in which the epigenetic change occurred. In one commonly cited example of epigenetic inheritance, Hugh Morgan et al. [93] found related inherited variations in coat color, diabetes, and other abnormalities in an inbred line of mice (agouti) to variations in the methylation patterns of an inserted retrotransposon. In this instance, the variations are transmitted through female meiosis. In other cases, epigenetic changes are environmentally induced. More recently, long-term epigenetic and behavioral effects of pain exposure in preterm infants were associated with a methylation pattern in the serotonin transporter SLC6A4 [94] or GR [95] promoters.

A second epigenetic mechanism, posttranslational modifications of histone tails, includes acetylation, methylation, phosphorylation, and ubiquitination. These modifications variably alter chromatin structure and, thereby, DNA accessibility to transcription factors and RNA polymerases. Functional effects depend on the specific amino acid modified and on the specific covalently attached group, e.g., acetylation commonly results in the loosening of chromatin and increased gene transcription, whereas methylated histones tightly bind DNA, restrict access to various enzymes, and reduce transcription. An example of how variation in maternal diet during early pregnancy can cause persistent changes in DNA methylation in offspring comes from The Gambia, where Dominguez-Salas et al. [96] examined DNA methylation in blood leukocyte and hair follicle samples collected from children born either during the rainy (hungry) or the dry (harvest) season. Children born in the rainy season, and hence exposed to higher maternal methyl-donor nutrient intake around the time of conception, showed increased methylation of six metastable alleles [96].

MicroRNAs (miRNAs) are single-stranded RNAs of ~21–23 nucleotides transcribed from DNA but not translated into proteins (noncoding RNAs). Their functional role includes gene expression regulation mediated by control of messenger RNA (mRNA) stability or translation. Mature miRNAs can be totally complementary to the mRNA; miRNA:mRNA pairing leads to mRNA degradation. Other miRNAs are only partially complementary to an mRNA, and their regulatory function is mediated by blocking mRNA translation. A single miRNA may regulate expression of many target genes; conversely, one gene may be regulated by numerous miRNAs. MicroRNAs play a key role in diverse developmental epigenetic processes, including morphogenesis, cell proliferation, differentiation, and apoptosis. MicroRNA changes may be sensitive indicators of the effects of acute and chronic environmental exposures [97], perhaps opening an opportunity to describe miRNA signatures for specific physical and psychosocial environmental exposures.

At first, epigenetic programming studies focused on maternally transmitted or influenced mechanisms, but, more recently, paternal factors have been shown to have transgenerational epigenetic effects. Even in the absence of contact with offspring, fathers can transmit environmentally induced effects, apparently through inherited epigenetic variation within the patriline [98]. Exposure of mice to undernutrition during fetal development leads to changes in the DNA methylome of germ cells in male offspring, even when the males are nourished normally from weaning [99]. These phenotypic differences were transmitted through the paternal line to the F2 offspring [99].

HEALTH VULNERABILITIES CAN BE HERITABLE

In the last decade, heritable environmentally induced epigenetic modifications are being shown to underlie *reversible* transgenerational alterations in phenotype [100,101]. In rats, in utero exposure to hydrocarbons, plastics, and dioxin can lead to early onset puberty and spermatogenic cell apoptosis in three subsequent generations. An examination of the sperm promoter epigenome has identified differentially methylated DNA regions in descendants of exposed animals [102]. One recent study linked maternal experience of trauma and consequent posttraumatic stress disorder symptoms to disrupted behavioral and physiological response (i.e., cortisol) in the infants [103].

A transgenerational data set of African Americans indicates that women who experienced early life impoverishment and who achieved upward economic mobility showed a decreased incidence of preterm birth compared with women with life-long impoverishment [104]. However, this effect was not seen in women with upward economic mobility who themselves had been of low birth weight. Although numerous interpretations of these findings are plausible, a possible explanation is that transgenerational epigenetic effect was not overcome despite a change in SES [104]. These lines of investigation advance understanding for transgenerational epigenetic effects of disparate environmental exposures as well as possible explanations for the increased risk of adverse health outcomes in minority individuals who otherwise would be considered low risk.

CONCLUSIONS

Sophisticated and integrative theoretical frameworks variously integrate critical developmental periods, evolutionary developmental theory, epigenetics, family and community ecology, maternal allostatic load, resilience, and local neighborhood and community level influences [105,106]. These comprehensive approaches are also applicable to low- and middle-income countries [107]. It is essential that these theoretical frameworks based on correlations be tested empirically in order to refine strategies that promote resilience and "stress inoculation" [108]. Too little is known how the social buffering of the HPA axis (that occurs in supportive environments despite mild everyday stressors) promotes health. Furthermore, as knowledge of epigenetic processes grows, so should the capacity to develop early life interventions to prevent and mitigate child health disparities. As noted by Darlene Francis [20], understanding how genes are differentially regulated by experience will play a profound role in how we conceptualize health inequalities by informing our concepts of the somatization or embodiment of social inequalities. Rather than engaging in nature-versus-nurture debates concerning race as a genetic or social construct, race may be viewed as an epigenomic construct in which genotype and the socially experienced world are perpetually entwined [20].

By highlighting sociocultural and socioeconomic correlates of newborn and childhood risk, allostatic load, weathering, and acculturation models suggest policy and interventions grounded in longitudinal and contextual change [40,109,110]. An understanding of economic theory about inequalities similarly should inform analysis of maternal and childhood health inequities. Priority areas identified for maternal child health research in order to close these knowledge gaps include: epigenetic mechanisms and their potential mutability; periconception as a critical and sensitive period for environmental exposures; maternal health prior to pregnancy; the role of the placenta as an important regulator of the intrauterine environment; and ways to strengthen early mother–child interactions [97]. Identification of environmentally modifiable epigenetic marks, such as gene-specific and global DNA methylation, is proceeding at a fast pace. These endeavors should advance sorting out causal relationships between loci-specific programmed epigenetic alterations in response to adverse early environments and metabolic and neurodevelopmental disease phenotypes later in life.

References

[1] Gluckman P, Hanson M. Developmental origins of health and disease. Cambridge: Cambridge Univ. Press; 2006.

[2] Barboza Solís C, Kelly-Irving M, Fantin R, Darnaudéry M, Torrisani J, Lang T, et al. Adverse childhood experiences and physiological wear-and-tear in midlife: findings from the 1958 British birth cohort. Proc Natl Acad Sci USA 2015;112:E738–46.

[3] Widom CS, Horan J, Brzustowicz L. Childhood maltreatment predicts allostatic load in adulthood. Child Abuse Negl 2015:S0145–2134.

[4] Hales CN, Barker DJ, Clark PM, Cox LJ, Fall C, Osmond C, et al. Fetal and infant growth and impaired glucose tolerance at age 64. BMJ 1991;303:1019–22.

[5] Kundakovic M, Champagne FA. Early-life experience, epigenetics, and the developing brain. Neuropsychopharmacology 2015;40:141–53.

[6] Godfrey KM, Lillycrop KA, Burdge GC, Gluckman PD, Hanson MA. Epigenetic mechanisms and the mismatch concept of the developmental origins of health and disease. Pediatr Res 2007;61:5R–10R.
[7] Ben-Shlomo Y. Rising to the challenges and opportunities of life course epidemiology. Intl J Epidemiol 2007;36:481–3.
[8] Hales CN, Barker DJ. The thrifty phenotype hypothesis. Br Med Bull 2001;60:5–20.
[9] Widdowson EM, McCance RA. The effect of finite periods of undernutrition at different ages on the composition and subsequent development of the rat. Proc R Soc Lond Ser B 1963;158:329–42.
[10] Geronimus AT, Hicken M, Keene D, Bound J. "Weathering" and age patterns of allostatic load scores among blacks and whites in the United States. Am J Public Health 2006;96:826–33.
[11] Ruiz RJ, Pickler RH, Marti CN, Jallo N. Family cohesion, acculturation, maternal cortisol, and preterm birth in Mexican-American women. Int J Womens Health 2013;5:243–52.
[12] Wright RJ. Moving towards making social toxins mainstream in children's environmental health. Curr Opin Pediatr 2009;21:222–9.
[13] Brody GH, Lei MK, Chen E, Miller GE. Neighborhood poverty and allostatic load in African American youth. Pediatrics 2014;134:e1362–8.
[14] Gump BB, Stewart P, Reihman J, Lonky E, Darvill T, Parsons PJ, et al. Low-level prenatal and postnatal blood lead exposure and adrenocortical responses to acute stress in children. Environ Health Perspect 2008;116:249–55.
[15] Rissman EF, Adli M. Transgenerational epigenetic inheritance: focus on endocrine disrupting compounds. Endocrinology 2014;155:2770–80.
[16] Garbarino J. Raising children in a socially toxic environment. San Francisco: Jossey-Bass; 1995.
[17] Blumenshine P, Egerter S, Barclay CJ, Cubbin C, Braveman PA. Socioeconomic disparities in adverse birth outcomes: a systematic review. Am J Prev Med 2010;39:263–72.
[18] Bethell CD, Newacheck P, Hawes E, Halfon N. Adverse childhood experiences: assessing the impact on health and school engagement and the mitigating role of resilience. Health Aff 2014;33:2106–15.
[19] Braveman P, Barclay C. Health disparities beginning in childhood: a life-course perspective. Pediatrics 2009;124:S163–75.
[20] Francis DD. Conceptualizing child health disparities: a role for developmental neurogenomics. Pediatrics 2009;124:S196–202.
[21] Finch BK. Early origins of the gradient: the relationship between socioeconomic status and infant mortality in the United States. Demography 2003;40:675–99.
[22] Wilkinson RG, Pickett KE. The problems of relative deprivation: why some societies do better than others. Soc Sci Med 2007;65:1965–78.
[23] Adjaye-Gbewonyo K, Kawachi I. Use of the Yitzhaki Index as a test of relative deprivation for health outcomes: a review of recent literature. Soc Sci Med 2012;75:129–37.
[24] Gini C. On the measure of concentration with special reference to income and statistics, vol. 208. Colorado College Publication; 1936. p. 73–79.
[25] Smith CA. Effects of maternal malnutrition upon the newborn infant in Holland. J Pediatr 1944–1945;1947(30):229–43.
[26] Antonov AN. Children born during the siege of Leningrad in 1942. J Pediatr 1947;30:250–9.
[27] Veenendaal MV, Painter RC, de Rooij SR, Bossuyt PM, van der Post JA, Gluckman PD, et al. Transgenerational effects of prenatal exposure to the 1944–45 Dutch famine. BJOG 2013;120:548–53.
[28] Popham F. Deprivation is a relative concept? Absolutely! J Epidemiol Comm Health 2015;69:199–200.
[29] Margerison-Zilko C, Cubbin C, Jun J, Marchi K, Fingar K, Braveman P. Beyond the cross-sectional: neighborhood poverty histories and preterm birth. Am J Public Health 2015;105:1174–80.
[30] Korte SM, Koolhaas JM, Wingfield JC, McEwen BS. The Darwinian concept of stress: benefits of allostasis and costs of allostatic load and the trade-offs in health and disease. Neurosci Biobehav Rev 2005;29:3–38.
[31] McEwen BS. Brain on stress: how the social environment gets under the skin. Proc Natl Acad Sci USA 2012;109:17180–5.
[32] Tyrka AR, Burgers DE, Philip NS, Price LH, Carpenter LL. The neurobiological correlates of childhood adversity and implications for treatment. Acta Psychiatr Scand 2013;128:434–47.
[33] Philip NS, Valentine TR, Sweet LH, Tyrka AR, Price LH, Carpenter LL. Early life stress impacts dorsolateral prefrontal cortex functional connectivity in healthy adults: informing future studies of antidepressant treatments. J Psychiatr Res 2014;52:63–9.
[34] Scharf SH, Schmidt MV. Animal models of stress vulnerability and resilience in translational research. Curr Psychiatry Rep 2012;14:159–65.
[35] Seeman TE, Singer B, Rowe JW, Horwitz R, McEwen BS. Price of adaptation—allostatic load and its health consequences: MacArthur studies of successful aging. Arch Intern Med 1997;157:2259–68.
[36] Hux VJ, Catov JM, Roberts JM. Allostatic load in women with a history of low birth weight infants: the national health and nutrition examination survey. J Womens Health 2014;23:1039–45.
[37] Wallace ME, Harville EW. Allostatic load and birth outcomes among white and black women in New Orleans. Matern Child Health J 2013;17:1025–9.

[38] Morrison S, Shenassa ED, Mendola P, Wu T, Schoendorf K. Allostatic load may not be associated with chronic stress in pregnant women, NHANES 1999–2006. Ann Epidemiol 2013;23:294–7.
[39] Geronimus AT. The weathering hypothesis and the health of African-American women and infants: evidence and speculations. Ethn Dis 1992;2:207–21.
[40] Misra DP, Straughen JK, Slaughter-Acey JC. Allostatic load and health: can perinatal epidemiology lead the way forward? Paediatr Perinat Epidemiol 2013;27:507–8.
[41] Wise PH, Pursley DM. Infant mortality as a social mirror. New Engl J Med 1992;326:558.
[42] Lu MC, Helfon N. Racial and ethnic disparities in birth outcomes: a life-course perspective. Matern Child Health J 2003;7:13–30.
[43] Burchard EG, Ziv E, Coyle N, Gomez SL, Tang H, Karter AJ, et al. The importance of race and ethnic background in biomedical research and clinical practice. N Engl J Med 2003;348:1170–5.
[44] Morenoff JD. Neighborhood mechanisms and the spatial dynamics of birthweight. Am J Sociol 2003;108:976–1017.
[45] Buka SL, Brennan RT, Rich-Edwards JW, Raudenbush SW, Earls F. Neighborhood support and the birthweight of urban infants. Am J Epidemiol 2003; 157:1–8.
[46] Cerdá M, Buka SL, Rich-Edwards JW. Neighborhood influences on the association between maternal age and birthweight: a multilevel investigation of age-related disparities in health. Soc Sci Med 2008;66:2048–60.
[47] Deal SB, Bennett AC, Rankin KM, Collins JW. The relation of age to low birth weight rates among foreign-born black mothers: a population-based exploratory study. Ethn Dis 2014;24:413–7.
[48] Dennis JA, Mollborn S. Young maternal age and low birth weight risk: an exploration of racial/ethnic disparities in the birth outcomes of mothers in the United States. Soc Sci J 2013;50:625–34.
[49] Kramer MR, Hogue CR. What causes racial disparities in very preterm birth? A biosocial perspective. Epidemiol Rev 2009;31:84–98.
[50] Schoendorf KC, Hogue CJR, Kleinman JC, Rowley DR. Mortality among infants of black compared with white college-educated parents. N Engl J Med 1992;326:1522–6.
[51] Colen CG, Geronimus AT, Bound J, James SA. Maternal upward socioeconomic mobility and black-white disparities in infant birthweight. Am J Public Health 2006;96:2032–9.
[52] Acevedo-Garcia D, Sanchez-Vaznaugh EV, Viruell-Fuentes EA, Almeida J. Integrating social epidemiology into immigrant health research: a cross-national framework. Soc Sci Med 2012;75:2060–8.
[53] Passel JS. Demography of immigrant youth: past, present, and future. Future Child 2011;21:19–41.
[54] Markides KS, Coreil J. The health of Hispanics in the southwestern United States: an epidemiologic paradox. Public Health Rep 1986;101:253–65.
[55] Forbes D, Frisbie WP. Spanish surname and Anglo infant mortality: differentials over a half-century. Demography 1991;28:639–60.
[56] Gutmann MP, Haines MR, Frisbie WP, Blanchard KS. Intra-ethnic diversity in Hispanic child mortality, 1890–1910. Demography 2000;37:467–76.
[57] Williams RL, Binkin NJ, Clingman EJ. Pregnancy outcomes among Spanish-surname women in California. Am J Public Health 1986;76:387–91.
[58] Warner WL, Strole L. The social systems of American ethnic groups. New Haven, CT: Yale Univ. Press; 1945.
[59] Singh GK, Yu SM. Adverse pregnancy outcomes: differences between US and foreign-born women in major US racial and ethnic groups. Am J Public Health 1996;86:837–43.
[60] Kaestner R, Pearson JA, Keene D, Geronimus AT. Stress, allostatic load, and health of Mexican immigrants. Soc Sci Q 2009;90:1089–111.
[61] Albraido-Lanza AFD, Chao MT, Florez KR. Do healthy behaviors decline with greater acculturation? Implications for the Latino mortality paradox. Soc Sci Med 2005;61:1243–55.
[62] Himmelgreen DA, Perez-Escamilla R, Martinez D, Bretnall A, Eells B, Peng Y, et al. The longer you stay, the bigger you get: length of time and language use in the U.S. are associated with obesity in Puerto Rican women. Am J Phys Anthropol 2004;125:90–6.
[63] Crimmins EM, Kim JK, Alley DE, Karlamangla A, Seeman T. Hispanic paradox in biological risk profiles. Am J Public Health 2007;97:1305–10.
[64] Eschbach K, Stimpson J, Kuo Y-F, Goodwin JS. Mortality of foreign-born and US-born Hispanic adults at younger ages: a reexamination of recent patterns. Am J Public Health 2007;97:1297–304.
[65] Peek MK, Cutchin MP, Salinas JJ, Sheffield KM, Eschbach K, Stowe RP, et al. Allostatic load among non-Hispanic whites, non-Hispanic blacks, and people of Mexican origin: effects of ethnicity, nativity, and acculturation. Am J Public Health 2010;100:940–6.
[66] de Figueiredo JM. Explaining the 'immigration advantage' and the 'biculturalism paradox': an application of the theory of demoralization. Int J Soc Psychiatry 2014;60:175–7.
[67] Becerra JE, Hogue CJ, Atrash HK, Pérez N. Infant mortality among Hispanics. A portrait of heterogeneity. JAMA 1991;265:217–21.
[68] Hummer RA, Powers DA, Pullum SG, Gossman GL, Frisbie WP. Paradox found (again): infant mortality among the Mexican origin population in the United States. Demography 2007;44:441–57.

[69] Wojcicki JM, Schwartz N, Jiménez-Cruz A, Bacardi-Gascon M, Heyman MB. Acculturation, dietary practices and risk for childhood obesity in an ethnically heterogeneous population of Latino school children in the San Francisco bay area. J Immigr Minor Health 2012;14:533–9.
[70] Sangi-Haghpeykar H, Lam K, Raine SP. Gestational weight gain among Hispanic women. Matern Child Health J 2014;18:153–60.
[71] Hazuda HP, Haffner SM, Stern MP. Acculturation and assimilation among Mexican Americans: scales and population-based data. Soc Sci Q 1988;69:687–705.
[72] Franzini L, Ribble JC, Keddie AM. Understanding the Hispanic paradox. Ethn Dis 2001;11:496–518.
[73] Markides KS, Eschbach K. Aging, migration and mortality: current status of research on the Hispanic Paradox. J Gerontol B 2005;60B:68–75.
[74] Ayala GX, Carnethon M, Arredondo E, Delamater AM, Perreira K, Van Horn L. Theoretical foundations of the Study of Latino (SOL) Youth: implications for obesity and cardiometabolic risk. Ann Epidemiol 2014;24:36–43.
[75] Fox M, Entringer S, Buss C, DeHaene J, Wadhwa PD. Intergenerational transmission of the effects of acculturation on health in Hispanic Americans: a fetal programming perspective. Am J Public Health 2015;105:S409–23.
[76] Halfon N, Larson K, Lu M, Tullis E, Russ S. Lifecourse health development: past, present and future. Matern Child Health J 2014;18:344–65.
[77] Russ SA, Larson K, Tullis E, Halfon N. A lifecourse approach to health development: implications for the maternal and child health research agenda. Matern Child Health J 2014;18:497–510.
[78] Barker DJP. Fetal and infant origins of adult disease. London: British Medical Publishing Group; 1992.
[79] Barker DJP. Mothers, babies and health in later life. 2nd ed. Edinburgh: Churchill Livingstone; 1998.
[80] Barker DJ, Osmond C, Winter PD, Margetts B, Simmonds SJ. Weight in infancy and death from ischaemic heart disease. Lancet 1989;334:577–80.
[81] Hanson MA, Gluckman PD. Early developmental conditioning of later health and disease: physiology or pathophysiology. Physiol Rev 2014;94:1027–76.
[82] Willson AE, Shuey KM, Elder Jr GH. Cumulative advantage processes as mechanisms of inequality in life course health. Am J Sociol 2007;112:1886–924.
[83] McLaughlin KA, Sheridan MA, Winter W, Fox NA, Zeanah CH, Nelson CA. Widespread reductions in cortical thickness following severe early-life deprivation: a neurodevelopmental pathway to attention-deficit/hyperactivity disorder. Biol Psychiatry 2014;76:629–38.
[84] Keding TJ, Herringa RJ. Abnormal structure of fear circuitry in pediatric post-traumatic stress disorder. Neuropsychopharmacology 2015;40:537–45.
[85] Jablonka E, Lamb MJ, Avital E. 'Lamarckian'mechanisms in Darwinian evolution. Trends Ecol Evol 1998;13:206–10.
[86] Koestler A. The case of the midwife toad. NY: Vintage; 1973.
[87] Medvedev ZA. The rise and fall of T.D. Lysenko, Lerner M (translator). NY: Columbia Univ. Press; 1969.
[88] Waddington CH. Organizers and genes. Cambridge: Cambridge Univ. Press; 1940.
[89] Jablonka E, Lamb MJ. The inheritance of acquired epigenetic variations. Int J Epidemiol April 7, 2015. pii: dyv020. [Epub ahead of print].
[90] Holliday R, Pugh JE. DNA modification mechanisms and gene activity during development. Science 1975;187:226–32.
[91] Teh AL, Pan H, Chen L, Ong ML, Dogra S, Wong J, et al. The effect of genotype and in utero environment on interindividual variation in neonate DNA methylomes. Genome Res 2014;24:1064–74.
[92] Eggermann T, Leisten I, Binder G, Begemann M, Spengler S. Disturbed methylation at multiple imprinted loci: an increasing observation in imprinting disorders. Epigenomics 2011;3:625–37.
[93] Morgan HD, Sutherland HG, Martin DI, Whitelaw E. Epigenetic inheritance at the agouti locus in the mouse. Nat Genet 1999;23:314–8.
[94] Chau CM, Ranger M, Sulistyoningrum D, Devlin AM, Oberlander TF, Grunau RE. Neonatal pain and COMT Val158Met genotype in relation to serotonin transporter (SLC6A4) promoter methylation in very preterm children at school age. Front Behav Neurosci 2014;8:409.
[95] Kantake M, Yoshitake H, Ishikawa H, Araki Y, Shimizu T. Postnatal epigenetic modification of glucocorticoid receptor gene in preterm infants: a prospective cohort study. BMJ Open 2014;4:e005318.
[96] Dominguez-Salas P, Moore SE, Baker MS, Bergen AW, Cox SE, Dyer RA, et al. Maternal nutrition at conception modulates DNA methylation of human metastable epialleles. Nat Commun 2014;5:3746.
[97] Vrijens K, Bollati V, Nawrot TS. MicroRNAs as potential signatures of environmental exposure or effect: a systematic review. Environ Health Perspect 2015;123:399–411.
[98] Braun K, Champagne FA. Paternal influences on offspring development: behavioural and epigenetic pathways. J Neuroendocrinol 2014;26:697–706.
[99] Radford EJ, Ito M, Shi H, Corish JA, Yamazawa K, Isganaitis E, et al. In utero effects. In utero undernourishment perturbs the adult sperm methylome and intergenerational metabolism. Science 2014;345:1255903.
[100] Jirtle RL, Skinner MK. Environmental epigenomics and disease susceptibility. Nat Rev Genet 2007;8:253–62.
[101] Franklin TB, Linder N, Russig H, Thöny B, Mansuy IM. Influence of early stress on social abilities and serotonergic functions across generations in mice. PLoS One 2011;6:e21842.

[102] Manikkam M, Guerrero-Bosagna C, Tracey R, Haque MM, Skinner MK. Transgenerational actions of environmental compounds on reproductive disease and identification of epigenetic biomarkers of ancestral exposures. PLoS One 2012;7:e31901.

[103] Yehuda R, Bierer LM. Transgenerational transmission of cortisol and PTSD risk. Prog Brain Res 2008;167:121–34.

[104] Collins Jr JW, Rankin KM, David RJ. African American women's lifetime upward economic mobility and preterm birth: the effect of fetal programming. Am J Public Health 2011;101:714–9.

[105] Braveman PA, Egerter SA, Woolf SH, Marks JS. When do we know enough to recommend action on the social determinants of health? Am J Prev Med 2011;40:S58–66.

[106] Ellis BJ, Del Giudice M. Beyond allostatic load: rethinking the role of stress in regulating human development. Dev Psychopathol 2014;26:1–20.

[107] Premji S. Perinatal distress in women in low- and middle-income countries: allostatic load as a framework to examine the effect of perinatal distress on preterm birth and infant health. Matern Child Health J 2014;18:2393–407.

[108] Lyons DM, Parker KJ, Katz M, Schatzberg AF. Developmental cascades linking stress inoculation, arousal regulation, and resilience. Front Behav Neurosci 2009;3:32.

[109] Jutte DP, Miller JL, Erickson DJ. Neighborhood adversity, child health, and the role for community development. Pediatrics 2015;135:S48–57.

[110] Aizer A, Currie J. The intergenerational transmission of inequality: maternal disadvantage and health at birth. Science 2014;344:856–61.

CHAPTER

3

DOHaD and the Periconceptional Period, a Critical Window in Time

Congshan Sun, Miguel A. Velazquez, Tom P. Fleming

Centre for Biological Sciences, University of Southampton, Southampton General Hospital, Southampton, UK

OUTLINE

INTRODUCTION: DOHaD AND THE PERICONCEPTIONAL PERSPECTIVE

Eggs and early embryos are well known to interact with and respond to their environment. Within the mother, factors secreted by the oviduct and uterus act to support and regulate blastocyst morphogenesis during early pregnancy, particularly through provision of nutrients and signaling molecules that may modulate embryo proliferation, metabolic status, and cellular differentiation, all in preparation for implantation. These interactions, mediated through metabolites, hormones, growth factors, and cytokines controlling short-term developmental progression,

The Epigenome and Developmental Origins of Health and Disease
http://dx.doi.org/10.1016/B978-0-12-801383-0.00003-7

Copyright © 2016 Elsevier Inc. All rights reserved.

have been well reviewed [1]. However, there are also several environmental interactions that can have a lasting influence on oocyte and embryo potential long after implantation, throughout gestation and even during postnatal life. This chapter will be concerned with these more heritable interactions that may alter the offspring's homeostatic profile in a permanent way and ultimately contribute to health and disease risk into adulthood.

Intensive research in recent years has demonstrated what is loosely known as the **periconceptional** period (broadly covering oocyte maturation and early embryo development beyond implantation, but in many studies comprising either all or only part of this timeline) to be a sensitive window across species to diverse environmental conditions with long-term consequences. In vivo, maternal dietary composition, metabolism, and health can all induce periconceptional (PC) "programming," while in vitro conditions such as in vitro fertilization (IVF), embryo culture, cryopreservation, and other assisted conception treatments may do likewise. Indeed, both in vivo and in vitro PC environments in different species have been shown to change the growth trajectory and alter the risk of cardiovascular, metabolic, behavioral, and immunologic traits in adults. Similarly, recent work has shown that the paternal environment, including diet and metabolism, can influence development of the fetus and postnatal offspring.

In some cases, these interactions in mothers induce changes in oocyte and embryo potential in what appears to be a passive way, for example, female obesity and over-rich maternal diets may result in accumulation of lipids and metabolites within the follicle and oocyte, adversely perturbing metabolism and mitochondrial function and increasing cellular stress [2,3]. However, in other circumstances, a more proactive series of interactions may occur, suggesting that the embryo is using environmental conditions to "select" an optimal development program to best-fit prevailing conditions. Here, an example would be the consequences of a maternal protein-restricted diet, which leads to an alteration in the nutrient composition of the reproductive tract that can be sensed by early embryos through signaling mechanisms and used to evoke specific responses to target compensatory changes in the extraembryonic lineages [4–6]. Such responses appear designed to overcome nutrient deficiency through activation of mechanisms leading to increased maternal-conceptus nutrient delivery throughout gestation, thereby protecting fetal growth.

In what context can we best consider the legacy of the PC environment? We believe it represents a critical window within the wider framework of the "developmental origins of health and disease" (DOHaD) concept. This concept, relating reproductive environment with later disease in adulthood, was first proposed by Professor David Barker. He hypothesized that homeostatic changes occurring in early life, including the prenatal period, affect an individual's later life, health, and mortality. A very famous proof supporting the DOHaD hypothesis is the coherent study of the Dutch Hunger Winter, a period of 5 months of famine in the winter of 1944–1945 in Amsterdam. Surveys of children born to women who were pregnant during the famine revealed that the babies were born smaller and had an enhanced risk of coronary heart disease, glucose intolerance, disturbed blood coagulation, increased stress responsiveness, and obesity in later life [7]. Elsewhere, cohort epidemiologic studies using Helsinki and Hertfordshire datasets on recorded births between the 1930s and 1940s support the DOHaD hypothesis in showing poor fetal growth and low birth weight associated with increased risk of coronary artery disease, hypertension, and insulin resistance in adulthood in both female and male offspring [8,9].

The DOHaD hypothesis is based on the "developmental plasticity" concept, which proposes that an organism undergoing prenatal development or early life is able to change its phenotype to adapt to the environment, and this ability will be gradually lost as it grows. Therefore, stimuli during an organism's early life may lead to persistent change. Low birth weight may be explained by reduction in fetal growth as a strategy to adapt to an unfavorable prenatal environment [9]. The strategy to adapt to maternal environment changes the organism permanently; such plasticity will be an aid to survival and competitive fitness during the life course, assuming the unfavorable prenatal nutrient availability persists into adulthood. However, if nutritional conditions change, the metabolic phenotype may become unsuitable for the actual rather than "predicted" postnatal environment, a mismatch that may lead to disease in adult life [10].

Why should the PC period be particularly vulnerable to adverse developmental programming? This likely reflects a combination of circumstances: (1) the early embryo only consists of a few cells, around 50 in a blastocyst; consequently, this finite pool will be vulnerable owing to its limited supply and essential future contribution to all the tissues of the body; (2) the early embryo's major role to generate and segregate the future placental cell lineages from the fetal progenitor makes it vulnerable because environmental information is likely required as a "blueprint" to define the structural and functional characteristics of these extraembryonic tissues before construction begins; and (3) the early embryo is a period of significant epigenetic reorganization involving DNA methylation and histone modifications, the major routes to regulating a heritable pattern of gene expression [11].

In this chapter, we will consider specific in vivo and in vitro environments shown to be influential in DOHaD, with emphasis on the PC period. But first, we provide a brief summary of the PC period and what key events of reproduction and development occur during this critical stage of our life cycle, with key references to support further reading.

THE PERICONCEPTIONAL PERIOD: A SUMMARY OF KEY EVENTS

The maturation of male and female gametes within testis and ovary reflects a combination of genetic, cell biological, and endocrinal regulation, including meiotic reduction of chromosome number to haploid. These complex differentiation processes are managed by distinct mechanisms between the sexes. Spermatogenesis and the production of spermatozoa is mainly dependent on the dual control of pituitary follicle-stimulating hormone and testis-derived androgens within the microenvironment provided by the Sertoli cells lining the seminiferous tubules and the interstitial Leydig cells (reviewed in [12]). During oogenesis, early meiotic arrest allows for subtle interaction between follicular cells and the oocyte and between the pituitary and the follicle to regulate growth and differentiation [13]. Oocyte maturation and ovulation, coordinated by the luteinizing hormone surge from the pituitary and the reactivation of meiosis [14], lead to the oocyte reaching competence for fertilization.

Fertilization restores a diploid biallelic genome essential for successful development and can be viewed as an intercellular signaling event. This signal allows the egg to initiate cell cycling and the program of cleavage to begin, leading to blastocyst formation and the segregation of the embryonic and extra-embryonic cell lineages before implantation (Figure 1). The sperm contributes phospholipase C-zeta (PLCζ) to the egg cytoplasm at the time of fertilization, which activates the phosphoinositide signaling pathway leading to regulated Ca^{2+} oscillations [15]. Following the oscillations, exocytosis of hydrolytic cortical granules changes the chemistry of the zona pellucida to inhibit other sperm from reaching the egg surface, thereby blocking polyspermy. The Ca^{2+} oscillations also allow meiosis to complete with extrusion of the second polar body, the formation of male and female pronuclei, and

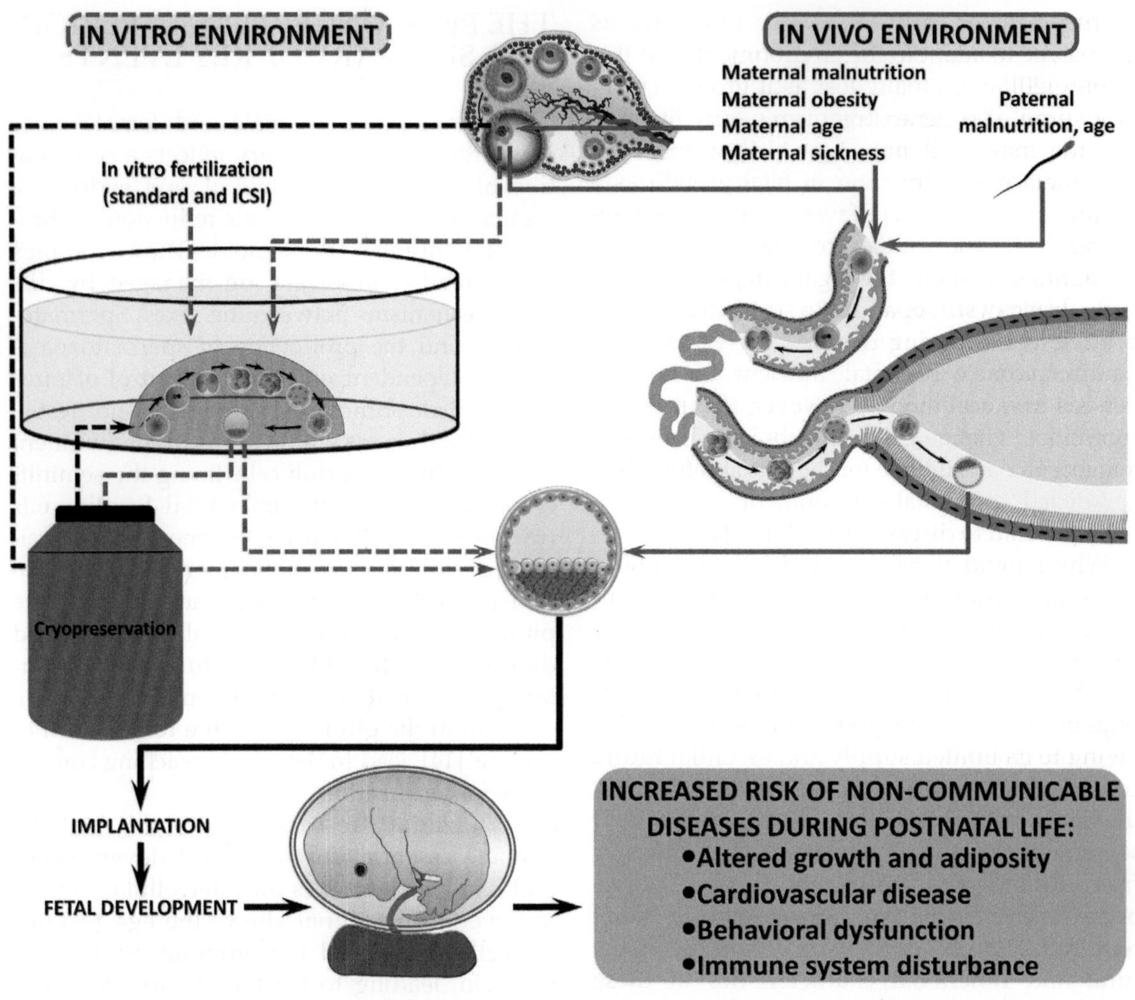

FIGURE 1 Schematic of the periconceptional period based on the mouse, from oocyte maturation through fertilization and cleavage to blastocyst formation and segregation of embryonic and extraembryonic cell lineages and their derivatives in later gestation.

subsequently the activation of the embryonic genome and resumption of cell cycling [15].

Cleavage in mammals is characteristically asynchronous with long cell cycles lasting some 12–24h and occurring in the absence of growth, resulting in smaller blastomeres at each division cycle. Signaling between blastomeres coordinates morphogenesis of the blastocyst by the 32-cell stage in mice and the formation of an outer trophectoderm epithelium (TE), which, through epithelial Na-K-ATPase-driven transport processes and construction of tight junctions, generates the blastocelic cavity [16,17]. TE differentiation begins at the 8-cell stage in mice when compaction occurs, a process by which E-cadherin-mediated intercellular adhesion initiates and blastomeres polarize along an apicobasal (outside–inside) orientation [18,19]. Compaction is mediated through several interacting signaling mechanisms, in particular involving PKC isoforms [17]. Cell polarity is generated through the apical localization of the Par3/Par6/aPKC complex [20].

Cleavage also generates an inner cluster of cells, the inner cell mass (ICM), eccentrically placed adjacent to the TE and blastocele. ICM cells are formed through differentiative division of outer polarized 8- and 16-cell blastomeres [16]. The ICM, during maturation and expansion of the blastocyst, separates into epiblast and primitive endoderm (PE) cell lineages, the latter adjacent to the blastocele. Collectively, these early cell lineages comprise the progenitors of all embryonic (epiblast) and extraembryonic (TE, giving rise to chorioallantoic placenta; PE, giving rise to parietal and visceral endoderm and yolk sac placenta) tissues of the conceptus (Figure 1). The segregation of early cell lineages is regulated both by cell-contact signaling and the expression of maternal and embryonic transcription factors that act in mutually antagonistic ways; these comprise the Caudal-related homeobox factor Cdx2, promoting TE differentiation on the outside of the embryo, and the pluripotency factors Oct4, Sox2, and Nanog, promoting ICM formation on the inside of the embryo [16,21]. The late blastocyst hatches from the zona pellucida some 4–5 days after fertilization, dependent upon species, followed by interaction between the TE and the uterine endometrium to initiate implantation.

IN VIVO MATERNAL NUTRITIONAL MODELS OF PERICONCEPTIONAL PROGRAMMING

Maternal Protein Deficiency in Rodents

Poor maternal nutrition has been considered a major factor leading to offspring metabolic syndrome. To simulate maternal undernutrition, a rodent low-protein diet (LPD) model has been developed using 9% casein, in contrast to control normal protein diet (NPD; 18% casein). In a pioneer study, rats born from LPD mothers demonstrated low birth weight and high blood pressure [22], mimicking the human metabolic syndrome. In subsequent mechanistic analyses of offspring hypertension, changes in the hypothalamic-pituitary-adrenal (HPA) axis and the renin-angiotensin system mediated through maternal glucocorticoid levels have been identified [23], together with altered kidney development comprising reduced glomerular number and reduced pancreatic islet size and beta-cell mass, all collectively contributing to the adult phenotype [24].

The importance of the PC period using the LPD model has been demonstrated in several studies using both mice and rats. Here, LPD was restricted to only the preimplantation period (Emb-LPD; up to E4.5 in rats and E3.5 in mice), resulting in compromised development of the embryo. This protocol revealed that maternal protein restriction limited to this short developmental window in rats was associated with hypertension in adult offspring [25]. Studies conducted on mouse offspring showed that maternal LPD throughout gestation and the shorter-duration Emb-LPD caused high blood pressure and anxiety-related behavior [4] even up to aging 1-year-old adults [26]. Further studies of adult cardiovascular disease and hypertension induced by LPD and Emb-LPD showed male mice had attenuated vasodilatation to isoprenaline and LPD female offspring showed elevated serum angiotensin-converting enzyme (ACE) activity, while Emb-LPD males had elevated lung ACE activity, all of which could be contributory to the hypertensive phenotype [27]. As anticipated for the metabolic syndrome–related phenotype, aging Emb-LPD female offspring show increased adiposity over control offspring, with over 20% increase in body fat recorded [26].

The relationship between PC Emb-LPD maternal diet and induction of adverse programming at the onset of pregnancy has been investigated. The serum and uterine fluid (UF) from Emb-LPD dams at the time of blastocyst formation show the amino acid profile to be modified with serum depletion of histidine, isoleucine, leucine, lysine, methionine, phenylalanine, tryptophan, tyrosine, and valine and UF depletion of the branched-chain amino acids (BCAAs) isoleucine, leucine, and valine. Insulin, as a growth activation hormone, was also studied and found to be reduced in Emb-LPD maternal serum [5]. Similar maternal serum reductions in insulin and BCAAs were also shown in the earlier rat model of Emb-LPD [25]. Our evidence suggests that blastocyst trophectoderm is able to sense the metabolite composition of bathing fluids of the external environment via the mTOR pathway [5]; this pathway regulates protein translation and cellular growth and is dependent upon extracellular signaling through insulin and BCAAs [28]. Thus, maternal Emb-LPD caused a reduction in mTOR signaling in blastocysts with decreased phosphorylated S6 mTOR effector compared to NPD but not the total amount of S6 or any effects on the other major mTOR effector, 4E-BP1[5]. Phosphorylated S6 (40S ribosomal protein) regulates the translation of terminal oligopyrimidine–dependent mRNAs that commonly encode translation factors including ribosomal proteins [29]. Blastocysts exhibiting this reduced mTOR signaling in response to Emb-LPD would therefore be expected to reduce protein translation and growth as a consequence. mTOR signaling has also been implicated in other models of dietary-induced developmental programming [30].

The sensing and induction of developmental programming in mouse blastocysts via mTOR signaling is rapidly followed by compensatory responses by extraembryonic lineages to protect future fetal growth. The outer TE, progenitor of the chorioallantoic placenta, exhibits increased proliferation and subsequent outgrowth expansion in response to Emb-LPD, suggesting a more invasive implantation as a compensatory mechanism [5]. In addition, the endocytosis system of TE cells was functionally stimulated by maternal Emb-LPD treatment, reflecting upregulated megalin receptor expression and increased ligand uptake and lysosome labeling [6]. This mechanism, mediated through RhoA GTPase activation and actin reorganization, increases the ability of TE cells to absorb nutrients and, from in vitro analyses, is directly caused by an environment depleted in BCAAs [6]. Moreover, the stimulation in nutrient uptake mediated by endocytosis upregulation is also observed in the primitive endoderm extraembryonic lineage and maintained in the derivative visceral yolk sac endoderm through late gestation [4,6]. Maternal LPD compensation also leads to stimulated placental transport in later gestation [31]. While these compensatory responses stimulate later fetal growth of Emb-LPD offspring despite poor maternal nutrition, they also associate with later adult-onset disease, since perinatal weight positively correlates with the disease-related conditions of hypertension, abnormal behavior, and excess body weight [4].

Maternal Undernutrition in Sheep

The PC undernutrition ovine model comprises reduced maintenance feed normally by 50% over the period of pre- and early postmating; however, the specific period, duration, and severity of undernutrition varies between studies. The physiological consequence of this model is similar to the rodent LPD model, with altered adult offspring blood pressure and cardiovascular function [32,33] and abnormal behavior [34]. These outcomes may in part be exacerbated by reduced fetal myogenesis, specifically decreased muscle fiber number, as a consequence of PC maternal undernutrition [35]. PC programming in this model has been proposed to derive from an acceleration in the maturation of the fetal HPA axis in late gestation with the precocious rise in fetal cortisol and prostaglandin inducing preterm delivery [36]. The advance in fetal HPA activity may be caused by increased growth of the fetal adrenal with attendant increase in steroid production [37]. This may be supplemented by changes in placental transport characteristics such that exposure to maternal glucocorticoids is increased [36].

The PC ovine undernutrition model, like the rodent LPD model, is characterized by early compensatory changes that may be contributory to later programming abnormalities. The ovine blastocyst responds to the dietary challenge by increasing TE cell number, suggesting a compensatory mechanism to increase supply of nutrient to the embryo [38]. Indeed, PC undernutrition causes increased vascular density in the ovine placentomes, activated by MAPK/ERK1/2 and/or PI3K/Akt signaling pathways [39]. In later gestation, further evidence of compensatory responses in the ovine undernutrition model include increased uterine blood flow [40].

Maternal Overnutrition and Obesity Models

Another form of malnutrition derives from maternal high fat and carbohydrate intake and resulting obesity, well recognized as increasing globally as a major health concern. Obesity and overweight are detrimental to maternal reproductive function, causing anovulation and delayed spontaneous conception [41] and reduced success in assisted reproductive treatment (ART), with lower clinical pregnancy and higher miscarriage rates [42]. Increased body mass index (BMI) associates with elevated triglycerides, free fatty acids, insulin, glucose, and lactate in serum and/or ovarian follicular fluid, which reduces oocyte potential [2]. Maternal obesity may induce excess maternal insulin and IGF-1 in blood and UF, which may act as negative factors for embryo quality [43,44].

We and others have used rodent maternal diet obesity models and shown harmful effects on oocyte quality and early embryo development. These include increased accumulation of cytoplasmic lipids, meiotic spindle abnormalities and aneuploidy, mitochondrial dysfunction, endoplasmic reticulum and oxidative stress, and increased apoptosis leading to slow embryo cleavage and increased degradation [3,43,45,46].

Maternal obesity and overweight not only affect fertility, but also lead to poor offspring health. Children of obese mothers tend to become obese themselves and develop hyperinsulinemia and glucose intolerance, leading to increased noncommunicable disease risk in later life [47]. Mouse models substantiate this: maternal high-fat (HF) diet leads to heavier offspring with increased adiposity, fatty liver disease, and metabolic and cardiovascular dysfunction indicated by insulin resistance, hyperleptinemia, hyperuricemia, and hepatic steatosis [48–50]. Significantly, embryo transfer and developmental studies indicate the adult offspring disease phenotype derived from maternal obesity and diabetes can be traced back to poor fetal growth and ultimately the PC period with reduced-quality oocytes/embryos [45,51].

IN VITRO MODELS OF PERICONCEPTIONAL PROGRAMMING RELATED TO ASSISTED REPRODUCTIVE TREATMENTS

The relatively low success of ART in the conventional measure of "take home baby" rate has remained relatively flat at ~28% of ART cycles annually, with male and female gamete quality, treatment limitations, and patient medical condition all contributory. Critically, the low birth rate does not factor the lifelong health of children born. Is there a legacy of ART for long-term health within the DOHaD context? ART children have a small increased risk of adverse perinatal outcomes (low birth weight, prematurity, birth defects) compared with spontaneously conceived peers [52] and exhibit higher rates of normally rare imprinting disorders, including Beckwith–Wiedemann and Angelman syndromes [53], and potentially an enhanced chance of congenital birth defects, but this is not consistent across studies [54]. A direct relationship has been identified between commercial ART culture medium selection and baby birth weight [55], although this is disputed [56]. It has been reported that ART children have, beyond childhood, elevated blood pressure, altered pubertal maturation, altered body fat distribution, impaired glucose tolerance, and vascular dysfunction [57–60].

Reported detrimental long-term effects of ART should not be overestimated, since generally they represent a small increased risk with no health concern in the vast majority of children born. Also, they relate to an infertile subpopulation, potentially older than naturally conceiving individuals, and may derive from diverse ART treatments that may be improving over time. The context should be to identify specific factors of concern so that improved conditions can be developed, using animal models where appropriate. In vitro culture conditions, the use of cryopreservation techniques, and increased maternal age are all ART-associated factors considered below in the context of DOHaD.

Culture Medium and Duration

Preimplantation embryo culture in animal models (rodents; domestic mammals) have indicated increased risk of epigenetic disorders with both fetal and placental tissues affected, leading to increased disease state preserved into adulthood [61–65]. For example, our own studies in mice reveal sustained hypertension throughout adulthood coupled with cardiovascular and metabolic dysfunction induced by embryo culture [66]. Mouse models have further demonstrated that suboptimal embryo culture may induce reduced fertility and glucose intolerance in male offspring [67]. The duration of human embryo culture in ART is also important in long-term outcomes. Direct comparison between cleavage stage (day 3) and blastocyst stage (day 5) singleton transfers have shown the latter to increase the risk of monozygotic twinning, fetal malformations, preterm birth, and larger-for-gestational-age births [68]. Larger birth weight in response to longer embryo culture period has also been identified in a Finnish study [69]. Further analysis of the effect of in vitro culture and its duration are required to fully evaluate long-term health risks.

Cryopreservation

The use of frozen and thawed embryos has increased since it was first introduced to ART such that it now can commonly exceed the number of fresh embryos used. Embryo cryopreservation has the logistical advantages of minimizing surgical intervention and lowering the risk of ovarian hyperstimulation syndrome; therefore it promotes maternal safety in ART. It also enhances fertility preservation for patients with cancer and other conditions. Vitrification is gradually becoming the main cryopreservation method, replacing slow freezing; it is particularly beneficial in oocyte and blastocyst freezing by reducing cellular damage [70] and has been shown to increase pregnancy rate over slow freezing in ART [71].

Cryopreservation, including vitrification, in animal models is known to decrease embryo survival and developmental viability in the short term compared with controls [72]. It is recognized that more studies on effects of cryopreservation on embryo metabolism and potential are required [73]. However, in broad terms, pregnancy rates with frozen–thawed embryos are comparable to those from fresh embryos [74], and indeed there is evidence that IVF obstetric and perinatal outcomes may be improved by using frozen versus fresh embryo transfer, possibly attributable to improved embryo–endometrium synchrony [75]. Moreover, oocyte vitrification does not appear to diminish rates of fertilization, embryo cleavage, or ongoing pregnancy [73]. Despite this, there are conflicting data on the effect of cryopreservation on human oocyte transcriptomics, with one study showing no effect [76], while another reports significant and distinct changes following slow-freezing and vitrification [77]. Slow-freezing of mouse blastocysts caused a significant change in the transcriptome but which was not caused by rapid vitrification [78]. Lastly, comparison of gene expression of fresh or frozen–thawed human embryos also showed differences, but variation within groups was also apparent [79].

While the above data available for cryopreservation are broadly reassuring, children born following transfer of vitrified embryos have been reported to have a higher birth weight and increased risk of large for gestational age births due at least partly to the cryopreservation technique rather than maternal factors [80]. Moreover, the cryopreservation technique appears to influence childhood (up to 11 years) height, plasma IGF and binding protein levels, and the lipid profile [81]. These data need to be extended but also implicate the need for rigorous controlled animal studies to investigate effects of cryopreservation on offspring health. Reassuringly, mouse oocyte vitrification has recently been reported to have no effect on adult offspring motor activity and behavioral characterization [82]. Although the above data available for cryopreservation are encouraging, there is no detailed analysis of long-term health of offspring born; such analysis should be addressed and supported through animal models.

Maternal Age

Maternal aging has been well established as a negative factor in fertility and can lead to an increased rate of subsequent spontaneous abortion and stillbirth, as well as increased perinatal death, preterm birth, and low birth weight [83]. These negative factors may also associate with suboptimal maternal reproductive endocrinology, defects in the reproductive tract, and increased rates of maternal hypertension and pregnancy complications [83]. In terms of reduced fertility, this has been shown to associate with reduced follicle number; a reduced quality and potential of the oocyte, especially an increase in aneuploidy; compromised signaling with cumulus cells; and mitochondrial dysfunction [83–85]. Oocyte effects may then lead to adverse decidual and placental development, likely contributed to by repeated sex steroid exposure during aging [83,85]. However, while a growing insight into the detrimental effects of maternal aging on ovarian reserve, uterine functionality, and perinatal complications has been forthcoming, we have less and rather conflicting understanding of longer-term consequences on offspring health into adulthood. A meta-analysis has shown an association between increasing maternal age and childhood autism [86]. Encouragingly, although older mothers have higher pregnancy-related morbidities, low-birth-weight babies from older mothers have higher survival chances [87]. However, children from older mothers tend to be at higher risk of being overweight, although this associates with increased maternal weight at older age [88]. Thus, from this limited literature, potential adverse effects on offspring health mediated through advancing maternal age may derive from PC effects associated with poorer oocyte competence and fertility or increased maternal BMI with age. Further animal studies are required to probe in more detail the relationship between maternal age and offspring health.

THE PATERNAL INFLUENCE ON PERICONCEPTIONAL DEVELOPMENTAL PROGRAMMING

The importance of paternal condition in reproductive programming of long-term health has been clearly demonstrated. High paternal BMI associates with reduced embryo development, pregnancy rate, and live birth outcomes in human ART [89]. Moreover, epidemiologic data suggest that a range of paternal conditions including obesity promote offspring obesity and metabolic dysfunction, potentially mediated through epigenetic changes such as hypomethylation at the imprinted IGF2 gene [90]. Again, animal models have been critical in identifying the relationship between paternal condition and offspring health. Thus, in mice, in vitro culture-produced males exhibit altered testis gene expression and reduced sperm production, motility, and fertility, and transmit glucose intolerance and altered organ sizing to F1 and F2 generations only through the male germline [67]. Mouse studies have further shown that paternal obesity induced by high-fat diet (HFD) associates with reduced sperm quality and increased DNA damage, leading to reduced fertility and embryo potential [91], outcomes that can be reversed by improved diet and exercise [92]. Moreover, first- and second-generation offspring display increased adiposity with reduced gamete quality and fertility [93]. Again, these transgenerational effects appear epigenetic in nature, with founders showing changes in testes gene expression, microRNA content, and germ cell DNA global methylation levels [94]. Similar paternal HFD programming of offspring adiposity, insulin sensitivity, and glucose intolerance with altered pancreatic islet gene expression and DNA methylation has been demonstrated in rats [95]. These paternal effects may also be mediated through the seminal plasma, which may modulate the trophic role of the maternal tract, affecting offspring growth and health [96]. Related outcomes have been reported for paternal dietary protein restriction [97].

In addition to paternal metabolism, paternal age has been shown to associate with adverse programming. There is good evidence that sperm fertility declines with paternal age as testicular histology deteriorates, leading to reduced sperm production, reduced motility, and increased sperm DNA damage, which leads to pregnancy complications including risk of stillbirth, prematurity, and rare autosomal pathologies [98,99]. In addition, paternal aging, often coupled with other age-related conditions such as obesity, results in altered gene expression patterns in offspring associated with neurocognitive and metabolic disorders [99,100]. The extent to which heritable effects from paternal aging result from epigenetic mechanisms has not been studied in detail; however, global DNA methylation changes in newborns have been shown to relate to paternal age [100]. Moreover, an epigenetic origin of paternal aging-related schizophrenia has been postulated [100].

CONCLUSIONS

The periconceptional environment is clearly shown to be one of vulnerability in the potential induction of developmental programming affecting long-term growth, physiology, and disease risk of offspring (Figure 2). This is demonstrated in human and animal models affecting nutritional environment in vivo as well as the in vitro environment and conditions associated with ART. Collectively, such studies indicate the relevance of animal studies, with similar outcomes being identified in the human. Most studies illustrate the sensitivity of the PC window with respect to postnatal and adult phenotype, but only gradually are underlying mechanisms being probed and identified. Given the critical biomedical and health consequences of this field, more well-controlled animal studies are called for to expand our understanding of mechanisms covering epigenetic, signaling, and metabolic criteria, and from this to devise preventative strategies. Lastly, our understanding of the type of environmental conditions inducing PC developmental programming may be limited. For example, our recent work has shown that maternal sickness and inflammatory response in vivo following bacterial endotoxin intraperitoneal injection during the PC period can also program the developing embryo [101]. Here, rather than affecting the cardiometabolic phenotype of offspring, the maternal sickness environment programs suppression of the offspring innate immune system [101]. The breadth of conditions inducing PC programming

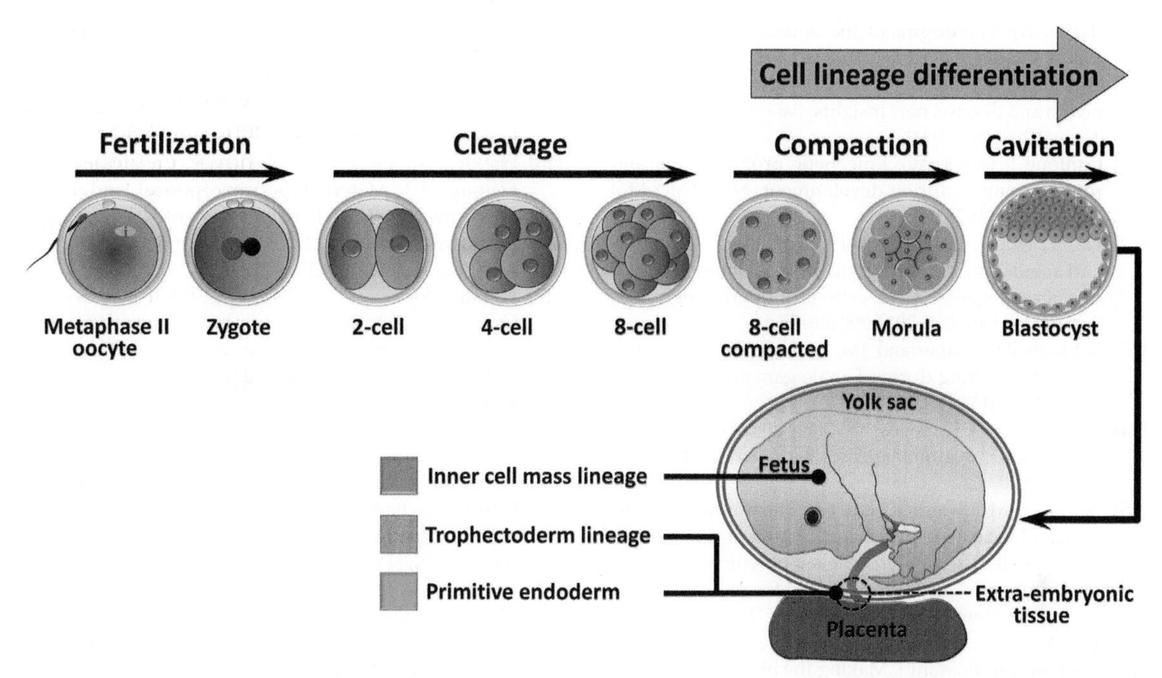

FIGURE 2 Schematic of a range of in vivo and in vitro environmental conditions that can affect the periconceptional stages of development with lasting consequences for adult health and disease risk.

and the range of bodily systems affected and responding to such conditions may be wider than we had ever anticipated and further argues for more intensive research.

Acknowledgments

We are grateful for research funding from BBSRC (BB/I001840/1; BB/F007450/1) and EU-FP7 EpiHealth program to TPF; CS was in receipt of a University of Southampton postgraduate scholarship bursary.

References

[1] Guzeloglu-Kayisli O, Kayisli UA, Taylor HS. The role of growth factors and cytokines during implantation: endocrine and paracrine interactions. Semin Reprod Med 2009;27(1):62–79.

[2] Robker RL, Akison LK, Bennett BD, Thrupp PN, Chura LR, Russell DL, et al. Obese women exhibit differences in ovarian metabolites, hormones, and gene expression compared with moderate-weight women. J Clin Endocrinol Metab 2009;94(5):1533–40.

[3] Igosheva N, Abramov AY, Poston L, Eckert JJ, Fleming TP, Duchen MR, et al. Maternal diet-induced obesity alters mitochondrial activity and redox status in mouse oocytes and zygotes. PLoS One 2010;5(4):e10074.

[4] Watkins AJ, Ursell E, Panton R, Papenbrock T, Hollis L, Cunningham C, et al. Adaptive responses by mouse early embryos to maternal diet protect fetal growth but predispose to adult onset disease. Biol Reprod 2008;78(2):299–306.

[5] Eckert JJ, Porter R, Watkins AJ, Burt E, Brooks S, Leese HJ, et al. Metabolic induction and early responses of mouse blastocyst developmental programming following maternal low protein diet affecting life-long health. PLoS One 2012;7(12):e52791.

[6] Sun C, Velazquez MA, Marfy-Smith S, Sheth B, Cox A, Johnston DA, et al. Mouse early extra-embryonic lineages activate compensatory endocytosis in response to poor maternal nutrition. Development 2014;141(5):1140–50.

[7] Roseboom T, de Rooij S, Painter R. The Dutch famine and its long-term consequences for adult health. Early Hum Dev 2006;82(8):485–91.

[8] Barker DJ, Osmond C, Kajantie E, Eriksson JG. Growth and chronic disease: findings in the Helsinki Birth Cohort. Ann Hum Biol 2009;36(5):445–58.

[9] Barker DJ. The origins of the developmental origins theory. J Intern Med 2007;261(5):412–7.
[10] Hanson MA, Gluckman PD. Developmental origins of health and disease: new insights. Basic Clin Pharmacol Toxicol 2008;102(2):90–3.
[11] Cantone I, Fisher AG. Epigenetic programming and reprogramming during development. Nat Struct Mol Biol 2013;20(3):282–9.
[12] Hai Y, Hou J, Liu Y, Yang H, Li Z, He Z. The roles and regulation of Sertoli cells in fate determinations of spermatogonial stem cells and spermatogenesis. Semin Cell Dev Biol 2014;29C:66–75.
[13] Sobinoff AP, Sutherland JM, McLaughlin EA. Intracellular signalling during female gametogenesis. Mol Hum Reprod 2013;19(5):265–78.
[14] Holt JE, Lane SI, Jones KT. The control of meiotic maturation in mammalian oocytes. Curr Top Dev Biol 2013;102:207–26.
[15] Nomikos M, Swann K, Lai FA. Starting a new life: sperm PLC-zeta mobilizes the Ca^{2+} signal that induces egg activation and embryo development: an essential phospholipase C with implications for male infertility. BioEssays News Rev Mol Cell Dev Biol 2012;34(2):126–34.
[16] Cockburn K, Rossant J. Making the blastocyst: lessons from the mouse. J Clin Invest 2010;120(4):995–1003.
[17] Eckert JJ, Fleming TP. Tight junction biogenesis during early development. Biochim Biophys Acta 2008;1778(3):717–28.
[18] Fleming TP, Sheth B, Fesenko I. Cell adhesion in the preimplantation mammalian embryo and its role in trophectoderm differentiation and blastocyst morphogenesis. Front Biosci 2001;6:D1000–7.
[19] Fierro-Gonzalez JC, White MD, Silva JC, Plachta N. Cadherin-dependent filopodia control preimplantation embryo compaction. Nat Cell Biol 2013;15(12):1424–33.
[20] Liu H, Wu Z, Shi X, Li W, Liu C, Wang D, et al. Atypical PKC, regulated by Rho GTPases and Mek/Erk, phosphorylates Ezrin during eight-cell embryocompaction. Dev Biol 2013;375(1):13–22.
[21] Saiz N, Plusa B. Early cell fate decisions in the mouse embryo. Reproduction 2013;145(3):R65–80.
[22] Langley-Evans SC. Maternal carbenoxolone treatment lowers birthweight and induces hypertension in the offspring of rats fed a protein-replete diet. Clin Sci (Lond) 1997;93(5):423–9.
[23] McMullen S, Langley-Evans SC. Sex-specific effects of prenatal low-protein and carbenoxolone exposure on renal angiotensin receptor expression in rats. Hypertension 2005;46(6):1374–80.
[24] Langley-Evans SC. Fetal programming of CVD and renal disease: animal models and mechanistic considerations. Proc Nutr Soc 2013;72(3):317–25.
[25] Kwong WY, Wild AE, Roberts P, Willis AC, Fleming TP. Maternal undernutrition during the preimplantation period of rat development causes blastocyst abnormalities and programming of postnatal hypertension. Development 2000;127(19):4195–202.
[26] Watkins AJ, Lucas ES, Wilkins A, Cagampang FR, Fleming TP. Maternal periconceptional and gestational low protein diet affects mouse offspring growth, cardiovascular and adipose phenotype at 1 year of age. PLoS One 2011;6(12):e28745.
[27] Watkins AJ, Lucas ES, Torrens C, Cleal JK, Green L, Osmond C, et al. Maternal low-protein diet during mouse pre-implantation development induces vascular dysfunction and altered renin-angiotensin-system homeostasis in the offspring. Br J Nutr 2010;103(12):1762–70.
[28] Wang X, Proud CG. Nutrient control of TORC1, a cell-cycle regulator. Trends Cell Biol 2009;19(6):260–7.
[29] Kim E. Mechanisms of amino acid sensing in mTOR signaling pathway. Nutr Res Pract 2009;3(1):64–71.
[30] Jansson N, Pettersson J, Haafiz A, Ericsson A, Palmberg I, Tranberg M, et al. Down-regulation of placental transport of amino acids precedes the development of intrauterine growth restriction in rats fed a low protein diet. J Physiol 2006;576(Pt 3):935–46.
[31] Coan PM, Vaughan OR, McCarthy J, Mactier C, Burton GJ, Constancia M, et al. Dietary composition programmes placental phenotype in mice. J Physiol 2011;589(Pt 14):3659–70.
[32] Torrens C, Snelling TH, Chau R, Shanmuganathan M, Cleal JK, Poore KR, et al. Effects of pre- and periconceptional undernutrition on arterial function in adult female sheep are vascular bed dependent. Exp Physiol 2009;94(9):1024–33.
[33] Gardner DS, Pearce S, Dandrea J, Walker R, Ramsay MM, Stephenson T, et al. Peri-implantation undernutrition programs blunted angiotensin II evoked baroreflex responses in young adult sheep. Hypertension 2004;43(6):1290–6.
[34] Hernandez CE, Matthews LR, Oliver MH, Bloomfield FH, Harding JE. Effects of sex, litter size and periconceptional ewe nutrition on offspring behavioural and physiological response to isolation. Physiol Behav 2010;101(5):588–94.
[35] Quigley SP, Kleemann DO, Kakar MA, Owens JA, Nattrass GS, Maddocks S, et al. Myogenesis in sheep is altered by maternal feed intake during the periconception period. Anim Reprod Sci 2005;87(3–4):241–51.
[36] Bloomfield FH, Oliver MH, Hawkins P, Holloway AC, Campbell M, Gluckman PD, et al. Periconceptional undernutrition in sheep accelerates maturation of the fetal hypothalamic-pituitary-adrenal axis in late gestation. Endocrinology 2004;145(9):4278–85.

[37] Zhang S, Rattanatray L, MacLaughlin SM, Cropley JE, Suter CM, Molloy L, et al. Periconceptional undernutrition in normal and overweight ewes leads to increased adrenal growth and epigenetic changes in adrenal IGF2/H19 gene in offspring. FASEB J 2010; 24(8):2772–82.

[38] Kakar MA, Maddocks S, Lorimer MF, Kleemann DO, Rudiger SR, Hartwich KM, et al. The effect of periconception nutrition on embryo quality in the superovulated ewe. Theriogenology 2005;64(5):1090–103.

[39] Zhu MJ, Du M, Hess BW, Nathanielsz PW, Ford SP. Periconceptional nutrient restriction in the ewe alters MAPK/ERK1/2 and PI3K/Akt growth signaling pathways and vascularity in the placentome. Placenta 2007;28(11–12):1192–9.

[40] Rumball CW, Bloomfield FH, Harding JE. Cardiovascular adaptations to pregnancy in sheep and effects of periconceptional undernutrition. Placenta 2008;29(1):89–94.

[41] van der Steeg JW, Steures P, Eijkemans MJ, Habbema JD, Hompes PG, Burggraaff JM, et al. Obesity affects spontaneous pregnancy chances in subfertile, ovulatory women. Hum Reprod 2008;23(2):324–8.

[42] Pinborg A, Petersen GL, Schmidt L. Recent insights into the influence of female bodyweight on assisted reproductive technology outcomes. Womens Health (Lond Engl) 2013;9(1):1–4.

[43] Minge CE, Bennett BD, Norman RJ, Robker RL. Peroxisome proliferator-activated receptor-gamma agonist rosiglitazone reverses the adverse effects of diet-induced obesity on oocyte quality. Endocrinology 2008;149(5):2646–56.

[44] Chi MM, Schlein AL, Moley KH. High insulin-like growth factor 1 (IGF-1) and insulin concentrations trigger apoptosis in the mouse blastocyst via downregulation of the IGF-1 receptor. Endocrinology 2000;141(12):4784–92.

[45] Jungheim ES, Schoeller EL, Marquard KL, Louden ED, Schaffer JE, Moley KH. Diet-induced obesity model: abnormal oocytes and persistent growth abnormalities in the offspring. Endocrinology 2010;151(8):4039–46.

[46] Luzzo KM, Wang Q, Purcell SH, Chi M, Jimenez PT, Grindler N, et al. High fat diet induced developmental defects in the mouse: oocyte meiotic aneuploidy and fetal growth retardation/brain defects. PLoS One 2012;7(11):e49217.

[47] Pasternak Y, Aviram A, Poraz I, Hod M. Maternal nutrition and offspring's adulthood NCD's: a review. J Maternal-Fetal Neonatal Med 2013;26(5):439–44.

[48] Elahi MM, Cagampang FR, Mukhtar D, Anthony FW, Ohri SK, Hanson MA. Long-term maternal high-fat feeding from weaning through pregnancy and lactation predisposes offspring to hypertension, raised plasma lipids and fatty liver in mice. Br J Nutr 2009; 102(4):514–9.

[49] Samuelsson AM, Matthews PA, Argenton M, Christie MR, McConnell JM, Jansen EH, et al. Diet-induced obesity in female mice leads to offspring hyperphagia, adiposity, hypertension, and insulin resistance: a novel murine model of developmental programming. Hypertension 2008;51(2):383–92.

[50] Dahlhoff M, Pfister S, Blutke A, Rozman J, Klingenspor M, Deutsch MJ, et al. Peri-conceptional obesogenic exposure induces sex-specific programming of disease susceptibilities in adult mouse offspring. Biochim Biophys Acta 2014;1842(2):304–17.

[51] Wyman A, Pinto AB, Sheridan R, Moley KH. One-cell zygote transfer from diabetic to nondiabetic mouse results in congenital malformations and growth retardation in offspring. Endocrinology 2008;149(2):466–9.

[52] Hansen M, Bower C. Impact of assisted reproductive technologies on intrauterine growth and birth defects in singletons. Semin Fetal Neonatal Med 2014;19(4):228–33.

[53] Chiba H, Hiura H, Okae H, Miyauchi N, Sato F, Sato A, et al. DNA methylation errors in imprinting disorders and assisted reproductive technology. Pediatr Int 2013;55(5):542–9.

[54] Davies MJ, Moore VM, Willson KJ, Van Essen P, Priest K, Scott H, et al. Reproductive technologies and the risk of birth defects. N Engl J Med 2012;366(19):1803–13.

[55] Nelissen EC, Van Montfoort AP, Smits LJ, Menheere PP, Evers JL, Coonen E, et al. IVF culture medium affects human intrauterine growth as early as the second trimester of pregnancy. Hum Reprod 2013;28(8): 2067–74.

[56] Vergouw CG, Kostelijk EH, Doejaaren E, Hompes PG, Lambalk CB, Schats R. The influence of the type of embryo culture medium on neonatal birthweight after single embryo transfer in IVF. Hum Reprod 2012;27(9):2619–26.

[57] Ceelen M, van Weissenbruch MM, Vermeiden JP, van Leeuwen FE, Delemarre-van de Waal HA. Cardiometabolic differences in children born after in vitro fertilization: follow-up study. J Clin Endocrinol Metab 2008;93(5):1682–8.

[58] Ceelen M, van Weissenbruch MM, Prein J, Smit JJ, Vermeiden JP, Spreeuwenberg M, et al. Growth during infancy and early childhood in relation to blood pressure and body fat measures at age 8-18 years of IVF children and spontaneously conceived controls born to subfertile parents. Hum Reprod 2009;24(11):2788–95.

[59] Sakka SD, Malamitsi-Puchner A, Loutradis D, Chrousos GP, Kanaka-Gantenbein C. Euthyroid hyperthyrotropinemia in children born after in vitro fertilization. J Clin Endocrinol Metab 2009;94(4):1338–41.

[60] Scherrer U, Rimoldi SF, Rexhaj E, Stuber T, Duplain H, Garcin S, et al. Systemic and pulmonary vascular dysfunction in children conceived by assisted reproductive technologies. Circulation 2012;125(15):1890–6.

[61] Fernandez-Gonzalez R, Moreira P, Bilbao A, Jimenez A, Perez-Crespo M, Ramirez MA, et al. Long-term effect of in vitro culture of mouse embryos with serum on mRNA expression of imprinting genes, development, and behavior. Proc Natl Acad Sci USA 2004;101(16):5880–5.

[62] Mann MR, Lee SS, Doherty AS, Verona RI, Nolen LD, Schultz RM, et al. Selective loss of imprinting in the placenta following preimplantation development in culture. Development 2004;131(15):3727–35.

[63] Morgan HD, Jin XL, Li A, Whitelaw E, O'Neill C. The culture of zygotes to the blastocyst stage changes the postnatal expression of an epigentically labile allele, agouti viable yellow, in mice. Biol Reprod 2008;79(4): 618–23.

[64] Rivera RM, Stein P, Weaver JR, Mager J, Schultz RM, Bartolomei MS. Manipulations of mouse embryos prior to implantation result in aberrant expression of imprinted genes on day 9.5 of development. Hum Mol Genet 2008;17(1):1–14.

[65] Calle A, Fernandez-Gonzalez R, Ramos-Ibeas P, Laguna-Barraza R, Perez-Cerezales S, Bermejo-Alvarez P, et al. Long-term and transgenerational effects of in vitro culture on mouse embryos. Theriogenology 2012;77(4):785–93.

[66] Watkins AJ, Platt D, Papenbrock T, Wilkins A, Eckert JJ, Kwong WY, et al. Mouse embryo culture induces changes in postnatal phenotype including raised systolic blood pressure. Proc Natl Acad Sci USA 2007;104(13):5449–54.

[67] Calle A, Miranda A, Fernandez-Gonzalez R, Pericuesta E, Laguna R, Gutierrez-Adan A. Male mice produced by in vitro culture have reduced fertility and transmit organomegaly and glucose intolerance to their male offspring. Biol Reprod 2012;87(2):34.

[68] Dar S, Joanne G, Librach CL. Reply: extended culture and the risk of preterm delivery in singletons: confounding by indication? Hum Reprod 2013;28(7): 2021–2.

[69] Makinen S, Soderstrom-Anttila V, Vainio J, Suikkari AM, Tuuri T. Does long in vitro culture promote large for gestational age babies? Hum Reprod 2013;28(3): 828–34.

[70] Edgar DH, Gook DA. A critical appraisal of cryopreservation (slow cooling versus vitrification) of human oocytes and embryos. Hum Reprod Update 2012;18(5):536–54.

[71] AbdelHafez FF, Desai N, Abou-Setta AM, Falcone T, Goldfarb J. Slow freezing, vitrification and ultra-rapid freezing of human embryos: a systematic review and meta-analysis. Reprod Biomed Online 2010;20(2):209–22.

[72] Manno III FA. Cryopreservation of mouse embryos by vitrification: a meta-analysis. Theriogenology 2010;74(2):165–72.

[73] Cobo A, Diaz C. Clinical application of oocyte vitrification: a systematic review and meta-analysis of randomized controlled trials. Fertil Steril 2011;96(2): 277–85.

[74] Herrero L, Martinez M, Garcia-Velasco JA. Current status of human oocyte and embryo cryopreservation. Curr Opin Obstet Gynecol 2011;23(4):245–50.

[75] Roque M, Lattes K, Serra S, Sola I, Geber S, Carreras R, et al. Fresh embryo transfer versus frozen embryo transfer in in vitro fertilization cycles: a systematic review and meta-analysis. Fertil Steril 2013;99(1): 156–62.

[76] Di Pietro C, Vento M, Guglielmino MR, Borzi P, Santonocito M, Ragusa M, et al. Molecular profiling of human oocytes after vitrification strongly suggests that they are biologically comparable with freshly isolated gametes. Fertil Steril 2010;94(7):2804–7.

[77] Monzo C, Haouzi D, Roman K, Assou S, Dechaud H, Hamamah S. Slow freezing and vitrification differentially modify the gene expression profile of human metaphase II oocytes. Hum Reprod 2012;27(7): 2160–8.

[78] Larman MG, Katz-Jaffe MG, McCallie B, Filipovits JA, Gardner DK. Analysis of global gene expression following mouse blastocyst cryopreservation. Hum Reprod 2011;26(10):2672–80.

[79] Shaw L, Sneddon SF, Brison DR, Kimber SJ. Comparison of gene expression in fresh and frozen-thawed human preimplantation embryos. Reproduction 2012; 144(5):569–82.

[80] Pinborg A, Henningsen AA, Loft A, Malchau SS, Forman J, Andersen AN. Large baby syndrome in singletons born after frozen embryo transfer (FET): is it due to maternal factors or the cryotechnique? Hum Reprod 2014;29(3):618–27.

[81] Green MP, Mouat F, Miles HL, Hopkins SA, Derraik JG, Hofman PL, et al. Phenotypic differences in children conceived from fresh and thawed embryos in in vitro fertilization compared with naturally conceived children. Fertil Steril 2013;99(7):1898–904.

[82] Liu XJ, Jia GH, Zhang G, Tian KY, Wang HX, Zhong WY, et al. Effect of vitrification of mouse oocyte on the behavior of adult offspring. Eur J Obstetrics Gynecol Reprod Biol 2013;169(2):279–82.

[83] Qiao J, Wang ZB, Feng HL, Miao YL, Wang Q, Yu Y, et al. The root of reduced fertility in aged women and possible therapentic options: current status and future perspects. Mol Aspects Med 2013;38:54–85.

[84] Handyside AH. Molecular origin of female meiotic aneuploidies. Biochim Biophys Acta 2012;1822(12): 1913–20.

[85] Nelson SM, Telfer EE, Anderson RA. The ageing ovary and uterus: new biological insights. Hum Reprod Update 2013;19(1):67–83.

[86] Sandin S, Hultman CM, Kolevzon A, Gross R, MacCabe JH, Reichenberg A. Advancing maternal age is associated with increasing risk for autism: a review and meta-analysis. J Am Acad Child Adolesc Psychiatry 2012;51(5):477–86. e1.

[87] Vohr BR, Tyson JE, Wright LL, Perritt RL, Li L, Poole WK. Maternal age, multiple birth, and extremely low birth weight infants. J Pediatr 2009;154(4):498–503. e2.

[88] Sutcliffe AG, Barnes J, Belsky J, Gardiner J, Melhuish E. The health and development of children born to older mothers in the United Kingdom: observational study using longitudinal cohort data. BMJ 2012;345:e5116.

[89] Bakos HW, Henshaw RC, Mitchell M, Lane M. Paternal body mass index is associated with decreased blastocyst development and reduced live birth rates following assisted reproductive technology. Fertil Steril 2011;95(5):1700–4.

[90] Soubry A, Hoyo C, Jirtle RL, Murphy SK. A paternal environmental legacy: evidence for epigenetic inheritance through the male germ line. BioEssays 2014;36(4):359–71.

[91] Mitchell M, Bakos HW, Lane M. Paternal diet-induced obesity impairs embryo development and implantation in the mouse. Fertil Steril 2011;95(4):1349–53.

[92] McPherson NO, Bakos HW, Owens JA, Setchell BP, Lane M. Improving metabolic health in obese male mice via diet and exercise restores embryo development and fetal growth. PLoS One 2013;8(8):e71459.

[93] Fullston T, Palmer NO, Owens JA, Mitchell M, Bakos HW, Lane M. Diet-induced paternal obesity in the absence of diabetes diminishes the reproductive health of two subsequent generations of mice. Hum Reprod 2012;27(5):1391–400.

[94] Fullston T, Ohlsson Teague EM, Palmer NO, DeBlasio MJ, Mitchell M, Corbett M, et al. Paternal obesity initiates metabolic disturbances in two generations of mice with incomplete penetrance to the F2 generation and alters the transcriptional profile of testis and sperm microRNA content. FASEB J 2013;27(10):4226–43.

[95] Ng SF, Lin RC, Laybutt DR, Barres R, Owens JA, Morris MJ. Chronic high-fat diet in fathers programs beta-cell dysfunction in female rat offspring. Nature 2010;467(7318):963–6.

[96] Bromfield JJ, Schjenken JE, Chin PY, Care AS, Jasper MJ, Robertson SA. Maternal tract factors contribute to paternal seminal fluid impact on metabolic phenotype in offspring. Proc Natl Acad Sci USA 2014;111(6):2200–5.

[97] Watkins AJ, Sinclair KD. Paternal low protein diet affects adult offspring cardiovascular and metabolic function in mice. Am J Physiol Heart Circ Physiol 2014;306(10):H1444–52.

[98] Alio AP, Salihu HM, McIntosh C, August EM, Weldeselasse H, Sanchez E, et al. The effect of paternal age on fetal birth outcomes. Am J Men's Health 2012; 6(5):427–35.

[99] Zitzmann M. Effects of age on male fertility. Best Pract Res Clin Endocrinol Metab 2013;27(4):617–28.

[100] Wiener-Megnazi Z, Auslender R, Dirnfeld M. Advanced paternal age and reproductive outcome. Asian J Androl 2012;14(1):69–76.

[101] Williams CL, Teeling JL, Perry VH, Fleming TP. Mouse maternal systemic inflammation at the zygote stage causes blunted cytokine responsiveness in lipopolysaccharide-challenged adult offspring. BMC Biol 2011;9:49.

CHAPTER

4

Introduction to Epigenetic Mechanisms: The Probable Common Thread for Various Developmental Origins of Health and Diseases Effects

Martha Susiarjo

Department of Cell and Developmental Biology, Perelman School of Medicine, University of Pennsylvania, Philadelphia, PA, USA

OUTLINE

INTRODUCTION

All cells in the human body are derived from one single cell during conception and are therefore genetically identical. Each cell, however, has a unique set of epigenetic instructions that serve to ensure that the genome is expressed in a lineage- or developmental-specific manner. Waddington first used the term "epigenetics" in 1942 to describe mechanisms in gene regulation leading to phenotypic effects [1]. The contemporary definition of epigenetics is the study of covalent and noncovalent modifications of the DNA and histone modifications, and the various

The Epigenome and Developmental Origins of Health and Disease
http://dx.doi.org/10.1016/B978-0-12-801383-0.00004-9

Copyright © 2016 Elsevier Inc. All rights reserved.

mechanisms that alter these modifications that ultimately affect gene expression, leading to phenotypic changes [2]. Note that changes in gene expression in the definition occur without genetic or nucleotide changes; rather, they involve modifications to the DNA that alter gene expression patterns through the modulation of transcription factors.

The complex interplay between various epigenetic mechanisms and transcription factors is crucial during mammalian development. Throughout the development of the zygote into the adult mammal, there are about 200 cell types and various tissues that become established; these developmental events are facilitated by the divergent processes dependent upon epigenetic mechanisms and transcription factors. Additionally, epigenetic mechanisms are implicated in natural phenomena in mammals including genomic imprinting and dosage compensation [3]. Deviations from developmentally regulated patterns of epigenetic modifications are reflected in processes such as aging and diseases [3].

Accumulating data have suggested that the environmental conditions experienced during early life can significantly affect human biology and long-term health. Early-life nutrition and stress, for instance, are among the best-documented examples of such perturbations because of their association with the risk for developing metabolic diseases (e.g., type 2 diabetes and cardiovascular diseases) as adults [4]. Increasing evidence suggest that epigenetic mechanisms play a crucial role in mediating these environmental-induced phenotypic changes. Unlike the genome, the mammalian epigenome is responsive to environmental cues and can therefore be modifiable during development. Cellular epigenetic instructions can memorize specific gene expression programs even after events such as replication or differentiation. Experimental data in animal models suggest that early-life exposure to inappropriate environmental conditions can create distinct epigenetic marks and that these marks induce long-term changes in gene expression, potentially leading to diseases in later life; these theories are known as the developmental origin of adult diseases (DOHaD). Researchers in the DOHaD field are interested in elucidating the relevant mechanisms of fetal programming in the event of environmental perturbations. Epigenetic mechanisms have been proposed to be attractive candidates for mediating gene–environment interaction relevant to fetal reprogramming events that may lead to adult diseases. It is thought that the mechanisms are used to hold memory for cellular gene regulation programs, i.e., something experienced during the fetal development is remembered until adulthood or even across generations (i.e., transgenerational).

In this chapter, the various epigenetic mechanisms involved in developmental gene regulation and the genome-wide epigenetic reprogramming events during embryonic and primordial germ cell (PGC) development will be discussed, as well as the potential that these events may be perturbed in response to environmental changes.

DNA METHYLATION

The best-studied epigenetic mechanism is DNA methylation; it refers to the modification of DNA involving cytosine methylation at the cytosine–guanine dinucleotides (CpG). This epigenetic modification converts a cytosine into a 5-methyl cytosine (5-mC) base of the DNA. It is thought that the presence of the methyl group disrupts the binding of transcription factors, recruits proteins containing methyl-binding domains (MBDs) that are typically associated with repressive histone modifications, recruits heterochromatin proteins, or leads to the transcription of noncoding RNA [5], so therefore, DNA methylation is commonly associated with transcriptional repression. The establishment and maintenance of DNA methylation depend on the DNA methyltransferase (*Dnmt*) family: the *Dnmt3a and 3b* genes encode for the de novo methyltransferases that initially lay down DNA methylation marks, and the *Dnmt1* gene for the maintenance methyltransferase that is responsible for maintaining methylation status [6]. The latter enzyme works by converting newly replicated, hemimethylated DNA into symmetrically methylated DNA through localization to the replication fork and interaction with proliferating cell nuclear antigen and NP95 (also known as UHRF1) that bind to hemimethylated DNA [7–9]. The importance of DNA methylation and the *Dnmt* gene family is illustrated by the fact that targeted gene deletion leads to early embryonic arrest in the mouse [10,11], and that defective DNA methylation is related to several human disorders [6].

CpG sites represent a very small percentage (approximately 1%) of the mammalian genome and are distributed nonrandomly, concentrated in regions with high frequency of CpGs, known as CpG islands [12]. CpG islands range in sizes from 200 base pairs to several kilobases. Approximately 85% of these CpG sites are highly methylated and 15% unmethylated. This "bimodal" distribution of DNA methylation pattern is maintained in all tissues except during the embryonic and PGC reprogramming events (discussed below in the *Embryonic Epigenetic Reprogramming* and *PGC Reprogramming* sections). Although unmethylated CpG islands are typically located within gene promoters of actively transcribed housekeeping genes and tumor suppressor genes, and methylated CpG islands within silent genes, more recent data illustrate that the association between CpG islands methylation and transcriptional status for many genes is not always direct. Weber et al. described the impact of DNA methylation status on gene expression using the methylated DNA immunoprecipitation (MeDIP) approach [13]. They defined three classes of promoters—poor, moderate, and high CpG contents—and found that DNA methylation of CpG-poor promoters was not linked to repressive gene transcription. Additionally, most promoters within CpG islands remain unmethylated even when genes are inactive. The study concluded that patterns of DNA methylation and its consequences on transcriptional activity were mostly relevant to key developmental genes. Moreover, a recent study by Shen and Waterland indicated that DNA methylation is developmentally programmed and these regulations are tissue-specific [14]. Additionally, nonpromoter CpG islands that are normally unmethylated become methylated in a tissue-specific manner during development and have the potential to alter gene expression. Based on these observations, it can be concluded that DNA methylation status of transcriptional regulatory elements, in addition to DNA methylation in gene body, can also be a source of transcriptional regulation.

In the context of gene regulation, states of DNA methylation are often linked to distinct posttranslational histone modifications. An example is the interactions of DNA methylation with MBDs. These proteins are associated with other chromatin-modifying complexes that target the histone tails. In general, MBD proteins have been biochemically purified along with a number of different histone deacetylase (HDAC) repressor complexes [15,16]. The recruitment of HDACs by MBD proteins facilitates a repressive state in chromatin via deacetylation of histone tails. In addition to this combination (i.e., methylated DNA, MBD, and repressor complex), another common combination of DNA methylation states and histone modification marks are the trimethylation of lysine 4 of histone H3 (H3K4me3, a mark for active transcription) and hypomethylated CpG islands at transcriptional start sites. This combination of marks has been shown to involve the mixed lineage leukemia (MLL) family Cis-X-X-Cis (CXXC) domains that target the MLL histone H3 methyltransferase complexes to unmethylated DNA [17,18]. Similarly, histone demethylase enzymes, such as Jumonji domain-containing histone demethylase 1a (JHDM1a, also known as KDM2a or

CXXC8), can also be recruited to unmethylated CpG-dense regions of the genome for demethylation of histone H3 lysine 36 [19,20].

In addition to CpG modifications, non-CpG methylation occurs in significant levels in embryonic stem (ES) cells and mouse oocytes [21–23], and with relatively high frequency in the mouse and human brain [24,25]. Non-CpG methylation has been detected in all dinucleotide contents (CpA, CpT, and CpC), although the most common is CpA, and it appears to be dependent on *Dnmt3a-3l*. Both gene activation and repression are associated with non-CpG methylation. Consistent with a functional epigenetic role for non-CpG methylation, its deposition in both ES cells and brain is developmentally regulated and linked to gene expression levels [21,22,25]. Furthermore, although 5-mC is the most characterized DNA methylation, recent data have begun to reveal another modification, 5 hydroxymethylcytosine (5-hmC). 5-mC marks can be oxidized to become 5-hmC marks through the action of the 10–11 translocation (TET) family of the methylcytosine hydroxidase [26,27]. Genome-wide maps in mouse and human reveal that 5-hmC occurs exclusively in the CpGs. Interestingly, 5-hmC occurs mostly at CpG-poor distal-regulatory regions corresponding to enhancers, CTCF binding sites, or DNAse-hypersensitive sites [28–30].

There are currently three main molecular biology-based techniques that are available to measure DNA methylation. The first method employs a chemical conversion by the use of bisulfite sequencing. It is the gold standard for mapping DNA methylation and involves a chemical conversion method that directly estimates the methylation state of each cytosine in a binary manner (i.e., unmethylated vs methylated). Treatment of genomic DNA with sodium bisulfite chemically converts unmethylated cytosine into uracil, while methylated cytosine is unaffected. After polymerase chain reaction (PCR), all unmethylated cytosines become thymidines and the remaining cytosines correspond to 5-mC. A traditional approach then assayed individual loci from bisulfite-treated genomic DNA with locus-specific PCR using the Sanger sequencing method. More recently, methods to increase whole genomic coverage have been created. The reduced representation bisulfite sequencing method combines restriction digestion with bisulfite sequencing for detailed analysis of high-CpG-density regions including CpG islands [31]. Bisulfite treatment followed by whole genome sequencing (methyl C-seq, BS-seq, or whole genome bisulfite sequencing) features nucleotide resolution and quantitated rate of methylation at all four nucleotides. It should be noted that both 5mC and 5hmC are resistant to bisulfite conversion, so therefore cannot be distinguished from each other in this chemical conversion–based method of DNA methylation analysis. The second method of measuring DNA methylation is digestion of genomic DNA with methyl-sensitive restriction enzymes. This method determines CpG methylation at base resolution; however, the genome-wide coverage is limited by the cutting frequency of the restriction enzymes and is biased toward enzyme-specific sequence recognition. The third method is an affinity-based method in which methylated fragments from sonicated DNA are captured with an antibody to 5-mC (MeDIP) or MBD followed by sequencing. The limitation of the method is the inaccuracy in directly determining the exact site of methylated cytosine, as the resolution of these affinity-based assays is highly dependent on the DNA fragment size, CpG density, and immunoprecipitation quality of the reagent. The results of both restriction enzyme- and affinity-based sequencing methods (second and third methods, respectively) are qualitative levels of enrichment rather than absolute. On the other hand, because affinity- and restriction enzyme-based methylation assays enrich or capture methylated DNA regions, the sequencing costs are moderate.

Increasing data suggest that DNA methylation changes are involved in linking fetal reprogramming events and the development of adult metabolic diseases. In 2004, a study by Weaver et al. studied the effects of maternal behavior on stress response development in the offspring using the rat model [32]. They found that differences in nurturing behavior in the mothers can influence stress responses in the offspring; the relevant mechanisms that were described included hypermethylation of specific CpG dinucleotides within the promoter of the glucocorticoid receptor (GR) gene in the hippocampus of the offspring. In addition to maternal behavior, maternal nutrition can significantly affect DNA methylation patterns, potentially leading to changes in gene expression of metabolic regulators. Changes in various epigenetic modifications have been shown in various rat models of maternal dietary restrictions. Feeding a protein-restricted maternal diet leads to lower body weight, cardiovascular diseases, and hypertension in the offspring. These changes are linked to reduced DNA methylation of the peroxisome proliferator-activated receptor (PPAR) alpha and GR promoters [33]. In humans, maternal starvation during the Dutch famine of 1944–1945 is linked to permanent DNA methylation alterations of *IGF2* in the offspring [34]. More recent evidence has also suggested the role of DNA methylation in carrying out transgenerational inheritance of gene expression instructions. Although it is known that most methylation status of germ cells and somatic cells are the result of developmental genomic programming, it is thought that additional changes in methylation can arise from biochemical environmental influences and this information may be inherited across multiple generations. In the protein-restricted model in rats, it has been found that feeding the protein-restricted diet in the F0 generation leads to adverse effects in glucose metabolism in the F1, F2, and F3 generations that are related to DNA methylation changes in the PPAR alpha and GR promoters [35]. Readers are referred to Chapters 6, 7, and 22 in this book for more information on the involvement of various environmental factors on DOHaD as well as evidence on transgenerational inheritance of relevant phenotypes.

POSTTRANSLATIONAL HISTONE MODIFICATIONS

Another type of epigenetic mechanism is posttranslational histone modifications. Nuclear chromatin is organized into nucleosomes, with each unit consisting of 146 base pairs of DNA wrapped twice around eight core histone proteins (i.e., two copies each of H2A, H2B, H3, and H4). Nucleosomes are assembled in a long array with each nucleosome connected by 10–70 bp of linker DNA. The compaction of these histone cores around the DNA as well as the folding of the nucleosomal arrays provide constraint to nuclear processes or enzymes that need to access the DNA. The majority of histone modifications occur in the 10–30 amino acids (aa) of the N-terminal domains of the histone proteins, often referred to as the histone tails. The "tail" of these histones contains subunits (e.g., arginine, lysine, serine, and threonine) that are frequently modifiable (e.g., methylated, phosphorylated, or acetylated). Each of these modifications alters the accessibility of DNA to transcription factors and may lead to the activation or repression of gene activity [36]. Common modifications include lysine acetylation; mono-, di-, or trimethylation of lysines; mono- or dimethylation of arginines; and phosphorylation of serines, threonines, or tyrosines, as well as lysine ubiquitylation and SUMOylation. Except for a few histone marks that drastically lead to major chromatin changes, the majority of histone marks alter the binding or activity of regulatory factors including ATP-dependent chromatin remodeling complexes (see *Nucleosome Remodeling* section).

Each posttranslational modification of the histone leads to either activation or repressive transcription. The acetylation of histone lysines, especially those in the H3 and H4 tails (i.e., H3Kac or H4Kac), has been described as a hallmark of transcriptionally active chromatin. This type of modification is believed to neutralize the positive charge of the lysine residue, therefore allowing the acetylated lysine to be bound more loosely around DNA and therefore leading to a more "open" chromatin configuration allowing transcription factors to access the DNA. Another potential mechanism of transcriptional activation is that the acetylated lysine may attract ATP-dependent chromatin remodeling complexes that harbor acetyl-lysine recognition domains known as the bromodomains. The specificity of this recruitment is thought to depend on insertion of the acetylated lysine to the hydrophobic binding pocket of the ~70 kDa bromodomains and the sequence-specific interaction of the residue on the surface of this binding pocket with nearby residues on the histone tails [37]. In contrast to acetylated lysine, the mono-, di-, or trimethylation of histone is not associated with neutralization of lysine residues and appears to be associated with both passive and active chromatin domains. Developmentally silenced loci are enriched in dimethylated lysine 27 histone 2 (H2K27me2), while heterochromatin domains are packed with H3K9me2 and me3 [38]. In contrast, actively transcribed genes contain nucleosomes that are di- and trimethylated at H3-K4, K36, and K79 [38]. H3-K4me2/3 is enriched at the 5′ end of transcribed genes, whereas H3-K36me2/3 and K79me2/3 are enriched in gene bodies [38,39]. As for acetylated lysine, both the "active" and "inactive" methyl marks are targets for chromatin remodeling enzymes.

A complex of enzymes (called histone-modifying enzymes) modify these histone proteins in spatial- and temporal-specific patterns. Lysine (K) methyltransferases (MTs) are key enzymes responsible for the mono-, di-, or trimethylation of the lysine residue regulation. Most KMTs harbor a predicted SET domain, which catalyzes the transfer of a methyl group from S-adenosylmethionine to the ε-amine on the side chain of lysine residue. KMTs that modify H3K4 include the MLL, SETD1A, and SET7 [40]. For H3K9, the first identified methyltransferase was SETDB1 [41]; others have been identified including SUV39H1, SUV39H2, G9A, and PRDM2 [42–45]. G9A, GLP, SETDB1, and SUV39H1 are known to form an enzymatic complex. For H3K27me, EZH1 and EZH2 are the enzymes linked to H3K27 methylation; H3K27me is enriched in actively transcribed promoters and H3K27me3 with repressive transcriptional promoter activity [46,47]. How histone-modifying enzymes recognize histones at specific loci is an ongoing research interest, but both DNA sequences and long noncoding RNA have been proposed to play critical roles.

To study posttranslational histone modification, chromatin immunoprecipitation followed by sequencing (ChIP-seq) is used to map the genome-wide binding pattern of chromatin-associated proteins. Alternatively, the same principle can be used to investigate patterns of binding of histone modification at specific loci by gene-specific quantitative PCR analysis. In the genome-wide method, chromatin is cross-linked and sonicated and DNA–protein complexes containing specific proteins of interest are immunoprecipitated. The DNA is subsequently purified and adaptor molecules are attached for PCR and sequencing. The digital sequences of enriched DNA, called reads, are computationally aligned to the reference genome to define punctate peaks or broad blocks of modified histones or protein occupancy. Since its development in 2007, researchers have used ChIP-seq extensively to survey the genomic profiles of histones and their modifications, transcription factors, DNA- and histone-modifying enzymes, transcriptional machinery, and other chromatin-associated proteins [48–51].

Changes in histone modification patterns have been observed in a limited number of studies linking environmental changes and fetal reprogramming events. In a rat model of maternal low-protein diet, it is known that offspring from mothers fed restricted protein during pregnancy are more likely to develop metabolic disorders as adults, including hypertension, obesity, and cardiovascular diseases. In these models, changes in posttranslational histone modifications are accompanied by relevant DNA methylation alterations. For instance, in this model, in addition to DNA methylation changes observed at PPAR alpha and GR promoters [33], further characterizations reveal that the offspring have altered histone modifications profiles at $GR1_{10}$ with an increase in H3K9 acetylation and a decrease in H3K9me2 and me3 [52].

NONCODING RNA

Only approximately 2% of mammalian transcripts are translated into functional proteins. The rest of the remaining transcripts were at first regarded as "junk"; however, increasing data have demonstrated that many of these untranslated RNAs can act as epigenetic modifiers and play important roles in gene regulation [53]. These noncoding RNAs (ncRNAs) can be long (more than 200 nucleotides) or short (less than 200 nucleotides). The long and short ncRNAs can be further classified based on their genomic origins and relevant biogenic processes.

Long ncRNAs (lncRNAs) are highly diverse in sequence, structure, and function. They are frequently processed similarly to mRNAs and are found in nuclear or cytosolic fractions. Often, lncRNAs are transcribed from protein-coding loci; however, there are exceptions, as some originate from introns and are composed of transposable genomic elements, or some from "natural antisense transcript" (NAT)/sense–antisense pairs. While function of lncRNA varies, there are increasing evidence that they serve as intermediate molecules or facilitate or block cellular events related to nucleic acids or proteins. One mechanism leading to lcRNA-mediated gene suppression is the recruitment of polycomb group (PcG) proteins and formation of heterochromatin. A classic example of lncRNA-mediated gene repression is the KvDMR1/LIT1/KCNQ1OT1 lncRNA at the 11p15.5 that is associated with human Beckwith–Wiedemann Syndrome. This 91-kb long transcript mediates bidirectional silencing by interacting with chromatin in a tissue lineage-specific manner and recruits histone methyltransferase members of the PRC2 complex, as well as the enzyme G9a, resulting in an increase in repressive posttranslational histone modifications in the *KCNQ1* domain [3]. Other examples of PcG-mediated repression for lncRNA include the *ANRIL*, located in the *p14/p15/INK4* locus, and the XIST in the context of X chromosome inactivation [54], a biological process to ensure equal X chromosome-linked gene dosage in female mammals. In the latter example, a short-repeat RNA (RepA) found within XIST can recruit PRC2 to the locus to be silenced, leading to repressive transcriptional activity via H3K27me3-dependent or independent pathways [55,56]. In addition to these *cis*-acting PcG complex-mediated lncRNA interactions, *trans*-acting mechanisms have been observed in which PcG complexes are used to silence expression of distant loci. The lncRNA *H19*, originally described as a key player in the H19/IGF2 genomic imprinting, has been shown to derepress the *Wnt* signaling pathway *in trans* through PRC2-dependent mechanism in bladder cancer metastasis [57]. HOTAIR, a 2.2kb lncRNA transcribed on human chromosome 12, recruits PRC2 and LSD1 and represses a 40-kb long homeobox gene on human chromosome 2 through H3K27me3 and loss of H3K4me3 [58]. In addition to transcriptional repression through posttranslational histone modification, DNA methylation-mediated gene repression can occur. For instance, at the *Kcnq1* domain, in addition to interacting with PRC2 and G9a, the lncRNA *Kcnq1ot1* can recruit *Dnmt1* to the CpG islands of imprinted genes [59].

In contrast to lncRNAs, short (s)ncRNAs have been extensively classified according to mechanisms of action or genomic origins. The best-studied type is microRNAs (miRNAs) that are

20–23 nucleotides in length and usually target mRNA in sequence complementarity to 2–7 nt of the 3′ untranslated region (UTR), although they can sometimes target the 5′UTR as well [60]. The complementary binding activates the miRNA-induced silencing complex; the binding leads to posttranslational transcriptional repression through the recruitment of the endonuclease Dicer that cleaves double-stranded RNA (dsRNA). Other popular sncRNAs are the PIWI interacting RNAs (piRNAs), which are 26–30 nt in length, single-stranded, and Dicer-independent. piRNA precursor transcripts undergo primary and secondary biogenesis that involve loading of the transcript onto the Aubergine and Argonaute 3 complexes, respectively. piRNAs silence transcription of target RNAs through perfect or mismatch base pairing, leading to H3K9me3 formation [61].

miRNAs regulate expression of several components of PcG complexes that subsequently interfere with downstream events. In several tumorigenesis cases, miRNAs downregulate the histone methyltransferase EZH2 [62,63]. In addition to directly donwregulating repressive histone methyltransferase, miRNA can also affect expression of HDAC, histone deacetylases that remove activating marks. miRNA-449a has been shown to normally repress HDAC1 and when this repression fails, it can lead to overexpression of HDAC1 and cancers [64]. The opposite patterns of expression (i.e., overexpressing of HDAC1–3 and decreased miRNA 4491 signaling), however, have been described in hepatocellular carcinoma [65], illustrating highly tissue-specific miRNA-dependent gene regulation. miRNA can also downregulate DNA methylation, leading to activated transcriptional activity. In lung cancer and leukemia cells, the miR-29 family has been shown to target DNMT1 and 3, and restored expression of these miRNAs is associated with decreased tumorigenicity in vitro and in vivo [66]. miR-132, which is dysregulated in schizophrenia and is involved in antiviral innate immunity and circadian timing, has been shown to regulate gene expression by directly inhibiting DNMT3A, the transcription factor GATA2, EP300 (a transcriptional co-activator), and several members of histone-modifying complexes in vitro and in vivo [67]. In some cases of human breast cancer, downregulation related to hypermethylation of a group of miRNAs has been noted [68]. Several miRNAs, including miR-128a, appear to be highly expressed in adult brain, and their downregulation is associated with cerebellar tumor/medullablastoma [69].

NUCLEOSOME REMODELING

The eukaryotic genome is organized into chromatin, consisting of DNA wrapped in histone and nonhistone chromatin components. To encode the genetic code in the context of development or in response to environmental cues, transcriptional machineries and regulatory elements need to gain access to the DNA despite the compact and protective chromatin organization. Nucleosome remodeling enzymes perform this critical task by modifying DNA–histone interactions so that nucleosomes can be moved, assembled, or disrupted, creating a local and potentially temporal-specific gene activation. Because nucleosomes are stable entities owing to the various DNA–histone interactions, nucleosome remodeling requires biochemical reactions coupled to ATP hydrolysis that are carried out by very specific remodeling enzymes and their cofactors.

The best known of nucleosome remodeling complexes is the Swi/Snf complex, originally identified genetically in yeast. Purified Swi/Snf complex could disrupt nucleosome structure in an ATP-dependent manner and stimulate the binding of transcription factors to nucleosomal DNA in vitro [70]. Several versions of Swi/Snf remodeling complex have been found in eukaryotic cells. For instance, in *Drosophila melanogaster*, *Brahma* (*Brm*), a homolog of Swi/Snf, was identified in a genetic screen for the suppressor of the transcriptional repressor Polycomb [71]. In mammals, *Brm* is involved in cellular homeostasis and is a component of Swi/Snf-like complexes [72]. Recent evidence has demonstrated that nucleosome-remodeling ATPases related to Swi/Snf (called the SNF1 family, as they are related to the nucleic acid helicase superfamily SNF) are conserved from yeast to human and there are approximately 1300 SNF2 family members [73]. *Snf2* family helicases commonly bind a double-stranded nucleic acid, move along one strand in one direction, and separate the two strands in doing so. Studies on nucleosome remodelers reveal that they are also DNA translocases and that they move along one strand of nucleosomal DNA, although they do not separate the two strands.

Gene promoters can be classified according to the content of nucleosomes and activity of remodeling complexes. "Housekeeping" or constitutively active genes are most likely depleted nucleosomes and rely less on nucleosomal remodeling, as compared to tightly regulated genes [74,75]. Nucleosomal remodelers are often recruited to genes through sequence-specific instructions and participate in the initiation and elongation of transcription by serving as a co-activator or co-suppressor. In mammals, for example, the Swi/Snf complex interacts with transcription factors including steroid receptors, tumor genes, and oncogenes [76]. Once recruited, they alter local chromatin organization by nucleosome movement or displacement that leads to activation or suppression of genes. Together with histone chaperones, the Swi/Snf complexes can displace nucleosomes *in trans* by forcing the histone octamer onto another site of DNA or by moving the histone onto the chaperone molecule. The displacement of nucleosomes by Swi/Snf is mediated through acetylation of histone and interaction with transcription factors.

Nucleosome remodelers can also interact with other epigenetic modifications, including DNA methylation and posttranslational histone modifications. As mentioned above, DNA methylation is often linked to gene repression owing to its association with MBD proteins. MBD proteins associate with a number of transcriptional co-repressor complexes, including histone methylases and histone deacetylases. The best-characterized nucleosome remodeling complex with known association with DNA methylation is the Mi-2/NuRD complex, which contains MBD 2 or MBD 3 as subunits and preferentially remodels nucleosomes that contain methylated DNA [77,78]. In addition to recognizing DNA methylation, chromatin remodelers also contain domains that mediate their specific interactions with histone marks; this interaction may play a regulatory role depending on the repressive or active nature of the histone marks. For instance, the CHD remodeler in humans has tandem chromodomains that recognize a methylated lysine 4 residue of the histone H3 tail (H3K4me); the interaction with active histone marks is believed to modulate the chromodomain activity at the gene promoter [76]. Additionally, several remodeling ATPases contain acetyl-lysine-binding bromodomains that have the potential to recruit, retain, or modulate remodeling activity on acetylated nucleosomes.

Owing to the large numbers of nucleosome remodelers (and subunits), combined with their versatile activities, nucleosomal remodeling affects many important aspects of genome function by forming the basis for higher-order chromatin organizations. During development, they may serve as critical co-regulators of transcriptional machinery and help to establish and maintain stable cell lineage-specific or temporal-specific gene expression programs. Nucleosome remodelers often contain core subunits that drive their fundamental function; however, their subunits are highly diverse and versatile. The BAF complex, for instance, contains diverse, cell-specific subunits that allow it to be involved in various developmental processes from embryonic stem-cell differentiation to postmitotic neuronal functions [76,79]. The cell-specific subunits' diversity mediates interactions with cell-specific transcriptional factors that recruit remodeler co-activators to different alternative sets of genes.

EMBRYONIC EPIGENETIC REPROGRAMMING

At fertilization, the male and female gametes fuse to produce the zygote. Immediately after fertilization, the two gametes are physically separated and undergo distinct chromatin changes. DNA compaction in the paternal genome switches from protamine-rich to maternally provided histones. In the paternal genome, histone acetylation initially dominates, including the acetylation of H4K2 and H4K4; later on, histone methylation starts appearing in a stage-specific manner [80,81]. The hyperacetylation and hypomethylation states of the chromatin alter the accessibility of the paternal genome for additional chromatin remodeling. In contrast, the maternal genome retains the same histone modification patterns from the oocyte (H3K9 or H3K27) in the zygote and subsequent cellular divisions [81,82]. This creates asymmetry between the male and female pronuclei. For instance, the distinct H3K9me3 mark is present until the four-cell stage, whereas others, including H3K20me3, H3K64me3, and H3K4me3, are present until the two-cell stage. The male and female pronuclei also have very distinctive DNA methylation profiles, as the paternal genome rapidly loses its methylation but methylation in the maternal genome is maintained [83]. The asymmetrical distribution of chromatin states in the male versus female gametes reflects the epigenetic profile of the previous generations. This epigenetic profile must then be reprogrammed to accommodate proper development in the embryo of the present generation. The functional consequences of the epigenetic reprogramming of the male and female gametes are unclear; however, it coincides with the genome activation of the gamete. Recent evidence shows that the two events are intertwined and that key developmental gene activation occurs in the context of these epigenetic reprogramming events. In the male genome, loss of early demethylation perturbs the activation kinetics of paternal alleles of several pluripotency genes and impairs development [84].

Using the mouse model, it has been shown that inheritance of the 5-mC through cell division happens through the action of *Dnmt1*. In the absence of *Dnmt1*, cells lose methylation each time they divide, a process known as passive demethylation. Passive, cell divison-dependent demethylation is thought to occur in the maternal genome. In contrast, the paternal genome is thought to undergo an "active," DNA replication-independent demethylation event. Recent studies have shown that 5-mC can be oxidized into 5-hmC, 5-formyl, or 5-carboxylmethyl by TET dioxygenases 1–3. *Tet3* is highly expressed in the oocyte and zygote, but is downregulated at the two-cell stage; when *Tet3* is knocked out, the conversion is perturbed [85]. In the maternal genome, *Tet3* oxidation is prevented owing to action of *Stella* or PGC7 [85]. Zygotes lacking *Stella* lose the asymmetrical DNA methylation status and exhibit global loss of 5-mC and acquisition of 5-hmc from both the paternal and maternal genome. Recent data show that *Stella* binds H3K9me2 and prevents *Tet*-mediated oxidation. Although loss of 5-mC is global, several genomic regions (e.g., imprinted loci, *Iap* retrotransposon, and centromeric heterochromatin) escape this demethylation event. Maintenance of high levels of DNA methylation at these genomic regions is essential for development, as it regulates chromosomal stability, maintains imprinted gene expression, and prevents retrotransposon activation [86,87].

Epigenetic reprogramming of the zygote is critical for the developmental program of the embryo. The earliest differentiation event, at the time when embryo and extraembryonic lineages are established, occurs at the 8- to 16-cell morula stage and is initiated as early as the 4-cell stage when the blastomeres are associated with high levels of H3R26me2 [88,89]. Perturbing these marks can alter the cell fate. So, histone modifications and DNA methylation are thought to be critical for lineage induction. When normal patterns of DNA methylation and histone modification are disrupted, differentiation of ES cells becomes impaired. PRCs, for instance, are required to maintain key developmental genes in the silent but transcriptionally poised state, as evidenced by the bivalency of key developmental genes marked for both repressive and active histone marks as a tool to be developmentally flexible [90].

PRIMORDIAL GERM CELL REPROGRAMMING

Another genome-wide reprogramming event during embryonic development that may be susceptible to environmental perturbation occurs in the PGCs. PGCs are derived from a subset of the epiblast cells. Because epiblast cells are originally destined to become somatic cells, PGCs must be reprogrammed to ensure totipotency and germ cell-specific developmental ability for the next generations. The earliest epigenetic changes observed in the PGCs occur during their migration to the genital ridge between embryonic day (E)7.5 and 10.5 in the mouse when loss of H3K9me2 and concomitant increase in H3K27me3 are observed [91,92]. Why PGCs switch from one repressive histone mark to another still remains a mystery; however, observation from ES cells has demonstrated the latter mark to be more associated with a repressive but poised differentiation state, so this could be a mechanism to be developmentally flexible. Recent data have suggested that the switch is important for expression of pluripotency genes (*Oct4, Nanog, Sall4, and SSEA1*) in vivo and derivation of embryonic germ cells in vitro [93].

From approximately E8.5, global reduction in 5-mC immunofluorescence staining occurs [91]. At imprinted loci, transposon elements and germ cell-specific genes are fully or partially demethylated by E10.5 in the mouse [94]. Based on the kinetics of the loss of DNA methylation, it is proposed that the early stage of DNA demethylation is DNA replication-dependent/passive and is then followed by rapid, active DNA demethylation. This hypothesis is consistent with observations in the mouse: Initially at E8.5, loss of DNA methylation is gradual; subsequently, between E11.5 and 12.5, it becomes accelerated. By E13.5, complete demethylation in gene bodies and intergenic regions, including repeat elements and imprinted loci, has occurred. Genomic regions excluded from this programming event are transposable elements of the IAP and LTR-ERV1 [94]. The global reprogramming in the PGCs is hypothesized to correct any error or epimutation from the previous generation. The global demethylation events coincide with transient lost of several histone modifications, including linker histone H1, H3K27me3, and H3K9me3, and permanent loss of H3K9ac and H2/H4R3me2 [91].

CONCLUSION

The various epigenetic mechanisms that are potentially linked to DOHaD are discussed in this chapter. Most evidence linking epigenetics and DOHaD originates from studies analyzing gene-specific changes in DNA methylation and posttranslational histone modifications. As various genome-wide approaches in studying epigenetic regulation improve in techniques and cost efficiency, the field of DOHaD will expand the relevant research in a more global context. Moreover, the involvement of other epigenetic mechanisms, including noncoding RNA and nucleosome remodeling in fetal reprogramming events leading to DOHaD, will, it is hoped, become more apparent in the near future.

References

[1] Waddington CH. The epigenotype. Int J Epidemiol 2012.

[2] Allis CD, Jenuwein T, Reinberg D. Epigenetics. CSHL Press; 2007.

[3] Lee JT, Bartolomei MS. X-inactivation, imprinting, and long noncoding RNAs in health and disease. Cell 2013;152:1308–23.

[4] Gluckman PD, Hanson MA, Bateson P, et al. Towards a new developmental synthesis: adaptive developmental plasticity and human disease. Lancet 2009;373:1654–7.

[5] Bird AP, Wolffe AP. Methylation-induced repression—belts, braces, and chromatin. Cell 1999;99:451–4.

[6] Gopalakrishnan S, Van Emburgh BO, Robertson KD. DNA methylation in development and human disease. Mutat Res 2008;647:30–8.

[7] Bostick M, Kim JK, Estève P-O, Clark A, Pradhan S, Jacobsen SE. UHRF1 plays a role in maintaining DNA methylation in mammalian cells. Science 2007; 317:1760–4.

[8] Leonhardt H, Page AW, Weier HU, Bestor TH. A targeting sequence directs DNA methyltransferase to sites of DNA replication in mammalian nuclei. Cell 1992;71:865–73.

[9] Sharif J, Muto M, Takebayashi S-I, et al. The SRA protein Np95 mediates epigenetic inheritance by recruiting Dnmt1 to methylated DNA. Nature 2007;450:908–12.

[10] Li E, Bestor TH, Jaenisch R. Targeted mutation of the DNA methyltransferase gene results in embryonic lethality. Cell 1992;69:915–26.

[11] Okano M, Bell DW, Haber DA, Li E. DNA methyltransferases Dnmt3a and Dnmt3b are essential for de novo methylation and mammalian development. Cell 1999;99:247–57.

[12] Gardiner-Garden M, Frommer M. CpG islands in vertebrate genomes. J Mol Biol 1987;196:261–82.

[13] Weber M, Hellmann I, Stadler MB, et al. Distribution, silencing potential and evolutionary impact of promoter DNA methylation in the human genome. Nat Genet 2007;39:457–66.

[14] Shen L, Waterland RA. Methods of DNA methylation analysis. Curr Opin Clin Nutr Metab Care 2007;10:576–81.

[15] Jones PL, Veenstra GJ, Wade PA, et al. Methylated DNA and MeCP2 recruit histone deacetylase to repress transcription. Nat Genet 1998;19:187–91.

[16] Nan X, Ng HH, Johnson CA, et al. Transcriptional repression by the methyl-CpG-binding protein MeCP2 involves a histone deacetylase complex. Nature 1998;393:386–9.

[17] Ayton PM, Chen EH, Cleary ML. Binding to nonmethylated CpG DNA is essential for target recognition, transactivation, and myeloid transformation by an MLL oncoprotein. Mol Cell Biol 2004;24:10470–8.

[18] Milne TA, Briggs SD, Brock HW, et al. MLL targets SET domain methyltransferase activity to Hox gene promoters. Mol Cell 2002;10:1107–17.

[19] Blackledge NP, Zhou JC, Tolstorukov MY, Farcas AM, Park PJ, Klose RJ. CpG islands recruit a histone H3 lysine 36 demethylase. Mol Cell 2010;38:179–90.

[20] Tsukada Y-I, Fang J, Erdjument-Bromage H, et al. Histone demethylation by a family of JmjC domain-containing proteins. Nature 2006;439:811–6.

[21] Lister R, Pelizzola M, Kida YS, et al. Hotspots of aberrant epigenomic reprogramming in human induced pluripotent stem cells. Nature 2011;471:68–73.

[22] Lister R, Pelizzola M, Dowen RH, et al. Human DNA methylomes at base resolution show widespread epigenomic differences. Nature 2009;462:315–22.

[23] Shirane K, Toh H, Kobayashi H, et al. Mouse oocyte methylomes at base resolution reveal genome-wide accumulation of non-CpG methylation and role of DNA methyltransferases. PLoS Genet 2013;9:e1003439.

[24] Varley KE, Gertz J, Bowling KM, et al. Dynamic DNA methylation across diverse human cell lines and tissues. Genome Res 2013;23:555–67.

[25] Lister R, Mukamel EA, Nery JR, et al. Global epigenomic reconfiguration during mammalian brain development. Science 2013;341:1237905.

[26] Tahiliani M, Koh KP, Shen Y, et al. Conversion of 5-methylcytosine to 5-hydroxymethylcytosine in mammalian DNA by MLL partner TET1. Science 2009;324:930–5.

[27] Ito S, D'Alessio AC, Taranova OV, Hong K, Sowers LC, Zhang Y. Role of Tet proteins in 5mC to 5hmC conversion, ES-cell self-renewal and inner cell mass specification. Nature 2010;466:1129–33.

[28] Liu Y, Liu P, Yang C, Cowley AW, Liang M. Base-resolution maps of 5-methylcytosine and 5-hydroxymethylcytosine in Dahl S rats: effect of salt and genomic sequence. Hypertension 2014;63:827–38.

[29] Song C-X, Szulwach KE, Dai Q, et al. Genome-wide profiling of 5-formylcytosine reveals its roles in epigenetic priming. Cell 2013;153:678–91.

[30] Shen L, Wu H, Diep D, et al. Genome-wide analysis reveals TET- and TDG-dependent 5-methylcytosine oxidation dynamics. Cell 2013;153:692–706.

[31] Meissner A, Gnirke A, Bell GW, Ramsahoye B, Lander ES, Jaenisch R. Reduced representation bisulfite sequencing for comparative high-resolution DNA methylation analysis. Nucleic Acids Res 2005;33:5868–77.

[32] Weaver ICG, Cervoni N, Champagne FA, et al. Epigenetic programming by maternal behavior. Nat Neurosci 2004;7:847–54.

[33] Lillycrop KA, Phillips ES, Jackson AA, Hanson MA, Burdge GC. Dietary protein restriction of pregnant rats induces and folic acid supplementation prevents epigenetic modification of hepatic gene expression in the offspring. J Nutr 2005;135:1382–6.

[34] Heijmans BT, Tobi EW, Stein AD, et al. Persistent epigenetic differences associated with prenatal exposure to famine in humans. Proc Natl Acad Sci USA 2008;105:17046–9.
[35] Benyshek DC, Johnston CS, Martin JF. Glucose metabolism is altered in the adequately-nourished grand-offspring (F3 generation) of rats malnourished during gestation and perinatal life. Diabetologia 2006;49:1117–9.
[36] Bannister AJ, Kouzarides T. Regulation of chromatin by histone modifications. Cell Res 2011;21:381–95.
[37] Josling GA, Selvarajah SA, Petter M, Duffy MF. The role of bromodomain proteins in regulating gene expression. Genes (Basel) 2012;3:320–43.
[38] Sims RJ, Reinberg D. Histone H3 Lys 4 methylation: caught in a bind? Genes Dev 2006;20:2779–86.
[39] Petty E, Pillus L. Balancing chromatin remodeling and histone modifications in transcription. Trends Genet 2013;29:621–9.
[40] Wang H, Cao R, Xia L, et al. Purification and functional characterization of a histone H3-lysine 4-specific methyltransferase. Mol Cell 2001;8:1207–17.
[41] Schultz DC, Ayyanathan K, Negorev D, Maul GG, Rauscher FJ. SETDB1: a novel KAP-1-associated histone H3, lysine 9-specific methyltransferase that contributes to HP1-mediated silencing of euchromatic genes by KRAB zinc-finger proteins. Genes Dev 2002;16:919–32.
[42] Rea S, Eisenhaber F, O'Carroll D, et al. Regulation of chromatin structure by site-specific histone H3 methyltransferases. Nature 2000;406:593–9.
[43] O'Carroll D, Scherthan H, Peters AH, et al. Isolation and characterization of Suv39h2, a second histone H3 methyltransferase gene that displays testis-specific expression. Mol Cell Biol 2000;20:9423–33.
[44] Tachibana M, Sugimoto K, Fukushima T, Shinkai Y. Set domain-containing protein, G9a, is a novel lysine-preferring mammalian histone methyltransferase with hyperactivity and specific selectivity to lysines 9 and 27 of histone H3. J Biol Chem 2001;276:25309–17.
[45] Kim K-C, Geng L, Huang S. Inactivation of a histone methyltransferase by mutations in human cancers. Cancer Res 2003;63:7619–23.
[46] Shen X, Liu Y, Hsu Y-J, et al. EZH1 mediates methylation on histone H3 lysine 27 and complements EZH2 in maintaining stem cell identity and executing pluripotency. Mol Cell 2008;32:491–502.
[47] Cao R, Wang L, Wang H, et al. Role of histone H3 lysine 27 methylation in Polycomb-group silencing. Science 2002;298:1039–43.
[48] Barski A, Cuddapah S, Cui K, et al. High-resolution profiling of histone methylations in the human genome. Cell 2007;129:823–37.
[49] Johnson DS, Mortazavi A, Myers RM, Wold B. Genome-wide mapping of in vivo protein-DNA interactions. Science 2007;316:1497–502.
[50] Mikkelsen TS, Ku M, Jaffe DB, et al. Genome-wide maps of chromatin state in pluripotent and lineage-committed cells. Nature 2007;448:553–60.
[51] Robertson G, Hirst M, Bainbridge M, et al. Genome-wide profiles of STAT1 DNA association using chromatin immunoprecipitation and massively parallel sequencing. Nat Methods 2007;4:651–7.
[52] Lillycrop KA, Slater-Jefferies JL, Hanson MA, Godfrey KM, Jackson AA, Burdge GC. Induction of altered epigenetic regulation of the hepatic glucocorticoid receptor in the offspring of rats fed a protein-restricted diet during pregnancy suggests that reduced DNA methyltransferase-1 expression is involved in impaired DNA methylation and changes in histone modifications. Br J Nutr 2007;97:1064–73.
[53] Mattick JS. Non-coding RNAs: the architects of eukaryotic complexity. EMBO Rep 2001;2:986–91.
[54] Lee JT. Lessons from X-chromosome inactivation: long ncRNA as guides and tethers to the epigenome. Genes Dev 2009;23:1831–42.
[55] Wang J, Mager J, Chen Y, et al. Imprinted X inactivation maintained by a mouse Polycomb group gene. Nat Genet 2001;28:371–5.
[56] Gieni RS, Hendzel MJ. Polycomb group protein gene silencing, non-coding RNA, stem cells, and cancer. Biochem Cell Biol 2009;87:711–46.
[57] Luo M, Li Z, Wang W, Zeng Y, Liu Z, Qiu J. Long non-coding RNA H19 increases bladder cancer metastasis by associating with EZH2 and inhibiting E-cadherin expression. Cancer Lett 2013;333:213–21.
[58] Rinn JL, Kertesz M, Wang JK, et al. Functional demarcation of active and silent chromatin domains in human HOX loci by noncoding RNAs. Cell 2007;129:1311–23.
[59] Mohammad FF, Mondal TT, Guseva NN, Pandey GKG, Kanduri CC. Kcnq1ot1 noncoding RNA mediates transcriptional gene silencing by interacting with Dnmt1. Development 2010;137:2493–9.
[60] Tollefsbol T. (ed). Handbook of Epigenetics: The New Molecular and Medical Genetics, Chapter 4, Elsevier, Oxford, UK.
[61] Siomi MC, Miyoshi T, Siomi H. piRNA-mediated silencing in *Drosophila* germlines. Semin Cell Dev Biol 2010;21:754–9.
[62] Lei Q, Shen F, Wu J, Zhang W, Wang J, Zhang L. miR-101, downregulated in retinoblastoma, functions as a tumor suppressor in human retinoblastoma cells by targeting EZH2. Oncol Rep 2014;32:261–9.
[63] Liang J, Zhang Y, Jiang G, et al. MiR-138 induces renal carcinoma cell senescence by targeting EZH2 and is downregulated in human clear cell renal cell carcinoma. Oncol Res 2014;21:83–91.
[64] Noonan EJ, Place RF, Pookot D, et al. miR-449a targets HDAC-1 and induces growth arrest in prostate cancer. Oncogene 2009;28:1714–24.

[65] Buurman R, Gürlevik E, Schäffer V, et al. Histone deacetylases activate hepatocyte growth factor signaling by repressing microRNA-449 in hepatocellular carcinoma cells. Gastroenterology 2012;143:811–5.
[66] Fang J, Hao Q, Liu L, et al. Epigenetic changes mediated by microRNA miR29 activate cyclooxygenase 2 and lambda-1 interferon production during viral infection. J Virol 2012;86:1010–20.
[67] Miller BH, Zeier Z, Xi L, et al. MicroRNA-132 dysregulation in schizophrenia has implications for both neurodevelopment and adult brain function. Proc Natl Acad Sci USA 2012;109:3125–30.
[68] Zhang K, Zhang Y, Liu C, Xiong Y, Zhang J. MicroRNAs in the diagnosis and prognosis of breast cancer and their therapeutic potential (Review). Int J Oncol 2014. http://dx.doi.org/10.3892/ijo.2014.2487.
[69] Vidal DO, Marques MMC, Lopes LF, Reis RM. The role of microRNAs in medulloblastoma. Pediatr Hematol Oncol 2013;30:367–78.
[70] Vignali M, Hassan AH, Neely KE, Workman JL. ATP-dependent chromatin-remodeling complexes. Mol Cell Biol 2000;20:1899–910.
[71] Dingwall AK, Beek SJ, McCallum CM, et al. The *Drosophila* snr1 and brm proteins are related to yeast SWI/SNF proteins and are components of a large protein complex. Mol Biol Cell 1995;6:777–91.
[72] Muchardt C, Yaniv M. The mammalian SWI/SNF complex and the control of cell growth. Semin Cell Dev Biol 1999;10:189–95.
[73] Ryan DP, Owen-Hughes T. Snf2-family proteins: chromatin remodellers for any occasion. Curr Opin Chem Biol 2011;15:649–56.
[74] Cairns BR. The logic of chromatin architecture and remodelling at promoters. Nature 2009;461:193–8.
[75] Rach EA, Winter DR, Benjamin AM, et al. Transcription initiation patterns indicate divergent strategies for gene regulation at the chromatin level. PLoS Genet 2011;7:e1001274.
[76] Hargreaves DC, Crabtree GR. ATP-dependent chromatin remodeling: genetics, genomics and mechanisms. Cell Res 2011;21:396–420.
[77] Clouaire T, Stancheva I. Methyl-CpG binding proteins: specialized transcriptional repressors or structural components of chromatin? Cell Mol Life Sci 2008;65:1509–22.
[78] Klose RJ, Bird AP. Genomic DNA methylation: the mark and its mediators. Trends Biochem Sci 2006;31:89–97.
[79] Yoo AS, Crabtree GR. ATP-dependent chromatin remodeling in neural development. Curr Opin Neurobiol 2009;19:120–6.
[80] Adenot PG, Mercier Y, Renard JP, Thompson EM. Differential H4 acetylation of paternal and maternal chromatin precedes DNA replication and differential transcriptional activity in pronuclei of 1-cell mouse embryos. Development 1997;124:4615–25.
[81] Santos F, Peters AH, Otte AP, Reik W, Dean W. Dynamic chromatin modifications characterise the first cell cycle in mouse embryos. Dev Biol 2005;280:225–36.
[82] Arney KL, Bao S, Bannister AJ, Kouzarides T, Surani MA. Histone methylation defines epigenetic asymmetry in the mouse zygote. Int J Dev Biol 2002;46:317–20.
[83] Mayer W, Niveleau A, Walter J, Fundele R, Haaf T. Demethylation of the zygotic paternal genome. Nature 2000;403:501–2.
[84] Gu T-P, Guo F, Yang H, et al. The role of Tet3 DNA dioxygenase in epigenetic reprogramming by oocytes. Nature 2011;477:606–10.
[85] Wossidlo M, Nakamura T, Lepikhov K, et al. 5-Hydroxymethylcytosine in the mammalian zygote is linked with epigenetic reprogramming. Nat Commun 2011;2:241.
[86] Lane N, Dean W, Erhardt S, et al. Resistance of IAPs to methylation reprogramming may provide a mechanism for epigenetic inheritance in the mouse. Genesis 2003;35:88–93.
[87] Olek A, Walter J. The pre-implantation ontogeny of the H19 methylation imprint. Nat Genet 1997;17:275–6.
[88] Johnson MH, Ziomek CA. The foundation of two distinct cell lineages within the mouse morula. Cell 1981;24:71–80.
[89] Torres-Padilla M-E, Parfitt D-E, Kouzarides T, Zernicka-Goetz M. Histone arginine methylation regulates pluripotency in the early mouse embryo. Nature 2007;445:214–8.
[90] Fisher CL, Fisher AG. Chromatin states in pluripotent, differentiated, and reprogrammed cells. Curr Opin Genet Dev 2011;21:140–6.
[91] Hajkova P, Ancelin K, Waldmann T, et al. Chromatin dynamics during epigenetic reprogramming in the mouse germ line. Nature 2008;452:877–81.
[92] Seki Y, Hayashi K, Itoh K, Mizugaki M, Saitou M, Matsui Y. Extensive and orderly reprogramming of genome-wide chromatin modifications associated with specification and early development of germ cells in mice. Dev Biol 2005;278:440–58.
[93] Mansour AA, Gafni O, Weinberger L, et al. The H3K27 demethylase Utx regulates somatic and germ cell epigenetic reprogramming. Nature 2012;488:409–13.
[94] Guibert S, Forné T, Weber M. Global profiling of DNA methylation erasure in mouse primordial germ cells. Genome Res 2012;22:633–41.

CHAPTER

5

Perinatal Neurohormonal Programming and Endocrine Disruption

Cheryl S. Rosenfeld

Department of Bond Life Sciences Center, Department of Biomedical Sciences, Genetics Area Program, Thompson Center for Autism and Neurobehavioral Disorders, University of Missouri, Columbia, MO, USA

OUTLINE

The Epigenome and Developmental Origins of Health and Disease
http://dx.doi.org/10.1016/B978-0-12-801383-0.00005-0

Copyright © 2016 Elsevier Inc. All rights reserved.

KEY CONCEPTS

- Sexual differentiation of the brain is tightly orchestrated by perinatal increase in the sex hormones, testosterone and estrogen, and their binding to neural androgen receptor (AR) and estrogen receptors (ESR1 and ESR2), respectively.
- Depending on the brain region, the effects may be largely testosterone mediated or require the conversion into estradiol. In males, the net effect is brain masculinization and defeminization, which results in them exclusively demonstrating male-typical sexual behaviors.
- In the absence of masculinization and defeminization responses, the female brain is instead programmed.
- Full manifestation of adult sexual behaviors requires the initial organization or programming of the brain during the in utero/postnatal period and later adult surge of the sex hormones, which is considered the activational period.
- Epigenetic changes likely modulate the organizational-activational effects of sex steroids and their cognate receptors.
- Maternal stress during the perinatal period and endogenous/exogenous glucocorticoids can also disrupt normal brain programming responses.
- Deficits in neurobehavioral programming can originate via changes in placental structure and function, suppression of enzymes regulating cortisol (corticosterone in rodents) production, glucocorticoid disturbances to the hypothalamic-pituitary-adrenal (HPA) axis, and via epigenetic changes.
- Perinatal exposure to endocrine-disrupting chemicals (EDC), such as bisphenol A (BPA) and phthalates, may also interrupt brain hormonal programming and later elaboration of adult sexual behaviors.
- These chemicals may induce their effects through disturbing steroidogenesis, via the sex steroid receptors, HPA axis, and epimutations.
- This chapter will consider how sex steroids and glucocorticoids shape perinatal programming of the brain, and the potential injurious effects of the EDC, BPA, and phthalates on these processes.

INTRODUCTION

Males and females of various species, including humans, demonstrate unique neurobehavioral traits. A precise cascade of events during the in utero, postnatal, and possibly even into adolescent periods is required to shape the various brain regions needed to support these later sexual behavioral differences. This initial brain programming is referred to as the "organizational" period [1–3]. Full manifestation of later adult sexual behaviors requires a later surge in sex hormones, especially testosterone, which is considered the "activational" period. Improper production of sex hormones or exposure to exogenous hormones that mimic estrogen and testosterone can cause disruptions in these processes. Further, as there is a lag between when the genital ridge (future male or female gonad) differentiates into a testis or ovary and hormonal sexual differentiation of the brain, an individual may possess a testis, but in the absence of masculinization and defeminization of the brain, the brain will become feminized [4]. Conversely, early exposure to androgenic or estrogenic chemicals may induce masculinization and defeminization of the brain in individuals classified as female based on the presence of ovaries. In this chapter, we will explore the early sex hormone programming of the brain and how various epigenetic mechanisms may shape various brain regions to support later male and female sexual behaviors. Such mechanisms include DNA methylation, histone protein modifications, and microRNA (miR) changes [5–7].

Early exposure to glucocorticoids (a stress hormone), such as cortisol (corticosterone in rodents), can also impact brain function, which may occur in a sex-dependent manner [8,9]. While in utero, the fetus may be subjected to glucocorticoids produced by the mother and synthetic forms administered to her. The placenta and brain express 11β-hydroxysteroid dehydrogenase 2 (11β-HSD2) enzyme that inactivates glucocorticoids and may therefore protect the fetus against a certain amount of maternal stress/hypercortisolemia [10]. However, elevated glucocorticoid concentrations can overwhelm and even suppress this enzyme. In rodents, placental production of this enzyme ceases prior to parturition, which may increase fetal vulnerability to maternal stress during the later stages of gestation [11]. Placental transfer of high amounts of endogenous or exogenous glucocorticoids can impact the hypothalamic-pituitary-adrenal (HPA) axis, which is tightly regulated under physiological conditions and controls the production of cortisol (corticosterone) by the adrenal cortex [12–14]. Negative feedback inhibition of cortisol on the hypothalamic (cortisol-releasing hormone, CRH) and pituitary (adrenocorticotropin, ACTH) hormones maintains normal homeostasis. Maternal glucocorticoids and synthetic glucocorticoid (sGC), however, can stimulate either hyposensitivity or hypersensitivity of the fetal HPA axis [14–17]. Further, these hormones/maternal stress may also disrupt normal perinatal brain programming via effects on the epigenome [18]. The various alterations in the fetal brain induced by maternal stress and glucocorticoids will be discussed.

Lastly, we will examine how two common endocrine-disrupting chemicals (EDC), bisphenol (BPA) and phthalates, may impede normal brain sexual differentiation [19]. These chemicals may impact various steroidogenic pathways that generate testosterone and estrogen. BPA and phthalates may alter brain expression of estrogen receptors (ESR1 and 2) and androgen receptor (AR). Similar to endogenous hormones, these EDC may cause dysfunction in the HPA axis and induce deleterious epigenetic changes [19–22]. These various pathways and mechanisms will thus be considered.

EFFECTS OF SEX STEROID HORMONES ON BRAIN SEXUAL DIFFERENTIATION

One of the most noteworthy neuroendocrinology discoveries of the past century was that sex steroid hormones, testosterone and estrogen, are required to program in a sex-dependent manner perinatal brain development [1–3]. This early brain programming is termed the "organizational effects" of these steroid hormones. However, steroid-dependent organization of other behaviors may also occur during adolescence [23,24]. Full elaboration of many sex-dependent behaviors requires a later surge of testosterone in adult males ("activational effects") [1–3]. The organizational-activational effects of testosterone in many brain regions, including the hippocampus, are due to aromatization of androgens to estrogen [25–28]. Developmental exposure to endogenous testosterone and estrogen may also underpin critical epigenetic changes in the brain [20–22]. This section will discuss the current evidence of how the sex hormones, testosterone and estrogen, induce sex-dependent differences in brain programming.

While it is presumed that gonad and brain sexual differentiation parallel each other, this is often not the case in humans and animal models. In humans, gonadal sexual differentiation occurs in the first few months of pregnancy; whereas brain sexual differentiation is initiated in the second half of pregnancy [4]. Therefore, hormonal fluctuations or deficiencies (discussed below) during this intermediate time can lead to discordance between the gonadal sex and masculinization/defeminization of the brain, which may give rise to transsexuality and sexual orientation differences in humans [4]. Even in the nineteenth century, it was recognized that a disconnect may exist between gonadal sex and gender identity. In 1864, the scientist Karl Heinrich Ulrichs coined the term, *Uranian*, to describe an individual with the mind of a woman who is trapped in a man's body. The name originates from the Greek goddess Aphrodite Urania, who was created from the testicles of the god Uranus (reviewed in [29]).

Transsexuality may also originate in animal models. For instance, partner preference in rams appears to be determined by the size of the sexually dimorphic nucleus (SDN) in the preoptic/anterior hypothalamus, which is generally larger in rams than ewes. However, this brain region is smaller in rams that prefer other males. Androgen binding to androgen receptor (AR) during the prenatal period stimulates these sexually dimorphic differences, as female lambs exposed to exogenous testosterone from days 30–90 of gestation exhibit a masculinized or enlarged SDN. Conversely, inhibition of aromatase activity in ram fetuses during this critical window had no effect [30,31]. Follow-up studies with an AR antagonist, flutamide, and dihydrotestosterone (DHT) confirm the role of AR in programming the SDN in sheep [32].

While genes expressed on the sex chromosomes are required for gonadal sexual differentiation, neural sexual differentiation is orchestrated by tight control of hormone synthesis and conversions during the critical perinatal period. In males, it is characterized by two processes: defeminization and masculinization [33]. The former entails permanent elimination of the "female brain" with corresponding inability to express female-typical sexual behaviors as adults. The latter involves coordinated development of brain regions responsible for mediating male sexual behaviors. In many rodent species, this critical window spans from embryonic day 18 (E18) to postnatal day 10 (PND 10) [34,35]. At this time, the testis produces testosterone, and upon reaching the brain, some of this hormone undergoes P450 aromatization and conversion to estradiol. Binding of estradiol to its cognate estrogen receptors, ESR1 and ESR2, results in increased gene transcription and eventual defeminization and masculinization [36,37]. Administration of testosterone to females during this developmental period triggers brain defeminization and masculinization, and these females will be rendered infertile due to neuroendocrine disruption [38,39]. As adults, they will instead exhibit male pattern of sexual behaviors. In rodents and other litter-bearing species, females may be exposed to a surfeit of testosterone if they are situated in utero between two male fetuses (2M). Hormonal transfer from male to female siblings can result in reprogramming and masculinization of the brain region, such as the sexually dimorphic nucleus of the preoptic area (SDN-POA) [40]. On the other hand, suppression of aromatase conversion of testosterone or treatment of ESR antagonists to males hinders normal male brain patterning [41,42]. The critical window for hormones to sculpt brain sexual differentiation thus accounts for potential incongruence in gonadal sex compared to brain sexual expression.

Marsupials also exhibit sexually dimorphic differences in central nervous system programming. The predominant regions include the lateral septal nucleus of the hypothalamus, the spinal nucleus of the bulbocavernous, and dorsolateral nucleus in the spinal cord. In contrast to eutherian animals, no sex-dependent differences in ESR and AR expression are observed in the brain of marsupial animals. Aromatase activity is, however, elevated in the brain of male compared to female opossums [43].

SEX STEROID-INDUCED EPIGENETIC REGULATION OF BRAIN SEXUAL DIFFERENTIATION

Epigenetic mechanisms may guide sex hormone programming of the brain [5–7]. Such modifications include those that affect gene expression without altering the DNA sequence. Consequently, the epigenome might serve as a bridge by which intrinsic (sex hormones) and extrinsic factors (environmental factors) can impact the genome. The best-characterized epigenetic changes in the brain include DNA methylation and histone protein modifications. More recent attention though has turned to brain microRNAs (miRs) [44].

DNA methylation: DNA methylation of CpG islands can impact brain function and likely governs developmental sex differences in the brain [6,7,45]. DNA methyl transferases (DNMTs) add methyl groups to cytosines in CpG-enriched regions or islands. Sex hormone stimulation of DNMTs may be required for brain sexual differentiation. In the amygdala, *Dnmt3a* is expressed in greater amounts in PND 1 females compared to males [46]. However, treatment of females with DHT and estradiol blunts these sex differences in *Dnmt3a* expression, and the female brain phenotype becomes masculinized. A recent report shows that sex steroid hormones suppress *Dnmt* expression in the sexually dimorphic preoptic area (POA) [47]. The subsequent decreased DNA methylation profile may relieve the repression of genes essential for brain masculinization. Female rats treated with *Dnmt* inhibitors exhibit increased number of masculinized neuronal markers and display male-typical sexual behavior as adults. By the same token, female mice who conditionally lack *Dnmt3a* demonstrate masculinized behavioral patterns. Early exposure to steroids may also lead to late-emerging DNA methylation and transcriptomic changes. Neonatal testosterone treatment of male and female mice only led to slight DNA methylation and gene expression changes early on [48]. However, upon reaching adulthood, a considerable number of sexually dimorphic CpG sites are differentially methylated in various brain regions, including the sexually dimorphic forebrain region—the bed nucleus of the stria terminalis (BNST)/POA and striatum, of neonatally androgenized animals. The changes reflect a masculinized signature response. Neonatal testosterone treatment also stimulates greater number of gene expression changes in the striatum of adults.

Methyl-CpG binding proteins, such as methyl-CpG-binding protein 2 (MeCP2), exert pleiotropic effects. Besides binding to methylated DNA, these proteins may recruit additional corepressor proteins and histone deacetylase enzymes (HDAC, discussed below). The net effect is chromatin modification in favor of gene silencing [49]. At PND 1, females possess greater amounts of *Mecp2* in the ventromedial hypothalamus (VMH) and amygdala than males, but by PND 10 this sex difference is attenuated [50]. At PND 10, *Mecp2* expression escalates in the POA of males compared to females. Neonatal treatment of males with siRNAs against *Mecp2* causes later deficits in juvenile social play behavior [51]. Taken together, the findings suggest that MeCP2 may play a key role in silencing genes involved in brain sexual differentiation, and these effects are precisely coordinated in a time, sex, and regional manner. For instance, the neonatal expression of *Mecp2* in the VMH and amygdala of females may prevent estrogen from inducing premature effects in these areas that are critical for mediating later female sexual behaviors [52].

Sex steroids may induce DNA methylation changes in ESR1 and ESR2 that are required for brain masculinization and defeminization. Female rodent adult offspring subjected to excessive maternal licking and grooming (high licking and grooming, LG) downregulate *Esr1* promoter methylation and possess greater amounts of this gene in the medial preoptic area (MPOA) compared to females whose mothers did not treat them in this manner or received low LG [53]. The decreased expression of *Esr1* in the low LG females may also be attributed to DNA methylation inhibition of Stat5b binding to the *Esr1* promoter site. Another study, however, reported that manual stimulation of the anogenital region in PND 5–7 females results in reduced expression of *Esr1*, but increased methylation of this gene in the POA region that approximates the expression level of control male siblings [54]. These effects are independent of circulating concentrations of estrogen and testosterone. The conflicting findings may be explained by several factors, including age of assessment (adult versus postnatal), brain region examined, and varying external stimuli.

Sexually dimorphic patterns of *Esr1* expression are evident in the MPOA region. *Esr1* is elevated in postnatal, postweaning, and adult females compared to males [54–56]. The *Esr1* promoter is hypermethylated, though, in males relative to females. A masculinized or hypermethylated pattern for the *Esr1* promoter and corresponding decreased gene expression occurs in neonatal female pups exposed to estradiol benzoate [54]. The data support the notion that sex hormones regulate sexually dimorphic differences in methylation status of the *Esr1* promoter that in turn may contribute to later sex differences in behavioral profiles.

Esr1 shapes other brain regions involved in brain sexual differentiation [57]. Heightened methylation at two CpG sites in the intron 1 region of *Esr1* is observed in the medial basal hypothalamus of newborn females, but these effects vanish with age [58]. Treatment of female rats with an ESR1 agonist, propyl-pyrazole triol (PPT), promotes apoptosis in the anteroventral periventricular (AVPV) nucleus but enhanced cell survival in the SDN-POA, which culminates in masculinization of the hypothalamus and a loss of female sexual receptivity at adulthood [59]. ESR1 knockout mice demonstrate impairments in male [60] and female sex behaviors [61].

Other nonclassical sexually dimorphic brain regions are influenced by estrogen at various life stages. For example, the neonatal cortex expresses high amounts of *Esr1*, but the expression declines by PND 10 and is absent by adulthood [62,63]. Elevation in DNMT1 and DNA methylation of the *Esr1* 5′ untranslated region coordinates this suppression in the adult cortex [64].

2M females may become androgenized because of in utero hormonal transfer from adjacent male conceptuses [40]. In the VMH, ESR1 protein and its mRNA (*Esr1*) are increased in 2M females [65]. Correspondingly, DNA methylation of the *Esr1b* promoter site is reduced in 2M compared to females who resided in utero between two other females (2F). The findings indicate that even subtle changes in the in utero environment during the critically sensitive hormonal period may lead to an imprint on the DNA methylation profile for *Esr1*, which manifests in later adult sexual behavioral disturbances.

ESR2 is abundant in the hippocampus and may regulate learning and memory, stress, and learned helplessness responses [66–69]. This ER form is presumably required for defeminization of the brain [70]. Neonatal exposure to estrogenic chemicals induces hypermethylation and decreased expression of *Esr2* in the ovary [71], but it is uncertain if similar effects are observed in the brain. The DNA methylation pattern of ESR2 exon 2 is not altered by sex or neonatal estradiol treatment to newborn or 3-week-old rats [58]. However, this hormone results in increased methylation at one CpG site in adults. Female mice exposed during this crucial perinatal window to an ESR2 agonist, diarylpropionitrile (DPN), fail to show normal lordosis behavior upon reaching sexual maturity [57]. Transgenic male mice deficient in *Esr2* display a feminized brain phenotype, as exemplified by increased lordosis behavior [70]. The two ESR forms appear to show temporal patterns in steroid-induced DNA methylation with *Esr1* vulnerable during the neonatal period, while *Esr2* is susceptible to sex-dependent DNA methylation after brain sexual differentiation, i.e., from weaning (PND 20) through adulthood (PND 60) [58,70]. Neonatal females primed with exogenous estradiol upregulate *Esr2* methylation in the hippocampus compared to both control males and nontreated females [70].

DNA methylation state of androgen receptor (*Ar*) and androgen induction of DNA methylation changes in other neural genes may be important in programming later sex-specific behavioral responses. Male rats and mice show higher amounts of *Ar* mRNA than females in the brain cortical region [72]. Coactivators, such as jumonji domain-containing 1c, JMJD1C, can affect *Ar* expression in the brain [73]. This protein also acts as a potential demethylase. Collectively, the findings suggest sex steroids might induce defeminization and masculinization by impacting DNA methylation status for promoters of key regulatory genes, including their own cognate receptors.

Histone Protein Modifications and Corepressor Proteins: Chromatin is comprised of DNA and histone proteins. This organization allows packaging of DNA into structural units called nucleosomes and DNA to be efficiently contained within the cell nucleus. This arrangement also facilitates mitosis, meiosis, and transcription [74]. There are five types of histone proteins: H1/H5, H2A, H2B, H3, and H4 [75–77]. H1/H5 are considered linker proteins, and the remainder are core proteins. Of these, considerable attention has been paid to H3 because it is one of the most highly modified histone proteins, and there is well-documented evidence of regulating gene transcription [78]. Two copies of the core histone proteins form an octamer that wraps around 147 DNA base pairs, which is equivalent to the nucleosome [79]. Chromatin is generally classified into two types: euchromatin, where the chromatin is loosened and active gene transcription can proceed, and heterochromatin, where the chromatin is tightly coiled and gene transcription is suppressed. There is increasing evidence though that the situation might be more complex with intermediate chromatin states [80].

The N-terminal region of histone proteins projects from the nucleosome and thus can undergo posttranslational modifications that in turn affect the chromatin state and ability for gene transcription to proceed. Such modifications include methylation, phosphorylation, acetylation, ubiquitination, and sumoylation [81,82]. These changes impact gene transcription by either altering the electrostatic charge between the DNA and histone protein or hindering histone–histone interactions [83,84]. The former relaxes the binding between the histone protein and DNA and increases the likelihood that transcriptional factors can access the DNA promoter region and induce gene transcription. The latter change affects posttranscriptional chromatin reassembly [85]. Depending on the type of modification, histone protein, and amino acid residue, some marks might be repressive and induce the histone protein to bind more tightly to DNA; whereas, other marks, namely acetylation and phosphorylation, pull the histone protein away from DNA and increase gene transcriptional activity [86,87]. Acetylation of histone proteins is regulated by two competing classes of enzymes: histone acetyl transferases (HATs) and HDACs [88]. HATs transfer a negatively charged acetyl group onto the lysine residues of the histone tail leading to electrostatic repulsion to the negatively charged phosphate backbone of DNA and subsequent binding of the genetic transcriptional machinery to DNA promoter regions. In contrast, HDACs remove the acetyl group from lysines, which restores the basic charge on these amino acid sites, thereby allowing for resumption of the electrostatic link between histone protein and DNA. There is increasing evidence that various histone proteins and marks might regulate sexual differentiation in various brain regions.

The earliest study to examine for such sex differences determined that H1 (0) was differentially expressed in the arcuate area of the hypothalamus [89]. Within this region, increased number of immunoreactive neurons were identified in postnatal females; whereas, expression remained low in all ages of males studied. Androgenization of postnatal females likewise reduced the number of positive neurons in the dorsolateral and ventromedial portions of the arcuate region. Another study showed that late gestational (embryonic day 18) males express greater amounts of H3K9/14Ac and H3K9Me3 in the combined amygdala, cortex, and hippocampal regions [21].

The BNST, which is part of the limbic system, is one neural region where sex steroids might influence histone proteins [45,90]. The principal nucleus (BNSTp) is larger in males of a variety of species, including humans [91–94]. Neonatal testosterone is thought to affect neuronal cell number in the BNSTp region by impacting histone protein acetylation. Treatment of male and female mice with the HDAC inhibitor, valproic acid, on PND 1 and 2 induces H3 acetylation in the brain and abolished the normal sex differences in the BNSTp [95]. Testosterone-stimulated HDAC activity is postulated to increase histone protein binding to DNA encoding pro-apoptotic transcripts. Thus, the expression of such genes is attenuated in males but upregulated in females where the histone proteins remain acetylated and loosened from the DNA encoding such transcripts. Additional evidence implicating a role for HDAC activity during the postnatal period is illustrated by the fact that knockdown of steroid hormone receptor coactivator-1 (SRC-1), which has HAT activity, during the postnatal period, compromises later adult sexual behaviors [96].

Androgenization of prenatal females through testosterone treatment of the dam increased histone acetylation but not DNA methylation expression compared to males. A genome-wide comparison of H3K4Me3 (a histone mark associated with peaks around transcriptional initiation sites in active genes) in the BNST and POA regions revealed that adult female mice exhibited larger amounts of genes associated with this histone protein mark (71%) [97]. Of these, many of these genes are integral for normal synaptic function.

In the MPOA, sex differences are detected in acetylated levels of H3 and H4 proteins that associate with *Esr1* and *Arom* genes [98]. The MPOA is a well-classified sexually dimorphic brain region that is important in male sexual behaviors [99,100]. Both *Esr1* and *Cyp19A1* (P450 aromatase) are elevated in the MPOA region of males relative to females [55,60,101–103]. Acetylation levels are reduced in PND 3 males, suggestive that there is sex-dependent regulation of histone acetylation levels during this critical period [98]. Intracerebroventricular infusion of the HDAC inhibitor, trichostatin A, or antisense oligonucleotides against mRNA for HDAC2 or 4 to newborn (PND 0 and 1) male rats results in selective suppression of later adult male sexual behaviors but does not affect other behaviors [98]. The treatments do not alter serum testosterone concentrations or HDAC2 and 4 gene expression in the MPOA. However, greater amounts of HDAC2 bind to the *Esr1b* promoter and more HDAC2 and 4 binding to the Arom II is evident in males than females. The results support the hypothesis that HDACs are critical for proper masculinization of the brain during the postnatal period.

It has been hypothesized that estrogen may induce histone modifications that are required for sexual differentiation [104]. Males may exhibit increased histone acetylation but reduced methylation in the preoptic nerve cells that modulate sexual behaviors; whereas the opposite effects are predicted to occur in females. Females may instead demonstrate increased histone acetylation and reduced methylation in the VMH neurons, which are important for female sexual behaviors.

Nuclear receptor corepressor proteins, such as NCOR1, inhibit neural expression of ESRs and AR [105–107]. NCOR1 binds and alters the activity of HDAC, methyl-binding proteins (MBP), including ZBTB33 (formerly called KAISO) [105–107] and MECP2 [108,109]. Estradiol may impact the sexually dimorphic expression of NCOR1 in the developing amygdala [110]. Further, small interfering RNA against *Ncor1* diminishes sex differences in juvenile social play behavior. Reduction of *Ncor1* also leads to anxiogenic effects in juvenile males and females.

MicroRNAs: Hypothalamic expression of select microRNAs (miRs), miR-132, miR-145, and miR-9, and coding genes affected by these biomolecules (c-Myc, Lin28, and Lin28b) fluctuate throughout the lifespan in male and female rats and monkeys. Neonatal treatment with estrogen or androgen disrupts brain sexual differentiation and pubertal expression ratios of Lin28/let-7 [111]. In Atlantic halibut fish (*Hippoglossus hippoglossus*), several miRs show age- and sex-dependent differences in the gonad and brain. Both miR-9 and miR-202 are abundant in the brain and gonad. Contrasting miR signature profiles between fish masculinized due to treatment with either cytochrome P450 aromatase inhibitor, fadrozole, or 17α-methyltestosterone compared to untreated controls support the notion that miRs are involved in gonadal and brain sexual differentiation in teleost fish [112]. In zebra finches (*Taeniopygia guttata*), sexually dimorphic differences in response to song exposure may be mediated by auditory forebrain expression of miRs with tgu-miR-2954-3p (located on the Z sex chromosome) proposed to be one especially integral in regulating eight song-responsive coding RNAs and exhibiting direct effects on cellular proliferation and neuronal differentiation [113]. In birds, males are the homogametic sex (ZZ), whereas females are heterogametic (ZW). Accordingly, hearing a bird singing stimulates greater amounts of this miR in males than females.

Gonadal sex hormones and sex chromosomes may lead to sex-dependent differences in neural miR patterns. As reviewed in [44], sex hormone effects may be at the transcriptional or the posttranscriptional level. Current evidences for sex hormones, such as estradiol, governing miR expression changes are primarily in hormone-responsive tumors, such as breast cancers (e.g., [114]). Posttranscriptional regulation of miRs by estradiol and ESR1 has been reported [115–117]. When miRs bind to coding RNAs, double-stranded RNA is formed. Drosha and Dicer are two components of a larger double-stranded RNA-protein complex that can activate or repress these complexes. These effects may target select miRs or indiscriminately impact a range of miRs [118–120]. Estradiol/ESR1 has been implicated in controlling Dicer, Argonaute, and Drosha expression in other cells and tissues [117,121,122].

Sex chromosomal regulation of miRs in the brain and other organs has to be X chromosome mediated as no miRs are seemingly present on the Y chromosome [123]. In mammalian females, one X chromosome is subjected in X inactivation due to the X-linked noncoding RNA, Xist. However, ~25% of coding RNAs on X chromosome destined for inactivation escape this process, resulting in greater expression of these transcripts in females [124]. X-linked miRs escaping inactivation, however, have yet to be discovered. Even so, many (~86%) X-linked miRs resist meiotic sex chromosome inactivation, which occurs prior to and after meiotic spermatogenesis [125]. One study identified miRs with sex-dependent differences in expression in the neonatal mouse brain [126]. The investigators postulated that the sex-biased miR profiles are attributed to sex chromosome–dependent mechanisms as they persisted after inhibition of gonadal sex hormones. Of the 47 miRs identified, only seven reside on the X chromosome, suggestive that transcription factors expressed from the sex chromosomes may influence the expression of miRs residing on autosomal chromosomes. Predictably, six of the seven X-linked miRs were expressed in greater amounts in the neonatal brain of females.

GLUCOCORTICOID AND MATERNAL STRESS EFFECTS ON BRAIN DEVELOPMENT AND SEXUAL DIFFERENTIATION

Glucocorticoids induce pleiotropic effects in a wide range of systems. Maternal glucocorticoid exposure that occurs with stress can dramatically impede normal brain development and adult neurobehavioral functions. Excessive amounts of maternal glucocorticoids may lead to offspring disruptions in neuroendocrine activities; cognition; sleep and emotional disorders; anxiety; learning; sociosexual behaviors; hypothalamo-pituitary-adrenal (HPA) axis; serotonergic, catecholamine, and cardiometabolic pathways; glucose intolerance; psychotic disorders; and brain size to list a few examples [14,15,127,128]. These effects may occur in a sex-dependent manner with some studies suggesting that males may be at greater risk for maternal stress (increased endogenous or exogenous glucocorticoid concentrations) [8,9].

11β-Hydroxysteroid Dehydrogenase Enzymes and P-glycoprotein: Glucocorticoids are regulated by two 11β-hydroxysteroid dehydrogenase (11β-HSD) enzymes, 11β-HSD1 and 11β-HSD2 [129]. 11β-HSD1 acts as both a dehydrogenase by converting cortisol to cortisone and a reductase with conversion of cortisone to cortisol. 11β-HSD1 has been identified in vertebrates and sharks but is absent in fish [130]. This enzyme is most prominently expressed by the liver, but it is also widely distributed throughout the adult brain, including the cerebellum, hippocampus, cortex, and to a lesser extent in the hypothalamus [131,132]. Elevated hippocampal and neocortical expression of 11β-HSD1 occurs with aging and is associated with cognitive decline [129]. Targeted deletion or pharmacological/surgical suppression of this enzyme improves various metabolic disorders and age-onset cognitive decline in rodent models and human clinical trials [10,133–136].

The 11β-HSD2 enzyme solely serves as a dehydrogenase and inactivates glucocorticoids (cortisol) to cortisone. It is expressed in high amounts in the fetal brain and placenta [10]. 11β-HSD2 has greater affinity for glucocorticoids than 11β-HSD1. Placental inactivation of maternal glucocorticoids presumably buffers the fetus against the harmful effects of maternal cortisol (corticosterone). Yet, maternal stress is associated with increased amounts of glucocorticoids penetrating into the fetal circulation [137]. Further, 11β-HSD2 ceases to be produced in the rodent placenta around mid-gestation, which may then result in increased transfer of glucocorticoids from the mother to the fetus [11]. While 11β-HSD2 production declines in the late-gestational rodent placenta, the human placenta continues to produce this enzyme up until parturition [138]. Thus, rodents and humans may demonstrate differing vulnerabilities to elevated maternal glucocorticoids in late gestation. In humans, maternal betamethasone (a synthetic glucocorticoid) treatment induces a rapid increase in 11β-HSD2 in daughters compared to sons [139]. Thus, rapid placental production of this enzyme in girls may confer protection against potentially deleterious effects of glucocorticoids.

Placental 11β-HSD2 tends to have low affinity to many sGCs [15]. The placenta may suppress the transfer of these forms via P-glycoprotein, a member of the ABC transporter family [140,141]. The syncytiotrophoblast cells of the placenta produce high levels of this glycoprotein that can bind to various hormones and pharmacological agents, including sGC, and inhibit their transfer into the fetal circulation. Predictably, P-glycoprotein is also abundant in the fetal blood–brain barrier, where it presumably serves a similar role to reduce transfer of sGC into the fetal brain parenchyma [140,142].

While placental expression of 11β-HSD2 is important in protecting the fetal brain against excessive concentrations of maternal glucocorticoids, expression of this enzyme in the brain is also important. This is exemplified by the recent creation of transgenic mice where 11β-HSD2 has been conditionally deleted in the brain [143]. These mice demonstrate normal placental and fetal growth and circulating concentrations of glucocorticoids. Fetal brain concentrations of glucocorticoids though are raised by mid-gestation. Adult mice lacking brain expression of this enzyme exhibit depressive-like behaviors and cognitive deficits. While mice lacking fetoplacental 11β-HSD2 show anxiogenic behaviors [144,145], this effect is not observed in the neural 11β-HSD2-deficient mice [143]. Prenatal stress in Long–Evans rats causes increase DNA methylation and corresponding decrease gene expression of placental *HSD11b2* [146]. While this environmental change results in hypermethylation of this gene in the brain, gene expression is not altered. Other animal studies suggest that prenatal stress alters the activity of 11β-HSD2 [147]. By epigenetically suppressing this enzyme, maternal stress may further increase fetal vulnerability to elevated glucocorticoid concentrations.

Maternal Stress, Placenta, and Central Nervous System Disruptions: Maternal stress may trigger other placental disturbances that contribute to later neurobehavioral disorders. Endogenous glucocorticoids and sGC may increase corticotropin-releasing hormone (CRH) production by the human placenta that stimulates downstream activation of the fetal and maternal HPA axes [148]. However, only the placenta of primates expresses CRH, which may further account for species divergence in vulnerability to endogenous and synthetic glucocorticoids. Other ways in which glucocorticoids can alter placental function and lead to secondary neural effects include suppression of placental growth, placental-specific hormones, such as placental lactogens in rodents, placental vascularization, and by compromising nutrient transfer from the mother to the fetus [149]. There may also be sex differences in placental vulnerability to maternal stress. In mice, prenatal stress escalates expression of *Pppara, Igfbp1, Hifa,* and *Glut4* in male but not female placentae [9]. These males go on to show heightened stress responses, anhedonia (inability to experience pleasure), and deficits in the HPA axis. Prenatally stressed mice show increased placental expression of proinflammatory cytokines (*Il6* and *Il1b*), hyperlocomotion, and disruptions in brain expression of dopamine D1 and D2 receptors [150].

Hypothalamic-Pituitary-Axis/Glucocorticoid and Mineralocorticoid Receptors in the Brain: Endogenous corticosteroid synthesis and release is tightly regulated by the HPA axis. The hypothalamus acts as the master conductor by producing and releasing CRH and vasopressin from the paraventricular nucleus. CRHs, and to a lesser extent vasopressin, are stored as Herring bodies in the median eminence of the neurohypophysis (posterior pituitary gland) [12–14]. These hormones will then enter the hypophyseal portal system, which allows efficient and rapid delivery of these hormones to the pars distalis region of the adenohypophysis (anterior pituitary gland). The two peptide hormones stimulate the basophil cells in this region to release adrenocorticotropic hormone. Under physiological conditions, ACTH will stimulate the release of cortisol (corticosterone in rodents) from the zona fasciculata region and sex steroids by the zona reticularis of the adrenal cortex. A negative feedback loop exists where increased cortisol (corticosterone) concentrations directly inhibit CRH and ACTH. Both mineralocorticoid receptor (MR) in the hippocampus and glucocorticoid receptor (GR) in the PVN and anterior pituitary regulate the negative feedback system. Antagonism of MR and GR blunts this loop causing increased HPA activity; whereas, upregulation of both receptors increases the responsiveness and thus facilitates decreased HPA activity [12,13].

Maternal stress and early fetal exposure to endogenous/exogenous glucocorticoids can disrupt the HPA axis in several ways. Fetal exposure to increased maternal or sGCs can directly disrupt the HPA axis and sensitivity to glucocorticoid-induced negative feedback [14,15]. Glucocorticoid exposure may either increase or decrease HPA function. In the former, sensitivity to glucocorticoid inhibition is reduced, especially in the hippocampus [16,17]. Glucocorticoid-induced suppression of HPA function may paradoxically result in increased glucocorticoid sensitivity by peripheral systems.

Endogenous versus exogenous glucocorticoids display differing binding affinities to neural GR and MR. For example, sGCs, such as betamethasone and dexamethasone, exclusively bind to GR. In contrast, endogenous glucocorticoids (cortisol and corticosterone) bind both receptors [151]. This differential binding likely explains the contrasting prenatal programming effects on the HPA axis. Exogenous glucocorticoid engagement of GR in the brain and pituitary gland can inhibit both the fetal and maternal HPA axes with decreased production of cortisol (corticosterone) occurring. In the absence of a ligand, MR is upregulated in the hippocampus [152]. On the other hand, prenatal stress upregulates maternal and fetal cortisol that is free to bind and activates both GR and MR.

The perinatal HPA axis responses to sGC depend also on a myriad of other factors. These include species-specific glucocorticoid vulnerability, timing of exposure (prenatal versus postnatal), dose, type of sGC, offspring sex, placentation form, and length of gestation [14,139]. A few examples will be illustrated. In humans, the fetal HPA axis is functional and can respond to stress by 20 weeks of gestation [153–155]. Treatment of mothers with a single or multiple doses of sGC dramatically reduces maternal and fetal plasma ACTH and cortisol concentrations [156–158]. Administration of a single or multiple doses of sGC to the mother prior to birth also suppresses cortisol production in premature babies, and this decreased responsiveness of the HPA axis and hypocortisolemia persists through the first week of life [159,160]. In term neonates, a single dose of sGC prior to birth, however, elicits elevated stress responses [161].

Contrasting sensitivities to glucocorticoids are observed across animal taxa. Rodents are considered some of the most highly sensitive species [14]. Humans, Old World primates (baboons, rhesus monkeys, and vervet monkeys), and sheep are moderately sensitive. In contrast, guinea pigs followed by the common marmoset (New World primate) are significantly resistant to the effects of glucocorticoids. Based on this inherent sensitivity, varying doses and times of gestational/neonatal exposure to sGC have been tested in different animal models. Thus, the disparate results may also reflect these differences. Multiple doses of a high concentration of a sGC (1 mg/kg) to late-stage pregnant guinea pigs decreases circulating plasma cortisol concentrations in their fetuses [152,162]. Additionally, this treatment results in higher levels of hippocampal expression of *Mr* but reduce expression of *Crh* in the PVN of hypothalamus and proopiomelanocortin mRNA (*Pomc*, which is the parental molecule for ACTH) in the adenohypophysis. However, exposure of late gestational sheep to single and multiple treatments of a 10-fold less sGC dosage (0.11 mg/kg) and who are then stressed show elevations in ACTH and cortisol [163–165]. Perinatal exposure of baboons to the sGC, betamethasone, suppresses the HPA axis via inhibition of *POMC* expression in the anterior pituitary gland and *MC2R* (ACTH receptor) in the adrenal cortex [166].

The effects of prenatal and postnatal stress and dexamethasone have been tested in rhesus macaques and marmosets [167]. Rhesus macaques subjected to prenatal stress display increased cortisol levels and inhibition of neurogenesis. While prenatal dexamethasone exposure generates deficits in social play behaviors and skilled motor responses in marmosets, these effects are independent of HPA dysfunction. In contrast, postnatal social stress generates hypercortisolemia, decreases social play behaviors, and reduces expression of *Gr* and *Mr* in the hippocampus.

Female rat offspring exposed to prenatal environmental stress go on to develop a range of hormonal and behavioral alterations [168]. These include deficits in maternal behavior, which resemble those of males, increased plasma concentrations of corticosterone, decreased circulating estradiol concentrations, and masculinization of the bed nucleus of the accessory olfactory tract (BAOT). Male rats born to dams treated with sGC (betamethasone) or who experienced restraint stress also show hormonal and sexual behavior deficits, which may be attributed to maternal hypocortisolemia. Decreased interest in female sexual partners, fecundity, and hypoandrogenemia are observed in male rat offspring derived from dams treated with a sGC [169]. Similarly, sons of stressed rat dams demonstrate low testosterone concentrations, sexual behavior deficits, and elevations in dopamine and serotonin in the striatum [170]. However, neonatal testosterone supplementation to prenatally stressed males mitigates the sexual behavior deficits, suggestive that there is some plasticity in the glucocorticoid-induced programming responses in the brain [171].

Epigenetic Effects of Maternal Stress and Glucocorticoids: Similar to testosterone and estrogen, maternal stress/glucocorticoids can trigger epigenetic modifications in various brain regions of their offspring [18]. HPA hypersensitivity is created in pre-natally stressed male but not female mice [9]. Additionally, stress-reported changes in males include DNA hypomethylation of the *Crh* promoter in the hypothalamus but hypermethylation of the *Gr* promoter in the hippocampus. In rats, maternal stress upregulates various miRs in the neonatal whole brain [172]. Another study reveals that prenatal stress causes a later surge in miRs in the hippocampus and prefrontal cortex of juvenile and adult rat offspring [173].

Maternal hypercortisolemia or sGC administration to dams in late pregnancy facilitates global DNA methylation and transcriptomic changes in the hippocampus [174,175]. In general, promoter DNA methylation tilts to demethylation by early exposure to sGC. Placental transfer of endogenous and exogenous glucocorticoids causes stage-dependent effects in the hippocampal epigenome of offspring. H3K9 acetylation patterns are also altered in the hippocampus in guinea pigs prenatally exposed to sGC [175], although the specific epimutations transform with time. Other studies with guinea pigs provide evidence that the sGC treatment results in acute and chronic changes in enzymes and proteins regulating DNA methylation patterns, such as *Dnmt1*, *Dnmt3a*, *Dnmt3b*, *Mbd2*, and *Mecp2*, and miRs in the brain and hippocampus [174,176].

ENDOCRINE DISRUPTION OF NORMAL BRAIN PROGRAMMING

In this section, we will consider how two EDCs, bisphenol A (BPA) and phthalates, may disrupt normal brain programming. BPA is a widely prevalent EDC [177,178], with current production estimated to be about 15 billion pounds per year [179,180]. The reason for this dramatic escalation in BPA is that it is used to make a wide range of common household items, such as polycarbonate plastic items, lining of metal food cans, dental sealants, thermal receipt paper, and food containers to list a few examples. The ubiquitous nature of BPA [178] has ensured continual exposure of animals and humans, including pregnant women [180–182]. Di(2-ethylhexyl) phthalate (DEHP) is another common EDC used to produce a large assortment of items, including personal care products (such as makeup), paints, textiles, pharmaceuticals, and food [183]. While it can be metabolized, the primary metabolite, mono-ethylhexyl-phthalate (MEHP) has also been ascribed to induce neuroendocrine disturbances [184]. The potential mechanisms by which these chemicals can disrupt normal brain programming will be considered.

Steroidogenic Pathway Disruptions: Steroidogenic pathways can be disrupted by both BPA and phthalates. Cholesterol is the backbone precursor for all steroid hormones, and it has to be transported into the cell by the mitochondrial protein, steroidogenic acute regulatory protein (StAR, STARD1). *StAR* mRNA expression is inhibited in male and female gonads of rodents and fish exposed to BPA [185–189]. Steroidogenic enzymes are essential in production of testosterone and estrogen. Of these, BPA inhibits several key enzymes, especially *Cyp11, Hsd3b, Hsd17b, Cyp17*, and *Cyp19* [185,187–190].

Suppression of testosterone (T) and estrogen (E2) production by male and female gonads is reported after BPA exposure [185,190–192]. However, one study suggests elevated urinary BPA concentrations in girls with premature puberty positively correlates with increased T, E2, and pregnenolone levels [193]. Reduced StAR protein, steroidogenic enzymes, and sex hormones are observed in the testis after phthalate exposure [194–201]. Likewise, increased phthalate exposure in men is linked with decreased serum concentrations of T and E2 [202].

Endocrine Disruption of Neural ESR and AR: Normal brain programming may also be hampered by BPA and phthalates directly altering brain expression of ESR1 and 2 and AR. There is not a consensus across studies though as to the directionality of these changes. The disparate results likely reflect the brain region examined, dose and time of exposure, and there may be species-dependent differences [203–211]. There is a paucity of data though on how phthalates affect the brain expression of these steroid receptors. Rats treated with a phthalate form, benzyl butyl phthalate (BBP, 5 mg/L, 0.16 μM), upregulate *Esr1* expression in the amygdala of rats. Addition of BPA (285.4 mg/kg) and DBP (285.4 mg/kg) to the diet of juvenile rats amplifies *Ar* expression in the brain [212].

Endocrine Disruption of the HPA axis: BPA, phthalates, and other EDC exposure may interrupt normal HPA axis function. Prenatal BPA exposure of rats via the dam (oral administration of 40 μg/kg bw/day to the dam throughout gestation and lactation) leads to increased serum corticosterone concentrations and reduced hypothalamic *Gr* expression in daughters [213,214]. BPA-exposed sons show suppression of hippocampal expression of *Gr*. Male but not female rats, whose mothers are treated with 2 μg BPA/kg bw/day from gestational day 10 to lactational day 7, develop elevations in basal concentrations of serum corticosterone and ACTH and *Crh* expression in the hypothalamus [215]. Application of a mild stressor potentiates further surges in corticosterone and ACTH concentrations in perinatally BPA-exposed males, but the opposite hormonal effects are observed in BPA + stressed females. Nonstressed perinatally BPA-exposed females exhibit upregulation *Gr* in the hippocampus and PVN, which are instead suppressed in males. Taken together, the data suggest that BPA exposure might affect the HPA axis in a sex- and dose-dependent manner. The HPA axis is also affected by exposure to phthalates. Current rodent and in vitro culture studies hint that this chemical causes elevations in circulating concentrations of ACTH and corticosterone, and treated adrenocortical cells may be hypersensitized to the effects of ACTH [216].

EDC-induced Epimutations: Similar to endogenous estrogen, testosterone, and glucocorticoids, EDC, such as BPA and phthalates, may also generate epimutations. Such alterations are another mechanism by which these chemicals can disrupt normal brain programming responses (as illustrated in Figure 1). Exposure to BPA, phthalates, or other EDCs can affect the DNA methylome in several systems, including the central nervous system [217–222]. DNA methylation patterns in the mouse forebrain are shifted by prenatal exposure to BPA [222]. Similarly, prenatal exposure to BPA causes persistent DNA methylation changes in one of the promoters of brain-derived neural factor (*Bdnf*) in the hippocampus and cord blood, but the direction depends on sex with hypermethylation evident

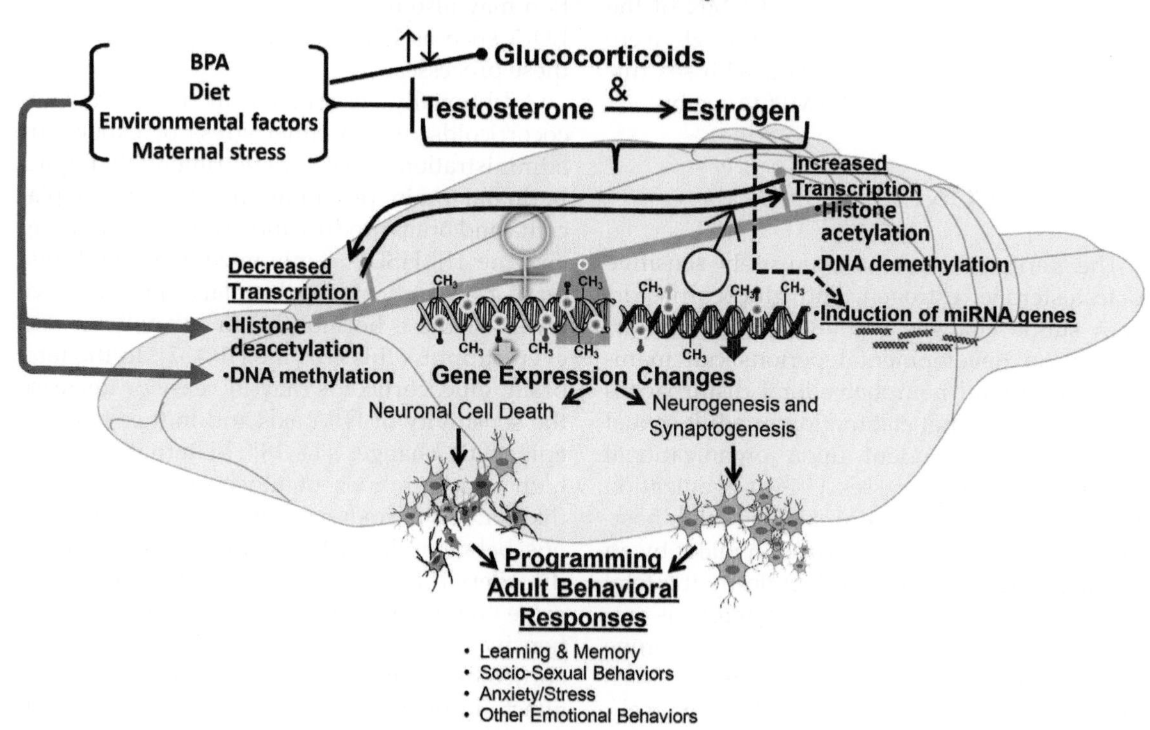

FIGURE 1 Hypothetical model of how estrogen, testosterone, and glucocorticoids program normal behavioral responses. The diagram also shows how EDCs, such as BPA and phthalates, may epigenetically disrupt these process steroid hormones, including estrogen testosterone, and glucocorticoids can modulate various epigenetic changes, including in DNA methylation patterns, histone protein modification, and miR changes in the brain. The landscape and cumulative changes can either lead to normal masculinization/defeminization or feminization of the brain or disrupt these pathways. *Modified from Ref. [22].*

in females; whereas, hypomethylation occurs in males [221].

Other studies have explored whether BPA simultaneously targets DNA methylation and histone proteins. BPA exposure of rats disturbs brain cortical expression of the ion transporter (*Kcc2*) through DNA methylation changes mediated by MECP2 and histone protein (H3K9) association with this gene [223]. Phthalates can also induce histone protein modifications in neuronal cells, including deacetylation (via HDAC4), and polyubiquination [224].

BPA may impact epigenetic regulating enzymes *Dnmt1*, *Dntm3a*, *Dnmt3b*, and *Mecp2* in several brain regions, such as the basolateral amygdala, cortex, and hypothalamus [205,225,226]. Adult males exposed to BPA possess greater amounts of *Hdac2* in the hippocampus [227].

No study to date has considered whether BPA exposure affects noncoding (nc) RNAs (including miRs) in the brain or neuronal cells. However, this chemical can affect expression of miRs in other cells, such as MCF7 [228], ovarian [229], and placental cells [230]. The effects of phthalate exposure on miRs in the brain or other body regions have not been examined to date. A recent study has shown that prenatal exposure of male and female rats to estradiol benzoate or a mixture of polychlorinated biphenyls alters postnatal miRs in the MPOA and VMN of the hypothalamus [231]. The effects though were region-, sex-, and age-dependent, with sex differences only observed in the MPOA region.

CONCLUSIONS

The perinatal brain is exquisitely sensitive to testosterone, estrogen, and glucocorticoids. Even subtle changes in these hormones during critical developmental periods can manifest as profound neurobehavioral disturbances later in life. Normal elaboration of adult sexual behaviors is dependent upon organizational and activational processes [1-3]. Organization of the brain generally occurs in the fetal/postnatal period and is characterized in males by an increase in testosterone production by the fetal testis. In some brain regions, testosterone is aromatized to estradiol. Collectively, testosterone and estradiol induce defeminization and masculinization of the male brain [4]. Conversely, the female brain is the default pathway that occurs in the absence of rising testosterone/estradiol concentrations in the brain during this critical time. However, females exposed in utero to endogenous/exogenous testosterone will undergo brain defeminization/masculinization and will demonstrate male-typical sexual behaviors as adults. Improper brain programming of males can result in them showing female-typical sexual behaviors as adults. Epigenetics may provide the link by which endogenous sex hormones and EDC, such as BPA and phthalates, can either result in normal brain programming or disrupt these vital processes [5–7,19–22]. Such changes include those to the DNA methylome, histone protein modifications, and alteration in miR patterns. Future studies though are needed to examine how endogenous and exogenous estrogens and testosterone interact with fetal sex to shape the brain epigenome. A better understanding of the normal cascade epigenetic events culminating in brain sexual differentiation may also reveal how perinatal exposure to EDC, such as BPA and phthalate, impinges on these processes.

Additionally, elevations in maternal glucocorticoids due to either stress or exogenous administration of sGC can detrimentally impact perinatal brain programming. While the placenta and brain produce the cortisol-inactivating enzyme 11β-HSD2 to minimize maternal transfer of glucocorticoids to the fetal brain, excessive levels of these hormones can overwhelm and even suppress this enzyme [10,137]. In the fetal brain, glucocorticoids may increase or decrease the sensitivity of HPA axis and induce harmful epigenetic changes [14–18]. Maternal stress or high concentrations of glucocorticoids during the in utero period may ultimately lead to various behavioral, cardiovascular, and metabolic disorders [14,15,127,128]. Further, sons may be more vulnerable than daughters [8,9]. A comprehensive analysis of how maternal stress/glucocorticoids results in a wide range of pathological effects in offspring is thus critically important. As with sex steroids, the epigenome may be the key in understanding how perinatal exposure to glucocorticoids underpins later neurobehavioral and other diseases.

Glossary

2F A female fetus situated in utero between female fetuses
2M A female fetus situated in utero between two male fetuses
11β-HSD1 11β-Hydroxysteroid dehydrogenase 1
11β-HSD2 11β-Hydroxysteroid dehydrogenase 2
ACTH Adrenocorticotropic hormone
AR Androgen receptor protein
Ar Androgen receptor mRNA
AVPV Anteroventral periventricular
Bdnf Brain-derived neural factor mRNA
BNST Bed nucleus of the stria terminalis
BNSTp Principal nucleus of the BNST
BPA Bisphenol A
BPP Benzyl butyl phthalate
CRH Corticotropin-releasing hormone protein

Crh Corticotropin-releasing hormone mRNA
DEHP Di(2-ethylhexyl) phthalate
DHT Dihydrotestosterone
DNMTs DNA methyl transferases
DNMT1 DNA methyl transferase 1 protein
Dnmt1 DNA methyl transferase 1 mRNA
DNMT3a DNA methyl transferase 3a protein
Dnmt3a DNA methyl transferase 3a mRNA
DNMT3b DNA methyl transferase 3b protein
Dnmt3b DNA methyl transferase 3b mRNA
DPN Diarylpropionitrile
EDC Endocrine-disrupting chemical
ESR1 Estrogen receptor 1 protein
Esr1 Estrogen receptor 1 mRNA
ESR2 Estrogen receptor 2 protein
Esr2 Estrogen receptor 2 mRNA
GR Glucocorticoid receptor protein
Gr Glucocorticoid receptor mRNA
HAT Histone acetyl transferases
HDAC Histone deacetylase enzymes
HPA Hypothalamic-pituitary-adrenal axis
HSD11b2 Gene-encoding 11β-HSD2
LG Licking and grooming
MBP Methyl-binding proteins
MeCP2 Methyl-CpG-binding protein 2
Mecp2 Methyl-CpG-binding protein 2 encoding mRNA
MEHP Monoethylhexyl-phthalate
miRs microRNAs
MPOA Medial preoptic area
MR Mineralocorticoid receptor protein
Mr Mineralocorticoid receptor mRNA
NCOR1 Nuclear receptor corepressor proteins
Ncor1 Nuclear receptor corepressor protein encoding mRNA
PND Postnatal day
POMC Proopiomelanocortin protein
Pomc Proopiomelanocortin mRNA
PPT Propyl-pyrazole triol
POA Preoptic area
SDN Sexually dimorphic nucleus
SDN-POA Sexually dimorphic nucleus of the preoptic area
sGC Synthetic glucocorticoids
SRC-1 Steroid hormone receptor coactivator-1
StAR Steroidogenic acute regulatory protein
VMH Ventromedial hypothalamus

References

[1] Arnold AP, Breedlove SM. Organizational and activational effects of sex steroids on brain and behavior: a reanalysis. Hormones Behav 1985;19:469–98.

[2] Morris JA, Jordan CL, Breedlove SM. Sexual differentiation of the vertebrate nervous system. Nat Neurosci 2004;7:1034–9.

[3] Phoenix CH, Goy RW, Gerall AA, Young WC. Organizing action of prenatally administered testosterone propionate on the tissues mediating mating behavior in the female guinea pig. Endocrinology 1959;65:369–82.

[4] Savic I, Garcia-Falgueras A, Swaab DF. Sexual differentiation of the human brain in relation to gender identity and sexual orientation. Prog Brain Res 2010;186:41–62.

[5] Menger Y, Bettscheider M, Murgatroyd C, Spengler D. Sex differences in brain epigenetics. Epigenomics 2010;2:807–21.

[6] Matsuda KI, Mori H, Kawata M. Epigenetic mechanisms are involved in sexual differentiation of the brain. Rev Endocr Metab Disord 2012;13:163–71.

[7] Nugent BM, McCarthy MM. Epigenetic underpinnings of developmental sex differences in the brain. Neuroendocrinology 2011;93:150–8.

[8] Bale TL. Sex differences in prenatal epigenetic programming of stress pathways. Stress 2011;14:348–56.

[9] Mueller BR, Bale TL. Sex-specific programming of offspring emotionality after stress early in pregnancy. J Neurosci 2008;28:9055–65.

[10] Chapman K, Holmes M, Seckl J. 11beta-hydroxysteroid dehydrogenases: intracellular gate-keepers of tissue glucocorticoid action. Physiol Rev 2013;93:1139–206.

[11] Brown RW, Diaz R, Robson AC, Kotelevtsev YV, Mullins JJ, Kaufman MH, et al. The ontogeny of 11 beta-hydroxysteroid dehydrogenase type 2 and mineralocorticoid receptor gene expression reveal intricate control of glucocorticoid action in development. Endocrinology 1996;137:794–7.

[12] Ulrich-Lai YM, Herman JP. Neural regulation of endocrine and autonomic stress responses. Nat Rev Neurosci 2009;10:397–409.

[13] de Kloet ER, Joels M, Holsboer F. Stress and the brain: from adaptation to disease. Nat Rev Neurosci 2005;6:463–75.

[14] Moisiadis VG, Matthews SG. Glucocorticoids and fetal programming part 1: outcomes. Nat Rev Endocrinol 2014;10:391–402.

[15] Moisiadis VG, Matthews SG. Glucocorticoids and fetal programming part 2: mechanisms. Nat Rev Endocrinol 2014;10:403–11.

[16] Kapoor A, Matthews SG. Short periods of prenatal stress affect growth, behaviour and hypothalamo-pituitary-adrenal axis activity in male guinea pig offspring. J Physiol 2005;566:967–77.

[17] Kapoor A, Leen J, Matthews SG. Molecular regulation of the hypothalamic-pituitary-adrenal axis in adult male guinea pigs after prenatal stress at different stages of gestation. J Physiol 2008;586:4317–26.

[18] Babenko O, Kovalchuk I, Metz GA. Stress-induced perinatal and transgenerational epigenetic programming of brain development and mental health. Neurosci Biobehav Rev 2015;48:70–91.

[19] Rosenfeld CS. Bisphenol A and phthalate endocrine disruption of parental and social behaviors. Front Neurosci 2015;9:1–15.
[20] Chung WC, Auger AP. Gender differences in neurodevelopment and epigenetics. Pflugers Arch 2013;465:573–84.
[21] Tsai HW, Grant PA, Rissman EF. Sex differences in histone modifications in the neonatal mouse brain. Epigenetics 2009;4:47–53.
[22] Jasarevic E, Geary DC, Rosenfeld CS. Sexually selected traits: a fundamental framework for studies on behavioral epigenetics. ILAR J 2012;53:253–69.
[23] Schulz KM, Molenda-Figueira HA, Sisk CL. Back to the future: the organizational-activational hypothesis adapted to puberty and adolescence. Horm Behav 2009;55:597–604.
[24] Schulz KM, Zehr JL, Salas-Ramirez KY, Sisk CL. Testosterone programs adult social behavior before and during, but not after, adolescence. Endocrinology 2009;150:3690–8.
[25] Watson J, Adkins-Regan E. Activation of sexual behavior by implantation of testosterone propionate and estradiol benzoate into the preoptic area of the male Japanese quail (Coturnix japonica). Horm Behav 1989;23:251–68.
[26] Watson J, Adkins-Regan E. Testosterone implanted in the preoptic area of male Japanese quail must be aromatized to activate copulation. Horm Behav 1989;23:432–47.
[27] Bowers JM, Waddell J, McCarthy MM. A developmental sex difference in hippocampal neurogenesis is mediated by endogenous oestradiol. Biol Sex Differ 2010;1:8.
[28] Konkle AT, McCarthy MM. Developmental time course of estradiol, testosterone, and dihydrotestosterone levels in discrete regions of male and female rat brain. Endocrinology 2011;152:223–35.
[29] Money J. Androgyne becomes bisexual in sexological theory: Plato to Freud and neuroscience. J Am Acad Psychoanal 1990;18:392–413.
[30] Roselli CE, Stormshak F. Prenatal programming of sexual partner preference: the ram model. J Neuroendocrinol 2009;21:359–64.
[31] Roselli CE, Stormshak F. The ovine sexually dimorphic nucleus, aromatase, and sexual partner preferences in sheep. J Steroid Biochem Mol Biol 2010;118:252–6.
[32] Roselli CE, Reddy RC, Estill CT, Scheldrup M, Meaker M, Stormshak F, et al. Prenatal influence of an androgen agonist and antagonist on the differentiation of the ovine sexually dimorphic nucleus in male and female lamb fetuses. Endocrinology 2014;155:5000–10.
[33] McCarthy MM. How it's made: organisational effects of hormones on the developing brain. J Neuroendocrinol 2010;22:736–42.
[34] Rhees RW, Shryne JE, Gorski RA. Onset of the hormone-sensitive perinatal period for sexual differentiation of the sexually dimorphic nucleus of the preoptic area in female rats. J Neurobiol 1990;21:781–6.
[35] Arnold AP, Gorski RA. Gonadal steroid induction of structural sex differences in the central nervous system. Annu Rev Neurosci 1984;7:413–42.
[36] Lephart ED. A review of brain aromatase cytochrome P450. Brain Res Brain Res Rev 1996;22:1–26.
[37] Naftolin F, Ryan KJ, Davies IJ, Reddy VV, Flores F, Petro Z, et al. The formation of estrogens by central neuroendocrine tissues. Recent Prog Horm Res 1975;31:295–319.
[38] Barraclough CA. Production of anovulatory, sterile rats by single injections of testosterone propionate. Endocrinology 1961;68:62–7.
[39] Baum MJ. Differentiation of coital behavior in mammals: a comparative analysis. Neurosci Biobehav Rev 1979;3:265–84.
[40] Pei M, Matsuda K, Sakamoto H, Kawata M. Intrauterine proximity to male fetuses affects the morphology of the sexually dimorphic nucleus of the preoptic area in the adult rat brain. Eur J Neurosci 2006;23:1234–40.
[41] Vreeburg JT, van der Vaart PD, van der Schoot P. Prevention of central defeminization but not masculinization in male rats by inhibition neonatally of oestrogen biosynthesis. J Endocrinol 1977;74:375–82.
[42] Bakker J, Honda S, Harada N, Balthazart J. Sexual partner preference requires a functional aromatase (cyp19) gene in male mice. Horm Behav 2002;42:158–71.
[43] Gilmore DP. Sexual dimorphism in the central nervous system of marsupials. Int Rev Cytol 2002;214:193–224.
[44] Morgan CP, Bale TL. Sex differences in microRNA regulation of gene expression: no smoke, just miRs. Biol Sex Differ 2012;3:22.
[45] McCarthy MM, Auger AP, Bale TL, De Vries GJ, Dunn GA, Forger NG, et al. The epigenetics of sex differences in the brain. J Neurosci 2009;29:12815–23.
[46] Kolodkin MH, Auger AP. Sex difference in the expression of DNA methyltransferase 3a in the rat amygdala during development. J Neuroendocrinol 2011;23:577–83.
[47] Nugent BM, Wright CL, Shetty AC, Hodes GE, Lenz KM, Mahurkar A, et al. Brain feminization requires active repression of masculinization via DNA methylation. Nat Neurosci 2015;18:690–7.
[48] Ghahramani NM, Ngun TC, Chen PY, Tian Y, Krishnan S, Muir S, et al. The effects of perinatal testosterone exposure on the DNA methylome of the mouse brain are late-emerging. Biol Sex Differ 2014;5:8.
[49] Nan X, Ng HH, Johnson CA, Laherty CD, Turner BM, Eisenman RN, et al. Transcriptional repression by the methyl-CpG-binding protein MeCP2 involves a histone deacetylase complex. Nature 1998;393:386–9.

[50] Kurian JR, Forbes-Lorman RM, Auger AP. Sex difference in mecp2 expression during a critical period of rat brain development. Epigenetics 2007;2:173–8.
[51] Kurian JR, Bychowski ME, Forbes-Lorman RM, Auger CJ, Auger AP. Mecp2 organizes juvenile social behavior in a sex-specific manner. J Neurosci 2008;28:7137–42.
[52] Kawata M. Roles of steroid hormones and their receptors in structural organization in the nervous system. Neurosci Res 1995;24:1–46.
[53] Champagne FA, Weaver IC, Diorio J, Dymov S, Szyf M, Meaney MJ. Maternal care associated with methylation of the estrogen receptor-alpha1b promoter and estrogen receptor-alpha expression in the medial preoptic area of female offspring. Endocrinology 2006;147:2909–15.
[54] Kurian JR, Olesen KM, Auger AP. Sex differences in epigenetic regulation of the estrogen receptor-alpha promoter within the developing preoptic area. Endocrinology 2010;151:2297–305.
[55] DonCarlos LL, Handa RJ. Developmental profile of estrogen receptor mRNA in the preoptic area of male and female neonatal rats. Brain Res Dev Brain Res 1994;79:283–9.
[56] Yokosuka M, Okamura H, Hayashi S. Postnatal development and sex difference in neurons containing estrogen receptor-alpha immunoreactivity in the preoptic brain, the diencephalon, and the amygdala in the rat. J Comp Neurol 1997;389:81–93.
[57] Kudwa AE, Michopoulos V, Gatewood JD, Rissman EF. Roles of estrogen receptors alpha and beta in differentiation of mouse sexual behavior. Neuroscience 2006;138:921–8.
[58] Schwarz JM, Nugent BM, McCarthy MM. Developmental and hormone-induced epigenetic changes to estrogen and progesterone receptor genes in brain are dynamic across the life span. Endocrinology 2010;151:4871–81.
[59] Patchev AV, Gotz F, Rohde W. Differential role of estrogen receptor isoforms in sex-specific brain organization. FASEB J 2004;18:1568–70.
[60] Ogawa S, Lubahn DB, Korach KS, Pfaff DW. Behavioral effects of estrogen receptor gene disruption in male mice. Proc Natl Acad Sci USA 1997;94:1476–81.
[61] Ogawa S, Taylor JA, Lubahn DB, Korach KS, Pfaff DW. Reversal of sex roles in genetic female mice by disruption of estrogen receptor gene. Neuroendocrinology 1996;64:467–70.
[62] Prewitt AK, Wilson ME. Changes in estrogen receptor-alpha mRNA in the mouse cortex during development. Brain Res 2007;1134:62–9.
[63] Miranda RC, Toran-Allerand CD. Developmental expression of estrogen receptor mRNA in the rat cerebral cortex: a nonisotopic in situ hybridization histochemistry study. Cereb Cortex 1992;2:1–15.
[64] Westberry JM, Prewitt AK, Wilson ME. Epigenetic regulation of the estrogen receptor alpha promoter in the cerebral cortex following ischemia in male and female rats. Neuroscience 2008;152:982–9.
[65] Mori H, Matsuda KI, Tsukahara S, Kawata M. Intrauterine position affects estrogen receptor alpha expression in the ventromedial nucleus of the hypothalamus via promoter DNA methylation. Endocrinology 2010;151:5775–81.
[66] Shughrue PJ, Lane MV, Merchenthaler I. Comparative distribution of estrogen receptor-alpha and -beta mRNA in the rat central nervous system. J Comp Neurol 1997;388:507–25.
[67] Krezel W, Dupont S, Krust A, Chambon P, Chapman PF. Increased anxiety and synaptic plasticity in estrogen receptor beta-deficient mice. Proc Natl Acad Sci USA 2001;98:12278–82.
[68] Imwalle DB, Gustafsson JA, Rissman EF. Lack of functional estrogen receptor beta influences anxiety behavior and serotonin content in female mice. Physiol Behav 2005;84:157–63.
[69] Rocha BA, Fleischer R, Schaeffer JM, Rohrer SP, Hickey GJ. 17 Beta-estradiol-induced antidepressant-like effect in the forced swim test is absent in estrogen receptor-beta knockout (BERKO) mice. Psychopharmacol Berl 2005;179:637–43.
[70] Kudwa AE, Bodo C, Gustafsson JA, Rissman EF. A previously uncharacterized role for estrogen receptor beta: defeminization of male brain and behavior. Proc Natl Acad Sci USA 2005;102:4608–12.
[71] Zama AM, Uzumcu M. Fetal and neonatal exposure to the endocrine disruptor methoxychlor causes epigenetic alterations in adult ovarian genes. Endocrinology 2009;150:4681–91.
[72] Kumar RC, Thakur MK. Androgen receptor mRNA is inversely regulated by testosterone and estradiol in adult mouse brain. Neurobiol Aging 2004;25:925–33.
[73] Wolf SS, Patchev VK, Obendorf M. A novel variant of the putative demethylase gene, s-JMJD1C, is a coactivator of the AR. Arch Biochem Biophys 2007;460:56–66.
[74] Widom J. Structure, dynamics, and function of chromatin in vitro. Annu Rev Biophys Biomol Struct 1998;27:285–327.
[75] Biterge B, Schneider R. Histone variants: key players of chromatin. Cell Tissue Res 2014;356:457–66.
[76] Hayes JJ, Hansen JC. Nucleosomes and the chromatin fiber. Curr Opin Genet Dev 2001;11:124–9.
[77] McBryant SJ, Adams VH, Hansen JC. Chromatin architectural proteins. Chromosome Res 2006;14:39–51.
[78] Felsenfeld G, Groudine M. Controlling the double helix. Nature 2003;421:448–53.
[79] Kornberg RD, Lorch Y. Twenty-five years of the nucleosome, fundamental particle of the eukaryote chromosome. Cell 1999;98:285–94.

[80] Ost A, Lempradl A, Casas E, Weigert M, Tiko T, Deniz M, et al. Paternal diet defines offspring chromatin state and intergenerational obesity. Cell 2014;159:1352–64.
[81] Mehler MF. Epigenetic principles and mechanisms underlying nervous system functions in health and disease. Prog Neurobiol 2008;86:305–41.
[82] Mehler MF. Epigenetics and the nervous system. Ann Neurol 2008;64:602–17.
[83] Hansen JC, Tse C, Wolffe AP. Structure and function of the core histone N-termini: more than meets the eye. Biochemistry 1998;37:17637–41.
[84] Wolffe AP, Hayes JJ. Chromatin disruption and modification. Nucleic Acids Res 1999;27:711–20.
[85] Ye J, Ai X, Eugeni EE, Zhang L, Carpenter LR, Jelinek MA, et al. Histone H4 lysine 91 acetylation a core domain modification associated with chromatin assembly. Mol Cell 2005;18:123–30.
[86] Grunstein M. Histone acetylation in chromatin structure and transcription. Nature 1997;389:349–52.
[87] Lee DY, Hayes JJ, Pruss D, Wolffe AP. A positive role for histone acetylation in transcription factor access to nucleosomal DNA. Cell 1993;72:73–84.
[88] Kuo MH, Allis CD. Roles of histone acetyltransferases and deacetylases in gene regulation. Bioessays 1998;20:615–26.
[89] Garcia-Segura LM, Luquin S, Martinez P, Casas MT, Suau P. Differential expression and gonadal hormone regulation of histone H1(0) in the developing and adult rat brain. Brain Res Dev Brain Res 1993;73:63–70.
[90] Qureshi IA, Mehler MF. Genetic and epigenetic underpinnings of sex differences in the brain and in neurological and psychiatric disease susceptibility. Prog Brain Res 2010;186:77–95.
[91] Forger NG, Rosen GJ, Waters EM, Jacob D, Simerly RB, de Vries GJ. Deletion of Bax eliminates sex differences in the mouse forebrain. Proc Natl Acad Sci USA 2004;101:13666–71.
[92] Guillamon A, Segovia S, del Abril A. Early effects of gonadal steroids on the neuron number in the medial posterior region and the lateral division of the bed nucleus of the stria terminalis in the rat. Brain Res Dev Brain Res 1988;44:281–90.
[93] Hines M, Allen LS, Gorski RA. Sex differences in subregions of the medial nucleus of the amygdala and the bed nucleus of the stria terminalis of the rat. Brain Res 1992;579:321–6.
[94] Hines M, Davis FC, Coquelin A, Goy RW, Gorski RA. Sexually dimorphic regions in the medial preoptic area and the bed nucleus of the stria terminalis of the guinea pig brain: a description and an investigation of their relationship to gonadal steroids in adulthood. J Neurosci 1985;5:40–7.
[95] Murray EK, Hien A, de Vries GJ, Forger NG. Epigenetic control of sexual differentiation of the bed nucleus of the stria terminalis. Endocrinology 2009;150:4241–7.
[96] Auger AP, Tetel MJ, McCarthy MM. Steroid receptor coactivator-1 (SRC-1) mediates the development of sex-specific brain morphology and behavior. Proc Natl Acad Sci USA 2000;97:7551–5.
[97] Shen EY, Ahern TH, Cheung I, Straubhaar J, Dincer A, Houston I, et al. Epigenetics and sex differences in the brain: a genome-wide comparison of histone-3 lysine-4 trimethylation (H3K4me3) in male and female mice. Exp Neurol 2015;268:21–9.
[98] Matsuda KI, Mori H, Nugent BM, Pfaff DW, McCarthy MM, Kawata M. Histone deacetylation during brain development is essential for permanent masculinization of sexual behavior. Endocrinology 2011;152:2760–7.
[99] Gorski RA. Sexual dimorphisms of the brain. J Anim Sci 1985;61(Suppl. 3):38–61.
[100] Larsson K, Heimer L. Mating behaviour of male rats after lesions in the preoptic area. Nature 1964;202:413–4.
[101] Matsumoto T, Honda S, Harada N. Alteration in sex-specific behaviors in male mice lacking the aromatase gene. Neuroendocrinology 2003;77:416–24.
[102] Colciago A, Celotti F, Pravettoni A, Mornati O, Martini L, Negri-Cesi P. Dimorphic expression of testosterone metabolizing enzymes in the hypothalamic area of developing rats. Brain Res Dev Brain Res 2005;155:107–16.
[103] DonCarlos LL. Developmental profile and regulation of estrogen receptor (ER) mRNA expression in the preoptic area of prenatal rats. Brain Res Dev Brain Res 1996;94:224–33.
[104] Gagnidze K, Weil ZM, Pfaff DW. Histone modifications proposed to regulate sexual differentiation of brain and behavior. Bioessays 2010;32:932–9.
[105] Horlein AJ, Naar AM, Heinzel T, Torchia J, Gloss B, Kurokawa R, et al. Ligand-independent repression by the thyroid hormone receptor mediated by a nuclear receptor co-repressor. Nature 1995;377:397–404.
[106] Lavinsky RM, Jepsen K, Heinzel T, Torchia J, Mullen TM, Schiff R, et al. Diverse signaling pathways modulate nuclear receptor recruitment of N-CoR and SMRT complexes. Proc Natl Acad Sci USA 1998;95:2920–5.
[107] Yoon HG, Wong J. The corepressors silencing mediator of retinoid and thyroid hormone receptor and nuclear receptor corepressor are involved in agonist- and antagonist-regulated transcription by androgen receptor. Mol Endocrinol 2006;20:1048–60.
[108] Cukier HN, Perez AM, Collins AL, Zhou Z, Zoghbi HY, Botas J. Genetic modifiers of MeCP2 function in Drosophila. PLoS Genet 2008;4:e1000179.

[109] Kokura K, Kaul SC, Wadhwa R, Nomura T, Khan MM, Shinagawa T, et al. The Ski protein family is required for MeCP2-mediated transcriptional repression. J Biol Chem 2001;276:34115–21.
[110] Jessen HM, Kolodkin MH, Bychowski ME, Auger CJ, Auger AP. The nuclear receptor corepressor has organizational effects within the developing amygdala on juvenile social play and anxiety-like behavior. Endocrinology 2010;151:1212–20.
[111] Sangiao-Alvarellos S, Manfredi-Lozano M, Ruiz-Pino F, Navarro VM, Sanchez-Garrido MA, Leon S, et al. Changes in hypothalamic expression of the Lin28/let-7 system and related microRNAs during postnatal maturation and after experimental manipulations of puberty. Endocrinology 2013;154:942–55.
[112] Bizuayehu TT, Babiak J, Norberg B, Fernandes JM, Johansen SD, Babiak I. Sex-biased miRNA expression in Atlantic halibut (*Hippoglossus hippoglossus*) brain and gonads. Sex Dev 2012;6:257–66.
[113] Gunaratne PH, Lin YC, Benham AL, Drnevich J, Coarfa C, Tennakoon JB, et al. Song exposure regulates known and novel microRNAs in the zebra finch auditory forebrain. BMC Genomics 2011;12:277.
[114] Hah N, Danko CG, Core L, Waterfall JJ, Siepel A, Lis JT, et al. A rapid, extensive, and transient transcriptional response to estrogen signaling in breast cancer cells. Cell 2011;145:622–34.
[115] Pawlicki JM, Steitz JA. Nuclear networking fashions pre-messenger RNA and primary microRNA transcripts for function. Trends Cell Biol 2010;20:52–61.
[116] Siomi H, Siomi MC. Posttranscriptional regulation of microRNA biogenesis in animals. Mol Cell 2010;38:323–32.
[117] Yamagata K, Fujiyama S, Ito S, Ueda T, Murata T, Naitou M, et al. Maturation of microRNA is hormonally regulated by a nuclear receptor. Mol Cell 2009;36:340–7.
[118] Viswanathan SR, Daley GQ, Gregory RI. Selective blockade of microRNA processing by Lin28. Science 2008;320:97–100.
[119] Michlewski G, Guil S, Semple CA, Caceres JF. Posttranscriptional regulation of miRNAs harboring conserved terminal loops. Mol Cell 2008;32:383–93.
[120] Gregory RI, Yan KP, Amuthan G, Chendrimada T, Doratotaj B, Cooch N, et al. The microprocessor complex mediates the genesis of microRNAs. Nature 2004;432:235–40.
[121] Bhat-Nakshatri P, Wang G, Collins NR, Thomson MJ, Geistlinger TR, Carroll JS, et al. Estradiol-regulated microRNAs control estradiol response in breast cancer cells. Nucleic Acids Res 2009;37:4850–61.
[122] Adams BD, Claffey KP, White BA. Argonaute-2 expression is regulated by epidermal growth factor receptor and mitogen-activated protein kinase signaling and correlates with a transformed phenotype in breast cancer cells. Endocrinology 2009;150:14–23.
[123] Kozomara A, Griffiths-Jones S. miRBase: integrating microRNA annotation and deep-sequencing data. Nucleic Acids Res 2011;39:D152–7.
[124] Carrel L, Willard HF. X-inactivation profile reveals extensive variability in X-linked gene expression in females. Nature 2005;434:400–4.
[125] Song R, Ro S, Michaels JD, Park C, McCarrey JR, Yan W. Many X-linked microRNAs escape meiotic sex chromosome inactivation. Nat Genet 2009;41:488–93.
[126] Morgan CP, Bale TL. Early prenatal stress epigenetically programs dysmasculinization in second-generation offspring via the paternal lineage. J Neurosci 2011;31:11748–55.
[127] Wyrwoll CS, Holmes MC. Prenatal excess glucocorticoid exposure and adult affective disorders: a role for serotonergic and catecholamine pathways. Neuroendocrinology 2012;95:47–55.
[128] Brunton PJ. Resetting the dynamic range of hypothalamic-pituitary-adrenal axis stress responses through pregnancy. J Neuroendocrinol 2010;22:1198–213.
[129] Wyrwoll CS, Holmes MC, Seckl JR. 11beta-hydroxysteroid dehydrogenases and the brain: from zero to hero, a decade of progress. Front Neuroendocrinol 2011;32:265–86.
[130] Baker ME. Evolutionary analysis of 11beta-hydroxysteroid dehydrogenase-type 1, -type 2, -type 3 and 17beta-hydroxysteroid dehydrogenase-type 2 in fish. FEBS Lett 2004;574:167–70.
[131] Moisan MP, Seckl JR, Brett LP, Monder C, Agarwal AK, White PC, et al. 11Beta-hydroxysteroid dehydrogenase messenger ribonucleic acid expression, bioactivity and immunoreactivity in rat cerebellum. J Neuroendocrinol 1990;2:853–8.
[132] Moisan MP, Seckl JR, Edwards CR. 11 beta-hydroxysteroid dehydrogenase bioactivity and messenger RNA expression in rat forebrain: localization in hypothalamus, hippocampus, and cortex. Endocrinology 1990;127:1450–5.
[133] Yau JL, Noble J, Hibberd C, Rowe WB, Meaney MJ, Morris RG, et al. Chronic treatment with the antidepressant amitriptyline prevents impairments in water maze learning in aging rats. J Neurosci 2002;22:1436–42.
[134] Landfield PW, Baskin RK, Pitler TA. Brain aging correlates: retardation by hormonal-pharmacological treatments. Science 1981;214:581–4.
[135] Meaney MJ, Aitken DH, van Berkel C, Bhatnagar S, Sapolsky RM. Effect of neonatal handling on age-related impairments associated with the hippocampus. Science 1988;239:766–8.

[136] Yau JL, McNair KM, Noble J, Brownstein D, Hibberd C, Morton N, et al. Enhanced hippocampal long-term potentiation and spatial learning in aged 11beta-hydroxysteroid dehydrogenase type 1 knockout mice. J Neurosci 2007;27:10487–96.

[137] Dauprat P, Monin G, Dalle M, Delost P. The effects of psychosomatic stress at the end of pregnancy on maternal and fetal plasma cortisol levels and liver glycogen in guinea-pigs. Reprod Nutr Dev 1984;24:45–51.

[138] McTernan CL, Draper N, Nicholson H, Chalder SM, Driver P, Hewison M, et al. Reduced placental 11beta-hydroxysteroid dehydrogenase type 2 mRNA levels in human pregnancies complicated by intrauterine growth restriction: an analysis of possible mechanisms. J Clin Endocrinol Metab 2001;86:4979–83.

[139] Stark MJ, Wright IM, Clifton VL. Sex-specific alterations in placental 11beta-hydroxysteroid dehydrogenase 2 activity and early postnatal clinical course following antenatal betamethasone. Am J Physiol Regul Integr Comp Physiol 2009;297:R510–4.

[140] Sun M, Kingdom J, Baczyk D, Lye SJ, Matthews SG, Gibb W. Expression of the multidrug resistance P-glycoprotein, (ABCB1 glycoprotein) in the human placenta decreases with advancing gestation. Placenta 2006;27:602–9.

[141] Iqbal M, Audette MC, Petropoulos S, Gibb W, Matthews SG. Placental drug transporters and their role in fetal protection. Placenta 2012;33:137–42.

[142] Iqbal M, Gibb W, Matthews SG. Corticosteroid regulation of P-glycoprotein in the developing blood-brain barrier. Endocrinology 2011;152:1067–79.

[143] Wyrwoll C, Keith M, Noble J, Stevenson PL, Bombail V, Crombie S, et al. Fetal brain 11beta-hydroxysteroid dehydrogenase type 2 selectively determines programming of adult depressive-like behaviors and cognitive function, but not anxiety behaviors in male mice. Psychoneuroendocrinology 2015;59:59–70.

[144] Holmes MC, Abrahamsen CT, French KL, Paterson JM, Mullins JJ, Seckl JR. The mother or the fetus? 11beta-hydroxysteroid dehydrogenase type 2 null mice provide evidence for direct fetal programming of behavior by endogenous glucocorticoids. J Neurosci 2006;26:3840–4.

[145] Wyrwoll CS, Seckl JR, Holmes MC. Altered placental function of 11beta-hydroxysteroid dehydrogenase 2 knockout mice. Endocrinology 2009;150:1287–93.

[146] Jensen Pena C, Monk C, Champagne FA. Epigenetic effects of prenatal stress on 11beta-hydroxysteroid dehydrogenase-2 in the placenta and fetal brain. PLoS One 2012;7:e39791.

[147] O'Donnell K, O'Connor TG, Glover V. Prenatal stress and neurodevelopment of the child: focus on the HPA axis and role of the placenta. Dev Neurosci 2009;31:285–92.

[148] Torricelli M, Novembri R, Bloise E, De Bonis M, Challis JR, Petraglia F. Changes in placental CRH, urocortins, and CRH-receptor mRNA expression associated with preterm delivery and chorioamnionitis. J Clin Endocrinol Metab 2011;96:534–40.

[149] Braun T, Challis JR, Newnham JP, Sloboda DM. Early-life glucocorticoid exposure: the hypothalamic-pituitary-adrenal axis, placental function, and long-term disease risk. Endocr Rev 2013;34:885–916.

[150] Bronson SL, Bale TL. Prenatal stress-induced increases in placental inflammation and offspring hyperactivity are male-specific and ameliorated by maternal antiinflammatory treatment. Endocrinology 2014; 155:2635–46.

[151] Krozowski ZS, Funder JW. Renal mineralocorticoid receptors and hippocampal corticosterone-binding species have identical intrinsic steroid specificity. Proc Natl Acad Sci USA 1983;80:6056–60.

[152] McCabe L, Marash D, Li A, Matthews SG. Repeated antenatal glucocorticoid treatment decreases hypothalamic corticotropin releasing hormone mRNA but not corticosteroid receptor mRNA expression in the fetal guinea-pig brain. J Neuroendocrinol 2001;13:425–31.

[153] Mastorakos G, Ilias I. Maternal and fetal hypothalamic-pituitary-adrenal axes during pregnancy and postpartum. Ann N Y Acad Sci 2003;997:136–49.

[154] Noorlander CW, De Graan PN, Middeldorp J, Van Beers JJ, Visser GH. Ontogeny of hippocampal corticosteroid receptors: effects of antenatal glucocorticoids in human and mouse. J Comp Neurol 2006;499:924–32.

[155] Kosinska-Kaczynska K, Bartkowiak R, Kaczynski B, Szymusik I, Wielgos M. Autonomous adrenocorticotropin reaction to stress stimuli in human fetus. Early Hum Dev 2012;88:197–201.

[156] Ballard PL, Granberg P, Ballard RA. Glucocorticoid levels in maternal and cord serum after prenatal betamethasone therapy to prevent respiratory distress syndrome. J Clin Invest 1975;56:1548–54.

[157] Korebrits C, Yu DH, Ramirez MM, Marinoni E, Bocking AD, Challis JR. Antenatal glucocorticoid administration increases corticotrophin-releasing hormone in maternal plasma. Br J Obstet Gynaecol 1998;105:556–61.

[158] Jeffray TM, Marinoni E, Ramirez MM, Bocking AD, Challis JR. Effect of prenatal betamethasone administration on maternal and fetal corticosteroid-binding globulin concentrations. Am J Obstet Gynecol 1999;181:1546–51.

[159] Karlsson R, Kallio J, Toppari J, Scheinin M, Kero P. Antenatal and early postnatal dexamethasone treatment decreases cortisol secretion in preterm infants. Horm Res 2000;53:170–6.

[160] Terrone DA, Rinehart BK, Rhodes PG, Roberts WE, Miller RC, Martin Jr JN. Multiple courses of betamethasone to enhance fetal lung maturation do not suppress neonatal adrenal response. Am J Obstet Gynecol 1999;180:1349–53.

[161] Davis EP, Waffarn F, Sandman CA. Prenatal treatment with glucocorticoids sensitizes the hpa axis response to stress among full-term infants. Dev Psychobiol 2011;53:175–83.

[162] Owen D, Matthews SG. Glucocorticoids and sex-dependent development of brain glucocorticoid and mineralocorticoid receptors. Endocrinology 2003; 144:2775–84.

[163] Rakers F, Frauendorf V, Rupprecht S, Schiffner R, Bischoff SJ, Kiehntopf M, et al. Effects of early- and late-gestational maternal stress and synthetic glucocorticoid on development of the fetal hypothalamus-pituitary-adrenal axis in sheep. Stress 2013; 16:122–9.

[164] Schwab M, Coksaygan T, Rakers F, Nathanielsz PW. Glucocorticoid exposure of sheep at 0.7 to 0.75 gestation augments late-gestation fetal stress responses. Am J Obstet Gynecol 2012;206:e16–22.

[165] Fletcher AJ, Ma XH, Wu WX, Nathanielsz PW, McGarrigle HH, Fowden AL, et al. Antenatal glucocorticoids reset the level of baseline and hypoxemia-induced pituitary-adrenal activity in the sheep fetus during late gestation. Am J Physiol Endocrinol Metab 2004;286:E311–9.

[166] Leavitt MG, Aberdeen GW, Burch MG, Albrecht ED, Pepe GJ. Inhibition of fetal adrenal adrenocorticotropin receptor messenger ribonucleic acid expression by betamethasone administration to the baboon fetus in late gestation. Endocrinology 1997;138:2705–12.

[167] Pryce CR, Aubert Y, Maier C, Pearce PC, Fuchs E. The developmental impact of prenatal stress, prenatal dexamethasone and postnatal social stress on physiology, behaviour and neuroanatomy of primate offspring: studies in rhesus macaque and common marmoset. Psychopharmacol Berl 2011;214:33–53.

[168] Del Cerro MC, Ortega E, Gomez F, Segovia S, Perez-Laso C. Environmental prenatal stress eliminates brain and maternal behavioral sex differences and alters hormone levels in female rats. Horm Behav 2015;73:142–7.

[169] Piffer RC, Garcia PC, Pereira OC. Adult partner preference and sexual behavior of male rats exposed prenatally to betamethasone. Physiol Behav 2009;98:163–7.

[170] Gerardin DC, Pereira OC, Kempinas WG, Florio JC, Moreira EG, Bernardi MM. Sexual behavior, neuroendocrine, and neurochemical aspects in male rats exposed prenatally to stress. Physiol Behav 2005;84:97–104.

[171] Pereira OC, Bernardi MM, Gerardin DC. Could neonatal testosterone replacement prevent alterations induced by prenatal stress in male rats? Life Sci 2006;78:2767–71.

[172] Zucchi FC, Yao Y, Ward ID, Ilnytskyy Y, Olson DM, Benzies K, et al. Maternal stress induces epigenetic signatures of psychiatric and neurological diseases in the offspring. PLoS One 2013;8:e56967.

[173] Monteleone MC, Adrover E, Pallares ME, Antonelli MC, Frasch AC, Brocco MA. Prenatal stress changes the glycoprotein GPM6A gene expression and induces epigenetic changes in rat offspring brain. Epigenetics 2014;9:152–60.

[174] Crudo A, Petropoulos S, Suderman M, Moisiadis VG, Kostaki A, Hallett M, et al. Effects of antenatal synthetic glucocorticoid on glucocorticoid receptor binding, DNA methylation, and genome-wide mRNA levels in the fetal male hippocampus. Endocrinology 2013;154:4170–81.

[175] Crudo A, Suderman M, Moisiadis VG, Petropoulos S, Kostaki A, Hallett M, et al. Glucocorticoid programming of the fetal male hippocampal epigenome. Endocrinology 2013;154:1168–80.

[176] Crudo A, Petropoulos S, Moisiadis VG, Iqbal M, Kostaki A, Machnes Z, et al. Prenatal synthetic glucocorticoid treatment changes DNA methylation states in male organ systems: multigenerational effects. Endocrinology 2012;153:3269–83.

[177] Galloway T, Cipelli R, Guralnick J, Ferrucci L, Bandinelli S, Corsi AM, et al. Daily bisphenol A excretion and associations with sex hormone concentrations: results from the InCHIANTI adult population study. Environ Health Perspect 2010;118:1603–8.

[178] Environment Canada. Screening Assessment for the Challenge Phenol, 4,4′-(1-methylethylidene)bis-(Bisphenol A) Chemical Abstracts Service Registry Number 80-05-7. In: Health MotEao. 2008. p. 1–107.

[179] GrandViewResearch. Global bisphenol A (BPA) market by appliation (appliances, automotive, consumer, construction, electrical & electronics) expected to reach USD 20.03 billion by 2020. 2014. http://www.digitaljournal.com/pr/2009287.

[180] Vandenberg LN, Ehrlich S, Belcher SM, Ben-Jonathan N, Dolinoy DC, Hugo ES, et al. Low dose effects of bisphenol A: an integrated review of in vitro, laboratory animal and epidemiology studies. Endocr Disrupt 2013;1:E1–20.

[181] Calafat AM, Kuklenyik Z, Reidy JA, Caudill SP, Ekong J, Needham LL. Urinary concentrations of bisphenol A and 4-nonylphenol in a human reference population. Environ Health Perspect 2005;113:391–5.

[182] Braun JM, Kalkbrenner AE, Calafat AM, Yolton K, Ye X, Dietrich KN, et al. Impact of early-life bisphenol a exposure on behavior and executive function in children. Pediatrics 2011;128:873–82.

[183] Latini G, De Felice C, Presta G, Del Vecchio A, Paris I, Ruggieri F, et al. In utero exposure to di-(2-ethylhexyl) phthalate and duration of human pregnancy. Environ Health Perspect 2003;111:1783–5.

[184] Leon-Olea M, Martyniuk CJ, Orlando EF, Ottinger MA, Rosenfeld CS, Wolstenholme JT, et al. Current concepts in neuroendocrine disruption. Gen Comp Endocrinol 2014;203:158–73.

[185] D'Cruz SC, Jubendradass R, Jayakanthan M, Rani SJ, Mathur PP. Bisphenol A impairs insulin signaling and glucose homeostasis and decreases steroidogenesis in rat testis: an in vivo and in silico study. Food Chem Toxicol 2012;50:1124–33.

[186] Horstman KA, Naciff JM, Overmann GJ, Foertsch LM, Richardson BD, Daston GP. Effects of transplacental 17-alpha-ethynyl estradiol or bisphenol A on the developmental profile of steroidogenic acute regulatory protein in the rat testis. Birth Defects Res B Dev Reprod Toxicol 2012;95:318–25.

[187] Liu S, Qin F, Wang H, Wu T, Zhang Y, Zheng Y, et al. Effects of 17alpha-ethinylestradiol and bisphenol A on steroidogenic messenger ribonucleic acid levels in the rare minnow gonads. Aquat Toxicol 2012;122–123:19–27.

[188] Savchuk I, Soder O, Svechnikov K. Mouse leydig cells with different androgen production potential are resistant to estrogenic stimuli but responsive to bisphenol a which attenuates testosterone metabolism. PLoS One 2013;8:e71722.

[189] Peretz J, Flaws JA. Bisphenol A down-regulates rate-limiting Cyp11a1 to acutely inhibit steroidogenesis in cultured mouse antral follicles. Toxicol Appl Pharmacol 2013;271:249–56.

[190] Nanjappa MK, Simon L, Akingbemi BT. The industrial chemical bisphenol A (BPA) interferes with proliferative activity and development of steroidogenic capacity in rat Leydig cells. Biol Reproduction 2012;86:1–12.

[191] Akingbemi BT, Sottas CM, Koulova AI, Klinefelter GR, Hardy MP. Inhibition of testicular steroidogenesis by the xenoestrogen bisphenol A is associated with reduced pituitary luteinizing hormone secretion and decreased steroidogenic enzyme gene expression in rat Leydig cells. Endocrinology 2004;145:592–603.

[192] Peretz J, Gupta RK, Singh J, Hernandez-Ochoa I, Flaws JA. Bisphenol A impairs follicle growth, inhibits steroidogenesis, and downregulates rate-limiting enzymes in the estradiol biosynthesis pathway. Toxicol Sci 2011;119:209–17.

[193] Lee SH, Kang SM, Choi MH, Lee J, Park MJ, Kim SH, et al. Changes in steroid metabolism among girls with precocious puberty may not be associated with urinary levels of bisphenol A. Reprod Toxicol 2014;44:1–6.

[194] Akingbemi BT, Ge R, Klinefelter GR, Zirkin BR, Hardy MP. Phthalate-induced Leydig cell hyperplasia is associated with multiple endocrine disturbances. Proc Natl Acad Sci USA 2004;101:775–80.

[195] Akingbemi BT, Youker RT, Sottas CM, Ge R, Katz E, Klinefelter GR, et al. Modulation of rat Leydig cell steroidogenic function by di(2-ethylhexyl)phthalate. Biol Reprod 2001;65:1252–9.

[196] Svechnikov K, Svechnikova I, Soder O. Inhibitory effects of mono-ethylhexyl phthalate on steroidogenesis in immature and adult rat Leydig cells in vitro. Reprod Toxicol 2008;25:485–90.

[197] Botelho GG, Golin M, Bufalo AC, Morais RN, Dalsenter PR, Martino-Andrade AJ. Reproductive effects of di(2-ethylhexyl)phthalate in immature male rats and its relation to cholesterol, testosterone, and thyroxin levels. Arch Environ Contam Toxicol 2009;57:777–84.

[198] Desdoits-Lethimonier C, Albert O, Le Bizec B, Perdu E, Zalko D, Courant F, et al. Human testis steroidogenesis is inhibited by phthalates. Hum Reprod 2012;27:1451–9.

[199] Chauvigne F, Plummer S, Lesne L, Cravedi JP, Dejucq-Rainsford N, Fostier A, et al. Mono-(2-ethylhexyl) phthalate directly alters the expression of Leydig cell genes and CYP17 lyase activity in cultured rat fetal testis. PLoS One 2011;6:e27172.

[200] Beverly BE, Lambright CS, Furr JR, Sampson H, Wilson VS, McIntyre BS, et al. Simvastatin and dipentyl phthalate lower ex vivo testicular testosterone production and exhibit additive effects on testicular testosterone and gene expression via distinct mechanistic pathways in the fetal rat. Toxicol Sci 2014;141:524–37.

[201] Saillenfait AM, Sabate JP, Robert A, Rouiller-Fabre V, Roudot AC, Moison D, et al. Dose-dependent alterations in gene expression and testosterone production in fetal rat testis after exposure to di-n-hexyl phthalate. J Appl Toxicol 2013;33:1027–35.

[202] Meeker JD, Calafat AM, Hauser R. Urinary metabolites of di(2-ethylhexyl) phthalate are associated with decreased steroid hormone levels in adult men. J Androl 2009;30:287–97.

[203] Cao J, Joyner L, Mickens JA, Leyrer SM, Patisaul HB. Sex-specific Esr2 mRNA expression in the rat hypothalamus and amygdala is altered by neonatal bisphenol A exposure. Reproduction 2014;147:537–54.

[204] Cao J, Mickens JA, McCaffrey KA, Leyrer SM, Patisaul HB. Neonatal bisphenol A exposure alters sexually dimorphic gene expression in the postnatal rat hypothalamus. Neurotoxicology 2012;33:23–36.

[205] Kundakovic M, Gudsnuk K, Franks B, Madrid J, Miller RL, Perera FP, et al. Sex-specific epigenetic disruption and behavioral changes following low-dose in utero bisphenol A exposure. Proc Natl Acad Sci USA 2013;110:9956–61.

[206] Cao J, Rebuli ME, Rogers J, Todd KL, Leyrer SM, Ferguson SA, et al. Prenatal bisphenol A exposure alters sex-specific estrogen receptor expression in the neonatal rat hypothalamus and amygdala. Toxicol Sci 2013;133:157–73.

[207] Patisaul HB, Sullivan AW, Radford ME, Walker DM, Adewale HB, Winnik B, et al. Anxiogenic effects of developmental bisphenol A exposure are associated with gene expression changes in the juvenile rat amygdala and mitigated by soy. PLoS One 2012;7:e43890.

[208] Mahoney MM, Padmanabhan V. Developmental programming: impact of fetal exposure to endocrine-disrupting chemicals on gonadotropin-releasing hormone and estrogen receptor mRNA in sheep hypothalamus. Toxicol Appl Pharmacol 2010;247:98–104.
[209] Monje L, Varayoud J, Luque EH, Ramos JG. Neonatal exposure to bisphenol A modifies the abundance of estrogen receptor alpha transcripts with alternative 5′-untranslated regions in the female rat preoptic area. J Endocrinol 2007;194:201–12.
[210] Ceccarelli I, Della Seta D, Fiorenzani P, Farabollini F, Aloisi AM. Estrogenic chemicals at puberty change ERalpha in the hypothalamus of male and female rats. Neurotoxicol Teratol 2007;29:108–15.
[211] Ramos JG, Varayoud J, Kass L, Rodriguez H, Costabel L, Munoz-De-Toro M, et al. Bisphenol a induces both transient and permanent histofunctional alterations of the hypothalamic-pituitary-gonadal axis in prenatally exposed male rats. Endocrinology 2003; 144:3206–15.
[212] Zhang WZ, Yong L, Jia XD, Li N, Fan YX. Combined subchronic toxicity of bisphenol A and dibutyl phthalate on male rats. Biomed Environ Sci 2013;26:63–9.
[213] Poimenova A, Markaki E, Rahiotis C, Kitraki E. Corticosterone-regulated actions in the rat brain are affected by perinatal exposure to low dose of bisphenol A. Neuroscience 2010;167:741–9.
[214] Panagiotidou E, Zerva S, Mitsiou DJ, Alexis MN, Kitraki E. Perinatal exposure to low-dose bisphenol A affects the neuroendocrine stress response in rats. J Endocrinol 2014;220:207–18.
[215] Chen F, Zhou L, Bai Y, Zhou R, Chen L. Sex differences in the adult HPA axis and affective behaviors are altered by perinatal exposure to a low dose of bisphenol A. Brain Res 2014;1571:12–24.
[216] Supornsilchai V, Soder O, Svechnikov K. Stimulation of the pituitary-adrenal axis and of adrenocortical steroidogenesis ex vivo by administration of di-2-ethylhexyl phthalate to prepubertal male rats. J Endocrinol 2007;192:33–9.
[217] Jang YJ, Park HR, Kim TH, Yang WJ, Lee JJ, Choi SY, et al. High dose bisphenol A impairs hippocampal neurogenesis in female mice across generations. Toxicology 2012;296:73–82.
[218] Tang WY, Morey LM, Cheung YY, Birch L, Prins GS, Ho SM. Neonatal exposure to estradiol/bisphenol A alters promoter methylation and expression of Nsbp1 and Hpcal1 genes and transcriptional programs of Dnmt3a/b and Mbd2/4 in the rat prostate gland throughout life. Endocrinology 2012;153:42–55.
[219] Martinez-Arguelles D, Papadopoulos V. Identification of hot spots of DNA methylation in the adult male adrenal in response to in utero exposure to the ubiquitous endocrine disruptor plasticizer di-(2-ethylhexyl) phthalate. Endocrinology 2015;156:124–33.
[220] Zhao Y, Shi HJ, Xie CM, Chen J, Laue H, Zhang YH. Prenatal phthalate exposure, infant growth, and global DNA methylation of human placenta. Environ Mol Mutagen 2015;56:286–92.
[221] Kundakovic M, Gudsnuk K, Herbstman JB, Tang D, Perera FP, Champagne FA. DNA methylation of BDNF as a biomarker of early-life adversity. Proc Natl Acad Sci USA 2015;112:6807–13.
[222] Yaoi T, Itoh K, Nakamura K, Ogi H, Fujiwara Y, Fushiki S. Genome-wide analysis of epigenomic alterations in fetal mouse forebrain after exposure to low doses of bisphenol A. Biochem Biophys Res Commun 2008;376:563–7.
[223] Yeo M, Berglund K, Hanna M, Guo JU, Kittur J, Torres MD, et al. Bisphenol A delays the perinatal chloride shift in cortical neurons by epigenetic effects on the Kcc2 promoter. Proc Natl Acad Sci USA 2013;110:4315–20.
[224] Guida N, Laudati G, Galgani M, Santopaolo M, Montuori P, Triassi M, et al. Histone deacetylase 4 promotes ubiquitin-dependent proteasomal degradation of Sp3 in SH-SY5Y cells treated with di(2-ethylhexyl) phthalate (DEHP), determining neuronal death. Toxicol Appl Pharmacol 2014;280:190–8.
[225] Zhou R, Chen F, Chang F, Bai Y, Chen L. Persistent overexpression of DNA methyltransferase 1 attenuating GABAergic inhibition in basolateral amygdala accounts for anxiety in rat offspring exposed perinatally to low-dose bisphenol A. J Psychiatr Res 2013;47:1535–44.
[226] Warita K, Mitsuhashi T, Ohta K, Suzuki S, Hoshi N, Miki T, et al. Gene expression of epigenetic regulatory factors related to primary silencing mechanism is less susceptible to lower doses of bisphenol A in embryonic hypothalamic cells. J Toxicol Sci 2013;38:285–9.
[227] Zhang Q, Xu X, Li T, Lu Y, Ruan Q, Lu Y, et al. Exposure to bisphenol-A affects fear memory and histone acetylation of the hippocampus in adult mice. Horm Behav 2014;65:106–13.
[228] Tilghman SL, Bratton MR, Segar HC, Martin EC, Rhodes LV, Li M, et al. Endocrine disruptor regulation of microRNA expression in breast carcinoma cells. PLoS One 2012;7:e32754.
[229] Veiga-Lopez A, Luense LJ, Christenson LK, Padmanabhan V. Developmental programming: gestational bisphenol-A treatment alters trajectory of fetal ovarian gene expression. Endocrinology 2013;154:1873–84.
[230] Avissar-Whiting M, Veiga KR, Uhl KM, Maccani MA, Gagne LA, Moen EL, et al. Bisphenol A exposure leads to specific microRNA alterations in placental cells. Reprod Toxicol 2010;29:401–6.
[231] Topper VY, Walker DM, Gore AC. Sexually dimorphic effects of gestational endocrine-disrupting chemicals on microRNA expression in the developing rat hypothalamus. Mol Cell Endocrinol 2015;414:42–52.

CHAPTER 6

Parental Nutrition and Developmental Origins of Health and Disease

Miguel A. Velazquez, Congshan Sun, Tom P. Fleming

Centre for Biological Sciences, University of Southampton, Southampton General Hospital, Southampton, UK

OUTLINE

INTRODUCTION

Noncommunicable diseases (NCDs) are leading causes of death in middle- and high-income countries [1]. The notion that NCDs are not the sole product of postnatal lifestyle is well established nowadays. Experimental research in animal models and human epidemiologic studies have provided strong evidence indicating that environmental conditions during the prenatal period can program the development of NCDs in postnatal life [2–4]. This unfavorable programming is the basis for the "developmental origins of health and disease" (DOHaD) hypothesis [5]. The DOHaD concept relies on the premise that disease arises as a result of a mismatch between the prenatal (i.e., predicted environment) and postnatal environment [3]. Several lines of evidence have demonstrated that malnutrition during prenatal development is a major contributor of adverse developmental programming [2,6]. Malnutrition can arise from a deficient (undernutrition) or excessive (overnutrition) bioavailability of one or more macro- and micronutrients [7]. Undernutrition can be caused by reduced

The Epigenome and Developmental Origins of Health and Disease
http://dx.doi.org/10.1016/B978-0-12-801383-0.00006-2

Copyright © 2016 Elsevier Inc. All rights reserved.

dietary intake, increased nutritional requirements or losses, or impaired ability to absorb or utilize nutrients [7]. On the other hand, overnutrition is the result of increased nutrient intake that exceeds the amounts of normal physiological activity and metabolism, which commonly leads to overweight or obesity [7]. There are several prenatal periods of development that can be affected by adverse programming, including the preconception phase and both preimplantation and postimplantation development [8]. The preconception period involves important molecular and cellular events that take place during folliculogenesis. During this developmental phase, ovarian cells (i.e., granulosa and theca cells) grow and differentiate along with the synthesis and storage of mRNAs and proteins by the oocyte, which are critical for oocyte maturation (i.e., cytoplasmic and nuclear maturation) [4,7]. The preimplantation period begins with the formation of a one-cell embryo (i.e., fertilization) and is characterized by important milestones of development, including embryonic genome activation, the formation of adhesive cell contacts, and the establishment of founder cell lineages at the blastocyst stage [9–11]. Following attachment of the blastocyst to the uterine wall (i.e., implantation) [12], several key developmental events also take place during the postimplantation period, including placentation, organogenesis, and fetal growth [13–15]. Animal models have shown that nutritional challenges during these developmental stages can affect postnatal health [2,3,7,16–20] (Figure 1). Furthermore, nutritional challenges during intrauterine development can affect the resultant offspring beyond the F1 generation [4,21]. In this chapter, we provide an overview of the role of nutritional programming during the prenatal period on the development of adverse phenotypes in adulthood, with emphasis on animal models.

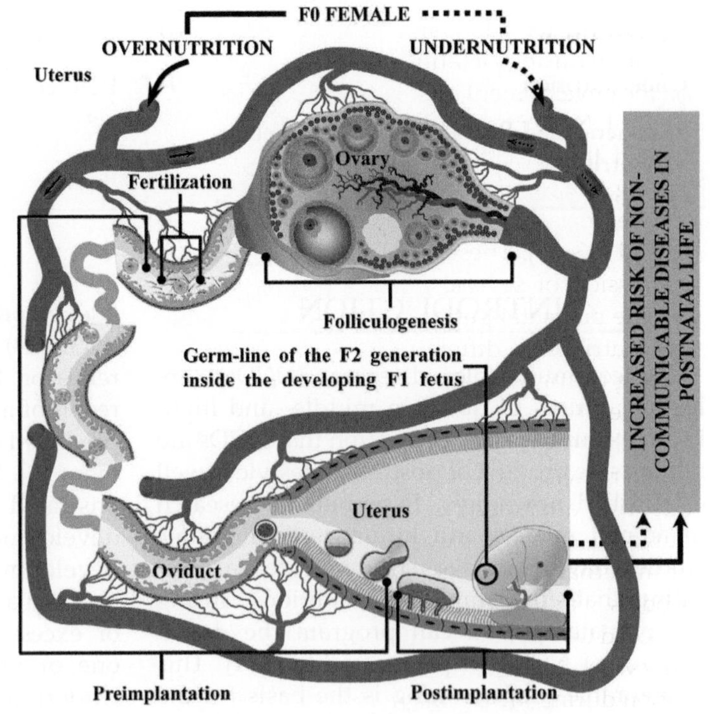

FIGURE 1 Nutritional challenges imposed to F0 females during the preconception phase (i.e., folliculogenesis) and throughout pregnancy, during intraoviductal (i.e., during fertilization and first cell divisions of the embryo) and intrauterine life (i.e., during blastocyst formation, implantation, placentation, and fetal development), can program the development of noncommunicable diseases during the postnatal period of the resultant offspring. This nutritional programming can be induced at various developmental windows that will impact directly both F1 offspring and F2 germline, with potential for transgenerational epigenetic inheritance (i.e., F3 generation). *Model based on animal models of undernutrition and overnutrition [2–4,7,16,17].*

OVERNUTRITION

Data from human cohort studies have revealed that offspring from mothers that are obese or gain excessive weight during pregnancy are at a higher risk of developing obesity and cardiovascular (e.g., hypertension) and metabolic dysfunction later in life (e.g., diabetes) [18,22]. However, with human studies, it is difficult to distinguish between the effects induced by prenatal nutritional challenges and those exerted by nutritional habits during postnatal life. Animal models are therefore required to elucidate the mechanism behind the prenatal programming induced by malnutrition. Most of the experimental settings investigating the role of overnutrition on developmental programming have used dietary regimes based on high-fat and/or high-sugar diets, or "cafeteria diet," in an effort to mimic Western diet habits [18–20]. However, most animal models published so far included offspring that were exposed to overnutrition not only during the prenatal period, but also during lactation, and it is known that overnutrition during lactation can have added effects on the development of diseases during postnatal life [18]. Nevertheless, some studies have investigated the effects of overnutrition exclusively during the prenatal period in offspring exposed to well-balanced diets during the postnatal period [23–33]. These studies have revealed the expression of several altered phenotypes during the postnatal period in offspring exposed to overnutrition at different developmental stages during the prenatal period (Table 1).

Several physiological and molecular alterations have been observed in offspring from mothers with diet-induced obesity, but these offspring alterations caused by maternal overnutrition can also be induced in the absence of maternal obesity [18]. Interestingly, in some models, adverse phenotypes in offspring exposed to prenatal overnutrition were observed only when the offspring was exposed to overnutrition during the postnatal period. For instance, in a cross-fostering experiment, offspring from obese female mice fed with a diet high in fat and sugar during pregnancy showed increases in blood pressure only when they were fostered by obese mothers [34]. Still, adverse programming (e.g., increased fat mass, hyperleptinemia) has been observed even when offspring exposed to intrauterine overnutrition are fostered by well-fed dams [35]. Nevertheless, at least in mice, cross-fostering per se can exert changes in offspring cardiovascular and metabolic function in adulthood [36], indicating that interpretation of data from this type of experiment should be done with caution. In this regard, a statistical consideration usually overlooked in the analysis of offspring data in litter-bearing animals is the variation in litter size. This is relevant since metabolic alterations can be induced by adjustment of litter size [37]. Litter size should be considered in statistical models in order to avoid false-positive results [38].

The studies of prenatal overnutrition suggest that alterations in cells and tissues during prenatal development take place in response to overnutrition exposure. Indeed, several phenotypic alterations have been detected during folliculogenesis [39–41], preimplantation embryo formation [42–44], and fetal development [45–51] in animal models of overnutrition (Figures 2 and 3). These prenatal changes induced by overnutrition will impact not only the F1 generation, but also the germline of the F2 generation developing inside the F1 fetus, increasing the risk for adverse transgenerational developmental programming (Figure 1). Indeed, alterations in F1 and F2 generations have been reported in offspring derived from F0 female mice fed with a high-fat diet before and during gestation [26]. Furthermore, in some experimental settings, the adverse phenotype caused by overnutrition can be more noticeable in the F2 generation [52]. The reason(s) behind this phenomenon is currently unknown.

An increasing number of studies strongly indicate that the altered phenotypes observed in

TABLE 1 Selected Animal Models Illustrating the Induction of Altered Phenotypes in Offspring Well Fed during the Postnatal Period but Exposed to Maternal Overnutrition during the Prenatal Period

Species	Time of Exposure in F0 Dam	Phenotype Induced in the Postnatal Period-sex	Generation Affected	References
Mice	8 weeks before mating and during gestation	Impaired glucose tolerance-♀♂	F1–F2	[26]
Mice	9 weeks before mating and during gestation	Increased fat mass-♀ Hyperleptinemia-♀ Hyperinsulinemia-♀	F1	[35]
Mice	4 weeks before mating and during gestation	Increased fat mass-♀♂ Increased blood pressure-♀♂ Impaired glucose tolerance-♀♂ Hyperleptinemia-♀♂	F1	[30]
Rats	During gestation	Hyperglycemia-♀♂ ↓ β-cell volume and number (pancreas)-♀♂ Impaired insulin release-♀♂	F1	[24,25]
Rats	4 weeks before mating and during gestation	Increased fat mass-♂ ↓ Insulin receptor β subunit protein (liver)-♂ ↓ IRS1 protein (liver)-♂ ↑ PKCζ protein (liver)-♂ ↑ Insulin receptor β subunit protein (skeletal muscle)-♂ ↑ p85 subunit of PI3-kinase protein (skeletal muscle)-♂	F1	[23]
Sheep	4 months before mating	Increased fat mass-♀	F1	[27]
Sheep	1 month before and 7 days after mating	↓ SLC2A4 protein (skeletal muscle)-♀ ↓ Ser 9 phospho-GSK3α protein (skeletal muscle)-♀♂ ↑ GSK3α protein (skeletal muscle) -♀♂ ↑ AT1R protein (adrenal cortex) -♀♂	F1	[28,29]
Sheep	2 months before mating and during gestation	Hyperglycaemia-♀♂ Hyperleptinemia-♀♂ Impaired glucose tolerance-♀♂	F1	[31,32]
Sheep	Mid- to late-gestation	Impaired insulin release-♀	F1	[33]

offspring during the postnatal period are not only induced by overnourished mothers, but also by paternal overnutrition [20,53,54]. These paternal effects could be transmitted not only via sperm, but also through seminal fluid [55]. In a recent study carried in mice, Bromfield et al. [56] reported that seminal fluid per se can play an important role in the induction of altered offspring phenotypes (e.g., high blood pressure). The study also suggested that programming induced by seminal fluid could be exerted through indirect effects on preimplantation embryos via oviduct expression of embryotrophic cytokines [56]. Literature on the effects of overnutrition on seminal fluid and how this may affect oviductal and uterine physiology is currently unavailable. The relevance of the paternal line during overnutrition is of further significance when considering that transmission of some

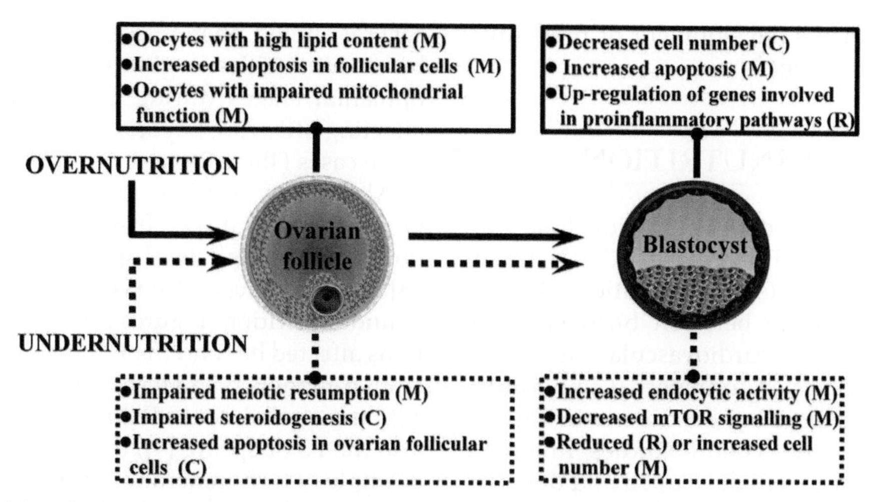

FIGURE 2 Selected examples illustrating the possible phenotypic alterations induced by maternal overnutrition [39–44] and undernutrition [58,69–73] in ovarian follicles and their respective oocyte (i.e., during folliculogenesis) and late preimplantation embryos (i.e., blastocysts before hatching and implantation). mTOR = mammalian target of rapamycin. *Data derived from animal models. M = mouse, R = rat, C = cattle.*

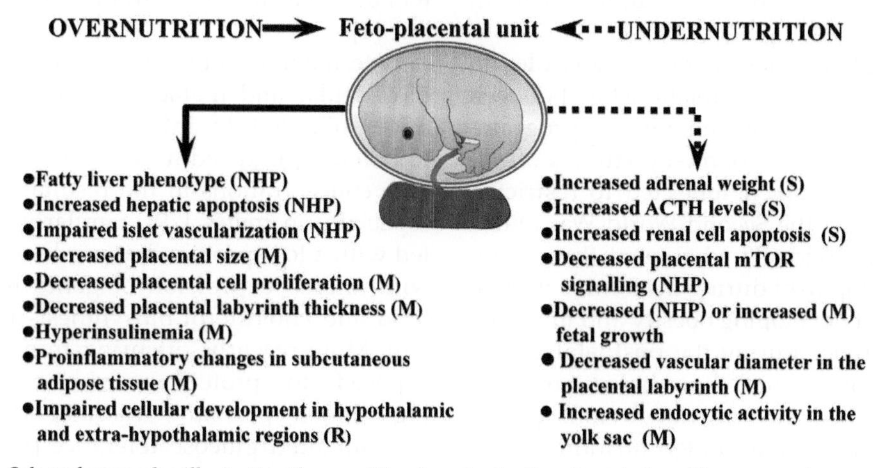

FIGURE 3 Selected examples illustrating the possible phenotypic alterations induced by maternal overnutrition [45–51] and undernutrition [62,74–78] during intrauterine development on both the fetus and the placenta. ACTH = adrenocorticotropic hormone. *Data derived from animal models. M = mouse, R = rat, S = sheep, NHP = nonhuman primate.*

phenotypic characteristics (e.g., increased body size) induced by maternal overnutrition to third-generation females takes place via the paternal line [57]. The transgenerational effect of overnutrition through both parental lines highlights the complexity of the current diet-induced obesity epidemic in human populations.

Although the above-discussed animal models of overnutrition do not replicate entirely the clinical expression of NCDs as seen in humans (i.e., obese mice do not present sudden cardiac death), they provide important insights for the elucidation of mechanisms behind the onset of NCDs, and therefore they are critical for the development of preventive and therapeutic measures.

UNDERNUTRITION

The DOHaD hypothesis, previously known as fetal origins of adult disease (FOAD), has its origins in epidemiologic observations describing the relationship between birth size and subsequent risk of cardiovascular disease in individuals subjected to maternal undernutrition due to poor living conditions in the UK [3,5]. Retrospective cohort studies involving women and men from developed and developing countries have confirmed the association between weight at birth and risk of cardiovascular and metabolic disease [3]. These studies strongly suggested an association between adverse prenatal events (e.g., undernutrition) and eventual development of disease in adulthood. One of the most relevant examples in humans illustrating the relationship between prenatal undernutrition and adverse programming comes from the follow-up studies of offspring born to women subjected to restricted nutrition during the Second World War Dutch famine [3]. Individuals who spent their intrauterine development during the famine were at greater risk of developing obesity and impaired glucose tolerance during the postnatal period compared to individuals born before and after the famine [3]. However, for obvious ethical reasons, experimental undernutrition is not possible in humans and the elucidation of the mechanism behind this adverse developmental reprogramming has to rely to a certain extent on animal models. Animal models of undernutrition usually employ low-protein diets or global caloric restriction regimes, and in some instances fasting approaches [3,6,16,17]. These types of studies have shown that metabolic and cardiovascular dysfunction during the postnatal period can be induced by prenatal undernutrition [58–68]. Indeed, in a number of species several altered phenotypes have been induced by undernutrition regimes at different developmental stages during the prenatal period, affecting offspring beyond the F1 generation in some cases (Table 2).

Alterations during the periconceptional period [58,69–73] (i.e., from folliculogenesis to implantation) and fetal growth phase [62,74–78] have also been observed in animal models of undernutrition (Figures 2 and 3). The generations affected beyond the F1 generation in some animal models of maternal prenatal undernutrition included not only the F2 generation, but also the F3 offspring [4], the latter being a true example of transgenerational epigenetic inheritance. This is because the F3 offspring will be the generation where progenitor cells are not exposed directly to an adverse environment, and therefore any phenotypic change observed will be the result of programming of the germline [4].

Paternal undernutrition can also induce cardiovascular and metabolic dysfunction. Male mice subjected to intermittent fasting prior to breeding induced a decrease in serum glucose concentrations in the resultant offspring, especially in males [79]. Similarly, male mice fed with a low-protein diet produced offspring with altered expression of genes involved in lipid and cholesterol biosynthesis in the liver [80]. More recently, offspring from male mice exposed to protein restriction developed high blood pressure, vascular dysfunction, and impaired glucose tolerance [81]. Paternal effects could be transmitted not only through the sperm, but also via seminal fluid; but as with overnutrition, information is lacking on the effects of undernutrition on the composition of seminal fluid and its influence on reproductive tract physiology.

From the above-discussed information, it is clear that altered phenotypes induced by undernutrition are transmitted via both maternal and paternal lineages. However, in some animal models of undernutrition, it has been observed

TABLE 2 Selected Animal Models Illustrating the Induction of Altered Phenotypes in Offspring Well Fed during the Postnatal Period but Exposed to Maternal Undernutrition during the Prenatal Period

Species	Time of Exposure in F0 Dam	Phenotype Induced in the Postnatal Period-sex	Generation Affected	References
Mice	During gestation	Increased fat mass-♂ Impaired glucose tolerance-♂ Increased adipocyte size-♂ Hyperleptinemia-♂	F1-F2	[66]
Mice	3.5 days before mating	Increased blood pressure-♀♂ Small kidneys with increased nephron number-♀	F1	[63]
Mice	0–3.5 days after mating	Increased blood pressure-♀♂ Impaired vasodilation of mesenteric arteries-♂	F1	[62,65]
Rats	0–4.25 days after mating	Increased blood pressure-♂	F1	[58]
Rats	During gestation	Increased blood pressure-♀(F1), ♀♂ (F2) Impaired vascular function-♀(F1), ♀♂(F2)	F1-F2	[60,61]
Sheep	−15 to 15 days after mating	Impaired vasodilation of coronary and renal arteries-♀	F1	[64]
Sheep	Late gestation	Increased fat mass-♀♂ Impaired glucose tolerance-♀♂ ↑ Insulin receptor β subunit protein (perirenal fat)-♀♂ ↓ SLC2A4 protein (perirenal fat)- ♀♂	F1	[59]
Sheep	1 month before and 7 days after mating	↓ SLC2A4 protein (skeletal muscle)-♀ ↓ IRS1 protein (skeletal muscle, liver)-♀♂ ↓ aPKCζ protein (skeletal muscle, liver)-♀♂ ↑ AT1R protein (adrenal cortex)-♀♂	F1	[28,29,68]
Sheep	−61 to 30 days after mating	Increased fat mass-♂ Decreased growth of lungs, adrenals, and heart-♂	F1	[67]

that parental gender-specific transmission of altered phenotypes may occur. For instance, in a mouse model of maternal caloric undernutrition, reduced birth weight progressed to the F2 generation through the paternal line and obesity through the maternal line, while impaired glucose tolerance was transmitted via both parental lineages [82]. Furthermore, some phenotypes (e.g., impaired glucose tolerance) were more severe in the F2 offspring exposed to both maternal and paternal undernutrition [82].

Some models of malnutrition have shown that undernutrition can interact with overnutrition in the induction of adverse phenotypes. In sheep, increased hepatic lipid accumulation [83], hyperinsulinemia [84], and increased myocardial lipid deposition [85] were exacerbated in diet-induced obese offspring that underwent in utero undernutrition between early and midgestation. In a sheep model of periconceptional nutrition, it was shown that obese females subjected to weight loss with an energy-restricted diet for 1 month and 7 days after mating produced offspring with alterations in the expression of members of the insulin signaling pathway in skeletal muscle [28] and the renin-angiotensin system in adrenal glands [29]. This latter model of malnutrition may have important clinical implications for humans, as it is generally believed that obese women who lose weight before conception can improve their chances of pregnancy success. However, current evidence strongly suggests that weight loss in obese women will improve their health and fertility at the cost of adverse programming in their children [86,87]. Therefore, it is critical to evaluate the benefits and risks of interventional weight loss during the periconceptional period in obese females. Data generated in animal models of periconceptional malnutrition will be pivotal to develop optimal weight loss approaches in obese women that will improve their fertility and well-being without compromising the future health of their children.

EPIGENETICS AND PRENATAL PROGRAMMING BY MALNUTRITION

Epigenetics can be defined as mitotically and/or meiotically heritable changes in genome function that cannot be explained by changes in DNA sequence [88]. The main epigenetic mechanisms include DNA methylation, histone modifications, and noncoding RNAs [89]. Epigenetic regulation is crucial for the establishment and maintenance of cell type-specific gene expression profiles necessary for the development and differentiation of various cell types [89]. Although epigenetic modifications are relatively stable and remembered across millions of cell divisions, they must undergo reprogramming in order to induce cellular identity. In general, the mammalian life cycle has two major epigenetic reprogramming events. The first genome-wide reprogramming event takes place immediately after fertilization and continues until the blastocyst stage. The other major genome-wide reprogramming occurs in primordial germ cells during early embryo development [88,89]. Nevertheless, extensive epigenetic reprogramming has also been observed in somatic tissue (i.e., kidney, lung, and brain) during normal human fetal development in the second trimester [90]. These natural epigenetic changes experienced throughout development create "windows of opportunity" where the establishment of epigenetic states can be altered by the environment [88]. Indeed, increasing evidence strongly indicates that the mechanism by which an organism can produce different phenotypes from a single genome in response to the environment is through the altered epigenetic regulation of genes [91]. It seems that epigenetic modifications can induce a cellular memory of perturbed environment that will maintain the molecular effects of an early-life experience until adulthood [88]. Hence maternal and paternal nutrition is able to impact future generations via relatively stable programming of germ cells in order to ensure that offspring receive important adaptive information [18]. This is the rationale behind the emerging consensus that epigenetic modifications play a critical role in developmental programming, including nutritional programming [3,18–20,88,89,91].

A link between epigenetic modifications and prenatal programming has started to emerge in human studies. For example, methylation variation in the promoter of the transcription factor retinoid X receptor alpha gene (RXRα) in umbilical cord tissue was found to explain up to 26% of the variation in childhood adiposity [92]. In a recent study, it was found that newborns from obese fathers showed a significant decrease in methylation at the insulin-like growth factor-2 (IGF-2) differentially methylated regions in DNA extracted from cord blood leukocytes [93]. It is interesting to note that this latter epigenetic modification has also been found in studies of prenatal undernutrition in humans. For instance, hypomethylation of the IGF-2 gene has been detected in genomic DNA from blood in individuals exposed in utero to the Dutch famine [94]. More recently, in a study with women exposed to marked annual variations (i.e., dry and rainy season) in nutrient intake in rural Gambia, it was found that children born in the dry season (i.e., children from mothers subjected to some degree of undernourishment) displayed hypomethylation of metastable epialleles (MEs) in peripheral blood lymphocytes and hair follicle samples [94]. MEs are genomic regions where DNA methylation is established stochastically in the early embryo, then stably maintained in differentiated tissues, leading to interindividual epigenetic variations that influence several cell types [95]. These changes in DNA methylation were related to periconceptional maternal plasma variations of methyl donor nutrients involved in one-carbon metabolism (i.e., methionine, betaine, folate, vitamin B2, homocysteine, dimethyl glycine, and S-adenosylhomocysteine) [95]. The relevance of methyl donors in nutritional programming is further illustrated by their potential use as interventional therapies to prevent or reverse epigenetic and phenotypic changes. Accordingly, the altered phenotypes induced in offspring derived from rats fed with a low-protein diet can be prevented by supplementation of the low-protein diet with glycine or folic acid [91]. However, in some instances, folic acid supplementation can cause additional epigenetic changes that can have detrimental effects [91]. Any prospective intervention to reset the epigenome (sum of total epigenetic modifications) to a healthy status should be developed with minimum side effects.

The above-discussed studies are just a minute sample of the information available, as the number of studies in animals [96] and humans [97] linking malnutrition with epigenetic programming is rapidly increasing. However, although most studies have shown significant associations between altered phenotypes induced by malnutrition and epigenetic changes, a truly cause–effect relationship has not been demonstrated that could be used as a starting point to develop effective interventional therapies. The development of effective interventional strategies would require the identification of epigenetic changes before the onset of disease, which is something that few studies have addressed [98]. Different epigenetic modifications should be analyzed in the same platform. This is because rather than working in isolation, different classes of epigenetic modifications work in concert to regulate the activity state of underlying DNA [98]. For instance, regions of highly dense DNA methylation localized to the start of genes are almost universally associated with a distinct profile of histone modifications that together silence associated genes [98]. Another important issue in epigenetic research in nutritional studies is the sample used for analysis, especially in humans, where peripheral blood constituents and tissues discarded at birth (i.e., umbilical cord tissue or blood, placenta, and extraembryonic tissue) are the preferred sample for practical reasons [97,98]. The assumption in this case is that all germ layers during early embryo development (i.e., at the time of major epigenetic reprogramming) are affected by malnutrition and therefore the epigenetic alterations will be detectable in all tissues [91]. However, if the nutritional challenge takes place at a later stage during gestation, it is possible that only tissue-specific epigenetic effects will be observed [91]. The latter highlights the relevance of analyzing epigenetic modifications across different tissues, which is challenging to carry out in humans, but quite feasible in animal models.

Finally, the development of knockout models will be pivotal for validation of the role of specific molecules or cellular pathways in the induction of altered phenotypes during prenatal programming by nutrition. Accordingly, oocytes exposed to hyperinsulinemia showed a decreased glycogen synthase kinase 3 (GSK-3) α and β, and oocyte-specific GSK-3α and β knockout mice showed cardiovascular dysfunction, indicating that GSK-3 plays a critical role in the induction of altered phenotypes during periconceptional hyperinsulinemia [99]. Animal models of malnutrition are crucial for the creation of conceptual frameworks that ultimately need to be explored in human populations. The goal is to identify epigenetic marks at periconception that will assist in the development of preconceptional and gestational interventions capable of manipulating the epigenome in order to prevent or correct prenatal adverse nutritional programming that could lead to development of NCDs in the postnatal life. The development of perinatal screening tools (i.e., epigenetic marks) is also required in order to identify individuals with susceptibility for NCDs later in life. Nevertheless, as suggested by Saffery and Novakovic [98], several decades will pass before clear associations between prenatal programming of adult disease and epigenetic changes are established.

CONCLUSIONS

Several epidemiologic studies in humans and experimental models in animals have proved that overnutrition and undernutrition during the prenatal period can induce the development of NCDs during postnatal life. Several animal models have also shown that both undernutrition and overnutrition during the prenatal period can induce similar altered phenotypes (e.g., impaired glucose tolerance, high blood pressure) in the resultant offspring. This adverse developmental programming can be induced by paternal and maternal lineages, and the effect can go beyond the F1 generation. In recent years, an increasing number of studies have shown that altered phenotypes induced by prenatal nutrition are associated with epigenetic modifications. However, although some associations seem to be strong, clear cause–effect relationships between disease occurrence and epigenetic changes are not available in the specialized literature. Research is needed to identify epigenetic marks that will help to develop preventive (i.e., preconception) and interventional (i.e., during gestation) strategies against adverse developmental programming induced by nutrition. Also, it is critical to identify early markers (i.e., epigenetic marks at birth) for disease risk during postnatal life in individuals exposed to malnutrition during the prenatal period.

Acknowledgments

We are grateful for research funding from BBSRC (BB/I001840/1; BB/F007450/1) and EU-FP7 EpiHealth program to TPF; CS was in receipt of a University of Southampton postgraduate scholarship bursary.

References

[1] Mathers CD, Boerma T, Ma Fat D. Global and regional causes of death. Br Med Bull 2009;92:7–32.
[2] McMillen IC, MacLaughlin SM, Muhlhausler BS, Gentili S, Duffield JL, Morrison JL. Developmental origins of adult health and disease: the role of periconceptional and foetal nutrition. Basic Clin Pharmacol Toxicol 2008;102:82–9.
[3] Langley-Evans SC, McMullen S. Developmental origins of adult disease. Med Princ Pract 2010;19:87–98.
[4] Velazquez MA, Fleming TP. Transgenerational risks by exposure in utero. In: Trounson A, Gosden R, Eichenlaub-Ritter U, editors. Biology and pathology of the oocyte. 2nd ed. Cambridge University Press; 2013. p. 353–61.
[5] Barker DJ. The origins of the developmental origins theory. J Intern Med 2007;261:412–7.
[6] Ojha S, Robinson L, Symonds ME, Budge H. Suboptimal maternal nutrition affects offspring health in adult life. Early Hum Dev 2013;89:909–13.
[7] Velazquez MA, Fleming TP. Maternal diet, oocyte nutrition and metabolism, and offspring health. In: Coticchio G, Albertini DF, De Santis L, editors. Oogenesis. 1st ed. Springer-Verlag; 2013. p. 329–51.
[8] Fowden AL, Giussani A, Forhead AJ. Intrauterine programming of physiological systems. Causes and consequences. Physiology 2006;21:29–37.
[9] Cockburn K, Rossant J. Making the blastocyst: lessons from the mouse. J Clin Invest 2010;120:995–1003.
[10] Li L, Zheng P, Dean J. Maternal control of early mouse development. Development 2010;137:859–70.
[11] Martinez Arias A, Nichols J, Schröter C. A molecular basis for developmental plasticity in early mammalian embryos. Development 2013;140:3499–510.
[12] Zhang S, Lin H, Kong S, Wang S, Wang H, Wang H, et al. Physiological and molecular determinants of embryo implantation. Mol Asp Med 2013;34:939–80.
[13] Koopman P. Organogenesis in development. Curr Top Dev Biol 2010;90:1–408.
[14] John R, Hemberger M. A placenta for life. Reprod Biomed Online 2012;25:5–11.
[15] van Uitert EM, Exalto N, Burton GJ, Willemsen SP, Koning AH, Eilers PH, et al. Human embryonic growth trajectories and associations with fetal growth and birthweight. Hum Reprod 2013;28:1753–61.

[16] Fleming TP, Velazquez MA, Eckert JJ, Lucas ES, Watkins AJ. Nutrition of females during the peri-conceptional period and effects on foetal programming and health of offspring. Anim Reprod Sci 2012;130:193–7.

[17] Fleming TP, Lucas ES, Watkins AJ, Eckert JJ. Adaptive responses of the embryo to maternal diet and consequences for post-implantation development. Reprod Fertil Dev 2012;24:35–44.

[18] Alfaradhi MZ, Ozanne SE. Developmental programming in response to maternal overnutrition. Front Genet 2011;2:27.

[19] Williams L, Seki Y, Vuguin PM, Charron MJ. Animal models of in utero exposure to a high fat diet: a review. Biochim Biophys Acta 2014;1842:507–19.

[20] Li M, Sloboda DM, Vickers MH. Maternal obesity and developmental programming of metabolic disorders in offspring: evidence from animal models. Exp Diabetes Res 2011;2011:592408.

[21] Aiken CE, Ozanne SE. Transgenerational developmental programming. Hum Reprod Update 2014;20: 63–75.

[22] Poston L. Maternal obesity, gestational weight gain and diet as determinants of offspring long term health. Best Pract Res Clin Endocrinol Metab 2012;26:627–39.

[23] Buckley AJ, Keserü B, Briody J, Thompson M, Ozanne SE, Thompson CH. Altered body composition and metabolism in the male offspring of high fat-fed rats. Metabolism 2005;54:500–7.

[24] Cerf ME, Chapman CS, Muller CJ, Louw J. Gestational high-fat programming impairs insulin release and reduces Pdx-1 and glucokinase immunoreactivity in neonatal Wistar rats. Metabolism 2009;58:1787–92.

[25] Cerf ME, Williams K, Nkomo XI, Muller CJ, Du Toit DF, Louw J, et al. Islet cell response in the neonatal rat after exposure to a high-fat diet during pregnancy. Am J Physiol Regul Integr Comp Physiol 2005;288: R1122–8.

[26] Gniuli D, Calcagno A, Caristo ME, Mancuso A, Macchi V, Mingrone G, et al. Effects of high-fat diet exposure during fetal life on type 2 diabetes development in the progeny. J Lipid Res 2008;49:1936–45.

[27] Rattanatray L, MacLaughlin SM, Kleemann DO, Walker SK, Muhlhausler BS, McMillen IC. Impact of maternal periconceptional overnutrition on fat mass and expression of adipogenic and lipogenic genes in visceral and subcutaneous fat depots in the postnatal lamb. Endocrinology 2010;151:5195–205.

[28] Nicholas LM, Morrison JL, Rattanatray L, Ozanne SE, Kleemann DO, Walker SK, et al. Differential effects of exposure to maternal obesity or maternal weight loss during the periconceptional period in the sheep on insulin signalling molecules in skeletal muscle of the offspring at 4 months of age. PLoS One 2013; 8:e84594.

[29] Zhang S, Morrison JL, Gill A, Rattanatray L, MacLaughlin SM, Kleemann D, et al. Dietary restriction in the periconceptional period in normal-weight or obese ewes results in increased abundance of angiotensin-converting enzyme (ACE) and angiotensin type 1 receptor (AT1R) in the absence of changes in ACE or AT1R methylation in the adrenal of the offspring. Reproduction 2013;146:443–54.

[30] Masuyama H, Hiramatsu Y. Additive effects of maternal high fat diet during lactation on mouse offspring. PLoS One 2014;9:e92805.

[31] Long NM, George LA, Uthlaut AB, Smith DT, Nijland MJ, Nathanielsz PW, et al. Maternal obesity and increased nutrient intake before and during gestation in the ewe results in altered growth, adiposity, and glucose tolerance in adult offspring. J Anim Sci 2010;88:3546–53.

[32] Zhang L, Long NM, Hein SM, Ma Y, Nathanielsz PW, Ford SP. Maternal obesity in ewes results in reduced fetal pancreatic β-cell numbers in late gestation and decreased circulating insulin concentration at term. Domest Anim Endocrinol 2011;40:30–9.

[33] Vonnahme KA, Luther JS, Reynolds LP, Hammer CJ, Carlson DB, Redmer DA, et al. Impacts of maternal selenium and nutritional level on growth, adiposity, and glucose tolerance in female offspring in sheep. Domest Anim Endocrinol 2010;39:240–8.

[34] Oben JA, Patel T, Mouralidarane A, Samuelsson AM, Matthews P, Pombo J, et al. Maternal obesity programmes offspring development of non-alcoholic fatty pancreas disease. Biochem Biophys Res Commun 2010;394:24–8.

[35] Dahlhoff M, Pfister S, Blutke A, Rozman J, Klingenspor M, Deutsch MJ, et al. Peri-conceptional obesogenic exposure induces sex-specific programming of disease susceptibilities in adult mouse offspring. Biochim Biophys Acta 2014;1842:304–17.

[36] Matthews PA, Samuelsson AM, Seed P, Pombo J, Oben JA, Poston L, et al. Fostering in mice induces cardiovascular and metabolic dysfunction in adulthood. J Physiol 2011;589:3969–81.

[37] Patel MS, Srinivasan M. Metabolic programming in the immediate postnatal life. Ann Nutr Metab 2011; 58(Suppl. 2):18–28.

[38] Lazic SE, Essioux L. Improving basic and translational science by accounting for litter-to-litter variation in animal models. BMC Neurosci 2013;14:37.

[39] Igosheva N, Abramov AY, Poston L, Eckert JJ, Fleming TP, Duchen MR, et al. Maternal diet-induced obesity alters mitochondria activity and redox status in mouse oocytes and zygotes. PLoS One 2010;5:e10074.

[40] Jungheim ES, Schoeller EL, Marquard KL, Louden ED, Schaffer JE, Moley KH. Diet-induced obesity model: abnormal oocytes and persistent growth abnormalities in the offspring. Endocrinology 2010;151:4039–46.

[41] Wu LL, Dunning KR, Yang X, Russell DL, Lane M, Norman RJ, et al. High-fat diet causes lipotoxicity responses in cumulus-oocyte complexes and decreased fertilization rates. Endocrinology 2010;151:5438–45.

[42] Kubandová J, Cikoš S, Burkuš J, Czikková S, Koppel J, Fabian D. Amount of maternal body fat significantly affected the quality of isolated mouse preimplantation embryos and slowed down their development. Theriogenology 2014;81:187–95.

[43] Shankar K, Zhong Y, Kang P, Lau F, Blackburn ML, Chen JR, et al. Maternal obesity promotes a proinflammatory signature in rat uterus and blastocyst. Endocrinology 2011;152:4158–70.

[44] Velazquez MA, Hadeler KG, Herrmann D, Kues WA, Ulbrich SE, Meyer HH, et al. In vivo oocyte developmental competence is reduced in lean but not in obese superovulated dairy cows after intraovarian administration of IGF1. Reproduction 2011;142:41–52.

[45] McCurdy CE, Bishop JM, Williams SM, Grayson BE, Smith MS, Friedman JE, et al. Maternal high-fat diet triggers lipotoxicity in the fetal livers of nonhuman primates. J Clin Invest 2009;119:323–35.

[46] Grant WF, Gillingham MB, Batra AK, Fewkes NM, Comstock SM, Takahashi D, et al. Maternal high fat diet is associated with decreased plasma n-3 fatty acids and fetal hepatic apoptosis in nonhuman primates. PLoS One 2011;6:e17261.

[47] Murabayashi N, Sugiyama T, Zhang L, Kamimoto Y, Umekawa T, Ma N, et al. Maternal high-fat diets cause insulin resistance through inflammatory changes in fetal adipose tissue. Eur J Obstet Gynecol Reprod Biol 2013;169:39–44.

[48] Sferruzzi-Perri AN, Vaughan OR, Haro M, Cooper WN, Musial B, Charalambous M, et al. An obesogenic diet during mouse pregnancy modifies maternal nutrient partitioning and the fetal growth trajectory. FASEB J 2013;27:3928–37.

[49] Stachowiak EK, Srinivasan M, Stachowiak MK, Patel MS. Maternal obesity induced by a high fat diet causes altered cellular development in fetal brains suggestive of a predisposition of offspring to neurological disorders in later life. Metab Brain Dis 2013;28:721–5.

[50] Kim DW, Young SL, Grattan DR, Jasoni CL. Obesity during pregnancy disrupts placental morphology, cell proliferation, and inflammation in a sex-specific manner across gestation in the mouse. Biol Reprod 2014;90:130.

[51] Pound LD, Comstock SM, Grove KL. Consumption of a Western-style diet during pregnancy impairs offspring islet vascularization in a Japanese macaque model. Am J Physiol Endocrinol Metab 2014;307:E115–23.

[52] King V, Dakin RS, Liu L, Hadoke PW, Walker BR, Seckl JR, et al. Maternal obesity has little effect on the immediate offspring but impacts on the next generation. Endocrinology 2013;154:2514–24.

[53] Fullston T, Ohlsson Teague EM, Palmer NO, DeBlasio MJ, Mitchell M, Corbett M, et al. Paternal obesity initiates metabolic disturbances in two generations of mice with incomplete penetrance to the F2 generation and alters the transcriptional profile of testis and sperm microRNA content. FASEB J 2013;27:4226–43.

[54] Ng SF, Lin RC, Maloney CA, Youngson NA, Owens JA, Morris MJ. Paternal high-fat diet consumption induces common changes in the transcriptomes of retroperitoneal adipose and pancreatic islet tissues in female rat offspring. FASEB J 2014;28:1830–41.

[55] Rando OJ. Daddy issues: paternal effects on phenotype. Cell 2012;151:702–8.

[56] Bromfield JJ, Schjenken JE, Chin PY, Care AS, Jasper MJ, Robertson SA. Maternal tract factors contribute to paternal seminal fluid impact on metabolic phenotype in offspring. Proc Natl Acad Sci USA 2014;111:2200–5.

[57] Dunn GA, Bale TL. Maternal high-fat diet effects on third-generation female body size via the paternal lineage. Endocrinology 2011;152:2228–36.

[58] Kwong WY, Wild AE, Roberts P, Willis AC, Fleming TP. Maternal undernutrition during the preimplantation period of rat development causes blastocyst abnormalities and programming of postnatal hypertension. Development 2000;127:4195–202.

[59] Gardner DS, Tingey K, Van Bon BW, Ozanne SE, Wilson V, Dandrea J, et al. Programming of glucose-insulin metabolism in adult sheep after maternal undernutrition. Am J Physiol Regul Integr Comp Physiol 2005;289:R947–54.

[60] Torrens C, Brawley L, Anthony FW, Dance CS, Dunn R, Jackson AA, et al. Folate supplementation during pregnancy improves offspring cardiovascular dysfunction induced by protein restriction. Hypertension 2006;47:982–7.

[61] Torrens C, Poston L, Hanson MA. Transmission of raised blood pressure and endothelial dysfunction to the F2 generation induced by maternal protein restriction in the F0, in the absence of dietary challenge in the F1 generation. Br J Nutr 2008;100:760–6.

[62] Watkins AJ, Ursell E, Panton R, Papenbrock T, Hollis L, Cunningham C, et al. Adaptive responses by mouse early embryos to maternal diet protect fetal growth but predispose to adult onset disease. Biol Reprod 2008;78:299–306.

[63] Watkins AJ, Wilkins A, Cunningham C, Perry VH, Seet MJ, Osmond C, et al. Low protein diet fed exclusively during mouse oocyte maturation leads to behavioural and cardiovascular abnormalities in offspring. J Physiol 2008;586:2231–44.

[64] Torrens C, Snelling TH, Chau R, Shanmuganathan M, Cleal JK, Poore KR, et al. Effects of pre- and periconceptional undernutrition on arterial function in adult female sheep are vascular bed dependent. Exp Physiol 2009;94:1024–33.
[65] Watkins AJ, Lucas ES, Torrens C, Cleal JK, Green L, Osmond C, et al. Maternal low-protein diet during mouse pre-implantation development induces vascular dysfunction and altered renin-angiotensin-system homeostasis in the offspring. Br J Nutr 2010;103:1762–70.
[66] Peixoto-Silva N, Frantz ED, Mandarim-de-Lacerda CA, Pinheiro-Mulder A. Maternal protein restriction in mice causes adverse metabolic and hypothalamic effects in the F1 and F2 generations. Br J Nutr 2011;106:1364–73.
[67] Jaquiery AL, Oliver MH, Honeyfield-Ross M, Harding JE, Bloomfield FH. Periconceptional undernutrition in sheep affects adult phenotype only in males. J Nutr Metab 2012;2012:123610.
[68] Nicholas LM, Rattanatray L, MacLaughlin SM, Ozanne SE, Kleemann DO, Walker SK, et al. Differential effects of maternal obesity and weight loss in the periconceptional period on the epigenetic regulation of hepatic insulin-signaling pathways in the offspring. FASEB J 2013;27:3786–96.
[69] Leroy JL, Vanholder T, Mateusen B, Christophe A, Opsomer G, de Kruif A, et al. Non-esterified fatty acids in follicular fluid of dairy cows and their effect on developmental capacity of bovine oocytes in vitro. Reproduction 2005;130:485–95.
[70] Eckert JJ, Porter R, Watkins AJ, Burt E, Brooks S, Leese HJ, et al. Metabolic induction and early responses of mouse blastocyst developmental programming following maternal low protein diet affecting life-long health. PLoS One 2012;7:e52791.
[71] Walsh SW, Mehta JP, McGettigan PA, Browne JA, Forde N, Alibrahim RM, et al. Effect of the metabolic environment at key stages of follicle development in cattle: focus on steroid biosynthesis. Physiol Genomics 2012;44:504–17.
[72] Tian X, Diaz FJ. Zinc depletion causes multiple defects in ovarian function during the periovulatory period in mice. Endocrinology 2013;153:873–86.
[73] Sun C, Velazquez MA, Marfy-Smith S, Sheth B, Cox A, Johnson D, et al. Mouse early extra-embryonic lineages activate compensatory endocytosis in response to poor maternal nutrition. Development 2014;141:1140–50.
[74] Bloomfield FH, Oliver MH, Hawkins P, Campbell M, Phillips DJ, Gluckman PD, et al. A periconceptional nutritional origin for noninfectious preterm birth. Science 2003;300:606.
[75] Lloyd LJ, Foster T, Rhodes P, Rhind SM, Gardner DS. Protein-energy malnutrition during early gestation in sheep blunts fetal renal vascular and nephron development and compromises adult renal function. J Physiol 2012;590:377–93.
[76] Liu NQ, Ouyang Y, Bulut Y, Lagishetty V, Chan SY, Hollis BW, et al. Dietary vitamin D restriction in pregnant female mice is associated with maternal hypertension and altered placental and fetal development. Endocrinology 2013;154:2270–80.
[77] Kavitha JV, Rosario FJ, Nijland MJ, McDonald TJ, Wu G, Kanai Y, et al. Down-regulation of placental mTOR, insulin/IGF-I signaling, and nutrient transporters in response to maternal nutrient restriction in the baboon. FASEB J 2014;28:1294–305.
[78] Williams-Wyss O, Zhang S, MacLaughlin SM, Kleemann DO, Walker SK, Suter CM, et al. Embryo number and periconceptional undernutrition in the sheep have differential effects on adrenal epigenotype, growth and development. Am J Physiol Endocrinol Metab. 2014;307:E141–50.
[79] Anderson LM, Riffle L, Wilson R, Travlos GS, Lubomirski MS, Alvord WG. Preconceptional fasting of fathers alters serum glucose in offspring of mice. Nutrition 2006;22:327–31.
[80] Carone BR, Fauquier L, Habib N, Shea JM, Hart CE, Li R, et al. Paternally induced transgenerational environmental reprogramming of metabolic gene expression in mammals. Cell 2010;143:1084–96.
[81] Watkins AJ, Sinclair KD. Paternal low protein diet affects adult offspring cardiovascular and metabolic function in mice. Am J Physiol Heart Circ Physiol 2014;306:H1444–52.
[82] Jimenez-Chillaron JC, Isganaitis E, Charalambous M, Gesta S, Pentinat-Pelegrin T, Faucette RR, et al. Intergenerational transmission of glucose intolerance and obesity by in utero undernutrition in mice. Diabetes 2009;58:460–8.
[83] Hyatt MA, Gardner DS, Sebert S, Wilson V, Davidson N, Nigmatullina Y, et al. Suboptimal maternal nutrition, during early fetal liver development, promotes lipid accumulation in the liver of obese offspring. Reproduction 2011;141:119–26.
[84] Sébert SP, Hyatt MA, Chan LL, Patel N, Bell RC, Keisler D, et al. Maternal nutrient restriction between early and midgestation and its impact upon appetite regulation after juvenile obesity. Endocrinology 2009;150:634–41.
[85] Chan LL, Sébert SP, Hyatt MA, Stephenson T, Budge H, Symonds ME, et al. Effect of maternal nutrient restriction from early to midgestation on cardiac function and metabolism after adolescent-onset obesity. Am J Physiol Regul Integr Comp Physiol 2009;296:R1455–63.
[86] Matusiak K, Barret HL, Callaway LK, Niter MD. Periconception weight loss: common sense for mothers, but what about for babies? J Obes 2014;2014:204295.
[87] S Zhang, Rattanatray L, Morrison JL, Nicholas LM, Lie S, McMillen IC. Maternal obesity and the early origins of childhood obesity: weighing up the benefits and costs of maternal weight loss in the periconceptional period for the offspring. Exp Diabetes Res 2011;2011:585749.

[88] Youngson NA, Morris MJ. What obesity research tells us about epigenetic mechanisms. Philos Trans R Soc Lond B Biol Sci 2013;368:20110337.

[89] Niculescu MD. Nutritional epigenetics. ILAR J 2012;53: 270–8.

[90] Yuen RK, Neumann SM, Fok AK, Peñaherrera MS, McFadden DE, Robinson WP, et al. Extensive epigenetic reprogramming in human somatic tissues between fetus and adult. Epigenetics Chromatin 2011;4:7.

[91] Lillycrop KA, Burdge GC. Epigenetic mechanisms linking early nutrition to long term health. Best Pract Res Clin Endocrinol Metab 2012;26:667–76.

[92] Godfrey KM, Sheppard A, Gluckman PD, Lillycrop KA, Burdge GC, McLean C, et al. Epigenetic gene promoter methylation at birth is associated with child's later adiposity. Diabetes 2011;60:1528–34.

[93] Soubry A, Schildkraut JM, Murtha A, Wang F, Huang Z, Bernal A, et al. Paternal obesity is associated with IGF2 hypomethylation in newborns: results from a Newborn Epigenetics Study (NEST) cohort. BMC Med 2013;11:29.

[94] Heijmans BT, Tobi EW, Stein AD, Putter H, Blauw GJ, Susser ES, et al. Persistent epigenetic differences associated with prenatal exposure to famine in humans. Proc Natl Acad Sci USA 2008;105:17046–9.

[95] Dominguez-Salas P, Moore SE, Baker MS, Bergen AW, Cox SE, Dyer RA, et al. Maternal nutrition at conception modulates DNA methylation of human metastable epialleles. Nat Commun 2014;5:3746.

[96] Seki Y, Williams L, Vuguin PM, Charron MJ. Minireview: epigenetic programming of diabetes and obesity: animal models. Endocrinology March 2012;153(3):1031–8.

[97] van Dijk SJ, Molloy PL, Varinli H, Morrison JL, Muhlhausler BS. Members of EpiSCOPE. Epigenetics and human obesity. Int J Obes (Lond) 2015;39:85–97.

[98] R1 Saffery, Novakovic B. Epigenetics as the mediator of fetal programming of adult onset disease: what is the evidence? Acta Obstet Gynecol Scand 2014;93: 1090–8.

[99] da Rocha AM, Ding J, Wolfe AM, Slawny N, Converso-Baran K, Smith GD. Chronic hyperinsulinemia actions on the oocyte and offspring cardiovascular defects. Fertil Steril 2013;100:S113–4.

CHAPTER

7

Maternal Prenatal Stress and the Developmental Origins of Mental Health: The Role of Epigenetics

Kieran J. O'Donnell[1], *Vivette Glover*[2]

[1]The Ludmer Centre for Neuroinformatics and Mental Health, Douglas Mental University Institute, McGill University, Montreal, QC, Canada; [2]Institute of Reproductive and Development Biology, Imperial College London, London, UK

OUTLINE

INTRODUCTION

There are profound developmental origins for individual differences in mental health that extend over the lifespan. This association is well established for severe early life trauma such as neglect, physical or sexual abuse [1]. Indeed, child maltreatment not only predicts the risk for multiple mental disorders, but, at least in the case of depression, also the severity

The Epigenome and Developmental Origins of Health and Disease
http://dx.doi.org/10.1016/B978-0-12-801383-0.00007-4

Copyright © 2016 Elsevier Inc. All rights reserved.

and treatment response [2]. Strong evidence has emerged that exposure to *prenatal* adversity predicts a range of mental health outcomes (see Refs [3–5]). Prospective, longitudinal analyses suggest that these effects persist through adolescence [6,7] into early adulthood [8] and are of clinical relevance [6].

An obvious question concerns the mechanism(s) by which the effects of the in utero environment are embedded in the molecular machinery that regulates genomic transcription. In this chapter we examine the evidence for the environmental epigenetics hypothesis, which proposes that the sustained effects of environmental conditions on gene expression are mediated by effects on the epigenetics mechanisms that regulate the transcriptional machinery.

Studies with a remarkable range of species have shown maternal prenatal influences on the phenotype of the offspring. Within evolutionary biology and ecology, such prenatal effects are defined as instances in which the phenotype of the offspring is affected by maternal signals, independent of genetic inheritance. Such effects are considered as a primary mechanism for phenotypic plasticity (i.e., variation in genotype–phenotype relations). Such parentally induced variation in phenotype has been suggested to be adaptive (i.e., when the resulting phenotype enhances survival and reproduction), referred to as the predictive adaptive response [9]. In the meadow vole, offspring growth, fur thickness, and sexual maturation are influenced by the length of daylight experienced by the mother during the perinatal period, preparing her offspring for the approaching winter [10,11]. In the red squirrel, maternal exposure to increased population density predicts greater offspring weight gain in the early postnatal period, which may enhance survival. These maternal prenatal effects on offspring phenotype are mediated at least in part by maternal neuroendocrine function and can be manipulated by administering melatonin to the pregnant meadow vole or cortisol to the pregnant red squirrel. These examples serve to demonstrate the importance of environment-dependent maternal signaling for offspring development, fitness, and survival. The question arises as to the extent and mode of transmission of prenatal effects in humans and their relevance for child development and vulnerability for later mental disorder.

Defining the Exposure

Stress is a generic term and has several different definitions. Many different types of prenatal stress have been shown to be associated with altered outcome for the child. These include maternal symptoms of anxiety and depression [12], daily hassles [13], pregnancy-specific anxiety [13,14], and a poor relationship with the partner [15]. The studies of maternal anxiety and depression in general used continuous measures. The associations between maternal mood in pregnancy and altered child outcome are certainly not only found in those with a clinical diagnosis but can be observed within the normal variation in a population (e.g., Ref. [12]). Other studies have included experience of war [16,17], famine [18–20], and acute disasters such as an earthquake [21], a Canadian ice storm [22], a hurricane in Louisiana [23], the Chernobyl nuclear disaster [24], and 9/11 [25]. It is clear that it is not just very extreme or "toxic stress" that can alter outcome. Exposures that can have an effect vary from the very severe, such as the death of a first-degree relative [26], to quite mild stresses, such as daily hassles. We do not yet know whether different forms of stress have different effects.

For some outcomes there is some evidence of a nonlinear relationship. For example, DiPietro et al. found that increased mild prenatal stress was associated with better cognition and motor development at two years [27]. With these outcomes it is possible that a moderate exposure to prenatal stress is optimal. This notion has some support from certain models of early life stress in nonhuman primates (NHPs) (see Ref. [28]).

However, with emotional and behavioral measures in the child the association with maternal prenatal stress appears linear [12]. This is an area where more research is needed.

Most of these studies are prospective, but those examining the effects of acute disasters are of necessity retrospective and are less able to allow for possible confounding factors. However, they have the advantage that the effects on the child are more clearly prenatal and unlikely to be due to genetic continuity.

All these studies have shown associations with adverse child outcomes. It is obviously harder to prove that these associations are, at least in part, causal. If the mother is anxious or depressed while pregnant, she may be affected postnatally also, and this may affect her parenting. There may be associated factors such as smoking or alcohol consumption (often underreported in pregnancy) [29], changes in diet (see Ref. [30]), lower maternal education, or lower socioeconomic status. These may all affect child outcome and the cumulative effect of these exposures on child outcome certainly requires further study. The evidence for thinking that there is a prenatal causal component comes from different types of evidence.

First, there are well-controlled animal studies, which employ cross-fostering designs on the first postnatal day or nursery rearing in the case of NHPs. These experiments show long-term effects of the prenatal stress and provide evidence for a prenatal rather than a postnatal or other confounding effect [31]. Secondly, there are large epidemiological human studies that have examined associations with prenatal anxiety or depression after allowing for a wide range of potential confounders including paternal mood, postnatal maternal mood, parenting behavior, maternal education, and maternal smoking and drinking alcohol during pregnancy, and still find a clinically significant relationship between prenatal maternal mood and child outcome [6]. Thirdly, there are studies that have found associations between prenatal maternal mood and aspects of child outcomes at birth, thus showing effects independent of postnatal maternal mood or parenting [14,32,33]. Finally, one in vitro fertilization study has shown some effect of maternal prenatal stress on conduct problems in children born from donor eggs, and therefore genetically unrelated to their birth mother, arguing against a simple genetic transmission of phenotype [34].

Effects on the Child

Many different outcomes have been shown to be changed in association with various types of maternal prenatal stress. The effects can be quite subtle and often have been measured on continuous scales rather than by diagnostic categories. However, they are also often of clinical significance [6]. Here we provide selected examples to show the range of outcomes found to be altered by maternal prenatal stress. The studies almost all date from the last 10–15 years. Although this type of research has been conducted in animals since the 1950s, it is relatively recent that similar research has been carried out in humans. Some studies have examined large population cohorts such as the Avon Longitudinal Study of Parents and Children (ALSPAC) cohort in Bristol, UK [6,35,36], and used maternal reports of offspring outcome. Other studies have been smaller observational cohorts, such as those of Van den Bergh et al. [37,38] and that of Bergman et al. [15,39].

The wide range of outcomes, which have been found to be altered by maternal prenatal stress, include those from birth until adulthood. At birth, an increase in congenital malformations has been found to be associated with very severe stress in the first trimester, such as the death of an older child [40]. Many studies have shown that less severe stress is associated with somewhat lower birth weight and reduced gestational age [34,41]. Another finding at birth is an altered sex ratio, with fewer males to females being born than in an unstressed population [42–44].

Many studies have looked at neurodevelopmental and psychopathological development. Some investigators have examined the newborn and found a poorer performance on the Neonatal Behavioral Assessment Scale [45], showing that adverse behavioral outcomes are observable from birth. Studies of infants and toddlers have shown more difficult temperament [46,47], more sleep problems [48], lower cognitive performance, and increased fearfulness [15]. More recent studies have shown that effects can persist from childhood into adolescence and early adulthood [6–8], emphasizing the influence of maternal prenatal stress on mental health across different developmental stages.

Others have examined the association between prenatal stress and neurodevelopmental outcomes in children, rather than babies, infants, or adults. Many independent groups have shown increases in child emotional problems, especially anxiety and depression, and symptoms of attention deficit hyperactivity disorder (ADHD) and conduct disorder [6,24,34–36,38,49] as well as increased vulnerability to victimization [50,51]. Others have shown a reduction in cognitive performance [22,52] and differences in learning strategies in adulthood [53].

Several studies have found an association between prenatal stress and increased risk of autism or autistic traits [23,54,55], although one population study failed to replicate this finding [56]. Two studies have found an increased risk of schizophrenia in adults. Both showed associations with severe stress, the death of a relative [26] or exposure to extreme adversity, the invasion of the Netherlands in 1940 [17], and both showed that the sensitive period of exposure was during the first trimester. Maternal famine exposure at the time of conception [57] or during pregnancy [20] has also been reported to be associated with an increased risk of schizophrenia. However, within the context of famine exposure it is impossible to disentangle the effect of maternal stress from that of nutritional deprivation.

Associations have also been found between prenatal stress and a range of altered physical and physiological outcomes in children. Alterations in brain structure have been observed, including reduced brain gray matter density [58,59] and altered limbic-prefrontal white matter circuitry [60]. An altered fingerprint pattern [61] and more mixed handedness [62,63] are physical changes observed to be associated with prenatal stress and are known to be determined in utero. Prenatal stress has also been shown to be associated with an altered diurnal pattern or altered function of the hypothalamic–pituitary–adrenal (HPA) axis, although the pattern of alteration is quite complex and may vary across developmental stages [64,65]. Finally, recent studies have shown that prenatal stress is associated with reduced telomere length in cord blood cells [66]. This is an intriguing finding as reduced telomere length is associated with a cellular senescence. This suggests that maternal prenatal stress may accelerate cellular aging; it will be important to determine the long-term implications of such findings.

Timing of Exposure

Studies of teratogens such as thalidomide and rubella show that the timing of exposure can influence the severity of the resulting phenotype. However, there is little consistency in the literature as to the most sensitive time in gestation for the influence of prenatal stress. It is likely that there are different periods of sensitivity dependent on the outcome studied, and the stage of development of the relevant brain structures. Studies of schizophrenia suggest the most sensitive period is the first trimester [17,26]. This is when neuronal cells are migrating to their eventual site in the brain, a process previously suggested to be disrupted in schizophrenia. In contrast, one study of conduct disorder found the greatest association with stress was in late pregnancy [34], as did a study on the risk for autism [23]. However, there is inconsistency in the literature even when the same outcome is being considered. Understanding the times of sensitivity for different outcomes is an area where much more research is needed. Likewise more work is needed to determine how maternal life history may influence the association between prenatal stress and child outcome. For example, Plant et al. reported that the association between maternal prenatal depression and child antisocial behavior was evident only in women who had experienced childhood trauma [67]. No association was observed between maternal prenatal depression and child antisocial behavior in women who did not experience childhood trauma. Although this is a small study (n=125 families), which requires replication, the findings suggest that some of the effects of maternal prenatal stress may be contingent on other adverse experiences earlier in life.

Underlying Mechanisms

Much remains to be understood about the mechanisms that underlie fetal programming by prenatal stress in humans. One early suggestion was a decrease in blood flow to the fetus [68]. However, it is not clear if the decrease observed in that study would be clinically significant, and others have failed to replicate the original finding [69,70].

Another possible mediating factor is increased exposure of the fetus to cortisol [71]. Glucocorticoids (cortisol in humans and NHPs, corticosterone in most rodents) are known to have a range of effects on the developing fetus, including on the brain [72]. Whilst they are essential for fetal development and tissue maturation, overexposure of the fetus to glucocorticoids may have effects that predispose the child to ill health in later life [73]. The potentially widespread role for exposure to increased cortisol in human fetal brain development is demonstrated by a microarray analysis, which showed that increasing cortisol exposure affected the expression of over a thousand genes in fetal brain cells [74]. These effects may be mediated at least in part by glucocorticoid-induced changes in the organization of DNA, giving rise to a more open configuration favoring transcription [75,76]. Similarly, a study in guinea pigs indicates widespread epigenetic changes in the hippocampus following prenatal administration of a synthetic glucocorticoid betamethasone [77]. In line with this, human fetuses exposed to synthetic glucocorticoids in the womb, because of threatened preterm labor, had more mental health problems than matched controls later in childhood [78]. Davis and colleagues have shown that babies exposed to synthetic glucocorticoids in utero have altered brain structure, including a thinner cortex, as shown by magnetic resonance imaging scans [79]. Also, the children of mothers who had consumed high levels of liquorice during pregnancy (which contains a natural inhibitor of 11-β-hydroxysteroid dehydrogenase type II (11β-HSD2), the enzyme that converts cortisol to its inactive form cortisone in the placenta), and were thus exposed to higher levels of cortisol in utero, were more likely to have emotional and cognitive problems in childhood [80].

Fetal overexposure to glucocorticoids could occur through increases in maternal cortisol associated with anxiety and during periods of stress, which then crosses the placenta into the fetal environment. In animal models, this has been shown to be one possible mechanism. Administration of adrenocorticotropic hormone (ACTH) to pregnant rhesus monkeys resulted in increased maternal cortisol production and adverse offspring neurodevelopmental outcomes similar to those seen in response to prenatal stress [81]. The effects of prenatal stress in rats have been shown to be prevented by adrenalectomy and reinstated by corticosterone administration [82]. However, the human HPA axis functions differently in pregnancy from most animal models because of the placental production of corticotrophin-releasing hormone (CRH), which in turn causes an increase in maternal cortisol. The maternal HPA axis becomes gradually less responsive to certain stressors as pregnancy progresses [83], and there is only a weak, if any, association between maternal mood and her cortisol level, especially later in pregnancy [84–86].

It is possible that fetal programming, caused by prenatal stress, may be mediated by raised fetal exposure to cortisol without increases in maternal levels. Maternal prenatal depression and maternal prenatal cortisol levels have been found to be independent predictors of infant temperament [47]. Stress or anxiety may cause increased transplacental transfer of maternal cortisol to the fetal compartment without a rise in maternal levels. The placenta clearly plays a crucial role in moderating fetal exposure to maternal factors, and presumably in preparing the fetus for the environment in which it is going to find itself [85] as part of a predictive adaptive response [9]. Thus, another mechanism by which the fetus could become overexposed to glucocorticoids is through changes in placental function, especially in a downregulation of the enzyme 11β-HSD2, the barrier enzyme that converts cortisol to the inactive cortisone [85]. If there is less of this barrier enzyme in the placenta, then the fetus will be exposed to more maternal cortisol, independently of any change in the maternal cortisol level. If the mother has higher basal levels of cortisol, then the amount of fetal exposure will be higher too, as there is a strong correlation between maternal and fetal cortisol levels [87]. There is evidence in rat models that prenatal stress can affect placental 11β-HSD2. Restraint stress of pregnant rats in the last week of pregnancy has been shown to result in decreased placental 11β-HSD2 expression and activity [88]. Additionally, there is also evidence that reduced 11β-HSD2 causes an alteration in the behavior of the offspring. Prenatal administration of the carbenoxolone, which reduces 11β-HSD2 activity, resulted in an increase in anxiety in a rodent model, mirroring the phenotype of offspring exposed to prenatal stress. Glover et al. [89] reported that the correlation between maternal and amniotic fluid cortisol levels was greater in women with high anxiety compared to less anxious women. This suggests that prenatal anxiety in humans can increase the placental permeability to cortisol. More direct evidence comes from a study of women undergoing elective caesarean section where an inverse association was observed between maternal prenatal trait anxiety and 11β-HSD2 expression [90].

In utero cortisol has been shown to be inversely correlated with infant cognitive development at about 18 months old [39]. This study also measured infant attachment style [91,92], an index of the early care environment, and found that this inverse correlation was strong in insecurely attached infants but absent in the securely attached. This suggests that the quality of attachment may be able to buffer this effect of raised in utero cortisol. In line with this, McGoron et al. report data from the Bucharest Early Intervention Project and demonstrate that a secure infant attachment moderates the association between institutionalization and risk for later psychopathology [93].

The timing of fetal exposure to raised cortisol may also be important for its effect on cognitive development. Elevated maternal cortisol early in gestation has been shown to be associated with a slower rate of mental development, as measured using the Bayley Scales of Infant Development (BSID) over the first year of life. However, elevated levels of maternal cortisol late in gestation were associated with accelerated cognitive development and higher BSID scores at 12 months of age [94]. The authors propose that elevated cortisol levels early in pregnancy may disturb neurodevelopmental processes, but higher cortisol levels later in pregnancy may promote maturation of neural circuitry. It is unclear how such differential effects of cortisol on child neurodevelopment are mediated, but these data underscore the complex associations between prenatal maternal HPA axis function and child outcome.

Many other systems are likely to be involved as well as the HPA axis and cortisol [95].

Another possible mediator of prenatal stress-induced programming effects on offspring neurocognitive and behavioral development is 5-hydroxytryptamine (5-HT). During gestation, 5-HT acts as a trophic factor regulating cell division, differentiation, and synaptogenesis in the fetal brain [96]. Animal studies have shown that increased 5-HT exposure during gestation is associated with alterations in many neuronal processes and subsequent changes in offspring behavior. Recent work has identified an endogenous 5-HT biosynthetic pathway in the human placenta [97], suggesting a possible role for alterations in placental 5-HT in human fetal programming. Prenatal depression has been found to be associated with a downregulation of the expression of placental monoamine oxidase A, an enzyme that metabolizes 5-HT to 5-hydroxy-indol-acetaldehyde, which is further oxidized to an inactive metabolite [98]. Thus, if the mother is depressed prenatally, the placenta may produce more 5-HT, which in turn may result in the brain of her fetus being exposed to more 5-HT, with altered neurodevelopment.

Another potentially interesting mediating factor is brain-derived neurotrophic factor (BDNF). This is a trophic factor known to be important in synaptic plasticity [99]. It is decreased in depression and raised by antidepressants [100]. Prenatal stress in a rat model has been shown to inhibit the formation of mature BDNF in the offspring brain [99]. Uguz et al. report data from a small sample of women with generalized anxiety disorder (GAD) and find cord blood levels of BDNF in neonates born to mothers with GAD were about half those in controls [101].

Thus, maternal stress, anxiety, or depression may be associated with altered placental function, which results in increased fetal exposure to cortisol, 5-HT, or BDNF, among other possible factors, and these in turn may change fetal neurodevelopment. These causal pathways all still need to be shown directly. This hypothesis also raises the question as to what biological changes take place in the mother during prenatal stress, anxiety, or depression, which in turn alters the function of the placenta. As discussed above, the maternal changes in cortisol are not of sufficient magnitude to be a likely sole mediator. And, maternal cortisol levels during pregnancy can be a predictor of child outcome independent of maternal mood [94]. The maternal mediator, or mediators, between prenatal stress, anxiety, and depression and altered child outcome is currently not known.

One possible group of biological maternal mediators could be those associated with the immune system and inflammation, including the proinflammatory cytokines, such as interleukin-6 IL6 [102]. A mouse model has shown an association between prenatal stress and alterations of immune pathways within the placenta, specifically in male offspring [103]. Treating the pregnant mice with nonsteroidal anti-inflammatory drugs both reduced the placental cytokine levels and hyperactive locomotion in the male offspring.

There is a growing literature associating proinflammatory cytokines with depression in humans [104]. Increased cytokines have been associated with psychosocial stress during pregnancy [105]. Elevated maternal stress was related to higher serum IL6 both in early and late pregnancy. No relationships between stress and cytokines were apparent during the second trimester. However, elevated stress levels across pregnancy were predictive of elevated production of the proinflammatory cytokines IL1β and IL6 by stimulated lymphocytes in the third trimester, suggesting that stress during pregnancy affects the function of immune system cells. A recent study has confirmed that depressed pregnant women have higher levels of IL6 in the first trimester [106]. However, another study has failed to find any association between maternal symptoms of anxiety and depression during pregnancy and levels of IL6 [107] at 18 or 32 weeks gestation. This is clearly an area that needs further exploration especially given the reciprocal links between the immune system and epigenome [108–110].

EPIGENETICS

There is currently great interest in the potential role of epigenetics in underlying the long-term effects of prenatal stress on the development of the fetus and the child [111]. Epigenetics, stemming from the Greek "epi" meaning "upon" and "genetics," refers to a series of chemical modifications to chromatin that regulate genomic transcription. Indeed, epigenetics has become largely synonymous with the study of transcriptional regulation, that is, the degree to which a gene is expressed or repressed.

As described in Chapter 1, DNA is commonly organized in a form that resembles beads lying along a string. The beads are nucleosomes comprised of ~146 base pairs wrapped around a core of histone proteins [112]. The histones and DNA together are referred to as chromatin; the nucleosome is a single unit of chromatin (see Figure 1).

Under normal conditions there is a tight physical relationship between the histone proteins and the accompanying DNA, resulting in a closed nucleosome configuration. This restrictive configuration is maintained, in part, by electrostatic bonds between the positively charged histones and the negatively charged DNA. The closed configuration impedes transcription factor binding and is associated with a reduced level of gene expression. Epigenetic modifications essentially favor a closed or open state of chromatin that either increases or decreases the ability of

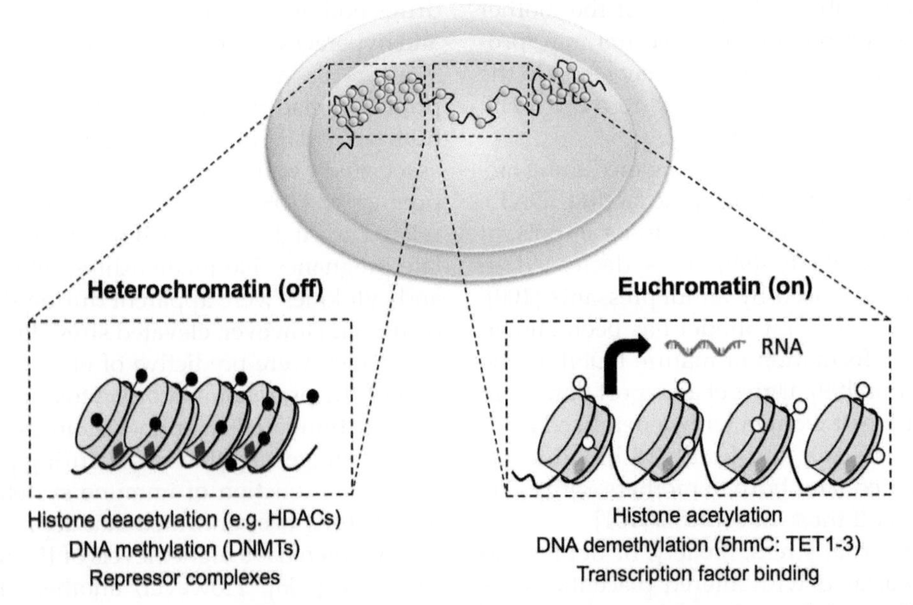

FIGURE 1 **Chromatin dynamics influence gene transcription.** DNA within a cell's nucleus coiled around histone proteins resembling beads on a string (upper panel: black lines/light blue circles). Increased repressive modifications to histone tails (red diamonds) and DNA methylation (black circles) promote nucleosome compaction and gene inactivation/repression (heterochromatin; lower left panel). Histone acetylation (green diamonds) and DNA demethylation/unmethylated cytosines (white circles) associate with an open chromatin configuration that permits transcription factor binding and active gene transcription (euchromatin; lower right panel). DNMTs, DNA methyltransferases; HDACs, histone deacetylases; 5hmC, hydroxymethylcytosine; TET, ten-eleven translocation family of enzymes.

transcription factors to access regulatory sites on the DNA, which control gene transcription.

The dynamic alteration of chromatin structure is achieved in part through covalent modifications to the histone proteins at the amino acids that form the histone protein tails extending out from the nucleosome. There are several examples of such modifications including, but not limited to, acetylation, phosphorylation, methylation, and ubiquitylation (see Ref. [113]). These modifications are achieved through a series of enzymes that bind to the histone tails and modify the local chemical properties of specific amino acids [114,115]. For example, the enzyme histone acetyltransferase "transfers" an acetyl group onto specific lysine amino acids on the histone tails. The addition of the acetyl group diminishes the positive charge, loosening the relation between the histones and DNA, opening the chromatin, and improving the ability of transcription factors to access the associated DNA. Thus, histone acetylation at specific lysine sites is commonly associated with active gene transcription. Nucleosomes in this open, permissive state are termed euchromatin (see Figure 1). The functional antagonists of the histone acetyltransferases are a class of enzymes known as histone deacetylases (HDACs). These enzymes remove acetyl groups and prevent further acetylation, thus maintaining a closed chromatin structure, decreasing transcription factor and gene expression. The acetylation and deacetylation of histones is a dynamic process that is regulated by environmental signals. Additional histone modifications, notably histone methylation, act less directly. Methylation can occur at multiple amino acid sites in the histone tails and then attracts protein complexes that mediate either the closure or opening of chromatin. Methylation or demethylation at individual sites is catalyzed by specific enzymes. Stability is an important feature of histone methylation; unlike acetylation, methylation marks can persist and have been shown to directly mediate the stable influence of environmental signals on gene transcription [116].

DNA Methylation

Another level of regulation occurs directly on the DNA. Indeed, the classic epigenetic alteration is that of DNA methylation, which involves the addition of a methyl group (CH_3) to cytosines linked by phosphate to guanines (CpGs) in the linear sequence of DNA, forming 5-methylcytosine (5mC) [117,118]. Methyl groups bound to cytosines occupy the major groove of DNA and may disrupt normal recognition of transcription factors [119] and influence transcription. Furthermore, methylated DNA attracts a class of proteins known as methylated-DNA binding proteins. These proteins, in turn, attract an entire cluster of proteins that form a repressor complex, which includes active mediators of gene silencing. The HDACs are a critical component of the repressor complex. HDACs prevent histone acetylation and favor a heterochromatin (closed) state that constrains transcription factor binding and gene expression [120].

More recently, non-CG methylation (mCH, where H = adenine, thymine, or cytosine), common in plants, has been described in rodents [121] and humans [122]. Both 5mC and mCH show dynamic change across human brain development and are relevant for the specification of neurons and glia [122]. Similarly, there is increasing interest in hydroxymethylcytosine (5hmC) as a novel epigenetic modification abundant in the human brain and of functional importance [122,123]. DNA demethylation can be achieved through iterative oxidation, catalyzed by the ten-eleven translocation (TET) family of enzymes (TET1-3). Oxidation of 5mC by TET enzymes gives rise to 5hmC, which can be further oxidized to 5-formylcytosine, then 5-carboxylcytosine both of which can be converted to unmodified cytosine. There is increasing evidence that 5hmC plays a functional role in gene transcription, in addition to facilitating DNA demethylation. Both positive and negative correlations between 5hmC and gene expression have been reported and 5hmC is enriched in enhancer regions, sites of alternate splicing, and gene promoters "poised" for transcription [122–124]. A recent study of prenatal restraint stress in rats found lower expression of BDNF transcripts in the hippocampus and cortex of prenatally stressed offspring. This covaried with increased expression of TET1 together with an enrichment of 5mC and 5hmC within the *BDNF* gene [125]. There is also some preliminary evidence from a small study of NHPs, which suggests an association between the postnatal rearing environment and 5hmC within the frontal cortex [126]. This is interesting as the rearing model used in this study is predictive of a range of adverse neurodevelopmental outcomes in maternally deprived offspring. Clearly, it will be important for future studies to determine the contribution of mCH and 5hmC to individual differences in vulnerability for mental disorders following prenatal stress. In contrast to the emerging literature on mCH and 5hmC, the influence of prenatal stress on 5mC has been studied in greater detail and is discussed below.

Early Life Stress and 5mC

Several studies have now shown epigenetic changes in the fetus or child after prenatal stress, in both animal models (e.g., Refs [127,128]) and in humans (see Ref. [129]). Hompes et al. have shown epigenetic changes in the promoter region of the glucocorticoid receptor (*NR3C1*) in the cord blood from mothers who suffered from pregnancy-related anxiety [14]. Oberlander et al. have also shown altered *NR3C1* promoter methylation in cord blood from neonates born to mothers with prenatal depression [130]. Similarly, Mulligan et al. report that maternal exposure to psychosocial stress predicts variation in *NR3C1* promoter methylation in cord blood from neonates in the Democratic Republic of Congo [131]. Appleton et al. investigated the association between methylation of the promoter region of *HSD11B2* in the placenta and markers of prenatal socioeconomic adversity, such as poverty, in 444 healthy newborns [132]. It was found that there was less methylation of *HSD11B2* in neonates whose mothers experienced the most socioeconomic adversity, particularly in the male infants.

Epigenetic changes have also been found in older children whose mothers experienced stress during pregnancy. For example, maternal prenatal stress, caused by violence by the partner, has been shown to be associated with increased methylation of the *NR3C1* promoter in the blood of their adolescent children [133]. While this finding is interesting and suggestive of transgenerational effects of domestic violence, these findings should be treated with some caution. The majority of adolescents examined in this study had undetectable levels of methylation within the *NR3C1* promoter (n=22), with low levels of methylation (<10% methylated) detected in a small subgroup (n=7). Eight women reported exposure to domestic violence during pregnancy; of these cases three adolescent offspring had undetectable levels of methylation in the *NR3C1* promoter while five adolescents had low levels of 5mC<6%, which presumably drive the observed association between maternal domestic violence exposure and adolescent 5mC.

An important consideration for all of these studies is that the changes in 5mC were small and the functional association with gene expression was not determined. Similarly, none of these studies corrected for the cellular heterogeneity of biosamples studied, which can contribute to group differences in 5mC (e.g., Ref. [134]). This makes it difficult to determine the origin of these epigenetic effects—active remodeling of the epigenome by prenatal stress or differences in the distribution of specific cell types, such as immune cells, which may have different methylation profiles.

In addition to candidate gene analyses, an increasing number of studies have employed genome-wide approaches to examine the effects of prenatal stress on 5mC. Currently, the Illumina Infinium HumanMethylation450 BeadChip (450K array) is one of the most popular platforms for genome-wide analyses. This array quantifies methylation at single-base resolution across 482,421 CpG sites and 3091 mCH sites [135] with probes enriched in gene promoters, proximal to the transcription start site but also located in the gene body and 3′ UTR. The array covers almost all (98.9%) of the University of California Santa Cruz (UCSC) RefGenes (~17 CpG sites/gene), however, the array has poor coverage of certain genomic regions of interest such as enhancers. This is an important limitation as enhancers contribute to the developmental regulation of gene expression, associated with levels of 5mC, and are increasingly implicated in neurodevelopmental disorders such as schizophrenia [136,137]. Likewise, it should be noted that the 450K arrays methylation at less than 2% of the 28 million CpGs within the human methylome. As such, 450K analyses providing estimates of CpGs associated with a particular phenotype of interest, including prenatal stress, should be treated as provisional and likely to change as coverage of the methylome increases.

Non et al. used the 450K array to survey genome-wide 5mC in a small sample of non-medicated depressive women (n = 13) compared to controls (n = 23). They observed 44 CpGs, which show differential methylation in neonatal cord blood samples as a function of maternal prenatal depression status. These differences were very small (~2%) but survived correction for the large number of multiple tests carried out [32]. Methylation levels were not significantly different between selective serotonin reuptake inhibitor (SSRI)-exposed neonates and controls. This contrasts with the findings of Schroeder et al., who found a significant difference in 5mC at a number of CpGs within *TNFRSF21* as a function of prenatal exposure to SSRIs [138] but no association between methylation and maternal depression. It should be noted that in their study, Schroeder and colleagues used a methylation platform with reduced coverage of the methylome (~27,000 CpGs) compared with the 450K used by Non et al. [32] and only included women with a lifetime history of a mood disorder. Neither study corrected for the cellular heterogeneity of cord blood, which may contribute to the discrepancies across studies.

Genetic and Gene-by-Environment Influences on the Epigenome

Genetic polymorphisms and epigenetic modifications both influence gene transcription. It is important to emphasize that these are coordinated and often interdependent processes. For example, a methylated cytosine is more readily mutated (deaminated) to thymine leading to mismatched guanine–thymine pairs, which are further modified to create an adenine–thymine bond. This process may, in part, explain the reduced frequency of CpGs across the human genome [120]. Similarly, genotype is an important determinant of the epigenetic landscape [117,136,139]. An analysis of 5mC in six individuals from a three-generation family illustrates that 5mC reflects genetic relatedness [140]. Likewise, a recent study of 614 people from 117 families reported high heritability for 5mC but observed a stronger correlation between maternal and child 5mC profiles than paternal and child comparisons [141], suggestive of maternal effects on the epigenome.

Genotype may influence 5mC in a number of ways. Genetic polymorphisms such as single nucleotide polymorphisms (SNPs), variable number tandem repeats, and copy number variants may introduce or remove CpGs, thereby adding (or removing) potential sites for methylation. Genetic polymorphisms that modulate transcription factor binding may also influence 5mC. Increased transcription factor binding protects CpG islands from acquiring 5mC [117,136,139] and predicts DNA demethylation, possibly via a 5hmC-dependent mechanism [142]. Both gene promoter and more distal regulatory regions appear to be influenced by transcription factor binding in this way. Importantly, transcription factor-mediated remodeling of the epigenome is a plausible candidate to explain the biological embedding of prenatal stress (Figure 2). Steroid receptors, such as the glucocorticoid receptor (GR), have the capacity to bind DNA and have been shown to alter the local methylation landscape at their consensus binding sites (Figure 2). This has been shown in liver [143] and neural stem cells [144], however, the magnitude of change in methylation and the specific sites affected may vary across different tissues.

Collectively, these findings point to a strong association between genotype and 5mC. Furthermore, there is emerging evidence that such genotype-methylation interdependency may be influenced by the early environment [33,144,145]. These allele-specific effects may provide a biological mechanism to explain statistical gene-by-environment (G × E) associations.

Work from the Growing Up in Singapore Towards healthy Outcomes (GUSTO) cohort has provided some of the first evidence that methylome variation may arise through G × E interactions [33]. Teh et al. [33] described methylome variation in umbilical cord–derived DNA in a sample of 237 Singaporean neonates. The authors characterize 1423 regions of the methylome that show marked interindividual variation, termed variably methylated regions (VMRs). Teh and colleagues observed that the majority of these VMRs (~75%) are best explained by interactions between genetic polymorphisms and prenatal environmental predictors, including (but not limited to) maternal depression. All these changes were found at birth, supporting an in utero effect of maternal depression on the epigenome. Interestingly, none of environmental predictors alone best

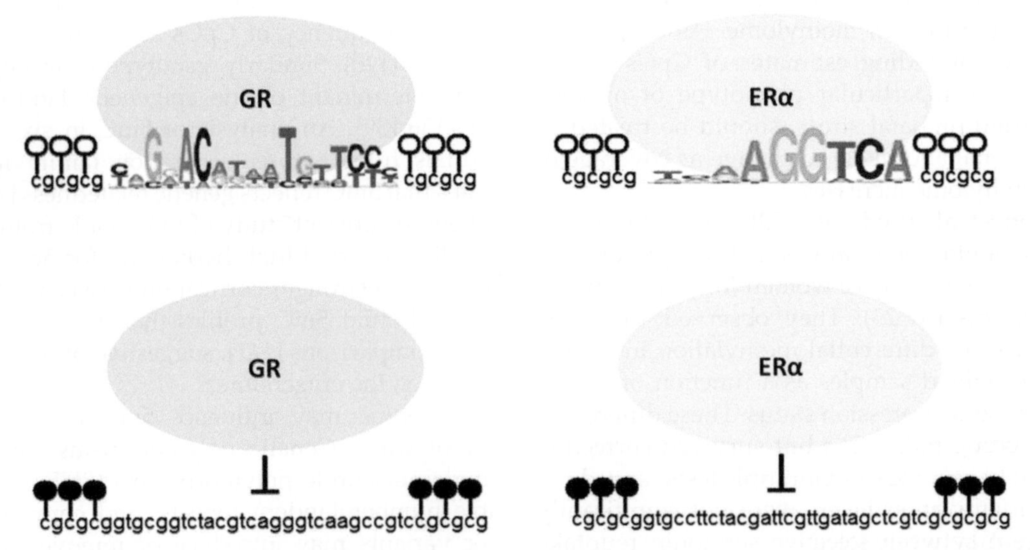

FIGURE 2 **Transcription factor remodeling of the DNA methylome.** Binding of the glucocorticoid receptor (GR; blue oval left panel), estrogen receptor alpha (ERα; orange oval right panel), and progesterone receptor (not shown) as well as other transcription factors, to their consensus DNA binding sites (colored text) promotes DNA demethylation (top row, white circles). Reduced transcription factor binding is association with increased DNA methylation (bottom row, black circles). Consensus binding sites from Motif Browser (Broad Institute). GRE/ERE, glucocorticoid/estrogen response element.

explained these VMRs, emphasizing the importance of multilevel, integrated analyses of the epigenome. Within the same data set, Chen et al. have recently expanded these analyses by integrating measures of neonatal neuroanatomy [146]. The authors focus their analysis on the *BDNF* val66met polymorphism (rs6265), often referred to as a "plasticity allele" (e.g., Ref. [147]). The authors found a much greater number of CpGs associated with maternal prenatal anxiety in neonates carrying the Met allele. Likewise, a greater number of CpGs were associated with right amygdala volume in carriers of the Met allele relative to other genotypes [147]. These findings suggest that child genotype moderates both the biological embedding of maternal prenatal anxiety and association between the methylome and measures of neuroanatomy. It will be important to determine the relevance of these findings for later mental health outcomes.

Collectively these studies highlight an important point for the DOHaD field. Often an attempt is made to dissect the "gene" from the "environment," however, given that most complex disorders, such as mental health disorders, are likely to arise from gene–environment interplay, this may prove a rather unfruitful endeavor.

Prenatal Stress and Posttranscriptional Regulation of Gene Expression

In addition to epigenetic modifications to DNA or histone proteins described above, microRNA (miRNA)-mediated gene repression is another epigenetic mechanism, which may mediate some of the effects of prenatal stress on offspring outcomes. miRNAs are short ~22 nucleotide noncoding RNAs that reduce the translation or stability of target messenger RNA transcripts [148]. Rodent studies indicate that prenatal stress may influence the abundance of specific miRNAs within the brain and the sperm of the offspring, which may, in part, mediate potential transgenerational effects of maternal prenatal stress on the subsequent generation [149]. Despite a growing appreciation for the role of miRNAs in health and disease, no clinical study to date has reported the effects of maternal prenatal stress on miRNA expression, which undoubtedly will be of interest for future studies in this area.

Transgenerational Effects of Prenatal Stress

The transgenerational transmission of prenatal stress effects is an area of clear relevance for the DOHaD field. Rodent studies of maternal prenatal undernutrition show transgenerational effects on gene expression and metabolism across two generations [150]. Radford and colleagues provide some evidence that epigenetic mechanisms, including DNA methylation, may influence the transmission of these transgenerational effects. However, the precise molecular mechanisms leading to changes in gene expression in the second generation are not well defined [150]. Likewise, Dias and Ressler [151] show transgenerational behavioral sensitivity to a specific odor when paired with a foot shock during fear conditioning experiments. Reduced methylation of *Olfr151*, which encodes an odorant receptor, was observed in sperm from F0 and F1 males suggesting a potential pathway for transgenerational transmission of this epigenetic state [151]. However, it should be noted that despite changes in 5mC in sperm, no changes in the epigenetic state of *Olfr151* were observed in the olfactory system. Collectively these two studies provide some evidence for a role of the epigenome in transgenerational transmission. However, it should be noted that transgenerational epigenetic inheritance is controversial [152,153] and the precise mediating mechanisms in higher order mammals are poorly defined. Indeed, some have proposed that a distinction should be drawn between inter- and transgenerational inheritance, the latter requiring effects to persist to the third generation (F3; see Ref. [154]). The timescale for a prospective analysis of such transgenerational effects in humans would be a logistic challenge.

Conversely, disentangling pre- from postnatal influences on the epigenome using retrospective human studies poses another distinct challenge (e.g., Ref. [133]).

Prenatal Stress and Other Regulators of the Epigenome

The existing literature on the biological pathways that mediate prenatal stress effects on child outcome has focused primarily on programming by maternal glucocorticoids. As discussed above, the GR can bind DNA and remodel the epigenome. Similar dynamic changes in 5mC have been observed as a result of estrogen receptor alpha (ERα) binding its consensus sequence. Interestingly, maternal estrogen signaling is sensitive to maternal perinatal mood and may influence child outcome. Metha and colleagues demonstrate that dynamic changes in abundance of estrogen-sensitive transcripts are predictive of maternal postnatal depression [155]. Likewise Guintivano et al. suggest that epigenetic regulation of specific estrogen-sensitive genes may be predictive of postnatal maternal mood [156]. Taken together these studies highlight estrogen signaling as of interest in predicting maternal mood, however, a direct link between altered maternal mood, estrogen signaling, and adverse child neurodevelopment outcomes has not been established. Indirect evidence in support of this pathway comes from animal studies of estrogenic endocrine disruptors, such as bisphenol A (BPA).

BPA is a selective estrogen receptor (ER) modulator that binds ERα and ERβ [157]. Prenatal exposure to BPA predicts long-lasting behavioral effects in exposed offspring [158]. These effects may be mediated by BPA-induced epigenetic changes in target gene expression. For example, prenatal BPA treatment has been shown to alter 5mC and expression of BDNF, an estrogen target gene [159]. Interestingly, the authors also show a correlation between prenatal BPA exposure and BDNF protein levels in cord blood from neonates, suggesting that these findings in a rodent model may be of relevance for clinical samples.

In addition to steroid-mediated remodeling of the epigenome there are other, interrelated pathways that may also mediate some of the effects of maternal prenatal stress on child neurodevelopment. There is an increasing appreciation of the importance of the "gut–brain axis" in the study of mental health [160]. Early life stress has been shown to produce alterations in the gut microbiome in both rodent [161] and NHP studies [162]. More recently, De Weerth and colleagues provided some of the first clinical data on the association between maternal prenatal stress and the diversity of the gut microbiome over the first 110 days of life [163]. The authors describe different trajectories of microbiome maturation between children exposed to high or low levels of maternal prenatal stress. The authors also report that this altered pattern of microbiome maturation correlates with increased maternal reports of child allergic reactions and gastrointestinal problems [163]. The capacity for prenatal stress to influence the gut microbiome is of great interest given the widespread influence of the gut microbiome on the HPA axis and immune system, both of which may regulate the epigenome. In line with this, germfree mice, born and raised in environments devoid of microbes, show exaggerated HPA axis stress responses, altered hippocampal BDNF and 5HT levels, and altered anxiety-like behavior and TNF-α response to LPS [164]. One study to date has reported an association between the gut microbiome and patterns of 5mC across the genome, however, this was a small study using whole blood [165]. Nonetheless, it provides preliminary evidence of an association of the microbiome and an aspect of the epigenome. Indeed the gut microbiome may interact with the epigenome at multiple levels, however, the implications of these associations for mental health are yet to be determined [166].

Methodological Considerations

The recent wave of epigenome-wide association studies reflects the growing interest in epigenetics in the field of developmental origins of mental health. These analyses are beginning to uncover disease risk markers (in peripheral tissue) in conditions often associated with exposure to various adverse experiences early in life, examples include major depression [167] and schizophrenia [168]. However, the application of conventional genome-wide association study (GWAS) designs may prove problematic for the field of epigenetic epidemiology. Chromatin modifications, such as 5mC, are dynamic and respond both to environmental influences but also to disease processes. Therefore, it remains a challenge to identify epigenetic modifications that are causal and not a consequence of disease processes [133]. In addition, unlike conventional genome-wide association studies in which the predictor (presence or absence of genetic risk factor) is invariant over time and tissue, epigenetic phenotypes are considerably more variable. An important consideration for all studies of the epigenome is the tissue specificity of epigenetic modifications. The epigenome is critically involved in establishing the identity of the 200 or more different cell types in the human body. As such cellular heterogeneity can be an important consideration when designing/analyzing epigenomic data [133,169]. Recently, statistical algorithms have been developed that enable the derivation of cell-type differences in both blood [170] and neural tissue [171]. This issue seems especially pertinent to studies of extremely heterogeneous tissues, such as the placenta, which contain trophoblast cells and components of mesenchyme, including fibroblasts, endothelial cells, macrophages, and hematopoietic cells, all of which may show different levels/patterns of 5mC [172,173]. Currently there is no reference-based method of correction available for tissues such as saliva or placenta, however, a reference-free approach has been described [164]. Acknowledging the potential role of tissue variability seems critical for DOHaD epigenomic analyses.

Additional considerations for epigenome-wide analyses are the issues of statistical power and stringency. Many of the studies described above employed microarray-based platforms for analysis of genome-wide 5mC including the 450K array. The ever-decreasing cost of sequencing technologies will undoubtedly increase the use of next-generation sequencing for whole-genome methylation analyses. While such approaches may reveal novel genomic regions of relevance to child vulnerability for mental disorder, increasing genome-wide coverage poses analytical challenges. For example, in the field of mental health few studies to date have identified between-group differences in 5mC that exceed 10%. These relatively small changes in 5mC, combined with the large number of comparisons (~485,000 for the Illumina 450K array) require adequately powered sample sizes to detect subtle differences in 5mC [174]. In the absence of large sample sizes, replication in independent samples or prospective longitudinal sampling may increase confidence in novel findings. Somewhat similar to early candidate gene association studies, the emerging field of epigenetic epidemiology is likely to identify several false positives if studies are not adequately powered and controlled.

Another important consideration for studies of prenatal stress and the epigenome is the timescale for establishing epigenetic changes and the stability of such effects. In vitro studies show that both steroid-mediated or activity-dependent DNA demethylation can occur rapidly (~100 mins) [175] or over 1–3 days [142,143,176] and that these effects can be sustained across neuronal differentiation [143]. There is some evidence that specific histone modifications may precede DNA methylation suggesting coordinated epigenetic changes [177,178]. It is likely the kinetics of such DNA (de)methylation and histone modifications will vary depending on the genomic context under study. With regard to the stability of the human methylome, recent estimates from whole genome bisulfite sequencing of diverse biological samples suggest a high degree of stability. Approximately 20–30% of the 28 million CpGs within the human methylome show dynamic change across development [179]. In infancy three longitudinal clinical studies have examined the stability of 5mC over time using commercially available arrays. Wang and colleagues examine 5mC in umbilical cord and venous blood samples collected at birth and during the first 24 months of life (average follow-up was 12 months of age) [180]. The authors report remarkable stability over time with significant changes occurring at <1% of methylation sites assessed. In contrast, Martino and colleagues show more dynamic change in mononuclear cell 5mC in their study of 0- to 5-year-old females. Profiles of 5mC from 1- and 5-year-old infants tended to form distinct clusters, while the two-and-a-half-year-old group clustered with both 1- and 5-year-olds [181]. Such differential developmental changes in 5mC may indicate that certain individuals may be more sensitive to early life factors, such as parenting behavior, which can influence 5mC. In a more recent study, the same group has characterized changes in 5mC across the first 18 months of life in DNA derived from buccal cells [182]. The authors report that ~30% of 450K probes show dynamic change over time, however, it should be noted that the average change in methylation was ~3%. When a more stringent definition of differential methylation was applied (>20%), ~1% of the probes assessed showed change over time. Finally, Spiers and colleagues describe the fetal brain methylome using tissue samples from early to mid-gestation [183]. The authors show that approximately 7% of sites investigated showed an association with gestational age and often overlap with steroid hormone binding sites [183]. Interestingly, the authors also report that sex differences in 5mC observed in adult brain tissue are also present in fetal brain samples, indicating that specific features of the brain methylome are established very early in development. Taken together these studies emphasize that there is likely to be variation in the stability of 5mC as a function of cell type and developmental stage. It is unknown if maternal prenatal stress influences the maturation of the fetal/neonatal methylome and how this contributes to child neurodevelopment.

Few clinical studies have determined if changes in 5mC mediate the association between maternal prenatal stress and child neurodevelopment. This may, in part, stem from the uncertainty in the field about the most appropriate surrogate tissue for brain and the appreciation that the direct correlation between 5mC in peripheral tissues and the brain is modest at best [184,185]. An alternative approach is to study 5mC within placental tissue as a potential mediator of the effects of prenatal stress on child neurodevelopment. One study has reported that the effects of maternal mood on aspects of infant neurobehavior vary as a function of 5mC within *NR3C1* and *HSD11B2* in the placenta [186]. The authors did not find any evidence to suggest that changes in 5mC of *NR3C1* and *HSD11B2* mediated the effect of maternal mood on neonatal neurobehavior but rather support both additive and interactive effects of maternal mood and placental methylation on neonatal neurobehavior. Such findings linking placental gene expression to infant neurobehavior are of interest given the literature on imprinted genes, which provide evidence of coexpression across placenta and certain neural structures [187]. It will be of interest to determine if other candidate genes show a similar pattern of coexpression and the relevance of such findings for child neurodevelopment.

Finally, as noted above, epigenetic modifications do not occur in isolation nor are they independent of the underlying DNA sequence. Genome-wide surveys of multiple epigenetic modifications, including different forms of 5mC, genotype, and paired gene expression will provide a more comprehensive understanding of how maternal prenatal stress shapes the epigenome and may help identify novel functional loci that contribute to the developmental origins of mental health.

CONCLUSIONS

It is clear that maternal prenatal stress is associated with a variety of adverse outcomes, which increase vulnerability for later mental disorder. These effects are evident in infancy and can persist into adulthood. There is emerging evidence in support of an environmental epigenetics hypothesis, with some evidence that maternal prenatal stress can influence epigenetic processes within her child. However, the evidence that these epigenetic changes mediate the effects of maternal prenatal stress on child outcome is rather limited, as is the evidence of transgenerational effects of maternal prenatal stress on the epigenome in humans. Thus, while epigenetics is an exciting and rapidly growing field, translating basic scientific findings to improve clinical care and mental health outcomes will require continued and cautious research efforts.

References

[1] Scott KM, Smith DR, Ellis PM. Prospectively ascertained child maltreatment and its association with DSM-IV mental disorders in young adults. Arch Gen Psychiatry 2010;67(7):712–9.

[2] Nanni V, Uher R, Danese A. Childhood maltreatment predicts unfavorable course of illness and treatment outcome in depression: a meta-analysis. Am J Psychiatry 2012;169(2):141–51.

[3] Glover V. Maternal depression, anxiety and stress during pregnancy and child outcome; what needs to be done. Best Pract Res Clin Obstet Gynaecol 2014;28(1):25–35.

[4] Talge NM, Neal C, Glover V. Antenatal maternal stress and long-term effects on child neurodevelopment: how and why? J Child Psychol Psychiatry 2007;48(3–4):245–61.

[5] Van den Bergh BR, Mulder EJ, Mennes M, Glover V. Antenatal maternal anxiety and stress and the neurobehavioural development of the fetus and child: links and possible mechanisms. A review. Neurosci Biobehav Rev 2005;29(2):237–58.

[6] O'Donnell KJ, Glover V, Barker ED, O'Connor TG. The persisting effect of maternal mood in pregnancy on childhood psychopathology. Dev Psychopathol 2014;26(2):393–403.

[7] Pearson RM, Evans J, Kounali D, Lewis G, Heron J, Ramchandani PG, et al. Maternal depression during pregnancy and the postnatal period: risks and possible mechanisms for offspring depression at age 18 years. JAMA Psychiatry 2013;70(12):1312–9.

[8] Betts KS, Williams GM, Najman JM, Alati R. The relationship between maternal depressive, anxious, and stress symptoms during pregnancy and adult offspring behavioral and emotional problems. Depress Anxiety 2015;32(2):82–90.

[9] Gluckman PD, Hanson MA, Spencer HG. Predictive adaptive responses and human evolution. Trends Ecol Evol 2005;20(10):527–33.

[10] Lee TM. Development of meadow voles is influenced postnatally by maternal photoperiodic history. Am J Physiol 1993;265(4):R749–55.

[11] Lee TM, Zucker I. Vole infant development is influenced perinatally by maternal photoperiodic history. Am J Physiol 1988;255(5):R831–8.

[12] O'Connor TG, Heron J, Golding J, Beveridge M, Glover V. Maternal antenatal anxiety and children's behavioural/emotional problems at 4 years. Report from the Avon Longitudinal Study of Parents and Children. Br J Psychiatry 2002;180:502–8.

[13] Huizink AC, Robles de Medina PG, Mulder EJ, Visser GH, Buitelaar JK. Stress during pregnancy is associated with developmental outcome in infancy. J Child Psychol Psychiatry 2003;44(6):810–8.

[14] Hompes T, Izzi B, Gellens E, Morreels M, Fieuws S, Pexsters A, et al. Investigating the influence of maternal cortisol and emotional state during pregnancy on the DNA methylation status of the glucocorticoid receptor gene (NR3C1) promoter region in cord blood. J Psychiatr Res 2013;47(7):880–91.

[15] Bergman K, Sarkar P, O'Connor TG, Modi N, Glover V. Maternal stress during pregnancy predicts cognitive ability and fearfulness in infancy. J Am Acad Child Adolesc Psychiatry 2007;46(11):1454–63.

[16] Meijer A. Child psychiatric sequelae of maternal war stress. Acta Psychiatr Scand 1985;72(6):505–11.

[17] van Os J, Selten JP. Prenatal exposure to maternal stress and subsequent schizophrenia. The May 1940 invasion of The Netherlands. Br J Psychiatry 1998;172:324–6.
[18] Brown AS, Susser ES, Lin SP, Neugebauer R, Gorman JM. Increased risk of affective disorders in males after second trimester prenatal exposure to the Dutch hunger winter of 1944–45. Br J Psychiatry 1995;166(5):601–6.
[19] Brown AS, van Os J, Driessens C, Hoek HW, Susser ES. Further evidence of relation between prenatal famine and major affective disorder. Am J Psychiatry 2000;157(2):190–5.
[20] St Clair D, Xu M, Wang P, Yu Y, Fang Y, Zhang F, et al. Rates of adult schizophrenia following prenatal exposure to the Chinese famine of 1959–1961. JAMA 2005;294(5):557–62.
[21] Glynn LM, Wadhwa PD, Dunkel-Schetter C, Chicz-Demet A, Sandman CA. When stress happens matters: effects of earthquake timing on stress responsivity in pregnancy. Am J Obstet Gynecol 2001;184(4):637–42.
[22] Laplante DP, Brunet A, Schmitz N, Ciampi A, King S. Project ice storm: prenatal maternal stress affects cognitive and linguistic functioning in 5 1/2-year-old children. J Am Acad Child Adolesc Psychiatry 2008;47(9):1063–72.
[23] Kinney DK, Miller AM, Crowley DJ, Huang E, Gerber E. Autism prevalence following prenatal exposure to hurricanes and tropical storms in Louisiana. J Autism Dev Disord 2008;38(3):481–8.
[24] Huizink AC, Dick DM, Sihvola E, Pulkkinen L, Rose RJ, Kaprio J. Chernobyl exposure as stressor during pregnancy and behaviour in adolescent offspring. Acta Psychiatr Scand 2007;116(6):438–46.
[25] Yehuda R, Engel SM, Brand SR, Seckl J, Marcus SM, Berkowitz GS. Transgenerational effects of posttraumatic stress disorder in babies of mothers exposed to the World Trade Center attacks during pregnancy. J Clin Endocrinol Metab 2005;90(7):4115–8.
[26] Khashan AS, Abel KM, McNamee R, Pedersen MG, Webb RT, Baker PN, et al. Higher risk of offspring schizophrenia following antenatal maternal exposure to severe adverse life events. Arch Gen Psychiatry 2008;65(2):146–52.
[27] DiPietro JA, Novak MF, Costigan KA, Atella LD, Reusing SP. Maternal psychological distress during pregnancy in relation to child development at age two. Child Dev 2006;77(3):573–87.
[28] Parker KJ, Maestripieri D. Identifying key features of early stressful experiences that produce stress vulnerability and resilience in primates. Neurosci Biobehav Rev 2011;35(7):1466–83.
[29] Spector LG, Murphy SE, Wickham KM, Lindgren B, Joseph AM. Prenatal tobacco exposure and cotinine in newborn dried blood spots. Pediatrics 2014;133(6).e1632–8.
[30] Monk C, Georgieff MK, Osterholm EA. Research review: maternal prenatal distress and poor nutrition – mutually influencing risk factors affecting infant neurocognitive development. J Child Psychol Psychiatry 2013;54(2):115–30.
[31] Weinstock M. The long-term behavioural consequences of prenatal stress. Neurosci Biobehav Rev 2008;32(6):1073–86.
[32] Non AL, Binder AM, Kubzansky LD, Michels KB. Genome-wide DNA methylation in neonates exposed to maternal depression, anxiety, or SSRI medication during pregnancy. Epigenetics 2014;9(7):964–72.
[33] Teh AL, Pan H, Chen L, Ong ML, Dogra S, Wong J, et al. The effect of genotype and in utero environment on inter-individual variation in neonate DNA methylomes. Genome Res 2014;24(7):1064–74.
[34] Rice F, Harold GT, Boivin J, van den Bree M, Hay DF, Thapar A. The links between prenatal stress and offspring development and psychopathology: disentangling environmental and inherited influences. Psychol Med 2010;40(2):335–45.
[35] O'Connor TG, Heron J, Glover V. Antenatal anxiety predicts child behavioral/emotional problems independently of postnatal depression. J Am Acad Child Adolesc Psychiatry 2002;41(12):1470–7.
[36] O'Connor TG, Heron J, Golding J, Glover V. Maternal antenatal anxiety and behavioural/emotional problems in children: a test of a programming hypothesis. J Child Psychol Psychiatry 2003;44(7):1025–36.
[37] Mennes M, Van den Bergh B, Lagae L, Stiers P. Developmental brain alterations in 17 year old boys are related to antenatal maternal anxiety. Clin Neurophysiol 2009;120(6):1116–22.
[38] Van den Bergh BR, Marcoen A. High antenatal maternal anxiety is related to ADHD symptoms, externalizing problems, and anxiety in 8- and 9-year-olds. Child Dev 2004;75(4):1085–97.
[39] Bergman K, Sarkar P, Glover V, O'Connor TG. Maternal prenatal cortisol and infant cognitive development: moderation by infant–mother attachment. Biol Psychiatry 2010;67(11):1026–32.
[40] Hansen D, Lou HC, Olsen J. Serious life events and congenital malformations: a national study with complete follow-up. Lancet 2000;356(9233):875–80.
[41] Wadhwa PD, Sandman CA, Porto M, Dunkel-Schetter C, Garite TJ. The association between prenatal stress and infant birth weight and gestational age at birth: a prospective investigation. Am J Obstet Gynecol 1993;169(4):858–65.
[42] Bruckner T, Catalano R. The sex ratio and age-specific male mortality: evidence for culling in utero. Am J Hum Biol 2007;19(6):763–73.
[43] Bruckner T, Catalano R, Ahern J. Male fetal loss in the U.S. following the terrorist attacks of September 11, 2001. BMC Public Health 2010;10(1):273.

[44] Obel C, Henriksen TB, Secher NJ, Eskenazi B, Hedegaard M. Psychological distress during early gestation and offspring sex ratio. Hum Reprod 2007;22(11):3009–12.
[45] Field T, Diego M, Hernandez-Reif M, Schanberg S, Kuhn C, Yando R, et al. Pregnancy anxiety and comorbid depression and anger: effects on the fetus and neonate. Depress Anxiety 2003;17(3):140–51.
[46] Buitelaar JK, Huizink AC, Mulder EJ, de Medina PG, Visser GH. Prenatal stress and cognitive development and temperament in infants. Neurobiol Aging 2003;24(Suppl. 1):S53–60. Discussion S7–8.
[47] Davis EP, Glynn LM, Schetter CD, Hobel C, Chicz-Demet A, Sandman CA. Prenatal exposure to maternal depression and cortisol influences infant temperament. J Am Acad Child Adolesc Psychiatry 2007;46(6):737–46.
[48] O'Connor TG, Caprariello P, Blackmore ER, Gregory AM, Glover V, Fleming P. Prenatal mood disturbance predicts sleep problems in infancy and toddlerhood. Early Hum Dev 2007;83(7):451–8.
[49] Rodriguez A, Bohlin G. Are maternal smoking and stress during pregnancy related to ADHD symptoms in children? J Child Psychol Psychiatry 2005;46(3):246–54.
[50] Pawlby S, Hay D, Sharp D, Waters CS, Pariante CM. Antenatal depression and offspring psychopathology: the influence of childhood maltreatment. Br J Psychiatry 2011;199(2):106–12.
[51] Lereya ST, Wolke D. Prenatal family adversity and maternal mental health and vulnerability to peer victimisation at school. J Child Psychol Psychiatry 2013;54(6):644–52.
[52] Mennes M, Stiers P, Lagae L, Van den Bergh B. Long-term cognitive sequelae of antenatal maternal anxiety: involvement of the orbitofrontal cortex. Neurosci Biobehav Rev 2006;30(8):1078–86.
[53] Schwabe L, Bohbot VD, Wolf OT. Prenatal stress changes learning strategies in adulthood. Hippocampus 2012;22(11):2136–43.
[54] Beversdorf DQ, Manning SE, Hillier A, Anderson SL, Nordgren RE, Walters SE, et al. Timing of prenatal stressors and autism. J Autism Dev Disord 2005;35(4):471–8.
[55] Walder DJ, Laplante DP, Sousa-Pires A, Veru F, Brunet A, King S. Prenatal maternal stress predicts autism traits in 6(1/2) year-old children: Project Ice Storm. Psychiatry Res 2014;219(2):353–60.
[56] Croen LA, Grether JK, Yoshida CK, Odouli R, Hendrick V. Antidepressant use during pregnancy and childhood autism spectrum disorders. Arch Gen Psychiatry 2011;68(11):1104–12.
[57] Susser E, Neugebauer R, Hoek HW, Brown AS, Lin S, Labovitz D, et al. Schizophrenia after prenatal famine. Further evidence. Arch Gen Psychiatry 1996;53(1):25–31.
[58] Buss C, Davis EP, Muftuler LT, Head K, Sandman CA. High pregnancy anxiety during mid-gestation is associated with decreased gray matter density in 6-9-year-old children. Psychoneuroendocrinology 2009;35(1):141–53.
[59] Sandman CA, Buss C, Head K, Davis EP. Fetal exposure to maternal depressive symptoms is associated with cortical thickness in late childhood. Biol Psychiatry 2015;77(4):324–34.
[60] Sarkar S, Craig MC, Dell'Acqua F, O'Connor TG, Catani M, Deeley Q, et al. Prenatal stress and limbic-prefrontal white matter microstructure in children aged 6–9 years: a preliminary diffusion tensor imaging study. World J Biol Psychiatry 2014;15(4):346–52.
[61] King S, Mancini-Marie A, Brunet A, Walker E, Meaney MJ, Laplante DP. Prenatal maternal stress from a natural disaster predicts dermatoglyphic asymmetry in humans. Dev Psychopathol 2009;21(2):343–53.
[62] Glover V, O'Connor TG, Heron J, Golding J. Antenatal maternal anxiety is linked with atypical handedness in the child. Early Hum Dev 2004;79(2):107–18.
[63] Rodriguez A, Waldenström U. Fetal origins of child non-right-handedness and mental health. J Child Psychol Psychiatry 2008;49(9):967–76.
[64] Glover V, O'Connor TG, O'Donnell K. Prenatal stress and the programming of the HPA axis. Neurosci Biobehav Rev 2010;35(1):17–22.
[65] O'Donnell KJ, Glover V, Jenkins J, Browne D, Ben-Shlomo Y, Golding J, et al. Prenatal maternal mood is associated with altered diurnal cortisol in adolescence. Psychoneuroendocrinology 2013;38(9):1630–8.
[66] Entringer S, Epel ES, Lin J, Buss C, Shahbaba B, Blackburn EH, et al. Maternal psychosocial stress during pregnancy is associated with newborn leukocyte telomere length. Am J Obstet Gynecol 2013;208(2):134.e1–7.
[67] Plant DT, Barker ED, Waters CS, Pawlby S, Pariante CM. Intergenerational transmission of maltreatment and psychopathology: the role of antenatal depression. Psychol Med 2013;43(3):519–28.
[68] Teixeira JMA, Fisk NM, Glover V. Association between maternal anxiety in pregnancy and increased uterine artery resistance index: cohort based study. BMJ 1999;318(7177):153–7.
[69] Kent A, Hughes P, Ormerod L, Jones G, Thilaganathan B. Uterine artery resistance and anxiety in the second trimester of pregnancy. Ultrasound Obstet Gynecol 2002;19(2):177–9.
[70] Monk C, Newport DJ, Korotkin JH, Long Q, Knight B, Stowe ZN. Uterine blood flow in a psychiatric population: impact of maternal depression, anxiety, and psychotropic medication. Biol Psychiatry 2012;72(6):483–90.
[71] Mina TH, Reynolds RM. Mechanisms linking in utero stress to altered offspring behaviour. Curr Top Behav Neurosci 2014;18:93–122.

[72] Herbert J. Cortisol and depression: three questions for psychiatry. Psychol Med 2013;43(3):449–69.

[73] Harris A, Seckl J. Glucocorticoids, prenatal stress and the programming of disease. Horm Behav 2011;59(3):279–89.

[74] Salaria S, Chana G, Caldara F, Feltrin E, Altieri M, Faggioni F, et al. Microarray analysis of cultured human brain aggregates following cortisol exposure: implications for cellular functions relevant to mood disorders. Neurobiol Dis 2006;23(3):630–6.

[75] Morsink MC, Joels M, Sarabdjitsingh RA, Meijer OC, De Kloet ER, Datson NA. The dynamic pattern of glucocorticoid receptor-mediated transcriptional responses in neuronal PC12 cells. J Neurochem 2006;99(4):1282–98.

[76] Stavreva DA, Coulon A, Baek S, Sung MH, John S, Stixova L, et al. Dynamics of chromatin accessibility and long-range interactions in response to glucocorticoid pulsing. Genome Res 2015;25(6).

[77] Crudo A, Petropoulos S, Suderman M, Moisiadis VG, Kostaki A, Hallett M, et al. Effects of antenatal synthetic glucocorticoid on glucocorticoid receptor binding, DNA methylation, and genome-wide mRNA levels in the fetal male hippocampus. Endocrinology 2013;154(11):4170–81.

[78] Khalife N, Glover V, Taanila A, Ebeling H, Jarvelin MR, Rodriguez A. Prenatal glucocorticoid treatment and later mental health in children and adolescents. PLoS One 2013;8(11):e81394.

[79] Davis EP, Sandman CA, Buss C, Wing DA, Head K. Fetal glucocorticoid exposure is associated with preadolescent brain development. Biol Psychiatry 2013;74(9):647–55.

[80] Räikkönen K, Pesonen A-K, Heinonen K, Lahti J, Komsi N, Eriksson JG, et al. Maternal licorice consumption and detrimental cognitive and psychiatric outcomes in children. Am J Epidemiol 2009;170(9):1137–46.

[81] Schneider ML, Moore CF, Kraemer GW, Roberts AD, DeJesus OT. The impact of prenatal stress, fetal alcohol exposure, or both on development: perspectives from a primate model. Psychoneuroendocrinology 2002;27(1–2):285–98.

[82] Barbazanges A, Piazza PV, Le Moal M, Maccari S. Maternal glucocorticoid secretion mediates long-term effects of prenatal stress. J Neurosci 1996;16(12):3943–9.

[83] Kammerer M, Adams D, Castelberg Bv B, Glover V. Pregnant women become insensitive to cold stress. BMC Pregnancy Childbirth 2002;2(1):8.

[84] Kammerer M, Taylor A, Glover V. The HPA axis and perinatal depression: a hypothesis. Arch Womens Ment Health 2006;9(4):187–96.

[85] O'Donnell K, O'Connor TG, Glover V. Prenatal stress and neurodevelopment of the child: focus on the HPA axis and role of the placenta. Dev Neurosci 2009;31(4):285–92.

[86] O'Keane V, Lightman S, Patrick K, Marsh M, Papadopoulos AS, Pawlby S, et al. Changes in the maternal hypothalamic–pituitary–adrenal axis during the early puerperium may be related to the postpartum "blues". J Neuroendocrinol 2011;23(11):1149–55.

[87] Gitau R, Cameron A, Fisk NM, Glover V. Fetal exposure to maternal cortisol. Lancet 1998;352(9129):707–8.

[88] Mairesse J, Lesage J, Breton C, Breant B, Hahn T, Darnaudery M, et al. Maternal stress alters endocrine function of the feto-placental unit in rats. Am J Physiol Endocrinol Metab 2007;292(6):E1526–33.

[89] Glover V, Bergman K, Sarkar P, O'Connor TG. Association between maternal and amniotic fluid cortisol is moderated by maternal anxiety. Psychoneuroendocrinology 2009;34(3):430–5.

[90] O'Donnell KJ, Bugge Jensen A, Freeman L, Khalife N, O'Connor TG, Glover V. Maternal prenatal anxiety and downregulation of placental 11beta-HSD2. Psychoneuroendocrinology 2012;37(6):818–26.

[91] Ainsworth MDS, Blehar MC, Waters E, Wall S. Patterns of attachment: a psychological study of the strange situation. Hillsdale, NJ: Erlbaum Associates; 1978.

[92] Bowlby J. Attachment theory and its therapeutic implications. Adolesc Psychiatry 1978;6:5–33.

[93] McGoron L, Gleason MM, Smyke AT, Drury SS, Nelson Iii CA, Gregas MC, et al. Recovering from early deprivation: attachment mediates effects of caregiving on psychopathology. J Am Acad Child Adolesc Psychiatry 2012;51(7):683–93.

[94] Davis EP, Sandman CA. The timing of prenatal exposure to maternal cortisol and psychosocial stress is associated with human infant cognitive development. Child Dev 2010;81(1):131–48.

[95] Beijers R, Buitelaar JK, de Weerth C. Mechanisms underlying the effects of prenatal psychosocial stress on child outcomes: beyond the HPA axis. Eur Child Adolesc Psychiatry 2014;23(10):943–56.

[96] Oberlander TF. Fetal serotonin signaling: setting pathways for early childhood development and behavior. J Adolesc Health 2012;51(Suppl. 2):S9–16.

[97] Bonnin A, Goeden N, Chen K, Wilson ML, King J, Shih JC, et al. A transient placental source of serotonin for the fetal forebrain. Nature 2011;472(7343):347–50.

[98] Blakeley PM, Capron LE, Jensen AB, O'Donnell KJ, Glover V. Maternal prenatal symptoms of depression and down regulation of placental monoamine oxidase A expression. J Psychosom Res 2013;75(4):341–5.

[99] Yeh C-M, Huang C-C, Hsu K-S. Prenatal stress alters hippocampal synaptic plasticity in young rat offspring through preventing the proteolytic conversion of pro-brain-derived neurotrophic factor (BDNF) to mature BDNF. J Physiol 2012;590(4):991–1010.

[100] Sen S, Duman R, Sanacora G. Serum brain-derived neurotrophic factor, depression, and antidepressant medications: meta-analyses and implications. Biol Psychiatry 2008;64(6):527–32.
[101] Uguz F, Sonmez EO, Sahingoz M, Gokmen Z, Basaran M, Gezginc K, et al. Maternal generalized anxiety disorder during pregnancy and fetal brain development: a comparative study on cord blood brain-derived neurotrophic factor levels. J Psychosom Res 2013;75(4):346–50.
[102] Buss C, Entringer S, Wadhwa PD. Fetal programming of brain development: intrauterine stress and susceptibility to psychopathology. Sci Signal 2012;5(245):pt7.
[103] Bronson SL, Bale TL. Prenatal stress-induced increases in placental inflammation and offspring hyperactivity are male-specific and ameliorated by maternal antiinflammatory treatment. Endocrinology 2014;155(7):2635–46.
[104] Hepgul N, Mondelli V, Pariante CM. Psychological and biological mechanisms of cytokine induced depression. Epidemiol Psichiatr Soc 2010;19(2):98–102.
[105] Coussons-Read ME, Okun ML, Nettles CD. Psychosocial stress increases inflammatory markers and alters cytokine production across pregnancy. Brain Behav Immun 2007;21(3):343–50.
[106] Haeri S, Baker AM, Ruano R. Do pregnant women with depression have a pro-inflammatory profile? J Obstet Gynaecol Res 2013;39(5):948–52.
[107] Blackmore ER, Moynihan JA, Rubinow DR, Pressman EK, Gilchrist M, O'Connor TG. Psychiatric symptoms and proinflammatory cytokines in pregnancy. Psychosom Med 2011;73(8):656–63.
[108] Lim PS, Li J, Holloway AF, Rao S. Epigenetic regulation of inducible gene expression in the immune system. Immunology 2013;139(3):285–93.
[109] Gasche JA, Hoffmann J, Boland CR, Goel A. Interleukin-6 promotes tumorigenesis by altering DNA methylation in oral cancer cells. Int J Cancer 2011;129(5):1053–63.
[110] Li C, Ebert PJ, Li QJ. T cell receptor (TCR) and transforming growth factor beta (TGF-beta) signaling converge on DNA (cytosine-5)-methyltransferase to control forkhead box protein 3 (foxp3) locus methylation and inducible regulatory T cell differentiation. J Biol Chem 2013;288(26):19127–39.
[111] Monk C, Spicer J, Champagne FA. Linking prenatal maternal adversity to developmental outcomes in infants: the role of epigenetic pathways. Dev Psychopathol 2012;24(4):1361–76.
[112] Turner BM. Chromatin and gene regulation: mechanisms in epigenetics. Cambridge, MA: Blackwell Science; 2001.
[113] Maze I, Noh K-M, Allis CD. Histone regulation in the CNS: basic principles of epigenetic plasticity. Neuropsychopharmacology 2013;38(1):3–22.
[114] Grunstein M. Histone acetylation in chromatin structure and transcription. Nature 1997;389(6649):349–52.
[115] Jenuwein T, Allis CD. Translating the histone code. Science 2001;293(5532):1074–80.
[116] Sun H, Kennedy PJ, Nestler EJ. Epigenetics of the depressed brain: role of histone acetylation and methylation. Neuropsychopharmacology 2013;38(1):124–37.
[117] Bird AP. CpG-rich islands and the function of DNA methylation. Nature 1986;321(6067):209–13.
[118] Razin A, Riggs AD. DNA methylation and gene function. Science 1980;210(4470):604–10.
[119] Lazarovici A, Zhou T, Shafer A, Dantas Machado AC, Riley TR, Sandstrom R, et al. Probing DNA shape and methylation state on a genomic scale with DNase I. Proc Natl Acad Sci USA 2013;110(16):6376–81.
[120] Deaton AM, Bird A. CpG islands and the regulation of transcription. Genes Dev 2011;25(10):1010–22.
[121] Ito S, D'Alessio AC, Taranova OV, Hong K, Sowers LC, Zhang Y. Role of Tet proteins in 5mC to 5hmC conversion, ES-cell self-renewal and inner cell mass specification. Nature 2010;466(7310):1129–33.
[122] Lister R, Mukamel EA, Nery JR, Urich M, Puddifoot CA, Johnson ND, et al. Global epigenomic reconfiguration during mammalian brain development. Science 2013;341(6146).
[123] Wang T, Pan Q, Lin L, Szulwach KE, Song CX, He C, et al. Genome-wide DNA hydroxymethylation changes are associated with neurodevelopmental genes in the developing human cerebellum. Hum Mol Genet 2012;21(26):5500–10.
[124] Wen L, Li X, Yan L, Tan Y, Li R, Zhao Y, et al. Whole-genome analysis of 5-hydroxymethylcytosine and 5-methylcytosine at base resolution in the human brain. Genome Biol 2014;15(3):R49.
[125] Dong E, Dzitoyeva SG, Matrisciano F, Tueting P, Grayson DR, Guidotti A. Brain-derived neurotrophic factor epigenetic modifications associated with schizophrenia-like phenotype induced by prenatal stress in mice. Biol Psychiatry 2015;77(6):589–96.
[126] Massart R, Suderman M, Provencal N, Yi C, Bennett AJ, Suomi S, et al. Hydroxymethylation and DNA methylation profiles in the prefrontal cortex of the non-human primate rhesus macaque and the impact of maternal deprivation on hydroxymethylation. Neuroscience 2014;268:139–48.
[127] Mueller BR, Bale TL. Sex-specific programming of offspring emotionality after stress early in pregnancy. J Neurosci 2008;28(36):9055–65.
[128] Jensen Pena C, Monk C, Champagne FA. Epigenetic effects of prenatal stress on 11beta-hydroxysteroid dehydrogenase-2 in the placenta and fetal brain. PLoS One 2012;7(6):e39791.
[129] Provencal N, Binder EB. The effects of early life stress on the epigenome: from the womb to adulthood and even before. Exp Neurol 2014;268:10–20.

[130] Oberlander TF, Weinberg J, Papsdorf M, Grunau R, Misri S, Devlin AM. Prenatal exposure to maternal depression, neonatal methylation of human glucocorticoid receptor gene (NR3C1) and infant cortisol stress responses. Epigenetics 2008;3(2):97–106.
[131] Mulligan CJ, D'Errico NC, Stees J, Hughes DA. Methylation changes at NR3C1 in newborns associate with maternal prenatal stress exposure and newborn birth weight. Epigenetics 2012;7(8):853–7.
[132] Appleton AA, Armstrong DA, Lesseur C, Lee J, Padbury JF, Lester BM, et al. Patterning in placental 11-B hydroxysteroid dehydrogenase methylation according to prenatal socioeconomic adversity. PLoS One 2013;8(9): e74691.
[133] Radtke KM, Ruf M, Gunter HM, Dohrmann K, Schauer M, Meyer A, et al. Transgenerational impact of intimate partner violence on methylation in the promoter of the glucocorticoid receptor. Transl Psychiatry 2011;1:e21.
[134] Liu Y, Aryee MJ, Padyukov L, Fallin MD, Hesselberg E, Runarsson A, et al. Epigenome-wide association data implicate DNA methylation as an intermediary of genetic risk in rheumatoid arthritis. Nat Biotechnol 2013;31(2):142–7.
[135] Pan H, Chen L, Dogra S, Ling Teh A, Tan JH, Lim YI, et al. Measuring the methylome in clinical samples: improved processing of the Infinium Human Methylation450 BeadChip Array. Epigenetics 2012;7(10):1173–87.
[136] Stadler MB, Murr R, Burger L, Ivanek R, Lienert F, Scholer A, et al. DNA-binding factors shape the mouse methylome at distal regulatory regions. Nature 2011;480(7378):490–5.
[137] Roussos P, Mitchell AC, Voloudakis G, Fullard JF, Pothula VM, Tsang J, et al. A role for noncoding variation in schizophrenia. Cell Rep 2014;9(4):1417–29.
[138] Schroeder JW, Smith AK, Brennan PA, Conneely KN, Kilaru V, Knight BT, et al. DNA methylation in neonates born to women receiving psychiatric care. Epigenetics 2012;7(4):409–14.
[139] Lienert F, Wirbelauer C, Som I, Dean A, Mohn F, Schubeler D. Identification of genetic elements that autonomously determine DNA methylation states. Nat Genet 2011;43(11):1091–7.
[140] Gertz J, Varley KE, Reddy TE, Bowling KM, Pauli F, Parker SL, et al. Analysis of DNA methylation in a three-generation family reveals widespread genetic influence on epigenetic regulation. PLoS Genet 2011; 7(8):e1002228.
[141] McRae AF, Powell JE, Henders AK, Bowdler L, Hemani G, Shah S, et al. Contribution of genetic variation to transgenerational inheritance of DNA methylation. Genome Biol 2014;15(5):R73.
[142] Feldmann A, Ivanek R, Murr R, Gaidatzis D, Burger L, Schubeler D. Transcription factor occupancy can mediate active turnover of DNA methylation at regulatory regions. PLoS Genet 2013;9(12):e1003994.
[143] Thomassin H, Flavin M, Espinas ML, Grange T. Glucocorticoid-induced DNA demethylation and gene memory during development. EMBO J 2001;20(8):1974–83.
[144] Klengel T, Mehta D, Anacker C, Rex-Haffner M, Pruessner JC, Pariante CM, et al. Allele-specific FKBP5 DNA demethylation mediates gene–childhood trauma interactions. Nat Neurosci 2013;16(1):33–41.
[145] Ursini G, Bollati V, Fazio L, Porcelli A, Iacovelli L, Catalani A, et al. Stress-related methylation of the catechol-*O*-methyltransferase Val158 allele predicts human prefrontal cognition and activity. J Neurosci 2011;31(18):6692–8.
[146] Chen L, Pan H, Tuan TA, Teh AL, MacIsaac JL, Mah SM, et al. Brain-derived neurotrophic factor (BDNF) Val66Met polymorphism influences the association of the methylome with maternal anxiety and neonatal brain volumes. Dev Psychopathol 2015;27(1):37–50.
[147] Drury SS, Gleason MM, Theall KP, Smyke AT, Nelson CA, Fox NA, et al. Genetic sensitivity to the caregiving context: the influence of 5httlpr and BDNF val66met on indiscriminate social behavior. Physiol Behav 2012;106(5):728–35.
[148] Issler O, Chen A. Determining the role of microRNAs in psychiatric disorders. Nat Rev Neurosci 2015;16(4):201–12.
[149] Morgan CP, Bale TL. Early prenatal stress epigenetically programs dysmasculinization in second-generation offspring via the paternal lineage. J Neurosci 2011;31(33):11748–55.
[150] Radford EJ, Ito M, Shi H, Corish JA, Yamazawa K, Isganaitis E, et al. In utero effects. In utero undernourishment perturbs the adult sperm methylome and intergenerational metabolism. Science 2014;345(6198):1255903.
[151] Dias BG, Ressler KJ. Parental olfactory experience influences behavior and neural structure in subsequent generations. Nat Neurosci 2014;17(1):89–96.
[152] Heard E, Martienssen RA. Transgenerational epigenetic inheritance: myths and mechanisms. Cell 2014;157(1):95–109.
[153] Grossniklaus U, Kelly WG, Ferguson-Smith AC, Pembrey M, Lindquist S. Transgenerational epigenetic inheritance: how important is it? Nat Rev Genet 2013;14(3):228–35.
[154] Martos SN, Tang WY, Wang Z. Elusive inheritance: transgenerational effects and epigenetic inheritance in human environmental disease. Prog Biophys Mol Biol 2015;118(1–2):44–54.
[155] Mehta D, Newport DJ, Frishman G, Kraus L, Rex-Haffner M, Ritchie JC, et al. Early predictive biomarkers for postpartum depression point to a role for estrogen receptor signaling. Psychol Med 2014:1–14.

[156] Guintivano J, Arad M, Gould TD, Payne JL, Kaminsky ZA. Antenatal prediction of postpartum depression with blood DNA methylation biomarkers. Mol Psychiatry 2013;19(5):560.

[157] Wetherill YB, Akingbemi BT, Kanno J, McLachlan JA, Nadal A, Sonnenschein C, et al. In vitro molecular mechanisms of bisphenol A action. Reprod Toxicol 2007;24(2):178–98.

[158] Kundakovic M, Gudsnuk K, Franks B, Madrid J, Miller RL, Perera FP, et al. Sex-specific epigenetic disruption and behavioral changes following low-dose in utero bisphenol A exposure. Proc Natl Acad Sci USA 2013;110(24):9956–61.

[159] Kundakovic M, Gudsnuk K, Herbstman JB, Tang D, Perera FP, Champagne FA. DNA methylation of BDNF as a biomarker of early-life adversity. Proc Natl Acad Sci USA 2014;112(22):6807–13.

[160] Borre YE, O'Keeffe GW, Clarke G, Stanton C, Dinan TG, Cryan JF. Microbiota and neurodevelopmental windows: implications for brain disorders. Trends Mol Med 2014;20(9):509–18.

[161] O'Mahony SM, Marchesi JR, Scully P, Codling C, Ceolho AM, Quigley EM, et al. Early life stress alters behavior, immunity, and microbiota in rats: implications for irritable bowel syndrome and psychiatric illnesses. Biol Psychiatry 2009;65(3):263–7.

[162] Bailey MT, Lubach GR, Coe CL. Prenatal stress alters bacterial colonization of the gut in infant monkeys. J Pediatr Gastroenterol Nutr 2004;38(4):414–21.

[163] Zijlmans MA, Korpela K, Riksen-Walraven JM, de Vos WM, de Weerth C. Maternal prenatal stress is associated with the infant intestinal microbiota. Psychoneuroendocrinology 2015;53:233–45.

[164] Hsiao EY, McBride SW, Hsien S, Sharon G, Hyde ER, McCue T, et al. Microbiota modulate behavioral and physiological abnormalities associated with neurodevelopmental disorders. Cell 2013;155(7):1451–63.

[165] Kumar H, Lund R, Laiho A, Lundelin K, Ley RE, Isolauri E, et al. Gut microbiota as an epigenetic regulator: pilot study based on whole-genome methylation analysis. mBio 2014;5(6).

[166] Stilling RM, Dinan TG, Cryan JF. Microbial genes, brain & behaviour—epigenetic regulation of the gut-brain axis. Genes Brain Behav 2014;13(1):69–86.

[167] Davies M, Krause L, Bell J, Gao F, Ward K, Wu H, et al. Hypermethylation in the ZBTB20 gene is associated with major depressive disorder. Genome Biol 2014;15(4):R56.

[168] Dempster EL, Pidsley R, Schalkwyk LC, Owens S, Georgiades A, Kane F, et al. Disease-associated epigenetic changes in monozygotic twins discordant for schizophrenia and bipolar disorder. Hum Mol Genet 2011;20(24):4786–96.

[169] Lam LL, Emberly E, Fraser HB, Neumann SM, Chen E, Miller GE, et al. Factors underlying variable DNA methylation in a human community cohort. Proc Natl Acad Sci USA 2012;109(Suppl. 2):17253–60.

[170] Houseman EA, Accomando WP, Koestler DC, Christensen BC, Marsit CJ, Nelson HH, et al. DNA methylation arrays as surrogate measures of cell mixture distribution. BMC Bioinf 2012;13:86.

[171] Guintivano J, Aryee MJ, Kaminsky ZA. A cell epigenotype specific model for the correction of brain cellular heterogeneity bias and its application to age, brain region and major depression. Epigenetics 2013;8(3):290–302.

[172] Grigoriu A, Ferreira JC, Choufani S, Baczyk D, Kingdom J, Weksberg R. Cell specific patterns of methylation in the human placenta. Epigenetics 2011;6(3):368–79.

[173] Hogg K, Price EM, Robinson WP. Improved reporting of DNA methylation data derived from studies of the human placenta. Epigenetics 2014;9(3):333–7.

[174] Mill J, Heijmans BT. From promises to practical strategies in epigenetic epidemiology. Nat Rev Genet 2013;14(8):585–94.

[175] Kangaspeska S, Stride B, Metivier R, Polycarpou-Schwarz M, Ibberson D, Carmouche RP, et al. Transient cyclical methylation of promoter DNA. Nature 2008;452(7183):112–5.

[176] Martinowich K, Hattori D, Wu H, Fouse S, He F, Hu Y, et al. DNA methylation-related chromatin remodeling in activity-dependent Bdnf gene regulation. Science 2003;302(5646):890–3.

[177] D'Alessio AC, Weaver IC, Szyf M. Acetylation-induced transcription is required for active DNA demethylation in methylation-silenced genes. Mol Cell Biol 2007;27(21):7462–74.

[178] Weaver IC, D'Alessio AC, Brown SE, Hellstrom IC, Dymov S, Sharma S, et al. The transcription factor nerve growth factor-inducible protein a mediates epigenetic programming: altering epigenetic marks by immediate-early genes. J Neurosci 2007;27(7):1756–68.

[179] Ziller MJ, Gu H, Muller F, Donaghey J, Tsai LT, Kohlbacher O, et al. Charting a dynamic DNA methylation landscape of the human genome. Nature 2013;500(7463):477–81.

[180] Wang D, Liu X, Zhou Y, Xie H, Hong X, Tsai HJ, et al. Individual variation and longitudinal pattern of genome-wide DNA methylation from birth to the first two years of life. Epigenetics 2012;7(6):594–605.

[181] Martino DJ, Tulic MK, Gordon L, Hodder M, Richman TR, Metcalfe J, et al. Evidence for age-related and individual-specific changes in DNA methylation profile of mononuclear cells during early immune development in humans. Epigenetics 2011;6(9):1085–94.

[182] Martino D, Loke YJ, Gordon L, Ollikainen M, Cruickshank MN, Saffery R, et al. Longitudinal, genome-scale analysis of DNA methylation in twins from birth to 18 months of age reveals rapid epigenetic change in early life and pair-specific effects of discordance. Genome Biol 2013;14(5):R42.
[183] Spiers H, Hannon E, Schalkwyk LC, Smith R, Wong CC, O'Donovan MC, et al. Methylomic trajectories across human fetal brain development. Genome Res 2015;25(3):338–52.
[184] Davies MN, Volta M, Pidsley R, Lunnon K, Dixit A, Lovestone S, et al. Functional annotation of the human brain methylome identifies tissue-specific epigenetic variation across brain and blood. Genome Biol 2012;13(6):R43.
[185] Farre P, Jones MJ, Meaney MJ, Emberly E, Turecki G, Kobor MS. Concordant and discordant DNA methylation signatures of aging in human blood and brain. Epigenetics Chromatin 2015;8:19.
[186] Conradt E, Lester BM, Appleton AA, Armstrong DA, Marsit CJ. The roles of DNA methylation of NR3C1 and 11beta-HSD2 and exposure to maternal mood disorder in utero on newborn neurobehavior. Epigenetics 2013;8(12):1321–9.
[187] Broad KD, Keverne EB. Placental protection of the fetal brain during short-term food deprivation. Proc Natl Acad Sci USA 2011;108(37):15237–41.

CHAPTER

8

Epigenetics in the Developmental Origin of Cardiovascular Disorders

Jeffrey S. Gilbert

Department of Biomedical Sciences, University of Minnesota Medical School, Duluth, MN, USA

OUTLINE

INTRODUCTION

Pregnancy is a dynamic stressor that relies upon numerous tightly choreographed adaptations to the maternal physiological state. In addition to homeostatic regulation of maternal energy, fluid, and electrolyte balance, the metabolic needs of a rapidly developing and growing fetus must also be met. When homeostasis fails and results in a suboptimal intrauterine environment, fetal growth and development are altered in ways that may confer short-term benefit but

The Epigenome and Developmental Origins of Health and Disease
http://dx.doi.org/10.1016/B978-0-12-801383-0.00008-6

Copyright © 2016 Elsevier Inc. All rights reserved.

result in deleterious long-term outcomes. Indeed, the contemporary literature is replete with evidence from clinical and experimental studies showing that a variety of stressors during pregnancy are transmitted to the fetus through an altered intrauterine environment that impacts long-term health outcomes [1–5].

Viewed in concert with the unrelenting worldwide increase in noncommunicable diseases, there has been a clear need to reevaluate many hypotheses regarding the primary underlying causes of chronic disease states [6]. This is especially evident in research regarding congenital deficits of cardiovascular development that have traditionally been viewed as genetic in origin [7]. Collectively these studies have become recognized as the "developmental origins of health and disease" (DOHaD) hypothesis. While much of this work has focused on embryonic and fetal development in humans and the analogous developmental periods in model species such as rats, mice, sheep, and nonhuman primates, it is becoming increasingly clear that the critical windows of programming may extend well beyond the gestation period and into adulthood [8,9].

Several themes have emerged from the many clinical and experimental investigations into the DOHaD hypothesis. These include but are not limited to the importance of biological sex in determining the effects of programming events, the existence of discrete (and often organ-specific) critical periods during which programming events occur, and that many factors including social (e.g., behavioral stress), environmental (e.g., diet, toxins, endocrine disruptors), clinical (e.g., idiopathic intrauterine growth restriction [IUGR], preeclampsia), and pharmacologic (e.g., inhibition of angiotensin II) can act as programming agents during development [2,8–10]. Despite significant progress in recent years toward identifying programming stressors and developing models for the study of DOHaD, many uncertainties have remained with respect to the exact mechanisms underlying these programming events. To that end, recent evidence strongly points to a significant role for epigenetic modifications as an important mechanism mediating events throughout the lifespan that culminate in the development and progression of cardiovascular disease (CVD) [11–13]. Thus, the goal of this chapter is to place into perspective the diverse and rapidly expanding body of literature that forms the basis of our current understanding of the mechanisms underlying the developmental programming of CVDs.

PERIODS OF SUSCEPTIBILITY TO DEVELOPMENTAL PROGRAMMING

Developmental programming is recognized as an important determinant of adult health and can be defined as the response by the developing organism to stimuli during critical periods of organogenesis that results in persistent effects on the adult phenotype. The initial acceptance and understanding of this concept was derived from human epidemiologic studies suggesting that many chronic conditions such as CVD [14,15] were associated with low birth weight. Since developing organisms pass numerous physiological milestones prior to birth, weight at birth was used as a proxy for adequate fetal development in many of the early experiments that generated this area of study. While we now understand birth weight has some limitations in this regard, it does remain an easily accessible and useful data point in both clinical and experimental studies. In addition, it is increasingly recognized that secondary stressors such as high-fat diet, toxin exposure, psychological stress, and the like, encountered in later life, may play an important role in the development and progression of health and disease.

Periconceptional and Embryonic

Susceptibility to programming events begins very early in the reproductive cycle, and recent studies suggest that in utero exposure to endocrine disruptors such as vinclozolin and methoxychlor as well as physiological stressors like undernutrition modifies primordial germ cell development and methylation patterns in mice [16,17]. Similarly, epigenetic processes play a prominent role by regulating the transcriptome of male and female preimplantation embryos such that several genes located on the X chromosome are more expressed in bovine and human female versus male embryos [18,19]. Likewise, several autosomal genes in trophoblast cells, such as those for interferon-γ [20], human choriogonadotropin hormone [21], and other imprinted genes [22], are also not expressed or methylated the same in male and female embryos. There are early morphologic differences evident as well, and these include different rates of development and cell division between male and female embryos [23]. Similarly, hyperglycemia has been shown to modify gene expression profiles in bovine blastocysts [24].

Fetal and Early Postnatal

The fetal period is a critical time for organogenesis, but it is important to recognize that several organ systems (e.g., lungs, mammary, vasculature) have important developmental milestones that occur postpartum [8,9,25]. The heart and cardiovascular system begin development in the early fetal period and are subject to programming from early pregnancy events such as hyperglycemia associated with maternal diabetes mellitus (DM) [26]. Since aberrant function of other organ systems and tissues may significantly contribute to the development and persistence of CVD, it bears mentioning these at least briefly in this context. Impaired lung and cardiopulmonary development, as often observed in preterm infants with bronchopulmonary dysplasia, results in increased risk for pulmonary hypertension [27,28]. Likewise, alterations in organs such as the liver and tissues such as adipose that have significant endocrine and metabolic regulatory functions may also contribute to adverse cardiovascular function in later life.

Adolescence and Adulthood

Concepts of developmental programming traditionally center around events associated with pregnancy and the early postnatal period, but it has long been recognized that the endocrine milieu of gestation modifies the incidence of mammary cancer and may have lasting impacts on the cardiovascular health in the mother [29–32]. There is accumulating evidence indicating that pregnancies complicated with preeclampsia increase the risk of CVD, DM, and early-stage renal disease in those mothers later in life [8,33–37]. Recent work also suggests that not only are alterations in circulating factors and inflammatory markers (e.g., sFlt-1, sEng, C3a, C5a) likely responsible for mediating pregnancy complications such as preeclampsia, HELLP syndrome, and fetal growth restriction, they may also predispose the maternal cardiovascular system to subsequent endothelial dysfunction as the mother ages [33–41]. Likewise, early-life exposures to programming stimuli such as gestational diabetes mellitus (GDM) have been shown to increase cardiovascular responses to stress in adolescence [42]. This in turn may adversely influence cardiovascular function in adulthood. While the exact mechanisms underlying these long-term effects of preeclampsia on later maternal health remain unclear, epigenetic processes seem likely to play a significant role.

PROGRAMMING STIMULI

As mentioned before, a wide range of factors are reported to program developmental outcomes. In general, these factors exert an influence on the organism during a critical period of development. These factors range from under- and overnutrition to exposure to clinical conditions or disease states (preeclampsia, DM, GDM), exposure to therapeutics (glucocorticoids), endocrine disruptor compounds (EDCs), and various physiological and psychological stressors (e.g., exercise, stress); even age at onset of pregnancy can be included in this list [8,43]. Importantly, while most of these are associated with deleterious outcomes, this is not always the case, as there is some evidence that outcomes can be enhanced by interventions (e.g., exercise) that mitigate the effects of the stressor during development [8,43]. In the sections below, these factors will be discussed briefly in relation to their developmental consequences for CVD and health.

Diet and Nutritional Factors

The influence of diet and nutrition on fetal development has been widely studied in regard to the DOHaD hypothesis, so much so that it is almost synonymous with this area of study [43]. In short, nutritional manipulations include low-protein diets, global calorie restriction, calorie excess, and various macro- (e.g., fat, carbohydrate) and micronutrient (e.g., specific vitamins) alterations. In addition, there are issues regarding timing of nutritional insult, and diets may be altered during various phases of gestation, often depending on the organism under study (long or short gestation) and the organ system of interest. It should also be noted that nutrition is typically modified on the maternal side (maternal nutrient restriction [MNR]), with the expectation that effects will be transmitted to the embryo or fetus but some clinical conditions may also represent a nutritional stress to the fetus if reduction of nutrient transport is impaired, as may be the case with preeclampsia or idiopathic IUGR [43,44].

Studies investigating MNR generally show decreased nephron endowment and altered expression of components of the intrarenal and cardiac renin–angiotensin system (RAS) [45–47]. Likewise, alterations in cardiovascular function have also been reported; these include impaired myogenic response in the mesenteric vascular bed of pregnant adult females exposed to MNR during development [48], and increased arterial pressure, which is often only observed in the male offspring [49,50]. Restriction of specific micronutrients also results in abnormalities as exemplified by studies evaluating maternal low-sodium diet in rats that generates IUGR, increased arterial pressure, and reduced creatinine clearance in female, but not male, offspring [51]. Maternal water deprivation has been reported to disrupt expression and function of various parts of the RAS [52], and epigenetic regulation of various osmolality- and furosimide-sensitive transport proteins has been proposed as a possible mechanism [53]. In addition, data from several groups indicate that protein restriction in the rat results in a deficiency of methyl donors and may be a critical factor altering gene methylation patterns and gene expression [54–57].

The role of overnutrition is an important aspect of study, considering the rapidly increasing rates of obesity worldwide. In general, these studies focus on a high-fat diet as a means to study overnutrition, but many of these animals become overweight during the studies; hence it is difficult to tease apart the effects of the diet versus obesity. The finer points of those studies are clearly beyond the scope of the present chapter and have been reviewed extensively elsewhere [58]. In short, previous studies show that male rat offspring, but not female, develop hypertension after exposure to a maternal diet high in saturated fat and low in linoleic acid [59]. In contrast, Elahi et al. reported that mice fed high-fat diets long before the onset of gestation are hypercholesterolemic and hypertensive and produce hypertensive, hypercholesterolemic female offspring [60]. Others have demonstrated that a diet rich in fat fed to pregnant rats results in male offspring gaining more body weight and presenting with decreased renal renin activity when compared to females [61]. In addition, offspring from this model reportedly are hypertensive and exhibit increased aortic stiffness, decreased aortic smooth muscle cell number, endothelial dysfunction, and decreased renal Na^+, K^+-ATPase activity [62,63]. Further, Khan et al. reported female offspring have reduced locomotor activity at 180 days of age compared to male offspring of pregnant rats fed a high-fat diet during pregnancy [62]. In addition, cross-fostering techniques after birth show the hypertension in females occurs whether exposure to maternal high-fat diet occurs before and during pregnancy or during the suckling period [64]. While the exact mechanisms responsible for programming due to high-fat diets remain under investigation, the report that statin treatment has beneficial effects on the offspring highlights at least one potential mechanism, alterations in lipid metabolism [60]. Because of the numerous pleiotropic effects of statins, the mechanisms for these effects remain unclear; nevertheless, these observations provide insights for further studies. In addition, it has been suggested that high levels of butyric acid that may result from a high-fat diet could lead to changes in chromatin structure and result in epigenetic alterations [65]. Taken together, these observations highlight the important role for dietary fats and intermediate metabolites in developmental programming.

Xenobiotics and Endocrine Disruptor Compounds

While EDCs are traditionally considered to be environmental pollutants, there are numerous examples of naturally occurring compounds that have similar deleterious effects on organisms if ingestion or exposure is sufficient. Examples include factors derived from *Pinus ponderosa* needles (isocupressic acid), leaves from *Veratrum californicum*, and carbenoxolone, an active ingredient of licorice [66], with the first two affecting grazing livestock species and the last humans.

Exposure to a number of environmental factors may interfere with endocrine systems through mimetic or antagonistic activity. Several known exogenous estrogen antagonists include polychlorinated and polybrominated biphenyls, dichlorodiphenyltrichloroethane, methoxychlor, diethylstilbestrol, 17β-estradiol, and bisphenol-A [67,68]. Xenoestrogens may exist in the system at levels that do not elicit strong or detectable estrogenic effects individually, but it has been shown that these have additive effects and several xenoestrogens in the system can act together to induce estrogenic activity [67]. With this in mind, fetal overexposure to estrogens is suggested to stunt growth and alter bone development and it commonly takes many years (i.e., into adulthood) for the manifestation of congenital endocrine complications [67,68]. In contrast to the large number of estrogen disruptors, only several androgen antagonists have been identified and studied. These are insecticide ingredients such as kepone and procymidone, dichlorodiphenyldichloroethylene, vinclosolin, and 2,3,7,8-tetrachlorodibenzodioxin [67,69]. Although the exact epigenetic mechanisms by which early-life exposures to androgen and estrogen mimetics continue to be investigated, one exciting possibility is that modifications occur in the developing resident stem/progenitor cells and these changes can go on to alter subsequent development in later life [70].

Interference in prostaglandin pathways has been associated with the development of several types of cancer and cardiovascular disorders and altered prostaglandin synthesis has been shown to disrupt endocrine processes. Several phthalates that are similar to pharmaceutical COX inhibitors have been found to disrupt levels of prostaglandin synthesis [71].

Chronic inhibition of COX activity is known to have deleterious effects on renal and cardiovascular function, resulting in mild to moderate hypertension and even renal failure. Developmental sex differences in this system have been reported and show that renal COX-2 expression is higher in female fetuses at gestational day 21 than in age-matched male fetuses [72]. Prostaglandin synthesis pathways are relatively understudied but may provide important insights into the development of sex differences in adult disease.

Clinical Conditions

Human Studies

There are a wide range of clinical conditions that may affect pregnant women and, as a consequence, result in a variety of situations leading to a suboptimal intrauterine environment for the gestating fetus. These may include placental insufficiency, preeclampsia, and hyperemesis gravidarum, all of which may be viewed as fetal undernutrition. In addition, complications such as GDM and maternal obesity may be sensed by the fetus as a combination of over- and undernutrition owing to the comorbidities often present in those patients. Maternal obesity is associated with a variety of conditions, including maternal hypertension, hypertriglyceridemia, hyperglycemia, and insulin resistance [73], that are independently correlated with a suboptimal in utero environment and consequently linked to DOHaD.

The role of clinical conditions in programming events can be viewed in two ways: first, as the effects of the condition itself (e.g., angiogenic imbalance in preeclampsia) on development; and second, as the effects of interventions (e.g., glucocorticoids for preterm birth) for treatment of the condition. Increasing evidence supports the idea that women who have had a preeclamptic pregnancy are at increased risk for metabolic and cardiovascular abnormalities in later life [33–37]. Further, a recent clinical study also shows women with a history of preeclampsia have increased sensitivity to Ang II and sodium eight months postpartum [74]. Although accumulating evidence suggests that persistent increases in circulating factors and inflammatory markers (e.g., sFlt-1, sEng, cytokines) may be responsible for the predisposition of the maternal cardiovascular system to subsequent dysfunction as the mother ages [34–36,38–41], the exact mechanisms underlying these observations (be they epigenetic or otherwise) remain unclear.

Animal Studies

Further evidence derives from experimental studies using well-characterized models of reduced uterine perfusion pressure that creates chronic placental ischemia and results in hypertension in the pregnant (but not virgin) rat and closely mimics many features of early-onset preeclampsia [75–78]. Recent studies in these models have shown that placental ischemia elevates levels of sFlt-1 in the placenta, amniotic fluid, and maternal plasma [79,80]. In the rat, this has been associated with decreased fetal growth and subsequent hypertension and glucose intolerance in the offspring [76–78]. In addition, studies by Gilbert et al. suggest rat dams have persistent deficits in endothelial function [81]. Recent studies in rodents have shown that elevated sFlt-1 levels alone (via adenoviral overexpression) results in fetal growth restriction and maternal hypertension [82]. The male mouse offspring of these pregnancies also present with increased arterial pressure in this model [83]. Taken together, it appears that the time of onset of preeclampsia (i.e., early vs late onset) may be an important factor in the long-term health risks associated with this syndrome [37,74]. This suggests that preterm preeclampsia (typically associated with reduced fetal growth) rather than late-term preeclampsia (not associated with reduced fetal growth) is associated with long-term health issues in the mother. Viewed in concert, these studies strongly suggest that in addition to the immediate well-being of the mother, a long-term outlook with regard to the well-being of the fetus must also be considered during complicated and/or high-risk pregnancies.

Treatments indicated for severe pregnancy complications have long been associated with DOHaD. Considering the well-known links between the kidney and long-term control of arterial pressure, altered renal organogenesis is recognized as an important potential contributor to hypertension and CVD. To this end, angiotensin II blockers, which comprise some of the most effective antihypertensive treatments, are contraindicated in conditions such as preeclampsia owing to deleterious effects on fetal renal development [84,85] and, ultimately, the development of hypertension in treated offspring. Likewise, treatment of preterm birth has long included use of synthetic glucocorticoids to improve neonatal outcomes by stimulating lung development and maturation. Studies clearly show that these near-term gains have long-term effects on many organ systems, including the cardiovascular system. Ortiz et al. have shown that antenatal dexamethasone elevates blood pressure in female offspring at three weeks of age while only male offspring had increased blood pressure at six months of age [86]. Interestingly, despite the observation that only male dexamethasone-treated rats were hypertensive at six months of age, both male and female offspring showed signs of glomerulosclerosis when compared to control rats [86]. Similar work has shown that a postnatal diet rich in ω-3 (n-3) fatty acids attenuates the effects of antenatal dexamethasone on blood pressure in the offspring [87] in a sex-independent manner. Recent work by Loria and colleagues has shown that early-life stress caused by maternal separation in rats results in increased sensitivity to angiotensin II and vascular inflammation [88]. With the wide-ranging effects reported in the various models, continued studies are required to tease out specific mechanisms of programming in these models.

MECHANISMS OF DOHaD IN CVD

General

There are several general mechanisms recognized to be at play in DOHaD. These include biological sex-related changes (some of which are epigenetic), constrained/accelerated fetal growth and development, and the interaction of these factors with critical windows of organogenesis [44,89]. These aspects of programming have been reviewed extensively elsewhere [2,43] and will not be further addressed in this chapter. Instead, the remaining focus will be on the emerging and recognized roles of epigenetic mechanisms and how they may play a role in the programming of CVD.

Epigenetic

The epigenome refers to various factors and modifications of DNA and its associated structures that are independent of the DNA sequence of a given gene. While the genome defines the complete set of genetic information contained in the DNA particular to an organism, the epigenome comprises the variety of modifications associated with the genomic DNA and that is often responsible for imparting unique cellular and developmental identity to individual organisms [90–92]. Further, the epigenome provides instructions for the unique gene expression program that defines specific cell types throughout the organism [92]. Epigenetic pathways are thought to be responsive to a variety of physiological and pathophysiological signals such as shear stress on the vascular endothelium, hypoxia, cytokines, and entry into the cell cycle [93]. Taken together, the epigenome could be viewed as the essence of an organism's ability to adapt to changes in how that organism interacts with its environment.

Epigenetic modifications appear to be central to the induction of persistent and heritable changes in gene expression that occur without alteration of DNA sequence [94–97]. While most cells in an organism contain the same DNA, gene expression varies widely across various tissues. Epigenetic mechanisms underlie this tissue- and cell type–specific gene expression [90] and include CpG methylation, histone modification (acetylation), the activity of autoregulatory DNA-binding proteins, and several RNA-dependent mechanisms such as long noncoding RNA (lncRNA) and short interfering (siRNA) or micro-RNA (miRNA) [98]. Since several of these mechanisms are implicated in the silencing of gene expression, X-inactivation, and X-linked dosage differences [99,100], one might posit that sex bias in differential gene expression linked to developmental programming may also have its roots in the epigenome. Indeed, these processes appear to have many sex-specific features, but further studies are needed to clarify these relationships.

DNA Methylation and Hydroxymethylation

DNA methylation is accomplished by DNA-methyltransferase (DNMT) enzymes, which may then be oxidized by ten-eleven translocation proteins to produce hydroxymethylation. While changes in methylation status have been the primary epigenetic mechanism studied in relation to the developmental programming of CVD, the role of hydroxymethylation is not as clear in this regard [101]. Indeed, some of the strongest evidence supporting the role of epigenetic mechanisms in the programming of cardiovascular dysfunction has come from nutrient restriction studies and supplementation of those pregnancies with folic acid or glycine to rescue the programming effects of the dietary insult [54,102–106]. While the majority of these studies have been in rat and mouse models, these short-gestation species have not been used exclusively to this end. One such example is a study by Sinclair et al. in which sheep exposed to a methyl-deficient diet during pregnancy produce hypertensive male offspring compared to females of similar rearing, as well as to male and female controls [106]. The authors then evaluated 1400 CpG sites (primarily gene promoter associated) in fetal liver at 90 days of gestation (term ~ 150 d) and reported that more than half of the affected loci were specific to males. These observations suggest male-specific hypomethylation that could provide a mechanistic basis for the phenotypic sex differences in the response to dietary stress during pregnancy [106].

Other studies have shown that promoters for genes of the RAS may also be responsive to a low-protein diet [107]. In particular, Bogdarina et al. reported that the proximal promoter for the AT1b gene was hypomethylated in the adrenal gland of mouse offspring fed a low-protein diet [107]. This effect was reversed by treatment with metapyrone, a glucocorticoid synthesis inhibitor suggesting that glucocorticoid exposure in early pregnancy contributes to the regulation of methylation patterns for genes in the RAS [108] and is consistent with a dysfunctional RAS that often plays an important role in hypertension. In contrast with the previous studies in rodents, recent work in the sheep suggests that changes in the expression of RAS proteins are not due to altered methylation patterns [109].

Another example of an important molecule for cardiovascular function that is regulated in part by chromatin accessibility is endothelial nitric oxide synthase (eNOS or NOS3). NOS3 is one of the genes that is endothelial restricted and expression of NOS3 helps define endothelial cell (EC) identity [93]. Studies by Chan et al. have shown that the NOS3 promoter in ECs is hypomethylated and enriched with activating histone modifications such as acetylated H3K9 and H4K12 [110]. Further, the authors also demonstrated the importance of these epigenetic modifications with pharmacologic inhibition studies on vascular smooth muscle cells (VSMCs) and ECs by using DNMT and histone deacetylase (HDAC) inhibitors to upregulate NOS3 in VSMCs and downregulate NOS3 in ECs [110].

Hypoxia is another stressor that is known to have major effects on ECs. Moreover, hypoxia has been recognized to induce a specific suite of epigenetic changes in tumor cells [111]. While the hypoxia inducible factor system for adapting to low oxygen supply that employs the *cis/trans* paradigm is well studied, the study of effects of hypoxia on chromatin-based pathways is in its early stages [93]. Recently, Dasgupta et al. reported that chronic hypoxia in the pregnant ewe increases estrogen receptor-alpha methylation, which is consistent with decreased protein expression and potentially increased vascular resistance in those animals [112]. The same group has also recently reported that pregnancy induced increases in Ca^{2+}-activated potassium channel β1 subunit expression due to hypomethylation and this was abrogated by chronic hypoxia in sheep uterine arteries [113]. These studies highlight possible mechanisms by which cardiovascular function may be programmed in the mother during pregnancies complicated by preeclampsia.

Histone Modifications

Histone modifications represent a large group of posttranslational changes on the tails of the histone protein. Examples of histone modifications include acetylation of lysine residues, methylation of lysine and arginine residues, phosphorylation of serine and threonine residues, and ubiquitination of lysine residues present on histone tails, as well as sumoylation and ADP ribosylation. All of these changes influence DNA transcription. Histone acetylation is carried out by enzymes called histone acetyltransferases, which are responsible for adding acetyl groups to lysine residues on histone tails, whereas HDACs are those that remove acetyl groups from acetylated lysines. Generally, the presence of acetylated lysine on histone tails leads to a relaxed chromatin state that promotes transcriptional activation of selected genes; in contrast, deacetylation of lysine residues leads to chromatin compaction and transcriptional inactivation [114].

Compared to CpG methylation status, histone modifications are not as thoroughly studied with respect to DOHaD. Nevertheless, recent work in this area suggests that histone code modifications such as histone 3 acetylation (H3Ac) play a role in regulating some factors in the RAS [115]. In a study comparing tissue-specific regulation of angiotensin-converting enzyme 1 (ACE1) in spontaneously hypertensive rats (SHR) compared to normotensive Wistar Kyoto (WKY) rats, Lee et al. reported that ACE1 was more abundant in SHR than in WKY and this was associated with H3Ac enrichment of ACE1 promoter regions [115]. While the SHR is not typically considered a model of programmed hypertension, the similarities between SHR and other DOHaD models along with a study by Woods have challenged this notion [116]. Taken together, it is likely that histone modifications play an important role in regulating expression of RAS proteins that regulate arterial pressure, but the exact roles of epigenetic modifications to stressors during development vis-à-vis adulthood remain to be elucidated.

RNA Mechanisms

RNA-based mechanisms are the most recently described and the least understood of the epigenetic regulatory systems, and research in this area currently focuses on noncoding RNAs that function to regulate gene transcription [100]. These noncoding RNAs are discriminated by size into lncRNA (>200 nucleotides) and short, micro- or small-interfering RNAs (miRNA or siRNA, respectively) [100].

The lncRNAs are thought to regulate transcription through the modulation of chromatin structure and often associate with activating or repressive chromatin remodeling complexes to modulate the structure of chromatin at target loci [100,117]. One ubiquitous example of lncRNA-mediated regulation of transcription via chromatin structure is Xist: an lncRNA that is involved in X-chromosome inactivation and systematic silencing of one of the two X chromosomes in each female somatic cell. Other regulatory mechanisms also include transcriptional co-repression, translational modulation, RNA splicing, and RNA degradation [99,100].

While lncRNAs appear to be one of the most predominant forms of RNA-based epigenetic regulation, and small noncoding RNAs function predominantly in the cytoplasm to control posttranscriptional processes (translation and RNA degradation), some small noncoding RNAs also play a role in chromatin-based silencing. Currently, little is known about the specific role(s) for many of these RNA mechanisms in the programming of CVD beyond the roles of the target proteins (e.g., vascular endothelial growth factor) in developmental processes.

CONCLUDING REMARKS

From a clinical and public health perspective it is hoped that increased understanding and awareness of developmental programming will lead to better diagnostic, preventative, and therapeutic measures. The persistence of programmed effects is likely due to covalent modifications of the genome resulting from changes in promoter methylation and histone acetylation. While epigenetic phenomena are central to the induction of persistent and heritable changes in gene expression that occur without alteration of DNA sequence, the exact mechanisms and specific contributions to the intensively studied area of developmental programming remain uncertain. While reversal of these molecular changes may be possible and improve long-term health outcomes if interventions are timed appropriately, loss of function due to hypoplastic development may be difficult to overcome. Nevertheless, continued investigation using hypothesis-driven mechanistic studies that incorporate epigenetic markers and processes into the models rather than solely evaluating phenotype is needed to identify targets and pathways for prevention, and not merely treatment, of chronic disease states such as CVD.

Acknowledgments

The author regrets that because of space limitations we have been unable to cite all the primary literature in the field. The authors' work has been supported in part by grants from the American Heart Association (10SDG2600040), the National Institutes of Health (HL109843, HD079547, HL114096), the American Physiological Society, and the University of Minnesota.

References

[1] Ravelli AC, van der Meulen JH, Michels RP, Osmond C, Barker DJ, Hales CN, et al. Glucose tolerance in adults after prenatal exposure to famine. Lancet January 17, 1998;351(9097):173–7.

[2] Gilbert JS, Nijland MJ. Sex differences in the developmental origins of hypertension and cardiorenal disease. Am J Physiol Regul Integr Comp Physiol December 1, 2008;295(6):R1941–52.

[3] Wentzel P, Gareskog M, Eriksson UJ. Folic acid supplementation diminishes diabetes- and glucose-induced dysmorphogenesis in rat embryos in vivo and in vitro. Diabetes February 1, 2005;54(2):546–53.

[4] Chang SY, Chen YW, Zhao XP, Chenier I, Tran S, Sauve A, et al. Catalase prevents maternal diabetes–induced perinatal programming via the Nrf2-HO-1 defense system. Diabetes June 25, 2012;10:2565–74.

[5] Roest PAM, Molin DGM, Schalkwijk CG, van Iperen L, Wentzel P, Eriksson UJ, et al. Specific local cardiovascular changes of N-(carboxymethyl)lysine, vascular endothelial growth factor, and Smad2 in the developing embryos coincide with maternal diabetes-induced congenital heart defects. Diabetes May 1, 2009;58(5):1222–8.

[6] Maher B. Personal genomes: the case of the missing heritability. Nature November 6, 2008;456(7218):18–21.

[7] Gilbert JS, Banek CT, Babcock SA, Dreyer HC. Diabetes in early pregnancy: getting to the heart of the matter. Diabetes January 2013;62(1):27–8.

[8] Gilbert JS, Nijland MJ, Knoblich P. Placental ischemia and cardiovascular dysfunction in preeclampsia and beyond: making the connections. Expert Rev Cardiovasc Ther November 1, 2008;6(10):1367–77.

[9] Gingery A, Bahe EL, Gilbert JS. Placental ischemia and breast cancer risk after preeclampsia: tying the knot. Expert Rev Anticancer Ther May 1, 2009;9(5):671–81.
[10] Loria AS, Ho DH, Pollock JS. A mechanistic look at the effects of adversity early in life on cardiovascular disease risk during adulthood. Acta Physiol February 1, 2014;210(2):277–87.
[11] Liang M, Cowley Jr AW, Mattson DL, Kotchen TA, Liu Y. Epigenomics of hypertension. Seminars Nephrol July 2013;33(4):392–9.
[12] Jirtle RL, Skinner MK. Environmental epigenomics and disease susceptibility. Nat Rev Genet April 2007;8(4):253–62.
[13] Dolinoy DC, Jirtle RL. Environmental epigenomics in human health and disease. Environ Mol Mutagen January 2008;49(1):4–8.
[14] Barker DJ. Fetal origins of coronary heart disease. Br Heart J March 1993;69(3):195–6.
[15] Roseboom TJ, van der Meulen JH, van Montfrans GA, Ravelli AC, Osmond C, Barker DJ, et al. Maternal nutrition during gestation and blood pressure in later life. J Hypertens January 2001;19(1):29–34.
[16] Radford EJ, Ito M, Shi H, Corish JA, Yamazawa K, Isganaitis E, et al. In utero undernourishment perturbs the adult sperm methylome and intergenerational metabolism. Science August 15, 2014;345(6198).
[17] Anway MD, Cupp AS, Uzumcu M, Skinner MK. Epigenetic transgenerational actions of endocrine disruptors and male fertility. Science June 3, 2005;308(5727): 1466–9.
[18] Wrenzycki C, Lucas-Hahn A, Herrmann D, Lemme E, Korsawe K, Niemann H. In vitro production and nuclear transfer affect dosage compensation of the X-linked gene transcripts G6PD, PGK, and Xist in preimplantation bovine embryos. Biol Reprod January 1, 2002;66(1):127–34.
[19] Peippo J, Farazmand A, Kurkilahti M, Markkula M, Basrur PK, King WA. Sex-chromosome linked gene expression in in-vitro produced bovine embryos. Mol Hum Reprod October 1, 2002;8(10):923–9.
[20] Larson MA, Kimura K, Kubisch HM, Roberts RM. Sexual dimorphism among bovine embryos in their ability to make the transition to expanded blastocyst and in the expression of the signaling molecule IFN–τ. Proc Natl Acad Sci USA August 14, 2001;98(17):9677–82.
[21] Haning Jr RV, Curet LB, Poole WK, Boehnlein LM, Kuzma DL, Meier SM. Effects of fetal sex and dexamethasone on preterm maternal serum concentrations of human chorionic gonadotropin, progesterone, estrone, estradiol, and estriol. Am J Obstet Gynecol December 1989;161(6 Pt 1):1549–53.
[22] Durcova-Hills G, Burgoyne P, McLaren A. Analysis of sex differences in EGC imprinting. Dev Biol April 1, 2004;268(1):105–10.
[23] Bernardi ML, Delouis C. Sex-related differences in the developmental rate of in-vitro matured/in-vitro fertilized ovine embryos. Hum Reprod March 1996;11(3):621–6.
[24] Cagnone GL, Dufort I, Vigneault C, Sirard MA. Differential gene expression profile in bovine blastocysts resulting from hyperglycemia exposure during early cleavage stages. Biol Reprod February 2012;86(2):50.
[25] Narayanan M, Owers-Bradley J, Beardsmore CS, Mada M, Ball I, Garipov R, et al. Alveolarization continues during childhood and adolescence. Am J Respir Crit Care Med January 15, 2012;185(2):186–91.
[26] Eriksson UJ. Congenital anomalies in diabetic pregnancy. Semin Fetal Neonatal Med April 2009;14(2):85–93.
[27] Khemani E, McElhinney DB, Rhein L, Andrade O, Lacro RV, Thomas KC, et al. Pulmonary artery hypertension in formerly premature infants with bronchopulmonary dysplasia: clinical features and outcomes in the surfactant era. Pediatrics December 1, 2007;120(6):1260–9.
[28] Xu XF, Lv Y, Gu WZ, Tang LL, Wei JK, Zhang LY, et al. Epigenetics of hypoxic pulmonary arterial hypertension following intrauterine growth retardation rat: epigenetics in PAH following IUGR. Respir Res 2013;14(1):20.
[29] Lawlor DA, Emberson JR, Ebrahim S, Whincup PH, Wannamethee SG, Walker M, et al. Is the association between parity and coronary heart disease due to biological effects of pregnancy or adverse lifestyle risk factors associated with child-rearing?: findings from the British Women's Heart and Health Study and the British Regional Heart Study. Circulation March 11, 2003;107(9):1260–4.
[30] Humphries KH, Westendorp ICD, Bots ML, Spinelli JJ, Carere RG, Hofman A, et al. Parity and carotid artery atherosclerosis in elderly women: the Rotterdam study. Stroke October 1, 2001;32(10):2259–64.
[31] Beral V. Long term effects of childbearing on health. J Epidemiol Community Health December 1, 1985;39(4):343–6.
[32] Green A, Beral V, Moser K. Mortality in women in relation to their childbearing history. BMJ August 6, 1988;297(6645):391–5.
[33] Chambers JC, Fusi L, Malik IS, Haskard DO, De Swiet M, Kooner JS. Association of maternal endothelial dysfunction with preeclampsia. J Am Med Assoc March 28, 2001;285(12):1607–12.
[34] Hubel CA, Snaedal S, Ness RB, Weissfeld LA, Geirsson RT, Roberts JM, et al. Dyslipoproteinaemia in postmenopausal women with a history of eclampsia. BJOG June 2000;107(6):776–84.
[35] Wikstrom AK, Haglund B, Olovsson M, Lindeberg SN. The risk of maternal ischaemic heart disease after gestational hypertensive disease. BJOG November 30, 2005;112(11):1486–91.

[36] Bellamy L, Casas JP, Hingorani AD, Williams DJ. Pre-eclampsia and risk of cardiovascular disease and cancer in later life: systematic review and meta-analysis. BMJ November 10, 2007;335(7627):974.

[37] Vikse BE, Irgens LM, Leivestad T, Skjaerven R, Iversen BM. Preeclampsia and the risk of end-stage renal disease. N Engl J Med August 21, 2008;359(8):800–9.

[38] Freeman DJ, McManus F, Brown EA, Cherry L, Norrie J, Ramsay JE, et al. Short- and long-term changes in plasma inflammatory markers associated with preeclampsia. Hypertension November 1, 2004;44(5):708–14.

[39] Hubel CA, Wallukat G, Wolf M, Herse F, Rajakumar A, Roberts JM, et al. Agonistic angiotensin II type 1 receptor autoantibodies in postpartum women with a history of preeclampsia. Hypertension March 2007;49(3):612–7.

[40] Hubel CA, Powers RW, Snaedal S, Gammill HS, Ness RB, Roberts JM, et al. C-Reactive protein is elevated 30 years after eclamptic pregnancy. Hypertension June 1, 2008;51(6):1499–505.

[41] Wolf M, Hubel CA, Lam C, Sampson M, Ecker JL, Ness RB, et al. Preeclampsia and future cardiovascular disease: potential role of altered angiogenesis and insulin resistance. J Clin Endocrinol Metab December 1, 2004;89(12):6239–43.

[42] Krishnaveni GV, Veena SR, Jones A, Srinivasan K, Osmond C, Karat SC, et al. Exposure to maternal gestational diabetes is associated with higher cardiovascular responses to stress in adolescent Indians. J Clin Endocrinol Metab December 5, 2014;100(3):986–93. [jc].

[43] Mcmillen IC, Robinson JS. Developmental origins of the metabolic syndrome: prediction, plasticity, and programming. Physiol Rev April 1, 2005;85(2):571–633.

[44] Gilbert JS, Cox LA, Mitchell G, Nijland MJ. Nutrient-restricted fetus and the cardio-renal connection in hypertensive offspring. Expert Rev Cardiovasc Ther March 2006;4(2):227–37.

[45] Ozaki T, Nishina H, Hanson MA, Poston L. Dietary restriction in pregnant rats causes gender-related hypertension and vascular dysfunction in offspring. J Physiol January 1, 2001;530(Pt 1):141–52.

[46] McMullen S, Langley-Evans SC. Maternal low-protein diet in rat pregnancy programs blood pressure through sex-specific mechanisms. Am J Physiol Regul Integr Comp Physiol January 1, 2005;288(1):R85–90.

[47] Gilbert JS, Lang AL, Nijland MJ. Maternal nutrient restriction and the fetal left ventricle: decreased angiotensin receptor expression. Reprod Biol Endocrinol July 14, 2005;3(1):27.

[48] Hemmings DG, Veerareddy S, Baker PN, Davidge ST. Increased myogenic responses in uterine but not mesenteric arteries from pregnant offspring of diet-restricted rat dams. Biol Reprod April 1, 2005;72(4):997–1003.

[49] Gilbert JS, Lang AL, Grant AR, Nijland MJ. Maternal nutrient restriction in sheep: hypertension and decreased nephron number in offspring at 9 months of age. J Physiol May 15, 2005;565(Pt 1):137–47.

[50] Kwong WY, Wild AE, Roberts P, Willis AC, Fleming TP. Maternal undernutrition during the preimplantation period of rat development causes blastocyst abnormalities and programming of postnatal hypertension. Development October 2000;127(19):4195–202.

[51] Battista MC, Oligny LL, St Louis J, Brochu M. Intrauterine growth restriction in rats is associated with hypertension and renal dysfunction in adulthood. Am J Physiol Endocrinol Metab July 1, 2002;283(1):E124–31.

[52] Guan J, Mao C, Xu F, Geng C, Zhu L, Wang A, et al. Prenatal dehydration alters renin-angiotensin system associated with angiotensin-increased blood pressure in young offspring. Hypertens Res September 25, 2009;32(12):1104–11.

[53] Raftopoulos L, Katsi V, Makris T, Tousoulis D, Stefanadis C, Kallikazaros I. Epigenetics, the missing link in hypertension. Life Sci 2015;129:22–6.

[54] Lillycrop KA, Phillips ES, Jackson AA, Hanson MA, Burdge GC. Dietary protein restriction of pregnant rats induces and folic acid supplementation prevents epigenetic modification of hepatic gene expression in the offspring. J Nutr June 1, 2005;135(6):1382–6.

[55] Burdge GC, Slater-Jefferies J, Torrens C, Phillips ES, Hanson MA, Lillycrop KA. Dietary protein restriction of pregnant rats in the F_0 generation induces altered methylation of hepatic gene promoters in the adult male offspring in the F_1 and F_2 generations. Br J Nutr March 2007;97(3):435–9.

[56] Lillycrop KA, Slater-Jefferies JL, Hanson MA, Godfrey KM, Jackson AA, Burdge GC. Induction of altered epigenetic regulation of the hepatic glucocorticoid receptor in the offspring of rats fed a protein-restricted diet during pregnancy suggests that reduced DNA methyltransferase-1 expression is involved in impaired DNA methylation and changes in histone modifications. Br J Nutr June 2007;97(6):1064–73.

[57] Lillycrop KA, Phillips ES, Torrens C, Hanson MA, Jackson AA, Burdge GC. Feeding pregnant rats a protein-restricted diet persistently alters the methylation of specific cytosines in the hepatic PPAR alpha promoter of the offspring. Br J Nutr August 2008;100(2):278–82.

[58] Dyer JS, Rosenfeld CR. Metabolic imprinting by prenatal, perinatal, and postnatal overnutrition: a review. Semin Reprod Med 2011;29(03):266–76.

[59] Langley-Evans SC. Intrauterine programming of hypertension in the rat: nutrient interactions. Comp Biochem Physiol A Physiol August 1996;114(4):327–33.

[60] Elahi MM, Cagampang FR, Anthony FW, Curzen N, Ohri SK, Hanson MA. Statin treatment in hypercholesterolemic pregnant mice reduces cardiovascular risk factors in their offspring. Hypertension April 1, 2008;51(4):939–44.

[61] Armitage JA, Lakasing L, Taylor PD, Balachandran AA, Jensen RI, Dekou V, et al. Developmental programming of aortic and renal structure in offspring of rats fed fat-rich diets in pregnancy. J Physiol (Lond) May 15, 2005;565(1):171–84.

[62] Khan IY, Taylor PD, Dekou V, Seed PT, Lakasing L, Graham D, et al. Gender-linked hypertension in offspring of lard-fed pregnant rats. Hypertension January 1, 2003;41(1):168–75.

[63] Samuelsson AM, Matthews PA, Argenton M, Christie MR, McConnell JM, Jansen EHJ, et al. Diet-induced obesity in female mice leads to offspring hyperphagia, adiposity, hypertension, and insulin resistance: a novel murine model of developmental programming. Hypertension February 1, 2008;51(2):383–92.

[64] Khan IY, Dekou V, Douglas G, Jensen R, Hanson MA, Poston L, et al. A high-fat diet during rat pregnancy or suckling induces cardiovascular dysfunction in adult offspring. Am J Physiol Regul Integr Comp Physiol January 1, 2005;288(1):R127–33.

[65] Junien C. Impact of diets and nutrients/drugs on early epigenetic programming. J Inherit Metab Dis April 2006;29(2–3):359–65.

[66] Challis JRG, Matthews SG, Gibb W, Lye SJ. Endocrine and paracrine regulation of birth at term and preterm. Endocr Rev October 1, 2000;21(5):514–50.

[67] Sonnenschein C, Soto AM. An updated review of environmental estrogen and androgen mimics and antagonists. J Steroid Biochem Mol Biol April 1998;65(1–6):143–50.

[68] Derfoul A, Lin FJ, Awumey EM, Kolodzeski T, Hall DJ, Tuan RS. Estrogenic endocrine disruptive components interfere with calcium handling and differentiation of human trophoblast cells. J Cell Biochem July 1, 2003;89(4):755–70.

[69] Hotchkiss AK, Rider CV, Furr J, Howdeshell KL, Blystone CR, Wilson VS, et al. In utero exposure to an AR antagonist plus an inhibitor of fetal testosterone synthesis induces cumulative effects on F1 male rats. Reprod Toxicol September 2010;30(2):261–70.

[70] Kilcoyne KR, Smith LB, Atanassova N, Macpherson S, McKinnell C, van den Driesche S, et al. Fetal programming of adult Leydig cell function by androgenic effects on stem/progenitor cells. Proc Natl Acad Sci May 6, 2014;111(18):E1924–32.

[71] Kristensen DM, Skalkam ML, Audouze K, Lesne L, Desdoits-Lethimonier C, Frederiksen H, et al. Many putative endocrine disruptors inhibit prostaglandin synthesis. Environ Health Perspect April 2011;119(4):534–41.

[72] Baserga M, Hale MA, Wang ZM, Yu X, Callaway CW, McKnight RA, et al. Uteroplacental insufficiency alters nephrogenesis and downregulates cyclooxygenase-2 expression in a model of IUGR with adult-onset hypertension. Am J Physiol Regul Integr Comp Physiol May 1, 2007;292(5):R1943–55.

[73] Wilson PWF, Grundy SM. The metabolic syndrome: practical guide to origins and treatment: Part I. Circulation September 23, 2003;108(12):1422–4.

[74] Saxena AR, Karumanchi SA, Brown NJ, Royle CM, McElrath TF, Seely EW. Increased sensitivity to angiotensin II is present postpartum in women with a history of hypertensive pregnancy. Hypertension May 1, 2010;55(5):1239–45.

[75] George EM, Granger JP. Recent insights into the pathophysiology of preeclampsia. Expert Rev Obstet Gynecol September 1, 2010;5(5):557–66.

[76] Heltemes A, Gingery A, Soldner ELB, Bozadjieva N, Jahr K, Johnson B, et al. Chronic placental ischemia alters amniotic fluid milieu and results in impaired glucose tolerance, insulin resistance and hyperleptinemia in young rats. Exp Biol Med 2010;235(7): 892–9.

[77] Alexander BT. Placental insufficiency leads to development of hypertension in growth-restricted offspring. Hypertension March 1, 2003;41(3):457–62.

[78] Ojeda NB, Grigore D, Alexander BT. Intrauterine growth restriction: fetal programming of hypertension and kidney disease. Adv Chronic Kidney Dis April 2008;15(2):101–6.

[79] Gilbert JS, Babcock SA, Granger JP. Hypertension produced by reduced uterine perfusion in pregnant rats is associated with increased soluble fms-like tyrosine kinase-1 expression. Hypertension October 8, 2007;50:1142–7.

[80] Makris A, Thornton C, Thompson J, Thomson S, Martin R, Ogle R, et al. Uteroplacental ischemia results in proteinuric hypertension and elevated sFLT-1. Kidney Int March 21, 2007;71(10):977–84.

[81] Gilbert JS, Bauer AJ, Gilbert SA, Banek CT. The opposing roles of anti-angiogenic factors in cancer and preeclampsia. Front Biosci (Elite Ed) 2012;4: 2752–69.

[82] Lu F, Longo M, Tamayo E, Maner W, Al-Hendy A, Anderson GD, et al. The effect of over-expression of sFlt-1 on blood pressure and the occurrence of other manifestations of preeclampsia in unrestrained conscious pregnant mice. Am J Obstet Gynecol April 2007;196(4):396.

[83] Lu F, Bytautiene E, Tamayo E, Gamble P, Anderson GD, Hankins GD, et al. Gender-specific effect of overexpression of sFlt-1 in pregnant mice on fetal programming of blood pressure in the offspring later in life. Am J Obstet Gynecol October 2007;197(4). 418.e1–5.

[84] Loria A, Reverte V, Salazar F, Saez F, Llinas MT, Salazar FJ. Sex and age differences of renal function in rats with reduced ANG II activity during the nephrogenic period. Am J Physiol Ren Physiol August 1, 2007;293(2):F506–10.

[85] Saez F, Castells MT, Zuasti A, Salazar F, Reverte V, Loria A, et al. Sex differences in the renal changes elicited by angiotensin II blockade during the nephrogenic period. Hypertension June 1, 2007;49(6):1429–35.

[86] Ortiz LA, Quan A, Zarzar F, Weinberg A, Baum M. Prenatal dexamethasone programs hypertension and renal injury in the rat. Hypertension February 1, 2003;41(2):328–34.

[87] Wyrwoll CS, Mark PJ, Mori TA, Puddey IB, Waddell BJ. Prevention of programmed hyperleptinemia and hypertension by postnatal dietary {ω}-3 fatty acids. Endocrinology January 1, 2006;147(1):599–606.

[88] Loria AS, Kang KT, Pollock DM, Pollock JS. Early life stress enhances angiotensin II- mediated vasoconstriction by reduced endothelial nitric oxide buffering capacity. Hypertension October 2011;58(4):619–26.

[89] McMillen IC, Adams MB, Ross JT, Coulter CL, Simonetta G, Owens JA, et al. Fetal growth restriction: adaptations and consequences. Reproduction August 2001;122(2):195–204.

[90] Waterland RA, Michels KB. Epigenetic epidemiology of the developmental origins hypothesis. Annu Rev Nutr August 1, 2007;27(1):363–88.

[91] Burdge GC, Lillycrop KA. Nutrition, epigenetics, and developmental plasticity: implications for understanding human disease. Annu Rev Nutr August 21, 2010;30:315–39.

[92] Csoka AB, Kanherkar RR, Bhatia-Dey N. Epigenetics across the human lifespan. Front Cell Dev Biol 2014;2.

[93] Yan MS-C, Matouk CC, Marsden PA. Epigenetics of the vascular endothelium. J Appl Physiol September 9, 2010;109(3):916–26.

[94] Akintola AD, Crislip ZL, Catania JM, Chen G, Zimmer WE, Burghardt RC, et al. Promoter methylation is associated with the age-dependent loss of N-cadherin in the rat kidney. Am J Physiol Ren Physiol January 2008;294(1):F170–6.

[95] Bird AP. CpG-rich islands and the function of DNA methylation. Nature May 15, 1986;321(6067):209–13.

[96] Holliday R, Ho T. DNA methylation and epigenetic inheritance. Methods June 2002;27(2):179–83.

[97] Wyrwoll CS, Mark PJ, Waddell BJ. Developmental programming of renal glucocorticoid sensitivity and the renin-angiotensin system. Hypertension September 2007;50(3):579–84.

[98] Kelly TL, Trasler JM. Reproductive epigenetics. Clin Genet April 2004;65(4):247–60.

[99] Gutschner T, Diederichs S. The hallmarks of cancer. RNA Biol June 1, 2012;9(6):703–19.

[100] Rinn JL, Chang HY. Genome regulation by long noncoding RNAs. Annu Rev Biochem June 4, 2012;81(1): 145–66.

[101] Ficz G. New insights into mechanisms that regulate DNA methylation patterning. J Exp Biol January 1, 2015;218(1):14–20.

[102] Brawley L, Torrens C, Anthony FW, Itoh S, Wheeler T, Jackson AA, et al. Glycine rectifies vascular dysfunction induced by dietary protein imbalance during pregnancy. J Physiol January 15, 2004;554(2): 497–504.

[103] Jackson AA, Dunn RL, Marchand MC, Langley-Evans SC. Increased systolic blood pressure in rats induced by a maternal low-protein diet is reversed by dietary supplementation with glycine. Clin Sci (Lond) December 2002;103(6):633–9.

[104] Pham TD, MacLennan NK, Chiu CT, Laksana GS, Hsu JL, Lane RH. Uteroplacental insufficiency increases apoptosis and alters p53 gene methylation in the full-term IUGR rat kidney. Am J Physiol Regul Integr Comp Physiol November 2003;285(5):R962–70.

[105] Torrens C, Brawley L, Anthony FW, Dance CS, Dunn R, Jackson AA, et al. Folate supplementation during pregnancy improves offspring cardiovascular dysfunction induced by protein restriction. Hypertension May 1, 2006;47(5):982–7.

[106] Sinclair KD, Allegrucci C, Singh R, Gardner DS, Sebastian S, Bispham J, et al. DNA methylation, insulin resistance, and blood pressure in offspring determined by maternal periconceptional B vitamin and methionine status. Proc Natl Acad Sci December 4, 2007;104(49):19351–6.

[107] Bogdarina I, Welham S, King PJ, Burns SP, Clark AJL. Epigenetic modification of the renin-angiotensin system in the fetal programming of hypertension. Circ Res March 2, 2007;100(4):520–6.

[108] Bogdarina I, Haase A, Langley-Evans S, Clark AJL. Glucocorticoid effects on the programming of AT1b angiotensin receptor gene methylation and expression in the rat. PLoS One February 16, 2010;5(2): e9237.

[109] Zhang S, Morrison JL, Gill A, Rattanatray L, MacLaughlin SM, Kleemann D, et al. Dietary restriction in the periconceptional period in normal-weight or obese ewes results in increased abundance of angiotensin-converting enzyme (ACE) and angiotensin type 1 receptor (AT1R) in the absence of changes in ACE or AT1R methylation in the adrenal of the offspring. Reproduction November 1, 2013;146(5): 443–54.

[110] Chan Y, Fish JE, D'Abreo C, Lin S, Robb GB, Teichert AM, et al. The cell-specific expression of endothelial nitric-oxide synthase: a role for DNA methylation. J Biol Chem August 13, 2004;279(33):35087–100.

[111] Johnson AB, Denko N, Barton MC. Hypoxia induces a novel signature of chromatin modifications and global repression of transcription. Mutat Res April 2, 2008;640(1–2):174–9.

[112] Dasgupta C, Chen M, Zhang H, Yang S, Zhang L. Chronic hypoxia during gestation causes epigenetic repression of the estrogen receptor-α gene in ovine uterine arteries via heightened promoter methylation. Hypertension September 1, 2012;60(3):697–704.

[113] Chen M, Dasgupta C, Xiong F, Zhang L. Epigenetic upregulation of large-conductance Ca^{2+}-activated K^{+} channel expression in uterine vascular adaptation to pregnancy. Hypertension September 1, 2014;64(3):610–8.

[114] Lennartsson A, Ekwall K. Histone modification patterns and epigenetic codes. Biochim Biophys Acta 2009;1790(9):863–8.

[115] Lee HA, Cho HM, Lee DY, Kim KC, Han HS, Kim IK. Tissue-specific upregulation of angiotensin-converting enzyme 1 in spontaneously hypertensive rats through histone code modifications. Hypertension March 1, 2012;59(3):621–6.

[116] Woods LL, Weeks DA. Naturally occurring intrauterine growth retardation and adult blood pressure in rats. Pediatr Res November 2004;56(5):763–7.

[117] Saxena A, Carninci P. Long non-coding RNA modifies chromatin. Bioessays November 1, 2011;33(11):830–9.

CHAPTER

9

Developmental Effects of Endocrine-Disrupting Chemicals in the Ovary and on Female Fertility

Mehmet Uzumcu, Aparna Mahakali Zama

Department of Animal Sciences, Rutgers, The State University of New Jersey, New Brunswick, NJ, USA

OUTLINE

The Epigenome and Developmental Origins of Health and Disease
http://dx.doi.org/10.1016/B978-0-12-801383-0.00009-8

Copyright © 2016 Elsevier Inc. All rights reserved.

INTRODUCTION

Widespread use of synthetic chemicals in the environment, including endocrine-disrupting chemicals (EDCs), shows a strong parallelism with the increasing prevalence of female reproductive problems such as early puberty and subfertility. Girls in various countries throughout the world are attaining puberty at an increasingly earlier age [1]. It is hypothesized that chemicals in the developmental environment are *in part* responsible for this trend [1]. The impaired fecundity rate in the United States (US) increased from 11% to 15% between 1982 and 2002 [2]. Although lifestyle changes such as planned delay in having a family [3] and the obesity epidemic can contribute to this decline, the role of the environment is strongly suspected. (See Effects of Endocrine-Disrupting Chemicals on Female Reproduction and Ovary for specific human association studies.) Importantly, the role of EDCs in the obesity epidemic is also seriously considered [4]. In addition, although overall cancer incidence and mortality have declined in the United States, the increased incidence of childhood cancers is most likely related to environmental causes, including EDCs (President's Cancer Panel Report, 2010).

The developmental origins of health and disease (DOHaD) hypothesis was originally proposed by the late David Barker [5] based on his observations that undernutrition during mid- to late gestation in humans leads to cardiovascular and associated pathologies during adulthood [6] (described in Chapter 1). This concept, which applies to other environmental stressors, such as improper maternal care [7] and nutrition [8] as well as exposure to EDCs [9], suggests that the environment to which an organism is exposed during its critical period of development programs an organ, tissue, or specific anatomical structure in a permanent manner. This "programming" then leads to pathophysiological outcomes during adulthood if the developmental environment is not "typical."

The critical roles of endogenous hormones in the differentiation and development of organs, especially the brain and reproductive tract in both sexes (described in Chapter 5), are well known, which are in line with the DOHaD hypothesis. It has long been known that endogenous gonadal sex steroids (i.e., testosterone and estradiol) or lack thereof during a critical developmental window determines the direction of the differentiation in the brain, and subsequent physiology and behaviors [10]. Exposure to exogenous testosterone (or estradiol) interferes with the normal differentiation of the female brain and directs it to a male pathway, leading to display of male-like behaviors and a lack of regular reproductive cycles [11]. Similarly, developmental exposure to estrogens or estrogenic compounds also alters normal formation of the uterus and oviduct, in severe cases leading to infertility [9]. In addition, more recent studies have shown that developmental exposure to estrogenic EDCs causes serious adverse effects on ovarian follicular development, leading to reduced fertility or infertility [12].

Evidence is accumulating that developmental exposure to estradiol or estrogenic compounds is associated with epigenetic alterations [13–16]. Developmental exposure to adverse nutritional conditions during early development is also associated with epigenetic changes [17]. Epigenetic mechanisms include DNA methylation, histone modifications, chromatin rearrangement, and noncoding RNA. The details of these mechanisms and their roles in embryonic epigenetic programming are discussed in Chapter 4. However, it is important to point out here that the long-lasting consequences of exposure to adverse environmental conditions during critical developmental windows are believed to be mediated by alterations in epigenetic marks in relevant loci in the genome. There appear to be several critical windows during which the organism is especially vulnerable to epigenetic influence of environmental factors, which include periconception, fertilization, peri-implantation embryonic development, and placental development, which are discussed in Chapter 3.

It should be apparent from the above discussion that exposure to EDCs during development can affect all major components of the female reproductive system, namely hypothalamus and pituitary, reproductive tract (e.g., uterus and oviduct), and ovary. In this review, we will focus on the ovary as well as female pathologies that can be attributed to developmental exposure to EDCs, most of which are estrogenic, antiestrogenic, and/or antiandrogenic.

OVARIAN DEVELOPMENT AND FUNCTION

The main function of the ovary is the production, maturation, and release of female gametes (ova). The ovary is also responsible for the biosynthesis and secretion of hormones that regulate most systems in the body. These two functions are intertwined in the process of folliculogenesis. To better understand ovarian functions, we will summarize the major stages of its development and folliculogenesis, emphasizing those stages that are affected by estrogens and androgens and susceptible to alterations by EDCs. Although we will provide select examples, for a comprehensive listing of the molecular players (including transcription and paracrine factors), we refer the readers to a recent review [18].

Mice and rats have extensively been used to understand important aspects of ovarian development and function in humans. Although the fundamental aspects of development and function are similar between these rodents and larger mammals, including humans, there are at least two main differences in folliculogenesis that provide advantages for using rodent species for ovarian research: (1) formation of the primordial follicular pool that takes place in utero in humans occurs neonatally in mice and rats, which makes the rodent ovary more accessible for experimentation; and (2) early folliculogenesis in rodents is more homogeneous, that is, rodent ovaries have a relatively uniform follicle population at the early stages of development, which makes data interpretation simpler. As relates to the first difference mentioned above (i.e., in utero development in humans vs neonatal development in rodents), it is important to recognize that there can be differences in toxicodynamics and toxicokinetics of EDCs during in utero and neonatal life. Additional differences exist between rodent and primate ovaries. For example, the effects of estradiol on early follicular events are somewhat opposite between rodents and primates (see Oocyte Nest Breakdown and Primordial Follicle Formation and Primordial-to-Primary Follicular Transition for more details). In addition, the rodents are polyovulators and the humans are monoovulators. All of these differences should be taken into account during interspecies data extrapolation. The timeline of main events for human and mouse ovaries is presented in Table 1.

Primordial Germ Cells and Gonadal Differentiation

The embryonic gonad in mammals develops from the genital ridge and is composed of somatic and primordial germ cells (PGCs). While somatic cells are derived from the recruitment and differentiation of local tissues, namely coelomic epithelium and mesonephros, the PGCs originate from extraembryonic tissues around embryonic day (E) 7.25 in mice. At E9, primordial germ cells start to migrate via the hindgut to the bipotential gonad, where they rapidly proliferate and form germ cell nests. Also known as oocyte nests, germ cell nests are composed of synchronously dividing germ cells [19], surrounded by precursors of the pregranulosa cells. Following gonadal sex determination at E12.5, proliferating germ cells, now called oogonia, enter meiosis (E12.5–13.5) in an anterior–posterior wave-like manner, and progress until prophase

TABLE 1 The Timeline for Ovarian Developmental Events in Mice and Humans

Developmental Events:	PGC Specification	PGC Migration to Hindgut	Genital Ridge Formation	PGCs Arrive at Bipotential Gonad	Gonadal Sex Determination	Proliferation of PGCs/ Oogonia	PGCs Enter into Meiosis	Cyst Breakdown/ Primordial Follicle Formation	Follicular Growth	Puberty	Permanent Anestrus/ Menopause
Mouse (days):	E7.25	E9–10.5	E10.5	E10.5	E12.5	E10.5–13.5	E12.5 – birth	E17.5 – PND4	PND3 – death	~PND40	~18 mo.
Human (weeks):	E3–4	E4–5	E5	E6–8	E8	E7.5–20	E8 – 22	E17/20 – birth	E22 – menopause	~12 yr.	~51 yr.

PGC = primordial germ cells; E = embryonic or in utero day (mouse) weeks (human); postnatal day (PND).
References used for preparing the table: [217-224].

I and arrest in the diplotene stage [20]. Depending on the species, the germ cells, now called oocytes, can stay arrested for weeks or months (mice) or years or decades (humans) and resume meiosis during follicular growth at puberty or the adult life.

During oogenesis, one of the events that is considered susceptible to long-lasting effects of EDCs is the epigenetic reprogramming in the germ cells, the process that erases and reestablishes the DNA methylation patterns of the germ cell genome [21]. The germ cell genome that is normally methylated prior to migration to the genital ridge [22] is demethylated, including at imprinted loci following the migration. Subsequently, remethylation, which is mediated through an interaction between somatic and germ cells, occurs in a sex-specific manner at different developmental stages [23]. Therefore, the epigenetic effects of EDCs on this event can be sex specific. In mice, while male germ cell remethylation starts between E14.5 and 16.5 and is mostly complete by birth [24], female germ cell remethylation starts during the early postnatal period, postnatal day (PND) 1–5, and continues throughout the growth of the oocyte. The progress in imprinting process is based on the size of the oocyte rather than the age of the animal [25].

Oocyte Nest Breakdown and Primordial Follicle Formation and Primordial-to-Primary Follicular Transition

In mice, a majority of oocytes in a nest are eliminated via apoptosis starting at E17.5 [26], a process called oocyte nest breakdown. The remaining oocytes are surrounded by a single layer of flattened pregranulosa cells and form the primordial follicles. Primordial follicles start appearing postnatally, and this process is mostly completed by PND3-4. Most of the primordial follicles remain quiescent, but some begin growing and transition to the next stage, primary follicles. Both of these early processes, primordial follicle formation (also known as follicular assembly) and the primordial-to-primary follicle transition, are tightly regulated by interactions between paracrine factors, transcription factors, and steroid hormones. Because these processes determine the size of the follicular pool in the ovary, which in turn determines the female reproductive lifespan, their disturbance can lead to early depletion of the pool and therefore may result in early reproductive senescence.

The sudden drop in circulating estradiol levels following birth is believed to be one of the major factors permitting oocyte nest breakdown and primordial follicle formation in mice and rats. A similar mechanism (e.g., a temporary but dramatic drop in estradiol production between E80 and 90 just prior to initiation of follicle formation) may exist in cattle ovary [27]. Using in vivo and in vitro approaches, several studies have demonstrated the inhibitory role of estradiol in these processes [27–29]. Such inhibition by exogenous estradiol leads to multioocyte follicles (MOFs; presence of more than one oocyte in a single follicle) in the postnatal ovary [29]. However, the role of estrogens in follicular assembly in primates appears to be opposite of rodents and cattle (i.e., estradiol stimulates the process of follicular assembly in primate ovary) [30,31].

The primordial-to-primary follicle transition, which is also known as "initial recruitment" (as opposed to "cyclic recruitment"; for details, see below and [32]), is one of the most highly regulated stages of folliculogenesis. Once a primordial follicle is activated, the process does not stop or reverse. Therefore, excessive follicular activation can prematurely deplete the follicular reserve. The factors that are critical for this process include KIT ligand (KL) and its receptor, KIT [33]; granulosa cell-specific anti-Mullerian hormone (AMH) [34]; and oocyte-specific forkhead box O3 (FOXO3) transcription factor [35]. KL and its downstream PI3K-signaling pathway stimulate initial recruitment by phosphorylating and inhibiting FOXO3, which is the major inhibitor of follicular activation. Thus, elimination of FOXO3 in the mouse ovary releases this inhibition and leads to massive activation of the primordial follicles and early depletion of the follicular reserve [35]. The paracrine factor AMH also suppresses this process, and, similar to FOXO3, the deletion of the gene encoding AMH leads to increased recruitment of primordial follicles and premature reproductive senescence [34].

Inhibitory roles of estradiol (and progesterone) for the initial recruitment (similar to their roles in primordial follicle formation) have been shown, using in vitro and in vivo experimental approaches [28]. Interestingly, if endogenous estradiol biosynthesis is prevented by deleting the gene encoding the P450 aromatase enzyme, the number of primordial and primary follicles is reduced [36]. Thus, it appears that while estradiol is needed for early folliculogenesis, excessive estradiol inhibits follicular progression.

Androgens have dual roles in folliculogenesis, acting as ligands for the androgen receptor (AR) and as substrates for the aromatase enzyme, and may play a stimulatory role in the recruitment of primordial follicles. In neonatal mouse ovary culture, testosterone or dihydrotestosterone (nonaromatizable androgen) increased the number of primary follicles but reduced the number of secondary follicles, suggesting that androgens stimulate the primordial-to-primary follicle transition but inhibit follicle growth beyond the primary stage [37]. Androgens may accomplish this by inhibiting FOXO3 activity and the expression of growth differentiation factor 9 (GDF9), a well-known stimulator of follicle development beyond the primary stage. As a result, exposure to androgens causes an accumulation of preantral-stage follicles. In sum, initial recruitment is inhibited by estrogens and stimulated by androgens.

Preantral and Antral Follicle Growth and Follicular Recruitment, Selection, and Maturation or Atresia

Preantral follicle development includes the proliferation of granulosa cells to multiple layers, the growth of the oocyte in size, and the establishment of a second layer of somatic cells, the thecal cells [38]. Preantral growth is mediated primarily by oocyte-derived factors, most notably GDF9 and bone morphogenic protein 15 (BMP15). Additional factors that are involved in the regulation of preantral follicle development include inhibins, KL, and neurotropins [39]. Evidence from studies with juvenile rats suggests that gonadotropins, luteinizing hormone (LH), and follicle-stimulating hormone (FSH) have some roles during preantral follicle growth [40] and their receptors are expressed starting at PND7 [41]. Although the gonadotropins stimulate growth and differentiation of preantral follicles, follicles can develop up to the antral stage in the absence of gonadotropins [42].

During antral follicle growth, proliferation of granulosa cells intensifies, and the oocyte continues to grow. Once granulosa layer becomes about five layers thick, an antrum (fluid-filled space) begins to develop forming antral follicles [43]. The antrum gradually expands and helps create two groups of granulosa cells with different locations in reference to the oocyte and distinct roles: cumulus granulosa cells (closely associated with oocytes) and mural granulosa cells (lining the follicular wall) [44].

Cyclic recruitment of antral follicles is dependent on gonadotropin (primarily FSH) stimulation, which begins at puberty and continues throughout adult reproductive life. One or more follicles within the recruited cohort are selected to complete the full course of folliculogenesis and ovulation: one follicle in monoovulators (e.g., humans) or multiple follicles in polyovulators (e.g., mice) [32]. Each selected follicle produces higher levels of estradiol secretion, which is accomplished by cooperation of thecal cells and granulosa cells. Under LH regulation, thecal cells produce androgen, which is transported to neighboring granulosa cells, which convert the androgen into estrogens under the regulation of FSH. This further elevates serum estrogen levels, suppressing the gonadotropin secretion from the pituitary. In addition, antral follicles produce inhibin, which further suppresses FSH secretion. It appears that each selected follicle remains FSH responsive, due to higher FSHR expression, while the remaining (nonselected) follicles within the growing cohort are *FSH starved* due to fewer FSHR [32,45]. In effect, this condition of FSH withdrawal favors the onset of apoptosis in granulosa cells, ultimately leading to atresia of nonselected follicles. Fine-tuning of this process is mediated by local growth factors such as insulin-like growth factors (IGF), activins, transforming growth factor (TGF) α and β, hepatocyte growth factor, and fibroblast growth factor 7 [39].

Ovulation and Corpus Luteum Formation

In the preovulatory stage of folliculogenesis, estradiol production reaches its peak levels, which exerts a positive-feedback effect on gonadotropin secretion. The rise in FSH and LH levels stimulates meiotic maturation of the oocyte(s) that become(s) competent for fertilization [46]. Increased gonadotropins also initiate luteinization, whereby granulosa cells switch from the almost exclusive production of estradiol to the production of both estradiol and progesterone. This process upregulates multiple genes, including the LH receptor (LHR) in the granulosa cell, to prepare for the upcoming ovulation. The feedback dynamics within the hypothalamic-pituitary-gonadal (HPG) axis continue and culminate with the preovulatory LH surge that stimulates ovulation (reviewed in Ref. [44]). Elevated LH, and to a lesser extent FSH, stimulate the terminal differentiation of granulosa cells, leading to their further luteinization. Multiple factors play roles in ovulation, including ESR2, progesterone receptor (PGR), proteases, epidermal growth factor-like proteins, and prostaglandin synthase-2. Following ovulation, the remnants of the ovulated follicle are stimulated by LH to terminally differentiate into the corpus luteum (CL). The CL, as the primary source of progesterone, is essential for enabling the initiation and maintenance of pregnancy (reviewed in Ref. [47]).

Endocrine-disrupting Chemicals Can Affect the Ovary Several Ways

Actions of EDCs in the ovary can be mediated by sex-steroid-nuclear receptors ERs (ESR1 and ESR2) and AR. ESR1 and ESR2 are expressed during early folliculogenesis in the ovary in a cell- and stage-specific manner in several species, including humans [48], rats [49], and mice [50]. ESR1 is expressed primarily in theca cells, and ESR2 is expressed in granulosa cells. Gene-deletion studies in mice show that ESR2 is essential for FSH-directed granulosa cell differentiation and LH responsiveness [51]. Data also show that ESR2 plays roles in primordial follicle formation [52] as well as follicle maturation from the early antral to the preovulatory stages [53]. In contrast, although ESR1 plays critical roles in the regulation of theca cell steroidogenesis in the ovary, its main function is to mediate estrogen-regulated feedback in the hypothalamus and pituitary [54]. The androgen receptor is expressed in multiple ovarian cell types (granulosa cells, thecal cells, and oocytes) in several species, including rats [55] and humans [56,57]. The expression of AR starts with the primordial-stage follicles and its level is elevated in the preantral stage and starts declining as folliculogenesis progresses [58]. Although androgens modulate follicular development, they may not be essential for female fertility as it is seen in testicular feminized mice (Tfm), in which the AR is dysfunctional, due to frame-shift mutation in the AR mRNA, results in a stop codon in the amino terminus. Despite the lack of functional ARs, Tfm females remain fertile with reduced fecundity [59]. Similarly, *Ar*-null mice, although initially fertile with a similar ovarian histology as the wild type, reach reproductive senescence earlier due to progressive follicular loss [60].

In addition, EDCs can act on the ovary through other receptors, including peroxisome proliferator-activated receptors (PPARs) and aryl hydrocarbon receptor (Ahr), as these are expressed in the ovary. All three subtypes of PPAR (α, β/δ, and γ) are expressed in rodent ovaries [61]. In addition, human granulosa cells express PPARγ [62]. Aryl hydrocarbon receptor is also expressed in rodents [63] and human ovarian tissues [64]. Furthermore, EDCs can act on the ovary indirectly by altering serum levels of hormones that regulate the ovarian function, such as pituitary gonadotropins, which can be considered as nonreceptor-mediated actions of EDCs.

In addition to their receptor-mediated and nonreceptor-mediated actions, EDCs can affect the epigenetic mechanisms that govern the highly dynamic cellular differentiation processes that take place throughout the ovarian development as well as cyclic folliculogenesis during adulthood [65–67]. It is well known that developmental exposure to environmentally relevant levels of EDCs, some of which are too low to act on steroid receptors, can influence adult functions, including female reproduction. Alterations of epigenetic mechanisms may explain the irreversible long-lasting effects of exposure to such low doses of EDCs during early development [68]. However, these alterations should correlate with phenotypic alterations, including gene expression.

Reproductive Senescence

Reproductive aging, the gradual loss of the ability to reproduce, at the backdrop of general aging, has unique aspects and has been examined closely on its own (reviewed in Ref. [69–71]). The end of the reproductive life in women is marked by menopause (cessation of menstrual cycles), which typically occurs at the age of 51 [72]. Although menopause per se does not occur in rats and mice, persistent anestrus (cessation of estrous cycles, or acyclicity) typically occurs between 12 and 18 months of age [73–75], but in some strains occurs as early as 6–8 months. In women, there is a menopausal transitioning period, characterized by irregular cycles, including a shorter follicular phase and anovulatory cycles, as well as a decline in serum estradiol level and an increase in serum FSH level [69]. In rodents, a similar transition to acyclicity occurs that is characterized by prolonged cycles, which in some cases is followed persistent (vaginal) estrus and recurring pseudopregnancy, and an alteration in serum gonadotropin levels, leading to persistent diestrus or anestrous [71].

In humans (and other primates) and rodents, there is a significant decline in primordial follicular reserve in the ovary during reproductive aging, with a concomitant decline in the hypothalamic-pituitary function, including an increase in FSH and a reduction in the preovulatory LH surge. Whether the depletion of the follicular reserve or decline in neuroendocrine function is the primary cause of reproductive senescence has been debated, and the answer appears to be species-specific. While in humans the depletion of the ovarian reserve is the primary driving force, alterations in the neuroendocrine axis are the initial signs of transition to reproductive senescence in rodents [69,71]. However, there is a strong interplay and reciprocity between the changes in ovarian and neuroendocrine components during the transition to reproductive senescence. For example, removal of either system (e.g., hypophysectomy or ovariectomy) delays the aging of the other. It is important to note that there is significant individual variation in the age of reproductive senescence even within inbred strains, emphasizing the role of the environment [75].

There is a trend among US women to delay their first pregnancy [3]. In addition, studies show that their fecundity significantly declines starting at age 32 [76]. Combined with these scenarios is the possibility that developmental exposure to EDCs shortens the female reproductive lifespan [77]. Therefore, it becomes even more critical to understand and prevent the adverse effects of environmental EDCs on female reproduction.

EFFECTS OF ENDOCRINE-DISRUPTING CHEMICALS ON FEMALE REPRODUCTION AND OVARY

EDCs are synthetic and natural chemicals in the environment that can interfere with endocrine functions. The EDCs, some of which are described in Chapter 7, include pesticides, plasticizers, industrial by-products, pharmaceuticals, flame retardants, phytoestrogens, parabens, perfluorooctanoic acid, and heavy metals. In this section, we will describe the effects of several EDCs, namely phthalates, organochlorine pesticides (DDT, methoxychlor, and their metabolites), dioxins, and diethylstilbestrol (DES). We will refer the reader to other recent reviews about another well-studied EDC, bisphenol A (BPA) [78,79], even though it is well known for its adverse effects in the ovary of multiple species [80–83].

For each chemical, we will first review epidemiological studies examining associations between potential exposures and female reproductive abnormalities, and then we will summarize in vivo studies with laboratory animals and related mechanistic studies (primarily in vitro approaches).

Phthalates

Phthalates are alkyl esters of *o*-phthalic acid and are named based on their alkyl chains. According to their alkyl chain length, phthalates can be divided into two general categories: (1) long-chain or high-molecular-weight phthalates (high-MWPs; e.g., di(2-ethylhexyl) phthalate [DEHP]); and (2) short-chain or low-MWPs (e.g., diethyl phthalate [DEP] or dibutyl phthalate [DBP]). High-MWPs are used to make polyvinylchloride (PVC) more flexible and are found in building and construction materials, cables and wires, car interiors, toys, and food containers [84]. Low-MWPs are usually found in non-PVC products such as personal care products, paints, and enteric-coated tablets [84]. Both groups of phthalates are noncovalently attached to the materials in which they are used and are readily released to the environment. Therefore, they are very common in the environment, and humans are exposed to them extensively. Among the high-MWPs, DEHP has been one of the most commonly used phthalates for many years. The reference dose (RfD) and tolerable daily intake (TDI) for DEHP are 20 and 50 μg/kg/day, respectively [85]. The daily adult human exposure to DEHP ranges from 3 to 30 μg/kg/day, and infant and children are exposed to several times higher doses [86]. In fact, DEHP and its metabolites have been found in breast milk, serum, amniotic fluid, sweat, and in urine of mothers and infants [87–90]. One of the most vulnerable populations is infants in neonatal intensive care units, or NICUs, whose daily exposure reaches 22.6 mg/kg [91]. Thus, the developmental exposure to DEHP is of special concern. A separate but important issue is related to the analytical chemistry of phthalates. Since phthalates are ubiquitous in the environment, if a mother compound (e.g., DEHP) were measured in a biological specimen, this could present a problem as the measurements can easily be tempered with environmental sample contamination. Therefore, its metabolites (in this case metabolites of DEHP), rather than the DEHP itself, must be measured to have an accurate assessment.

Phthalates—Epidemiological Studies

For more than a decade, researchers have studied the correlation between concentration of phthalates (or their metabolites) in various body fluids and various female reproductive disorders, including early puberty, infertility/subfertility, and adverse pregnancy outcomes [84].

The results of studies examining the effect of phthalates on pubertal age are mixed. Two studies, one in Puerto Rico [92] and one in Taiwan [93], found higher levels of phthalates or phthalate metabolites in the serum or urine of early thelarche or precocious pubertal patients. In contrast, two separate groups, one in Denmark [94] and one in the United States [95], found that urinary levels of phthalate metabolites are significantly correlated with delayed pubarche, which is attributed to the antiandrogenic effects of phthalates. The US group confirmed their initial findings with a longitudinal study that followed the same girls over 7 years and found that highest quintiles reached the same pubertal level at least 8.5 months later than the lowest quintiles [96]. Still, some studies found no correlation between phthalates and pubertal age. There was no significant difference in the concentrations of nine phthalates in the urine of girls ($n=28$) with central precocious puberty and age- and race-matched normal girls [97]. All of the aforementioned studies investigated phthalate concentrations close to the time of the measurement of the pubertal parameters. In contrast, a study examined potential long-term effects in adolescents of exposure to large quantities of DEHP as neonates undergoing extracorporeal membrane oxygenation and found no significant effects on physical growth and pubertal maturity [98]. The lack of effect can be due to small sample size (six females). To resolve whether phthalate exposure alters the pubertal age in humans, longitudinal studies involving larger cohorts are needed.

Effects of phthalates on female fertility parameters, such as time to pregnancy (TTP) and loss of pregnancy, were studied. In a self-reporting study, Burdorf and coworkers found that exposure to phthalates was significantly associated with prolonged TTP [99]. In a prospective study that examined the fecundity score of 501 couples as a function of TTP, males', but not females' phthalate exposure was associated with an approximately 20% reduction in fecundity, underscoring the importance of assessing exposure of both partners to minimize erroneous conclusions [100]. A limited number of studies investigated the association between the loss of pregnancy and exposure to phthalates. A prospective fecundity study that measured the metabolites of various phthalates (e.g., DEP, DBP, DEHP) in urine 10 days after the first day of the last menstrual period before conception found that the average level of exposure to mono(2-ethylhexyl)phthalate (MEHP)—a more active metabolite of DEHP—was significantly higher ($P<0.01$) in women who experienced a pregnancy loss than those who did not (23.4 ng/ml [range < level of detection [LOD] – 84.0]) vs 16.2 ng/ml (range < LOD – 64.0); other phthalate metabolites did not differ significantly [101].

The association between phthalate exposure and preterm birth and gestational age in humans has been studied with mixed results. Some of the earlier studies examining the relationship between increased levels of phthalates in maternal urine or cord blood and preterm birth found a positive correlation [102–104], while other studies found a negative correlation [105,106] or no correlation at all [99]. Conflicting results can be due to various reasons, such as differences in the population under investigation or methodology. For example, these studies measured the phthalates or metabolites only one time in the urine during the pregnancy or relied on self-reporting for the exposure [99]. In contrast, more recent longitudinal studies that measured phthalate metabolites multiple times, either in maternal urine throughout the pregnancy (three to four times) [107,108] or in the urine of infants (nine times during the first 14 months of the life) [109], found a significant correlation between phthalate levels and preterm birth. Although results from studies differ, overall they suggest that exposure to phthalates can result in preterm birth.

Phthalates—Animal Studies

In this section, we will focus on DEHP, along with it metabolite MEHP, because it is one of the most commonly used and studied phthalate esters. Animals that are exposed to various doses of DEHP (from 20 μg/kg/day to 2 g/kg/day) during various times in their lives, including peripubertal period and/or adulthood, show adverse effects in multiple reproductive parameters, including pubertal age, estrous cyclicity, and litter size as well as alterations in serum hormone levels and ovarian morphology [110–113]. Transient daily oral exposure to 2 g/kg of DEHP in female rats results in prolonged estrous cycle and alterations in natural ovulation time that often results in no ovulation and hence absence of CL. In addition, DEHP-exposed animals have suppressed levels of estradiol, progesterone, and LH. The primary cause of these disruptions appears to be the low levels of estradiol, insufficient to induce the preovulatory LH surge [113,114]. Reduced serum estradiol and progesterone levels have also been reported following prepubertal exposure to 500 mg/kg/day DEHP for 10 days [115].

Early in vitro studies elucidated the molecular basis of disrupted ovarian steroidogenesis [114]. Studies with cultured rat granulosa cells suggest that MEHP inhibits FSH-stimulated cyclic adenosine monophophate (cAMP) production, thereby preventing activation of the enzymes for progesterone production, and suppresses levels of aromatase mRNA (*Cyp19a1*) levels. MEHP appears to be acting also via activation of PPARs, involving both PPARα and PPARγ [114]. The role of PPAR signaling was recently confirmed in an in vivo three-generation study in mice [116]. The ovarian effects of exposure to ~80 mg/kg/day DEHP (via feed) were present only in wild-type mice, but not in mice lacking the *Ppara* gene, suggesting that PPARs play roles in mediating actions of DEHP in the ovary. Studies with exposure to DEHP as low as 0.05 mg/kg/day during in utero and postnatal life (E0.5 to PND21) resulted in reduced expression of *Cyp17a1*, *Cyp19a1*, *Pgr*, *Lhcgr*, and *Fshr* in the adult ovary (PND42) of the CD-1 mice, all of which may affect ovarian steroidogenesis [117]. In addition, the same treatment led to increased ovarian weight, impaired maturation, and developmental ability of oocytes, suggesting problems with follicular dynamics.

More recent studies investigating the developmental effects of DEHP or MEHP suggest that DEHP can affect all major stages of folliculogenesis, including primordial follicle formation [118], primordial to the primary follicle transition [77,112], and follicular maturation [119], as well as ovulation and CL function [113] (see also Figures 1–3). Zhang and coworkers using neonatal mouse ovary cultures showed that

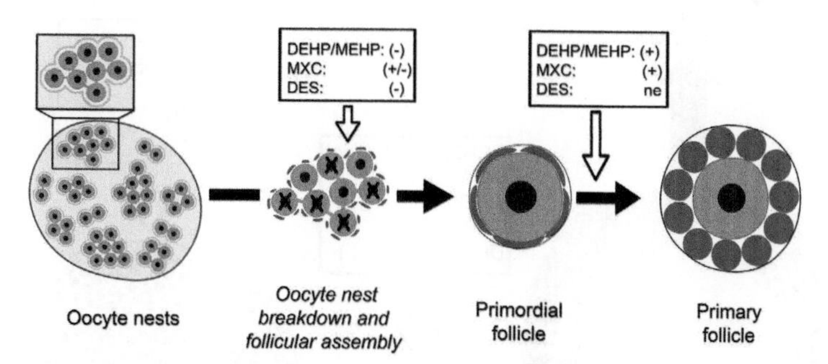

FIGURE 1 **EDCs affect oocyte nest breakdown, primordial follicle formation, and/or primordial-to-primary follicle transition.** In female embryonic gonad, primordial germ cells proliferate and form oocyte nests. Starting 17.5 in mice, majority of oocytes in each nest dies via apoptosis, a process called oocyte nest breakdown. The remaining oocytes are surrounded by precursor of granulosa cells (GCs, red), forming primordial follicles, a process also known as follicular assembly. These processes can be affected by DEHP [118], MXC [14], and DES [12,202–205]. Following their formation, which is complete by PND3–4 in mice, while most primordial follicles remain quiescent, waiting for their recruitment for growth for the duration of the female's reproductive life, some immediately transition into primary follicular stage via a process known as follicular activation. The follicular activation, which is characterized by growth of oocyte and GC in size and proliferation of GC, is affected by DEHP [112] and MXC [14]. For Figures 1–3, (−) = inhibits; (+/−) = mixed results (in some studies or doses simulates and in some inhibits); (+) = stimulates; (ne) = no effect observed in reported studies.

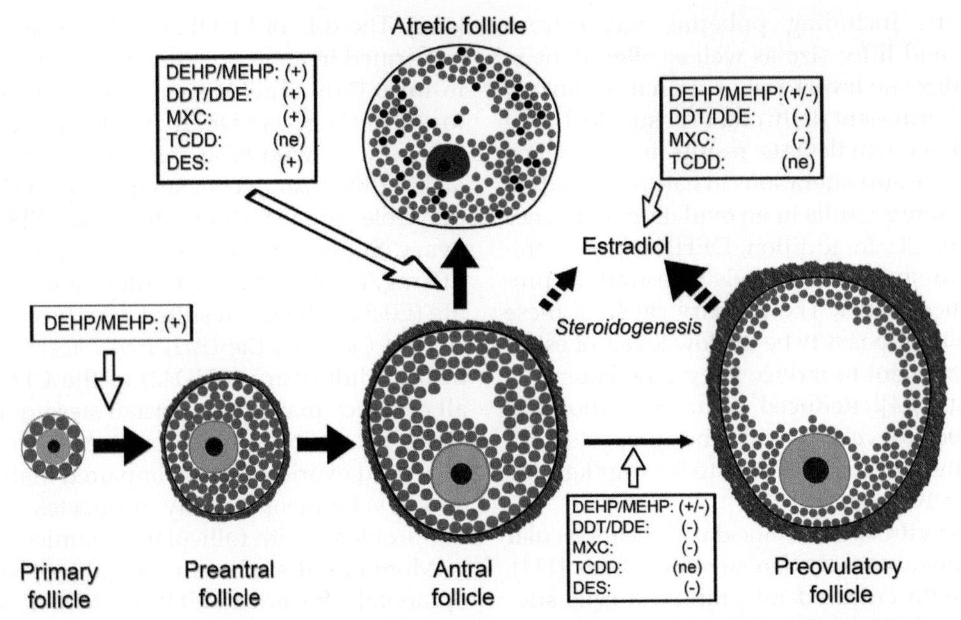

FIGURE 2 **EDCs affect preantral and antral follicle growth and follicular selection and maturation.** Primary follicle undergoes rapid growth, initially stimulated by local growth factors. Its growth is also stimulated by gonadotropins, especially after antral follicular stage. These processes can be affected by DEHP [77,112]. With the involvement of second cell type (thecal cells; tan color), postprimary follicular stage, the follicle begins producing progressively increased amount of estradiol via steroidogenic pathway, a process that is affected by DEHP [113,214], DDT/DDE [149,150], and MXC [163]. At antral follicle stage, selection of follicles that are destined to further growth and maturation takes place, which can be affected by DEHP [77,119] and MXC [163]. Unselected follicles, which constitute 99% of follicles, undergo atresia, a process that is induced by EDCs such as DEHP [121] and MXC [215].

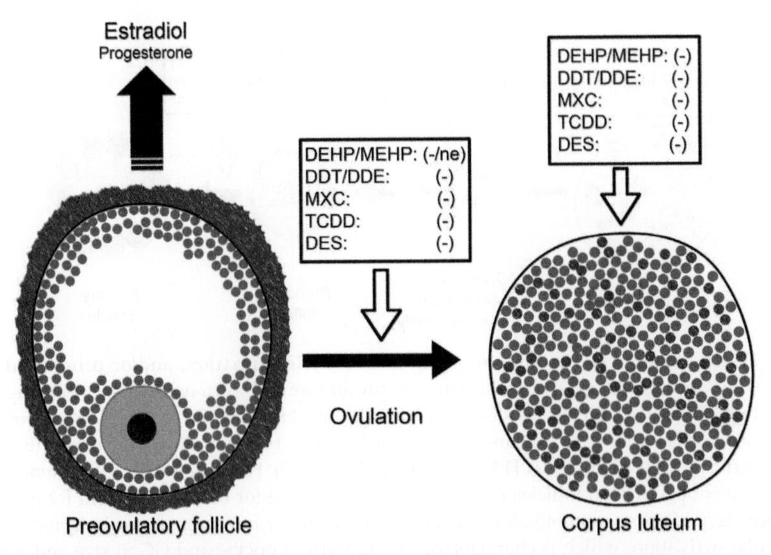

FIGURE 3 **EDCs affect the ovulation and corpus luteum formation.** The preovulatory follicle secretes even larger quantities of estradiol (and also begins secreting progesterone), which can be altered by EDCs as depicted in Figure 2 (not shown here). Following ovulation, a process that is adversely affected by all of the EDCs described in this chapter (DEHP [77,113], MXC [163], dioxin [180,182], DDT [145,146], and DES [199]), remaining somatic cells, GC (red), and thecal cells (tan color) form corpus luteum. Corpus luteum functions can also be adversely affected by DEHP [216], MXC [163], dioxin [180,182], DDT [145,146], and DES [199].

10 and 100 μM DEHP increase the number of oocyte nests while reducing the number of follicles, suggesting that DEHP impairs primordial follicle formation [118]. The same study also showed that DEHP exposure inhibits the expression of multiple oocyte-specific transcription factors including *Figla*, a key molecule in follicular assembly.

In addition, DEHP appears to accelerate the recruitment of primordial follicles into the growing follicle pool. Early postnatal (PND5-20) exposure of mice to relatively low levels of DEHP (20 or 40 μg/kg/day) depletes primordial follicles while increasing the number of secondary and antral follicles [112], which is associated with altered patterns of imprinted genes and increased metaphase II spindle abnormalities. Follicular recruitment was similarly accelerated in adult mice that were transiently (10–30 days) exposed to DEHP (200 μg/kg to 700 mg/kg), which was associated with alterations of the PI3K-signaling pathway [120]. It is possible that the impacts on primordial follicles can be indirect through the decline in estradiol concentrations, as well as direct. It is also important to note that although accelerated recruitment reduces the number of primordial follicles and increases the secondary and early antral follicular pools [112], further growth and maturation may be impaired. Thus, DEHP-exposed antral follicles, rather than going through the final stages of maturation and ovulation, undergo atresia [121,122]. Using antral follicle cultures, it was shown that DEHP inhibits follicular growth, primarily due to increased oxidative stress, which was prevented by *N*-acetyl cysteine, an antioxidant chemical [119]. DEHP treatment also increases granulosa cell apoptosis, which is supported by in vivo studies that investigated prolonged exposure (16 weeks; 6 days/week) via oral gavage to 125–2000 mg/kg/day of adult ICR mice. Granulosa cells from exposed mice show reduced cell growth and increased apoptosis [123]. Impairments in folliculogenesis that are described above can cause early exhaustion of follicles, leading to early reproductive senescence. Indeed, mice exposed to 100–1000 mg/kg/day of MEHP between E17 and E19 had a 1-month-shorter reproductive life as well as delayed puberty and altered estrous cyclicity [77].

Organochlorine Pesticides

DDT and Its Metabolites

DDT (1,1,1-trichloro-2,2-bis[4-chlorophenyl] ethane) is an organochlorine pesticide that was used extensively in the United States from the 1940s to 1970s, particularly as an aerial spray to control populations of mosquitoes and insects. It was the first EDC that was shown, as far back as the 1960s, to have detrimental effects on human and animal health [124]. Many studies have since shown that DDT is highly effective in bioaccumulating in lipids and can cause widespread deleterious effects on multiple organ systems. Even though DDT has been banned in the United States since the 1970s, it is still heavily used in some parts of the world for malaria control [125], making possible cross-contamination across large geographical areas.

DDT—EPIDEMIOLOGICAL STUDIES

DDT is dechlorinated during metabolism to 1,1-dichloro-2,2-bis(4-chloro-phenyl)ethylene (DDE), an antiandrogenic compound. DDT is also found in mixtures of its isomers, *p,p′-DDT*, *o,p′-DDT*, and *o,o′*-DDT, which are predominantly estrogenic [126]. These three isomers and DDE have been shown to have endocrine-disrupting effects. DDE has been detected in multiple tissues of humans and was present in 90% of the population in North America many years after DDT was banned [127]. The DDT isomers and/or DDE have been found in serum, follicular fluid, as well as umbilical cord samples [128]. Interestingly, age, body mass index, education, and income levels of pregnant women were positively correlated with their intake of beef, poultry, and seafood and also positively correlated with higher *p,p′-DDE* levels [129–132]. Although DDT use is restricted in most countries and overall values are declining (at least in some parts of the world), the DDT isomers and DDE continue to be a public health problem because of their presence in human tissues [133].

In a study of Chinese textile workers, serum DDT levels were associated with early menarche (~1 year, 10 ng/l increments = 0.20 years) and shortened menstrual cycle (<21 days) [134]. A similar study spanning two generations of women (1973–1991) using data from two cohorts in a Michigan fisherman population showed a correlation between in utero DDE levels in the serum and early menarche (15 μg/l increments = 1 year earlier menarche). The same study measured other persistent organic pollutants such as polychlorinated biphenyls (PCBs) and found no association with onset of puberty [135].

DDE has been found in human follicular fluid and has been correlated with failure in IVF procedures [136] and with an increased incidence of spontaneous abortions [137]. In some instances, DDE and DDT isomers were found in follicular fluid but were not associated with any detrimental effects in pregnancy [138].

A population of Latina agricultural workers residing in California had preterm births associated with increased levels of DDT isomers and DDE in serum [139]. In the same population, it was found that the nonpregnant serum levels of these women were much higher than the average population in the United States [139]. Maternal serum and umbilical cord blood levels of DDT and/or DDE have been related, in some studies, to preterm birth [140], decreased birth weight [141], or intrauterine growth retardation [142]. The study by Longnecker and colleagues had the largest population size (2380 children), among the highest DDE levels, and found a dose–response relationship for preterm delivery and small for gestational age (SGA) [140]. Presence of *p,p′*-DDT in the mothers' serum 1–3 days after their daughters' birth is associated with a longer time to pregnancy, a reduced probability of pregnancy, and high infertility [143].

DDT—ANIMAL STUDIES

In vivo studies have shown that *o,p′*-DDT has estrogen-like activity and causes disruption in follicular composition, fewer ovulations, and smaller litter sizes in rats. It was also associated with follicular cysts akin to polycystic ovary syndrome (PCOS) [144]. The *o,p′*-isomer of DDT, which comprises approximately 20% of technical-grade DDT, is several times more uterotrophic than *p,p′*-DDT. Furthermore, in utero exposure to a 10 mg *o,p′*-DDT dose between E15 and E19 in rats induced delayed puberty and persistent estrus with associated anovulations [145]. A more controlled study recapitulated these findings, although it was found that the higher dose in the studies (150 ppm) induced PCOS-like cysts but not the 75-ppm dose, which is more related to typical DDT exposure levels [146].

In vitro studies have shown that DDE has effects on progesterone synthesis in porcine granulosa cells—its level was inversely proportional to the dosage of DDE (10 ng/ml vs 0.3–320 μg/ml) used [147,148]. On the other hand, *p,p′*-DDT decreased estradiol production while DDE and *o,p′*-DDT increased estradiol production (an estrogenic response). Both ESR2 and CYP19A1 levels were affected by these isomers. In addition, it was found that members of the steroidogenic pathway, especially Cyp11a1, were downregulated when progesterone and estradiol levels were reduced [149].

Methoxychlor

Methoxychlor (MXC; 1,1,1-trichloro-2,2-bis(4-methoxyphenyl)) is a well-studied organochlorine pesticide that is used as a replacement for DDT. It is restricted in use in the United States but is used in other parts of the world. It is an estrogenic compound that demonstrates low-affinity binding for estrogen receptors [150]. The major MXC metabolites, HPTE and mono-OH MXC, can function as estrogenic, antiestrogenic, or antiandrogenic compounds [151], therefore it has been used as a model compound for commonly occurring EDCs in the environment [152].

MXC—EPIDEMIOLOGICAL STUDIES

There is a strong association between developmental exposure to MXC and underdeveloped fetuses and subsequent female fertility problems [153]. A two- to three-fold increase in risk of prolonged TTP and spontaneous abortion among female greenhouse workers [154] and increased infertility in women with agricultural work histories that involved exposures to organochlorine pesticides have also been reported [155].

MXC—ANIMAL STUDIES

Adverse effects that were observed in these association studies are similar to the effects observed in experimental animals exposed to MXC during adulthood. Exposure of rats to MXC (2500 or 5000 ppm) interfered with their normal estrous cycle and reduced mating rate and litter size [156]. However, when the exposure was withdrawn, these animals reverted to regular estrous cycles. The reversibility of the effects was also suggested in a mice study, which reported that immature follicles appeared not to be affected by 2–4 weeks of MXC (~50–200 mg/kg/day) exposure via oral gavage while larger follicles were atretic [157]. In addition, studies demonstrated that adult mice and rats that were exposed to MXC showed persistent vaginal estrus [157], direct inhibition of embryonic growth, implantation failure [158], pregnancy loss [159], and ovarian atrophy due to inhibition of folliculogenesis leading to atretic follicles, and reduced ovulation and fewer CLs [157]. It was later shown that MXC exposure in adult mice selectively affects the antral follicles and induces atresia using the Bcl2/Bax signaling pathway, without affecting the HPG axis [160].

In contrast, when rats were exposed to MXC during in utero and early postnatal development, the effects lasted into adulthood with more severe outcomes on reproductive parameters. These included acceleration of the vaginal opening (sign of puberty), acceleration of the onset of the first estrus, irregular cycles with persistent vaginal estrus, reduced pregnancy rate and litter size despite apparent mating, and early reproductive senescence [161,162]. Serum estradiol and progesterone levels were altered with increased FSH levels [162]. The effects on the ovary itself were dramatic, with both folliculogenesis and ovulation being inhibited.

In a more recent study in which female rats were treated during fetal and neonatal development (E19-PND7) with a dose of MXC that is comparable to the dose used in the above studies (100 mg/kg/day), the exposed females displayed similar abnormalities in reproductive parameters and ovarian morphology by adulthood [163]. Follicular composition analysis showed that developmental MXC treatment did not affect the total number of follicles in adult females or the number of follicles at primary and secondary stages. However, the number of preantral and early antral follicles was increased and the number of CLs was reduced, with numerous large cystic follicles. Immunohistochemical staining and quantification of expression patterns of important regulators of ovarian functions revealed that while LHR, CYP11A1, and CYP19A1 levels were reduced, levels of AMH and AR were increased and levels of steroidogenic acute regulatory protein and ESR1 were unchanged [163]. Especially noteworthy was that the ESR2 levels were unchanged in both primary and secondary follicles but decreased dramatically in periantral stage follicles, which are responsive to gonadotropins. These observations suggest that gonadotropin-dependent follicles are most affected by EDC exposure.

Epigenetic analyses have shown that DNA methylation levels are altered in an age- and treatment-dependent manner, which has also been shown in the uterine and other tissues with other EDCs (e.g., DES, genistein) [14,164]. A more recent targeted genome-wide DNA methylation array study has revealed that members of essential signaling pathways are hypermethylated and their gene expression is downregulated in

MXC-treated ovaries. IGF-1 signaling was the most significantly affected pathway wherein several members of the family—*Igf1r*, *insulin receptor* (*Insr*), *Pik3r1*, *Hras*, and *Foxo3*—were hypermethylated [165]. These data suggested that the initial DNA methylation patterns soon after exposure were representative of the gene expression patterns responsive to the EDC exposure and not the adult hypermethylation events. Furthermore, the long-lasting effects observed by adulthood could be due to histone modifications.

Dioxins

Polychlorinated aromatic hydrocarbons such as dioxins and PCBs are usually industrial effluents, by-products of herbicide production, chlorination, and combustion such as brush burning. They are ubiquitous in the environment and tend to bioaccumulate with detrimental effects on both plants and animals. In fact, the public health impact of dioxins may rival the impact that DDT had on public health in the 1960s. According to an EPA report, the lowest-observed-adverse-effect level or LOAEL is 20 ng/kg, but levels of dioxin and dioxin-like chemicals have been found in the general US population that are "at or near levels associated with adverse health effects" [166]. Some groups of people may be exposed to higher levels of dioxins because of their diet (e.g., high consumers of fish in certain parts of the world) or their occupation (e.g., workers in the pulp and paper industry, in incineration plants, and at hazardous waste sites); others may be exposed in industrial accidents or war (http://www.epa.gov/ncea/pdfs/dioxin/nas-review/).

Dioxin—Epidemiological Studies

2,3,7,8-Tetrachlorodibenzo-*p*-dioxin (TCDD) is a potent EDC and the most toxic of all the known dioxins. Exposure to TCDD in adults causes obvious symptoms such as skin lesions and chloracne as exemplified by the 2004 dioxin poisoning case of Viktor Yushchenko, former president of Ukraine. Long-term exposure is linked to impairment of the immune system, the liver, the developing nervous system, the endocrine system, and reproductive functions. One of the most notable incidents of population exposure to TCDD is that of Seveso, Italy, where a large industrial accident in 1976 caused release of dioxins approximately over seven square miles of surrounding area. Subsequent to the accident, a women's health study was set up and the health status of the women exposed in the accident was followed over many years. Findings of the study have revealed that there have been long-term detrimental effects on the women as well as their offspring, as discussed below [167,168].

Another egregious use of dioxins has been Agent Orange—an herbicide used as a defoliant during the Vietnam War. It consisted of a 50:50 mixture by weight of the *n*-butyl esters of two phenoxy acids: 2,4-dichlorophenoxyacetic acid (2,4-D) and 2,4,5-trichlorophenoxyacetic acid (2,4,5-T). A synthetic contaminant of 2,4,5-T is TCDD [169]. Levels of TCDD contamination in Agent Orange ranged from less than 0.05 to almost 50 ppm, with a mean value of about 2 ppm [170]. Millions of pounds of dioxin were sprayed in Vietnam over a 6-year period [169] with very serious consequences to the Vietnamese as well as the war veterans involved in the handling of Agent Orange.

Epidemiological studies have shown that there is a negative correlation between the levels of serum dioxins and pubertal age—higher levels of prenatal exposure to dioxin-like compounds were associated with delayed thelarche and menarche [171]. The Seveso Women's Health study showed mixed results as to the effects of dioxins on menarche. Depending on the age at exposure, girls had an earlier menarche, although there was not a significant correlation between the onset of menarche and serum dioxin levels in girls who were premenarcheal at the time of the accident.

Interestingly, the adult women in these studies had an increased incidence of endometriosis, breast and uterine cancers, and other conditions [167,168,172].

A cytogenetic analysis of maternal, placental, and aborted (induced) fetal tissue from Seveso [173] noted a higher frequency of chromosomal aberrations in fetal tissue from TCDD-exposed women than from the unexposed comparison group.

Several studies of spontaneous abortion have been conducted by Vietnamese researchers reviewed in Ref. [174], which reported an increase in spontaneous abortions among people living in Agent Orange–sprayed areas.

Other studies conducted in Vietnam have examined the risk for hydatidiform moles and found them to be of higher incidence. This disorder, associated with the death of a fetus, produces a mass of degenerated placental tissue in the uterus. It is benign but has been considered a precursor of choriocarcinoma, a malignant tumor of embryonic tissue [174].

Dioxin—Animal Studies

Numerous cancers have been associated with TCDD exposures. The LD50 for TCDD is 0.04 mg/kg for rats, while for other dioxins the LD50 values are as high as 100 mg/kg [175]. Interestingly, rats are more sensitive to TCDD reproductive toxicity than mice [176].

TCDD has been found to have antiestrogenic effects as an AhR agonist and has direct effects on the ovaries in rodent studies. Its toxic effects are mediated via the AhR. The AhR is complexed with two heat shock proteins and X-associated protein 2 (XAP2) in the cytoplasm, and binding to TCDD causes translocation to the nucleus where XAP2 and the heat shock proteins are released. The ligand-AhR complex interacts with the aryl hydrocarbon nuclear translocator (ARNT), and the heterodimer binds to DNA to transactivate or repress target genes, thus leading to various cell signaling events [177].

Exposure of adult rats to TCDD either for very short periods (just prior to a pregnant mare serum gonadotropin [PMSG] treatment) or for prolonged periods (weeks) caused lower ovarian weights, anovulations, and disruption of follicular maturation [178,179]. Furthermore, there was a loss of estrous cyclicity. The dioxin-treated ovaries were also found to have fewer CLs and a number of antral follicles that did not progress to ovulation [180]. Interestingly, this anovulatory effect occurred even in hypophysectomized rats that were primed with PMSG, suggesting that there are direct effects on the ovary. It was also found that steroidogenesis per se was unaffected in these animals and that the TCDD-induced defects were primarily in the follicle maturation and/or ovulation processes [181]. In addition, TCDD-exposed rats underwent premature reproductive senescence [182].

Studies with exposure to TCDD in utero have demonstrated that the DOHaD concept applies to ovary reprogramming. Doses as low as 1 μg/kg of TCDD given to rats, in utero, at E8 or E15 caused long-term disruption in reproductive function. However, certain rat strains such as Long Evans and Sprague–Dawley [183] were more responsive to TCDD than the Holtzman strain. In the affected strains, adult or aged animals had a significant disruption in estrous cyclicity (persistent estrus or reduced number of days spent in estrus), reduction in ovarian weights, and reduced number of litters. Importantly, the follicular composition of TCDD-exposed animals was not altered yet the number of ovulations decreased. Exposure at E15 led to reduced ovarian weights and more ovarian neoplasms by PND70 than exposure at E8 in rats [184]. Cystic follicles were also increased in number. Therefore, the timing of the in utero exposure was the major determinant in the observed abnormalities in adult and aged animals, suggesting that one of the more sensitive periods in the ovary for dioxin exposure is during or right after gonadal sex differentiation.

Diethylstilbestrol

DES is a nonsteroidal synthetic estrogen that was prescribed from the 1940s to the 1970s to pregnant women at doses of 5–150 mg/day to prevent miscarriages [185]. Even though early on DES was shown to be largely ineffective, its use continued until the 1970s [186]. Given that DES is a potent estrogen, numerous abnormalities in the reproductive, cardiovascular, and immune systems have since been reported in both male and female offspring of women treated with DES; similar effects appeared in animal models [187].

It is well known that DES caused T-shaped uteri and clear cell adenocarcinoma of the uterus, cervix, and vagina in women whose mothers were exposed to DES during pregnancy [188]. There are numerous animal studies validating these human reports. For example, progeny of DES-treated mice have shown malformations of the uterus, squamous metaplasia of the luminar and glandular epithelium, endometrial hyperplasia and leiomyomas, and oviductal proliferative lesions [189,190]. Over 95% of the DES-exposed women have reported various problems with other reproductive functions as well.

DES—Epidemiological Studies

Some studies have shown that menarche is achieved at a similar age by DES-exposed offspring as unexposed girls (~12.6 years of age). However, menstrual regularization took significantly longer (16.2 years vs 15.8 years) [191]. In addition, compared to unexposed women, DES-exposed women had greater difficulty in achieving pregnancy, which could be associated with both primary and secondary infertility since DES-exposed women have more frequent uterine or tubal problems [192]. Furthermore, DES exposure was associated with an increase in preterm birth and a higher risk of SGA. Both first- and second-trimester miscarriages were also more common in DES-exposed women compared to controls, and DES-exposed women had at least one ectopic pregnancy more than unexposed women, suggesting that uterine abnormalities occurred [191]. These studies provide evidence that prenatal DES exposure is associated with fetal growth and gestational length, which may mediate associations between DES and other health outcomes in later life.

There are limited reports that these effects are being observed in the granddaughters of DES-treated women as well [194]. Since there is evidence of multigenerational effects, epigenetic mechanisms could play an important role and are therefore being further investigated [195,196].

DES—Animal Studies

Mice injected with a single dose of 10 μg/kg DES on E15 and examined at 7 months of age had no CL and numerous atretic follicles [197]. Other studies with varying doses of DES (5–100 μg/kg), which were administered either in utero (E9–E16) [198] or neonatally (PND1–PND5) [199], demonstrated that adult DES ovaries developed similar hypertrophy and vacuolation of interstitial tissue, hemorrhagic cysts, and lack of CL. These animals also had high levels of testosterone [198]. There was a dose-dependent reduction in the number of litters as well as the number of oocytes ovulated after stimulation with exogenous gonadotropins [200]. The oocytes derived from DES-treated ovaries showed lower levels of fertilizability during IVF, suggesting reduced oocyte quality [201].

DES can bind to both ESRs with more than two-fold higher affinities than estradiol [52]. Multiple studies from Iguchi and colleagues showed that in utero (E15–18) and neonatally (PND1–5) DES-treated mice had ovaries containing excessive number of multioocyte follicles by adulthood [12,201]. MOFs were also observed in ovaries that were treated in vitro at PND1–5, following their transplantation to untreated mice, suggesting a direct effect of DES in the ovary [12]. Recent studies showed that neonatal exposure to 3 μg/kg DES induced MOFs, a process mediated by ESR2 but not ESR1 [203]. DES exposure reduced oocyte apoptosis (potentially suppressing oocyte nest breakdown) via ESR2 signaling mechanisms. Furthermore, it has been hypothesized that such alterations in the germ cell and somatic cell populations may affect the invasion of pregranulosa cells and basement membrane remodeling during primordial follicle formation [204]. Interestingly, the incidence of MOFs has also been reported after exposures to other estrogenic EDC such as genistein and BPA [205].

Ovariectomized animals when supplemented with estradiol are able to respond with a transient increase in gene expression and concomitant uterine proliferation and growth [206,207]. When such a stimulus is removed, the uterus returns to its unstimulated state. However, when DES or estradiol is administered during neonatal development, expression of immediate early genes such as *lactoferrin*, *EGF*, and proto-oncogenes such as *c-fos*, *c-jun*, and *c-myc* is upregulated even into adulthood [206]. Inversely, expression of genes that are necessary for uterine development, such as the *Abdominal B* (*AbdB*) *Hox* gene, *Hoxa-10* (known to be controlled by estradiol and progesterone) [208], *Wnt7a* as well as *Msx2* are repressed, leading to structural abnormalities of the reproductive tract [209]. Numerous studies have been conducted to assess the methylation patterns of promoters of several of these estrogen-responsive genes associated with uterine development.

Couse and colleagues have shown that ESR1 is essential for the mediation of DES effects in the uterus; *Esr1*-null female mice exhibited a complete resistance to the effects of DES, while *Esr2*-null mice did not [210]. Additionally, as mentioned earlier, ESR1 induction is necessary for activation of estrogen responsive gene expression including that of the *lactoferrin* and *c-fos* genes [206]. Since these genes are all downstream of ESR1 signaling, it is imperative to thoroughly examine the potential role of epigenetic mechanisms in the regulation of ESR1 expression after EDC exposure. Interesting new studies by Bredfeldt and colleagues have now provided a link between ESR signaling and regulation of histone modifications. They found that rapid PI3K/AKT signaling downstream of membrane-associated ESR, in response to estradiol as well as DES, caused reduction in trimethylation of H3K27. More interestingly, activation of this nongenomic signaling caused reprogramming of the uterine gene expression profile [211]. Rapid estrogen signaling in ovaries is also possible, and our recent work with MXC showed that developmental exposure to MXC causes epigenetic reprogramming of the genes encoding signaling molecules in IGF-1 and PI3K/AKT signaling [14]. Therefore, we need to test the hypothesis that similar signaling pathways may be epigenetically modified in the ovaries and uteri.

CONCLUSIONS

Developmental effects of EDCs on the ovary and female reproduction is a serious concern. Human epidemiological studies suggest that there are developmental effects of EDCs on female reproduction. However, except in certain cases, such as the pharmaceutical (e.g., DES) or accidental (e.g., dioxins) exposures, the case to establish the link between the developmental exposure and human adult disease outcome is relatively weak, with some inconclusive or inconsistent data. There are multiple reasons for the relatively weak link [212]. For example, there is a big delay, often decades, between the time of the exposure and the emergence of the symptoms. In addition, some chemicals such as phthalates have a very short half-life [109]. Therefore, establishing cause-and-effect relationships between exposures and pathologies is more challenging. Furthermore, there is genetic heterogeneity within a human population, which can alter sensitivity to EDCs. In addition, there can be a large variation in the degree of exposure among individuals.

Since controlled studies with humans are not ethically possible, animal studies are performed to assess the risks and to understand the mechanisms of action of EDCs. Animal studies using environmentally relevant doses of EDCs clearly show that these chemicals can have long-lasting effects on the female reproductive system, including the ovary [112,117,120,175]. Combined with the reported effects of EDCs on wildlife and the potential transgenerational effects of EDCs, these results remind us of the seriousness of the EDC problem. Moreover, unlike laboratory animals in EDCs studies, humans are exposed to a mixture of numerous potential EDCs, with additive or synergistic effects in a given time, which makes the problem even more serious.

As basic researchers with interests in female reproduction and the ovary, here we offer our point of view. First, animal studies that are built on findings from human epidemiological studies are essential. These in vivo studies should be environmentally and physiologically relevant. In addition, identifying phenotypic and gene expression alterations should be one of the main priorities of these studies. Second, as mentioned above, "mixture studies" are important because humans are exposed daily to myriads of EDCs simultaneously. Even the limited research conducted so far shows that EDCs often act synergistically, amplifying each other's impact. Thus, in order to get the full picture of the exposure to multiple EDCs, mixture studies should be expanded and multiple chemicals, especially those that are likely to come in contact with humans at the same developmental period, should be tested together. Third, analyses of alterations of various gene-specific and genome-wide epigenetic mechanisms (epigenome), in parallel with gene expression and phenotypic analyses, are needed. Results from epigenetic analyses can offer alternative mechanisms to explain the developmental effects of EDCs. DNA methylation has been more extensively studied in this field, but other mechanisms are being considered more often nowadays, which should be continued. In addition, cell- and follicle-stage-specific epigenetic analyses are critical. The epigenetic patterns in the ovary are not only cell specific, but also specific to the follicular stages. Thus, experimental designs should take this into the account.

To make a long-term prediction regarding the role of epigenetics: although the "field of epigenetics" is vast and past progress, especially in the EDC area, has been slow, there have been major advances in the field, such as the availability of large data bases through the ENCODE Project (https://www.encodeproject.org/) and the recent release of integrative analysis of over 100 reference human epigenomes by the NIH Roadmap Epigenomics Consortium [213]. Combined with the rapid progress in sequencing and bioinformatics technologies, these advances can provide reference points to better understanding of the epigenome of our interest. This can make it possible in the future to decipher an individual's epigenome, just like the genome today, but with much more functional information. The epigenome then can be monitored with routine intervals, which can provide a more comprehensive way to assess the state of the individual's health, including impacts of environmental EDCs.

Acknowledgments

The authors thank Dr Kathy Manger for editing the manuscript and Mr Aaron DeLaRosa for his help making the figures. The studies cited from the authors' laboratories were supported in part by National Institute of Environmental Health Science grants (ES013854, ES017059, and ES017847) and NIEHS Center grant ES005022.

References

[1] Cesario SK, Hughes LA. Precocious puberty: a comprehensive review of literature. J Obstet Gynecol Neonatal Nurs 2007;36(3):263–74.

[2] Guzick DS, Swan S. The decline of infertility: apparent or real? Fertil Steril 2006;86(3):524–6. discussion 34.

[3] Ventura SJ, Curtin SC, Abma JC, Henshaw SK. Estimated pregnancy rates and rates of pregnancy outcomes for the United States, 1990–2008. Natl Vital Stat Rep 2012;60(7):1–21.

[4] Chamorro-Garcia R, Sahu M, Abbey RJ, Laude J, Pham N, Blumberg B. Transgenerational inheritance of increased fat depot size, stem cell reprogramming, and hepatic steatosis elicited by prenatal exposure to the obesogen tributyltin in mice. Environ Health Perspect 2013;121(3):359–66.
[5] Barker DJ. Fetal origins of coronary heart disease. BMJ 1995;311(6998):171–4.
[6] Barker DJ, Winter PD, Osmond C, Margetts B, Simmonds SJ. Weight in infancy and death from ischaemic heart disease. Lancet 1989;2(8663):577–80.
[7] Weaver IC, Meaney MJ, Szyf M. Maternal care effects on the hippocampal transcriptome and anxiety-mediated behaviors in the offspring that are reversible in adulthood. Proc Natl Acad Sci USA 2006;103(9):3480–5.
[8] de Rooij SR, Wouters H, Yonker JE, Painter RC, Roseboom TJ. Prenatal undernutrition and cognitive function in late adulthood. Proc Natl Acad Sci USA 2010;107(39):16881–6.
[9] Newbold RR, Tyrey S, Haney AF, McLachlan JA. Developmentally arrested oviduct: a structural and functional defect in mice following prenatal exposure to diethylstilbestrol. Teratology 1983;27(3):417–26.
[10] Phoenix CH, Goy RW, Gerall AA, Young WC. Organizing action of prenatally administered testosterone propionate on the tissues mediating mating behavior in the female guinea pig. Endocrinology 1959;65:369–82.
[11] Barraclough CA. Production of anovulatory, sterile rats by single injections of testosterone propionate. Endocrinology 1961;68:62–7.
[12] Iguchi T, Fukazawa Y, Uesugi Y, Takasugi N. Polyovular follicles in mouse ovaries exposed neonatally to diethylstilbestrol in vivo and in vitro. Biol Reprod 1990;43(3):478–84.
[13] Bromer JG, Wu J, Zhou Y, Taylor HS. Hypermethylation of homeobox A10 by in utero diethylstilbestrol exposure: an epigenetic mechanism for altered developmental programming. Endocrinology 2009;150(7):3376–82.
[14] Zama AM, Uzumcu M. Fetal and neonatal exposure to the endocrine disruptor methoxychlor causes epigenetic alterations in adult ovarian genes. Endocrinology 2009;150(10):4681–91.
[15] Schwarz JM, Nugent BM, McCarthy MM. Developmental and hormone-induced epigenetic changes to estrogen and progesterone receptor genes in brain are dynamic across the life span. Endocrinology 2010;151(10):4871–81.
[16] Manikkam M, Haque MM, Guerrero-Bosagna C, Nilsson EE, Skinner MK. Pesticide methoxychlor promotes the epigenetic transgenerational inheritance of adult-onset disease through the female germline. PLoS One 2014;9(7):e102091.
[17] Heijmans BT, Tobi EW, Stein AD, et al. Persistent epigenetic differences associated with prenatal exposure to famine in humans. Proc Natl Acad Sci USA 2008;105(44):17046–9.
[18] Sanchez F, Smitz J. Molecular control of oogenesis. Biochim Biophys Acta 2012;1822(12):1896–912.
[19] Pepling ME, Spradling AC. Female mouse germ cells form synchronously dividing cysts. Development 1998;125(17):3323–8.
[20] Menke DB, Koubova J, Page DC. Sexual differentiation of germ cells in XX mouse gonads occurs in an anterior-to-posterior wave. Dev Biol 2003;262(2):303–12.
[21] McCarrey JR. Distinctions between transgenerational and non-transgenerational epimutations. Mol Cell Endocrinol 2014;398.
[22] Reik W, Dean W, Walter J. Epigenetic reprogramming in mammalian development. Science 2001;293(5532):1089–93.
[23] Hajkova P, Erhardt S, Lane N, et al. Epigenetic reprogramming in mouse primordial germ cells. Mech Dev 2002;117(1–2):15–23.
[24] Kato Y, Kaneda M, Hata K, et al. Role of the Dnmt3 family in de novo methylation of imprinted and repetitive sequences during male germ cell development in the mouse. Hum Mol Genet 2007;16(19):2272–80.
[25] Hiura H, Obata Y, Komiyama J, Shirai M, Kono T. Oocyte growth-dependent progression of maternal imprinting in mice. Genes Cells 2006;11(4):353–61.
[26] Pepling ME, Spradling AC. Mouse ovarian germ cell cysts undergo programmed breakdown to form primordial follicles. Dev Biol 2001;234(2):339–51.
[27] Fortune JE, Yang MY, Muruvi W. In vitro and in vivo regulation of follicular formation and activation in cattle. Reprod Fertil Dev 2011;23(1):15–22.
[28] Kezele P, Skinner MK. Regulation of ovarian primordial follicle assembly and development by estrogen and progesterone: endocrine model of follicle assembly. Endocrinology 2003;144(8):3329–37.
[29] Chen Y, Jefferson WN, Newbold RR, Padilla-Banks E, Pepling ME. Estradiol, progesterone, and genistein inhibit oocyte nest breakdown and primordial follicle assembly in the neonatal mouse ovary in vitro and in vivo. Endocrinology 2007;148(8):3580–90.
[30] Billiar RB, Zachos NC, Burch MG, Albrecht ED, Pepe GJ. Up-regulation of alpha-inhibin expression in the fetal ovary of estrogen-suppressed baboons is associated with impaired fetal ovarian folliculogenesis. Biol Reprod 2003;68(6):1989–96.
[31] Zachos NC, Billiar RB, Albrecht ED, Pepe GJ. Developmental regulation of baboon fetal ovarian maturation by estrogen. Biol Reprod 2002;67(4):1148–56.
[32] McGee EA, Hsueh AJ. Initial and cyclic recruitment of ovarian follicles. Endocr Rev 2000;21(2):200–14.
[33] Packer AI, Hsu YC, Besmer P, Bachvarova RF. The ligand of the c-kit receptor promotes oocyte growth. Dev Biol 1994;161(1):194–205.
[34] Durlinger AL, Kramer P, Karels B, et al. Control of primordial follicle recruitment by anti-Mullerian hormone in the mouse ovary. Endocrinology 1999;140(12):5789–96.

[35] Castrillon DH, Miao L, Kollipara R, Horner JW, DePinho RA. Suppression of ovarian follicle activation in mice by the transcription factor Foxo3a. Science 2003;301(5630):215–8.

[36] Britt KL, Saunders PK, McPherson SJ, Misso ML, Simpson ER, Findlay JK. Estrogen actions on follicle formation and early follicle development. Biol Reprod 2004;71(5):1712–23.

[37] Yang JL, Zhang CP, Li L, et al. Testosterone induces redistribution of forkhead box-3a and down-regulation of growth and differentiation factor 9 messenger ribonucleic acid expression at early stage of mouse folliculogenesis. Endocrinology 2009;151(2):774–82.

[38] Hirshfield AN. Theca cells may be present at the outset of follicular growth. Biol Reprod 1991;44(6):1157–62.

[39] Skinner MK. Regulation of primordial follicle assembly and development. Hum Reprod Update 2005;11(5): 461–71.

[40] McGee EA, Perlas E, LaPolt PS, Tsafriri A, Hsueh AJ. Follicle-stimulating hormone enhances the development of preantral follicles in juvenile rats. Biol Reprod 1997;57(5):990–8.

[41] Sokka T, Huhtaniemi I. Ontogeny of gonadotrophin receptors and gonadotrophin-stimulated cyclic AMP production in the neonatal rat ovary. J Endocrinol 1990;127(2):297–303.

[42] Kumar TR. What have we learned about gonadotropin function from gonadotropin subunit and receptor knockout mice? Reproduction 2005;130(3):293–302.

[43] Oakberg EF. Follicular growth and atresia in the mouse. In Vitro 1979;15(1):41–9.

[44] Richards JS, Pangas SA. The ovary: basic biology and clinical implications. J Clin Invest 2010;120(4):963–72.

[45] diZerega GS, Hodgen GD. Folliculogenesis in the primate ovarian cycle. Endocr Rev 1981;2(1):27–49.

[46] Matzuk MM, Burns KH, Viveiros MM, Eppig JJ. Intercellular communication in the mammalian ovary: oocytes carry the conversation. Science 2002;296(5576): 2178–80.

[47] Stocco C, Telleria C, Gibori G. The molecular control of corpus luteum formation, function, and regression. Endocr Rev 2007;28(1):117–49.

[48] Jakimiuk AJ, Weitsman SR, Yen HW, Bogusiewicz M, Magoffin DA. Estrogen receptor alpha and beta expression in theca and granulosa cells from women with polycystic ovary syndrome. J Clin Endocrinol Metab 2002;87(12):5532–8.

[49] Sar M, Welsch F. Differential expression of estrogen receptor-beta and estrogen receptor-alpha in the rat ovary. Endocrinology 1999;140(2):963–71.

[50] Chen Y, Breen K, Pepling ME. Estrogen can signal through multiple pathways to regulate oocyte cyst breakdown and primordial follicle assembly in the neonatal mouse ovary. J Endocrinol 2009;202(3):407–17.

[51] Couse JF, Yates MM, Deroo BJ, Korach KS. Estrogen receptor-beta is critical to granulosa cell differentiation and the ovulatory response to gonadotropins. Endocrinology 2005;146(8):3247–62.

[52] Kuiper GG, Lemmen JG, Carlsson B, et al. Interaction of estrogenic chemicals and phytoestrogens with estrogen receptor beta. Endocrinology 1998;139(10):4252–63.

[53] Emmen JM, Couse JF, Elmore SA, Yates MM, Kissling GE, Korach KS. In vitro growth and ovulation of follicles from ovaries of estrogen receptor (ER) {alpha} and ER{beta} null mice indicate a role for ER{beta} in follicular maturation. Endocrinology 2005;146(6):2817–26.

[54] Woodruff TK, Mayo KE. To beta or not to beta: estrogen receptors and ovarian function. Endocrinology 2005;146(8):3244–6.

[55] Szoltys M, Slomczynska M. Changes in distribution of androgen receptor during maturation of rat ovarian follicles. Exp Clin Endocrinol Diabetes 2000;108(3): 228–34.

[56] Sajjad Y, Quenby SM, Nickson P, Lewis-Jones DI, Vince G. Expression of androgen receptors in upper human fetal reproductive tract. Hum Reprod 2004;19(7):1659–65.

[57] Rice S, Ojha K, Whitehead S, Mason H. Stage-specific expression of androgen receptor, follicle-stimulating hormone receptor, and anti-Mullerian hormone type II receptor in single, isolated, human preantral follicles: relevance to polycystic ovaries. J Clin Endocrinol metabolism 2007;92(3):1034–40.

[58] Tetsuka M, Hillier SG. Androgen receptor gene expression in rat granulosa cells: the role of follicle-stimulating hormone and steroid hormones. Endocrinology 1996;137(10):4392–7.

[59] Lyon MF, Glenister PH. Reduced reproductive performance in androgen-resistant Tfm/Tfm female mice. Proc R Soc Lond B Biol Sci 1980;208(1170): 1–12.

[60] Shiina H, Matsumoto T, Sato T, et al. Premature ovarian failure in androgen receptor-deficient mice. Proc Natl Acad Sci USA 2006;103(1):224–9.

[61] Komar CM, Braissant O, Wahli W, Curry Jr TE. Expression and localization of PPARs in the rat ovary during follicular development and the periovulatory period. Endocrinology 2001;142(11):4831–8.

[62] Mu YM, Yanase T, Nishi Y, et al. Insulin sensitizer, troglitazone, directly inhibits aromatase activity in human ovarian granulosa cells. Biochem Biophys Res Commun 2000;271(3):710–3.

[63] Baba T, Mimura J, Nakamura N, et al. Intrinsic function of the aryl hydrocarbon (dioxin) receptor as a key factor in female reproduction. Mol Cell Biol 2005;25(22):10040–51.

[64] Khorram O, Garthwaite M, Golos T. Uterine and ovarian aryl hydrocarbon receptor (AHR) and aryl hydrocarbon receptor nuclear translocator (ARNT) mRNA expression in benign and malignant gynaecological conditions. Mol Hum Reprod 2002;8(1):75–80.

[65] Lee L, Asada H, Kizuka F, et al. Changes in histone modification and DNA methylation of the StAR and Cyp19a1 promoter regions in granulosa cells undergoing luteinization during ovulation in rats. Endocrinology 2013;154(1):458–70.

[66] Meldi KM, Gaconnet GA, Mayo KE. DNA methylation and histone modifications are associated with repression of the inhibin alpha promoter in the rat corpus luteum. Endocrinology 2012;153(10):4905–17.

[67] Hoivik EA, Aumo L, Aesoy R, et al. Deoxyribonucleic acid methylation controls cell type-specific expression of steroidogenic factor 1. Endocrinology 2008;149(11): 5599–609.

[68] Cruz G, Foster W, Paredes A, Yi KD, Uzumcu M. Long-term effects of early-life exposure to environmental oestrogens on ovarian function: role of epigenetics. J Neuroendocrinol 2014;26(9):613–24.

[69] Neal-Perry G, Santoro NF. Aging in the hypothalamic-pituitary-ovarian Axis. In: Neill JD, editor. Knobil and Neill's pysiology of reproduction. 3rd ed. New York: Elsevier Academic Press; 2006. p. 2729–55.

[70] Kermath BA, Gore AC. Neuroendocrine control of the transition to reproductive senescence: lessons learned from the female rodent model. Neuroendocrinology 2012;96(1):1–12.

[71] vom Saal FS, Finch CE. Reproductive senescence: phenomena and mechanisms in mammals and seleceted vertebrates. In: Knobil E, Neill JD, editors. The physiology of reproduction. 1st ed. New York: Raven Press; 1988. p. 2351–413.

[72] NIA. AgePage. NIA/NIH; 2013. http://www.nia.nih.gov/health/publication/menopause - menopause [accessed 03.15.15].

[73] Sone K, Yamamoto-Sawamura T, Kuwahara S, et al. Changes of estrous cycles with aging in female F344/n rats. Exp Anim 2007;56(2):139–48.

[74] Felicio LS, Nelson JF, Finch CE. Longitudinal studies of estrous cyclicity in aging C57BL/6J mice: II. Cessation of cyclicity and the duration of persistent vaginal cornification. Biol Reprod 1984;31(3):446–53.

[75] Nelson JF, Felicio LS, Randall PK, Sims C, Finch CE. A longitudinal study of estrous cyclicity in aging C57BL/6J mice: I. Cycle frequency, length and vaginal cytology. Biol Reprod 1982;27(2):327–39.

[76] American College of Obstetricians and Gynecologists Committee on Gynecologic Practice and Practice Committee. Female age-related fertility decline. Committee Opinion No. 589. Fertil Steril 2014;101(3): 633–4.

[77] Moyer B, Hixon ML. Reproductive effects in F1 adult females exposed in utero to moderate to high doses of mono-2-ethylhexylphthalate (MEHP). Reprod Toxicol 2012;34(1):43–50.

[78] Peretz J, Vrooman L, Ricke WA, et al. Bisphenol a and reproductive health: update of experimental and human evidence, 2007–2013. Environ Health Perspect 2014;122(8):775–86.

[79] Rochester JR. Bisphenol A and human health: a review of the literature. Reprod Toxicol 2013;42:132–55.

[80] Wang W, Hafner KS, Flaws JA. In utero bisphenol A exposure disrupts germ cell nest breakdown and reduces fertility with age in the mouse. Toxicol Appl Pharmacol 2014;276(2):157–64.

[81] Hunt PA, Lawson C, Gieske M, et al. Bisphenol A alters early oogenesis and follicle formation in the fetal ovary of the rhesus monkey. Proc Natl Acad Sci USA 2012;109(43):17525–30.

[82] Adewale HB, Jefferson WN, Newbold RR, Patisaul HB. Neonatal bisphenol-a exposure alters rat reproductive development and ovarian morphology without impairing activation of gonadotropin-releasing hormone neurons. Biol Reprod 2009;81(4):690–9.

[83] Veiga-Lopez A, Luense LJ, Christenson LK, Padmanabhan V. Developmental programming: gestational bisphenol-a treatment alters trajectory of fetal ovarian gene expression. Endocrinology 2013;154(5):1873–84.

[84] Kay VR, Chambers C, Foster WG. Reproductive and developmental effects of phthalate diesters in females. Crit Rev Toxicol 2013;43(3):200–19.

[85] Koch HM, Preuss R, Angerer J. Di(2-ethylhexyl) phthalate (DEHP): human metabolism and internal exposure—an update and latest results. Int J Androl 2006;29(1):155–65. discussion 81–5.

[86] Shelby MD. NTP-CERHR monograph on the potential human reproductive and developmental effects of di (2-ethylhexyl) phthalate (DEHP). NTP CERHR MON 2006;18. v, vii-7, II-iii-xiii passim.

[87] Genuis SJ, Beesoon S, Lobo RA, Birkholz D. Human elimination of phthalate compounds: blood, urine, and sweat (BUS) study. Sci World J 2012;2012:615068.

[88] Main KM, Mortensen GK, Kaleva MM, et al. Human breast milk contamination with phthalates and alterations of endogenous reproductive hormones in infants threemonthsofage.EnvironHealthPerspect2006;114(2): 270–6.

[89] Wittassek M, Angerer J, Kolossa-Gehring M, et al. Fetal exposure to phthalates—a pilot study. Int J Hyg Environ Health 2009;212(5):492–8.

[90] Volkel W, Kiranoglu M, Schuster R, Fromme H, Hbmnet. Phthalate intake by infants calculated from biomonitoring data. Toxicol Lett 2014;225(2):222–9.

[91] NTP. Di(2-ethylhexyl) phthalate. Rep Carcinog 2011;12: 156–9.

[92] Colon I, Caro D, Bourdony CJ, Rosario O. Identification of phthalate esters in the serum of young Puerto Rican girls with premature breast development. Environ Health Perspect 2000;108(9):895–900.

[93] Chen CY, Chou YY, Wu YM, Lin CC, Lin SJ, Lee CC. Phthalates may promote female puberty by increasing kisspeptin activity. Hum Reprod 2013;28(10):2765–73.

[94] Frederiksen H, Sorensen K, Mouritsen A, et al. High urinary phthalate concentration associated with delayed pubarche in girls. Int J Androl 2012;35(3):216–26.

[95] Wolff MS, Teitelbaum SL, Pinney SM, et al. Investigation of relationships between urinary biomarkers of phytoestrogens, phthalates, and phenols and pubertal stages in girls. Environ Health Perspect 2010;118(7):1039–46.

[96] Wolff MS, Teitelbaum SL, McGovern K, et al. Phthalate exposure and pubertal development in a longitudinal study of US girls. Hum Reprod 2014;29.

[97] Lomenick JP, Calafat AM, Melguizo Castro MS, et al. Phthalate exposure and precocious puberty in females. J Pediatr 2010;156(2):221–5.

[98] Rais-Bahrami K, Nunez S, Revenis ME, Luban NL, Short BL. Follow-up study of adolescents exposed to di(2-ethylhexyl) phthalate (DEHP) as neonates on extracorporeal membrane oxygenation (ECMO) support. Environ Health Perspect 2004;112(13):1339–40.

[99] Burdorf A, Brand T, Jaddoe VW, Hofman A, Mackenbach JP, Steegers EA. The effects of work-related maternal risk factors on time to pregnancy, preterm birth and birth weight: the Generation R Study. Occup Environ Med 2011;68(3):197–204.

[100] Buck Louis GM, Sundaram R, Sweeney AM, Schisterman EF, Maisog J, Kannan K. Urinary bisphenol A, phthalates, and couple fecundity: the Longitudinal Investigation of Fertility and the Environment (LIFE) Study. Fertil Steril 2014;101(5):1359–66.

[101] Toft G, Jonsson BA, Lindh CH, et al. Association between pregnancy loss and urinary phthalate levels around the time of conception. Environ Health Perspect 2012;120(3):458–63.

[102] Latini G, De Felice C, Presta G, et al. In utero exposure to di-(2-ethylhexyl)phthalate and duration of human pregnancy. Environ Health Perspect 2003;111(14):1783–5.

[103] Meeker JD, Hu H, Cantonwine DE, et al. Urinary phthalate metabolites in relation to preterm birth in Mexico city. Environ Health Perspect 2009;117(10):1587–92.

[104] Whyatt RM, Adibi JJ, Calafat AM, et al. Prenatal di(2-ethylhexyl)phthalate exposure and length of gestation among an inner-city cohort. Pediatrics 2009;124(6):e1213–20.

[105] Adibi JJ, Hauser R, Williams PL, et al. Maternal urinary metabolites of di-(2-ethylhexyl) phthalate in relation to the timing of labor in a US multicenter pregnancy cohort study. Am J Epidemiol 2009;169(8):1015–24.

[106] Wolff MS, Engel SM, Berkowitz GS, et al. Prenatal phenol and phthalate exposures and birth outcomes. Environ Health Perspect 2008;116(8):1092–7.

[107] Ferguson KK, McElrath TF, Meeker JD. Environmental phthalate exposure and preterm birth. JAMA Pediatr 2014;168(1):61–7.

[108] Ferguson KK, McElrath TF, Ko YA, Mukherjee B, Meeker JD. Variability in urinary phthalate metabolite levels across pregnancy and sensitive windows of exposure for the risk of preterm birth. Environ Int 2014;70:118–24.

[109] Frederiksen H, Kuiri-Hanninen T, Main KM, Dunkel L, Sankilampi U. A longitudinal study of urinary phthalate excretion in 58 full-term and 67 preterm infants from birth through 14 months. Environ Health Perspect 2014;122.

[110] Takai R, Hayashi S, Kiyokawa J, et al. Collaborative work on evaluation of ovarian toxicity. 10) Two- or four-week repeated dose studies and fertility study of di-(2-ethylhexyl) phthalate (DEHP) in female rats. J Toxicol Sci 2009;34(Suppl. 1):SP111–9.

[111] Ma M, Kondo T, Ban S, et al. Exposure of prepubertal female rats to inhaled di(2-ethylhexyl)phthalate affects the onset of puberty and postpubertal reproductive functions. Toxicol Sci 2006;93(1):164–71.

[112] Zhang XF, Zhang LJ, Li L, et al. Diethylhexyl phthalate exposure impairs follicular development and affects oocyte maturation in the mouse. Environ Mol Mutagen 2013;54(5):354–61.

[113] Davis BJ, Maronpot RR, Heindel JJ. Di-(2-ethylhexyl) phthalate suppresses estradiol and ovulation in cycling rats. Toxicol Appl Pharmacol 1994;128(2):216–23.

[114] Lovekamp-Swan T, Davis BJ. Mechanisms of phthalate ester toxicity in the female reproductive system. Environ Health Perspect 2003;111(2):139–45.

[115] Svechnikova I, Svechnikov K, Soder O. The influence of di-(2-ethylhexyl) phthalate on steroidogenesis by the ovarian granulosa cells of immature female rats. J Endocrinol 2007;194(3):603–9.

[116] Kawano M, Qin XY, Yoshida M, et al. Peroxisome proliferator-activated receptor alpha mediates di-(2-ethylhexyl) phthalate transgenerational repression of ovarian Esr1 expression in female mice. Toxicol Lett 2014;228(3):235–40.

[117] Pocar P, Fiandanese N, Secchi C, et al. Exposure to di(2-ethyl-hexyl) phthalate (DEHP) in utero and during lactation causes long-term pituitary-gonadal axis disruption in male and female mouse offspring. Endocrinology 2012;153(2):937–48.

[118] Zhang T, Li L, Qin XS, et al. Di-(2-ethylhexyl) phthalate and bisphenol A exposure impairs mouse primordial follicle assembly in vitro. Environ Mol Mutagen 2014;55(4):343–53.
[119] Wang W, Craig ZR, Basavarajappa MS, Gupta RK, Flaws JA. Di(2-ethylhexyl) phthalate inhibits growth of mouse ovarian antral follicles through an oxidative stress pathway. Toxicol Appl Pharmacol 2012;258(2): 288–95.
[120] Hannon PR, Peretz J, Flaws JA. Daily exposure to di(2-ethylhexyl) phthalate alters estrous cyclicity and accelerates primordial follicle recruitment potentially via dysregulation of the phosphatidylinositol 3-kinase signaling pathway in adult mice. Biol Reprod 2014;90(6):136.
[121] Grande SW, Andrade AJ, Talsness CE, et al. A dose-response study following in utero and lactational exposure to di-(2-ethylhexyl) phthalate (DEHP): reproductive effects on adult female offspring rats. Toxicology 2007;229(1–2):114–22.
[122] Xu C, Chen JA, Qiu Z, et al. Ovotoxicity and PPAR-mediated aromatase downregulation in female Sprague–Dawley rats following combined oral exposure to benzo[a]pyrene and di-(2-ethylhexyl) phthalate. Toxicol Lett 2010;199(3):323–32.
[123] Li N, Liu T, Zhou L, He J, Ye L. Di-(2-ethylhcxyl) phthalate reduces progesterone levels and induces apoptosis of ovarian granulosa cell in adult female ICR mice. Environ Toxicol Pharmacol 2012;34(3):869–75.
[124] Carsen R. Silent spring. 1962.
[125] Turusov V, Rakitsky V, Tomatis L. Dichlorodiphenyltrichloroethane (DDT): ubiquity, persistence, and risks. Environ Health Perspect 2002;110(2):125–8.
[126] Smeets JM, van Holsteijn I, Giesy JP, Seinen W, van den Berg M. Estrogenic potencies of several environmental pollutants, as determined by vitellogenin induction in a carp hepatocyte assay. Toxicol Sci 1999;50(2):206–13.
[127] Gladen BC, Ragan NB, Rogan WJ. Pubertal growth and development and prenatal and lactational exposure to polychlorinated biphenyls and dichlorodiphenyl dichloroethene. J Pediatr 2000;136(4):490–6.
[128] De Felip E, di Domenico A, Miniero R, Silvestroni L. Polychlorobiphenyls and other organochlorine compounds in human follicular fluid. Chemosphere 2004;54(10):1445–9.
[129] Kamarianos A, Karamanlis X, Goulas P, Theodosiadou E, Smokovitis A. The presence of environmental pollutants in the follicular fluid of farm animals (cattle, sheep, goats, and pigs). Reprod Toxicol 2003;17(2):185–90.
[130] Furusawa N. Transferring and distributing profiles of p,p′-(DDT) in egg-forming tissues and eggs of laying hens following a single oral administration. J Vet Med A Physiol Pathol Clin Med 2002;49(6):334–6.
[131] Cao LL, Yan CH, Yu XD, et al. Relationship between serum concentrations of polychlorinated biphenyls and organochlorine pesticides and dietary habits of pregnant women in Shanghai. Sci Total Environ 2011;409(16):2997–3002.
[132] Ibarluzea J, Alvarez-Pedrerol M, Guxens M, et al. Sociodemographic, reproductive and dietary predictors of organochlorine compounds levels in pregnant women in Spain. Chemosphere 2011;82(1): 114–20.
[133] Hardell E, Carlberg M, Nordstrom M, van Bavel B. Time trends of persistent organic pollutants in Sweden during 1993–2007 and relation to age, gender, body mass index, breast-feeding and parity. Sci Total Environ 2010;408(20):4412–9.
[134] Ouyang F, Perry MJ, Venners SA, et al. Serum DDT, age at menarche, and abnormal menstrual cycle length. Occup Environ Med 2005;62(12):878–84.
[135] Vasiliu O, Muttineni J, Karmaus W. In utero exposure to organochlorines and age at menarche. Hum Reprod 2004;19(7):1506–12.
[136] Younglai EV, Foster WG, Hughes EG, Trim K, Jarrell JF. Levels of environmental contaminants in human follicular fluid, serum, and seminal plasma of couples undergoing in vitro fertilization. Arch Environ Contam Toxicol 2002;43(1):121–6.
[137] Korrick SA, Chen C, Damokosh AI, et al. Association of DDT with spontaneous abortion: a case-control study. Ann Epidemiol 2001;11(7):491–6.
[138] Jarrell JF, Villeneuve D, Franklin C, et al. Contamination of human ovarian follicular fluid and serum by chlorinated organic compounds in three Canadian cities. CMAJ 1993;148(8):1321–7.
[139] Bradman AS, Schwartz JM, Fenster L, Barr DB, Holland NT, Eskenazi B. Factors predicting organochlorine pesticide levels in pregnant Latina women living in a United States agricultural area. J Expo Sci Environ Epidemiol 2007;17(4):388–99.
[140] Longnecker MP, Klebanoff MA, Zhou H, Brock JW. Association between maternal serum concentration of the DDT metabolite DDE and preterm and small-for-gestational-age babies at birth. Lancet 2001;358(9276): 110–4.
[141] Weisskopf MG, Anderson HA, Hanrahan LP, et al. Maternal exposure to Great Lakes sport-caught fish and dichlorodiphenyl dichloroethylene, but not polychlorinated biphenyls, is associated with reduced birth weight. Environ Res 2005;97(2):149–62.
[142] Siddiqui MK, Srivastava S, Srivastava SP, Mehrotra PK, Mathur N, Tandon I. Persistent chlorinated pesticides and intra-uterine foetal growth retardation: a possible association. Int Arch Occup Environ Health 2003;76(1):75–80.

[143] Cohn BA, Cirillo PM, Wolff MS, et al. DDT and DDE exposure in mothers and time to pregnancy in daughters. Lancet 2003;361(9376):2205–6.
[144] Heinrichs WL, Gellert RJ, Bakke JL, Lawrence NL. DDT administered to neonatal rats induces persistent estrus syndrome. Science 1971;173(997):642–3.
[145] Gellert RJ, Heinrichs WL. Effects of ddt homologs administered to female rats during the perinatal period. Biol Neonate 1975;26(3–4):283–90.
[146] Jonsson Jr HT, Keil JE, Gaddy RG, Loadholt CB, Hennigar GR, Walker Jr EM. Prolonged ingestion of commercial DDT and PCB: effects on progesterone levels and reproduction in the mature female rat. Arch Environ Contam Toxicol 1975;3(4):479–90.
[147] Chedrese PJ, Feyles F. The diverse mechanism of action of dichlorodiphenyldichloroethylene (DDE) and methoxychlor in ovarian cells in vitro. Reprod Toxicol 2001;15(6):693–8.
[148] Crellin NK, Kang HG, Swan CL, Chedrese PJ. Inhibition of basal and stimulated progesterone synthesis by dichlorodiphenyldichloroethylene and methoxychlor in a stable pig granulosa cell line. Reproduction 2001;121(3):485–92.
[149] Wojtowicz AK, Augustowska K, Gregoraszczuk EL. The short- and long-term effects of two isomers of DDT and their metabolite DDE on hormone secretion and survival of human choriocarcinoma JEG-3 cells. Pharmacol Rep 2007;59(2):224–32.
[150] Cummings AM. Methoxychlor as a model for environmental estrogens. Crit Rev Toxicol 1997;27(4):367–79.
[151] Gaido KW, Maness SC, McDonnell DP, Dehal SS, Kupfer D, Safe S. Interaction of methoxychlor and related compounds with estrogen receptor alpha and beta, and androgen receptor: structure–activity studies. Mol Pharmacol 2000;58(4):852–8.
[152] Uzumcu M, Zachow R. Developmental exposure to environmental endocrine disruptors: consequences within the ovary and on female reproductive function. Reprod Toxicol 2007;23(3):337–52.
[153] Axmon A, Rylander L, Stromberg U, Hagmar L. Miscarriages and stillbirths in women with a high intake of fish contaminated with persistent organochlorine compounds. Int Arch Occup Environ Health 2000;73(3):204–8.
[154] Bretveld RW, Hooiveld M, Zielhuis GA, Pellegrino A, van Rooij IA, Roeleveld N. Reproductive disorders among male and female greenhouse workers. Reprod Toxicol 2008;25(1):107–14.
[155] Fuortes L, Clark MK, Kirchner HL, Smith EM. Association between female infertility and agricultural work history. Am J Ind Med 1997;31(4):445–51.
[156] Harris SJ, Cecil HC, Bitman J. Effect of several dietary levels of technical methoxychlor on reproduction in rats. J Agric Food Chem 1974;22(6):969–73.
[157] Martinez EM, Swartz WJ. Effects of methoxychlor on the reproductive system of the adult female mouse. 1. Gross and histologic observations. Reprod Toxicol 1991;5(2):139–47.
[158] Hall DL, Payne LA, Putnam JM, Huet-Hudson YM. Effect of methoxychlor on implantation and embryo development in the mouse. Reprod Toxicol 1997;11(5):703–8.
[159] Cummings AM, Gray Jr LE. Antifertility effect of methoxychlor in female rats: dose- and time-dependent blockade of pregnancy. Toxicol Appl Pharmacol 1989;97(3):454–62.
[160] Borgeest C, Symonds D, Mayer LP, Hoyer PB, Flaws JA. Methoxychlor may cause ovarian follicular atresia and proliferation of the ovarian epithelium in the mouse. Toxicol Sci 2002;68(2):473–8.
[161] Gray Jr LE, Ostby J, Ferrell J, et al. A dose-response analysis of methoxychlor-induced alterations of reproductive development and function in the rat. Fundam Appl Toxicol 1989;12(1):92–108.
[162] Chapin RE, Harris MW, Davis BJ, et al. The effects of perinatal/juvenile methoxychlor exposure on adult rat nervous, immune, and reproductive system function. Fundam Appl Toxicol 1997;40(1):138–57.
[163] Armenti AE, Zama AM, Passantino L, Uzumcu M. Developmental methoxychlor exposure affects multiple reproductive parameters and ovarian folliculogenesis and gene expression in adult rats. Toxicol Appl Pharmacol 2008;233(2):286–96.
[164] Tang WY, Newbold R, Mardilovich K, et al. Persistent hypomethylation in the promoter of nucleosomal binding protein 1 (Nsbp1) correlates with overexpression of Nsbp1 in mouse uteri neonatally exposed to diethylstilbestrol or genistein. Endocrinology 2008; 149(12):5922–31.
[165] Zama AM, Uzumcu M. Targeted genome-wide methylation and gene expression analyses reveal signaling pathways involved in ovarian dysfunction after developmental EDC exposure in rats. Biol Reprod 2013;88(2):52.
[166] Agency EP. Estimating exposure to dioxin-like compounds. 1994.
[167] Warner M, Eskenazi B, Mocarelli P, et al. Serum dioxin concentrations and breast cancer risk in the Seveso Women's Health Study. Environ health Perspect 2002;110(7):625–8.
[168] Eskenazi B, Warner M, Samuels S, et al. Serum dioxin concentrations and risk of uterine leiomyoma in the Seveso Women's Health Study. Am J Epidemiol 2007;166(1):79–87.
[169] Gough M. Dioxin, agent orange: the facts. Springer; 1986.
[170] Herbicides IoMUCtRtHEiVVoEt. Veterans and agent orange: health effects of herbicides used in vietnam. Washington, DC: Institute of Medicine (US) Committee to Review the Health Effects in Vietnam Veterans of Exposure to Herbicides; 2012.

[171] Leijs MM, Koppe JG, Olie K, et al. Delayed initiation of breast development in girls with higher prenatal dioxin exposure; a longitudinal cohort study. Chemosphere 2008;73(6):999–1004.

[172] Warner M, Eskenazi B, Olive DL, et al. Serum dioxin concentrations and quality of ovarian function in women of Seveso. Environ Health Perspect 2007;115(3):336–40.

[173] Tenchini ML, Crimaudo C, Pacchetti G, Mottura A, Agosti S, De Carli L. A comparative cytogenetic study on cases of induced abortions in TCDD-exposed and nonexposed women. Environ Mutagen 1983;5(1):73–85.

[174] Constable JD, Hatch MC. Reproductive effects of herbicide exposure in Vietnam: recent studies by the Vietnamese and others. Teratog, Carcinog Mutagen 1985;5(4):231–50.

[175] Huff J, Lucier G, Tritscher A. Carcinogenicity of TCDD: experimental, mechanistic, and epidemiologic evidence. Annu Rev Pharmacol Toxicol 1994;34:343–72.

[176] Theobald HM, Peterson RE. In utero and lactational exposure to 2,3,7,8-tetrachlorodibenzo-p-dioxin: effects on development of the male and female reproductive system of the mouse. Toxicol Appl Pharmacol 1997;145(1): 124–35.

[177] Petrulis JR, Perdew GH. The role of chaperone proteins in the aryl hydrocarbon receptor core complex. Chemico-Biological Interact 2002;141(1–2):25–40.

[178] Li X, Johnson DC, Rozman KK. Effects of 2,3,7,8-tetr achlorodibenzo-p-dioxin (TCDD) on estrous cyclicity and ovulation in female Sprague–Dawley rats. Toxicol Lett 1995;78(3):219–22.

[179] Son DS, Ushinohama K, Gao X, et al. 2,3,7,8-Tetrach lorodibenzo-p-dioxin (TCDD) blocks ovulation by a direct action on the ovary without alteration of ovarian steroidogenesis: lack of a direct effect on ovarian granulosa and thecal-interstitial cell steroidogenesis in vitro. Reprod Toxicol 1999;13(6):521–30.

[180] Petroff BK, Gao X, Rozman KK, Terranova PF. Interaction of estradiol and 2,3,7,8-tetrachlorodibenzo-p-dioxin (TCDD) in an ovulation model: evidence for systemic potentiation and local ovarian effects. Reprod Toxicol 2000;14(3):247–55.

[181] Li X, Johnson DC, Rozman KK. Reproductive effects of 2,3,7,8-tetrachlorodibenzo-p-dioxin (TCDD) in female rats: ovulation, hormonal regulation, and possible mechanism(s). Toxicol Appl Pharmacol 1995; 133(2):321–7.

[182] Franczak A, Nynca A, Valdez KE, Mizinga KM, Petroff BK. Effects of acute and chronic exposure to the aryl hydrocarbon receptor agonist 2,3,7,8-tetrachlorodibenzo-p-dioxin on the transition to reproductive senescence in female Sprague–Dawley rats. Biol Reprod 2006;74(1):125–30.

[183] Salisbury TB, Marcinkiewicz JL. In utero and lactational exposure to 2,3,7,8-tetrachlorodibenzo-p-dioxin and 2,3,4,7,8-pentachlorodibenzofuran reduces growth and disrupts reproductive parameters in female rats. Biol Reprod 2002;66(6):1621–6.

[184] Gray Jr LE, Ostby JS. In utero 2,3,7,8-tetrachlorodibenzo-p-dioxin (TCDD) alters reproductive morphology and function in female rat offspring. Toxicol Appl Pharmacol 1995;133(2):285–94.

[185] Smith OW, Gabbe SG. Diethylstilbestrol in the prevention and treatment of complications of pregnancy. 1948. Am J Obstet Gynecol 1999;181(6):1570–1.

[186] Dieckmann WJ, Davis ME, Rynkiewicz LM, Pottinger RE. Does the administration of diethylstilbestrol during pregnancy have therapeutic value? Am J Obstet Gynecol 1953;66(5):1062–81.

[187] Newbold RR. Lessons learned from perinatal exposure to diethylstilbestrol. Toxicol Appl Pharmacol 2004;199(2):142–50.

[188] Herbst AL, Ulfelder H, Poskanzer DC. Adenocarcinoma of the vagina. Association of maternal stilbestrol therapy with tumor appearance in young women. N Engl J Med 1971;284(15):878–81.

[189] Kitajewski J, Sassoon D. The emergence of molecular gynecology: homeobox and Wnt genes in the female reproductive tract. Bioessays 2000;22(10):902–10.

[190] McLachlan JA, Newbold RR, Bullock BC. Long-term effects on the female mouse genital tract associated with prenatal exposure to diethylstilbestrol. Cancer Res 1980;40(11):3988–99.

[191] Titus-Ernstoff L, Troisi R, Hatch EE, et al. Menstrual and reproductive characteristics of women whose mothers were exposed in utero to diethylstilbestrol (DES). Int J Epidemiol 2006;35(4):862–8.

[192] Goldberg JM, Falcone T. Effect of diethylstilbestrol on reproductive function. Fertil Steril 1999; 72(1):1–7.

[193] Hatch E, Herbst A, Hoover R, et al. Incidence of squamous neoplasia of the cervix and vagina in des-exposed daughters. Ann Epidemiol 2000;10(7):467.

[194] Blatt J, Van Le L, Weiner T, Sailer S. Ovarian carcinoma in an adolescent with transgenerational exposure to diethylstilbestrol. J Pediatr Hematol Oncol 2003;25(8):635–6.

[195] Turusov VS, Trukhanova LS, Parfenov Yu D, Tomatis L. Occurrence of tumours in the descendants of CBA male mice prenatally treated with diethylstilbestrol. Int J Cancer 1992;50(1):131–5.

[196] Newbold RR, Hanson RB, Jefferson WN, Bullock BC, Haseman J, McLachlan JA. Increased tumors but uncompromised fertility in the female descendants of mice exposed developmentally to diethylstilbestrol. Carcinogenesis 1998; 19(9):1655–63.

[197] Wordinger RJ, Highman B. Histology and ultrastructure of the adult mouse ovary following a single prenatal exposure to diethylstilbestrol. Virch Arch B Cell Pathol Incl Mol Pathol 1984;45(3):241–53.
[198] Haney AF, Newbold RR, McLachlan JA. Prenatal diethylstilbestrol exposure in the mouse: effects on ovarian histology and steroidogenesis in vitro. Biol Reprod 1984;30(2):471–8.
[199] Tenenbaum A, Forsberg JG. Structural and functional changes in ovaries from adult mice treated with diethylstilboestrol in the neonatal period. J Reprod Fertil 1985;73(2):465–77.
[200] McLachlan JA, Newbold RR, Shah HC, Hogan MD, Dixon RL. Reduced fertility in female mice exposed transplacentally to diethylstilbestrol (DES). Fertil Steril 1982;38(3):364–71.
[201] Halling A, Forsberg JG. Effects of neonatal exposure to diethylstilbestrol on early mouse embryo development in vivo and in vitro. Biol Reprod 1991;45(1):157–62.
[202] Iguchi T, Takasugi N. Polyovular follicles in the ovary of immature mice exposed prenatally to diethylstilbestrol. Anat Embryol Berl 1986;175(1):53–5.
[203] Kirigaya A, Kim H, Hayashi S, et al. Involvement of estrogen receptor beta in the induction of polyovular follicles in mouse ovaries exposed neonatally to diethylstilbestrol. Zool Sci 2009;26(10):704–12.
[204] Kim H, Nakajima T, Hayashi S, et al. Effects of diethylstilbestrol on programmed oocyte death and induction of polyovular follicles in neonatal mouse ovaries. Biol Reprod 2009;81(5):1002–9.
[205] Jefferson WN, Couse JF, Padilla-Banks E, Korach KS, Newbold RR. Neonatal exposure to genistein induces estrogen receptor (ER)alpha expression and multioocyte follicles in the maturing mouse ovary: evidence for ERbeta-mediated and nonestrogenic actions. Biol Reprod 2002;67(4):1285–96.
[206] Nelson KG, Sakai Y, Eitzman B, Steed T, McLachlan J. Exposure to diethylstilbestrol during a critical developmental period of the mouse reproductive tract leads to persistent induction of two estrogen-regulated genes. Cell Growth Differ 1994;5(6):595–606.
[207] Loose-Mitchell DS, Chiappetta C, Stancel GM. Estrogen regulation of c-fos messenger ribonucleic acid. Mol Endocrinol 1988;2(10):946–51.
[208] Ma L, Benson GV, Lim H, Dey SK, Maas RL. Abdominal B. (AbdB) Hoxa genes: regulation in adult uterus by estrogen and progesterone and repression in mullerian duct by the synthetic estrogen diethylstilbestrol (DES). Dev Biol 1998;197(2):141–54.
[209] Huang WW, Yin Y, Bi Q, et al. Developmental diethylstilbestrol exposure alters genetic pathways of uterine cytodifferentiation. Mol Endocrinol 2005;19(3):669–82.
[210] Couse JF, Korach KS. Estrogen receptor-alpha mediates the detrimental effects of neonatal diethylstilbestrol (DES) exposure in the murine reproductive tract. Toxicology 2004;205(1–2):55–63.
[211] Bredfeldt TG, Greathouse KL, Safe SH, Hung MC, Bedford MT, Walker CL. Xenoestrogen-induced regulation of EZH2 and histone methylation via estrogen receptor signaling to PI3K/AKT. Mol Endocrinol 2010;24(5):993–1006.
[212] Gore AC, Patisaul HB. Neuroendocrine disruption: historical roots, current progress, questions for the future. Front Neuroendocrinol 2010;31(4):395–9.
[213] Roadmap Epigenomics C, Kundaje A, Meuleman W, et al. Integrative analysis of 111 reference human epigenomes. Nature 2015;518(7539):317–30.
[214] Davis BJ, Weaver R, Gaines LJ, Heindel JJ. Mono-(2-ethylhexyl) phthalate suppresses estradiol production independent of FSH-cAMP stimulation in rat granulosa cells. Toxicol Appl Pharmacol 1994;128(2):224–8.
[215] Borgeest C, Miller KP, Gupta R, et al. Methoxychlor-induced atresia in the mouse involves Bcl-2 family members, but not gonadotropins or estradiol. Biol Reprod 2004;70(6):1828–35.
[216] Romani F, Tropea A, Scarinci E, et al. Endocrine disruptors and human reproductive failure: the in vitro effect of phthalates on human luteal cells. Fertil Steril 2014;102(3):831–7.
[217] Sarraj MA, Drummond AE. Mammalian foetal ovarian development: consequences for health and disease. Reproduction 2012;143(2):151–63.
[218] Hartshorne GM, Lyrakou S, Hamoda H, Oloto E, Ghafari F. Oogenesis and cell death in human prenatal ovaries: what are the criteria for oocyte selection? Mol Hum Reprod 2009;15(12):805–19.
[219] Pepling ME. From primordial germ cell to primordial follicle: mammalian female germ cell development. Genesis 2006;44(12):622–32.
[220] Ginsburg M, Snow MH, McLaren A. Primordial germ cells in the mouse embryo during gastrulation. Development 1990;110(2):521–8.
[221] Motta PM, Makabe S, Nottola SA. The ultrastructure of human reproduction. I. The natural history of the female germ cell: origin, migration and differentiation inside the developing ovary. Hum Reprod Update 1997;3(3):281–95.
[222] Kerr JB, Myers M, Anderson RA. The dynamics of the primordial follicle reserve. Reproduction 2013;146(6): R205–15.
[223] Monniaux D, Clement F, Dalbies-Tran R, et al. The ovarian reserve of primordial follicles and the dynamic reserve of antral growing follicles: what is the link? Biol Reprod 2014;90(4):85.
[224] Goto T, Adjaye J, Rodeck CH, Monk M. Identification of genes expressed in human primordial germ cells at the time of entry of the female germ line into meiosis. Mol Hum Reprod 1999;5(9):851–60.

CHAPTER

10

Developmental and Epigenetic Origins of Male Reproductive Pathologies

Carlos Guerrero-Bosagna

Avian Behavioral Genomics and Physiology Group, IFM Biology, Linköping University, Linköping, Sweden

OUTLINE

INTRODUCTION

Noncommunicable diseases, e.g., noninfectious and nontransmissible among people, are the most prevalent in human populations. The focus of discussion about the causes of noncommunicable diseases has shifted in recent years. If only a decade ago the main focus of attention about the etiology of noncommunicable diseases was on genetic predisposition, it has strongly switched to the role of early development and environmental exposures. Focusing on early developmental exposures when searching for the causes of diseases is becoming standard

The Epigenome and Developmental Origins of Health and Disease
http://dx.doi.org/10.1016/B978-0-12-801383-0.00010-4

Copyright © 2016 Elsevier Inc. All rights reserved.

practice [1,2], especially in the light of current human exposures to environmental toxicants derived from globalization and the industrialization of increasing parts of the planet [1,2].

This emphasis on early developmental exposures is not casual. Biologists have increasingly perceived a need for a more systemic approach to interrogate organisms in order to balance the dominating gene-centric view that has permeated biology for decades [3]. Although the technical complexities of analyzing organisms in their developmental and environmental context are high, even higher are the gains in the understanding of biological processes when a systemic view is applied. In that sense, focusing on the developmental origins of phenotypes (diseases included) requires understanding two enormous dimensions not properly considered previously: (1) developmental timing and (2) emergent properties. Interestingly, these are strongly interconnected because emergent properties arise from the integration of molecular processes during the development of organisms.

Reproductive pathologies appear to be particularly sensitive to environmental cues, and most of the time its emergence is explained by environmental exposures rather than by genetic factors. In this sense, the consideration of both *developmental timing* and *emergent properties* is paramount in order to understand the molecular processes involved in the emergence of reproductive pathologies. This is because reproduction is in itself highly dependent on environmental exposures. Indeed, reproduction can be defined as the conservation in the progeny of both structural (e.g., molecules, cellular compartments) and environmental features involved in the emergence of a living organization [4]. Therefore, reproduction is much more than the action of hormones, sexual coupling, or the transgenerational transfer of genes, and it is influenced by factors that are both intrinsic and extrinsic to organisms. It is from the interaction of these factors during reproductive and developmental processes that epigenetic marks emerge in the genome. These marks are accessory chemical modifications of the DNA that are mitotically stable [5] and are crucial for the regulation of gene expression. The list of known epigenetic modifications is constantly growing, but the best known include DNA methylation or hydroxymethylation of CG dinucleotides, chemical modifications of histones, interaction of DNA with small RNAs, or states of chromatin condensation [6,7].

Epigenetic patterns in the genome correlate to the environmental exposures that organisms have been subjected to during their ontogeny. Therefore, a "coupling" exists between epigenetic patterns and the environment where organisms live and/or develop prenatally, which generates those phenotypes that we recognize as "normal." Therefore, these "normal" phenotypes and associated epigenomes are contingent to the environment that is usually experienced by organisms. However, if either these intrinsic or extrinsic signals are in some way altered, epigenomic alterations and consequently phenotype "abnormalities" (e.g., disease states) can also emerge [6]. The study of male reproductive pathologies, as it will be seen, is an excellent approach to understand mechanisms through which early developmental exposures (fetal and perinatal) and resulting epigenetic modifications will generate adult onset diseases or dysfunctions.

This chapter delves into several aspects of the epigenetic and developmental processes involved in the etiology of male reproductive pathologies. Such aspects include: (1) how male fertility is affected by or correlates with environmental exposures in adulthood; (2) how environmental exposures can affect pregnant females and interfere with sensitive periods of offspring development (pre- and postnatally); (3) how this early developmental interference can lead to altered epigenetic patterns; (4) how certain developmental interferences can induce changes that last for several generations; and (5) how epigenetic marks can help to identify male reproductive pathologies.

ENVIRONMENTAL FACTORS KNOWN TO CORRELATE WITH MALE REPRODUCTIVE DYSFUNCTION

Trends in human populations showing decreasing male reproductive parameters in the last decades are well documented [8–10]. Examples of these trends include decline in sperm quality and count [9,11] and increases in the incidence of testicular cancer [10], cryptorchidism, or hypospadias [8]. A recent nationwide study in Sweden demonstrates an approximate twofold increased incidence in hypospadias from 1990 to 1999 that is not attributable to any previously known risk factors [12]. Because genetic context does not seem to explain these trends, the most plausible explanation is that environmental exposures to which humans have been subjected in the last decades are promoting epigenetic changes related to the observed impairments in male reproductive parameters [13].

Several factors in the environment can affect the quality and daily production of sperm. One of the most studied factors that correlates to reduced fertility is high temperatures on male genitals due to habitual exposure [9,14]. For example, studies indicate that sitting for prolonged periods is related to a decrease in the number and quality of sperm due to a reduced thermal exchange in this position that leads to increased temperature of the testes [9]. Other habits that work similarly are prolonged immersion in baths with temperatures higher than 37°C, the use of computers on top of the lap, or saunas set at very high temperatures for prolonged periods of time [9]. Testicular heating has even been studied as a contraception method [15].

The role of overweight and obesity on male reproductive parameters has recently gained great attention. Although studies show contradictory conclusions regarding associations of body mass index and sperm parameters [16,17], strong evidence supports a negative correlation between body mass index and testosterone or sex hormone-binding globulin levels [17]. Also, overweight and obese are associated with an increased prevalence of azoospermia or oligozoospermia [18], with obese individuals outnumbering nonobese by threefold among men with poor semen quality [19]. Also, a study evaluating men from subfertile couples found that obese individuals present higher sperm DNA damage than overweight and normal weight individuals [20].

In addition to the reproductive effects of unintended environmental exposures, the use of certain medications or drugs is also associated with a reduction in fertility parameters. For example, studies have shown that antidepressants are genotoxic for germ cells [21] and that certain natural compounds used as antidepressants, such as St. John's wort, are mutagenic to sperm cells [22] and induce abnormalities in testis chromosomes and in the morphology of spermatozoa [23]. There is evidence both in humans and laboratory animals indicating that the chronic use of drugs like marijuana, cocaine, methamphetamine, narcotics, or opioids produce anabolic androgenic effects associated with reduced fertility [24]. These effects include reduction in the sperm quality and number, and reduced levels of key fertility hormones, such as testosterone. Also, both cigarette smoking and alcohol consumption are shown to be associated with decreased sperm parameters [25].

DEVELOPMENTAL ENVIRONMENT AND ESTABLISHMENT OF EPIGENETIC PATTERNS

The environment that the embryo experiences during early development (fetal and perinatal) is crucial for determining the developmental path that is taken by the organism at later stages. Therefore, external influences that affect these early developmental stages can have enormous consequences on the adult phenotype. In mammals, in order to affect embryos these environmental influences have to cross physiological barriers imposed by organs such as the placenta and the uterus. Several factors are among the environmental influences reported to have effects in uterine and/or placental tissues with consequences for the developing embryos. These compounds include endocrine-disrupting chemicals such as phytoestrogens [26], diethylstilbestrol (DES) [27], organochlorines (dichlorodiphenyl dichloroethylene and polychlorinated biphenyls) [28], and components of plastics (bisphenol A [29] and phthalates [30]). Other compounds capable of crossing these barriers are inorganic environmental contaminants such as cadmium [31], constituents of tobacco smoking [32], or nutritional compounds such as folates [33]. Interestingly, even though the concentration of some compounds is reduced after crossing the placental barrier, other compounds can be completely transferred to the fetus [34].

Importantly, one of the main molecular mechanisms affected during these developmental exposures are epigenetic processes. The identification of crucial developmental periods of major epigenetic reprogramming in laboratory rodents was a fundamental step in epigenetic research [35]. For DNA methylation, two developmental periods are identified in which major resetting occurs in mammalian genomes [35,36]. One of these periods is postfertilization, during cleavage and blastocyst stages, where a global reduction in DNA methylation occurs followed by reestablishment of DNA methylation patterns by the time of embryo implantation [36,37]. This epigenetic reprogramming is crucial for the differentiation of somatic cells lines.

Another period of epigenetic reprogramming occurs during the migration of primordial germ cells (PGCs) toward their final establishment in the gonads [38]. During this migration, a major demethylation of the genome also occurs followed by remethylation [35,37,38]. In addition to this resetting in DNA methylation, major changes in other epigenetic mechanisms also occur during PGC development [39]. Both of these periods of resetting of DNA methylation patterns are windows of sensitivity to environmental exposure [5,6]. Interfering with the resetting period of PGCs, however, has different implications than interfering with the resetting period of preimplantation embryos. Because the germ line has the ability of transmitting epigenetic marks to next generations, altered DNA methylation patterns produced in the germ line after interferences with this epigenetic reprogramming can be transgenerationally perpetuated [5]. This is described as the phenomenon of *transgenerational epigenetic inheritance* [40] and will be later described in detail.

DEVELOPMENTAL EXPOSURES AND THE INCIDENCE OF MALE REPRODUCTIVE ABNORMALITIES

Developmental disruption of epigenetic processes can have drastic phenotypic consequences in animals, ranging from changes in coat color [41] to increased propensity to diseases [3]. The field of "developmental origins of health and diseases" (DOHaD) investigates developmental and epigenetic processes involved in the etiology of adult diseases [42]. Nowadays, several diseases have been found to correlate with early developmental exposures and epigenetic changes. These include allergies [43], hepatic cancer [44], gastric cancer [45], asthma [46], colorectal cancer [47], prostate cancer [48], HIV latency [49], metabolic diseases [50], and cardiovascular diseases [51]. Sociobiological aspects such as behavior [52], depression [53], and brain disorders [54] have also been found to have associations with environmental exposures and epigenetic alterations during early development.

Interestingly, reproductive abnormalities (in both males and females) seem to be particularly prone to emerge as a result of early developmental exposures that alter epigenetic patterns. Well-studied examples include reproductive abnormalities and epigenetic alterations after exposures to DES [55–57], phytoestrogens [58,59], the fungicide vinclozolin [60–63], or the plasticizer bisphenol A (BPA) [64]. The consideration of environmental exposures during early development would be fundamental for understanding the etiology of currently observed male reproductive pathologies.

Because the genetic background is essentially static in populations where increases in male disorders are occurring, early exposures to environmental toxicants during fetal development and associated epigenetic changes may be fundamental to explain the observed trends [65]. In an etiology that is similar to the phenotypes associated with the metabolic disease syndrome [66], male reproductive disorders are suggested to have a common developmental origin and to be physiologically linked [67,68]. These common attributes have led the authors to group these male reproductive disorders into the concept of "testicular dysgenesis syndrome" (TDS) [68,69]. TDS describes these male reproductive disorders as sharing the same pathophysiological etiology and as being caused by early developmental disruption during testicular development. It is therefore crucial to correlate environmental exposures with epigenetic variation in germ and somatic cells involved in male reproduction in order to address fertility issues in humans [70].

Many examples of prenatal exposures related to impaired adult male reproductive function exist. These include fetal exposures generated from maternal daily habits or resulting from maternal contact with environmental toxicants. Exposures related to maternal daily habits that are known to influence male reproductive parameters in the offspring include smoking, administration of analgesics, and alcohol consumption. Maternal smoking during pregnancy, for example, produces negative reproductive effects in their adult male descendants, which present reduced sperm number [71,72], earlier onset of puberty, lower final adult height, higher body mass index, smaller testicles, reduced spermatogenesis-related hormones, and higher free testosterone [72]. Reproductive effects on male offspring associated with maternal exposure to analgesics include congenital cryptorchidism in humans and reduction in the anogenital distance and testicular testosterone production in rats [73]. Regarding maternal alcohol consumption, some studies suggest increased risk of congenital cryptorchidism in newborns [74], while other researchers have found no effects [75].

Compounds categorized as environmental toxicants that negatively influence male reproduction or the development of the urogenital tract through a maternal exposure include plastic compounds, pesticides, the fungicide vinclozolin, dioxin, DES, and hydrocarbon fumes. Maternal exposure to phthalates, for example, is widely known to disrupt the offspring's male reproduction. The evidence for male offspring abnormalities derived from maternal exposure to phthalates is vast and has been reported in rodents and humans (reviewed in [76]). Such effects include reduced testicular testosterone synthesis; decreased anogenital distance; reduced penile length; occurrence of retained nipples; malformations of the epididymis, vas deferens, seminal vesicles, and prostate; incidence of hypospadias, cryptorchidism, and lesions of the gubernaculum [76]; disruption of Leydig cell development [77]; and altered testicular gene expression [78,79]. In humans, fetal exposure to phthalates has been associated with reduced anogenital distance in the offspring [80,81].

In addition to phthalates, other plastic compounds have been shown to alter male reproduction after an early developmental exposure. The effects of gestational exposure to the plasticizer BPA are well documented (reviewed in [82]) and include hormonal disturbances on the HPT axis, changes in testicular morphology, reduced sperm production, and impairments in sperm motility and morphology [82]. More recently, it has been shown that gestational exposure to BPA decreases anogenital distance in both the male and female offspring and induces nipple retention in males [83]. Also, increased pubertal abnormalities have been observed in the offspring of rat mothers treated with a mixture of BPA and phthalates [84].

Another well-characterized maternal exposure that results in male reproductive abnormalities is to the fungicide vinclozolin. Broad reproductive effects in the male offspring have been reported by different groups in response to gestational exposure to vinclozolin. Such effects include ejaculation failure, incidence of permanent nipples, hypospadias, cryptorchidism, reduced anogenital distance, and sperm count, in addition to reduced weights of the seminal vesicle, ventral prostate, and epididymis [85–87], increased spermatogenic cell apoptosis, and reduced sperm motility [88].

Exposure to pesticides is a major concern, since in vast regions of the world agro-workers and populations living nearby agricultural areas can be exposed to large amounts of pesticides. Fetal exposure to pesticides has been shown to correlate with the incidence of cryptorchidism [89,90]. A study focusing on a cohort of newborns in France found that maternal exposure to pesticides was the main risk factor associated with increased cryptorchidism [90]. Interestingly, mother-mediated exposure of the child to pesticides could persist after parturition, since higher amounts of pesticides are found in the breast milk of mothers of boys with cryptorchidism [89]. In rodents, gestational exposures to pesticides are also reported to generate male reproductive or urogenital abnormalities in the offspring. Maternal exposure to methoxychlor causes pubertal abnormalities [91], and maternal exposure to a mixture of permethrin and DEET induces kidney abnormalities [92].

DES is a powerful estrogenic compound formerly used to "treat" miscarriage in women. It was later observed that treatment of these women from the 1930s to 1970s with DES generated reproductive effects in their offspring (reviewed in [93]). The incidence of hypospadias in the sons of these exposed women has been correlated to a gestational exposure to DES [94–96]. It is also important to highlight that even after DES was banned for use by women it was still used to stimulate growth in cattle until 1979. Interestingly, a recent study observed a difference in sperm parameters in sons of women eating large amounts of beef during their pregnancy [97]. This difference was suggested to be due to contaminants in beef such as DES. Although these studies in humans are correlational, the casual effects of gestational exposure to DES have been defined in animal models. In male rodents, gestational exposure to DES has been shown to generate reproductive tract tumors [57] and testes of reduced size [98]. The DES-stimulated hormonal toxicity involves epigenetic modifications in the seminal vesicle [56].

An exposure to natural compounds that seems to also induce a maternally mediated negative effect for male fertility is to phytoestrogens. It has been suggested in humans that a maternal diet high in phytoestrogens could induce a higher male offspring propensity to hypospadias [99]. This has been shown in rodents, in a recent study that finds increased incidence of hypospadias in the male offspring of mothers exposed to high amounts of the phytoestrogen genistein alone or in combination with vinclozolin during gestation [100].

Interestingly, studies are scarce regarding the effects of maternal nutrition (one of the most important aspects of pregnancy) on offspring reproductive parameters. A recent study in rats, however, indicates that maternal consumption of a high energy content cafeteria diet before and/or during pregnancy had reproductive effects in the male offspring such as reduced plasmatic follicle stimulating hormone, luteinizing hormone, and testosterone in the male offspring, as well as lower frequencies of intromission and smaller percentage of animals with intromission behavior [101]. Also in rats, it has been shown that a high-fat maternal diet during pregnancy and lactation triggers early onset of puberty in both the male and female offspring [102]. A study in sheep found that maternal undernutrition during pregnancy reduces the number of Sertoli cells in the newborn lamb [103].

Other less studied maternal exposures that also affect male urogenital development and reproduction are to jet fuel 8, dioxin, and heavy metals. In rats, gestational exposure to jet fuel 8 generates prostate, kidney, and pubertal abnormalities in the male offspring [104], while gestational exposure to dioxin induces prostate and kidney abnormalities in the male offspring [105]. In humans, maternal exposure to heavy metals is associated with increased hypospadias in their sons [106,107].

ENVIRONMENTALLY INDUCED TRANSGENERATIONAL EPIGENETIC EFFECTS RELATED TO MALE REPRODUCTION

The first study that indicated that gestational exposure to endocrine disruptors could generate disease phenotypes that could be transmitted to future generations was performed by Newbold and collaborators in 2000 [57], while testing reproductive effects of a gestational exposure to DES. DES was administered to pregnant rats during early postimplantation development and generated an increased susceptibility for an otherwise rare proliferative lesion in the *rete* testis, and for tumor formation in other reproductive tract tissues, in the F1 and F2 generation male offspring [57]. This evidence suggested transmissible epigenetic alterations in the offspring due to this gestational exposure to DES.

The first description of an environmentally induced transgenerational epigenetic inheritance process was using an early developmental exposure to the endocrine disruptor vinclozolin [60]. Vinclozolin is a fungicide with anti-androgenic effects widely used in crops around the world [108]. It was previously reported in rats that a developmental exposure to vinclozolin via intraperitoneal injections on the mother produced increased apoptosis in spermatogenic cells in the offspring [88]. Interestingly, further experiments revealed that this increase in spermatogenic cell apoptosis occurred also in the next generations in the absence of any further exposure, being observed even four generations later [60,61,109]. The transmission of this phenotype across generations was attributed to an initial alteration of the sperm epigenome (DNA methylation) and the further transgenerational propagation of these epigenetic changes through the germ line, three generations after the developmental exposure to vinclozolin [60,62,110].

In addition to spermatogenic cell apoptosis, other male reproductive abnormalities have been shown to be transmitted through transgenerational epigenetic inheritance. Moreover, recent studies show that a variety of environmental toxicants can trigger this phenomenon. These transgenerationally transmitted altered phenotypes include kidney abnormalities after ancestral exposures to vinclozolin [61], dioxin [105], DDT [111], and methoxychlor [91]; prostate abnormalities after an ancestral exposure to vinclozolin [61]; pubertal abnormalities after ancestral exposures to dioxin [105]; and testis abnormalities after ancestral exposure to a mixture of pesticides (Permethrin and DEET) [92] and DDT [111]. Ancestral exposure to all these compounds induced sperm epimutations (i.e., DNA methylation changes) in the unexposed F3 generation, therefore opening the conceptual possibility that sperm epimutations can be indicators of both ancestral exposures and susceptibility to diseases [111,112].

Other groups reported similar transgenerational transmission of altered DNA methylation patterns in imprinted genes in the sperm induced by an ancestral developmental exposure to vinclozolin [113]. The same group reported that early developmental exposure to methoxychlor also altered DNA methylation in imprinted genes in the sperm, but this effect persisted for two generations and disappeared in the third generation after the exposure [114]. Transgenerational effects related to impairment of male fertility have also been shown in response to an early developmental exposure to di-(2-thylhexyl) phthalate, which was reported to induce transgenerationally transmitted disruption of testicular germ cell associations, reduced sperm count, and decreased sperm motility [115]. Recently, transgenerational male reproductive effects have also been reported in response to a developmental exposure to DDE [116], which is a degradation product of DDT. Early developmental exposure to DDE generated increased testicular germ line apoptosis from the F1 to the F3 generations [116]. Moreover, F3 generation males showed higher incidence of infertility and small testes, which was not observed in the F1 or F2 generations [116]. In the sperm, altered methylation patterns in the IGF2/H19 DMR, decreased expression of IGF2, and increased expression of H19 were observed from the F1 to the F3 generations [116]. Therefore, the phenomenon of transgenerational epigenetic inheritance has been currently reported by different groups, in several model organisms, and induced by a variety of compounds.

Not only developmental exposures but exposures after birth have also been evaluated for their transgenerational effects. For example, perinatal exposures to BPA have been shown to impair spermatogenesis and fertility [82] and to induce transgenerational alterations in the expression of steroid receptors and their co-regulators in testis [117]. Six-week-old male mice given oral administration of benzo(a) pyrene presented impairments in several fertility parameters up to the F2 generation [118]. These include testicular malformations, decreased sperm count, and reduced number of seminiferous tubes with elongated spermatids [118]. Although studies reporting transgenerational effects derived from an adult germ line disturbance are currently not many, they certainly raise attention and open interesting questions regarding the process of transgenerational epigenetic inheritance. These include which period of the adult germ line differentiation can be affected so as to generate epigenetic changes, and whether these epigenetic changes are similar in nature to those produced after a fetal developmental exposure.

GERM LINE TO SOMATIC EPIGENETIC EFFECTS: LESSON FROM THE VINCLOZOLIN TRANSGENERATIONAL MODEL

The germ line development has a crucial role in the phenomena of transgenerational epigenetic inheritance since these germ line epigenetic alterations induced by environmental exposures occurring early during development can be transmitted to next generations [5,121]. These epigenetic germ line alterations will subsequently affect gene expression and epigenetic patterns in somatic tissues [63,119]. These altered somatic epigenetic marks and gene expression are involved in generating altered phenotypes that are observed to be transgenerationally transmitted [5,120].

Given all the angles and variables involved in this process, expectedly there are gaps in knowledge regarding the biological mechanisms by which an environmental insult that interferes with the establishment of epigenetic patterns early in the germ line development can influence the phenotype of individuals several generations after this exposure. Further investigations attempted to address some of these issues. One important open question was whether the gestational exposure to vinclozolin indeed altered patterns of DNA methylation transgenerationally in the germ line development in the embryo. To address this question, PGCs were obtained at embryonic days E13 and E16, which encompass the period between the onset of gonadal sex determination and after testis cord formation [121].

We investigated patterns of DNA methylation in these two developmental times in PGCs in the F3 generation unexposed embryo from a lineage of vinclozolin exposure. Altered DNA methylation patterns were observed in both of these developmental times with regard to controls, which were also concomitant with alterations in gene expression [121].

A further step was to investigate whether DNA methylation patterns and gene expression were also transgenerationally altered in somatic cells involved in supporting germ line development. For this, Sertoli cells were selected from unexposed F3 generation males from a lineage of ancestral exposure to vinclozolin and were compared to controls. We observed DNA methylation changes in these Sertoli cells that had distinct degree of connections to the changes in gene expression in the same cells [63]. These connections included physical proximity in genomic regions as well as functional correlations. It is important to highlight that the changes in gene expression were not necessarily related to changes in DNA methylation because the latter takes place early in the development and the former are observed later on, and as a consequence of the DNA methylation alterations. In the same way, the DNA methylation changes in the germ line certainly influence the somatic DNA methylation changes but are also generally not the same. A schematic representation of this concept is shown in Figure 1.

Overall, this research indicates that epigenetic markers both in somatic tissues and in the germ line will be able to be used to detect ancestral exposures and propensity to diseases such as male reproductive pathologies. Not much research has been done on using germ line biomarkers to detect ancestral exposures, but a growing body of evidence indicates that epigenetic biomarkers can be used to identify propensity to diseases. The particular case for epigenetic marks of male reproductive pathologies is described in the next two sections.

EPIGENETIC MARKS IN FERTILITY: DNA METHYLATION AND HISTONE MODIFICATIONS

Epigenetic marks in the germ line can be useful both as indicators of ancestral exposures to specific environmental toxicants [112] or as indicators of propensity to diseases, particularly of male infertility [70]. Extensive natural sperm epigenomic variation has been observed between healthy men, with some regions being more variable than others [122]. Epigenetic disruptions have been documented in association with the incidence of different types of germ cell tumors [123] or with impaired fertility and spermatogenesis [124]. Epigenetic disruption is observed as abnormal histone (e.g., H3K4me and H3K27me) [125,126] or DNA methylation marks at imprinted and developmental loci [125–130] in the sperm DNA of infertile men. Genome-wide changes in histone marks would alter sperm DNA chromatin packaging during spermiogenesis, generating poor reproductive outcomes [125,131]. For DNA methylation, the changes associated with infertility also appear to be widespread in the sperm genome [132]. These include nonimprinted genes [132] as well as imprinted and developmental loci [125–130,132]. An association between oligospermia and aberrant imprinting in the sperm DNA has consistently been found in studies interrogating epigenetic marks of infertility [127–130]. Interestingly, such abnormal imprinting could even be transmitted to next generations if these sperm cells are used in assisted reproductive technologies [127,128].

Several specific DNA methylation changes in the germ line that are associated with infertility in either humans or rodents have been documented to date. These include DNA methylation alteration in the promoter region of *Mthfr* [133], hypomethylation in regions of the imprinted *Igf2-H19* locus [128–130,134], hypermethylation in the imprinted genes *Mest* [126,129], *Lit1*, *Snrpn*, *Peg3*, and *Zac* [126], as well as altered methylation in several other (imprinted and nonimprinted)

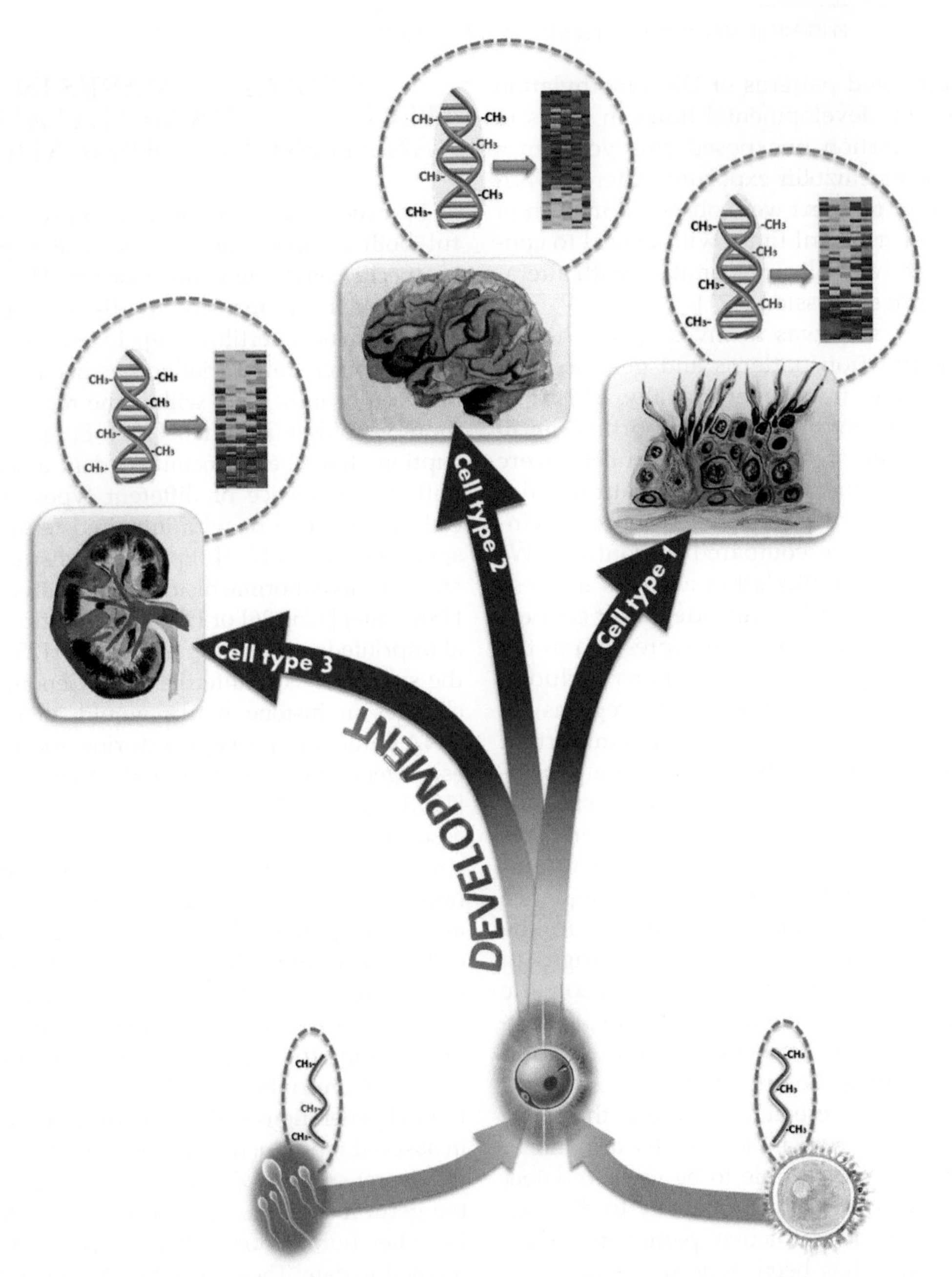

FIGURE 1 Schematic representation of the effects germ line alterations in DNA methylation will have after fertilization, on the further development of different cell types. Methyl groups in red indicate altered methylation, while methyl groups in black indicate a normal methylation pattern. Epigenetic germ line alterations will influence epigenetic patterns in somatic tissues (e.g., kidney, brain, or Sertoli cells, depicted in the figure), although the changes observed in either case are not necessarily the same. Somatic epigenetic changes can also be more numerous in comparison with the germ line changes. These somatic epigenetic modifications will in turn trigger changes in gene expression (shown as heat maps). In the same way, the somatic changes in DNA methylation do not necessarily produce changes in the expression of the same genes. An amplification effect can take place in gene expression, initially induced by the changes in DNA methylation triggered early in the development. In addition, changes in DNA methylation can regulate the activity of several genes in parallel, through epigenetic control regions.

genes such as *Hras, Nt3, Mt1a, Pax8, Diras3, Plagl1, Sfn,* and *Sat2chrm1* [132].

Histone marks in the sperm DNA have also been associated with infertility. These include increased H3K9 acetylation and H3K27 trimethylation in exonic regions of *Brdt* (which leads to its repression) [135], reduced H4 acetylation in spermatids of infertile men with qualitatively normal or abnormal spermatogenesis [136], loss of normal demethylation in H3K9 (which reduces expression of TNP1 and PRM1, required for histone replacement in spermiogenesis) [137], and aberrant histone acetylation in promoters of developmental genes (which leads to an insufficient sperm chromatin compaction that persist in the zygote) [131].

Besides the importance of epigenetic mechanisms in the germ line, epigenetic marks in somatic cells that surround or support germ cells can also be found altered in male reproductive pathologies. A good example is the *Rhox* homeobox gene cluster at the X chromosome. This gene cluster has recently appeared as important for androgen regulation of developmental reproductive processes in mammals, in which epigenetic marks play an important role [138]. In Sertoli cells, *Rhox5* is the founding member of the *Rhox* cluster, and its deletion associates with subfertility, increased germ cell apoptosis, and decreased sperm count and motility [138]. Interestingly, *Rhox5* expression is controlled by the independent regulation of two promoters, with their activity being tissue- and DNA-methylation dependent [138].

EPIGENETIC MARKS AND FERTILITY: SMALL NONCODING RNAs

It is also becoming increasingly evident that noncoding RNAs in somatic or germ cells involved in spermatogenesis have a fundamental role in spermatogenesis [139]. Small noncoding RNAs involved in sperm production include micro-RNAs (miRNAs), endogenous small interfering RNAs (endo-siRNAs), and PIWI-interacting RNAs (piRNAs) [139]. These small RNAs are categorized depending on the involvement of the RNase III endonuclease DICER in their biogenesis mechanisms [139]. DICER is essential for microRNA processing and in testicular tissues is involved in the haploid differentiation of germ cells [140]. Endo-siRNAs and miRNAs are DICER dependent, while piRNAs are DICER independent and expressed predominantly in tissues involved in male germ line development [139].

The disruption of the biogenesis or expression of small noncoding RNAs has important consequences from a fertility standpoint [139]. Removal of functional DICER in PGCs generates defects in proliferation, which affects further spermatogenesis [141,142]. Prebirth conditional deletion of DICER produces delayed progression of meiotic prophase I and increased apoptosis in spermatocytes [143,144]. Effects of postbirth DICER deletion are milder but include disruption in chromatin organization, abnormalities in the sperm head morphology, as well as reduction in the number of epididymal spermatozoa [140]. In Sertoli cells, ablation of DICER leads to infertility with testis degeneration and full absence of spermatozoa [145]. Moreover, dysregulation of microRNAs in Sertoli cells is shown to mediate androgen-induced spermiation failure and suppression of the follicle-stimulating hormone [146].

Several miRNAs have been implicated in the process of spermatogonia differentiation. These include miR-146 [147], miR-221, and miR-222 [148], and the clusters Mir-17-92 and Mir-106b-25 [149]. In mouse, the deletion of the Mir-17-92 cluster results in small testes and reduced numbers of epididymal sperm [149]. In cultured mouse spermatogonial stem cells, the transient inhibition of miR-21 increases the number of apoptotic germ cells [150]. On the other hand, the expression of miR-34 seems to induce germ cell apoptosis [151].

The small RNAs termed piRNAs are associated to PIWI proteins. These are a subset of the Argonaute proteins expressed predominantly in the germ line and have been found in a variety of organisms [152,153]. PIWI proteins are essential for germ cell maintenance and spermatogenesis [152,153]. Analysis of the expression profiles of piRNAs during mouse spermatogenesis has revealed that distinct classes of piRNAs are expressed during different developmental stages of spermatogenesis [152]. These classes are related to the genomic regions from which piRNAs can derive, which include intergenic regions or genomic regions coding for retrotransposons or mRNAs [152]. Disruption of the pathway involved in piRNA biogenesis has consequences in fertility. For example, in the testicular tissue of men with different fertility problems, two genes involved in the pathway of piRNAs, PIWIL2, and TDRD1 are hypermethylated [154]. Table 1 summarizes the epigenetic modifications that are known to have a developmental role in the etiology of male infertility.

In addition to small RNAs and DNA methylation, the activity of repeat elements such as B1 SINEs has also been proposed to have a role in transcriptional regulation of testis-specific genes [155]. Interestingly, DICER would have a fundamental role in the regulation of the activity of transposable elements [139]. One important known role of piRNAs is indeed repression of the activity of transposable elements [156,157]. It is important to highlight that repeat element activity is tightly regulated by epigenetic mechanisms, mainly DNA methylation [158].

CONCLUSIONS AND PERSPECTIVES

The study of developmental processes and early environmental exposures is helping researchers understand previously underestimated biological mechanisms in disease etiology. Reproductive biology is a special field in this sense because it was within the realm of reproductive biology that the importance of the action of endocrine-disrupting chemicals surfaced. Moreover, it was due to investigation of the early developmental effects of these endocrine disruptors that transgenerational epigenetic effects were also discovered.

Currently, the same research emphasis can be applied to several other common environmental exposures, whether or not they are endocrine disrupting chemicals. The number of papers investigating early environmental exposures and epigenetic effects have multiplied in recent years, alongside studies investigating transgenerational effects. This is certainly positive in biology because it opens a completely new dimension in the understanding of disease etiology and allows us to focus on the involvement of environmental exposures (ancestral or current) as an essential causal factor of diseases. This is in addition to the common focus on genetic causes of diseases. However, together with the opening of new research possibilities, this new focus also brings new challenges, since studies are becoming increasingly multifactorial and complex. Moreover, it also brings new responsibilities for researchers as communicators for the dangers of some environmental exposures and of compounds that are purposely being taken in pharmacological doses by human populations. The administration of these compounds can have unintended important side effects on health, particularly in fertility.

One more dimension is that exposures can affect future generations in their susceptibility to diseases, particularly disease related to reproduction. As for human fertility, it is becoming increasingly obvious that current trends and observed diseases have great relationship to what we have been exposed to as modern society. Moreover, due to the description of the process of transgenerational epigenetic inheritance, it is now feasible, in theory, to detect epigenetically based transgenerational susceptibility to

TABLE 1 Epigenetic Modifications Shown to Have a Developmental Role in the Etiology of Male Infertility

Epigenetic Modifications Related to Male Infertility		Cell Type	References
DNA methylation	Hypermethylation in *Lit1*, *Snrpn*, *Peg3*, and *Zac*	Germ cells	[125]
	Hypermethylation in *Mest*	Germ cells	[125,128]
	Hypomethylation in regions of the *Igf2-H19* locus	Germ cells	[127–129,133]
	Altered DNA methylation in *Hras*, *Nt3*, *Mt1a*, *Pax8*, *Diras3*, *Plagl1*, *Sfn*, and *Sat2chrm1*	Germ cells	[131]
	Altered DNA methylation in *Mthfr*	Germ cells	[132]
	Involvement of *Rhox5* DNA methylation	Sertoli cells	[137]
Histones	Aberrant histone acetylation in promoters of developmental genes	Germ cells	[130]
	Reduced H4 acetylation	Germ cells	[135]
	H3K9 acetylation and H3K27 trimethylation in *Brdt*	Germ cells	[134]
	Absence of normal H3K9 demethylation	Germ cells	[136]
Small RNAs	Impaired biogenesis of DICER-dependent small RNAs	Germ cells	[140,141]
	Micro-RNA involvement (specifically miR-23b, miR-30c, miR-30d, and miR-690)	Sertoli cells	[145]
	Involvement of miR-146	Germ cells	[146]
	Impaired function of miR-221 and miR-222	Germ cells	[147]
	Impaired biogenesis of micro-RNAs originating from the Mir-17-92 cluster	Testes and germ cells	[148]
	Inhibition of miR-21	Germ cells	[149]
	Expression of miR-34	Germ cells	[150]
	Impaired biogenesis of piRNAs	Testes	[153]

diseases in individuals and populations. Considering this technical possibility and the trends indicating progressive impairment of reproductive parameters in human males, the identification of epigenetic biomarkers related to infertility will be paramount as a diagnostic tool. Future research efforts should focus on identifying these biomarkers and on understanding the epigenetic mechanisms that link environmental exposures with impairment of fertility. The knowledge obtained will be important in order to provide the general public with information that could be used to change habits and minimize harmful exposures, as well as to address the biological causes of human fertility decline from a policy-making perspective.

Acknowledgments

The author greatly appreciates critical revision of the manuscript by Dr Per Jensen, Dr Heriberto Rodriguez-Martinez, and Dr Jordi Altimiras, as well as drawings of organ tissues by Dr Julia Uffenorde. The author is grateful for funding support by the European Research Council Advanced Research Grant Genewell 322206, held by Dr Per Jensen.

References

[1] Balbus JM, Barouki R, Birnbaum LS, et al. Early-life prevention of non-communicable diseases. Lancet 2013; 381(9860):3–4.

[2] Vineis P, Stringhini S, Porta M. The environmental roots of non-communicable diseases (NCDs) and the epigenetic impacts of globalization. Environ Res 2014;133:424–30.

[3] Guerrero-Bosagna C, Skinner MK. Environmentally induced epigenetic transgenerational inheritance of phenotype and disease. Mol Cell Endocrinol 2012; 354(1–2):3–8.

[4] Maturana-Romesín H, Mpodozis J. The origin of species by means of natural drift. Rev Chil Hist Nat 2000;73:261–300.

[5] Skinner MK, Manikkam M, Guerrero-Bosagna C. Epigenetic transgenerational actions of environmental factors in disease etiology. Trends Endocrinol Metab 2010;21(4):214–22.

[6] Feil R, Fraga MF. Epigenetics and the environment: emerging patterns and implications. Nat Rev Genet 2011;13(2):97–109.

[7] Teperek-Tkacz M, Pasque V, Gentsch G, Ferguson-Smith AC. Epigenetic reprogramming: is deamination key to active DNA demethylation? Reproduction 2011;142(5):621–32.

[8] Main KM, Skakkebaek NE, Virtanen HE, Toppari J. Genital anomalies in boys and the environment. Best Pract Res Clin Endocrinol Metab 2010;24(2):279–89.

[9] Sharpe RM. Environmental/lifestyle effects on spermatogenesis. Philos Trans R Soc Lond B Biol Sci 2010;365(1546):1697–712.

[10] Skakkebaek NE, Rajpert-De Meyts E, Jorgensen N, et al. Testicular cancer trends as 'whistle blowers' of testicular developmental problems in populations. Int J Androl 2007;30(4):198–204. discussion -5.

[11] Merzenich H, Zeeb H, Blettner M. Decreasing sperm quality: a global problem? BMC Public Health 2010; 10:24.

[12] Nordenvall AS, Frisen L, Nordenstrom A, Lichtenstein P, Nordenskjold A. Population based nationwide study of hypospadias in Sweden, 1973 to 2009: incidence and risk factors. J Urol 2014;191(3):783–9.

[13] Guerrero-Bosagna C, Skinner MK. Environmentally induced epigenetic transgenerational inheritance of male infertility. Curr Opin Genet Dev 2014;26C:79–88.

[14] Mieusset R, Bujan L. Testicular heating and its possible contributions to male infertility: a review. Int J Androl 1995;18(4):169–84.

[15] Mieusset R, Bujan L. The potential of mild testicular heating as a safe, effective and reversible contraceptive method for men. Int J Androl 1994;17(4):186–91.

[16] Belloc S, Cohen-Bacrie M, Amar E, et al. High body mass index has a deleterious effect on semen parameters except morphology: results from a large cohort study. Fertil Steril 2014;102(5):1268–73.

[17] MacDonald AA, Herbison GP, Showell M, Farquhar CM. The impact of body mass index on semen parameters and reproductive hormones in human males: a systematic review with meta-analysis. Hum Reprod Update 2010;16(3):293–311.

[18] Sermondade N, Faure C, Fezeu L, et al. BMI in relation to sperm count: an updated systematic review and collaborative meta-analysis. Hum Reprod Update 2013;19(3):221–31.

[19] Magnusdottir EV, Thorsteinsson T, Thorsteinsdottir S, Heimisdottir M, Olafsdottir K. Persistent organochlorines, sedentary occupation, obesity and human male subfertility. Hum Reprod 2005;20(1):208–15.

[20] Dupont C, Faure C, Sermondade N, et al. Obesity leads to higher risk of sperm DNA damage in infertile patients. Asian J Androl 2013;15(5):622–5.

[21] Attia SM, Bakheet SA. Citalopram at the recommended human doses after long-term treatment is genotoxic for male germ cell. Food Chem Toxicol 2013;53:281–5.

[22] Ondrizek RR, Chan PJ, Patton WC, King A. An alternative medicine study of herbal effects on the penetration of zona-free hamster oocytes and the integrity of sperm deoxyribonucleic acid. Fertil Steril 1999;71(3):517–22.

[23] Aleisa AM. Cytological and biochemical effects of St. John's Wort supplement (a complex mixture of St. John's Wort, Rosemary and Spirulina) on somatic and germ cells of Swiss Albino mice. Int J Environ Res Public Health 2008;5(5):408–17.

[24] Fronczak CM, Kim ED, Barqawi AB. The insults of illicit drug use on male fertility. J Androl 2012;33(4):515–28.

[25] Gaur DS, Talekar MS, Pathak VP. Alcohol intake and cigarette smoking: impact of two major lifestyle factors on male fertility. Indian J Pathol Microbiol 2010;53(1):35–40.

[26] Moller FJ, Zierau O, Hertrampf T, Bliedtner A, Diel P, Vollmer G. Long-term effects of dietary isoflavones on uterine gene expression profiles. J Steroid Biochem Mol Biol 2009;113(3–5):296–303.

[27] Newbold RR, Jefferson WN, Padilla-Banks E, Haseman J. Developmental exposure to diethylstilbestrol (DES) alters uterine response to estrogens in prepubescent mice: low versus high dose effects. Reprod Toxicol 2004;18(3):399–406.

[28] Hsu WW, Osuch JR, Todem D, et al. DDE and PCB serum concentration in maternal blood and their adult female offspring. Environ Res 2014;132:384–90.

[29] Moors S, Diel P, Degen GH. Toxicokinetics of bisphenol A in pregnant DA/Han rats after single i.v. application. Archiv Toxicol 2006;80(10):647–55.

[30] Latini G, De Felice C, Presta G, et al. In utero exposure to di-(2-ethylhexyl)phthalate and duration of human pregnancy. Environ Health Perspect 2003;111(14): 1783–5.

[31] Nakamura Y, Ohba K, Suzuki K, Ohta H. Health effects of low-level cadmium intake and the role of metallothionein on cadmium transport from mother rats to fetus. J Toxicol Sci 2012;37(1):149–56.

[32] Shiverick KT, Salafia C. Cigarette smoking and pregnancy I: ovarian, uterine and placental effects. Placenta 1999;20(4):265–72.

[33] Guay F, Jacques Matte J, Girard CL, Palin MF, Giguere A, Laforest JP. Effects of folic acid and vitamin B12 supplements on folate and homocysteine metabolism in pigs during early pregnancy. Br J Nutr 2002;88(3):253–63.

[34] Li LX, Chen L, Meng XZ, et al. Exposure levels of environmental endocrine disruptors in mother–newborn pairs in China and their placental transfer characteristics. PLoS One 2013;8(5):e62526.

[35] Hackett JA, Surani MA. Beyond DNA: programming and inheritance of parental methylomes. Cell 2013;153(4):737–9.

[36] Hackett JA, Surani MA. DNA methylation dynamics during the mammalian life cycle. Philos Trans R Soc Lond B Biol Sci 2013;368(1609):20110328.

[37] Reik W, Dean W, Walter J. Epigenetic reprogramming in mammalian development. Science 2001;293(5532): 1089–93.

[38] Lees-Murdock DJ, Walsh CP. DNA methylation reprogramming in the germ line. Epigenetics 2008;3(1): 5–13.

[39] Hackett JA, Zylicz JJ, Surani MA. Parallel mechanisms of epigenetic reprogramming in the germline. Trends Genet 2012;28(4):164–74.

[40] Daxinger L, Whitelaw E. Understanding transgenerational epigenetic inheritance via the gametes in mammals. Nat Rev Genet 2012;13(3):153–62.

[41] Dolinoy DC, Huang D, Jirtle RL. Maternal nutrient supplementation counteracts bisphenol A-induced DNA hypomethylation in early development. Proc Natl Acad Sci USA 2007;104(32):13056–61.

[42] Godfrey KM, Lillycrop KA, Burdge GC, Gluckman PD, Hanson MA. Epigenetic mechanisms and the mismatch concept of the developmental origins of health and disease. Pediatr Res 2007;61(5 Pt 2):5R–10R.

[43] Kuriakose JS, Miller RL. Environmental epigenetics and allergic diseases: recent advances. Clin Exp Allergy 2010;40(11):1602–10.

[44] Pogribny IP, Ross SA, Wise C, et al. Irreversible global DNA hypomethylation as a key step in hepatocarcinogenesis induced by dietary methyl deficiency. Mutat Res 2006;593(1–2):80–7.

[45] Nan HM, Song YJ, Yun HY, Park JS, Kim H. Effects of dietary intake and genetic factors on hypermethylation of the hMLH1 gene promoter in gastric cancer. World J Gastroenterol 2005;11(25):3834–41.

[46] Martino D, Prescott S. Epigenetics and prenatal influences on asthma and allergic airways disease. Chest 2011;139(3):640–7.

[47] Choong MK, Tsafnat G. Genetic and epigenetic biomarkers of colorectal cancer. Clin Gastroenterol Hepatol 2011;10(1):9–15.

[48] Perry AS, Watson RW, Lawler M, Hollywood D. The epigenome as a therapeutic target in prostate cancer. Nat Rev Urol 2010;7(12):668–80.

[49] Hakre S, Chavez L, Shirakawa K, Verdin E. Epigenetic regulation of HIV latency. Curr Opin HIV AIDS 2011;6(1):19–24.

[50] Lillycrop KA. Effect of maternal diet on the epigenome: implications for human metabolic disease. Proc Nutr Soc 2011;70(1):64–72.

[51] Bertram C, Khan O, Ohri S, Phillips DI, Matthews SG, Hanson MA. Transgenerational effects of prenatal nutrient restriction on cardiovascular and hypothalamic-pituitary-adrenal function. J Physiol 2008;586(8):2217–29.

[52] Champagne FA, Weaver IC, Diorio J, Dymov S, Szyf M, Meaney MJ. Maternal care associated with methylation of the estrogen receptor-alpha1b promoter and estrogen receptor-alpha expression in the medial preoptic area of female offspring. Endocrinology 2006;147(6):2909–15.

[53] Oberlander TF, Weinberg J, Papsdorf M, Grunau R, Misri S, Devlin AM. Prenatal exposure to maternal depression, neonatal methylation of human glucocorticoid receptor gene (NR3C1) and infant cortisol stress responses. Epigenetics 2008;3(2):97–106.

[54] Kaminsky Z, Tochigi M, Jia P, et al. A multi-tissue analysis identifies HLA complex group 9 gene methylation differences in bipolar disorder. Mol Psychiatry 2010;17(7):728–40.

[55] Jefferson WN, Chevalier DM, Phelps JY, et al. Persistently altered epigenetic marks in the mouse uterus after neonatal estrogen exposure. Mol Endocrinol 2013;27(10):1666–77.

[56] Li Y, Hamilton KJ, Lai AY, et al. Diethylstilbestrol (DES)-stimulated hormonal toxicity is mediated by ERalpha alteration of target gene methylation patterns and epigenetic modifiers (DNMT3A, MBD2, and HDAC2) in the mouse seminal vesicle. Environ Health Perspect 2014;122(3):262–8.

[57] Newbold RR, Hanson RB, Jefferson WN, Bullock BC, Haseman J, McLachlan JA. Proliferative lesions and reproductive tract tumors in male descendants of mice exposed developmentally to diethylstilbestrol. Carcinogenesis 2000;21(7):1355–63.

[58] Guerrero-Bosagna CM, Skinner MK. Environmental epigenetics and phytoestrogen/phytochemical exposures. J Steroid Biochem Mol Biol 2014;139:270–6.

[59] Jefferson WN, Padilla-Banks E, Goulding EH, Lao SP, Newbold RR, Williams CJ. Neonatal exposure to genistein disrupts ability of female mouse reproductive tract to support preimplantation embryo development and implantation. Biol Reprod 2009;80(3):425–31.

[60] Anway MD, Cupp AS, Uzumcu M, Skinner MK. Epigenetic transgenerational actions of endocrine disruptors and male fertility. Science 2005;308(5727):1466–9.

[61] Anway MD, Leathers C, Skinner MK. Endocrine disruptor vinclozolin induced epigenetic transgenerational adult-onset disease. Endocrinology 2006;147(12):5515–23.

[62] Guerrero-Bosagna C, Covert TR, Haque MM, et al. Epigenetic transgenerational inheritance of vinclozolin induced mouse adult onset disease and associated sperm epigenome biomarkers. Reprod Toxicol 2012;34(4):694–707.

[63] Guerrero-Bosagna C, Savenkova M, Haque MM, Nilsson E, Skinner MK. Environmentally induced epigenetic transgenerational inheritance of altered Sertoli cell transcriptome and epigenome: molecular etiology of male infertility. PLoS One 2013;8(3):e59922.

[64] Mileva G, Baker SL, Konkle AT, Bielajew C. Bisphenol-A: epigenetic reprogramming and effects on reproduction and behavior. Int J Environ Res Public Health 2014;11(7):7537–61.

[65] Hughes IA, Martin H, Jaaskelainen J. Genetic mechanisms of fetal male undermasculinization: a background to the role of endocrine disruptors. Environ Res 2006;100(1):44–9.

[66] Brenseke B, Prater MR, Bahamonde J, Gutierrez JC. Current thoughts on maternal nutrition and fetal programming of the metabolic syndrome. J Pregnancy 2013;2013:368461.

[67] Nordkap L, Joensen UN, Blomberg Jensen M, Jorgensen N. Regional differences and temporal trends in male reproductive health disorders: semen quality may be a sensitive marker of environmental exposures. Mol Cell Endocrinol 2012;355(2):221–30.

[68] Wohlfahrt-Veje C, Main KM, Skakkebaek NE. Testicular dysgenesis syndrome: foetal origin of adult reproductive problems. Clin Endocrinol 2009;71(4):459–65.

[69] Giwercman A, Giwercman YL. Environmental factors and testicular function. Best Pract Res Clin Endocrinol Metab 2011;25(2):391–402.

[70] Guerrero-Bosagna C, Skinner MK. Environmental epigenetics and effects on male fertility. Adv Exp Med Biol 2014;791:67–81.

[71] Ramlau-Hansen CH, Thulstrup AM, Storgaard L, Toft G, Olsen J, Bonde JP. Is prenatal exposure to tobacco smoking a cause of poor semen quality? A follow-up study. Am J Epidemiol 2007;165(12):1372–9.

[72] Ravnborg TL, Jensen TK, Andersson AM, Toppari J, Skakkebaek NE, Jorgensen N. Prenatal and adult exposures to smoking are associated with adverse effects on reproductive hormones, semen quality, final height and body mass index. Hum Reprod 2011;26(5):1000–11.

[73] Kristensen DM, Hass U, Lesne L, et al. Intrauterine exposure to mild analgesics is a risk factor for development of male reproductive disorders in human and rat. Hum Reprod 2011;26(1):235–44.

[74] Damgaard IN, Jensen TK, Petersen JH, Skakkebaek NE, Toppari J, Main KM. Cryptorchidism and maternal alcohol consumption during pregnancy. Environ Health Perspect 2007;115(2):272–7.

[75] Jensen MS, Bonde JP, Olsen J. Comment on maternal alcohol consumption–Re: "Cryptorchidism and hypospadias as a sign of testicular dysgenesis syndrome: environmental connection". Birth Defects Res Part A Clin Mol Teratol 2011;91(2):126. author reply 7.

[76] Albert O, Jegou B. A critical assessment of the endocrine susceptibility of the human testis to phthalates from fetal life to adulthood. Hum Reprod Update 2014;20(2):231–49.

[77] Ivell R, Heng K, Nicholson H, Anand-Ivell R. Brief maternal exposure of rats to the xenobiotics dibutyl phthalate or diethylstilbestrol alters adult-type Leydig cell development in male offspring. Asian J Androl 2013;15(2):261–8.

[78] Martinez-Arguelles DB, Culty M, Zirkin BR, Papadopoulos V. In utero exposure to di-(2-ethylhexyl) phthalate decreases mineralocorticoid receptor expression in the adult testis. Endocrinology 2009;150(12):5575–85.

[79] Saillenfait AM, Sabate JP, Robert A, et al. Dose-dependent alterations in gene expression and testosterone production in fetal rat testis after exposure to di-*n*-hexyl phthalate. J Appl Toxicol 2013;33(9):1027–35.

[80] Swan SH, Main KM, Liu F, et al. Decrease in anogenital distance among male infants with prenatal phthalate exposure. Environ Health Perspect 2005;113(8):1056–61.

[81] Suzuki Y, Yoshinaga J, Mizumoto Y, Serizawa S, Shiraishi H. Foetal exposure to phthalate esters and anogenital distance in male newborns. Int J Androl 2012;35(3):236–44.

[82] Salian S, Doshi T, Vanage G. Perinatal exposure of rats to bisphenol A affects fertility of male offspring–an overview. Reprod Toxicol 2011;31(3):359–62.

[83] Christiansen S, Axelstad M, Boberg J, Vinggaard AM, Pedersen GA, Hass U. Low-dose effects of bisphenol A on early sexual development in male and female rats. Reproduction 2014;147(4):477–87.

[84] Manikkam M, Tracey R, Guerrero-Bosagna C, Skinner MK. Plastics derived endocrine disruptors (BPA, DEHP and DBP) induce epigenetic transgenerational inheritance of adult-onset disease and sperm epimutations. PLoS One 2012;8(1):e55387.

[85] Gray Jr LE, Ostby J, Monosson E, Kelce WR. Environmental antiandrogens: low doses of the fungicide vinclozolin alter sexual differentiation of the male rat. Toxicol Ind Health 1999;15(1–2):48–64.
[86] Gray Jr LE, Ostby JS, Kelce WR. Developmental effects of an environmental antiandrogen: the fungicide vinclozolin alters sex differentiation of the male rat. Toxicol Appl Pharmacol 1994;129(1):46–52.
[87] Wolf CJ, LeBlanc GA, Ostby JS, Gray Jr LE. Characterization of the period of sensitivity of fetal male sexual development to vinclozolin. Toxicol Sci 2000;55(1):152–61.
[88] Uzumcu M, Suzuki H, Skinner MK. Effect of the anti-androgenic endocrine disruptor vinclozolin on embryonic testis cord formation and postnatal testis development and function. Reprod Toxicol 2004;18(6):765–74.
[89] Damgaard IN, Skakkebaek NE, Toppari J, et al. Persistent pesticides in human breast milk and cryptorchidism. Environ Health Perspect 2006;114(7):1133–8.
[90] Gaspari L, Paris F, Jandel C, et al. Prenatal environmental risk factors for genital malformations in a population of 1442 French male newborns: a nested case–control study. Hum Reprod 2011;26(11):3155–62.
[91] Manikkam M, Haque MM, Guerrero-Bosagna C, Nilsson EE, Skinner MK. Pesticide methoxychlor promotes the epigenetic transgenerational inheritance of adult-onset disease through the female germline. PLoS One 2014;9(7):e102091.
[92] Manikkam M, Tracey R, Guerrero-Bosagna C, Skinner MK. Pesticide and insect repellent mixture (Permethrin and DEET) induces epigenetic transgenerational inheritance of disease and sperm epimutations. Reprod Toxicol 2012;34(4):708–19.
[93] Giusti RM, Iwamoto K, Hatch EE. Diethylstilbestrol revisited: a review of the long-term health effects. Ann Intern Med 1995;122(10):778–88.
[94] Kalfa N, Paris F, Soyer-Gobillard MO, Daures JP, Sultan C. Prevalence of hypospadias in grandsons of women exposed to diethylstilbestrol during pregnancy: a multigenerational national cohort study. Fertil Steril 2011;95(8):2574–7.
[95] Klip H, Verloop J, van Gool JD, et al. Hypospadias in sons of women exposed to diethylstilbestrol in utero: a cohort study. Lancet 2002;359(9312):1102–7.
[96] Pons JC, Papiernik E, Billon A, Hessabi M, Duyme M. Hypospadias in sons of women exposed to diethylstilbestrol in utero. Prenat Diagn 2005;25(5):418–9.
[97] Swan SH, Liu F, Overstreet JW, Brazil C, Skakkebaek NE. Semen quality of fertile US males in relation to their mothers' beef consumption during pregnancy. Hum Reprod 2007;22(6):1497–502.
[98] Ikeda Y, Tanaka H, Esaki M. Effects of gestational diethylstilbestrol treatment on male and female gonads during early embryonic development. Endocrinology 2008;149(8):3970–9.
[99] North K, Golding J. A maternal vegetarian diet in pregnancy is associated with hypospadias. The ALSPAC Study Team. Avon Longitudinal Study of Pregnancy and Childhood. BJU Int 2000;85(1):107–13.
[100] Vilela ML, Willingham E, Buckley J, et al. Endocrine disruptors and hypospadias: role of genistein and the fungicide vinclozolin. Urology 2007;70(3):618–21.
[101] Jacobs S, Teixeira DS, Guilherme C, et al. The impact of maternal consumption of cafeteria diet on reproductive function in the offspring. Physiology Behav 2014;129:280–6.
[102] Connor KL, Vickers MH, Beltrand J, Meaney MJ, Sloboda DM. Nature, nurture or nutrition? Impact of maternal nutrition on maternal care, offspring development and reproductive function. J Physiol 2012;590(Pt 9):2167–80.
[103] Alejandro B, Perez R, Pedrana G, et al. Low maternal nutrition during pregnancy reduces the number of Sertoli cells in the newborn lamb. Reprod Fertil Dev 2002;14(5–6):333–7.
[104] Tracey R, Manikkam M, Guerrero-Bosagna C, Skinner MK. Hydrocarbon (Jet fuel JP-8) induces epigenetic transgenerational inheritance of adult-onset disease and sperm epimutations. Reprod Toxicol 2012; 36:104–16.
[105] Manikkam M, Tracey R, Guerrero-Bosagna C, Skinner MK. Dioxin (TCDD) induces epigenetic transgenerational inheritance of adult onset disease and sperm epimutations. PLoS One 2012;7(9):e46249.
[106] Nassar N, Abeywardana P, Barker A, Bower C. Parental occupational exposure to potential endocrine disrupting chemicals and risk of hypospadias in infants. Occup Environ Med 2010;67(9):585–9.
[107] Sharma T, Banerjee BD, Yadav CS, Gupta P, Sharma S. Heavy metal levels in adolescent and maternal blood: association with risk of hypospadias. ISRN Pediatr 2014;2014:714234.
[108] Wong C, Kelce WR, Sar M, Wilson EM. Androgen receptor antagonist versus agonist activities of the fungicide vinclozolin relative to hydroxyflutamide. J Biol Chem 1995;270(34):19998–20003.
[109] Anway MD, Memon MA, Uzumcu M, Skinner MK. Transgenerational effect of the endocrine disruptor vinclozolin on male spermatogenesis. J Androl 2006;27(6):868–79.
[110] Guerrero-Bosagna C, Settles M, Lucker B, Skinner MK. Epigenetic transgenerational actions of vinclozolin on promoter regions of the sperm epigenome. PLoS One 2010;5(9).
[111] Skinner MK, Manikkam M, Tracey R, Guerrero-Bosagna C, Haque MM, Nilsson E. Ancestral dichlorodiphenyltrichloroethane (DDT) exposure promotes epigenetic transgenerational inheritance of obesity. BMC Med 2013;11(228).

[112] Manikkam M, Guerrero-Bosagna C, Tracey R, Haque MM, Skinner MK. Transgenerational actions of environmental compounds on reproductive disease and epigenetic biomarkers of ancestral exposures. PLoS One 2012;7(2):e31901.
[113] Stouder C, Paoloni-Giacobino A. Transgenerational effects of the endocrine disruptor vinclozolin on the methylation pattern of imprinted genes in the mouse sperm. Reproduction 2010;139(2):373–9.
[114] Stouder C, Paoloni-Giacobino A. Specific transgenerational imprinting effects of the endocrine disruptor methoxychlor on male gametes. Reproduction 2011;141(2):207–16.
[115] Doyle TJ, Bowman JL, Windell VL, McLean DJ, Kim KH. Transgenerational effects of di-(2-ethylhexyl) phthalate on testicular germ cell associations and spermatogonial stem cells in mice. Biol Reprod 2013;88(5):112.
[116] Song Y, Wu N, Wang S, et al. Transgenerational impaired male fertility with an Igf2 epigenetic defect in the rat are induced by the endocrine disruptor p,p′-DDE. Hum Reprod 2014;29(11):2512–21.
[117] Salian S, Doshi T, Vanage G. Impairment in protein expression profile of testicular steroid receptor coregulators in male rat offspring perinatally exposed to bisphenol A. Life Sci 2009;85(1–2):11–8.
[118] Mohamed el SA, Song WH, Oh SA, et al. The transgenerational impact of benzo(a)pyrene on murine male fertility. Hum Reprod 2010;25(10):2427–33.
[119] Nilsson E, Larsen G, Manikkam M, Guerrero-Bosagna C, Savenkova MI, Skinner MK. Environmentally induced epigenetic transgenerational inheritance of ovarian disease. PLoS One 2012;7(5):e36129.
[120] Skinner MK. Environmental epigenetic transgenerational inheritance and somatic epigenetic mitotic stability. Epigenetics 2011;6(7):838–42.
[121] Skinner MK, Guerrero-Bosagna C, Haque MM, Nilsson E, Bhandari R, McCarrey JR. Environmentally induced transgenerational epigenetic reprogramming of primordial germ cells and the subsequent germ line. PLoS One 2013;8(7):e66318.
[122] Flanagan JM, Popendikyte V, Pozdniakovaite N, et al. Intra- and interindividual epigenetic variation in human germ cells. Am J Hum Genet 2006;79(1):67–84.
[123] Oosterhuis JW, Looijenga LH. Testicular germ-cell tumours in a broader perspective. Nat Rev Cancer 2005;5(3):210–22.
[124] Rajender S, Avery K, Agarwal A. Epigenetics, spermatogenesis and male infertility. Mutat Res 2011;727(3):62–71.
[125] Hammoud SS, Nix DA, Hammoud AO, Gibson M, Cairns BR, Carrell DT. Genome-wide analysis identifies changes in histone retention and epigenetic modifications at developmental and imprinted gene loci in the sperm of infertile men. Hum Reprod 2011;26(9):2558–69.
[126] Hammoud SS, Purwar J, Pflueger C, Cairns BR, Carrell DT. Alterations in sperm DNA methylation patterns at imprinted loci in two classes of infertility. Fertil Steril 2010;94(5):1728–33.
[127] Kobayashi H, Sato A, Otsu E, et al. Aberrant DNA methylation of imprinted loci in sperm from oligospermic patients. Hum Mol Genet 2007;16(21):2542–51.
[128] Marques CJ, Carvalho F, Sousa M, Barros A. Genomic imprinting in disruptive spermatogenesis. Lancet 2004;363(9422):1700–2.
[129] Marques CJ, Costa P, Vaz B, et al. Abnormal methylation of imprinted genes in human sperm is associated with oligozoospermia. Mol Hum Reprod 2008;14(2):67–74.
[130] Marques CJ, Francisco T, Sousa S, Carvalho F, Barros A, Sousa M. Methylation defects of imprinted genes in human testicular spermatozoa. Fertil Steril 2010;94(2):585–94.
[131] Paradowska AS, Miller D, Spiess AN, et al. Genome wide identification of promoter binding sites for H4K12ac in human sperm and its relevance for early embryonic development. Epigenetics 2012;7(9):1057–70.
[132] Houshdaran S, Cortessis VK, Siegmund K, Yang A, Laird PW, Sokol RZ. Widespread epigenetic abnormalities suggest a broad DNA methylation erasure defect in abnormal human sperm. PLoS One 2007;2(12):e1289.
[133] Wu W, Shen O, Qin Y, et al. Idiopathic male infertility is strongly associated with aberrant promoter methylation of methylenetetrahydrofolate reductase (MTHFR). PLoS One 2010;5(11):e13884.
[134] Boissonnas CC, Abdalaoui HE, Haelewyn V, et al. Specific epigenetic alterations of IGF2-H19 locus in spermatozoa from infertile men. Eur J Hum Genet 2010;18(1):73–80.
[135] Steilmann C, Cavalcanti MC, Bartkuhn M, et al. The interaction of modified histones with the bromodomain testis-specific (BRDT) gene and its mRNA level in sperm of fertile donors and subfertile men. Reproduction 2010;140(3):435–43.
[136] Sonnack V, Failing K, Bergmann M, Steger K. Expression of hyperacetylated histone H4 during normal and impaired human spermatogenesis. Andrologia 2002;34(6):384–90.
[137] Okada Y, Scott G, Ray MK, Mishina Y, Zhang Y. Histone demethylase JHDM2A is critical for Tnp1 and Prm1 transcription and spermatogenesis. Nature 2007;450(7166):119–23.
[138] Shanker S, Hu Z, Wilkinson MF. Epigenetic regulation and downstream targets of the Rhox5 homeobox gene. Int J Androl 2008;31(5):462–70.
[139] Yadav RP, Kotaja N. Small RNAs in spermatogenesis. Mol Cell Endocrinol 2014;382(1):498–508.

[140] Korhonen HM, Meikar O, Yadav RP, et al. Dicer is required for haploid male germ cell differentiation in mice. PLoS One 2011;6(9):e24821.

[141] Hayashi K, Chuva de Sousa Lopes SM, Kaneda M, et al. MicroRNA biogenesis is required for mouse primordial germ cell development and spermatogenesis. PLoS One 2008;3(3):e1738.

[142] Maatouk DM, Loveland KL, McManus MT, Moore K, Harfe BD. Dicer1 is required for differentiation of the mouse male germline. Biol Reprod 2008;79(4):696–703.

[143] Liu D, Li L, Fu H, Li S, Li J. Inactivation of Dicer1 has a severe cumulative impact on the formation of mature germ cells in mouse testes. Biochem Biophys Res Commun 2012;422(1):114–20.

[144] Romero Y, Meikar O, Papaioannou MD, et al. Dicer1 depletion in male germ cells leads to infertility due to cumulative meiotic and spermiogenic defects. PLoS One 2011;6(10):e25241.

[145] Papaioannou MD, Pitetti JL, Ro S, et al. Sertoli cell Dicer is essential for spermatogenesis in mice. Dev Biol 2009;326(1):250–9.

[146] Nicholls PK, Harrison CA, Walton KL, McLachlan RI, O'Donnell L, Stanton PG. Hormonal regulation of sertoli cell micro-RNAs at spermiation. Endocrinology 2011;152(4):1670–83.

[147] Huszar JM, Payne CJ. MicroRNA 146 (Mir146) modulates spermatogonial differentiation by retinoic acid in mice. Biol Reprod 2013;88(1):15.

[148] Yang QE, Racicot KE, Kaucher AV, Oatley MJ, Oatley JM. MicroRNAs 221 and 222 regulate the undifferentiated state in mammalian male germ cells. Development 2013;140(2):280–90.

[149] Tong MH, Mitchell DA, McGowan SD, Evanoff R, Griswold MD. Two miRNA clusters, Mir-17-92 (Mirc1) and Mir-106b-25 (Mirc3), are involved in the regulation of spermatogonial differentiation in mice. Biol Reprod 2012;86(3):72.

[150] Niu Z, Goodyear SM, Rao S, et al. MicroRNA-21 regulates the self-renewal of mouse spermatogonial stem cells. Proc Natl Acad Sci USA 2011;108(31):12740–5.

[151] Liang X, Zhou D, Wei C, et al. MicroRNA-34c enhances murine male germ cell apoptosis through targeting ATF1. PLoS One 2012;7(3):e33861.

[152] Gan H, Lin X, Zhang Z, et al. piRNA profiling during specific stages of mouse spermatogenesis. RNA 2011;17(7):1191–203.

[153] Thomson T, Lin H. The biogenesis and function of PIWI proteins and piRNAs: progress and prospect. Annu Rev Cell Dev Biol 2009;25:355–76.

[154] Heyn H, Ferreira HJ, Bassas L, et al. Epigenetic disruption of the PIWI pathway in human spermatogenic disorders. PLoS One 2012;7(10):e47892.

[155] Ichiyanagi K, Li Y, Watanabe T, et al. Locus- and domain-dependent control of DNA methylation at mouse B1 retrotransposons during male germ cell development. Genome Res 2011;21(12):2058–66.

[156] Brennecke J, Aravin AA, Stark A, et al. Discrete small RNA-generating loci as master regulators of transposon activity in Drosophila. Cell 2007; 128(6):1089–103.

[157] Houwing S, Kamminga LM, Berezikov E, et al. A role for Piwi and piRNAs in germ cell maintenance and transposon silencing in Zebrafish. Cell 2007;129(1): 69–82.

[158] Macia A, Munoz-Lopez M, Cortes JL, et al. Epigenetic control of retrotransposon expression in human embryonic stem cells. Mol Cell Biol 2011;31(2):300–16.

CHAPTER

11

Developmental Origins of Childhood Asthma and Allergic Conditions—Is There Evidence of Epigenetic Regulation?

Steve Turner

Child Health, Royal Aberdeen Children's Hospital, Aberdeen, UK

OUTLINE

The Epigenome and Developmental Origins of Health and Disease
http://dx.doi.org/10.1016/B978-0-12-801383-0.00011-6

Copyright © 2016 Elsevier Inc. All rights reserved.

INTRODUCTION

Asthma is a very common noncommunicable disease, affecting 1 million children and a total of 5.4 million individuals across the United Kingdom (UK) [1]. In addition to being a highly prevalent condition, asthma is a serious condition since, in the UK for example, each year there are approximately 80,000 hospital admissions and 1200 deaths due to asthma, and the National Health Service spends approximately £1 billion caring for people with asthma [1]. Asthma also impacts on quality of life due to symptoms brought on by exercise, sleep disruption due to nocturnal cough, and missed days at school and work. In contrast with many other noncommunicable diseases, for example coronary artery disease, type 2 diabetes mellitus, and chronic obstructive airways disease, there is no agreed definition of asthma and there is no diagnostic test. A definition of asthma might be "a condition characterized by episodic wheeze and other respiratory symptoms that responds to treatment with inhaled corticosteroids." From the outset, it is important for the reader to appreciate the rather ethereal nature of asthma diagnosis since this might explain some of the issues discussed later. Individuals with asthma are at increased risk for having atopy, that is, they produce IgE specific to at least one common environmental allergen, and asthma is commonly considered as an allergic condition alongside eczema (also known as atopic dermatitis), hay fever (also known as allergic rhinitis), and food allergies; but in this chapter, asthma will be considered distinct to other allergic conditions for reasons that will become apparent.

Eczema will be used in this chapter as the exemplar for allergic conditions. As with asthma, defining eczema is far from straightforward; the European Academy of Allergy and Clinical Immunology found that "reaching a consensus on the definitions of the dermatological aspects of allergy has been the most difficult issue for the nomenclature task force" [2]. The National Institute for Clinical Excellence [3] defines eczema as an itchy skin condition plus three of the following: visible dermatitis (skin inflammation) in the skin creases; a history of dermatitis in skin creases; dry skin in the previous year; a personal history of asthma or allergic rhinitis (or parental history in under four-year-olds); onset of signs by 2 years of age. As with asthma, eczema is not always associated with atopy, and nonatopic eczema is a recognized condition [4] that responds to the same treatment as atopic eczema but is much less on common in children compared to atopic eczema and is not associated with asthma. Notwithstanding the imprecise nature of diagnosing atopic eczema, approximately 50% of adults have a history of ever having had eczema according to a study of 9 million primary care records in the UK [5], making eczema a very common problem. Although hospitalizations and deaths due to eczema are uncommon, this skin condition does cause considerable morbidity due to incessant itching through the day and night.

Asthma and eczema are complex conditions where "developmental origins" are implicated in causation and gene–environmental interactions are generally considered important. By developmental origins, what is meant here is that there is a deviation from "normal" development of an organ, rather than normal development that then becomes secondarily "abnormal." The mechanism whereby environmental exposures cause asthma and eczema is widely considered thought to involve epigenetics. The term *epigenetics* embraces a number of biochemical modifications of genes that regulate genetic expression; the most commonly reported epigenetic modifications is DNA methylation, which occurs when a methyl group is added to a cytosine or guanine nucleic acid, and this generally "switches off" gene expression. A fuller description of epigenetic mechanisms is given later in this chapter. The focus of this chapter is on the epidemiology of asthma and eczema and the evidence for their developmental origins, and then concludes by placing epigenetics in the pathway(s) leading to asthma and allergic diseases.

What Does the Epidemiology Tell Us?

Historical records indicate that asthma and eczema symptoms have been present for millennia. The word *asthma* is derived from the Greek word *aazein* that means to pant or breathe with an open mouth. The word *eczema* is also from a Greek word, *ekzema*, which means to boil, that is, describing the intense burning/itchiness of the skin condition. Having established that asthma and eczema are not new conditions, what is clear is that the prevalence of these chronic conditions in children (and adults) is considerably higher in the early twenty-first century than in the 50 years previously. Since asthma and eczema were not particularly common conditions in the early 1900s, their prevalence was not measured until the early 1960s. In Aberdeen, the lifetime prevalence for asthma in 9–12-year-olds was 4% in 1964, and this increased to 10% in 1989 and further increased to 28% in 2004 with similar changes observed for eczema prevalence. There was a divergence in asthma and eczema prevalence in the Aberdeen population survey of 2009; lifetime prevalence for asthma had reduced to 22% whilst remaining at 31% for eczema [6], and this has been reported in a second population [7]. The International Study of Asthma and Allergy in Children undertook the first of its three surveys between 1991 and 1995 and since then has charted the "asthma epidemic." More recently, in 2002–2003, there was evidence of asthma and eczema prevalence in 13–14-year-olds still rising from 1991 to 1995 levels in some countries (e.g., Central America, Spain, Portugal, and Eastern Europe), having reached a plateau in some (e.g., New Zealand, Brazil, and Uruguay) and declining in others (e.g., the UK) [8]. Whilst it is likely that a proportion of the rise in asthma and eczema prevalence is due to lower thresholds for diagnosing these conditions in children ("diagnostic reflagging"), there is consistent evidence of genuine and marked rise in prevalence.

The divergence in asthma and eczema prevalence suggests different underlying mechanisms. Epidemiology studies have identified a number of inconsistencies in the relationship between asthma and eczema, which are summarized in Table 1. For example, although the majority of children with asthma have coexistent eczema or hay fever (estimated variously at approximately 60–80%), the majority of children with eczema do not have asthma (prevalence approximately 25–50%).

What are the Developmental Origins of Asthma and Eczema?

Identifying the origins of asthma and eczema is not straightforward since symptoms can present at any age; for example, some individuals present with respiratory or chest symptoms from shortly after birth whereas others have symptoms triggered by environmental exposures in later life, for example, occupational asthma. Birth cohort studies, where individuals were recruited before birth, have given insight into the natural history of asthma and eczema. Although findings differ slightly between different populations, what is apparent is that a predisposition to asthma and eczema can be identified before the onset of symptoms. What must be acknowledged is that since these birth cohort studies first recruited in the mid-1980s, the results are necessarily limited to outcomes in childhood and adolescence.

Birth weight has been linked to risk for asthma in some populations, and this suggests that antenatal factors might determine asthma risk. A review of the literature identified 22 studies published since 2000 of which 9 found an association between reduced birth weight and increased asthma risk, 10 found no association, and 3 found increased birth weight (i.e., greater than 4.5 kg) and increased risk for asthma [9]. Similarly, there are many studies that have linked fetal size at birth to risk for eczema.

TABLE 1 Some Consistencies and Inconsistencies in the Relationship between Asthma and Eczema

	Consistency	Inconsistency
Comorbidity	The majority of asthmatic patients have coexisting eczema or hay fever.	Not all asthmatics have eczema or hay fever, and not all individuals with eczema or hay fever have asthma.
Change in prevalence	There were parallel rises in asthma and eczema prevalence between the 1980s and 1990s.	Some studies find falling asthma prevalence in the 2000s but eczema prevalence remains high.
Fetal growth	Asthma and eczema risk is increased among babies with increased birth weight.	Small first-trimester fetal size is associated with asthma but not eczema. Changing fetal size is more clearly associated with risk for eczema compared to asthma. Reduced birth weight is associated with increased asthma risk.
Eczema predicts asthma outcome	Early onset eczema (atopy) is a risk factor for later asthma.	
Physiological outcomes at birth		Abnormalities of respiratory physiology in very early life are associated with asthma outcomes in later life, but cord blood immune function does not differ between those who do and do not develop atopy.
Intervention studies	Complex interventions are associated in some studies with reduced asthma and eczema risk.	"Single" interventions are associated with reduced eczema but not asthma risk.
Genetic and epigenetic studies	Twin studies demonstrate good concordance for asthma and eczema.	There are different candidate genes for asthma and eczema.

A recent systematic review found that a birth weight of less than 2.5kg was associated with reduced risk for eczema (odds ratio 0.7 (95% confidence interval 0.6, 0.8)), and a birth weight of more than 4kg was associated with increased risk for eczema (OR 1.1 (95% CI 1.0, 1.2)) [10]. Other birth measurements have been linked to asthma and eczema outcomes, albeit in much fewer studies than those where birth weight is the outcome, and increased birth length has been linked to reduced risk for asthma [9] and eczema [11]. Increased head circumference has been linked to increased risk for hay fever [12] and asthma [9] but not eczema risk. While all birth measurements are correlated within an individual, there are theories why some anthropometric measurements might be more relevant to outcomes, for example, increased head growth might be associated with enhanced thymic development and an atypical immune system (although there is no evidence to support this rather tenuous hypothesis). If birth weight at either end of the "normal" spectrum might lead to the same endpoint (i.e., asthma or eczema), it is not surprising that most studies find no association between birth weight and outcome, since the majority of infants are of "normal weight" and any effect of very heavy or light weight will cancel each other out. It is remarkable that positive or negative associations between birth weight and asthma and eczema are seen in any population, and reported association might reflect false positive findings, publication bias, or incomplete follow-up leading to a nonrepresentative population.

Several groups have explored the hypothesis that the physiological abnormalities associated with asthma and eczema are present before symptoms, and studies have included measurement of pulmonary mechanics and immune function in neonates and followed these individuals up to relate the physiological abnormalities to clinical outcomes. Studies that have measured abnormalities of pulmonary physiology in asymptomatic infants, for example, reduced expiratory flows [13] or shortened ratio of inspiratory to expiratory time [14], have demonstrated that these physiological differences increase the risk for later asthma symptoms in childhood and early adulthood. Given the challenge (and imprecision) of measuring lung function in infants aged between one day and one month, it is remarkable that associations are detectable. Abnormalities of lung function in early life also persist into childhood [13] and adulthood [15], suggesting that the physiological abnormalities in the lungs that are associated with asthma are determined before birth. Evidence for antenatal origins of postnatal respiratory morbidity is provided by one study that related fetal size at 10 weeks gestation to lung function and risk for asthma at 10 years of age [16]. In this study, small first-trimester fetal size was associated with increased risk for asthma symptoms (but not eczema) and for reduced lung function at age 10 years, independent of birth weight and the child's size at 10 years, suggesting that the mechanism is primarily one of abnormal respiratory development and not simply small fetuses becoming small babies and children with small lungs. Although tracking or lifetime trajectory of lung function, and risk for asthma symptoms, may be determined in early pregnancy, there are modifying factors that include genetic factors (as evidenced by parental history of asthma), early onset atopy, and the current environment [13]. Factors associated with changes in lung function during infancy and childhood include sex (boys initially have reduced lung function relative to girls, but this relationship inverts during puberty) and no longer being exposed to maternal smoking [17].

Although asthma symptoms in childhood generally resolve, the follow-up of young adults with asthma demonstrates that symptoms relapse in later life, for example, in the 30s, for at least 30% of individuals [18]; this proportion is likely to rise with increasing age due to the natural decline in lung function beyond 25 years of age, and this may explain a blurring between asthma and chronic obstructive pulmonary disease (COPD) seen in adult populations. One study has reported that children with severe asthma are at increased risk for COPD in adulthood [19]. Evidence for subclinical asthma in adolescents with apparent remission of asthma can be seen in the form of reduced lung function and evidence of allergic (eosinophilic) airway inflammation [20]. These studies indicate that the physiological abnormalities associated with asthma may persist, and although symptoms may appear to remit, these often relapse with extended follow-up.

Cord blood mononuclear cells (CBMC) have been used as a surrogate for the physiological manifestation of "atopy." A study of infants born at term and prematurely reported that CBMCs were able to respond to allergens regardless of gestation, indicating that the immune system was functional from an early gestation [21]. Experiments in the 1990s demonstrated variously that infants did not have substantially different CMBC responses to common environmental exposures regardless of risk for atopy (usually evidenced by at least one parent with asthma, eczema, or hay fever) [22] or that parental atopy was associated with enhanced CBMC responses [23], that is, there was no consistency for infants being predisposed at birth to atopy. Maternal smoking and maternal dietary factors have been, respectively, associated with exaggerated or blunted CBMC responses suggesting that immune system responses might be primed at birth by exposures that are modifiable during pregnancy [23]. During infancy, mononuclear cell responses to allergens in atopic and nonatopic children become divergent [24], and responsiveness in mid-childhood may be established between the ages of 1 and 6 years [25]. In summary, the evidence is that the immune system is responsive before term but the mononuclear cell changes associated with atopy become more apparent in the first year(s) of life.

In conclusion, although asthma and eczema symptoms may not present until adolescence or adulthood, there is a good level of evidence that physiological abnormalities associated with these outcomes can be detected before the onset of symptoms and these may be lifelong. The first 1000 days, that is, between conception to the second birthday, seems to be a particularly critical window when predisposition is determined; the antenatal period might be more important to asthma outcome compared to eczema.

What Factors are Associated with Asthma and Eczema?

There are many factors linked to both asthma and eczema causation, but the nature of many of these relationships is by association and not causation, that is, the relationship reflects a third factor, which is causally associated with both exposure and outcome. The previously described rise in childhood asthma and eczema prevalence during the 1980s and 1990s occurred too rapidly to be explained by genetic change, and a "Westernized" life style has been implicated and supported by subsequent studies. The fall of the Berlin Wall in 1989 was accompanied with a change in lifestyle for children living in East Berlin, and the prevalence of eczema rose marginally from 12% to 14% between 1991/1992 and 1995/1996, and hay fever increased from 2% to 5% but recurrent wheeze fell from 26% to 19% [26]. A second study comparing outcomes in Estonia between 1992/1993 and 1996/1997 found no difference in prevalence for hay fever but did observe, as in East Berlin, the prevalence of asthma and eczema was considerably lower compared to Western European countries, and with time this East–West difference may narrow [27]. Another approach to examine the impact of a "Western" lifestyle on asthma and eczema outcomes is to look at asthma risk for children who move to or from a Western environment in early life. One such study described a three-fold reduced risk for wheeze among children living in Hong Kong (the "Westernized" lifestyle) who had been born in mainland China but moved before the age of 3 years compared to their peers born in Hong Kong [28]. The hygiene hypothesis [29] gained much attention in the 1990s and was based on an apparent protective effect of older siblings on hay fever symptoms in younger siblings, and initially the proposed mechanism included increased infections in early life that "switched off" allergy. The hygiene hypothesis was supported by associations between day care attendance and reduced asthma risk [30]. More recent understanding is that it is not hygiene, as evidenced by hand washing and use of disinfectants, per se but encounters with an environment with more diverse bacterial, fungal, and viral matter that reduces risks for asthma and atopy.

Changes in diet have been associated with altered risk for asthma [31], and a Mediterranean (i.e., non-Westernized) diet has been associated with a lower prevalence of asthma but not atopy (eczema outcome was not reported) using data from 50,000 children who participated in ISAAC surveys across many different countries [32]. The mechanism for dietary exposures altering asthma risk is not clear, but theories include the antioxidant properties of dietary components [31] and the direct effect of factors in diet such as vitamin D on immune function. Diet is a complex exposure where one constituent is usually linked to others, for example, omega-3 fatty acids and fat-soluble vitamins A, D, E, and K are strongly correlated, and constituents vary between what is ostensibly the same food item, for example, the nutrient content of an apple depends on the characteristics of the soil in which its tree grew. Dietary habits are also confounded by socioeconomic factors, and it is possible that the associations described are confounded by economic factors.

Relating exposure and asthma or eczema outcome is complicated by different age at onset of symptoms, and cohort studies have tried to better understand causation by creating phenotypes stratified by age at onset and remission of symptoms. The classical study was carried out in a cohort from Tucson where individuals were grouped by wheezing status at ages 3 and 6 years. Compared to those who never wheezed, those with transient wheeze (i.e., wheeze only before 3 years) had reduced premorbid lung function and were more likely to have mothers who smoked; those with wheeze only between 3 and 6 years were more likely to have rhinitis, to be boys, and to have a history of maternal asthma. Children with wheeze before and after 3 years of age had all the risk factors of the early and later wheezers plus a trend for reduced lung function [33]. These findings have been replicated in other cohorts despite the rather arbitrary age cutoffs [13,34] and give insight into the concept of "multiple hits," that is, a combination of factors rather than a single factor has to be present for symptoms to arise.

A full description of the factors implicated in asthma and eczema is beyond the remit of this chapter, but there are reviews of this complex literature for the interested reader [35] and Table 2 summarizes exposures associated with asthma and eczema and whether epigenetic factors might be relevant. What is apparent is that the effect of single exposure is modest but interactions between exposures apparently have a greater effect size. Additionally, there are many risk

TABLE 2 A Summary of Some Factors Associated with Altered Risk for Asthma and/or Eczema and Whether There Might be Underlying Epigenetic Mechanisms

Exposure	Asthma	Eczema	Epigenetic mechanism?
Antenatal smoking	Increased risk [35]	Increased risk [96]	✓ [89]
Postnatal smoking	Increased risk [35]	Increased risk [96]	✓
Breast-feeding	Reduced risk [35] (increased in some populations)	Reduced risk [97]	?
Maternal diet during pregnancy	Increased vitamins D, E, and fatty acids associated with reduced risk [35]	Some foods, e.g., fish, may reduce risk [98]	✓ [79]
Respiratory infection	Increased risk with severe illness [35]	?	?
Obesity	Increased risk (particularly for nonatopic asthma) [99]	Increased risk [100]	✓ [101]
Living on a farm	Reduced risk [35]	X increased risk in nonaffluent countries [102]	✓ [91]
Damp housing/mold	Increased risk [35]	? [103]	?
Inhaled chemicals	Increased risk [35]	?	?
Poor quality ambient air	Increased risk [35]	?	?
Inhaled allergens (e.g., pets, grass)	Increased risk [35]	?	?
Ingested allergens in early life (e.g., egg, milk)	Increased risk [35]	Increased risk	?
Antenatal exposure to paracetamol	Increased risk [35]	X [104]	?

✓ = Yes; ? = Unknown; X = No.

factors for both asthma and eczema, but several factors associated with asthma are not linked to eczema, for example, respiratory virus infection. Furthermore, exposures associated with eczema risk are not associated with asthma, for example, the apparently protective effect of older siblings is seen for eczema but not asthma. Puberty is worthy of mention since asthma prevalence is higher in younger boys compared to girls, and this sex difference resolves during puberty before reversing in adulthood [36]. The inversion of the male:female preponderance for asthma symptoms is also seen in lung physiology where lung function is lower in boys compared to girls but higher in men compared to women [17]. The mechanism underlying the changing profile of asthma either side of puberty, which is seen to a lesser extent for eczema, may include epigenetic factors since these are particularly active during puberty and sex hormones are also likely to be important.

What Do Intervention Studies Tell Us?

Whilst observational studies can give insight into what exposures may be relevant to asthma and eczema, only intervention studies can prove that exposure causes the outcome. Many of the risk factors for asthma and eczema (and most noncommunicable diseases) cannot be modified, for example, age, gender, family history, and socioeconomic status, and intervention is not possible, however, studies have modified many of the exposures listed in Table 2 in the hope of preventing asthma and eczema. There is currently no cure for asthma and eczema (as is the case for most noncommunicable diseases), so there is a great appetite for preventative interventions, and the epidemiology points to interventions during the first 1000 days being most likely to be effective. The results from most interventions in early development have been disappointing, but these have given insight into the complexity and timing of asthma and eczema causation.

Three intervention studies have successfully reduced asthma (but not eczema) risk in children by changing multiple exposures, that is, they introduced a complex intervention. In a study carried out in 120 infants born in the Isle of Wight, UK, postnatal exposure to cow's milk protein (and other dietary allergens) and house dust mites were reduced, and the intervention group had 50% reduced risk for asthma at 18 years of age but not eczema [37]. A second study from Canada recruited 545 pregnant mothers and randomized them to standard care or an intervention that modified antenatal and postnatal exposures to house dust mites, pets, secondhand smoke (SHS), promoted breast feeding and delayed weaning, and reported reduced asthma but not eczema incidence at 7 years of age [38]. The third intervention advised about the benefits of antenatal and postnatal oily fish and the harm of SHS and dampness in 2860 mothers and observed a 30% reduction in asthma risk at 2 years, but no change in eczema risk [39]. An intervention study from Australia where exposure to house dust mites and cow's milk was modified in 616 infants found no difference in asthma symptoms between the groups at 5 years of age [40]. A fifth complex intervention reduced early life exposure to SHS, inhaled and ingested allergens, and promoted breastfeeding but found no difference in asthma outcome at age 6 years [41]. What these studies tell us is that complex interventions in early life might prevent asthma, but there are practical issues in delivering these interventions and changing lifestyles. These studies indicate that many risk factors contributing to asthma may be modifiable, which is beneficial in developing preventative strategies.

Other interventions have focused on reducing asthma and eczema risk by modifying maternal diet during pregnancy, promoting breast feeding, and modifying infant weaning diet. An intervention where maternal diet was modified by fish oil supplementation during pregnancy reported reduced risk for severe eczema among infants of mothers who received the supplement, but not eczema per se nor asthma symptoms [42]. In one clinical trial where mothers were randomized by center to receive an intervention that increased breast feeding uptake and duration demonstrated reduced eczema in infants randomized to the intervention but with follow-up to 6 years, this effect was lost and the prevalence of atopy was higher among the intervention group [43]. A review of the literature has concluded that interventions where infant diet was modified generally have no effect on eczema although exclusive and/or prolonged breast feeding may have a protective effect [44]. Dietary interventions with fish oil supplements in the first six months of life did not change risk for asthma symptoms in infancy [45]. A systematic review of two trials found no link between infant diet supplementation prebiotics and asthma risk [46], and a trial where infants were randomized to supplement with probiotic (+/−prebiotic) or placebo also found no difference in asthma risk [47].

There are some intervention studies that were designed to reduce asthma and/or eczema where the intervention has resulted in undesired effects. One example, the CAPS study, found increased eczema among five-year-olds who had received house dust mite reduction intervention as infants [40]. In a second study where antenatal and postnatal house dust mite exposure was very effectively reduced, the children in the intervention group had better lung function but were more likely to be sensitized to house dust mite at 3 years of age [48]. A third intervention study found no evidence that house dust mite avoidance in early life alters asthma risk in childhood [49]. Despite there being a good level of epidemiological evidence that exposure to house dust mite was relevant to asthma causation, the results of intervention studies have demonstrated that there is no causal relationship between early house dust mite exposure and asthma or eczema.

Overall, the results of intervention studies designed to prevent asthma have been disappointing. The interventions that are associated with reduced asthma risk have demonstrated that changes to several facets of the environment, for example, inhaled and ingested allergen exposure and breast feeding and reduced exposure to secondhand smoke, are more likely to be effective than interventions that alter one exposure, for example, house dust mite. Given the evidence linking antenatal environmental exposures to asthma outcome, intervention that did not modify the fetuses' exposures seem less likely to have an effect. Other reasons for interventions not working might include some or all of the following: the intervention was not effectively delivered, the study was underpowered to detect a small difference in risk, the wrong target population was used, and the outcome was poorly defined.

Asthma and Eczema on the Context of Other Noncommunicable Diseases

Whilst this chapter is focused on asthma and eczema, there are associations between these conditions and other noncommunicable diseases where developmental origins and epigenetic factors are implicated in causation. For example, there is an association between obesity and asthma, in particular nonatopic asthma [50]; the underlying mechanism(s) is unknown, and in addition to developmental origins and epigenetic factors, there may include the compressive effect of excessive chest wall tissues, a perception of reduced exercise capacity due to symptoms and/or proinflammatory cytokine release from adipocytes. What is clear is that weight loss in obese individuals with asthma leads to better asthma control [51], and there is a good level of evidence that obesity is causally related to asthma symptoms. Children with asthma are also more likely to have some of the subclinical features of the metabolic syndrome [52], characterized by obesity, hypertension, dyslipidemia, and insulin resistance. The mechanism linking asthma to metabolic syndrome is not fully understood, but these apparently different conditions may be the result of a common insult at a critical stage of development that predisposes the individuals to both respiratory and metabolic abnormalities. A common and early origin for many noncommunicable diseases is evidenced by the link with reduced birth weight and noncommunicable disease, and the next section will explore possible fetal origins of NCD.

Fetal Origins of Disease

Until the 1980s the fetus was considered to have a privileged environment where it was protected from harm by the maternoplacental unit. This opinion was supported by observations from those babies who survived to term whose mothers were starved during pregnancy; infants born during the Leningrad siege had apparently modest reductions in birth weight, typically less than 500–600 g [53]. What was not appreciated was that although these fetuses survived with a reduced birth weight, they were at increased risk for noncommunicable diseases in later life.

The first "developmental origins" hypothesis, which sought to explain associations between reduced birth weight and increased risk for insulin resistance, was the 1992 thrifty phenotype hypothesis, which proposed that "poor fetal and infant growth and the subsequent development of type 2 diabetes and the metabolic syndrome result from the effects of poor nutrition in early life, which produces permanent changes in glucose-insulin metabolism" [54]. The "fetal origins hypothesis" was proposed in a 1995 publication and stated that "fetal undernutrition in middle to late gestation, leads to disproportionate fetal growth, programs later [coronary heart] disease" [55]. The concept behind both the thrifty phenotype and fetal origins hypothesis was that reduced fetal nutrition set in motion a series of physiological adaptations that allowed survival to term but at the expense of later noncommunicable disease. From a "survival of the fittest point" perspective, this adaptation makes sense, and moreover, low birth weight is associated with early onset puberty and thus "disadvantaged fetuses" manage to survive and pass on their genes at an earlier age—the "fertility first hypothesis" [56]. Systematic reviews have confirmed associations between reduced birth weight and insulin resistance [57], hypertension [58], and minor increases in plasma lipids [59]. However, there is awareness that increased birth weight is also associated with increased risk for noncommunicable diseases, for example, increased birth weight (i.e., >4 kg or >90th centile) is associated with increased blood pressure [60] and asthma risk [9] in children, and these associations cannot be explained easily by the thrifty phenotype and fetal origins hypotheses since low birth weight was the "exposure" leading to increased risk for noncommunicable diseases. Additionally, the profile of maternal malnutrition has changed from undernutrition in Edwardian Britain to today's overnutrition (i.e., obesity), but noncommunicable disease prevalence (e.g., asthma) is not falling.

In the early 2000s, a series of hypotheses based on the principle of plasticity were put forward, including predictive adaptive responses [61] and developmental plasticity [62], and placed under the umbrella term "developmental origins of health and disease" [63]. These paradigms accommodated observations that both increased and reduced birth weight were associated with adverse outcomes and proposed that a mismatch between "cues" to the developing fetus in the antenatal environment and the postnatal environment lead to increased risk for noncommunicable disease. In this context, an individual might be born too small or too large for their environment depending on whether the antenatal cues were accurate. Testing these hypotheses is made difficult by the challenges of assessing antenatal fetal well-being and the necessary interval between birth and the onset of noncommunicable diseases such as type 2 diabetes and coronary artery disease, but there is proof of concept that both accelerating and faltering antenatal growth is associated with increased risk for "adverse" outcomes as measured in children [64].

Asthma and eczema are unusual as noncommunicable diseases since onset of symptoms can be from infancy and childhood, and these give insight into the developmental origins of noncommunicable diseases early in the life course. Ultrasound measurement of fetal size has been used as an index of fetal well-being and related to postnatal outcomes including asthma and eczema. There are at least two major factors that influence interpretation of fetal ultrasound measurements: first, ultrasound measurements have an intersubject variability of up to 10% [65]; and second, the fetus grows rapidly during pregnancy, and gestational accuracy, which is very important to interpreting fetal size, is uncertain in at least one-third of pregnancies. Despite these limitations to the measurement and interpretation of fetal ultrasound measurements, there are some highly consistent findings between studies (but also some inconsistencies).

Antenatal Fetal Measurements and Asthma Outcomes

Airway branching and relative caliber is determined by 18 weeks gestation, and alveoli start to develop from 24 weeks gestation continuing through to at least 3 years of age [66], and factors that affect the predominantly antenatal development of the respiratory system might be expected to have implications for postnatal respiratory outcomes. At the time of writing, associations between fetal size and asthma outcomes have been reported in three cohorts. The cohort with the longest follow-up (the SEATON study) has reported an association between small fetal size persisting from the first trimester and increased risk for asthma and also reduced lung function (a physiological abnormality associated with asthma) at 10 years of age [16]. Growth acceleration between the first and second trimesters and second and third trimesters was also associated with increased risk for asthma and reduced lung function [16]. These findings implicate factors acting both in very early pregnancy and throughout pregnancy in the development of asthma.

A second study (the Southampton Women's Study; SWS) related fetal growth between three trimesters, but not fetal size, to a history of wheeze ever at 3 years of age and found no association [67]. When children with wheeze were subcategorized by atopy (skin prick reactivity), growth acceleration for head size between the first and second trimesters was associated with reduced risk for atopic wheeze and growth acceleration for abdominal girth after the second trimester was associated with reduced risk for nonatopic wheeze [67]. In the third cohort (Generation R), fetal size was not linked to asthma risk at 4 years of age, but persistent small fetal size was associated with reduced risk for asthma [68].

There are some striking contrasts for asthma outcomes seen between these cohorts, and some differences might be due to differences in the ages at which the children were assessed. Growth acceleration was positively associated with symptoms in SEATON [16] but negatively for SWS [67], whilst persistent small size was positively associated with asthma risk in SEATON [16] but negatively for the Generation R cohort [68]. These apparent inconsistencies might be explained by the predictive adaptive responses concept. With extended follow-up of these cohorts and with results from other populations, the picture may become clearer.

Antenatal Fetal Measurements and Eczema Outcomes

The adaptive immune system is "primed" at birth but not "established" until postnatal life; nonetheless there are associations between antenatal fetal growth and eczema and atopy. Two of the cohorts previously described (i.e., SEATON and Generation R) have related fetal growth to eczema. Both found that growth acceleration was positively associated with eczema risk; first-second trimester growth acceleration and eczema at 10 years for the SEATON cohort [16] and second-third trimester growth acceleration and eczema at 4 years for Generation R [68]. Although eczema outcome was not reported for SWS, the risk for skin prick positivity was directly proportional to growth acceleration between the first and second and second and third trimesters [67]. Skin prick reactivity was not associated with fetal growth acceleration in the SEATON cohort, but here rhinitis risk was lower among the group with faltering growth between the first and second trimesters [16].

Together these findings suggest that factors leading to fetal growth acceleration may also be related to a process that places an individual at increased risk for eczema and allergy. When the results for asthma and eczema are viewed together, it might be speculated that very early fetal size and growth are important to asthma and lung function whereas accelerating fetal growth after the first trimester might be more relevant to eczema and allergy. In this paradigm, individuals with abnormalities in fetal growth (either high or low growth trajectory) during the first trimester that are maintained throughout pregnancy might be at risk for nonatopic asthma, those with early growth failure followed by growth acceleration might be predisposed to atopic asthma, and those with normal early growth followed by growth acceleration have nonasthmatic atopy. Although this remains to be proven, mechanisms leading to differences in antenatal growth trajectories might explain the inconsistent relationship between asthma and eczema; see Table 1.

Factors Associated with Altered Fetal Size

Maternal diet and plasma nutrients have been linked to altered fetal size in a number of populations, but what remains to be determined is whether these associations are causal and whether an individual component of the diet is important or alternatively whether a combination of nutrients is important. Maternal plasma vitamin E is positively associated with first-trimester fetal size [69]; hence, a maternal lifestyle in early pregnancy characterized by "healthy" vitamin E intake might lead to "good" fetal growth and reduce risk for asthma. In a cross-sectional analysis, maternal plasma polyunsaturated fatty acids in mid-pregnancy have been inversely associated with fetal size [70], and a further study has demonstrated associations between maternal vitamin D insufficiency and reduced fetal growth (in particular, limb growth) and increased dietary calcium intake and greater fetal growth [71]. The association between changing growth trajectory after the first trimester and risk for asthma suggests that factors acting during pregnancy might be modifying asthma risk and maternal smoking is associated with fetal growth failure, which is more obvious after the first trimester [72]. However, if maternal smoking during pregnancy leads to faltering fetal growth, then this would not explain how accelerating fetal growth was associated with increased risk for asthma. A further inhaled maternal exposure that might influence fetal growth is exposure to poor quality ambient air during pregnancy, which has been associated with reduced fetal growth [73]. Maternal obesity and weight gain during pregnancy are also modifiable fetal exposures that may influence the developing fetus; a systematic review has described associations between both maternal obesity and gestational weight gain and a twofold increased risk for childhood asthma symptoms [74]. In addition to maternal obesity, there are many exposures associated with reduced birth weight, but associations with antenatal size have not yet been explored, for example, endocrine disruptors, epigenetic modification.

Postnatal Weight Gain

There is consistent evidence that the first 1000 days post conception, that is, throughout pregnancy up to the second birthday, are important to the development of asthma and eczema.

There may be different factors driving antenatal and postnatal development, for example, placental sufficiency for the former and oral calorie intake for the latter, and complete understanding of developmental origins can be gained by studying both antenatal and postnatal development together (usually evidenced as growth). For example, accelerated postnatal growth might be an index of a calorie-rich postnatal diet but might reflect antenatal growth failure ("catch-up growth"); thus excessive growth might be due to antenatal or postnatal factors (or both). Postnatal weight gain is associated with increased childhood asthma symptoms, and this risk is greater after premature delivery [75]. Premature delivery per se is a risk factor for increased risk for asthma [76], and whilst both birth weight and gestation are determinants of asthma risk, the latter is the stronger determinant [75], and this adds weight to the developmental origins of asthma. In the Generation R cohort, increased asthma symptoms (but not eczema) were associated with accelerated postnatal growth independent of antenatal growth [68]. The mechanism underlying this association may be a slowing of lung function growth during infancy [77] relative to somatic growth; accelerated somatic growth is associated with a number of risk factors for asthma including male sex and not being breast fed [77]. Although accelerated growth in infancy might be associated with faltering infant lung function growth, there is an apparently inconsistent trend for this to be associated with improved lung function in childhood [77,78].

So How Does Epigenetics Fit into the Development of Asthma and Eczema?

Epigenetic Mechanisms

In 1942 Conrad Waddington defined epigenetics as "the branch of biology which studies the causal interactions between genes and their products which bring the phenotype into being." Over the intervening years an understanding of the complexity of epigenetic regulation has emerged, and at the time of writing there are two epigenetic mechanisms described [79]:

1. Methylation. Here a methyl group is added to a cytosine or adenine nucleotide within the DNA molecule. Usually clusters of nucleotides are methylated to prevent transcription of whole genes, and this process results in suppression of gene expression.
2. Chromatin remodeling. Chromatin is the complex of DNA structured around histone, and histone is the protein "spool" around which DNA is structured within the cell nucleus. Gene expression requires the relevant DNA to be moved from a compact to relaxed structure, and this is an enzyme-dependent process. Modification of histone by enzymatic activity including methylation, acetylation, and phosphorylation alters the properties of histone and thus DNA structure. Acetylation or the absence of methylation is associated with a more relaxed DNA structure. The effect of histone modification can be to either suppress or stimulate gene expression. In addition to histone modification, ATP-dependent enzymes (also known as complexes) can affect DNA expression by directly moving nucleosomes (a short section of DNA) to be more or less accessible to the transcription machinery within the cell nucleus.

Changes in epigenetic regulation over time would explain the rapidly changing prevalence of conditions, such as asthma and eczema, where it has been suggested that approximately 50% of causation is explained by inherited factors [80]. Epigenetic mechanisms would also explain the developmental origins of disease and in particular account for developmental plasticity where an individual might achieve one of a number of phenotypes.

At the time of writing, the importance of epigenetics to the development of asthma and eczema is uncertain due to a number of reasons that have already appeared in this chapter when exploring the epidemiology, and these include the imprecise nature of asthma and eczema diagnosis, confounders, reverse causation, and publication bias. Additionally, there are issues of difficulty in accessing respiratory tissues for studying asthma outcome and many methodological issues that are beyond the expertise of this author and include power, tissue specificity, biological plausibility of association, and functionality, and these are described elsewhere [81].

It is highly plausible that epigenetic factors are important to the etiology of both asthma and eczema since these conditions have such a compelling hereditary origin, but changes in DNA cannot have occurred swiftly enough to explain the rise in prevalence of these conditions; epigenetic changes in gene expression, perhaps triggered by changing environmental exposures, might at least partly explain the asthma epidemic. Furthermore, as previously discussed, there are developmental origins to both asthma and eczema, and epigenetic factors are thought to at least partly explain antenatal programming of noncommunicable diseases [82].

Associations between Asthma, Eczema, and Epigenetic Variation

Circulating lymphocytes are considerably easier to access in comparison with respiratory tissue, and not unsurprisingly the literature describing epigenetic variation in association with atopy or immune cell function is more comprehensive than the literature for asthma epigenetics. Individuals with atopy typically have type 2 T-helper lymphocytes (Th2), which release increased amounts of interleukin (IL)-4, IL-5, and IL-13 and reduced gamma interferon (IFN-g) after exposure to stimuli to allergens or microbial products in vitro, and in comparison, nonatopic individuals have predominantly type 1 T-helper cells (Th1). The normal neonatal immune system is suppressed in utero to reduce interactions between maternal and fetal immune systems, and this suppression, which is skewed toward a Th2 profile, particularly with suppressed IFN-g, involves epigenetic regulation [79]. The trajectories for the "normal" neonatal immune system to remain either atopic (Th2 skewed) or deviate to a nonatopic (Th1 skewed) phenotype can be likened to developmental plasticity where an individual can reach a number of phenotypes depending on environmental cues acting in early life. There is good evidence that atopy is associated with epigenetic regulation of genes coding for mediators associated with the Th1 or Th2 phenotype [83]. One of the key determinants of T cell differentiation is FOXP3, and there is a complex (and not completely understood) relationship between hyper- and hypomethylation of the genes expressing and suppressing FOXP3 activity [83]; epigenetic regulation of T cell phenotypes may be in a constant state of flux in recognition of new environmental encounters that are particularly frequent in early life. A complete review of epigenetic variation in genes coding for immune mediators is beyond the remit this chapter, and the reader is referred elsewhere [79,83]. Whilst there is a considerable literature linking epigenetic variants to lymphocyte function, there are relatively few associating epigenetics to diagnosed disease, but there are associations with asthma [84] and eczema [85] in children. With time, the current evidence gap(s) linking epigenetics to asthma and eczema will narrow.

Associations between Antenatal and Postnatal Environmental Exposures and Epigenetic Variation

There are many studies linking epigenetic factors to modifiable exposures during pregnancy that are implicated in asthma and eczema causation. Maternal nutrition during pregnancy and its effect on the unborn child in postnatal life has been a focus of many research groups, perhaps inspired by the findings from the Dutch Hunger Winter Cohort; here, adults who were in utero whilst their mothers were starving had less DNA methylation of the gene coding for insulin-like growth factor 2 [86]. Interventions in animal models have confirmed that maternal dietary exposures during pregnancy are associated with differences in the epigenetic profile of offspring [87]. Although no study has yet undertaken epigenetic analysis on respiratory tissue or lymphocytes and related these to antenatal dietary exposures, dietary folate is known to alter epigenetic expression and may be associated with increased risk for asthma, although a recent systematic review found no clear evidence for this [88]. Other antenatal dietary exposures factors that are linked to epigenetic changes include vitamin B12 and fish oils [79], although only fish oils have been consistently associated with asthma and eczema.

Exposures to maternal smoking and poor ambient air quality have also been linked to epigenetic changes and provide further evidence supporting the potential role for epigenetics in the mechanism between environmental exposures and noncommunicable diseases. Antenatal exposure to maternal smoking is associated with epigenetic changes in oral mucosal cells [89], but what remains to be determined is whether these changes are seen in respiratory tissues. Grandmaternal smoking during mother's gestation has been associated with increased risk for asthma in her grandchild, independent of mother's smoking, and this might reflect intergenerational epigenetic programming [90].

Exposure to farms during early life has been associated with reduced risk for asthma and eczema for mechanisms that are not fully understood but could include exposure to high concentrations of a wide range of inhaled and ingested microbial products. Epigenetic regulation during early development has been proposed as a mechanism linking the environmental exposure with disease outcome, and a pilot study has found evidence to support this theory. Exposed children in five European countries including Germany and Switzerland had increased DNA methylation in regions containing prespecified candidate genes for IgE compared to children who did not live on farms [91] but hypomethylation for genes associated with increased asthma risk. A second small study from Sweden linked farm exposure to epigenetic variation in placental tissue and observed that exposure was associated with less DNA methylation for the CD14 promotor [92]; CD14 is important in the immune response to microbial exposures.

Exposure to increasing concentrations of fine airborne particulates has been linked to increased exhaled nitric oxide (ENO; a biomarker for airway inflammation) and also to hypermethylation of the gene coding for one of the family of enzymes collectively termed *nitric oxide synthase* [93]. Over time, epigenetic associations with other exposures associated with asthma and eczema such as breast-feeding [94] are likely to emerge. It is likely that these associations will not be replicated in different populations, and heterogeneity between populations might give insight into epigenetic mechanisms. Additionally, standard statistical approaches are probably not well suited to understanding the likely complex interaction between epigenetic changes to multiple genes within the setting of the multiple exposures experienced by an individual, and novel approaches are required [95]. A further question that needs answering in the future is how durable are epigenetic effects? For example, are these intergenerational or short lived? The scientific community also needs to consider the prospect that epigenetics may not explain the translation from environmental exposure to pathology and therefore needs to continue to explore alternative mechanisms.

CONCLUSIONS

Asthma and eczema are common conditions where many environmental exposures are implicated in causation, and these causal pathways may be separate but with some intersections. Figure 1(a) and (b) outlines a "traditional" model

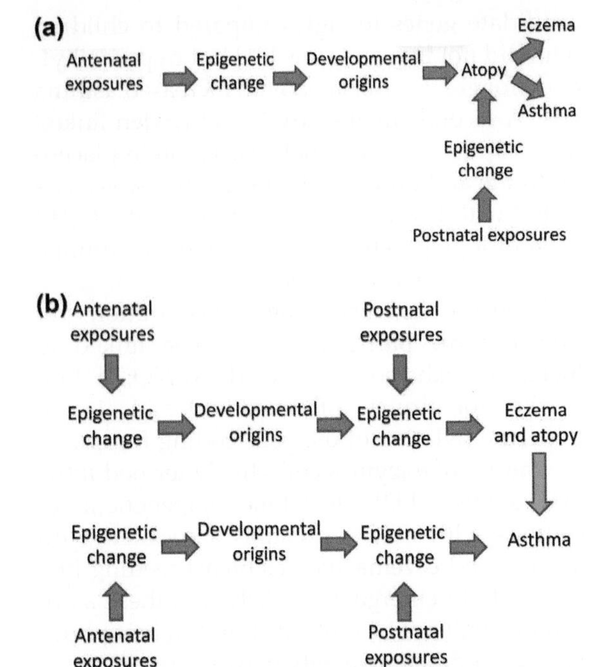

FIGURE 1 (a) A traditional schematic depicting how antenatal and postnatal factors lead to the development of asthma. (b) An alternative paradigm where separate pathway lead to both asthma and atopy and where atopy exaggerates coexisting asthma.

of the pathway toward asthma and eczema and one alternative. Developmental origins are also thought to be important to asthma and eczema etiology, and the first 1000 days postconception seem to be particularly important. At the time of writing there is an expectation that epigenetic variations may explain how fetal growth and environmental exposures are linked to asthma and eczema.

References

[1] Asthma UK. Asthma facts and FAQs. 2014. Available at: http://www.asthma.org.uk/asthma-facts-and-statistics. [accessed 11/15, 2014].

[2] Johansson SG, Hourihane JO, Bousquet J, Bruijnzeel-Koomen C, Dreborg S, Haahtela T, et al. A revised nomenclature for allergy. An EAACI position statement from the EAACI nomenclature task force. Allergy 2001;56(9):813–24.

[3] National Institute for Clinical Excellence. Atopic eczema in children: management of atopic eczema in children from birth up to the age of 12 years. 2014. Available at: http://www.nice.org.uk/guidance/CG57. [accessed 11/05, 2014].

[4] Brown S, Reynolds NJ. Atopic and non-atopic eczema. BMJ 2006;332(7541):584–8.

[5] Simpson CR, Newton J, Hippisley-Cox J, Sheikh A. Trends in the epidemiology and prescribing of medication for eczema in England. J R Soc Med 2009; 102(3):108–17.

[6] Malik G, Tagiyeva N, Aucott L, McNeill G, Turner SW. Changing trends in asthma in 9-12 year olds between 1964 and 2009. Arch Dis Child 2011;96(3):227–31.

[7] Ponsonby AL, Glasgow N, Pezic A, Dwyer T, Ciszek K, Kljakovic M. A temporal decline in asthma but not eczema prevalence from 2000 to 2005 at school entry in the Australian Capital Territory with further consideration of country of birth. Int J Epidemiol 2008;37(3):559–69.

[8] Asher MI, Montefort S, Bjorksten B, et al. Worldwide time trends in the prevalence of symptoms of asthma, allergic rhinoconjunctivitis, and eczema in childhood: ISAAC Phases One and Three repeat multicountry cross-sectional surveys. Lancet 2006;368(9537):733–43.

[9] Turner S. Perinatal programming of childhood asthma: early fetal size, growth trajectory during infancy, and childhood asthma outcomes. Clin Dev Immunol 2012;2012:962923.

[10] Panduru M, Salavastru CM, Panduru NM, Tiplica GS. Birth weight and atopic dermatitis; systematic review and meta-analysis. Acta Dermatovenerol Croat 2014;22:91–6.

[11] Bisgaard H, Halkjaer LB, Hinge R, Giwercman C, Palmer C, Silveira L, et al. Risk analysis of early childhood eczema. J Allergy Clin Immunol 2009; 123(6):1355–60. e5.

[12] Katz KA, Pocock SJ, Strachan DP. Neonatal head circumference, neonatal weight, and risk of hayfever, asthma and eczema in a large cohort of adolescents from Sheffield, England. Clin Exp Allergy 2003;33(6):737–45.

[13] Mullane D, Turner SW, Cox DW, Goldblatt J, Landau LI, le Souef PN. Reduced infant lung function, active smoking, and wheeze in 18-year-old individuals. JAMA Pediatr 2013;167(4):368–73.

[14] Haland G, Carlsen KC, Sandvik L, et al. Reduced lung function at birth and the risk of asthma at 10 years of age. N Engl J Med 2006;355(16):1682–9.

[15] Stern DA, Morgan WJ, Wright AL, Guerra S, Martinez FD. Poor airway function in early infancy and lung function by age 22 years: a non-selective longitudinal cohort study. Lancet 2007;370(9589):758–64.

[16] Turner S, Prabhu N, Danielan P, et al. First- and second-trimester fetal size and asthma outcomes at age 10 years. Am J Respir Crit Care Med 2011;184(4):407–13.
[17] Turner S, Fielding S, Mullane D, Cox D, Goldblatt J, Landau LI, et al. A longitudinal study of lung function from 1 month to 18 years of age. Thorax 2014;69:1015.
[18] Phelan PD, Robertson CF, Olinsky A. The Melbourne asthma study: 1964–1999. J Allergy Clin Immunol 2002;109(2):189–94.
[19] Tai A, Tran H, Roberts M, Clark N, Wilson J, Robertson CF. The association between childhood asthma and adult chronic obstructive pulmonary disease. Thorax 2014;9:805.
[20] van Den Toorn LM, Prins JB, Overbeek SE, Hoogsteden HC, de Jongste JC. Adolescents in clinical remission of atopic asthma have elevated exhaled nitric oxide levels and bronchial hyperresponsiveness. Am J Respir Crit Care Med 2000;162(3 Pt 1):953–7.
[21] Szepfalusi Z, Pichler J, Elsasser S, van Duren K, Ebner C, Bernaschek G, et al. Transplacental priming of the human immune system with environmental allergens can occur early in gestation. J Allergy Clin Immunol 2000;106(3):530–6.
[22] Prescott SL, Macaubas C, Holt BJ, Smallacombe TB, Loh R, Sly PD, et al. Transplacental priming of the human immune system to environmental allergens: universal skewing of initial T cell responses toward the Th2 cytokine profile. J Immunol 1998;160(10):4730–7.
[23] Devereux G, Barker RN, Seaton A. Antenatal determinants of neonatal immune responses to allergens. Clin Exp Allergy 2002;32(1):43–50.
[24] Prescott SL, Macaubas C, Smallacombe T, Holt BJ, Sly PD, Holt PG. Development of allergen-specific T-cell memory in atopic and normal children. Lancet 1999;353(9148):196–200.
[25] Turner SW, Heaton T, Rowe J, et al. Early-onset atopy is associated with enhanced lymphocyte cytokine responses in 11-year-old children. Clin Exp Allergy 2007;37(3):371–80.
[26] von Mutius E, Weiland SK, Fritzsch C, Duhme H, Keil U. Increasing prevalence of hay fever and atopy among children in Leipzig, East Germany. Lancet 1998;351(9106):862–6.
[27] Riikjarv MA, Annus T, Braback L, Rahu K, Bjorksten B. Similar prevalence of respiratory symptoms and atopy in Estonian schoolchildren with changing lifestyle over 4 years. Eur Respir J 2000;16(1):86–90.
[28] Wong GW, Leung TF, Ma Y, Liu EK, Yung E, Lai CK. Symptoms of asthma and atopic disorders in preschool children: prevalence and risk factors. Clin Exp Allergy 2007;37(2):174–9.
[29] Strachan DP. Hay fever, hygiene, and household size. BMJ 1989;299(6710):1259–60.
[30] Ball TM, Castro-Rodriguez JA, Griffith KA, Holberg CJ, Martinez FD, Wright AL. Siblings, day-care attendance, and the risk of asthma and wheezing during childhood. N Engl J Med 2000;343(8):538–43.
[31] Seaton A, Godden DJ, Brown K. Increase in asthma: a more toxic environment or a more susceptible population? Thorax 1994;49(2):171–4.
[32] Nagel G, Weinmayr G, Kleiner A, Garcia-Marcos L, Strachan DP, ISAAC Phase Two Study Group. Effect of diet on asthma and allergic sensitisation in the International Study on Allergies and Asthma in Childhood (ISAAC) Phase Two. Thorax 2010; 65(6):516–22.
[33] Martinez FD, Wright AL, Taussig LM, Holberg CJ, Halonen M, Morgan WJ. Asthma and wheezing in the first 6 years of life. The Group Health Medical Associates. N Engl J Med 1995;332(3):133–8.
[34] Henderson J, Granell R, Heron J, et al. Associations of wheezing phenotypes in the first 6 years of life with atopy, lung function and airway responsiveness in mid-childhood. Thorax 2008;63(11):974–80.
[35] Dick S, Friend A, Dynes K, et al. A systematic review of associations between environmental exposures and development of asthma in children aged up to 9 years. Brit Med J Open 2014;4(11).
[36] Chen W, Mempel M, Schober W, Behrendt H, Ring J. Gender difference, sex hormones, and immediate type hypersensitivity reactions. Allergy 2008;63(11): 1418–27.
[37] Scott M, Roberts G, Kurukulaaratchy RJ, Matthews S, Nove A, Arshad SH. Multifaceted allergen avoidance during infancy reduces asthma during childhood with the effect persisting until age 18 years. Thorax 2012;67(12):1046–51.
[38] Chan-Yeung M, Ferguson A, Watson W, et al. The Canadian childhood asthma primary prevention study: outcomes at 7 years of age. J Allergy Clin Immunol 2005;116(1):49–55.
[39] Dotterud CK, Storro O, Simpson MR, Johnsen R, Oien T. The impact of pre- and postnatal exposures on allergy related diseases in childhood: a controlled multicentre intervention study in primary health care. BMC Public Health 2013;13:123.
[40] Marks GB, Mihrshahi S, Kemp AS, et al. Prevention of asthma during the first 5 years of life: a randomized controlled trial. J Allergy Clin Immunol 2006;118(1):53–61.
[41] Maas T, Dompeling E, Muris JW, Wesseling G, Knottnerus JA, van Schayck OC. Prevention of asthma in genetically susceptible children: a multifaceted intervention trial focussed on feasibility in general practice. Pediatr Allergy Immunol 2011; 22(8):794–802.

[42] Dunstan JA, Mori TA, Barden A, Beilin LJ, Taylor AL, Holt PG, et al. Fish oil supplementation in pregnancy modifies neonatal allergen-specific immune responses and clinical outcomes in infants at high risk of atopy: a randomized, controlled trial. J Allergy Clin Immunol 2003;112(6):1178–84.
[43] Kramer MS, Matush L, Vanilovich I, Platt R, Bogdanovich N, Sevkovskaya Z, et al. Effect of prolonged and exclusive breast feeding on risk of allergy and asthma: cluster randomised trial. BMJ 2007; 335(7624):815.
[44] Foisy M, Boyle RJ, Chalmers JR, Simpson EL, Williams HC. Overview of reviews. The prevention of eczema in infants and children: an overview of Cochrane and non-Cochrane reviews. Evid Based Child Health 2011;6:1322–39.
[45] D'Vaz N, Meldrum SJ, Dunstan JA, et al. Postnatal fish oil supplementation in high-risk infants to prevent allergy: randomized controlled trial. Pediatrics 2012;130(4):674–82.
[46] Osborn DA, Sinn JK. Prebiotics in infants for prevention of allergy. Cochrane Database Syst Rev 2013;3:006474.
[47] Kukkonen AK, Kuitunen M, Savilahti E, Pelkonen A, Malmberg P, Makela M. Airway inflammation in probiotic-treated children at 5 years. Pediatr Allergy Immunol 2011;22(2):249–51.
[48] Woodcock A, Lowe LA, Murray CS, et al. Early life environmental control: effect on symptoms, sensitization, and lung function at age 3 years. Am J Respir Crit Care Med 2004;170(4):433–9.
[49] Gehring U, de Jongste JC, Kerkhof M, et al. The 8-year follow-up of the PIAMA intervention study assessing the effect of mite-impermeable mattress covers. Allergy 2012;67(2):248–56.
[50] Gibson PG. Obesity and asthma. Ann Am Thorac Soc 2013;10(Suppl.):S138–42.
[51] Eneli IU, Skybo T, Camargo Jr CA. Weight loss and asthma: a systematic review. Thorax 2008;63(8):671–6.
[52] Cottrell L, Neal WA, Ice C, Perez MK, Piedimonte G. Metabolic abnormalities in children with asthma. Am J Respir Crit Care Med 2011;183(4):441–8.
[53] Antonov AN. Children born during the Siege of Leningrad in 1942. J Pediatr 1947;30(3):250–9.
[54] Hales CN, Barker DJ. Type 2 (non-insulin-dependent) diabetes mellitus: the thrifty phenotype hypothesis. Diabetologia 1992;35(7):595–601.
[55] Barker DJ. Fetal origins of coronary heart disease. BMJ 1995;311(6998):171–4.
[56] Corbett SJ, McMichael AJ, Prentice AM. Type 2 diabetes, cardiovascular disease, and the evolutionary paradox of the polycystic ovary syndrome: a fertility first hypothesis. Am J Hum Biol 2009;21(5):587–98.
[57] Whincup PH, Kaye SJ, Owen CG, et al. Birth weight and risk of type 2 diabetes: a systematic review. JAMA 2008;300(24):2886–97.
[58] Mu M, Wang SF, Sheng J, Zhao Y, Li HZ, Hu CL, et al. Birth weight and subsequent blood pressure: a meta-analysis. Arch Cardiovasc Dis 2012;105(2):99–113.
[59] Owen CG, Whincup PH, Odoki K, Gilg JA, Cook DG. Birth weight and blood cholesterol level: a study in adolescents and systematic review. Pediatrics 2003;111(5 Pt 1):1081–9.
[60] Zhang Y, Li H, Liu SJ, et al. The associations of high birth weight with blood pressure and hypertension in later life: a systematic review and meta-analysis. Hypertens Res 2013;36(8):725–35.
[61] Gluckman PD, Hanson M, Spencer HG. Predictive adaptive responses and human evolution. Trends Ecol Evol 2005;20:527–33.
[62] Hanson M, Godfrey KM, Lillycrop KA, Burdge GC, Gluckman PD. Developmental plasticity and developmental origins of non-communicable disease: theoretical considerations and epigenetic mechanisms. Prog Biophys Mol Biol 2011;106(1):272–80.
[63] Gluckman PD, Hanson MA, Beedle AS. Early life events and their consequences for later disease: a life history and evolutionary perspective. Am J Hum Biol 2007;19(1):1–19.
[64] Alkandari F, Ellahi A, Aucott L, Devereux G, Turner S. Fetal ultrasound measurements and associations with postnatal outcomes in infancy and childhood: a systematic review of an emerging literature. J Epidemiol Comm Health 2015;69:41–8.
[65] Sarris I, Ioannou C, Chamberlain P, Ohuma E, Roseman F, Hoch L, et al. Intra- and interobserver variability in fetal ultrasound measurements. Ultrasound Obstetrics Gynecol 2012;39(3):266–73.
[66] Stick S. Pediatric origins of adult lung disease. 1. The contribution of airway development to paediatric and adult lung disease. Thorax 2000;55(7):587–94.
[67] Pike KC, Crozier SR, Lucas JS, et al. Patterns of fetal and infant growth are related to atopy and wheezing disorders at age 3 years. Thorax 2010;65(12):1099–106.
[68] Sonnenschein-van der Voort AM, Jaddoe VW, Raat H, Moll HA, Hofman A, de Jongste JC, et al. Fetal and infant growth and asthma symptoms in preschool children: the Generation R Study. Am J Respir Crit Care Med 2012;185(7):731–7.
[69] Turner SW, Campbell D, Smith N, et al. Associations between fetal size, maternal {alpha}-tocopherol and childhood asthma. Thorax 2010;65(5):391–7.
[70] Carlsen K, Pedersen L, Bonnelykke K, Stark KD, Lauritzen L, Bisgaard H. Association between whole-blood polyunsaturated fatty acids in pregnant women and early fetal weight. Eur J Clin Nutr 2013;67(9):978–83.

[71] Young BE, McNanley TJ, Cooper EM, McIntyre AW, Witter F, Harris ZL, et al. Maternal vitamin D status and calcium intake interact to affect fetal skeletal growth in utero in pregnant adolescents. Am J Clin Nutr 2012;95(5):1103–12.

[72] Prabhu N, Smith N, Campbell D, et al. First trimester maternal tobacco smoking habits and fetal growth. Thorax 2010;65(3):235–40.

[73] van den Hooven EH, Pierik FH, de Kluizenaar Y, et al. Air pollution exposure during pregnancy, ultrasound measures of fetal growth, and adverse birth outcomes: a prospective cohort study. Environ Health Perspect 2012;120(1):150–6.

[74] Forno E, Young OM, Kumar R, Simhan H, Celedon JC. Maternal obesity in pregnancy, gestational weight gain, and risk of childhood asthma. Pediatrics 2014;134(2):e535–46.

[75] Sonnenschein-van der Voort AM, Arends LR, de Jongste JC, et al. Preterm birth, infant weight gain, and childhood asthma risk: a meta-analysis of 147,000 European children. J Allergy Clin Immunol 2014; 133(5):1317–29.

[76] Jaakkola JJ, Ahmed P, Ieromnimon A, Goepfert P, Laiou E, Quansah R, et al. Preterm delivery and asthma: a systematic review and meta-analysis. J Allergy Clin Immunol 2006;118(4):823–30.

[77] Turner S, Zhang G, Young S, Cox M, Goldblatt J, Landau L, et al. Associations between postnatal weight gain, change in postnatal pulmonary function, formula feeding and early asthma. Thorax 2008;63(3):234–9.

[78] Kotecha SJ, Watkins WJ, Heron J, Henderson J, Dunstan FD, Kotecha S. Spirometric lung function in school-age children: effect of intrauterine growth retardation and catch-up growth. Am J Respir Crit Care Med 2010;181(9):969–74.

[79] Harb H, Renz H. Update on epigenetics in allergic disease. J Allergy Clin Immunol 2015;135:15–24.

[80] Carroll W. Asthma genetics: pitfalls and triumphs. Paediatr Respir Rev 2005;6(1):68–74.

[81] Gruzieva O, Merid SK, Melen E. An update on epigenetics and childhood respiratory diseases. Paediatr Respir Rev 2014. http://dx.doi.org/10.1016/j.prrv.2014.07.003.

[82] Gluckman PD, Hanson MA, Low FM. The role of developmental plasticity and epigenetics in human health. Birth Defects Res Part C Embryo Today Reviews 2011;93(1):12–8.

[83] Begin P, Nadeau KC. Epigenetic regulation of asthma and allergic disease. Allergy Asthma Clin Immunol 2014;10:27.

[84] Soto-Ramirez N, Arshad SH, Holloway JW, et al. The interaction of genetic variants and DNA methylation of the interleukin-4 receptor gene increase the risk of asthma at age 18 years. Clin Epigenetics 2013;5:1.

[85] Ziyab AH, Karmaus W, Yousefi M, et al. Interplay of filaggrin loss-of-function variants, allergic sensitization, and eczema in a longitudinal study covering infancy to 18 years of age. PLoS One 2012; 7(3):e32721.

[86] Heijmans BT, Tobi EW, Stein AD, et al. Persistent epigenetic differences associated with prenatal exposure to famine in humans. Proc Natl Acad Sci USA 2008;105(44):17046–9.

[87] Levin BE. Epigenetic influences on food intake and physical activity level: review of animal studies. Obesity 2008;16(Suppl. 3):S51–4.

[88] Crider KS, Cordero AM, Qi YP, Mulinare J, Dowling NF, Berry RJ. Prenatal folic acid and risk of asthma in children: a systematic review and meta-analysis. Am J Clin Nutr 2013;98(5):1272–81.

[89] Breton CV, Byun HM, Wenten M, Pan F, Yang A, Gilliland FD. Prenatal tobacco smoke exposure affects global and gene-specific DNA methylation. Am J Respir Crit Care Med 2009;180(5):462–7.

[90] Miller LL, Henderson J, Northstone K, Pembrey M, Golding J. Do grandmaternal smoking patterns influence the etiology of childhood asthma? Chest 2014;145(6):1213–8.

[91] Michel S, Busato F, Genuneit J, et al. Farm exposure and time trends in early childhood may influence DNA methylation in genes related to asthma and allergy. Allergy 2013;68(3):355–64.

[92] Slaats GG, Reinius LE, Alm J, Kere J, Scheynius A, Joerink M. DNA methylation levels within the CD14 promoter region are lower in placentas of mothers living on a farm. Allergy 2012;67(7):895–903.

[93] Salam MT, Byun HM, Lurmann F, Breton CV, Wang X, Eckel SP, et al. Genetic and epigenetic variations in inducible nitric oxide synthase promoter, particulate pollution, and exhaled nitric oxide levels in children. J Allergy Clin Immunol 2012;129(1):232–9.e1–7.

[94] Verduci E, Banderali G, Barberi S, et al. Epigenetic effects of human breast milk. Nutrients 2014;6:1711–24.

[95] Li J, Zhao Q, Bolund L. Computational methods for epigenetic analysis: the protocol of computational analysis for modified methylation-specific digital karyotyping based on massively parallel sequencing. Methods Mol Biol 2011;791:313–28.

[96] Hofhuis W, de Jongste JC, Merkus PJ. Adverse health effects of prenatal and postnatal tobacco smoke exposure on children. Arch Dis Child 2003;88(12):1086–90.

[97] Gdalevich M, Mimouni D, David M, Mimouni M. Breast-feeding and the onset of atopic dermatitis in childhood: a systematic review and meta-analysis of prospective studies. J Am Acad Dermatol 2001;45(4): 520–7.

[98] Sausenthaler S, Koletzko S, Schaaf B, et al. Maternal diet during pregnancy in relation to eczema and allergic sensitization in the offspring at 2 y of age. Am J Clin Nutr 2007;85(2):530–7.
[99] Papoutsakis C, Priftis KN, Drakouli M, et al. Childhood overweight/obesity and asthma: is there a link? A systematic review of recent epidemiologic evidence. J Acad Nutr Dietetics 2013;113(1):77–105.
[100] Sliverberg JI, Simpson EL. Association between obesity and eczema prevalence, severity and poorer health in US adolescents. Dermatitis 2014;25:172–81.
[101] van Dijk SJ, Molloy PL, Varinli H, Morrison J, Muhlhausler BS, Members of EpiSCOPE. Epigenetics and human obesity. Int J Obes 2015;39:85–97.
[102] Brunekreef B, Von Mutius E, Wong GK, Odhiambo JA, Clayton TO, ISAAC Phase Three Study Group. Early life exposure to farm animals and symptoms of asthma, rhinoconjunctivitis and eczema: an ISAAC Phase Three Study. Int J Epidemiol 2012;41(3):753–61.
[103] Tischer C, Chen CM, Heinrich J. Association between domestic mould and mould components, and asthma and allergy in children: a systematic review. Eur Respir J 2011;38(4):812–24.
[104] Shaheen SO, Newson RB, Henderson AJ, et al. Prenatal paracetamol exposure and risk of asthma and elevated immunoglobulin E in childhood. Clin Exp Allergy 2005;35(1):18–25.

CHAPTER

12

Immune Disorders, Epigenetics, and the Developmental Origins of Health and Disease

Rodney R. Dietert[1], *Amrie C. Grammer*[2]

[1]Department of Microbiology and Immunology, College of Veterinary Medicine, Cornell University, Ithaca, NY, USA; [2]University of Virginia Research Park, VA, USA

OUTLINE

The Epigenome and Developmental Origins of Health and Disease
http://dx.doi.org/10.1016/B978-0-12-801383-0.00012-8

Copyright © 2016 Elsevier Inc. All rights reserved.

INTRODUCTION

The paradigm of the developmental origins of adult health and disease (DOHaD) was recently reviewed by Barouki et al. [1]. Importantly, these investigators noted that the capacity of chemical exposure in early life to program the epigenome and affect later gene expression in any form of chemical exposure was not restricted by the somewhat arbitrary disciplinary boundary used to distinguish nutrition from toxicology. In effect, programming of the epigenome and the processes to accomplish it is unaffected by whether the agent is categorized as a nutrient, an environment chemical, or a drug. In fact, the definition of useful biomarkers for human toxicity is likely to change in the near future with the understanding that humans are really more appropriately defined as a symbiotic superorganism that is majority microbial. When chemical exposures capable of altering the microbial phenotypes and microbiome are considered as opposed to purely mammalian cell toxicity, early-life programming of disease takes on an entirely different light. For this reason, the potential impact of epigenetic alteration and the microbiome are included in this review.

DOHaD evolved over decades from the observations of Robert Barker and colleagues, eventually termed "the Barker hypothesis," who found that prenatal dietary conditions affected the risk of adult cardiovascular disease. Initially, these observations were restricted to maternal nutritional status and cardiovascular health versus disease [2]. But during the past two to three decades, the number of environmental risk factors affecting DOHaD has expanded well beyond diet as have the biological systems perturbed and the spectrum of later-life diseases connected to early-life conditions.

In fact, the pervasive nature through which early-life conditions program both the mammalian and microbial components of the infant and later adult has led to a greater focus on critical developmental windows of physiological vulnerability [3–6]. From the standpoint of health protection, strategies utilizing DOHaD provide opportunities to positively affect health across the entire lifespan. This is even more striking when the relationship of comorbid, noncommunicable diseases (NCDs) is considered. NCDs arise in predictable patterns of comorbidity [7], and preventing the onset of childhood diseases and conditions such as asthma, autism, celiac disease, and type 1 diabetes can reduce the risk of additional comorbid interrelated NCDs [7]. However, to be maximally effective, a deeper understanding of potentially common pathways and/or mechanisms connecting early-life risk factors to later-life diseases is needed. This would aid both risk reduction strategies as well as potential therapeutic interventions.

With the emerging understanding of epigenetics and its role in gene expression and phenotypes, there is an opportunity to search among known associations within disease models for early-life developmental programming to identify commonly shared pathways and/or mechanisms. This is likely to be useful in a reductionist approach to NCD prevention and treatment. Additionally, because there is evidence suggesting that some epigenetic alterations may be transmitted transgenerationally [8], it is important to recognize that some environmental epigenetic alteration resulting in later life disease could have origins in past generations. At present, there is no clear understanding of the extent to which present generational exposure versus that from prior decades and ancestry result in present-day NCD diagnoses. This chapter will consider the evidence for common pathways for DOHaD involving immune disorders with attention given to epigenetic routes, timing of exposure, and the potential for involvement of the microbiome.

The focus of this chapter is on immune system disorders. Because the immune system is represented in virtually every tissue and organ of the body, immune dysfunction-associated diseases can cover a wide spectrum of both systemic and tissue-directed pathologies. These include diseases that also fall into other categories (e.g., dermal, respiratory, metabolic, rheumatoid) and may be included in other chapters of this book as well. In the present chapter, we will focus on three categories of immune system disorders: allergic, autoimmune, and inflammatory. Table 1 provides examples of the range of diseases whose risk is elevated by specific environmental exposures and conditions via developmental immunotoxicity (DIT). Rather than discussing every disease among these categories, we focus on those examples of highly prevalent diseases and conditions where the most evidence exists for epigenetically programmed associations and common pathways. For one disease example, systemic lupus erythematosus (SLE), we provide a more detailed discussion of the suggested risk factors and biological processes.

ENVIRONMENTAL RISK FACTORS AFFECTING IMMUNE FUNCTION AND INDIVIDUAL VULNERABILITY

Knowledge of environmental insults that produce developmental immunotoxicity (DIT) and are associated with later-life immune dysfunction and disorders dates back to at least the 1970s [9]. A wide array of environmental exposures, physical conditions, and psychosocial experiences has been demonstrated to alter immune function and increase the risk of one or more immune disorders. For specific references for chemicals, drugs, dietary factors, and physical/psychosocial factors as well as approaches for DIT evaluation, see [10–15]. The list includes widely distributed environmental chemicals (e.g., arsenic, lead, atrazine, bisphenol A, dioxins) and widely consumed drugs (e.g., paracetamol). Developmental timing, genetic background, prior ancestral exposure, sex, and co-exposures are critical factors in determining risk of a given exposure producing an immune disorder in later life.

For the vast majority of risk factors, one of greatest vulnerabilities is the developmental timing of exposure. During the prenatal and early postnatal periods of development, the host is particularly sensitive to environmentally induced immunotoxicity leading to persistent immune dysfunction and elevated risk of both communicable and noncommunicable diseases. This age-based increase in sensitivity manifests in several different ways (see [16]).

1. Lower doses and/or durations of exposure are required to produce adverse immune outcomes in the young than in adults.
2. A broader spectrum of immune alterations is seen following early-life versus adult exposure.
3. The immune dysfunction is more persistent after early-life versus adult exposure.
4. Programming for adult-triggered immune dysfunction is more likely following early-life exposure.

Genetic background and/or ancestrally determined epigenetic patterns can be important in determining which subset of children is most likely to respond to a given environmental exposure with immune dysfunction-driven pathogenesis. Gene expression levels and/or products of allelic variants with different activities can alter handling or metabolism of compounds, the potential for immune insult, or the resiliency of the immune system to withstand immune-mediated tissue damage. One example in the last category appears to occur with the heavy metal lead. Lead is known to induce misregulated inflammation resulting in potential oxidative damage of tissues. One vulnerability factor appears to be the levels of available glutathione, which is affected by glutathione S-transferase gene polymorphisms [17]. Toll-like receptor (TLR) gene polymorphisms have been reported to affect the susceptibility to traffic-related, pollution-induced asthma. Likewise, polymorphisms of several immune-related genes appear to affect the vulnerability for alcohol-induced, cytokine-driven liver damage.

TABLE 1 Human Immune Disorders and Developmental Immunotoxicity[a]

Disease, Disorder or Susceptibility State	Disease Category	Suggested Early-Life Immune-Modulating Risk Factor	Category of Risk Factor
Acute myeloid leukemia	Cancer	Benzene	Aromatic hydrocarbon
Allergic sensitization	Allergic	Polychlorinated biphenyls	Organochlorine
Asthma	Allergic/Nonatopic	Maternal paracetamol use	Drug
Atherosclerosis	Inflammatory/ Cardiovascular	Maternal hypercholesterolemia	Dietary/Metabolic
Atopic dermatitis	Allergic	Maternal smoking	Environmental/Lifestyle
Allergic rhinitis	Allergic	Antibiotics in infancy	Drug
Autism spectrum disorders	Neurobehavioral	Maternal immune activation	Multiple/Infectious
Bipolar disorder	Neurobehavioral	Gestational influenza	Infectious
Cardiovascular disease	Cardiovascular	Childhood abuse	Psychosocial
Celiac disease	Allergic/Autoimmune	Elective Cesarean delivery	Medical/Lifestyle (when elective)
Crohn's disease	Autoimmune	Maternal smoking	Environmental/Lifestyle
Chronic obstructive pulmonary disease	Inflammatory	Smoke from biomass fuels	Environmental
Depression	Inflammatory	Childhood trauma	Psychosocial
Endometriosis	Reproductive	Environmental tobacco smoke	Environmental
Hypertension	Cardiovascular	Pesticides (DDT)	Environmental
Insulin resistance	Metabolic	Maternal diet	Dietary
Lack of protection against diphtheria and tetanus following childhood vaccination	Immune	Perfluorinated pollutants	Environmental
Multiple sclerosis	Autoimmune	Vitamin D insufficiency	Dietary/Environmental
Myalgic encephalomyelitis (chronic fatigue syndrome)	Systemic/Inflammatory	Childhood trauma	Psychosocial
Narcolepsy (specific subpopulation)	Autoimmune	H1N1 flu vaccination	Drug
Obesity/overweight risk	Metabolic	Cesarean delivery	Medical procedure
Otitis media	Infectious/Allergic	Maternal smoking/ETS	Environmental/Lifestyle
Parkinson's disease	Neurodegenerative	Pesticides	Environmental
Preeclampsia		Traffic-related air pollution	Environmental
Psoriasis	Inflammatory/ Autoimmune	Environmental tobacco smoke	Environmental

TABLE 1 Human Immune Disorders and Developmental Immunotoxicity[a]—cont'd

Disease, Disorder or Susceptibility State	Disease Category	Suggested Early-Life Immune-Modulating Risk Factor	Category of Risk Factor
Respiratory infections	Communicable	Polychlorinated biphenyls	Environmental
Rheumatoid arthritis	Autoimmune	Maternal smoking	Environmental/Lifestyle
Schizophrenia	Neurobehavioral	Prenatal immune activation	
Sudden infant death syndrome	Unclassified	Maternal smoking and alcohol consumption	Environmental/ Lifestyle/Drug
Type 1 diabetes	Autoimmune	Lack of or short duration breast-feeding	Dietary
Ulcerative colitis	Autoimmune	Urban living	Environmental

[a] *Adapted from: Dietert [15].*

Male versus female differences also affect the comparative risk of exposure-induced immune disorders. For example, in utero exposure to arsenic has been associated with immunotoxicity. However, the association of arsenic concentration and acute respiratory infections was only seen in male offspring in a study in rural Bangladesh [18].

Co-exposures to two or more environmental factors in early life can significantly alter the risk of later-life immune disorders. Depending upon which specific factors are involved, the overall immune-related risk of disease risk can be significantly elevated or reduced. For example, in rodents co-exposure to the heavy metals arsenic and lead result in macrophage immunotoxicity at synergistically lower concentrations than is seen with exposure to either single metal [19]. Pesticide use on a farm was reported to reduce the benefit of early-life exposure to farm animals relative to risk of asthma [20].

THE IMMUNE SYSTEM, DEVELOPMENTAL ORIGINS OF ADULT HEALTH AND DISEASE, AND PUBLIC HEALTH IMPLICATIONS

One of the striking patterns in public health has been the significant increase in noncommunicable disease during the past 40 years. Across the different categories of immune disorders, diseases such as asthma, food allergies, celiac disease, type 1 diabetes, inflammatory bowel disease, chronic obstructive pulmonary disease, as well as inflammation-driven obesity and cardiovascular disease have all experienced significant increases in prevalence. In fact, the Harvard School of Public Health and World Economic Forum in a joint study recently reported that NCDs already are the most significant cause of global mortality, and the costs associated with NCDs are expected to approximate almost half of global net worth by the year 2030 [21]. Because most NCDs have their origins in early life, it is important to identify modifiable risk factors and approach prevention with a greater urgency.

There are three recently identified factors that make it challenging to track the progress in avoidance of NCD risk factors. First, early-life programming of NCDs is increasingly seen as involving epigenetic alterations. Because some of these epigenetic marks may be transmitted transgenerationally for several generations, currently observed immune dysfunction and newly diagnosed immune disorders could be connected to problematic environmental exposures of prior decades, if not centuries. It is not entirely clear what portion of the current immune disease prevalences are due to current-generation exposures versus exposures that might date back to the 1970s or even the 1950s. For this reason, the population level response to avoidance of developmental immunotoxicants might require decades to be fully reflected in changes in disease prevalence.

A second challenge is the discovery that the microbiome dictates much of postnatal immune maturation, exerts a huge effect on human metabolism, alters the sensitivity for toxicant and drug exposures, and also affects the functional status of various organs. Status of the microbiome appears to affect the risk of some but not all immune disorders as is discussed under the different categories of disease. Yet, this is only beginning to be fully considered in terms of prevention of immune disorders starting in early life.

A final challenge is the observation that NCDs connected to immune dysfunction are far more interconnected in mechanisms and risk than had been previously thought. This is considered further in the following section.

IMMUNE PROGRAMMING BY THE MICROBIOTA

While environmental, physical, and psychological conditions can exert direct effects on the immune system during critical windows of prenatal and postnatal development [15,22], evidence suggests that a properly seeded, well-balanced microbiome is required for functionally effective immune maturation [23]. The concept that the microbiome and immune system co-mature as a unit served as the basis for the "completed self" paradigm in which newborns lacking an adequate diversity of microbiota are essentially biologically incomplete [24,25]. Even short periods of missing gut microbiota following birth appear to affect the levels of specialized immune cells [26].

Several different immune cell populations are affected by the state of the microbiome in early life. For example, the numbers and activity of T regulatory cells (Tregs), natural killer (NK) cells, and natural killer-like T cells (NKT cells) can be affected by gut microbiota, and these alterations appear to extend across the life course [27]. A recently reported epigenetic pathway for the regulation of colonic Tregs involves the DNA-methylation adaptor Uhrf1. When commensal microbes are lacking, reduced Uhrf1 expression and overexpression of a gene (Cdkn1a) encoding a cycle blocker in Tregs, the cyclin-dependent kinase inhibitor p21, produces a deficiency in both Treg cell numbers and maturation. The result is predisposition for colonic inflammation [28].

Additional controls of microbiota over immune cells exist. For example, it has been found that commensal microbes regulate granulopoiesis and are important in establishing a steady state via MyD88/TICAM signaling to innate immune cells [29]. Research also suggests that macrophage polarization can be regulated at a distance by gut microbiota. Antibiotic treatment and gut microbial dysbiosis (including fungal overgrowth) can promote distant M2 macrophage polarization in the airways with increased airway inflammation [30]. Finally, bone marrow–derived dendritic cell maturation and function can be shifted via specific alterations to the gut microbiota [31]. Adaptive immunity can be affected by microbiota as microbes can help to drive T helper cell differentiation [32]. In turn, there is extensive cross-talk, and host adaptive immunity helps to craft the gut microbiome [33].

INTERLINKAGE OF IMMUNE DISORDERS

Immune disorders used to be thought of as primarily a result of immunosuppression. This idea was only heightened during the era in which AIDS-related immunosuppression was emerging. However, it is now clear that any significant deviation from a balanced immune system and useful tissue homeostasis is likely to result in pathology and disease. In fact, inappropriately enhanced immune responses can be as devastating as immunosuppression [34], and targeted immunosuppression and immune enhancement can occur simultaneously as a result of early-life immunotoxic environmental conditions.

Among the striking features of immune dysfunction–driven NCDs is that they are: (1) highly interlinked by comorbid risk and (2) appear to require ongoing misregulated inflammation.

Co-occurrence of Immunosuppression and Inappropriate Immune Enhancement

An example of selected immunosuppression co-occurring with inappropriate immune enhancement can be found following early-life exposure to the heavy metal lead. Developmental lead exposure depresses Th1-driven, cell-mediated immune responses (immunosuppression of this function), but elevates Th2-driven allergic responses and causes innate immune cell inflammatory dysregulation. The result is a package of elevated health risks that can include increased risk of certain types of infections (elevated communicable disease risk) while at the same time elevating the risk of specific noncommunicable allergic, autoimmune, and inflammatory conditions. In fact, it appears to be more the rule than the exception that immunotoxic chemicals and drugs produce later-life immune dysregulation in which selected immunosuppression exists along with inappropriately enhanced immune function.

In terms of comorbidity, specific immune disorders do not tend to occur individually. That is, onset and diagnosis of one immune disorder is accompanied by an elevated risk of additional immune-related disorders. This was described in a series of papers illustrating the patterns of comorbid NCD risk [7,35,36]. While interlinkage of risk associated with various autoimmune conditions and allergic diseases had been suspected for years, the comorbid connections linking these diseases to obesity, cardiovascular disease, depression, sleep disorders, and behavioral disorders is comparatively new. This new understanding of multidisease risk makes prevention of a first NCD all the more valuable when viewed across the life course. For this reason, it is particularly important to consider potential, common, epigenetic-based pathways that could lead from initial environmental exposures and conditions to later-life immune disorders.

Among NCDs, it is clear that diseases within the same category of immune disorders (e.g., the allergic diseases) are often closely linked. That is, the onset of one allergic disease such as allergic rhinitis tends to be associated with an elevated risk of one or more additional allergic diseases (e.g., asthma). This also holds true among autoimmune diseases (e.g., type 1 diabetes and celiac disease). However, comorbidities are not restricted to the same category. For this reason, childhood asthma, obesity, and behavioral disorders are interlinked as are psoriasis, depression, and skin cancer [7]. Many of the common comorbidities associated with allergic, autoimmune, and inflammatory conditions such as depression, sleep disorders, sensory loss, and cardiovascular complications are among those conditions requiring significant, and often prolonged, pharmacological intervention.

IMMUNE DISORDERS AND COMMUNICABLE DISEASES

Immune disorders can affect host immune-mediated defenses and the results of challenge with infectious agents or vaccines. The immune dysfunction resulting from early-life immunotoxic exposures or conditions can influence: (1) the level of protection afforded by vaccination, (2) the overall susceptibility to infections, (3) the frequency of infections, (4) the severity of infections as measured by either disease duration or tissue pathology, and/or (5) the consequences of infection that could include risk of mortality, susceptibility to secondary infections, loss of tissue function, and/or risk of noncommunicable diseases (such as asthma or colitis).

A wide range of environmental factors has been reported to affect susceptibility to communicable diseases including exposure to endocrine-disrupting chemicals and drugs, heavy metals, plasticizers, organic solvents, pesticides and herbicides, naturally occurring toxicants (e.g., fungal toxins), and physical and psychological stressors. In some cases deficiencies in certain metals (e.g., zinc) or micronutrients (e.g., vitamin D) can also program the immune system for targeted immunosuppression and result in increased susceptibility to infections. For example, Thorton et al. [37] recently reported that vitamin A concentrations are inversely related to apparent infection-related respiratory and gastrointestinal morbidity in school-age children in Bogotá, Columbia. Inadequate zinc in maternal and infant diet can increase the risk of infections, and zinc supplementation can protect again both respiratory and gastrointestinal infections [38]. Toxicant-associated increases in childhood respiratory infections have been reported in the case of polychlorinated biphenyls (PCBs) [39] and airborne fine particulate matter [40]. Metzger et al. [41] found that maternal smoking was associated with increased risk of infant infectious disease (both respiratory and nonrespiratory) and that this outcome extended to even full-term, normal-weight babies.

As more detailed evaluations have been performed, it is clear that developmental immunotoxicants rarely affect only the risk of communicable diseases or noncommunicable diseases. Instead, the reported effects are often limited by the range of assessment. Earlier DIT studies emphasized assessment of risk of infectious diseases while more recent studies have focused on noncommunicable diseases. It is rare that both disease categories are included within the same evaluation although the study of PCBs and the Faroe Island children are a welcomed exception where effects on both vaccine failure and risk of allergic sensitization have been noted. A second example where allergic disease and elevated risk of infections are joint outcomes after early-life exposure is in the case of metal (nickel) allergy. Rosato et al. [42] reported that children with systemic nickel allergy syndrome (a contact dermatitis allergy) had recurrent respiratory infections and an elevated prevalence of otitis media.

IMMUNE DISORDERS AND NONCOMMUNICABLE DISEASES

Allergic Disorders

Allergies represent one of the earliest-onset categories of NCDs [43,44]. For this reason there is considerable interest in both early-life risk factors as well as prominent pathways for allergy-associated immune disorders. Four major types of allergic disorders have been examined at least to some extent for early-life origins as well as linkage to epigenetic alterations. In addition to food allergies, three disorders form what has been termed the "atopic triad": asthma, allergic rhinitis, and atopic dermatitis (eczema). The importance of both developmental origins of these diseases and epigenetic programming in disease pathogenesis has been noted in a review by North and Elis [45]. In general, elevated risk of one or more forms of allergy appears to be associated with persistent maintenance of a Th2 state originally established during pregnancy. Additionally, incomplete or skewed maturation of certain immune cell populations can result in deficient immune surveillance and improperly regulated inflammation, which appears to fit the profile of allergy-based immune disorders [46–48]. Not surprisingly, allergic disorders have an intrinsic connection to specific infections including vulnerability to and/or aberrant host responses against specific infections (Table 2).

While some environmental factors and conditions have extensive research supporting their connection with specific allergic diseases, others

TABLE 2 Co-occurrence of Immune Dysfunction-driven Infectious and Allergic Diseases[a]

Allergic Disease	Infection(s)	References
Allergic rhinitis	*Streptococcus pyogenes* upper respiratory infections	[143]
Asthma	Influenza A	[144]
Asthma	*Escherichia coli* bloodstream infections	[145]
Asthma	Otitis media	[146]
Atopy	Recurrent community-acquired pneumonia	[147]
Atopic dermatitis	Methicillin-resistant *Staphylococcus aureus* (MRSA)	[148]

[a] *See also Dietert [142].*

have been more generally characterized as causing an increased risk of allergic sensitization and/or loss of immune tolerance. Chief among these are the two categories of endocrine disruptors, the PCBs and the perfluorinated compounds such as perfluorooctanoic acid (PFOA). Developmental exposures to PCBs are associated with both immunosuppression as in the context of childhood vaccine responses, while at the same time increasing the risk of allergic sensitization. Early-life exposure to PFOA also reduced childhood vaccine antibody responses [49,50] and elevated allergic inflammation [51] while epigenetically reducing the levels of host cell antioxidant protection [52].

Among the factors potentially affecting several forms of allergic diseases during childhood is the introduction of furry pets into the household. Although timing of pet introduction and family history of allergic diseases is important, many studies show a modest protection with pets present in the household of nonatopic families at or near birth [53,54]. Part of the protection has been thought to be a result of the presence of potential pet-related allergens in the child's environment at the same time that postnatal immune maturation is occurring and the establishment of tolerance. But a second factor may be the recently identified role of pets in the transfer and spread of microbes resulting in a potential diversification of the household members' microbiomes [55,56]. Such activity during pregnancy and during the first year of an infant's life might aid microbiome establishment and early maturation as well as useful immune maturation.

Allergic Rhinitis

Allergic rhinitis (AR) is the upper respiratory condition historically referred to as hay fever. It is among the most prevalent of the allergic diseases and may well be the most extensively medicated (prescription and over-the-counter drugs) on a daily basis. AR overlaps with asthma in some of the associations with early-life environmental risk factors, immune dysfunction, and epigenetic alterations. For example, significant association of environmental exposure to polycyclic aromatic hydrocarbons (PAHs) producing reduced T regulatory cell activity has been found in children with AR [57]. Additionally, antibiotics during the first week of life have been associated with elevated risk of AR in school-age children [58]. In contrast, living on a farm was highly protective against AR. The protective effect of early-life exposure to certain animals appears to be based on more than just the timing of allergen exposure relative to immune maturation. Using a mouse animal model, Fujimura et al. [56] found that exposure to dog-associated house dust resulted in specific gut microbial restructuring, which, in turn, led to reduced Th2-driven airway responses and protection against airway allergen challenge. These combined observations suggest that microbial-originated signaling of the immune system in the newborn/infant

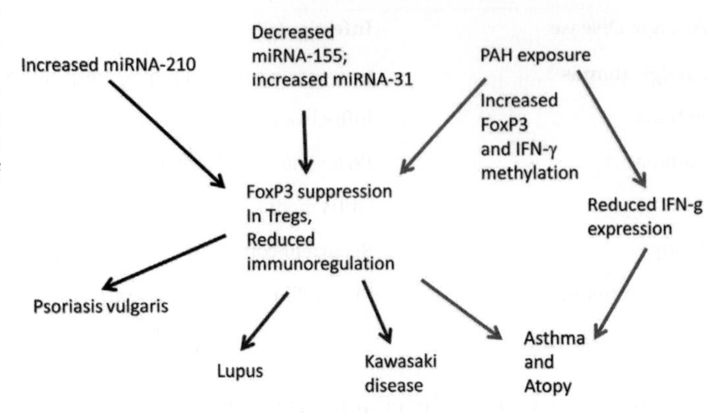

FIGURE 1 The figure illustrates the epigenetics regulation of FoxP3+ Tregs, which has been reported as a common pathway associated with several different immune disorders. Additionally, it has been suggested that exposure to polycyclic aromatic hydrocarbons, as occurs via air pollution, may increase the risk of asthma and atopy via early-life epigenetic programming of Treg activity.

is among the pathways involved in the risk of AR.

Atopic Dermatitis

Atopic dermatitis (AD), also known as eczema, is part of the atopic triad that includes both asthma and allergic rhinitis. It is estimated to have a lifetime prevalence of approximately 20% following what is often an early-life onset. The cohorts of children with this condition often progress to asthma later in life. In a longitudinal study within the Korean population, evidence suggests that genetic polymorphisms for the IL-13 gene, prenatal exposure to antibiotics, and Cesarean delivery are additive for elevated risk of atopic dermatitis [59]. Maternal intake of alcohol during the third trimester is also a risk factor for atopic dermatitis [60].

Genome-wide association studies (GWAS) have been conducted to look for potential gene targets of epigenetic regulation of AD. Candidate genes for AD are involved with innate-adaptive immunity, interleukin (IL)-1 family signaling, regulatory T cells, epidermal barrier functions, the vitamin D pathway, and the nerve growth factor pathway [61]. One of the potential routes to elevated risk of atopic dermatitis appears to involve pathways that overlap with risk of other allergic diseases. Maternal prenatal smoking has been associated with epigenetic regulation of CD4+ forkhead box P3-positive (FoxP3+) T regulatory cells (Tregs) via elevated production of microRNA 223 both in the mother as well as in cord blood [62]. Additionally, elevation of this microRNA is associated with reduced cord blood Treg cell numbers resulting in increased risk of atopic dermatitis by age three. Epigenetically programmed pathways involving Tregs and immune-driven health risks appear to represent one of the common pathways associated with immune disorders (Figure 1).

Asthma

Asthma is among the immune disorders that not only have numerous prenatal and postnatal environmental risk factors but also a significant emergence during childhood. There is a significant association of birthplace (US vs non-US) and prevalence of current asthma among both children and adults [63], suggesting that early-life environment may exert a significant influence on risk of this disease. Numerous perinatal environmental risk factors have been reported for later-life asthma, and these are listed in Table 3.

As with other allergic diseases, epigenetic modifications are thought to contribute to disease risk. Hew et al. [57] reported that FoxP3 inhibited expression via either rMicroRNA

TABLE 3 Reported Prenatal and Childhood Environmental Risk Factors for Later-life Asthma[a]

Traffic-related air pollution
Maternal smoking
Environmental tobacco smoke
Paracetamol use
Vitamin D insufficiency
Urban-type environment
Lower socioeconomic status
Maternal stress, anxiety, or depression
Childhood abuse
Maternal intimate partner violence
PCBs and certain pesticides
Heavy metals
Cesarean birth delivery
Neonatal antibiotic use
Earlier timing of childhood vaccinations
Certain respiratory infections (trigger)
Lack of or short duration of breast-feeding
Later rather than earlier introduction of household pets
Juvenile fast food diet

[a] *Adapted from Dietert [149].*

changes or methylation at the Treg-specific demethylated region (TSDR) site can result in inhibition of Treg function and uncontrolled inflammation. However, selective circumstances have been noted where TSDR methylation status is uncoupled from FoxP3 expression levels (e.g., within highly inflamed tissue) [64].

There is evidence suggesting that the microbiome may play a role in risk and severity of allergic asthma. For example, researchers have reported that perinatal antibiotics alter the microbiome in a manner that promotes Th2-driven allergic inflammatory disease [65]. Major microbiome-associated changes appear to involve both the polarization of macrophages to M2 and reduced Treg activity.

Food Allergies

Because food allergies often arise early in life, a focus has been placed on the early-life windows during which environmental exposure and conditions may predispose for this immune disorder. Candidate genes for potential association with occurrence and/or severity of food allergy include the MHC human leukocyte antigen (HLA class II gene family including HLA-DRB1, HLA-DQB1, and HLA-DPB1), cluster of differentiation 14 (CD14), FOXP3, signal transducer and activator of transcription 6 (STAT6), serine protease inhibitor Karzal type 5 (SPINK5), IL-10, and IL-13 [66].

Commensal gut microbes have emerged as significant players in developmental risk of food allergies. Specific anaerobe microbes are able to support the integrity of the epithelial barrier by affecting immune production of the cytokine IL-22. Barrier function prevents food allergens from reaching sites when sensitization could take place, particularly during childhood. A second effect is that gut microbes produce a numerous array of metabolites including short-chain fatty acids. Some of these appear to promote FoxP3-driven differentiation of Treg immune cells and suppress aberrant responses to food [67].

Using an epigenome-wide analysis, Martino et al. [68] reported that a cadre of epigenetic marks were present at birth that appear to be associated with subsequent food allergy in children and were longitudinally stable (also present at one year of age). A recent review by Hong and Wang [69] emphasized that candidate epigenetic marks appear promising in association with pathways to food allergy. However, currently the level of detail for the role of epigenetics in this immune disorder is less well defined in comparison with that for asthma. For example, 10 candidate genes have been identified of interest in risk of food allergy. However, linkage between environmental exposures and epigenetic marks associated with these genes remains to be investigated [70].

Some evidence suggests an overlap among pathways of immune dysfunction for asthma and food allergies. For example, Syed et al. [71] reported that oral immunotherapy for peanut allergy that resulted in clinically tolerant patients was associated with hypomethylation of the FoxP3 gene and increased Treg function. The investigators suggested that this may be a predictive marker for clinical tolerance during immunotherapy.

AUTOIMMUNE DISORDERS

When taken together, autoimmune disorders are comparable in prevalence to heart disease or cancer. However, no single autoimmune condition is in the 10 highest-ranking noncommunicable diseases in the United States. This seeming discrepancy is based on the fact that myriad autoimmune conditions (in excess of 100) have been defined. In some cases they are systemic such as with SLE, while in other cases they are targeted to a single organ such as with type 1 diabetes or autoimmune thyroiditis. Yet in all of these examples, underlying immune dysfunction exists, and it is this dysfunction that, at least in part, leaves patients facing an elevated risk of additional autoimmune and/or inflammatory diseases.

Risk factors for autoimmune diseases include both genetic and environmental factors. In general in contrast with the allergic diseases, identification of the early-life environmental factors contributing to specific autoimmune diseases has been challenging. In part this may be due to the exquisite gene-by-environment interactions that occur where specific environmental exposures and conditions can program specific genetic subpopulations of humans for the onset of later-life autoimmunity. For a few autoimmune conditions such as multiple sclerosis, SLE, and type 1 diabetes, significant information has been gathered on both early-life environmental risk factors as well as the involvement of epigenetic alterations in the programming of immune dysfunction. For others, such as systemic sclerosis [72] and Sjögren's syndrome [73], the role of epigenetics is just coming into focus. Many other autoimmune conditions await further study.

Despite the large number of genetic polymorphisms and environmental factors that may interact to produce a specific autoimmune condition, there are likely commonalities in vulnerable pathways. For example, one theme running through the autoimmune literature is that activity of Tregs and, in particular, the environmental vulnerability of FoxP3 expression levels may be key in tipping the immunoregulatory balance toward autoimmunity. Barbi et al. [74] proposed that dramatic shifts in FoxP3 expression in response to environmental changes make it the perfect target for epigenetic programming of autoimmunity.

Celiac Disease

Celiac disease has both allergic and autoimmune components. Among the early-life risk factors for pediatric celiac disease are Cesarean delivery, use of antibiotics during the first year of life, lack of breast-feeding, and infant gastrointestinal infections [75,76]. The disease is doubly significant as its prevalence has risen dramatically the past two decades [77] and particularly among children [78].

Furthermore, it is an entryway disease for additional autoimmune and inflammatory noncommunicable diseases and conditions as reviewed in Dietert [75]. These include small bowel adenocarcinoma and T cell lymphomas as well as depression (women), osteoporosis, sarcoidosis, eating disorders (women), COPD, cardiovascular disease, hearing loss, and restless leg syndrome [75]. Celiac disease and type 1 diabetes are comorbid with each other. A diagnosis of either disease increases the risk for the other, and one can readily precede the other. Occasionally, a dual diagnosis is made [79]. Ludvigsson et al. [80] found that celiac disease patients were at an elevated risk of psoriasis both before and after with celiac disease diagnosis. One of the tenets considered about these celiac disease comorbidities is that immune-driven imbalances found in even young children and reflected in overproduction of specific cytokines (e.g., IFN-γ and IL-12p70) create states of systemic inflammation that over time promote additional inflammation-driven diseases [81].

Multiple Sclerosis

Multiple sclerosis is an autoimmune neurological condition with both genetic and environmental risk factors as well as environment-by-genetics interactions. Among the putative environmental risk factors are insufficient vitamin D, cigarette smoking, alcohol consumption, heavy metals, and microbial infectious triggers. Several recent reviews have summarized the apparent role of epigenetic modification in the pathogenesis of multiple sclerosis (MS). Zhou et al. [82] point out that gene variants alone do not explain the heritability component of MS and that some combination of DNA methylation, histone modifications, and microRNA-mediated gene silencing are likely to play a role in the inflammation-initiated neurodegenerative condition. Küçükali et al. [83] argue that the greater influence of maternal over paternal intergenerational transmission is also suggestive of a strong role for epigenetic modifications in MS pathogenesis.

Most investigators have focused on the links between epigenetic modification and inappropriate neuroinflammation at the levels of both innate immune cells [84] (e.g., microglia or brain macrophages) and T regulatory cell populations (e.g., Tregs and Th17 cells).

Further suggestion of a link between epigenetic modification affecting inflammatory regulation and MS is reported. Researchers using two different mouse models of MS found that treatment with a hypomethylating agent, 5-aza-2′-deoxycytidine, ameliorated the neurological effects of the condition. A primary change in gene expression involved Tregs and FoxP3 expression. Tregs were elevated among the immune cell populations, while mRNA transcripts of anti-inflammatory cytokines were increased and those involved in the production of proinflammatory mediators were decreased.

The immune response controlling gene complex, the human major histocompatibility complex (MHC) known as the human leukocyte antigen (HLA) complex, has been implicated with risk of MS. But even here, epigenetic modification may play a role. In early studies, Chao et al. [85] reported evidence directing a focus on epigenetic modifications to human class II genes. Similarly, using genome-wide DNA methylation analysis, Graves et al. [86] reported that a significant association exists between the methylation levels of the HLA-DRB1 locus among 30 MS patients versus 28 healthy controls.

Systemic Lupus Erythematosus

SLE is a prototypic, chronic autoimmune disease that mostly affects women and emerges within the second or third decade of life, depending upon the load of genetic susceptibility loci as well as environmental exposure to cigarette smoking, ultraviolet light, and infection [87–90].

Many scientists have examined the role of toxins and the microbiota as potential initiating factors for SLE. There is a plethora of toxins that induce a SLE-like syndrome but upon close examination do not contribute to full-blown SLE. Moreover, unlike other autoimmune diseases such as rheumatoid arthritis and Crohn's disease, the evidence is less clear for the role of microbiota in SLE [91], although intestinal dysbiosis with a decrease of some Firmicutes family members and a decreased ratio of Firmicutes/Bacteroidetes has been observed in

SLE patients compared with controls [92]. This dysbiosis is hypothesized to be a consequence of altered immune function in SLE patients and presents as an overrepresentation of oxidative phosphorylation and glycan utilization pathways in SLE patient microbiota compared with normal. Moreover, a recent study has found that SLE-prone mice (B6/lpr) with lower colonization of the anti-inflammatory Lactobacilli in the gut is associated with higher disease activity as measured by lymphadenopathy and renal pathology [93].

Female Prevalence in SLE

Environmental factors contribute to the emergence of SLE in genetically prone individuals. The principal evidence for this conclusion comes from studies of disease-unaffected relatives of SLE patients [94]. For example, B cells in peripheral blood mononuclear cells (PBMC) isolated from female, but not male, blood relatives of SLE patients have been observed to secrete IgG anti-dsDNA autoantibodies spontaneously at levels significantly higher than controls, but less than SLE patients. In contrast, both male and female blood relatives of SLE patients secreted more IgG in response to the polyclonal memory B cell activator, pokeweed mitogen, when compared to controls [95]. Differences between the genders were apparent when antibodies to specific antigens were measured. Following stimulation of memory B cells with pokeweed mitogen, male relatives secreted significantly more IgM anti-ssDNA when compared to female relatives or male/female controls, whereas female relatives secreted more IgG to the common environmental antigen, influenza hemagglutinin when compared to male relatives or male/female controls [95].

These data suggest that relatives of SLE patients, like SLE patients themselves, have hyperactive B cells when compared to normal people in the control population. Finally, the finding that male SLE relatives secrete low-affinity, nonswitched IgM autoantibodies, whereas female SLE relatives secrete high-affinity, autoantibodies that have undergone switching to IgG, suggests that the female gender may contribute to initiation of SLE disease activity in humans.

Estrogen-containing hormone replacement therapy has been found to precipitate SLE or exacerbate the existing disease. Some studies have observed similar results with oral contraceptive use [96,97]. E2 induces its effects following diffusion across the plasma membrane and binding to cytoplasmic estrogen receptors, hER a and b, found in human B cells, T cells, and monocytes [98]. Functionally, E2 has been reported to rescue autoreactive B cells from tolerization to increase total spontaneous IgG secretion from PBMC isolated from SLE patients and total spontaneous IgG production from PBMC isolated from normal controls markedly [99]. Furthermore, E2 has been shown to increase spontaneous secretion of anti-dsDNA IgG antibody from PBMC isolated from active SLE patients markedly, but not from PBMC isolated from inactive SLE patients or normal controls [100].

An additional mechanism that may contribute to E2-mediated activation of B cells may be an autocrine feedback loop in which E2-induced prolactin produced by B cells may bind to prolactin receptors that have been described on human B cells, T cells, monocytes, and neutrophils [101]. Of note, prolactin has been detected in human secondary lymphoid tissues such as spleen, lymph node, and tonsil that provide the microenvironment necessary for B cell differentiation and selection resulting in secretion of specific, high affinity, class-switched Abs [102]. Estrogen-induced hyperprolactinemia, which may occur during pregnancy or postpartum breast-feeding at levels significantly higher than that constitutively present, with glomerulonephritis and the production of a variety of autoantibodies with specificities to ds DNA, cardiolipin, and endothelium [103]. Engagement of the prolactin receptor induces transcription of a variety of signaling molecules, expression of high-affinity IL-2R on B cells, and proliferation of lymphocytes. It should be noted that the potential role of E2 and prolactin as initiating factors in the development of SLE has been emphasized by studies with the SLE-prone mice [104].

Infectious Agents in Systemic Lupus Erythematosus

Viral and/or bacterial infections, or what has recently become known as the infectome, play a significant role as an initiating factor for SLE [105]. While multiple viruses have been indicated to be involved in SLE including CMB, parvovirus B19, and Epstein–Barr virus (EBV), the most is known about EBV infection as an initiator of SLE [106]. A comparison of the prevalence of infection with EBV in young SLE patients compared to normal age-matched controls of European American, African American, and Hispanic origin suggested that EBV infection may be an etiologic factor in SLE [107].

Microbial infection of SLE-susceptible individuals has been suggested to play a role in secretion of IgM anti-DNA Ab and to contribute to initiation of active disease. Polyclonal B cell activation leading to secretion of anti-DNA Ab has been observed following stimulation with staphylococcal and streptococcal B cell "superantigens" (sAgs), pneumococcal polysaccharides, and DNA in the form of phosphorothioate oligodeoxynucleotides (sODNs) [108,109]. Of importance, in light of the finding that individuals susceptible to SLE have an exaggerated response to sIg engagement, all of these stimuli utilize sIg ligation in a manner that is independent of the antigen-binding site for all or part of their stimulatory signal.

Staphylococcal sAgs, which have been shown to induce IgM anti-DNA Ab from B cells isolated from normal and SLE donors, include staphylococcal protein A (SPA), staphylococcal enterotoxin (SE) A, and SED. Whereas SPA and SEA bind sIg of B cells in the framework regions of VH3, SED binds sIg of B cells in the framework region of VH4 [110]. All of these sAgs initially induce polyclonal activation of B cells, as assayed by the ability to detect all available VH gene families that can be sustained in the presence of activated T cells. Of note, the TSST-1 staphylococcal sAg has been shown to stimulate T cells to express CD154 in a manner that was capable of facilitating Ig secretion and Ig heavy chain class switching of bystander B cells in a nonantigen-specific manner [111]. Products of microbial infection such as sAgs, polysaccharides, and DNA may contribute to the initiation of active SLE in susceptible individuals by polyclonally activating B cells, leading to expansion of autoreactive clones or induction of somatic hypermutation in normal clones that may change the nature of the Ag that the secreted Ab binds to from foreign to self.

Epigenetics in Systemic Lupus Erythematosus

A profound number of lymphoid and myeloid abnormalities that may either reflect the composite impact of genetic factors or secondary event to other primary immunologic abnormalities. SLE is characterized by the presence of hyperreactive lymphocytes and aberrant antibody responses to both nuclear and cytoplasmic antigens. End-organ damage from inflammation and autoantibodies affects the kidneys, lungs, joints, heart, and brain. Higher prevalence for SLE exists in persons of African American, Hispanic, or Asian ancestry compared with European Americans. There is a strong genetic bias with high sibling risk ratios and higher concordance among monozygotic twins compared with dizygotic twins or full siblings. However, the concordance rate of SLE in monozygotic twins is only 15–25%, indicating that other factors besides genetics play a role in disease pathogenesis. Genome-wide association studies (GWAS) have identified multiple genetic risk factors including TRAF1, MHC, ITGAM/CD11b, IRF5, STAT4, PTPN22, PRDM1/Blimp1, JAZF1/ZnF802, UHRF1BP1, IL10, ETS1, WDFY4, BLK, BANK1,

Ets-1, Csk-1, TNFSF4/OX40L, TNFAIP3/A20, TNIP1 (TNFAIP3-interacting protein 1), and REL [112].

Genetic susceptibility loci and environmental influences do not completely explain the etiology of SLE. Recent studies examining genome-wide analysis of DNA methylation patterns in lymphoid and myeloid subsets in the periphery of SLE patients have shown hypomethylation of IFN-alpha-regulated genes (that are a gene expression signature of active SLE patients) as well as composite remodeling of CD4+ T cell transcriptional activity especially in areas of transcription of cell cycle genes and MAPK signaling [113,114]. Of interest, drugs that inhibit DNA methylation induce T cell autoreactivity and SLE symptoms in mice [115], and drug-induced SLE in humans is associated with reduced DNA methylation due to repression of DNA methyltransferase transcription [116].

Epigenetic control of gene expression in utero can be heritable and is controlled by the DNA methyl-transferases DNMT -3A and/or -3B. In contrast, DNMT1 is a maintenance DNA methyltransferase that is expressed in an organism after birth. After association with DNMT3 proteins, DNMTL3 (which has not catalytic activity of its own) directs binding to specific loci and increases activity of DNMT3A/B. DNMTL3 focuses DNA methylation activity of DNMT3A/B by mediating specific binding to a subset of transcription factors. A subset of these DNMTL3-targeted transcription factors that includes RXRalpha, NFkB-p65, ATF3, CDX2, and E2F has been shown to play a role in induction and progression of SLE. DNMT3L recognizes unmethylated histone H3 lysine 4 (H3K4) residues and recruits DNMT3A/B for posttranslational histone modification [117].

Interestingly, the DNMT3B/DNMT3L/NFkB-p65 complex epigenetically negatively regulates TRAF1, an adaptor protein that modulates signaling through TNF-receptor family members [118]. In contrast, NFkB-p65 alone induces expression of TRAF1. Cyclical changes in methylation/demethylation status of gene promoters are crucial events in transcriptional regulation with methylation repressing and demethylation inducing transcriptional activity. Not only is TRAF1 an SLE-susceptibility locus but TRAF1 mediates NFkB activation through a number of TNF-receptor family members that are involved in the development and selection of lymphocytes into the functional nonautoimmune repertoire.

SLE is a complex, polygenic, chronic autoimmune disease characterized by the production of multiple autoantibodies. Initiating factors, such as female sex hormones or infection with bacteria or viruses, appear to contribute to the emergence of active disease in genetically prone individuals. Environmental toxins, ultraviolet light, and cigarette smoking contribute in some individuals. Microbiota may play a regulatory role in progression and/or severity of disease. Epigenetic control of genetic susceptibility loci is an emerging area of research in SLE and may occur as early as in utero. Hyperactivity of lymphocytes is a characteristic of SLE, but which cell abnormalities are intrinsic and which are acquired remains to be fully elucidated.

Type 1 Diabetes

Type 1 diabetes is an autoimmune disease largely arising during childhood or in the young adult. Hence, by age of onset early-life exposures and/or intergenerational alterations are involved in the environmental risk of and programming for type 1 diabetes. The pathogenesis involves the destruction of pancreatic B-cells through a specific autoimmune response. As a result, contributing environmental and epigenetics factors are by definition either resulting from prenatal or early postnatal environmental exposure-host responses or transgenerational alterations. While T cell-mediated destruction of pancreatic islets is known to play a major role in the pathogenesis, inappropriate chemotactic recruitment and subsequent responses among innate immune cells appear to initiate the cascade of organ-directed events. Inappropriate activation of Toll-like receptors (TLRs) on innate immune cells seems to play a role in the early stage of immune dysfunction-mediated disease. Among the environmental risk factors in early life suggested to increase the risk for type 1 diabetes are: Cesarean delivery, insufficient vitamin D, and the timing of introduction of solid foods. Vitamin D receptor levels also appear to be important in risk of type 1 diabetes, and these may be influenced via genetic polymorphisms and/or through epigenetic modifications.

Some evidence suggests that the microbiome may be important in risk of type 1 diabetes. Using the nonobese diabetic (NOD) mouse model of Type 1 diabetes, Markle et al. [119] found that microbial makeup of the microbiome affected both testosterone level and metabolic phenotypes needed for type 1 diabetes to manifest and display sex-specific differences in prevalence. In fact, disturbed microbiota phenotypes are among the earliest biomarkers at the onset of type 1 diabetes and de Goffau et al. [120] found that butyrate-producing bacterial strains were more prevalent in age-matched, nondiabetic versus diabetic children.

There is some evidence that specific epigenetic marks may play a role in programming of type 1 diabetes. Jayaraman et al. [121] examined gene expression differences and epigenetics marks associated with the murine model for type 1 diabetes (NOD strain of mice). They found that a cluster of related inflammatory genes was elevated in diabetic mice and that the expression of these genes was significantly reduced in mice with suppression of the diabetes. The authors propose that suppression of histone hyperacetylation and subsequent hyperacetylation of the IFN-γ gene and a transactivator, Tbx21/Tbet (upregulating IFN-γ expression without affecting expression levels of several other cytokines), provided amelioration of disease. Dang et al. [122] suggested a similar process linked with type 1 diabetes. Finally, a drug facilitating histone deacetylation has been proposed as a useful epigenetic-based therapeutic avenue in the treatment of type 1 diabetes [123].

As with other autoimmune conditions, type 1 diabetes has several comorbidities and prevalence complications. These include a host of vascular conditions including retinopathy, nephropathy, and hypertension [124] as well as depression, sleep problems, autoimmune thyroiditis, celiac disease, hearing loss, multiple sclerosis, and atherosclerosis (reviewed in [36]).

INFLAMMATORY DISORDERS

Obesity and cardiovascular disease (CVD) are now recognized as being intrinsically linked with immune-inflammatory dysfunction and could be viewed as inflammatory disorders with macrophages as key regulators of disease processes [125,126]. However, these diseases are covered extensively in other chapters of this book. As a result, only the subcategory of CDV directly linked with macrophage dysfunction (atherosclerosis) will be discussed here.

Atherosclerosis

Atherosclerosis is a multifactorial cardiovascular disease whose roots can be found in the misregulated inflammatory processes of atherogenesis but appear to include other misdirected immune-driven responses as well. For example, Wick et al. [127] recently reviewed the evidence of autoimmune responses directed at heat shock proteins, and lack of early-life tolerance against a target that shared homology with certain microbial heat shock proteins may contribute to the early-life changes and atherogenesis. Even though atherosclerosis is often thought of as an adult-onset disease, biomarkers of atherosclerosis are present during childhood [128]. Epigenetic alterations appear to play a significant role in atherogenesis and the macrophage epigenome has been suggested as a major target for therapeutic strategies [129].

Depression

While there may be several different routes involving the pathogenesis of depression, a significant component in some cases of depression appears to be misregulated inflammation and includes age-related consideration [130,131]. Depression is a common comorbidity with a majority of inflammation-driven immune-based disorders. Elevated levels of specific proinflammatory cytokines are associated with the occurrence of depressive symptoms [132]. Early-life stress as well as specific environmental exposures are thought to play a major role in inflammation-driven disorders such as major depressive disorder (MDD). Nieratschker et al. [133] identified a cross-species, early life, stress-induced epigenetic mark, methylation of MORC-1 (MORC family CW-type zinc finger 1), which is associated with cord blood cells, newborn T cells at birth, and adult brain cells and has been linked with MDD.

Psoriasis

Psoriasis is both an autoimmune skin disease and an inflammatory disorder. Risk of psoriasis appears to be affected by both genetic/epigenetic factors as well as by environmental risk factors. Among genetic factors reported are variables affecting FoxP3 expression levels among Tregs. Additional risk factors include environmental tobacco smoke, BMI/obesity, alcohol consumption, and early-life stressful events [134,135]. Trowbridge and Pittelkow [136] describe that epigenetic regulation of multiple genes appears to be involved with the pathogenesis of psoriasis. These include both the IFN-γ and TNF-α genes.

One of the complications of psoriasis is that a subset of patients develops psoriatic arthritis. Among the environmental risk factors reported to be associated with this complication are infections requiring the use of antibiotics [137].

Sleep Disorders

Sleep disorders represent one of the more common comorbid conditions arising from early-life programmed immune disorders. Sleep problems are very commonly associated with autoimmune and/or inflammatory conditions such as atherosclerosis, and epigenetic alterations are suspected in the programming of several types of sleep disorders such as sleep apnea [138,139].

Narcolepsy, one type of disorder, was recently recognized as an autoimmune condition where childhood onset has been associated with influenza A H1N1 infection and H1N1 vaccination [140]. In another manifestation, childhood obstructive sleep apnea syndrome has been associated with increased NF-*k*B activity and elevated levels of proinflammatory markers such as C-reactive protein and IL-1 [141].

CONCLUSIONS

The developing immune system is a sensitive target for programming of later-life immune dysfunction and disease. At least three environmentally associated, overlapping pathways can result in immune dysfunction: direct disruption of critical windows of immune development, epigenetic programming of altered gene expression, and microbiome-dysbiosis resulting in improper immune development. Increased risk of both infectious and noncommunicable diseases results from early-life programmed immune dysfunction. The vast majority of these involve tissues and organs other than the primary or secondary lymphoid organs (e.g., lung, skin, cardiovascular system). Many of the immune disorders are interlinked by comorbid risk suggesting that a single, developmentally programmed problem with the immune system could result in several different types of diseases. Misregulation of inflammation is among the most common outcome, although the nature of the inflammation (e.g., Th1- vs Th2-driven) may differ among disorders. Common pathways of programming appear to exist, and one of these is the epigenetic regulation of FoxP3 gene expression affecting T cell regulation and control of inflammation. A second example is represented by epigenetic- and/or microbiome-associated skewing for the polarization of M1 versus M2 macrophages.

References

[1] Barouki R, Gluckman PD, Grandjean P, Hanson M, Heindel JJ. Developmental origins of non-communicable disease: implications for research and public health. Environ Health 2012;11:42.

[2] Barker DJ, Gluckman PD, Godfrey KM, Harding JE, Owens JA, Robinson JS. Fetal nutrition and cardiovascular disease in adult life. Lancet 1993;341:938–41.

[3] Dietert RR, Etzel RA, Chen D, Halonen M, Holladay SD, Jarabek AM, et al. Workshop to identify critical windows of exposure for children's health: immune and respiratory systems work group summary. Environ Health Perspect 2000;108(Suppl 3):483–90.

[4] Makris SL, Thompson CM, Euling SY, Selevan SG, Sonawane B. A lifestage-specific approach to hazard and dose-response characterization for children's health risk assessment. Birth Defects Res B Dev Reprod Toxicol 2008;83:530–46.

[5] Borre YE, Moloney RD, Clarke G, Dinan TG, Cryan JF. The impact of microbiota on brain and behavior: mechanisms & therapeutic potential. Adv Exp Med Biol 2014;817:373–403.

[6] Haugen AC, Schug TT, Collman G, Heindel JJ. Evolution of DOHaD: the impact of environmental health sciences. J Dev Orig Health Dis 2014;4:1–10.

[7] Dietert RR, DeWitt JC, Germolec DR, Zelikoff JT. Breaking patterns of environmentally influenced disease for health risk reduction: immune perspectives. Environ Health Perspect 2010;118:1091–9.

[8] Dietert RR. Transgenerational epigenetics of endocrine-disrupting chemicals. [Chapter 18]. In: Tollefsbol T, editor. Transgenerational epigenetics: evidence and debate. Elsevier; 2014.

[9] Luster M, Faith RE, McLachlan JA. Alterations of the antibody response following in utero exposure to diethylstilbestrol. Bull Environ Contam Toxicol 1978;20:433–7.

[10] Holladay SD, editor. Developmental immunotoxicology. CRC Press; 1996.

[11] Van Loveren H, Piersma A. Immunotoxicological consequences of perinatal chemical exposures. Toxicol Lett 2004;149:141–5.

[12] Herzyk DJ, Jeanine L, Bussiere JL, editors. Immunotoxicology strategies for pharmaceutical safety assessment. New York: Wiley; 2008.

[13] Dietert RR, Luebke RK, editors. Immunotoxicity, immune dysfunction and chronic disease. New York: Springer Business + Media; 2012.

[14] Coelho R, Viola TW, Walss-Bass C, Brietzke E, Grassi-Oliveira R. Childhood maltreatment and inflammatory markers: a systematic review. Acta Psychiatr Scand 2014;129:180–92.

[15] Dietert RR. Developmental immunotoxicity, perinatal programming, and noncommunicable diseases: focus on human studies. Adv Med 2014. Article ID 867805. http://dx.doi.org/10.1155/2014/867805.

[16] Luebke RW, Chen DH, Dietert R, Yang Y, King M, Luster MI. Immunotoxicology Workgroup. The comparative immunotoxicity of five selected compounds following developmental or adult exposure. J Toxicol Environ Health B Crit Rev 2006;9:1–26.

[17] Sirivarasai J, Wananukul W, Kaojarern S, Chanprasertyothin S, Thongmung N, Ratanachaiwong W, et al. Association between inflammatory marker, environmental lead exposure, and glutathione S-transferase gene. Biomed Res Int 2013;2013:474963.

[18] Raqib R, Ahmed S, Sultana R, Wagatsuma Y, Mondal D, Hoque AM, et al. Effects of in utero arsenic exposure on child immunity and morbidity in rural Bangladesh. Toxicol Lett 2009;185:197–202.

[19] Sengupta M, Bishayi B. Effect of lead and arsenic on murine macrophage response. Drug Chem Toxicol 2002;25:459–72.

[20] Hoppin JA, Umbach DM, London SJ, Henneberger PK, Kullman GJ, Alavanja MC, et al. Pesticides and atopic and nonatopic asthma among farm women in the agricultural Health Study. Am J Respir Crit Care Med 2008;177:11–8.

[21] Bloom DE, Cafiero ET, Jane-Llopis E, Abrams-Gessel S, Bloom LR, Fathima S, et al. The global economic burden of noncommunicable diseases. Geneva: World Economic Forum; 2011.

[22] DeWitt JC, Peden-Adams MM, Keil DE, Dietert RR. Current status of developmental immunotoxicity: early-life patterns and testing. Toxicol Pathol 2012;40:230–6.

[23] Martin R, Nauta AJ, Ben Amor K, Knippels LM, Knol J, Garssen J. Early life: gut microbiota and immune development in infancy. Benef Microbes 2010;1:367–82.

[24] Dietert R, Dietert J. The completed self: an immunological view of the human-microbiome superorganism and risk of chronic diseases. Entropy 2012;14:2036–65.

[25] Dietert RR. The microbiome in early life: self-completion and microbiota protection as health priorities. Birth Defects Res B Dev Reprod Toxicol 2014;101:333–40.

[26] Olszak T, An D, Zeissig S, Vera MP, Richter J, Franke A, et al. Microbial exposure during early life has persistent effects on natural killer T cell function. Science 2012;336:489–93.

[27] Hansen CH, Nielsen DS, Kverka M, Zakostelska Z, Klimesova K, Hudcovic T, et al. Patterns of early gut colonization shape future immune responses of the host. PLoS One 2012;7:e34043.

[28] Obata Y, Furusawa Y, Endo TA, Sharif J, Takahashi D, Atarashi K, et al. The epigenetic regulator Uhrf1 facilitates the proliferation and maturation of colonic regulatory T cells. Nat Immunol 2014;15:571–9.

[29] Balmer ML, Schürch CM, Saito Y, Geuking MB, Li H, Cuenca M, Kovtonyuk LV, McCoy KD, Hapfelmeier S, Ochsenbein AF, Manz MG, Slack E, Macpherson AJ. Microbiota-derived compounds drive steady-state granulopoiesis via MyD88/TICAM signaling. J Immunol 2014;193:5273–83.

[30] Kim YG, Udayanga KG, Totsuka N, Weinberg JB, Núñez G, Shibuya A. Gut dysbiosis promotes M2 macrophage polarization and allergic airway inflammation via fungi-induced PGE_2. Cell Host Microbe 2014;15:95–102.

[31] Burgess SL, Buonomo E, Carey M, Cowardin C, Naylor C, Noor Z, et al. Bone marrow dendritic cells from mice with an altered microbiota provide interleukin 17A-dependent protection against Entamoeba histolytica colitis. MBio 2014;5:e01817.

[32] Zielinski CE. Microbe driven T-helper cell differentiation: lessons from Candida albicans and Staphylococcus aureus. Exp Dermatol 2014;23:795–8.

[33] Zhang H, Sparks JB, Karyala SV, Settlage R, Luo XM. Host adaptive immunity alters gut microbiota. ISME 2015;9(3):770–81.

[34] Oldstone MB, Rosen H. Cytokine storm plays a direct role in the morbidity and mortality from influenza virus infection and is chemically treatable with a single sphingosine-1-phosphate agonist molecule. Curr Top Microbiol Immunol 2014;378:129–47.

[35] Dietert RR, Zelikoff JT. Pediatric immune dysfunction and health risks following early-life immune insult. Curr Pediat Rev 2009;5:36–51.

[36] Dietert RR, Zelikoff JT. Identifying patterns of immune-related disease: use in disease prevention and management. World J Pediatr 2010;6:111–8.

[37] Thornton KA, Mora-Plazas M, Marín C, Villamor E. Vitamin A deficiency is associated with gastrointestinal and respiratory morbidity in school-age children. J Nutr 2014;144:496–503.

[38] Basnet S, Mathisen M, Strand TA. Oral zinc and common childhood infections-An update. J Trace Elem Med Biol 2015;31:163–6.

[39] Stølevik SB, Nygaard UC, Namork E, Haugen M, Meltzer HM, Alexander J, et al. Prenatal exposure to polychlorinated biphenyls and dioxins is associated with increased risk of wheeze and infections in infants. Food Chem Toxicol January 2013;51:165–72.

[40] Jedrychowski WA, Perera FP, Spengler JD, Mroz E, Stigter L, Flak E, et al. Intrauterine exposure to fine particulate matter as a risk factor for increased susceptibility to acute broncho-pulmonary infections in early childhood. Int J Hyg Environ Health 2013;216:395–401.

[41] Metzger MJ, Halperin AC, Manhart LE, Hawes SE. Association of maternal smoking during pregnancy with infant hospitalization and mortality due to infectious diseases. Pediatr Infect Dis J 2013;32:e1–7.

[42] Rosato E, Carello R, Gabriele I, Molinaro I, Galli E, Salsano F. Recurrent infections in children with nickel allergic contact dermatitis. J Biol Regul Homeost Agents 2011;25:661–5.

[43] Lyons JJ, Milner JD, Stone KD. Atopic dermatitis in children: clinical features, pathophysiology, and treatment. Immunol Allergy Clin North Am 2015;35:161–83.

[44] Amarasekera M, Prescott SL, Palmer DJ. Nutrition in early life, immune-programming and allergies: the role of epigenetics. Asian Pac J Allergy Immunol 2013;31:175–82.

[45] North ML, Ellis AK. The role of epigenetics in the developmental origins of allergic disease. Ann Allergy Asthma Immunol 2011;106:355–61.

[46] Holt PG, Strickland DH, Hales BJ, Sly PD. Defective respiratory tract immune surveillance in asthma: a primary causal factor in disease onset and progression. Chest February 2014;145(2):370–8.

[47] Larsen JM, Brix S, Thysen AH, Birch S, Rasmussen MA, Bisgaard H. Children with asthma by school age display aberrant immune responses to pathogenic airway bacteria as infants. J Allergy Clin Immunol 2014;133:1008–13.

[48] Noutsios GT, Floros J. Childhood asthma: causes, risks, and protective factors; a role of innate immunity. Swiss Med Wkly 2014;144:w14036.

[49] Grandjean P, Andersen EW, Budtz-Jørgensen E, Nielsen F, Mølbak K, Weihe P, et al. Serum vaccine antibody concentrations in children exposed to perfluorinated compounds. JAMA 2012;307:391–7.

[50] Granum B, Haug LS, Namork E, Stølevik SB, Thomsen C, Aaberge IS, et al. Pre-natal exposure to perfluoroalkyl substances may be associated with altered vaccine antibody levels and immune-related health outcomes in early childhood. J Immunotoxicol 2013;10:373–9.

[51] Singh TS, Lee S, Kim HH, Choi JK, Kim SH. Perfluorooctanoic acid induces mast cell-mediated allergic inflammation by the release of histamine and inflammatory mediators. Toxicol Lett 2012;210:64–70.

[52] Tian M, Peng S, Martin FL, Zhang J, Liu L, Wang Z, et al. Perfluorooctanoic acid induces gene promoter hypermethylation of glutathione-S-transferase Pi in human liver L02 cells. Toxicology 2012;296:48–55.
[53] Pohlabeln H, Jacobs S, Böhmann J. Exposure to pets and the risk of allergic symptoms during the first 2 years of life. J Investig Allergol Clin Immunol 2007;17:302–8.
[54] Lodge CJ, Lowe AJ, Gurrin LC, Matheson MC, Balloch A, Axelrad C, et al. Pets at birth do not increase allergic disease in at-risk children. Clin Exp Allergy 2012;42:1377–85.
[55] Lax S, Smith DP, Hampton-Marcell J, Owens SM, Handley KM, Scott NM, et al. Longitudinal analysis of microbial interaction between humans and the indoor environment. Science 2014;345:1048–52.
[56] Fujimura KE, Demoor T, Rauch M, Faruqi AA, Jang S, Johnson CC, et al. House dust exposure mediates gut microbiome Lactobacillus enrichment and airway immune defense against allergens and virus infection. Proc Natl Acad Sci USA 2014;111:805–10.
[57] Hew KM, Walker AI, Kohli A, Garcia M, Syed A, McDonald-Hyman C, et al. Childhood exposure to ambient polycyclic aromatic hydrocarbons is linked to epigenetic modifications and impaired systemic immunity in T cells. Clin Exp Allergy 2015;45(1):238–48.
[58] Alm B, Goksör E, Pettersson R, Möllborg P, Erdes L, Loid P, et al. Antibiotics in the first week of life is a risk factor for allergic rhinitis at school age. Pediatr Allergy Immunol 2014;25:468–72.
[59] Lee SY, Yu J, Ahn KM, Kim KW, Shin YH, Lee KS, et al. Additive effect between IL-13 polymorphism and cesarean section delivery/prenatal antibiotics use on atopic dermatitis: a birth cohort study (COCOA). PLoS One 2014;9:e96603.
[60] Carson CG, Halkjaer LB, Jensen SM, Bisgaard H. Alcohol intake in pregnancy increases the child's risk of atopic dermatitis. the COPSAC prospective birth cohort study of a high risk population. PLoS One 2012;7:e42710.
[61] Tamari M, Hirota T. Genome-wide association studies of atopic dermatitis. J Dermatol 2014;41:213–20.
[62] Herberth G, Bauer M, Gasch M, Hinz D, Röder S, Olek S, et al. Maternal and cord blood miR-223 expression associates with prenatal tobacco smoke exposure and low regulatory T-cell numbers. J Allergy Clin Immunol 2014;133:543–50.
[63] Iqbal S, Oraka E, Chew GL, Flanders WD. Association between birthplace and current asthma: the role of environment and acculturation. Am J Public Health 2014;104:S175–82.
[64] Bending D, Pesenacker AM, Ursu S, Wu Q, Lom H, Thirugnanabalan B, et al. Hypomethylation at the regulatory T cell-specific demethylated region in CD25hi T cells is decoupled from FOXP3 expression at the inflamed site in childhood arthritis. J Immunol August 4, 2014;193(6):2699–708.
[65] Russell SL, Gold MJ, Reynolds LA, Willing BP, Dimitriu P, Thorson L, et al. Perinatal antibiotic-induced shifts in gut microbiota have differential effects on inflammatory lung diseases. J Allergy Clin Immunol 2015;135(1):100–9.
[66] Tan TH, Ellis JA, Saffery R, Allen KJ. The role of genetics and environment in the rise of childhood food allergy. Clin Exp Allergy January 2012;42:20–9.
[67] Cao S, Feehley TJ, Nagler CR. The role of commensal bacteria in the regulation of sensitization to food allergens. FEBS Lett 2014;588(22):4258–66.
[68] Martino D, Joo JE, Sexton-Oates A, Dang T, Allen K, Saffery R, et al. Epigenome-wide association study reveals longitudinally stable DNA methylation differences in CD4+ T cells from children with IgE-mediated food allergy. Epigenetics 2014;9:998–1006.
[69] Hong X, Wang X. Epigenetics and development of food allergy (FA) in early childhood. Curr Allergy Asthma Rep 2014;14:460.
[70] Hong X, Wang X. Early life precursors, epigenetics, and the development of food allergy. Semin Immunopathol 2012;34:655–69.
[71] Syed A, Garcia MA, Lyu SC, Bucayu R, Kohli A, Ishida S, et al. Peanut oral immunotherapy results in increased antigen-induced regulatory T-cell function and hypomethylation of forkhead box protein 3 (FOXP3). J Allergy Clin Immunol 2014;133:500–10.
[72] Broen JC, Radstake TR, Rossato M. The role of genetics and epigenetics in the pathogenesis of systemic sclerosis. Nat Rev Rheumatol 2014;10(11):671–81.
[73] Konsta OD, Thabet Y, Le Dantec C, Brooks WH, Tzioufas AG, Pers JO, et al. The contribution of epigenetics in Sjögren's Syndrome. Front Genet 2014;5:71.
[74] Barbi J, Pardoll D, Pan F. Treg functional stability and its responsiveness to the microenvironment. Immunol Rev 2014;259:115–39.
[75] Dietert R. Inflammatory bowel disease and celiac disease. [Chapter 12]. In: Dietert RR, Luebke RK, editors. Immunotoxicity, immune dysfunction and chronic disease. New York: Springer Business + Media; 2012.
[76] Canova C, Zabeo V, Pitter G, Romor P, Baldovin T, Zanotti R, et al. Association of maternal education, early infections, and antibiotic use with celiac disease: a population-based birth cohort study in northeastern Italy. Am J Epidemiol 2014;180:76–85.
[77] Burger JP, Roovers EA, Drenth JP, Meijer JW, Wahab PJ. Rising incidence of celiac disease in the Netherlands; an analysis of temporal trends from 1995 to 2010. Scand J Gastroenterol 2014;49:933–41.
[78] White LE, Merrick VM, Bannerman E, Russell RK, Basude D, Henderson P, et al. The rising incidence of celiac disease in Scotland. Pediatrics 2013; 132:e924–31.

[79] Szaflarska-Popławska A. Coexistence of coeliac disease and type 1 diabetes. Prz Gastroenterol 2014;9:11–7.
[80] Ludvigsson JF, Lindelöf B, Zingone F, Ciacci C. Psoriasis in a nationwide cohort study of patients with celiac disease. J Invest Dermatol 2011;131:2010–6.
[81] Björck S, Lindehammer SR, Fex M, Agardh D. Serum cytokine pattern in young children with screening detected celiac disease. Clin Exp Immunol 2015;179(2):230–5.
[82] Zhou Y, Simpson Jr S, Holloway AF, Charlesworth J, van der Mei I, Taylor BV. The potential role of epigenetic modifications in the heritability of multiple sclerosis. Mult Scler 2014;20:135–40.
[83] Küçükali CI, Kürtüncü M, Coban A, Cebi M, Tüzün E. Epigenetics of multiple sclerosis: an updated review. Neuromolecular Med 2015;17(2):83–96.
[84] Garden GA. Epigenetics and the modulation of neuroinflammation. Neurotherapeutics 2013;10:782–8.
[85] Chao MJ, Ramagopalan SV, Herrera BM, Lincoln MR, Dyment DA, Sadovnick AD, et al. Epigenetics in multiple sclerosis susceptibility: difference in transgenerational risk localizes to the major histocompatibility complex. Hum Mol Genet 2009;18:261–6.
[86] Graves M, Benton M, Lea R, Boyle M, Tajouri L, Macartney-Coxson D, et al. Methylation differences at the HLA-DRB1 locus in CD4+ T-Cells are associated with multiple sclerosis. Mult Scler 2013;20:1033–41.
[87] Takvorian SU, Merola JF, Costenbader KH. Cigarette smoking, alcohol consumption and risk of systemic lupus erythematosus. Lupus 2014;23:537–44.
[88] Böckle BC, Sepp NT. Smoking is highly associated with discoid lupus erythematosus and lupus erythematosus tumidus: analysis of 405 patients. Lupus 2015;24(7):669–74.
[89] Pasoto SG, Ribeiro AC, Bonfa E. Update on infections and vaccinations in systemic lupus erythematosus and Sjögren's syndrome. Curr Opin Rheumatol 2014;26:528–37.
[90] Mak A, Tay SH. Environmental factors, toxicants and systemic lupus erythematosus. Int J Mol Sci 2014;15:16043–56.
[91] Vieira SM, Pagovich OE, Kriegel MA. Diet, microbiota and autoimmune diseases. Lupus 2014;23:518–26.
[92] Hevia A, Milani C, López P, Cuervo A, Arboleya S, Duranti S, et al. Intestinal dysbiosis associated with systemic lupus erythematosus. MBio 2014;5:e01548–14.
[93] Zhang H, Liao X, Sparks JB, Luo XM. Dynamics of gut microbiota in autoimmune lupus. Appl Environ Microbiol 2014;80(24):7551–60.
[94] Schwartzman-Morris J, Putterman C. Gender differences in the pathogenesis and outcome of lupus and of lupus nephritis. Clin Dev Immunol 2012;2012:604892.
[95] Clark J, Bourne T, Salaman MR, Seifert MH, Isenberg DA. B lymphocyte hyperactivity in families of patients with systemic lupus erythematosus. J Autoimmun 1996;9:59–65.
[96] Rojas-Villarraga A, Torres-Gonzalez JV, Ruiz-Sternberg ÁM. Safety of hormonal replacement therapy and oral contraceptives in systemic lupus erythematosus: a systematic review and meta-analysis. PLoS One 2014;9:e104303.
[97] Gilbert EL, Ryan MJ. Estrogen in cardiovascular disease during systemic lupus erythematosus. Clin Ther 2014;36:1901–12.
[98] Cunningham M, Gilkeson G. Estrogen receptors in immunity and autoimmunity. Clin Rev Allergy Immunol 2011;40:66–73.
[99] Karpuzoglu E, Zouali M. The multi-faceted influences of estrogen on lymphocytes: toward novel immunointerventions strategies for autoimmunity management. Clin Rev Allergy Immunol 2011;40:16–26.
[100] Kanda N, Tsuchida T, Tamaki K. Estrogen enhancement of anti-double-stranded DNA antibody and immunoglobulin G production in peripheral blood mononuclear cells from patients with systemic lupus erythematosus. Arthritis Rheum 1999;42:328–37.
[101] Jeganathan V, Peeva E, Diamond B. Hormonal milieu at time of B cell activation controls duration of autoantibody response. J Autoimmun 2014;53:46–54.
[102] Saha S, Tieng A, Pepeljugoski KP, Zandamn-Goddard G, Peeva E. Prolactin, systemic lupus erythematosus, and autoreactive B cells: lessons learnt from murine models. Clin Rev Allergy Immunol 2011;40:8–15.
[103] Orbach H, Shoenfeld Y. Hyperprolactinemia and autoimmune diseases. Autoimmun Rev 2007;6:537–42.
[104] Gonzalez J, Saha S, Peeva E. Prolactin rescues and primes autoreactive B cells directly and indirectly through dendritic cells in B6.Sle3 mice. Clin Exp Immunol 2013;172:311–20.
[105] Nelson P, Rylance P, Roden D, Trela M, Tugnet N. Viruses as potential pathogenic agents in systemic lupus erythematosus. Lupus 2014;23:596–605.
[106] James JA, Robertson JM. Lupus and Epstein-Barr. Curr Opin Rheumatol 2012;24:383–8.
[107] Draborg AH, Duus K, Houen G. Epstein-Barr virus and systemic lupus erythematosus. Clin Dev Immunol 2012;2012:370516.
[108] Kristiansen SV, Pascual V, Lipsky PE. Staphylococcal protein A induces biased production of Ig by VH3-expressing B lymphocytes. J Immunol 1994;153:2974–82.
[109] Liang H, Nishioka Y, Reich CF, Pisetsky DS, Lipsky PE. Activation of human B cells by phosphorothioate oligodeoxynucleotides. J Clin Invest 1996;98:1119–29.

[110] Domiati-Saad R, Lipsky PE. Staphylococcal enterotoxin A induces survival of VH3-expressing human B cells by binding to the VH region with low affinity. J Immunol 1998;161:1257–66.

[111] Jabara HH, Geha RS. The superantigen toxic shock syndrome toxin-1 induces CD40 ligand expression and modulates IgE isotype switching. Int Immunol 1996;8:1503–10.

[112] Crampton SP, Morawski PA, Bolland S. Linking susceptibility genes and pathogenesis mechanisms using mouse models of systemic lupus erythematosus. Dis Model Mech 2014;7:1033–46.

[113] Marion TN, Postlethwaite AE. Chance, genetics, and the heterogeneity of disease and pathogenesis in systemic lupus erythematosus. Semin Immunopathol 2014;36:495–517.

[114] Nestor CE, Barrenäs F, Wang H, Lentini A, Zhang H, Bruhn S, et al. DNA methylation changes separate allergic patients from healthy controls and may reflect altered CD4+ T-cell population structure. PLoS Genet 2014;10:e1004059.

[115] Quddus J, Johnson KJ, Gavalchin J, Amento EP, Chrisp CE, Yung RL, et al. Treating activated CD4+ T cells with either of two distinct DNA methyltransferase inhibitors, 5-azacytidine or procainamide, is sufficient to cause a lupus-like disease in syngeneic mice. J Clin Invest 1993;92:38–53.

[116] Chang C, Gershwin ME. Drugs and autoimmunity–a contemporary review and mechanistic approach. J Autoimmun 2010;34:J266–75.

[117] Pacaud R, Sery Q, Oliver L, Vallette FM, Tost J, Cartron PF. DNMT3L interacts with transcription factors to target DNMT3L/DNMT3B to specific DNA sequences: role of the DNMT3L/DNMT3B/p65-NFκB complex in the (de-)methylation of TRAF1. Biochimie 2014;104:36–49.

[118] Wang Y, Zhang P, Liu Y, Cheng G. TRAF-mediated regulation of immune and inflammatory responses. Sci China Life Sci 2010;53:159–68.

[119] Markle JG, Frank DN, Adeli K, von Bergen M, Danska JS. Microbiome manipulation modifies sex-specific risk for autoimmunity. Gut Microbes 2014;5(4):485–93.

[120] de Goffau MC, Fuentes S, van den Bogert B, Honkanen H, de Vos WM, Welling GW, et al. Aberrant gut microbiota composition at the onset of type 1 diabetes in young children. Diabetologia 2014;57:1569–77.

[121] Jayaraman S, Patel A, Jayaraman A, Patel V, Holterman M, Prabhakar B. Transcriptome analysis of epigenetically modulated genome indicates signature genes in manifestation of type 1 diabetes and its prevention in NOD mice. PLoS One 2013;8:e55074.

[122] Dang MN, Buzzetti R, Pozzilli P. Epigenetics in autoimmune diseases with focus on type 1 diabetes. Diabetes Metab Res Rev 2013;29:8–18.

[123] Jayaraman S. Novel methods of type 1 diabetes treatment. Discov Med 2014;17:347–55.

[124] James S, Gallagher R, Dunbabin J, Perry L. Prevalence of vascular complications and factors predictive of their development in young adults with type 1 diabetes: systematic literature review. BMC Res Notes 2014;7:593.

[125] Lim J, Iyer A, Liu L, Suen JY, Lohman RJ, Seow V, et al. Diet-induced obesity, adipose inflammation, and metabolic dysfunction correlating with PAR2 expression are attenuated by PAR2 antagonism. FASEB J 2013;27:4757–67.

[126] McNelis JC, Olefsky JM. Macrophages, immunity, and metabolic disease. Immunity 2014;41:36–48.

[127] Wick G, Jakic B, Buszko M, Wick MC, Grundtman C. The role of heat shock proteins in atherosclerosis. Nat Rev Cardiol 2014;11:516–29.

[128] Okur I, Tumer L, Ezgu FS, Yesilkaya E, Aral A, Oktar SO, et al. Oxidized low-density lipoprotein levels and carotid intima-media thickness as markers of early atherosclerosis in prepubertal obese children. J Pediatr Endocrinol Metab 2013;26:657–62.

[129] Hoeksema MA, Gijbels MJ, Van den Bossche J, van der Velden S, Sijm A, Neele AE, et al. Targeting macrophage Histone deacetylase 3 stabilizes atherosclerotic lesions. EMBO Mol Med 2014;6:1124–32.

[130] Berk M, Williams LJ, Jacka FN, O'Neil A, Pasco JA, Moylan S, et al. So depression is an inflammatory disease, but where does the inflammation come from? BMC Med 2013;11:200.

[131] Bay-Richter C, Linderholm KR, Lim CK, Samuelsson M, Träskman-Bendz L, Guillemin GJ, et al. A role for inflammatory metabolites as modulators of the glutamate N-methyl-d-aspartate receptor in depression and suicidality. Brain Behav Immun 2015;43:110–7.

[132] Dannehl K, Rief W, Schwarz MJ, Hennings A, Riemer S, Selberdinger V, et al. The predictive value of somatic and cognitive depressive symptoms for cytokine changes in patients with major depression. Neuropsychiatr Dis Treat 2014;10:1191–7.

[133] Nieratschker V, Massart R, Gilles M, Luoni A, Suderman MJ, Krumm B, et al. MORC1 exhibits cross-species differential methylation in association with early life stress as well as genome-wide association with MDD. Transl Psychiatry 2014;4:e429.

[134] Ozden MG, Tekin NS, Gürer MA, Akdemir D, Doğramacı C, Utaş S, et al. Environmental risk factors in pediatric psoriasis: a multicenter case-control study. Pediatr Dermatol 2011;28:306–12.

[135] Quan C, Zhu KJ, Zhang C, Liu Z, Liu H, Zhu CY, et al. Combined effects of the BDNF rs6265 (Val66Met) polymorphism and environment risk factors on psoriasis vulgaris. Mol Biol Rep 2014;41:7015–22.

[136] Trowbridge RM, Pittelkow MR. Epigenetics in the pathogenesis and pathophysiology of psoriasis vulgaris. J Drugs Dermatol 2014;13:111–8.

[137] Eder L, Law T, Chandran V, Shanmugarajah S, Shen H, Rosen CF, et al. Association between environmental factors and onset of psoriatic arthritis in patients with psoriasis. Arthritis Care Res Hob 2011;63:1091–7.

[138] Kim J, Bhattacharjee R, Khalyfa A, Kheirandish-Gozal L, Capdevila OS, Wang Y, et al. DNA methylation in inflammatory genes among children with obstructive sleep apnea. Am J Respir Crit Care Med 2012;185:330–8.

[139] Marin JM, Artal J, Martin T, Carrizo SJ, Andres M, Martin-Burriel I, et al. Epigenetics modifications and Subclinical atherosclerosis in obstructive sleep apnea: the EPIOSA study. BMC Pulm Med 2014;14:114.

[140] De la Herrán-Arita AK, García-García F. Narcolepsy as an immune-mediated disease. Sleep Disord 2014;2014:792687. http://dx.doi.org/10.1155/2014/792687.

[141] Israel LP, Benharoch D, Gopas J, Goldbart AD. A pro-inflammatory role for nuclear factor kappa B in childhood obstructive sleep apnea syndrome. Sleep 2013;36:1947–55.

[142] Dietert RR. Distinguishing environmental causes of immune dysfunction from pediatric triggers of disease. Open Pediatr Med J 2009;3:38–44.

[143] Juhn YJ, Frey D, Li X, Jacobson R. Streptococcus pyogenes upper respiratory infection and atopic conditions other than asthma: a retrospective cohort study. Prim Care Respir J 2012;21:153–8.

[144] Furuya Y, Roberts S, Hurteau GJ, Sanfilippo AM, Racine R, Metzger DW. Asthma increases susceptibility to heterologous but not homologous secondary influenza. J Virol 2014;88:9166–81.

[145] Bang DW, Yang HJ, Ryoo E, Al-Hasan MN, Lahr B, Baddour LM, et al. Asthma and risk of non-respiratory tract infection: a population-based case-control study. BMJ Open 2013;3:e003857.

[146] MacIntyre EA, Heinrich J. Otitis media in infancy and the development of asthma and atopic disease. Curr Allergy Asthma Rep 2012;12:547–50.

[147] Patria F, Longhi B, Tagliabue C, Tenconi R, Ballista P, Ricciardi G, et al. Clinical profile of recurrent community-acquired pneumonia in children. BMC Pulm Med 2013;13:60.

[148] Ong PY. Recurrent MRSA skin infections in atopic dermatitis. J Allergy Clin Immunol Pract 2014;2:396–9.

[149] Dietert RR. Maternal and childhood asthma: risk factors, interactions, and ramifications. Reprod Toxicol 2011;32:198–204.

CHAPTER

13

Neurobehavioral Disorders and Developmental Origins of Health and Disease

Curt A. Sandman[1], Quetzal A. Class[2], Laura M. Glynn[1,3], Elysia Poggi Davis[1,4]

[1]Department of Psychiatry and Human Behavior, University of California, Irvine, CA, USA; [2]Department of Psychological and Brain Sciences, Indiana University, Bloomington, IN, USA; [3]Department of Psychology, Chapman University, Orange, CA, USA; [4]Department of Psychology, University of Denver, Denver, CO, USA

OUTLINE

The Epigenome and Developmental Origins of Health and Disease
http://dx.doi.org/10.1016/B978-0-12-801383-0.00013-X

Copyright © 2016 Elsevier Inc. All rights reserved.

INTRODUCTION

The National Comorbidity Survey Replication study estimated that exposure to early-life adversity could account for 32.4% of psychiatric disorders [1]. Early-life stress and exposure to trauma have been linked to depression [2–4], posttraumatic stress disorder (PTSD), panic disorder, schizophrenia, substance abuse, attention-deficit disorder [3], eating disorders [5], an abnormal stress response [4,6], mental retardation [7], and other serious neurological disorders [7]. Evidence suggests that exposure to early-life adversity, including poor birth outcomes, may be *causally* related to neurological changes underlying the risk for psychopathology [8]. In recognition of the importance of early-life events for mental health, the National Institute of Mental Health (NIMH) advocated identifying and understanding periods of development that exhibit the most dramatic transitions in both humans and model organisms. It was argued that particular attention should be paid to altered trajectories during periods of rapid developmental transition for the identification of individuals at risk for mental illness. Because mental illnesses were defined by the NIMH as disorders of life trajectories that begin before birth, it is critical to examine the complex processes that shape early-life neurodevelopmental trajectories and determine how these processes lead to vulnerabilities for cognitive and emotional disorders [9]. The study of early-life stress includes the fetal period in the life cycle, which is unmatched in growth and development and is the period in the human life span that is most vulnerable to both organizing and disorganizing influences. Despite these facts, the consequences for mental health of fetal exposures to adversity is understudied and poorly understood.

There are two fundamental approaches to assess fetal programming effects on neurobehavioral outcomes. One approach is to evaluate the association between birth phenotype and measures of behavioral and neurological function. The important and fertile initial studies introducing the programming hypothesis [10] relied on birth phenotype, either birth weight or gestational length, as indirect markers of intrauterine exposures. In this way birth phenotype was viewed as a reflection of fetal adaptive processes to intrauterine life that shaped the structure and function of physiological systems that underlie health and disease risk in later life [11,12]. It is unlikely that birth phenotype alone, except in extreme cases, plays a causal role in the development of disease risk. Moreover, birth phenotype is associated with numerous environmental risks, such as poverty, that are themselves predictive of subsequent difficulties [13]. The second approach is to directly assess features (metabolic, inflammatory, epigenomic, and stress) of the intrauterine environment that may be associated with concurrent (fetal) or later health outcomes. One challenge inherent with this approach is that intrauterine exposures often are related and even may be causative factors for birth phenotype, however, there now exists a growing body of studies showing that fetal exposures influence health risk independently of birth phenotype. Our discussion on the neurobehavioral consequences of early-life exposures will consider these two pathways with a focus on early-life (fetal) exposure to psychological and biological stress signals.

There is a loose linkage between the choice of approach and the analytical and study design strategies to assess neurobehavioral consequences of early-life exposure. The vanguard studies of the influential Developmental Origins of Health and Disease (DOHaD) or Programming model [10] were primarily *retrospective* and relied on birth phenotype as predictors of health outcome. The approach of these seminal observational studies was to select individuals who suffered from various diseases and to ascertain their birth histories from past medical records. Although retrospective studies are efficient, heuristic, and relatively easy to conduct, there are several limitations with this approach. First, the archival medical/birth records may not be accurate or complete, and the longer the span between present and past, the more likely there will be data loss. Second, possible confounding variables may not be available for collection. Third, there is no ability to explore factors that are related to predictor variables. For instance, in a retrospective study the question cannot be answered of whether the association between birth weight and disease is a function of another factor such as prenatal exposures. Some of these concerns can be alleviated by the use of quasi-experimental methods including family-based cousin, sibling, and twin comparisons as well as random exposure to "natural experiments." These approaches make it possible to rule out plausible confounding by genetic and environmental selection factors and offer powerful options for examining causal inference [14–22]. In full recognition of the limitations associated with retrospective studies, the advantage of knowing an outcome that may not be expressed for decades compels us to include in this chapter neurobehavioral programming findings discovered with this approach.

Contemporary studies have included prospective evaluation of intrauterine exposures with strong confirmation of the DOHaD model. In contrast to retrospective studies, prospective studies are conceived and designed, subjects are recruited, and baseline data and predictor variables are collected before subjects have developed outcomes of interest. The studies are necessarily longitudinal because subjects are followed over time. The limitations of this approach are that they typically are long and expensive studies to conduct, and often direct outcomes of interest cannot be determined. For instance, the association between prenatal exposures and health outcomes in adults requires many years of follow-up. Often, intermediary or prodromal conditions are examined as surrogate measures of the outcomes of interest. Moreover, during extended study periods there is inevitable loss of subjects, creating potential cohort biases. The strength of the prospective strategy is that it is required to make precise estimates of either the incidence or the relative risk of an outcome based on exposure. Prospective designs uniquely present the ability to examine relevant and co-existing moderating and mediating variables. We will discuss prospective studies of intrauterine stress exposures and neurobehavioral outcomes in children and in adolescent and adult populations and include the recent reports relating intrauterine experience to brain structure and function.

THE RISK ASSOCIATED WITH EARLY-LIFE INSULTS

The global burden of neurodevelopmental consequences of intrauterine and neonatal insults was evaluated in a comprehensive review of the world literature [23]. Title search identified over 28,000 publications from which 1330 were reviewed. From these, only 153 articles met the reasonably liberal inclusion criteria. The primary reasons for exclusion included: "numbers of survivors with sequelae could not be extracted; no clear description or diagnosis of the neonatal insult was provided; less than 80% of survivors were followed up after definite exposure in the first 6 months; and no appropriate test to establish the nature and extent of the sequelae or impairment was done. Additional reports were excluded because they were reviews, case series, or commentaries." The median risk of at least one major consequence of early-life stress was 39.4%. Among a wide range of outcomes, the most common impairments were learning difficulties, cognition or developmental delay, and sensory deficiencies. This important article provides evidence that intrauterine and neonatal insults result in significant long-term neurological morbidity and have a high risk of affecting more than one domain. The authors caution that their assessment of neurobehavioral problems among survivors "is likely to be an underestimate." Not only did this analysis highlight the global significance of the consequences of early-life insults but it also revealed the relative infancy of quality studies in this area of research. This chapter will focus on the growing number of studies on the neurobehavioral consequences of early-life exposures to stress that meet rigorous peer-reviewed criteria.

CHANGES IN THE STRESS SYSTEM DURING PREGNANCY

The maternal endocrine stress system is profoundly altered during the course of human pregnancy [24–29]. As gestation advances, the pituitary gland doubles in size increasing the synthesis and release of stress peptides such as ACTH into the maternal/fetal circulation and a two- to four-fold increase in the production of cortisol from the adrenal gland. But it is the growth and development of a new fetal organ, the placenta, that is primarily responsible for the profound changes in the maternal/fetal stress systems.

The human placenta synthesizes and releases the body's main stress hormone, corticotropin-releasing hormone (CRH), into the maternal and fetal circulation. Placental CRH (pCRH) is identical to hypothalamic CRH in structure and activity [30–32]; however, in contrast to the inhibitory influence on the CRH gene in the hypothalamus, maternal stress signals (cortisol) from the adrenal glands stimulate the expression of hCRHmRNA and establishes a positive feedback loop that allows for the simultaneous increase of placental CRH (pCRH), ACTH, and cortisol over the course of gestation. All of these stress-related peptide/hormone levels rise as pregnancy advances, peak during labor, and fall to basal levels within 24 h after delivery [31,33–36]. The exponential increase of these stress signals (especially pCRH) over the latter part of human gestation plays a fundamental role in the organization of the fetal nervous system [37], influences the maturation of the fetal Hypothalamic-Pituitary-Adrenal (HPA) axis and other systems, and modifies birth phenotype (the timing of the onset of spontaneous labor and delivery) [38–42].

These changes in physiological responding are mirrored by changes in psychological responding. As gestation advances, women become less psychologically responsive to stressors in the environment and exhibit alterations in the appraisal of stress [19,29,43–45]. For instance, women rated their response to a major earthquake as more stressful when it occurred early in pregnancy compared with women who were exposed to the same event later in pregnancy [19]. Moreover, as pregnancy advances, there are predictable changes in generalized anxiety symptoms [29], in depressive symptoms, and in fears and anxieties specific to pregnancy [46]. For these reasons we have included in this chapter the neurobehavioral consequences of early-life exposure to *variations* in biological and psychological programming signals of maternal stress.

ORGANOGENESIS

A central tenet of the Programming model is that the fetus is an active participant in its own development [45,47,48]. Through intimate communications with the mother, the fetus receives information that is integrated into its neurodevelopmental program to prepare for life after birth. Because the fetal period in the life cycle is unmatched by any other in growth and development, it is the stage in the human life span that is most vulnerable to both organizing and disorganizing (programming) influences. Fetal organs develop from progenitor stem cells at precise times and in a specific sequence from conception to maturity, so the timing of information is a critical factor in determining the structure of the neurodevelopmental program. Disruption in the timing or sequence of organ development can result in tissue remodeling producing smaller organs or altered organ morphology [49]. Remodeled tissue modifies the function and physiological capacity of the organ throughout the life span and is a fundamental assumption of how fetal exposures influence health and disease.

TIMING OF INTRAUTERINE EXPOSURES

Signals from the maternal stress systems are continuous sources of information to the fetus over the course of gestation. These systems are undergoing massive but orderly changes as pregnancy advances. Deviations in these systems from normative patterns can be signals to the fetus that there is a threat to the host (mother). Depending on the severity and the timing of deviant (stressful) maternal signals, the sequence of neural development can be disrupted, resulting in programmed consequences for brain structure and behavior. Fetal exposure to psychobiological stress has been significantly associated with alterations in neurological structure. Moreover, the impact of the signals can change as gestation progresses. Women become progressively less sensitive both psychologically and physiologically to stress exposures as pregnancy advances, so exposure to stress may be less likely to shape neurobehavioral development late in pregnancy compared with earlier exposures. Exposures to maternal distress early in pregnancy have been linked to adverse birth outcome, less optimal language development, impaired learning and memory performance, greater risk for anxiety disorders, and gray matter reductions in the frontal cortex [50–52]. Similarly, elevated stress and placental hormones early in pregnancy have been associated with less optimal neurodevelopment, more irritable and fearful temperament, larger volumes in limbic areas of the brain, and slower behavioral recovery from pain [53–55]. Increases in these hormones occurring very late in gestation have been associated in some instances with enhanced cognitive development [56]. These influences occur against the background of the timing and sequence of fetal organ development, which is another significant factor in determining the consequences of stress exposures for brain development. Studies that have included serial or longitudinal measures are necessary to assess this critical variable.

EMOTIONAL REGULATION

Birth Phenotype

A substantial literature documents an increased risk for psychopathology among individuals who are born preterm. Increasing evidence from both retrospective studies using birth phenotype and from prospective studies of psychobiological stress signals suggests that anxiety and mood disorders have part of their origin in prenatal and early postnatal experiences [9,24].

Length of Gestation

Evaluation of existing literature indicates that one phenotypic characteristic of preterm/low birth weight individuals is increased risk for emotional problems [57]. A recent meta-analysis of studies using parent or teacher reports revealed that anxiety disorders were nearly twice as high among very preterm and very low birth weight children as compared to term/normal birth weight children [58].

In a prospective, longitudinal evaluation, very preterm children were three times more likely to meet criteria for any psychiatric disorder compared to their term counterparts at seven years of age. Almost 25% of the children in the very preterm group had a psychiatric disorder. The largest group differences were seen for anxiety disorders as well as autism spectrum disorder (ASD), and attention deficit hyperactivity disorder (ADHD) [59]. Those children who were very preterm, who had high social risk, brain abnormalities at term, and emotional problems at age five had the greatest risk of a psychiatric disorder at seven years. The consequences of preterm birth are apparent across the continuum and not only for very preterm children. Compared with school-aged children born full term, children born in the late preterm period are at an increased risk for anxiety disorders and have altered gray matter volume [60].

Birth Weight

One limitation of much of the existing literature is that birth weight and gestational length are confounded. More direct evidence for a role for fetal programming of anxiety disorders comes from a study that evaluated fetal growth. In a sample of 682 individuals, length and weight were assessed at birth and used to calculate a ponderal index standardized for gestational age at birth. Data indicated that individuals with a high ponderal index at birth were at a significantly decreased risk for a diagnosis of generalized anxiety disorder based on DSM criteria as compared with individuals with a lower ponderal index [61]. Consistent with these findings, adults who were small for gestational age (SGA) and born full term showed an increased risk for any psychiatric disorder, based on DSM criteria, as compared to individuals who were born at term and appropriate for gestational age (AGA) [62]. Anxiety disorders were the most frequently diagnosed in the SGA group followed by ADHD. These data provide evidence that fetal growth, independently from gestational length, is associated with risk for anxiety disorders.

Intrauterine Exposures

Animal models provide compelling experimental evidence that prenatal stress is associated with consequences for brain and behavior in the offspring. Although impairments are present across a variety of domains, one of the consistent effects of gestational exposure to stress and stress hormones in rodent models is an increase in anxiety-like behavior [63–65]. Similarly, in primate studies prenatal stress is associated with behavioral disturbances and decreased exploratory behavior [66] as well as smaller hippocampal volume and an inhibition of neurogenesis in the dentate gyrus [67,68].

Fetal Exposure to Maternal Psychological Distress

Perhaps one of the most consistent neurobehavioral findings is that fetal exposure to adversity is associated with an increased subsequent risk for stress and emotional problems. Studies evaluating outcomes during infancy and early childhood are consistent with this hypothesis. Fetal exposure to maternal psychological distress including anxiety, stress, and depression is associated with altered stress responses [69–72] as well as more fearful or reactive temperament, greater negative emotionality, and more internalizing problems during infancy and early childhood [73–79]. Infants who are easily aroused by varied stimulation are more likely to become behaviorally inhibited as young children [80], exhibit social anxiety during adolescence [81], and show greater amygdalar activation to novelty as adults [82]. Thus, this set of findings suggests that fetal exposure to maternal stress signals may program the development of vulnerability to later mood and anxiety problems that may emerge during adolescence.

Emerging evidence has shown that fetal exposure to maternal psychological stress signals is associated with higher levels of emotional problems during childhood and adolescence [51,83–85]. For example, in a prospectively evaluated cohort of 178 maternal–child pairs, fetal exposure to maternal psychological stress was associated with child affective problems at six to nine years of age [51]. Prenatal exposure to elevated maternal reports of depression, perceived stress, and pregnancy-specific anxiety (but not general anxiety) was associated with increased anxiety in children. These associations were not accounted for by potential confounding factors including sociodemographic risk, postnatal maternal anxiety, or depression or obstetric risk.

Consistent with these findings, in a large prospective sample of 7944 maternal–child pairs, O'Donnell and colleagues [86] found that elevated maternal anxiety and depression during the prenatal period was associated with child mental health problems. Specifically, children who were exposed to high maternal anxiety during gestation were at a significantly elevated risk of mental health problems during childhood and early adolescence. The children of anxious mothers had almost a two-fold increase in the risk of a mental health problem by the age of 13. These associations remained after considering a range of potential confounds including postnatal maternal anxiety, paternal anxiety, sociodemographic, and maternal health factors.

Neurobehavioral consequences of maternal distress are observed as early as the fetal period. Fetal exposure to maternal psychological distress are associated with greater fetal reactivity to a challenge [87,88]. DiPietro and colleagues have demonstrated that fetal responses are predictive of infant behavior. For example, fetuses who reacted more intensely to a manipulation of maternal stress were more irritable as infants [89]. These findings illustrate that maternal psychological stress signals program fetal behavior, before there are influences related to parenting or socialization, and that these early adaptations to stress may have consequences that persist into the postnatal period.

Fetal Exposure to Maternal HPA Axis and Placental Hormones

One of the primary pathways by which maternal psychological distress is thought to affect the fetus is through alteration of the functioning of the *maternal HPA and placental axis*. As early as the fetal period, exposure to elevated levels of stress responsive hormones (maternal cortisol and placental CRH) is associated with greater fetal reactivity to a challenge [90,91]. Several studies have demonstrated that elevated levels of both maternal cortisol and placental CRH during gestation are associated with more fearful or negative temperament during infancy [54,75,78,92]. This phenotype during infancy may be predictive of childhood risk for anxiety problems. Children with anxiety ratings within the borderline/clinically significant range were twice as likely to have been exposed to higher maternal cortisol during gestation, even after considering relevant confounding factors, compared to children with ratings in the normal range (odds ratio = 2.1, 95% confidence interval = 1.1–3.9, $p < 0.05$) [51]. The association between prenatal cortisol and child anxiety was observed primarily among females [93].

Prenatal stress exposures may contribute to child risk for emotional problems through modification or programming of the developing *child's HPA axis*. Exposure to elevated levels of endogenous maternal cortisol and the administration of synthetic glucocorticoids during pregnancy are associated with greater neonatal cortisol reactivity to a painful stressor [69,94] as well as altered infant responses to a laboratory challenge [71]. Alterations to child HPA axis functioning persist into childhood and adolescence. Evaluation of prenatal influences on the HPA axis regulation later in childhood primarily has focused on circadian regulation of cortisol production. In a healthy system, cortisol levels increase in response to morning awakening and then decline across the day. Fetal exposure to maternal anxiety is associated with a flattening of the cortisol awakening response at 10 years of age [95] and with a flattened diurnal profile in the offspring during adolescence [85,96]. This flattened diurnal profile is similar to the profile in children exposed to other types of early adversity [97]. In support of the possibility that HPA axis dysregulation may be a mechanism underlying the link between prenatal stress and child mental health, Van den Bergh and colleagues [85] showed that the diurnal cortisol profile mediated the association between prenatal maternal anxiety and depressive symptoms among adolescent girls.

COGNITIVE REGULATION

Birth Phenotype

There is unequivocal evidence that those born early or small are at risk for cognitive impairment ranging from quite mild to severe [98–101]. A quantitative review of studies published from 1970 to 1997 documented that among survivors of very low birth weight (<800 g) or very preterm delivery (<26 weeks) the average rates of disability for the most prevalent causes were: mental retardation, 14.3%; cerebral palsy, 7.5%; blindness, 7.8%; and deafness, 2.9% [102]. Additionally, estimation of the probability that an individual exhibited at least one clinically significant disability was 24.2% and 22.1% among those born very low birth weight and very preterm, respectively. Similarly, from an examination of nearly half a million British births of less than 28 weeks gestation, it was estimated that "one in ten of all survivors has a disability so profound that he or she is never likely to become independently mobile or to communicate effectively with others" [103]. It is worth noting that although there are significant advancements in neonatal care, this has not resulted in noticeable decreases in rates of disability among those born early or small. For example, in a meta-analysis of all publications on birth cohorts with a median year no earlier than 2000, is was estimated that among those born at <28 weeks, the rate of moderate to severe neurological disability was 24.5% [98].

Among those born preterm or low birth weight that escape severe disability, rates of less severe developmental delays also are prevalent. In studies examining only those children who are not developmentally disabled and apparently neurologically healthy, those born with adverse birth phenotypes have been found repeatedly to exhibit lower IQ scores (both performance and verbal) [104–107]. Those who escape obvious intellectual disabilities also show specific deficits across a range of cognitive processes including visual processing and visual memory [106], episodic memory [108], and spatial working memory [109]. Further, these more subtle impairments appear to have implications for achievement outcomes such as arithmetic [106] and verbal fluency performance [110] as well as school performance [111,112].

Length of Gestation

Very recent empirical findings point to the fact that variation in length of gestation within the term range (≥37 weeks gestation), also is associated with increased vulnerability to cognitive impairments. Comparison of gestational age at birth among children born at term has shown that gestating longer is associated with higher IQ [113], enhanced spatial working memory [109], better executive function [114], and increased school performance [115]. In addition, two recent prospective studies provide further support for the effects of variation in this range of the distribution. Rose et al. [116] demonstrated in a cohort of over 1500 one-year-old infants that for each additional week of gestation from 37 to 41 weeks, mental development scores increased in a linear manner. Similarly, in a cohort of over 200 infants, using the "gold standard" for determining gestational age at birth, Espel et al. [117] showed that longer gestation in the term range was consistently associated with higher scores on assessments of mental development at 3, 6, and 12 months of age.

Birth Weight

Although the evidence is clear that those at the lower levels of viability are at risk for less optimal cognitive outcomes, a few studies report that the effects of being born early or small are not restricted to the extreme low end of the distributions. There have been at least three studies examining the link between birth weight and later cognitive performance that examined the influences within the normal range. Schenkin et al. [107] found a positive relation between birth weight in the normal range (>2500 g) in 445 Scottish children. Richards et al. [118] found that birth weight within the normal range predicted cognitive ability at the ages of 8, 11, 15, and 25 among British males and females. Similarly, cognitive function in 4300 Danish men (mean age of 18) also was positively correlated with birth weight in the normal range [119]. This study further revealed that very high birth weight (>4200 g) was associated with decreased cognitive abilities, a finding that is consistent with the broader literature examining the links between birth weight and health outcomes. These studies reveal that very high birth weight children, like the extremes at the other end, are at risk for diseases including schizophrenia, diabetes, and polycystic ovary syndrome [120–123].

Intrauterine Exposures

Rodent models provide compelling evidence that prenatal maternal stress exerts persisting and pervasive effects on the brain and behavior of offspring [124–126]. In addition, among nonhuman primates, exposures to stressors during the prenatal period are associated with compromised neuromotor development [127,128] and poorer attention regulation [129]. Evidence from animal models further suggests that both maternal stress and placental hormones represent a plausible and likely pathway through which maternal stress exposures program the developing fetus [130,131].

Fetal Exposure to Maternal Psychological Distress

Relatively few prospective studies have examined the effects of prenatal maternal psychological distress and stress exposures on cognitive development in humans. Although there is evidence that exposures to large-scale negative life events and maternal self-report of elevated stress, depression and anxiety during the prenatal period are associated with delayed infant cognitive and neuromotor development [73,132,133], and that these deficits may persist into adolescence, the findings across studies are not fully consistent [79,134]. As discussed previously, one reason for these apparent inconsistencies may relate to the timing of assessment or to the timing of stress exposures. A second potential explanation is that measures of distress that are directly relevant to pregnancy (e.g., "I am fearful regarding the health of my baby," "I am concerned or worried about losing my baby," "I am concerned or worried about developing medical problems during my pregnancy") are superior to more generalized measures of psychological distress in predicting neurodevelopmental outcomes. Two recent studies strongly support this view [56,135]. In both studies, the investigators directly compared assessments of pregnancy-specific anxiety to more generalized measures (e.g., perceived stress or state anxiety), and in both cases the pregnancy-specific measures were more potent predictors of outcome. Specifically, generalized measures of prenatal distress were less predictive of both cognitive development in 12-month-olds [56] and of executive function in six- to nine-year-old children [135].

Fetal Exposure to the Maternal HPA Axis

Even fewer studies have examined exposures to maternal HPA axis or placental hormones on human cognitive development. However, one study has shown that beginning as early as 24h after birth, the effects of exposures are detectable. Fetal exposure to increased levels of maternal cortisol at 15 and at 19 weeks gestation and increased levels of placental CRH at 31 weeks were associated with significant decreases in newborn physical and neuromuscular maturation [136]. In 2010, two complementary studies were published that examined the influence of exposures to prenatal maternal cortisol on infant and toddler development. In a cohort of 125 pregnant women, cortisol levels were determined five times during pregnancy, and cognitive development was assessed with the Bayley Scales of Infant Development (BSID) at 3, 6, and 12 months of age [56]. Results revealed that elevated maternal salivary cortisol early in pregnancy (15 gestational weeks) was associated with a slower cognitive development trajectory across the first year, which resulted in lower scores on measures of mental development at 12 months of age. Similarly, in another prospective prenatal cohort of 125 mother–infant pairs, Bergman et al. [137] showed that cortisol levels assessed from amniotic fluid at 17 weeks gestation was associated with poorer mental development scores on the BSID at 18 months of age. A link between prenatal cortisol and child outcomes also was demonstrated by LeWinn et al. [138] who reported that maternal free cortisol levels sampled between 31 and 36 weeks gestation were inversely associated with IQ performance at seven years of age. Interestingly, it was also shown that within sibling pairs, those that were exposed to higher levels of cortisol were at greater risk for impaired IQ performance, which allows additional confidence that the effects were due to the intrauterine exposures as opposed to genetics or to the early-life environment.

Strong supporting evidence for the programming influences described above in infants and children is derived from studies that have examined the influence of maternal and placental hormones on fetal development, before the extrauterine environment could have an opportunity to exert influences. Examination of the fetal heart rate and motor response to acoustic or vibroacoustic stimulation has been a useful tool for the assessment of fetal neurological integrity and development [139–141]. Documentation of the developmental trajectories in these responses suggests that at 25 weeks gestation, some fetuses are able to mount a response to vibroacoustic stimulation while others are not [52]. Using this paradigm as an index, it has been demonstrated that exposures to elevated placental CRH and to maternal cortisol at 15 weeks gestation are associated with a failure to exhibit a response [90,91]. Also it has been shown that elevated placental CRH during the third trimester is associated with an impaired ability to habituate to repeated presentations of the same stimulus and to the ability to identify a novel acoustic stimulus [37].

EPIDEMIOLOGICAL EVIDENCE FOR PROGRAMMING NEUROBEHAVIORAL OUTCOMES

In this section, retrospective studies of programming effects on neurobehavioral outcomes are considered. Retrospective studies ascertain early-life experiences and birth histories, either by self-report of earlier experiences or by reviewing medical records of individuals who exhibit various diseases. Retrospective studies complement prospective studies by providing the longitudinal resources to examine later-onset and rare disorders. Additionally, genetically informed retrospective designs enhance the ability to draw causal inferences from the associations identified.

Birth Phenotype: Neurodevelopmental and Clinical Outcomes

Population-based epidemiological studies suggest that indicators of poor fetal growth, such as low birth weight (e.g., ≤2500 g), are linked with increased offspring risk for neurodevelopmental disorders such as ADHD and ASD [142–145] as well as nonclinical levels of academic and social difficulties [146,147]. Similarly, preterm birth is associated with offspring morbidity across the life span [148], including psychiatric disorders [149] and nonclinical academic and social difficulties [150–152].

Several researchers have utilized family and twin-based quasi-experimental designs to address the possibility of unmeasured confounds. Twin studies comparing neurodevelopmental disorders across birth weight of discordant twins have suggested that impaired fetal growth may act as an independent, environmental risk factor for ASD and ADHD [153–155]. Further, a recent non-twin sibling-comparison study found that lower birth weight predicted increased risk for ASD, ADHD, and academic problems [16]. Figure 1 presents the association between birth weight adjusted for gestational age and ADHD in a Swedish population–based cohort using both traditional cross-cohort

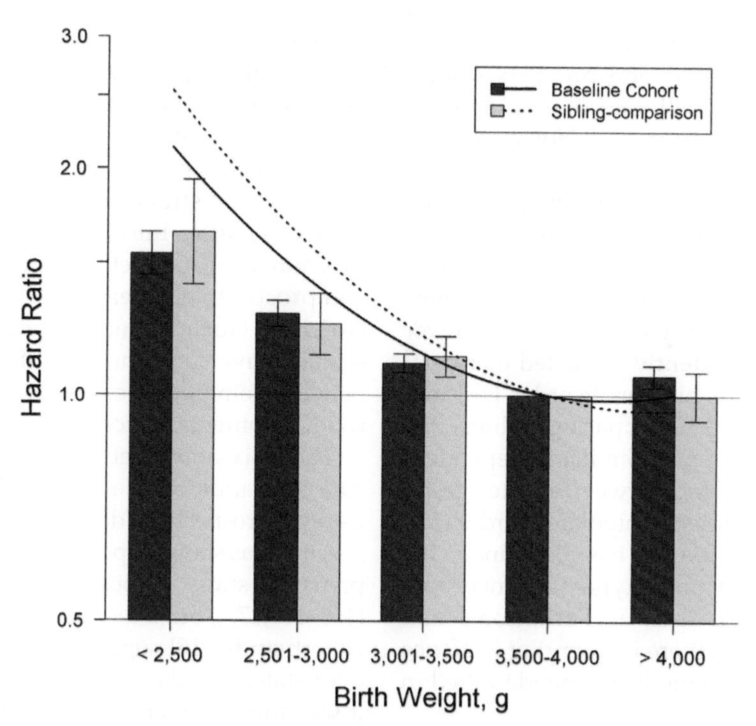

FIGURE 1 Swedish population-based cohort (N = 2,032,803) and sibling-comparison analysis illustrates the significant association between birth weight controlling for gestational age and ADHD. Associations are presented from continuous (line) and ordinal (bar with 95% CI) representation of birth weight controlling for gestational age when predicting ADHD. Solid line and dark bars baseline represent population-wide estimates. Sibling-comparison, fixed effects models are illustrated by the dotted lines and light bars. In the reference group are those born in the 3500–4000 g birth weight category. The maintenance of association magnitude across population and sibling-comparison models, consistent with a causal inference, can be noted when predicting ADHD. *Adapted from Class et al. [16].*

analyses and sibling-comparison analyses. The findings showed fetal growth at birth is consistently associated with ADHD independent of measured covariates and all genetic and environmental factors that make siblings similar. Another sibling comparison found commensurate results showing largely independent associations between shorter gestational age and ASD and ADHD [17]. Thus, findings across study designs support strong associations between birth phenotype and childhood neurodevelopmental disorders.

Weaker and more inconsistent associations have been found when examining adult-onset psychiatric outcomes including depression, bipolar disorder, and schizophrenia. A recent meta-analysis reported that associations between low birth weight and depression may be due to publication bias [156], and previous studies have been limited by self- and parent report of both risk and outcome [144,157]. However, large epidemiological studies have suggested associations do exist between adverse birth phenotype and severe mental illness including schizophrenia, substance use problems, and affective disorder [143]. A recent, large Swedish cohort sibling-comparison study also found that lower birth weight independently predicted increased risk for psychotic or bipolar disorder [16] that was present even when comparing siblings discordant for birth weight. Similar independent associations were found between shorter gestational age and psychotic or bipolar disorder [17].

Though little research has examined the links between birth phenotypes and neurodegenerative diseases, such as Parkinson's and Alzheimer's disease, there is emerging evidence that significant associations will be identified in future research [158].

Intrauterine Effects: Neurodevelopmental and Clinical Outcomes

Accumulating retrospective epidemiological evidence links prenatal maternal stress to increased risk of psychopathological morbidity in offspring [159–164]. Associations with adverse psychopathological outcomes have been reported following maternal exposure to physical stressors, such as famine [165,166], and psychological stressors, such as bereavement [160,161,167], trauma [168], war [164], and natural disaster [19,169,170].

Using self and other report of maternal experiences during pregnancy, many large epidemiological studies have identified associations with severe, impairing, and costly psychiatric disorders [160,162,171–174]. A large, population-based cohort study that utilized death of a family member as an indicator of maternal exposure to stress during pregnancy reported that death of a first-degree relative of the mother during the third trimester was associated with increased risk for offspring ASD and ADHD [15]. Similar positive associations with childhood neurodevelopmental outcomes have been reported in studies measuring potentially less severe but more chronic stress exposure from family discord [175], after hurricanes and tropical storm exposure [170], and retrospectively recalled [171] and prospectively measured stressful life events [162,163]. However, not all retrospective cohort studies have confirmed significant associations between retrospective reports of stress and neurodevelopmental outcomes [176].

The associations between prenatal stress exposure and adult-onset neuropsychiatric outcomes using retrospective designs are mixed. Some research has shown a positive association between prenatal stress exposure and schizophrenia [159–161,177], but a meta-analysis [178] and large cohort study [16] suggest that these effects were over stated. Studies also suggest that the strong association between prenatal stress and affective disorder observed in prospective studies is not apparent in retrospective studies of clinical cases [15,164]. Accurate ascertainment of prenatal stress and small base rates of both risk and outcome make these questions difficult to answer, even when large population-based data are available.

PROGRAMMING THE VULNERABLE HUMAN FETAL BRAIN

The human brain is the most complicated organ in living systems [179]. This elaborate structure originates from a simple neural tube, followed by a series of differentiation processes over a long period of time including the entire fetal period and into early adulthood. Radial glial cells differentiate from the neuroepithelial progenitor cells at the beginning of neurogenesis. The transition of neuroepithelial cells to radial glia is accompanied by a series of structural and functional changes, including the appearance of "glial" features, as well as the appearance of new signaling molecules and junctional proteins. Radial glia not only provide scaffolding for neuronal migration but also are the main source of neurons in several regions of the central nervous system including the cerebral cortex [180].

Radial neuronal cell migration begins in the human brain around 42 days gestational age [181,182]. There are critical periods of neural migration in early human development when the nervous system is especially vulnerable to disruptions. Between gestational age (GA) 8 and 16 weeks, migrating neurons form the subplate zone awaiting connections from afferent neurons originating in the thalamus, basal forebrain, and brain stem. Concurrently, cells accumulating in the outer cerebral wall form the cortical plate, which eventually will become the cerebral cortex. By gestational week 20, there is an exponential increase in cortical thickness (Figure 2) [179], axons form synapses with the cortical plate, and by gestational week 24, cortical circuits are organized [183,184]. By gestational week 28, the fetal brain is forming secondary and tertiary gyri, and exhibiting neuronal differentiation, dendritic arborization, axonal elongation, synapse formation and collateralization, and myelination

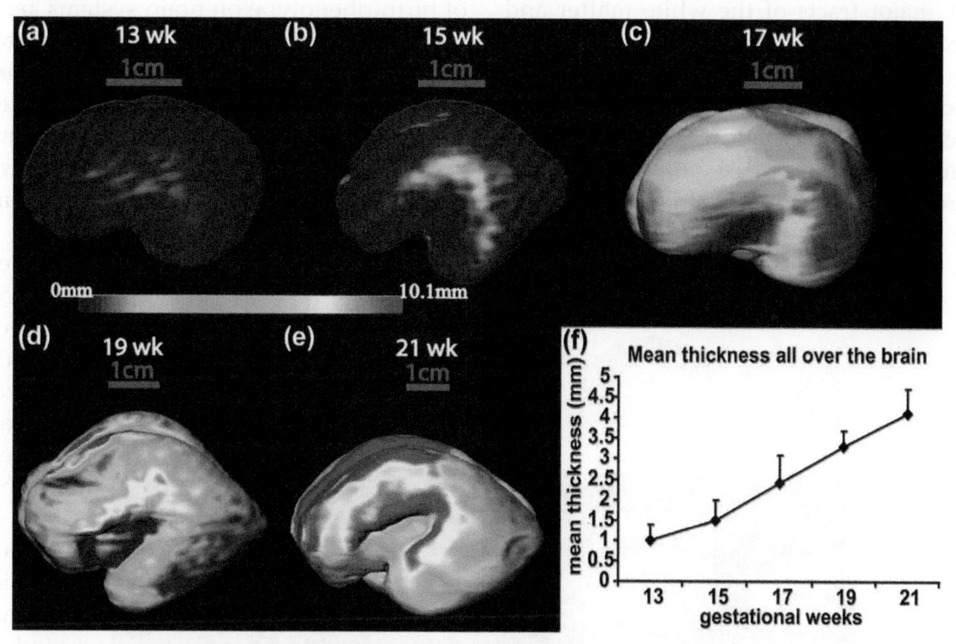

FIGURE 2 Averaged thickness profiles of the cortical plate plus the subplate across the cortical surface for 13–21-week gestation brains (a–e). Color bar indicates the thickness (hot colors represent greater thickness). A summary of the increase in cortical growth is shown in (f). *Reproduced with permission from [179].*

[185,186]. The fetal human brain contains billions of neurons and is 40% greater in number than in the adult [184,187–189]. The rate of synaptogensis reaches an astonishing peak so that at gestational week 34 through 24 months postpartum, there is an increase of 40,000 synapses per second [185,190]. The timing of disruptions in neurogenesis or in the sequence of neuronal events coupled with the timetable for brain development determines the nature of a programmed effect.

BIRTH PHENOTYPE AND NEURONAL CONSEQUENCES

There is a high rate of neurobehavioral problems in children and adults who survive extreme preterm birth and low birth weight. During the last third of human pregnancy, gray matter volume increases fourfold, the surface area of the cortex increases approximately eightfold [191,192], major tracts of the white matter and its connections increase in volume and complexity (Figure 3), and the majority of cortico–cortico connections are established [193]. In normal human development, this dramatic growth in brain connections takes place within the nurturing environment of the uterus. In many preterm and low birth weight infants, brain development during this sensitive period occurs within the hyperstimulating intensive care environment with unknown consequences.

Hundreds of excellent articles have examined and discovered that infants born at very early gestational ages or low birth weight experience pervasive insults to the developing brain [194–199]. Abbreviated gestation and low birth weight has been associated with thalamocortical abnormalities [200–202], compromised amygdala, basal ganglia and insula [203], cerebellar volume deficits [204], alteration of hemispheric connections [205], loss of gray/white matter [203,206–208], and abnormal cortical folding [209]. The causes for the majority of infants born early or small are unknown, but it is known that there are often medical complications (e.g., intraventricular hemorrhage) associated with poor birth outcomes that contribute to the neurological impairments that are observed. For this reason it is difficult to determine if the consequences of birth phenotype on brain systems are related specifically to birth outcomes or to factors that are causally related to fetal growth or gestational length at birth. Studies that have examined neurological integrity and function in low risk children who are born within the term (after 37 weeks) gestational age intervals or are born

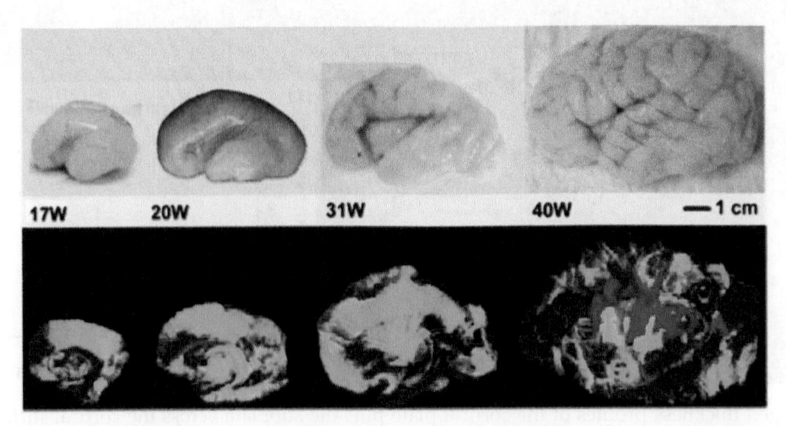

FIGURE 3 Human fetal brains (upper row) and examples of tractography pathways passing through a sagittal slice (lower row) from weeks 17 through week 40 of gestation. *Reproduced with permission from [181].*

small but at term offer an opportunity to explore birth phenotype as a programming influence on the nervous system.

Length of Gestation

In a study of 100 preadolescent healthy children, significant regionally specific associations were found between gestational length and brain morphology [69]. Children in this study had a stable neonatal course and did not experience common perinatal risk factors such as mechanical ventilation, intraventricular hemorrhage, or periventricular leukomalacia. Longer gestation, even within the normal GA interval (37–42 weeks) was associated with increased volume of the temporal cortex, one of the latest brain regions to mature [210]. In a related study, children born at term had larger volumes than children born late preterm (34–36 GA) in the parietal and temporal areas of the right hemisphere [60]. In a novel study of the association between neural network organization and gestational length, diffusion tensor imaging and fiber tractography were applied in a study of 147 preadolescent children [211]. Longer gestation was positively associated with higher network efficiency and increased local efficiency in the posteriormedial cortex, including the precuneus, cuneus, and superior parietal regions. Analysis of interconnected structural hubs (rich clubs) indicated that longer gestation conferred higher levels of structural connectivity and increased communication capacity of the brain network. These findings emphasize that there are highly significant neurological benefits of increased gestational length for the developing fetal brain.

Birth Weight/Size

The effects of growth, *independent from gestational age*, on brain morphometry and white matter microstructure were reported in a series of follow-up studies from teenage to 20-year-old adults [198,212,213]. Born small for gestation age (SGA) but at term was associated with smaller brains in 15-year-old subjects [212] and with deviations in white matter microstructure at 19 years (but not at 15 years) [213]. Among the 20-year-old cohort, total brain volume was reduced by almost 6% in the SGA group. The most widespread areas of reductions were observed in the frontal, temporal, and parietal lobes. The areas affected are similar to those observed for gestational influences and support the conclusion that birth phenotype even in individuals with minimal medical complications and within normal limits can exert persisting programming influences on the brain.

A recent retrospective study [214] reported that measures of birth size were significantly associated with brain size in a group of 1254 participants (mean age = 75 years). Lower birth weight, birth length, and ponderal index (at birth) were associated with smaller intracranial volumes (ICV), smaller absolute brain volumes, and larger absolute CSF volume determined by MRI late in life. After adjustments for ICV (head size), ponderal index (at birth) was the best predictor of total brain and white matter volume. The authors reported that although all measures of birth size were significantly associated with head size in the 75-year-old cohort, only ponderal index, "as an indicator of fetal wasting," was associated significantly with brain volumes. Lower ponderal index could be a selective reflection of growth inhibition in late pregnancy with consequences for third-trimester white and total brain volumes. Lower ponderal index at birth also was associated with poorer cognitive performance in these subjects, but it was unclear if brain volumes mediated this relationship.

NEURONAL CONSEQUENCES OF INTRAUTERINE EXPOSURES

As presented above, signals from the maternal stress systems are continuous sources of information to the fetus over the course of gestation.

Depending on the severity and the timing of deviant (stressful) maternal signals, the sequence of neural development can be disrupted resulting in programmed consequences for brain structure and behavior. Fetal exposure to psychobiological stress has been significantly associated with alterations in neurological structure.

There is a large animal literature that supports the conclusion that prenatal exposure to maternal stress at critical periods of development alters the programming of the fetal brain [130,215]. This literature has identified brain areas that are most susceptible to exposure to stress. Exposure to prenatal stress in animal models primarily affects development of the amygdala, hippocampus, and prefrontal cortex. Fetal exposure to the biological stress system alters the density of cortisol receptors [131], increases the production of CRH in the amygdala [65,216], and increases amygdala volume [217]. Perinatal stress, in rodents as well as nonhuman primates, produces behavioral abnormalities, such as an elevated and prolonged stress response, impaired learning and memory, deficits in attention, altered exploratory behavior, altered social and play behavior, and an increased preference for alcohol [130,218]. Reports in animal models have identified disrupted neuronal differentiation and axonal growth associated with activation of the maternal immune system [219]. Because there are significant differences in the brain and in the stress and reproductive systems, direct extrapolation across species is tenuous.

Most of the studies evaluating the long-term consequences of human fetal exposures have focused on environmental toxins and teratogenic agents [220]. Alterations in human brain development have been associated with fetal exposure to alcohol [221], drugs of abuse, smoking, and prescription medications [220]. There is also evidence in humans that prenatal exposures to maternal obesity [222] and immune activation or infection [223] have behavioral consequences for the fetus. Far fewer studies have evaluated more subtle and far more pervasive environmental signals including fetal exposure to maternal stress.

Exposure to Early-Life Stress

Four recent studies of brain structure or function in human subjects with different methods have each implicated the effects of early-life stress on prefrontal and limbic regions or connections. Results from a novel functional magnetic resonance imaging (fMRI) study of 16 children institutionalized from birth were compared with 10 control subjects [224]. Neuronal activity was examined in frontolimbic regions including the amygdala, ventromedial prefrontal cortex, and hippocampus during an aversive task requiring discriminating responses to pictures depicting human emotions. Only increased activity in the amygdala differentiated the early-life stress-exposed children from the comparison group. Similar findings were reported in a study of current resting state connectivity from MRI scans, psychiatric symptoms, and measures of early-life stress assessed in a group of 57 adolescents [225]. Between four and five years of age, basal levels of salivary cortisol had been collected and used in this analysis to assess an early biological marker of stress. The complex findings indicated that exposure to early-life stress during infancy was associated with higher levels of childhood cortisol *in females only*. Higher levels of childhood cortisol at this early age were associated with diminished connectivity between the ventromedial prefrontal cortex and amygdala. This inverse relation was associated with decreased current self-report of anxiety but increased report of depression even though there was a significant comorbidity between them.

Studies of fiber tracts and the default network also found consequences for the limbic and frontal systems associated with early-life stress. A retrospective study of *prenatal stress* and white matter microstructure was conducted in 22 healthy seven-year-old children [226]. Maternal reports of prenatal stressful life events were associated with increased fractional anisotropy in the right uncinate fasciculus, which connects the paralimbic regions, the amygdala, and hippocampus. In a small retrospective study (N = 19), normal healthy adults reporting early-life stress before the age of 18 years displayed significantly greater brain default network (DN) deactivation during a working memory task [227]. (The DN includes the posterior cingulate cortex, medial prefrontal cortex, middle frontal regions, and lateral parietal and medial temporal regions.) The association between early-life stress and deactivation of the DN, especially the prefrontal regions, may be evidence of a risk for developing psychiatric conditions later in life.

The findings for these studies support the conclusion that there are programming influences on brain structure, function, and networks related to a variety of exposures either experienced prenatally or during infancy. Despite the different assessments of early-life stress, and the varied methods and approaches for assessing the nervous system, each study found that the primary areas or circuits affected were associated with regulation of emotion. Formal mediational analysis confirmed this conclusion [225].

Fetal Exposure to Maternal Psychological Distress

As described earlier, maternal anxiety related to concerns and beliefs about the outcome of pregnancy (pregnancy-specific anxiety; PSA) has been associated with behavioral outcomes and with the trajectory of maternal cortisol levels over gestation [228]. In a sample of 35 mother–child pairs, PSA was evaluated at three gestational intervals and structural, MRI scans were conducted when the children were six- to nine-years-old [135]. Pregnancy anxiety at 19 weeks gestation, but not other intervals, was associated with gray matter volume reductions in the prefrontal cortex, the premotor cortex, the medial temporal lobe, the lateral temporal cortex, the postcentral gyrus, as well as the cerebellum extending to the middle occipital gyrus and the fusiform gyrus. These effects were significant after adjusting for total gray matter volume, age, gestational age at birth, handedness, and postpartum maternal stress. The pervasive alteration of brain systems in children exposed to PSA may moderate the association between maternal distress and behavior that has been reported.

The association between prenatal maternal generalized anxiety and the hippocampus was assessed in 175 neonates with a follow-up of 35 infants [229]. MRIs were collected at birth and then again in a subsample at six months of age. Infants of mothers reporting increased anxiety during pregnancy (all subjects were in the normal range for anxiety) showed slower growth of both the left and right hippocampus over the first six months of life. The effect of prenatal maternal anxiety on right hippocampal growth was stronger when adjusted for postnatal maternal anxiety. A complex association between pre- and postnatal maternal anxiety and bilateral hippocampal growth suggested that the left hippocampus was constrained by exposure to postnatal anxiety and the right hippocampus by prenatal anxiety. Although these bilateral findings are difficult to interpret, the general finding that hippocampal development is influenced by maternal signals of anxiety is consistent with a larger literature.

Maternal depression is one of the most common prenatal complications, and the consequences of fetal exposure to maternal depression are poorly understood. There is recent evidence that the stress of prenatal exposure to maternal depression is associated with risk for externalizing behavior problems extending into adolescence [230–232]. The consequences of fetal exposure to maternal symptoms of depression were observed in a study of 157, 6- to 10-day-old newborns [233]. Alterations in brain microstructure were determined with structural magnetic resonance imaging and diffusion tensor imaging in infants from mothers scoring high on depression at 26 weeks gestation (28 women in the clinically significant range). Fetal exposure to prenatal maternal depressive symptoms was associated with lower fractional anisotropy and axial diffusivity, but not volume in the right amygdala. The association between fetal exposure to maternal depression and cortical thickness (using FreeSurfer software) was assessed in a group of 81 children, ages six to nine years, who had been followed to term beginning at 19 gestational weeks [234]. Significant cortical thinning in children primarily in the right prefrontal area was associated with exposure to prenatal maternal depression (Figure 4). The strongest association was observed in children who were exposed to maternal depression at 25 gestational weeks. Fetal exposure to maternal depression at this sensitive period of brain development was associated with cortical thinning in 19% of the whole cortex and 24% of the frontal lobes, primarily in the right superior, medial orbital and frontal pole regions of the prefrontal cortex. Fetal vulnerability to maternal signals at 25 gestational weeks has been reported previously [24,48,54,233,235] and may result from the enormous growth and dramatic structural changes in the nervous system during this interval. The pattern of cortical thinning in children exposed to prenatal maternal depression is similar to patterns in depressed patients [236–238] and in individuals with risk for depression [239]. Exposure to prenatal maternal depression and cortical thinning in this cohort were both associated with child externalizing behavior. The relation between prenatal maternal depression and child externalizing behavior was mediated by thinning in the frontal areas.

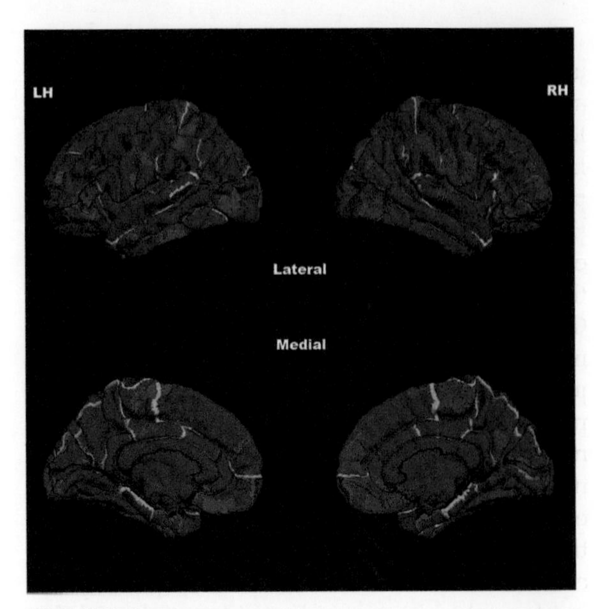

FIGURE 4 The areas in blue represent cortical regions that are significantly thinner in six- to nine-year-old children who were exposed to symptoms of maternal depression at 25 weeks gestation. *Adapted from Sandman et al. [234].*

Fetal Exposure to the Maternal HPA Axis

The consequences of fetal exposure to a biological marker (cortisol) of stress on manually traced volumes of the amygdala and hippocampus were assessed in a study of 65, typically developing six- to nine-year-old children [53]. Prenatal exposure to high maternal cortisol concentrations in early gestation was associated with a larger right amygdala volume in girls but not boys. A one standard deviation increase in maternal cortisol was associated with a 6.4% increase in the size of the right amygdala. Moreover, amygdala volume mediated the association between maternal cortisol levels and affective problems in girls. These findings in the right amygdala of girls with prenatal exposure to cortisol are nearly identical to those reported in girls with elevated cortisol at five years of age [225] and are consistent with the findings for early-life stress from Cohen et al. [224] of no association between maternal cortisol and children's hippocampus volume. Findings from these studies may provide an important clue to the origins of neuropsychiatric disorders. For instance, the magnitude of effect between prenatal cortisol exposure and amygdala size is comparable to the differences in amygdala volume between clinically depressed patients and healthy comparison volunteers [240] and consistent with findings that exposure to high levels of stress in early postnatal life or cortisol in early childhood is associated with altered (programmed) development and function of the amygdala [225,241–245].

Additional evidence that stress hormones exert neural programming influences comes from a study of fetal exposure to synthetic glucocorticoids (GCs) [246]. Structural brain images were compared between a group of 6- to 10-year-old typically developing children who were born full term with fetal exposure to GCs and children without prenatal GC exposure. Prenatal exposure to GCs was associated with significantly (8–9%) thinner cortices, primarily in the rostral anterior cingulate (rACC) than the unexposed children. Moreover, the GC-related reduction in rACC volume was associated with increased report of child affective disorder. The magnitude of the programming effect of synthetic GC exposure on rACC thinning in these normal healthy children is similar to the 10–14% reduction in the rACC among children with depressive symptomatology [247].

SEX DIFFERENCES AND THE DOHaD MODEL

Environmental pressures on population sex ratios, even prior to conception, exert unique patterns of developmental "programming" with consequences that confer differential risk for morbidity and mortality early in life and vulnerability to disease later in life. Historical records and current studies [248,249] confirm that more males than females are conceived (primary sex ratio) and born (secondary sex ratio). This suggests that under certain conditions sex ratio is associated with, and perhaps programmed by, preconceptional and maternal/fetal exposures to environmental events.

Sex differences have been observed in mammalian animal models as early as meiosis. When faced with adversity, male meiosis is interrupted resulting in infertility [250]. However, in females, a similar adversity does not interrupt meiosis resulting in greater chances of survival but with the possible risk of subsequent chromosomal abnormalities [250]. There is further evidence that within weeks of implantation the female placenta is more responsive than the male placenta to changes in stress signals including detection and response to maternal glucocorticoid concentration [251]. It has been argued that because male fetuses invest resources in growth, largely independent of maternal conditions, they do not adapt to maternal signals (such as glucocorticoid concentrations) and have limited ability to adjust to adversity placing them at greater risk for subsequent mortality and neurological morbidity. In contrast, the female fetus does not invest as heavily in growth but conserves resources and adjusts to prenatal adversity. This developmental/evolutionary "strategy" allows the female fetus to conserve its energy needs and increases the probability of survival when exposed to stress that reduces resources later in gestation.

Because of their failure to make adjustments to more subtle environmental signals, males are more vulnerable to severe consequences of early-life stress. For instance, more males than females are born preterm [248], have poorer neonatal and infant health outcomes [252], are less likely to survive in intensive care, and are at higher risk for developmental delay [253]. Females, with their adaptive flexibility early in gestation, escape the consequences of early-life exposure to adversity [93], but their escape from the risk of early mortality and morbidity has a price of increased vulnerability for effective and neurobehavioral problems expressed later in development [93]. For instance, preadolescent girls, but not boys, exposed to elevated levels of maternal cortisol early in pregnancy had significantly enlarged amygdale and increased levels of anxiety [53]. Six- to nine-year-old girls but not boys exposed to high levels of pregnancy-specific anxiety early in gestation additionally performed poorly on tests of executive function [254] and had decreased gray matter volume in areas of the brain critical for emotional regulation, inhibitory control, and memory [135]. Growing evidence supports the conclusion that sexually dimorphic neurobehavioral patterns in response to hostile conditions, or more specifically stress, may be programmed very early in gestation.

CONCLUSIONS

Evidence unequivocally supports the conclusion that exposure to early-life stress, especially during the fetal period, exerts profound and persisting consequences for human neurobehavioral development. However, there are alternatives to the fetal programming and developmental origins of disease models. The stress-inoculation model predicts that exposure to mild adversity during early development promotes *resilience* when confronted with stressful circumstances later in life [255]. The adaptive calibration model [256] proposes that early exposure to adversity may result in an *amplified response* to subsequent exposure to adversity but also may result in the resilience to benefit from supportive and protective features of the environment. The predictive adaptive response (PAR) model [12,257,258] predicts that optimal outcomes are related to the congruence or similarity between the prenatal and postnatal environments. This model predicts that a discordant transition between fetal and infant life places the child at risk for health complications. In a test of the PAR model, symptoms of maternal depression were assessed at regular intervals throughout the pregnancies in a healthy pregnant women and then again postpartum [25]. Infants whose mothers had concordant prenatal and early postnatal depressive symptoms (either high prenatal and postnatal symptoms or low prenatal and postnatal symptoms) had higher mental and motor development scores than infants exposed to discordant pre- and postnatal environments. It is possible that under some circumstances congruence between prenatal and postnatal environments prepares the fetus for postnatal life and confers an adaptive advantage for critical survival functions during early development.

Despite these alternative models, the evidence we reviewed from prospective and retrospective studies indicates that adverse birth phenotype and fetal exposures to maternal signals of psychobiological stress both result in increased risk for emotional and cognitive disorders and in alterations in brain structures in children and adults. Although there may be different consequences related to the timing of exposures to stress and there is some indication that prenatal stress exposure affects males and females differently [93], it is remarkable that a

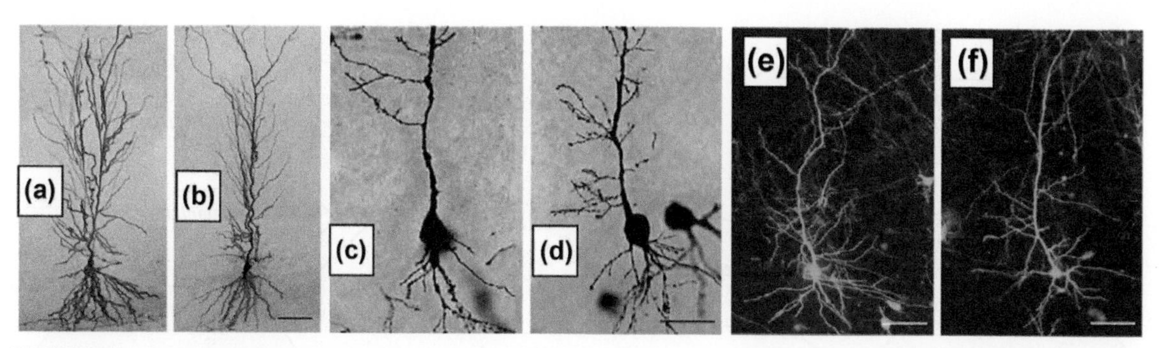

FIGURE 5 Dendritic impoverishment in pyramidal cells of adult rats exposed to behavioral and biological early-life stress: (a) control vs (b) behavioral stress (fragmented maternal care); (c) wild type vs (d) knockout of CRH receptor type 1 results in attenuation of detrimental effects of glucocorticoid system on dendritic arborization; (e) control tissue vs (f) tissue treated with CRH. *Adapted from [267].*

range of early-life exposures exerts relatively similar consequences on neurobehavioral development. This suggests that there is a common and critical neural mechanism that communicates between the source of stress and neuronal integrity and development.

Dendrites are treelike extensions that communicate critical information among neurons and transmit electrical stimulation to the soma. The shapes and sizes (complexity) of dendritic trees directly influence synaptic integration and neuronal excitability [259]. Increases in dendritic complexity result in (1) more synapses, (2) enlarged patterns of connectivity, (3) participation in larger communication networks [260], and (4) denser patterns of cortical thickness. In contrast, defects in dendritic arborization are associated with impaired neurological function and human neurodevelopmental disorders [261]. Although the molecular basis for the regulation of dendritic branching is only recently understood to be associated with inhibition of Cdc42 (controlling a protein) and activation of cofilin (actin regulation) during early development, it has been known for at least 50 years that structural components, including the microstructure of the brain, can be "powerfully shaped by experiences before birth" [262].

Convincing evidence indicates that prenatal exposure to stress decreases dendritic spine density in the anterior cingulate gyrus and orbitofrontal cortex of rats [263,264]. The existence or pattern of dendritic density and complexity in prenatally stressed animals may be sex specific [265,266]. Figure 5 illustrates in a rodent model[1] that early-life exposure to either behavioral or endocrine (CRH) stress results in a reduction of dendritic arborization [267]. This selective neuronal loss after early-life stress is associated with attenuated synaptic potentiation and memory deficits [268], and unlike the transient effects of stress on the brain and behavior of adults, the consequences of early-life stress permanently alter the construction of the nervous system. It is reasonable to assume that the programming influences of early-life stress on the brain and behavior have roots in the microstructure of the brain [217] (Figure 6).

[1] We have presented evidence [269] that the neurological milestones in first postnatal week of the rat are equivalent to the last trimester of the human fetus.

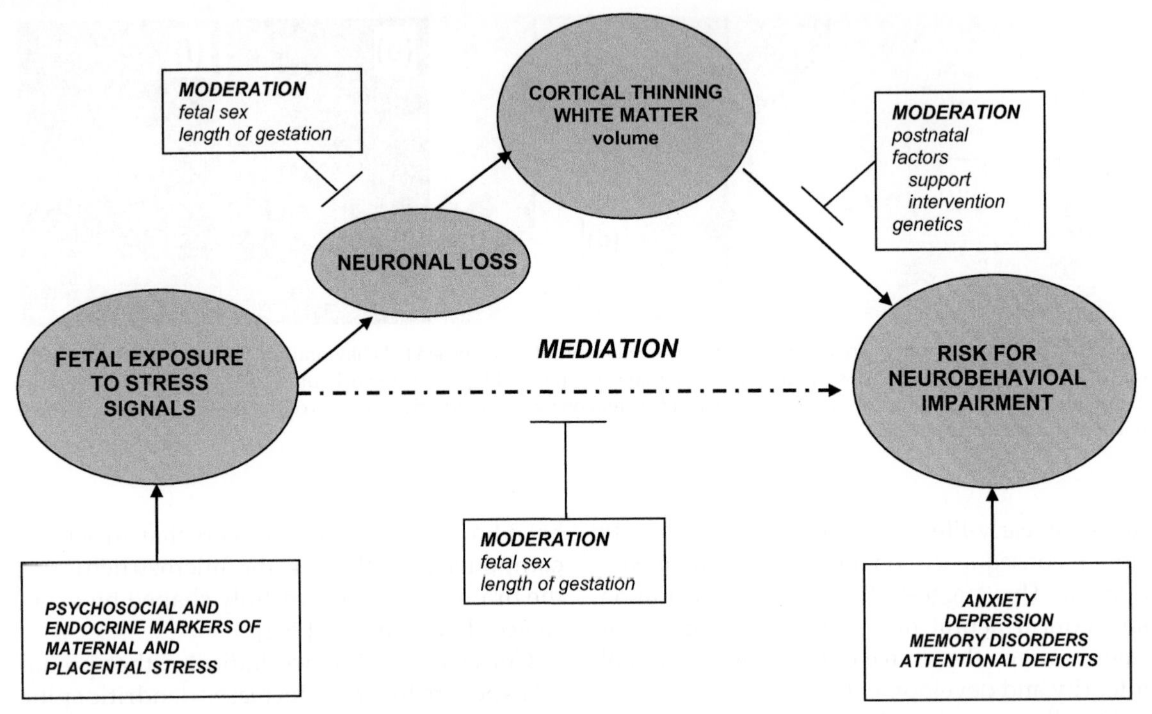

FIGURE 6 A model that illustrates the relations proposed in this chapter among exposure to early-life (fetal) stress, brain microstructure, brain volume, and behavioral consequences. The solid lines reflect hypothesized direct (causal) pathways connecting fetal exposure to neurological changes that mediate (dotted line) the association with behavioral risk. The model proposes that a significant association between fetal stress exposure and behavioral risk occurs as a function (or because) and there are alterations in brain structure (mediation). The model also recognizes that there are links sensitive to moderation (examples are listed, i.e., male vs female).

Acknowledgments

The research reported here was supported by National Institute of Health grants NS-41298, HD-51852, HD-28413 HD-40967, HD-50662, HD-65823 and Conte Center award MH-96889.

We are grateful for the expert contributions of Claudia Buss, PhD, Cheryl Crippen, PhD, Kevin Head, BA, Christina Canino-Brown, MA, Kendra Leak, BA, Megan (Blair) Faulkner, BA, Natalie Hernandez, BA, Mariann Howland, BA, Amanda Appel, MPH, Christine Dunkel-Schetter, PhD, and Cal Hobel, MD. Our program of research would not be possible without the participation of the families. Special gratitude is expressed especially to the families (mothers and children) who have continued in our longitudinal studies.

References

[1] Green JG, McLaughlin KA, Berglund PA, et al. Childhood adversities and adult psychiatric disorders in the national comorbidity survey replication I: associations with first onset of DSM-IV disorders. Arch Gen Psychiatry 2010;67(2):113–23.

[2] Coffino B. The role of childhood parent figure loss in the etiology of adult depression: findings from a prospective longitudinal study. Attachment Hum Dev 2009;11(5):445–70.

[3] Heim C, Nemeroff CB. The role of childhood trauma in the neurobiology of mood and anxiety disorders: preclinical and clinical studies. Biol Psychiatry 2001;49(12):1023–39.

[4] Kaufman J, Plotsky PM, Nemeroff CB, Charney DS. Effects of early adverse experiences on brain structure and function: clinical implications. Biol Psychiatry 2000;48(8):778–90.
[5] Gilbert R, Widom CS, Browne K, Fergusson D, Webb E, Janson S. Burden and consequences of child maltreatment in high-income countries. Lancet 2009;373(9657):68–81.
[6] Kajantie E, Raikkonen K. Early life predictors of the physiological stress response later in life. Neurosci Biobehav Rev 2010;35(1):23–32.
[7] Glynn LM, Sandman CA. The influence of prenatal stress and adverse birth outcomes on human cognitive and neurological development. Int Rev Res Ment Retard 2006:109–29. Academic Press.
[8] Woon FL, Hedges DW. Hippocampal and amygdala volumes in children and adults with childhood maltreatment-related posttraumatic stress disorder: a meta-analysis. Hippocampus 2008;18(8): 729–36.
[9] Baram TZ, Davis EP, Obenaus A, et al. Fragmentation and unpredictability of early life experience in mental disorders. Am J Psychiatry 2012;169(9):907–15.
[10] Barker DJ. In utero programming of chronic disease. Clin Sci 1998;95(2):115–28.
[11] Morley R, Owens J, Blair E, Dwyer T. Is birthweight a good marker for gestational exposures that increase the risk of adult disease? Paediatr Perinat Epidemiol 2002;16:194–9.
[12] Gluckman PD, Hanson MA. Living with the past: evolution, development, and patterns of disease. Science 2004;305(5691):1733–6.
[13] Weightman AL, Morgan HE, Shepherd MA, Kitcher H, Roberts C, Dunstan FD. Social inequality and infant health in the UK: systematic reviews and meta-analyses. Br Med J Open 2012;2(3):e000964.
[14] Group AoMSW. Identifying the environmental causes of disease: how should we decide what to believe and when to take action? London: Academy of Medical Sciences; 2007.
[15] Class QA, Abel KM, Khashan AS, et al. Offspring psychopathology following preconception, prenatal and postnatal maternal bereavement stress. Psychol Med 2014;44(1):71–84.
[16] Class QA, Rickert ME, Larsson H, Lichtenstein P, D'Onofrio BM. Fetal growth and psychiatric and socioeconomic problems: population-based sibling comparison. Br J Psychiatry 2014;205(5):355–61.
[17] D'Onofrio BM, Class QA, Rickert ME, Larsson H, Langstrom N, Lichtenstein P. Preterm birth and mortality and morbidity: a population-based quasi-experimental study. JAMA Psychiatry 2013;70(11):1231–40.
[18] D'Onofrio BM, Lahey B, Turkheimer E, Lichtenstein P. Critical need for family-based, quasi-experimental designs in integrating genetic and social science research. Am J Public Health 2013;103:S46–55.
[19] Glynn LM, Wadhwa PD, Dunkel-Schetter C, Chicz-Demet A, Sandman CA. When stress happens matters: effects of earthquake timing on stress responsivity in pregnancy. Am J Obstet Gynecol 2001;184(4):637–42.
[20] Kendler KS. Psychiatric genetics: a methodological critique. Am J Psychiatry 2005;162:3–11.
[21] Lahey BB, D'Onofrio BM, Waldman ID. Using epidemiologic methods to test hypotheses regarding causal influences on child and adolescent mental disorders. J Child Psychol Psychiatry 2009;50:53–62.
[22] Lindström K, Lindblad F, Hjern A. Preterm birth and attention-deficit/hyperactivity disorder in schoolchildren. Pediatrics 2011;127(5):858–65.
[23] Mwaniki MK, Atieno M, Lawn JE, Newton CR. Long-term neurodevelopmental outcomes after intrauterine and neonatal insults: a systematic review. Lancet 2012;379(9814):445–52.
[24] Sandman CA, Davis EP. Neurobehavioral risk is associated with gestational exposure to stress hormones. Expert Rev Endocrinol Metab 2012;7(4):445–59.
[25] Sandman CA, Davis EP, Glynn LM. Prescient human fetuses thrive. Psychol Sci 2012;23(1):93–100.
[26] Mairesse J, Lesage J, Breton C, et al. Maternal stress alters endocrine function of the feto-placental unit in rats. Am J Physiol Endocrinol Metab 2007;292(6):E1526–33.
[27] de Weerth C, Buitelaar JK. Physiological stress reactivity in human pregnancy–a review. Neurosci Biobehav Rev 2005;29(2):295–312.
[28] Entringer S, Kumsta R, Hellhammer DH, Wadhwa PD, Wust S. Prenatal exposure to maternal psychosocial stress and HPA axis regulation in young adults. Hormones Behav 2009;55(2):292–8.
[29] Glynn LM, Schetter CD, Hobel CJ, Sandman CA. Pattern of perceived stress and anxiety in pregnancy predicts preterm birth. Health Psychol 2008;27(1): 43–51.
[30] Petraglia F, Sutton S, Vale W. Neurotransmitters and peptides modulate the release of immunoreactive corticotropin-releasing factor from cultured human placental cells. Am J Obstet Gynecol 1989;160(1):247–51.
[31] Sasaki A, Tempst P, Liotta AS, et al. Isolation and characterization of a corticotropin-releasing hormone-like peptide from human-placenta. J Clin Endocr Metab 1988;67(4):768–73.
[32] Sasaki A, Shinkawa O, Yoshinaga K. Placental corticotropin-releasing hormone may be a stimulator of maternal pituitary adrenocorticotropic hormone-secretion in humans. J Clin Invest 1989;84(6):1997–2001.

[33] Campbell EA, Linton EA, Wolfe CD, Scraggs PR, Jones MT, Lowry PJ. Plasma corticotropin-releasing hormone concentrations during pregnancy and parturition. J Clin Endocrinol Metab 1987;64(5):1054–9.
[34] Chan EC, Smith R, Lewin T, et al. Plasma corticotropin-releasing hormone, beta-endorphin and cortisol inter-relationships during human pregnancy. Acta Endocrinol 1993;128(4):339–44.
[35] Goland RS, Conwell IM, Warren WB, Wardlaw SL. Placental corticotropin-releasing hormone and pituitary-adrenal function during pregnancy. Neuroendocrinology 1992;56(5):742–9.
[36] Wolfe CD, Patel SP, Campbell EA, et al. Plasma corticotrophin-releasing factor (CRF) in normal pregnancy. Br J Obstet Gynaecol 1988;95(10):997–1002.
[37] Sandman CA, Wadhwa PD, Chicz-DeMet A, Porto M, Garite TJ. Maternal corticotropin-releasing hormone and habituation in the human fetus. Dev Psychobiol 1999;34(3):163–73.
[38] McLean M, Bisits A, Davies J, Woods R, Lowry P, Smith R. A placental clock controlling the length of human pregnancy. Nat Med 1995;1(5):460–3.
[39] Sandman CA, Glynn L, Schetter CD, et al. Elevated maternal cortisol early in pregnancy predicts third trimester levels of placental corticotropin releasing hormone (CRH): priming the placental clock. Peptides 2006;27(6):1457–63.
[40] Smith R, Mesiano S, McGrath S. Hormone trajectories leading to human birth. Regul Pept 2002;108(2–3): 159–64.
[41] Smith R, Nicholson RC. Corticotrophin releasing hormone and the timing of birth. Front Biosci 2007;12: 912–8.
[42] Tyson EK, Smith R, Read M. Evidence that corticotropin-releasing hormone modulates myometrial contractility during human pregnancy. Endocrinology 2009;150(12): 5617–25.
[43] Glynn LM, Schetter CD, Wadhwa PD, Sandman CA. Pregnancy affects appraisal of negative life events. J Psychosom Res 2004;56(1):47–52.
[44] Glynn LM, Sandman CA. Prenatal origins of neurological development: a critical period for fetus and mother. Curr Dir Psychol Sci 2011;20(6):384–9.
[45] Sandman CA, Davis EP, Buss C, Glynn LM. Prenatal programming of human neurological function. Int J Pept 2011;2011:837596.
[46] Sandman CA, Glynn LM, Davis EP. Neurobehavioral consequences of fetal exposure to gestational stress. In: Kisilevsky BS, Reissland N, editors. Advancing research on fetal development. New York, NY: Springer, in press.
[47] Sandman CA, Davis EP. Gestational stress influences cognition and behavior. Future Neurol 2010;5(5): 675–90.
[48] Sandman CA, Davis EP, Buss C, Glynn LM. Exposure to prenatal psychobiological stress exerts programming influences on the mother and her fetus. Neuroendocrinology 2012;95(1):7–21.
[49] McMullen S, Langley-Evans SC, Gambling L, Lang C, Swali A, McArdle HJ. A common cause for a common phenotype: the gatekeeper hypothesis in fetal programming. Med Hypotheses 2012;78(1):88–94.
[50] Gutteling BM, de Weerth C, Zandbelt N, Mulder EJ, Visser GH, Buitelaar JK. Does maternal prenatal stress adversely affect the child's learning and memory at age six? J Abnorm Child Psychol 2006;34(6):789–98.
[51] Davis EP, Sandman CA. Prenatal psychobiological predictors of anxiety risk in preadolescent children. Psychoneuroendocrinology 2012;37(8):1224–33.
[52] Buss C, Davis EP, Class QA, et al. Maturation of the human fetal startle response: evidence for sex-specific maturation of the human fetus. Early Hum Dev 2009;85(10):633–8.
[53] Buss C, Davis EP, Shahbaba B, Pruessner JC, Head K, Sandman CA. Maternal cortisol over the course of pregnancy and subsequent child amygdala and hippocampus volumes and affective problems. Proc Natl Acad Sci USA 2012;109(20):E1312–9.
[54] Davis EP, Glynn LM, Dunkel Schetter C, Hobel C, Chicz-Demet A, Sandman CA. Corticotropin-releasing hormone during pregnancy is associated with infant temperament. Dev Neurosci 2005;27(5):299–305.
[55] Davis EP, Glynn LM, Dunkel Schetter C, Hobel C, Chicz-DeMet A, Sandman CA. Prenatal exposure to maternal depression and cortisol influences infant temperament. J Am Acad Child Adolesc Psychiatry 2007;46(6):737–46.
[56] Davis EP, Sandman CA. The timing of prenatal exposure to maternal cortisol and psychosocial stress is associated with human infant cognitive development. Child Dev 2010;81(1):131–48.
[57] Johnson S, Wolke D. Behavioural outcomes and psychopathology during adolescence. Early Hum Dev 2013;89(4):199–207.
[58] Somhovd MJ, Hansen BM, Brok J, Esbjorn BH, Greisen G. Anxiety in adolescents born preterm or with very low birthweight: a meta-analysis of case-control studies. Dev Med Child Neurol 2012;54(11):988–94.
[59] Treyvaud K, Ure A, Doyle LW, et al. Psychiatric outcomes at age seven for very preterm children: rates and predictors. J Child Psychol Psychiatry Allied Discip 2013;54(7):772–9.
[60] Rogers CE, Barch DM, Sylvester CM, et al. Altered gray matter volume and school age anxiety in children born late preterm. J Pediatr 2014;165(5):928–35.
[61] Vasiliadis HM, Buka SL, Martin LT, Gilman SE. Fetal growth and the lifetime risk of generalized anxiety disorder. Depress Anxiety 2010;27(11):1066–72.

[62] Lund LK, Vik T, Skranes J, Brubakk AM, Indredavik MS. Psychiatric morbidity in two low birth weight groups assessed by diagnostic interview in young adulthood. Acta Paediatr 2011;100(4):598–604.
[63] Roussel S, Boissy A, Montigny D, Hemsworth PH, Duvaux-Ponter C. Gender-specific effects of prenatal stress on emotional reactivity and stress physiology of goat kids. Hormones Behav 2005;47(3):256–66.
[64] Schulz KM, Pearson JN, Neeley EW, et al. Maternal stress during pregnancy causes sex-specific alterations in offspring memory performance, social interactions, indices of anxiety, and body mass. Physiol Behav 2011;104(2):340–7.
[65] Mueller BR, Bale TL. Sex-specific programming of offspring emotionality after stress early in pregnancy. J Neurosci 2008;28(36):9055–65.
[66] Schneider ML. Prenatal stress exposure alters postnatal behavioral expression under conditions of novelty challenge in rhesus-monkey infants. Dev Psychobiol 1992;25(7):529–40.
[67] Uno H, Eisele S, Sakai A, et al. Neurotoxicity of glucocorticoids in the primate brain. Hormones Behav 1994;28(4):336–48.
[68] Coe CL, Kramer M, Czeh B, et al. Prenatal stress diminishes neurogenesis in the dentate gyrus of juvenile rhesus monkeys. Biol Psychiatry 2003;54(10): 1025–34.
[69] Davis EP, Glynn LM, Waffarn F, Sandman CA. Prenatal maternal stress programs infant stress regulation. J Child Psychol Psychiatry Allied Discip 2011;52(2):119–29.
[70] Gutteling BM, De Weerth C, Buitelaar JK. Maternal prenatal stress and 4–6 year old children's salivary cortisol concentrations pre- and post-vaccination. Stress 2004;7(4):257–60.
[71] O'Connor TG, Bergman K, Sarkar P, Glover V. Prenatal cortisol exposure predicts infant cortisol response to acute stress. Dev Psychobiol 2013;55(2):145–55.
[72] Tollenaar MS, Beijers R, Jansen J, Riksen-Walraven JM, de Weerth C. Maternal prenatal stress and cortisol reactivity to stressors in human infants. Stress 2011;14(1):53–65.
[73] Bergman K, Sarkar P, O'Connor TG, Modi N, Glover V. Maternal stress during pregnancy predicts cognitive ability and fearfulness in infancy. J Am Acad Child Adolesc Psychiatry 2007;46(11):1454–63.
[74] Blair MM, Glynn LM, Sandman CA, Davis EP. Prenatal maternal anxiety and early childhood temperament. Stress 2011;14(6):644–51.
[75] Davis EP, Glynn LM, Schetter CD, Hobel C, Chicz-Demet A, Sandman CA. Prenatal exposure to maternal depression and cortisol influences infant temperament. J Am Acad Child Adolesc Psychiatry 2007;46(6): 737–46.
[76] Davis EP, Snidman N, Wadhwa PD, Dunkel Schetter C, Glynn L, Sandman CA. Prenatal maternal anxiety and depression predict negative behavioral reactivity in infancy. Infancy 2004;6(3):319–31.
[77] de Bruijn AT, van Bakel HJ, van Baar AL. Sex differences in the relation between prenatal maternal emotional complaints and child outcome. Early Hum Dev 2009;85(5):319–24.
[78] de Weerth C, van Hees Y, Buitelaar J. Prenatal maternal cortisol levels and infant behavior during the first 5 months. Early Hum Dev 2003;74:139–51.
[79] Van den Bergh BR. The influence of maternal emotions during pregnancy on fetal and neonatal behavior. J Prenat Perinat Psychol Health 1990;5(2): 119–30.
[80] Kagan J, Snidman N, Arcus D. Childhood derivatives of high and low reactivity in infancy. Child Dev 1998;69(6):1483–93.
[81] Schwartz CE, Snidman N, Kagan J. Adolescent social anxiety as an outcome of inhibited temperament in childhood. J Am Acad Child Adolesc Psychiatry 1999;38(8):1008–15.
[82] Schwartz CE, Wright CI, Shin LM, Kagan J, Rauch SL. Inhibited and uninhibited infants "grown up": adult amygdalar response to novelty. Science 2003;300(5627): 1952–3.
[83] O'Connor TG, Heron J, Golding J, Glover V. Maternal antenatal anxiety and behavioural/emotional problems in children: a test of a programming hypothesis. J Child Psychol Psychiatry 2003;44(7):1025–36.
[84] Van den Bergh BR, Marcoen A. High antenatal maternal anxiety is related to ADHD symptoms, externalizing problems, and anxiety in 8- and 9-year-olds. Child Dev 2004;75(4):1085–97.
[85] Van den Bergh BR, Van Calster B, Smits T, Van Huffel S, Lagae L. Antenatal maternal anxiety is related to HPA-axis dysregulation and self-reported depressive symptoms in adolescence: a prospective study on the fetal origins of depressed mood. Neuropsychopharmacology 2008;33(3):536–45.
[86] O'Donnell KJ, Glover V, Barker ED, O'Connor TG. The persisting effect of maternal mood in pregnancy on childhood psychopathology. Dev Psychopathol 2014;26(2):393–403.
[87] DiPietro JA, Kivlighan KT, Costigan KA, Laudenslager ML. Fetal motor activity and maternal cortisol. Dev Psychobiol 2009;51(6):505–12.
[88] Monk C, Fifer WP, Myers MM, Sloan RP, Trien L, Hurtado A. Maternal stress responses and anxiety during pregnancy: effects on fetal heart rate. Dev Psychobiol 2000;36(1):67–77.
[89] DiPietro JA, Ghera MM, Costigan KA. Prenatal origins of temperamental reactivity in early infancy. Early Hum Dev 2008;84(9):569–75.

[90] Class QA, Buss C, Davis EP, et al. Low levels of corticotropin-releasing hormone during early pregnancy are associated with precocious maturation of the human fetus. Dev Neurosci 2008;30(6):419–26.
[91] Glynn LM, Sandman CA. Sex moderates associations between prenatal glucocorticoid exposure and human fetal neurological development. Dev Sci 2012;15(5):601–10.
[92] Bergman K, Glover V, Sarkar P, Abbott DH, O'Connor TG. In utero cortisol and testosterone exposure and fear reactivity in infancy. Hormones Behav 2010;57(3):306–12.
[93] Sandman CA, Glynn LM, Davis EP. Is there a viability-vulnerability tradeoff? Sex differences in fetal programming. J Psychosom Res 2013;75(4):327–35.
[94] Davis EP, Waffarn F, Sandman CA. Prenatal treatment with glucocorticoids sensitizes the hpa axis response to stress among full-term infants. Dev Psychobiol 2011;53(2):175–83.
[95] O'Connor TG, Ben-Shlomo Y, Heron J, Golding J, Adams D, Glover V. Prenatal anxiety predicts individual differences in cortisol in pre-adolescent children. Biol Psychiatry 2005;58(3):211–7.
[96] O'Donnell KJ, Glover V, Jenkins J, et al. Prenatal maternal mood is associated with altered diurnal cortisol in adolescence. Psychoneuroendocrinology 2013;38(9):1630–8.
[97] Koss KJ, Hostinar CE, Donzella B, Gunnar MR. Social deprivation and the HPA axis in early development. Psychoneuroendocrinology 2014;50:1–13.
[98] Blencowe H, Lee ACC, Cousens S, et al. Preterm birth-associated neurodevelopmental impairment estimates at regional and global levels for 2010. Pediatr Res 2013;74:17–34.
[99] Anderson PJ, Doyle LW, Grp VIC. Executive functioning in school-aged children who were born very preterm or with extremely low birth weight in the 1990s. Pediatrics 2004;114(1):50–7.
[100] Cooke RWI, Foulder-Hughes L. Growth impairment in the very preterm and cognitive and motor performance at 7 years. Arch Dis Child 2003;88(6):482–7.
[101] O'Brien F, Roth S, Stewart A, Rifkin L, Rushe T, Wyatt J. The neurodevelopmental progress of infants less than 33 weeks into adolescence. Arch Dis Child 2004;89(3):207–11.
[102] Lorenz JM, Wooliever DE, Jetton JR, Paneth N. A quantitative review of mortality and developmental disability in extremely premature newborns. Arch Pediat Adol Med 1998;152(5):425–35.
[103] Tin W, Wariyar U, Hey E. Changing prognosis for babies of less than 28 weeks' gestation in the north of England between 1983 and 1994. Brit Med J 1997;314(7074):107–11.
[104] Levy-Shiff R, Einat G, Mogilner MB, Lerman M, Krikler R. Biological and environmental correlates of developmental outcome of prematurely born infants in early adolescence. J Pediatr Psychol 1994;19(1):63–78.
[105] Olsen P, Vainionpaa L, Paakko E, Korkman M, Pyhtinen J, Jarvelin MR. Psychological findings in preterm children related to neurologic status and magnetic resonance imaging. Pediatrics 1998;102(2 Pt 1): 329–36.
[106] Rickards AL, Kelly EA, Doyle LW, Callanan C. Cognition, academic progress, behavior and self-concept at 14 years of very low birth weight children. J Dev Behav Pediatr 2001;22(1):11–8.
[107] Shenkin SD, Starr JM, Pattie A, Rush MA, Whalley LJ, Deary IJ. Birth weight and cognitive function at age 11 years: the Scottish Mental Survey 1932. Arch Dis Child 2001;85(3):189–95.
[108] Isaacs EB, Lucas A, Chong WK, et al. Hippocampal volume and everyday memory in children of very low birth weight. Pediatr Res 2000;47(6):713–20.
[109] Luciana M, Lindeke L, Georgieff M, Mills M, Nelson CA. Neurobehavioral evidence for working-memory deficits in school-aged children with histories of prematurity. Dev Med Child Neurol 1999;41(8):521–33.
[110] Rushe TM, Rifkin L, Stewart AL, et al. Neuropsychological outcome at adolescence of very preterm birth and its relation to brain structure. Dev Med Child Neurol 2001;43(4):226–33.
[111] Williams BL, Dunlop AL, Kramer M, Dever BV, Hogue C, Jain L. Perinatal origins of first-grade academic failure: role of prematurity and maternal factors. Pediatrics 2013;131(4):693–700.
[112] Pharoah POD, Stevenson CJ, West CR. General certificate of secondary education performance in very low birthweight infants. Arch Dis Child 2003;88(4):295–8.
[113] Yang S, Platt RW, Kramer MS. Variation in child cognitive ability by week of gestation among healthy term births. Am J Epidemiol 2010;171(4):399–406.
[114] Phua DYL, Rifkin-Graboi A, Saw SM, Meaney MJ, Qiu AQ. Executive functions of six-year-old boys with normal birth weight and gestational age. Plos One 2012;7(4):e36502.
[115] Noble KG, Fifer WP, Rauh VA, Nomura Y, Andrews HF. Academic achievement varies with gestational age among children born at term. Pediatrics 2012;130(2): E257–64.
[116] Rose O, Blanco E, Martinez SM, et al. Developmental scores at 1 year with increasing gestational age, 37–41 weeks. Pediatrics 2013;131(5):E1475–81.
[117] Espel EV, Glynn LM, Sandman CA, Davis EP. Longer gestation among children born full term influences cognitive and motor development. Plos One 2014;9(11):e113758.

[118] Richards M, Hardy R, Kuh D, Wadsworth ME. Birth weight and cognitive function in the British 1946 birth cohort: longitudinal population based study. BMJ 2001;322(7280):199–203.
[119] Sorensen HT, Sabroe S, Olsen J, Rothman KJ, Gillman MW, Fischer P. Birth weight and cognitive function in young adult life: historical cohort study. BMJ 1997;315(7105):401–3.
[120] Cresswell JL, Barker DJ, Osmond C, Egger P, Phillips DI, Fraser RB. Fetal growth, length of gestation, and polycystic ovaries in adult life. Lancet 1997;350(9085): 1131–5.
[121] Gunnell D, Rasmussen F, Fouskakis D, Tynelius P, Harrison G. Patterns of fetal and childhood growth and the development of psychosis in young males: a cohort study. Am J Epidemiol 2003;158(4): 291–300.
[122] McCance DR, Pettitt DJ, Hanson RL, Jacobsson LT, Knowler WC, Bennett PH. Birth weight and non-insulin dependent diabetes: thrifty genotype, thrifty phenotype, or surviving small baby genotype? BMJ 1994;308(6934):942–5.
[123] Savona-Ventura C, Chircop M. Birth weight influence on the subsequent development of gestational diabetes mellitus. Acta Diabetol 2003;40(2):101–4.
[124] Provençal N, Binder EB. The effects of early-life stress on the epigenome: from the womb to adulthood and even before. Exp Neurol June 2015;268:10–20.
[125] Bale TL. Sex differences in prenatal epigenetic programing of stress pathways. Stress Int J Biol Stress 2011;14(4):348–56.
[126] Harris A, Seckl J. Glucocorticoids, prenatal stress and the programming of disease. Hormones Behav 2011;59(3):279–89.
[127] Schneider ML, Coe CL. Repeated social stress during pregnancy impairs neuromotor development of the primate infant. J Dev Behav Pediatr 1993;14(2): 81–7.
[128] Schneider ML, Coe CL, Lubach GR. Endocrine activation mimics the adverse effects of prenatal stress on the neuromotor development of the infant primate. Dev Psychobiol 1992;25(6):427–39.
[129] Schneider ML, Roughton EC, Koehler AJ, Lubach GR. Growth and development following prenatal stress exposure in primates: an examination of ontogenetic vulnerability. Child Dev 1999;70(2):263–74.
[130] Weinstock M. The long-term behavioural consequences of prenatal stress. Neurosci Biobehav Rev 2008;32(6): 1073–86.
[131] Kapoor A, Dunn E, Kostaki A, Andrews MH, Matthews SG. Fetal programming of hypothalamo-pituitary-adrenal function: prenatal stress and glucocorticoids. J Physiol 2006;572(Pt 1):31–44.
[132] Huizink AC, de Medina PGR, Mulder EJH, Visser GHA, Buitelaar JK. Stress during pregnancy is associated with developmental outcome in infancy. J Child Psychol Psyc 2003;44(6):810–8.
[133] Mennes M, Stiers P, Lagae L, Van den Bergh B. Long-term cognitive sequelae of antenatal maternal anxiety: involvement of the orbitofrontal cortex. Neurosci Biobehav Rev 2006;30(8):1078–86.
[134] DiPietro JA, Novak MFSX, Costigan KA, Atella LD, Reusing SP. Maternal psychological distress during pregnancy in relation to child development at age two. Child Dev 2006;77(3):573–87.
[135] Buss C, Davis EP, Muftuler LT, Head K, Sandman CA. High pregnancy anxiety during mid-gestation is associated with decreased gray matter density in 6-9-year-old children. Psychoneuroendocrinology 2010;35(1):141–53.
[136] Ellman LM, Schetter CD, Hobel CJ, Chicz-Demet A, Glynn LM, Sandman CA. Timing of fetal exposure to stress hormones: effects on newborn physical and neuromuscular maturation. Dev Psychobiol 2008;50(3):232–41.
[137] Bergman K, Sarkar P, Glover V, O'Connor TG. Maternal prenatal cortisol and infant cognitive development: moderation by infant-mother attachment. Biol Psychiatry 2010;67(11):1026–32.
[138] LeWinn KZ, Stroud LR, Molnar BE, Ware JH, Koenen KC, Buka SL. Elevated maternal cortisol levels during pregnancy are associated with reduced childhood IQ. Int J Epidemiol 2009;38(6):1700–10.
[139] DiPietro JA, Bornstein MH, Costigan KA, et al. What does fetal movement predict about behavior during the first two years of life? Dev Psychobiol 2002;40(4):358–71.
[140] Nijhuis JG. Behavioral states – concomitants, clinical implications and the assessment of the condition of the nervous-system. Eur J Obstet Gyn R B 1986;21(5–6):301–8.
[141] Kisilevsky BS, Muir DW, Low JA. Maturation of human fetal responses to vibroacoustic stimulation. Child Dev 1992;63(6):1497–508.
[142] Abel KM, Dalman C, Svensson AC, et al. Deviance in fetal growth and risk of autism spectrum disorder. Am J Psychiatry 2013;170(4):391–8.
[143] Abel KM, Wicks S, Susser ES, et al. Birth weight, schizophrenia, and adult mental disorder: is risk confined to the smallest babies? Archives General Psychiatry 2010;67(9):923–30.
[144] Hack M, Youngstrom EA, Cartar L, et al. Behavioral outcomes and evidence of psychopathology among very low birth weight infants at age 20 years. Pediatrics 2004;114:932–40.
[145] Losh M, Esserman D, Anckarsater H, Sullivan PF, Lichtenstein P. Lower birth weight indicates higher risk of autistic traits in discordant twin pairs. Psychol Med 2012;42:1091–102.

[146] Aarnoudse-Moens CSH, Weisglas-Kuperus N, van Goudoever B, Oosterlaan J. Meta-analysis of neurobehavioral outcomes in very preterm and/or very low birth weight children. Pediatrics 2009;124:717–28.

[147] Lawlor DA, Bor W, O'Callaghan MJ, Williams GM, Najman JM. Intrauterine growth and intelligence within sibling pairs: findings from Mater-University study of pregnancy and its outcomes. J Epidemiol Community Health 2005;59:279–82.

[148] McCormick MC, Litt JS, Smith VC, Zupancic JAF. Prematurity: an overview and public health implications. Annu Rev Public Health 2011;32:367–79.

[149] Crump C, Winkleby MA, Sundquist K, Sundquist J. Preterm birth and psychiatric medication prescription in young adulthood: a Swedish National Cohort study. Int J Epidemiol 2010;39:1522–30.

[150] Mathiasen R, Hansen BM, Anderson AN, Greisen G. Socio-economic achievements of individuals born very preterm at the age of 27 to 29 years: a nationwide cohort study. Dev Med Child Neurol 2009;51:901–8.

[151] McGowan JE, Alderdice FA, Holmes VA, Johnston L. Early childhood development of late-preterm infants: a systematic review. Pediatrics 2011;127:1111–24.

[152] Saigal S, Streiner D. Socio-economic achievements of individuals born very preterm at the age 0f 27 to 29. Dev Med Child Neurol 2009;51:845–50.

[153] Groen-Blokhuis MM, Middeldorp CM, van Beijsterveldt CEM, Boomsma DI. Evidence for a causal association of low birth weight and attention problems. J Am Acad Child Adolesc Psychiatry 2011;50:1247–54.

[154] Hultman CM, Torrang A, Tuvblad C, Cnattingius S, Larsson J, Lichtenstein P. Birth weight and attention-deficit/hyperactivity symptoms in childhood and early adolescence: a prospective Swedish Twin study. J Am Acad Child Adolesc Psychiatry 2007;46(3):370–7.

[155] Lehn H, Derks EM, Hudziak JJ, van Beijsterveldt TC, Boomsma DI. Attention problems and attention-deficit/hyperactivity disorder in discordant and concordant monozygotic twins: evidence of environmental mediators. J Am Acad Child Adolesc Psychiatry 2007;46:83–91.

[156] Wojcik W, Lee W, Colman I, Hardy R, Hotopf M. Foetal origins of depression? A systematic review and meta-analysis of low birth weight and later depression. Psychol Med 2013;43(1):1–12.

[157] Foley DL, Neale MC, Kendler KS. Does intra-uterine growth discordance predict differential risk for adult psychiatric disorder in a population-based sample of monozygotic twins? Psychiatr Genet 2000;10(1):1–8.

[158] Miller DB, O'Callaghan JP. Do early-life insults contribute to the late-life development of Parkinson and Alzheimer diseases? Metabolism Clin Exp 2008;57(S2):S44–9.

[159] Huttunen M, Niskanen P. Prenatal loss of father and psychiatric disorders. ArchGen Psychiatry 1978;35(4):429–31.

[160] Khashan AS, Abel KM, McNamee R, et al. Higher risk of offspring schizophrenia following antenatal maternal exposure to severe adverse life events. Arch Gen Psychiatry 2008;65(2):146–52.

[161] Khashan AS, McNamee R, Henriksen TB, et al. Risk of affective disorders following prenatal exposure to severe life events: a Danish population-based cohort study. J Psychiatr Res 2011;45:879–85.

[162] Rodriguez A, Bohlin G. Are maternal smoking and stress during pregnancy related to ADHD symptoms in children? J Child Psychol Psychiatry 2005;46(3):246–54.

[163] Ronald A, Pennell CE, Whitehouse AJO. Prenatal maternal stress associated with ADHD and autistic traits in early childhood. Front Psychol 2011;1:1–8.

[164] Van Os J, Selten JP. Prenatal exposure to maternal stress and subsequent schizophrenia. The May 1940 invasion of the Netherlands. Br J Psychiatry 1998;172(4):324–6.

[165] Brown AS, Susser ES, Lin SP, Neugebauer R, Gorman JM. Increased risk of affective disorder in males after second trimester prenatal exposure to the Dutch Hunger Winter of 1944-1945. Br J Psychiatry 1995;166:601–6.

[166] Brown AS, van Os J, Driessens C, Hoek HW, Susser ES. Further evidence of relation between prenatal famine and major affective disorder. Am J Psychiatry 2000;157(2):190–5.

[167] Li J, Olsen J, Vestergaard M, Obel C. Attention-deficit/hyperactivity disorder in the offspring following prenatal maternal bereavement: a nationwide follow-up study in Denmark. Eur Child Adolesc Psychiatry 2010;19(10):747–53.

[168] Brand SR, Engel SM, Canfield RL, Yehuda R. The effect of maternal PTSD following in utero trauma exposure on behavior and temperament in the 9-month-old infant. Ann NY Acad Sci 2006;1071:454–8.

[169] King S, Laplante DP. The effects of prenatal maternal stress on children's cognitive development: project ice storm. Stress 2005;8(1):35–45.

[170] Kinney DK, Miller AM, Crowley DJ, Huang E, Gerber E. Autism prevalence following prenatal exposure to hurricanes and tropical storms in Louisiana. J Autism Dev Disord 2008;38(3):481–8.

[171] Beversdorf DQ, Manning SE, Hillier A, et al. Timing of prenatal stressors and autism. J Autism Dev Disord 2005;35(4):471–8.

[172] de Bie HM, Oostrom KJ, Delemarre-van de Waal HA. Brain development, intelligence and cognitive outcome in children born small for gestational age. Paediatrics 2010;73:6–14.

[173] Pelham WE, Foster EM, Robb JA. The economic impact of attention-deficit/hyperactivity disorder in children and adolescents. J Pediatr Psychol 2007;32(6): 711–27.

[174] Polanczyk G, de Lima MS, Horta BL, Biederman J, Rohde LA. The worldwide prevalence of ADHD: a systematic review and metaregression analysis. Am J Psychiatry 2007;164:942–8.

[175] Ward AJ. A comparison and analysis of the presence of family problems during pregnancy of mothers of "autistic" children and mothers of typically developing children. Child Psychiatry Hum Dev 1990;20:279–88.

[176] Rai D, Golding J, Magnusson C, Steer C, Lewis G, Dalman C. Prenatal and early-life exposure to stressful life events and risk of autism spectrum disorders: population-based studies in Sweden and England. Plos One 2012;7(6):e38893.

[177] Beydoun H, Saftlas AF. Physical and mental health outcomes of prenatal maternal stress in human and animal studies: a review of recent evidence. Paediatr Perinat Epidemiol 2008;22:438–66.

[178] Selten JP, Cantor-Graae E, Nahon D, Levav I, Aleman A, Kahn RS. No relationship between risk of schiophrenia and prenatal exposure to stress during the Six-Day War or Yom Kippur War in Israel. Schizophrenia Res 2003;63:131–5.

[179] Huang H, Xue R, Zhang JY, et al. Anatomical characterization of human fetal brain development with diffusion tensor magnetic resonance imaging. J Neurosci 2009;29(13):4263–73.

[180] Malatesta P, Appolloni I, Calzolari F. Radial glia and neural stem cells. Cell Tissue Res 2008;331(1): 165–78.

[181] Takahashi E, Folkerth RD, Galaburda AM, Grant PE. Emerging cerebral connectivity in the human fetal brain: an MR tractography study. Cereb Cortex 2012;22(2):455–64.

[182] Stiles J, Jernigan TL. The basics of brain development. Neuropsychol Rev 2010;20(4):327–48.

[183] Kostovic I, Judas M, Rados M, Hrabac P. Laminar organization of the human fetal cerebrum revealed by histochemical markers and magnetic resonance imaging. Cereb Cortex 2002;12(5):536–44.

[184] Bourgeois JP, Goldmanrakic PS, Rakic P. Synaptogenesis in the prefrontal cortex of rhesus-monkeys. Cereb Cortex 1994;4(1):78–96.

[185] Bourgeois JP. Synaptogenesis, heterochrony and epigenesis in the mammalian neocortex. Acta Paediatr 1997;86:27–33.

[186] Volpe JJ. Neurology of the newborn. Philadelphia, PA: Elsevier; 2008.

[187] Huttenlocher PR, Dabholkar AS. Regional differences in synaptogenesis in human cerebral cortex. J Comp Neurol 1997;387(2):167–78.

[188] Huttenlocher PR, Decourten C, Garey LJ, Vanderloos H. Synaptogenesis in human visual-cortex: evidence for synapse elimination during normal development. Neurosci Lett 1982;33(3):247–52.

[189] Becker LE, Armstrong DL, Chan F, Wood MM. Dendritic development in human occipital cortical-neurons. Dev Brain Res 1984;13(1):117–24.

[190] Levitt P. Structural and functional maturation of the developing primate brain. J Pediatr 2003;143(4):S35–45.

[191] Adams-Chapman I. Neurodevelopmental outcome of the late preterm infant. Clin Perinatol 2006;33(4): 947–64.

[192] Kinney HC. The near-term (late preterm) human brain and risk for periventricular leukomalacia: a review. Semin Perinatol 2006;30(2):81–8.

[193] Wyatt JS. Mechanisms of brain injury in the newborn. Eye 2007;21(10):1261–3.

[194] Nosarti C, Al-Asady MHS, Frangou S, Stewart AL, Murray RM, Rifkin L. Reduction of hippocampal volume in very preterm adolescents: a model for schizophrenia. Schizophrenia Res 2000;41(1):119.

[195] Peterson BS, Anderson AW, Ehrenkranz R, et al. Regional brain volumes and their later neurodevelopmental correlates in term and preterm infants. Pediatrics 2003;111(5):939–48.

[196] Delobel-Ayoub M, Arnaud C, White-Koning M, et al. Behavioral problems and cognitive performance at 5 years of age after very preterm birth: the EPIPAGE study. Pediatrics 2009;123(6):1485–92.

[197] Grunewaldt KH, Fjortoft T, Bjuland KJ, et al. Follow-up at age 10years in ELBW children - functional outcome, brain morphology and results from motor assessments in infancy. Early Hum Dev 2014;90(10):571–8.

[198] Ostgard HF, Lohaugen GC, Bjuland KJ, et al. Brain morphometry and cognition in young adults born small for gestational age at term. J Pediatr 2014;165(5):921–7.e1.

[199] Bjuland KJ, Lohaugen GCC, Martinussen M, Skranes J. Cortical thickness and cognition in very-low-birth-weight late teenagers. Early Hum Dev 2013;89(6): 371–80.

[200] Ball G, Boardman JP, Rueckert D, et al. The effect of preterm birth on thalamic and cortical development. Cereb Cortex 2012;22(5):1016–24.

[201] Ball G, Boardman JP, Aljabar P, et al. The influence of preterm birth on the developing thalamocortical connectome. Cortex 2013;49(6):1711–21.

[202] Rose J, Vassar R, Cahill-Rowley K, et al. Neonatal physiological correlates of near-term brain development on MRI and DTI in very-low-birth-weight preterm infants. NeuroImage Clin 2014;5:169–77.

[203] Padilla N, Junque C, Figueras F, et al. Differential vulnerability of gray matter and white matter to intrauterine growth restriction in preterm infants at 12 months corrected age. Brain Res 2014;1545:1–11.

[204] Parker J, Mitchell A, Kalpakidou A, et al. Cerebellar growth and behavioural neuropsychological outcome in preterm adolescents. Brain 2008;131:1344–51.
[205] Peterson BS, Vohr B, Staib LH, et al. Regional brain volume abnormalities and long-term cognitive outcome in preterm infants. JAMA 2000;284(15):1939–47.
[206] Groeschel S, Tournier JD, Northam GB, et al. Identification and interpretation of microstructural abnormalities in motor pathways in adolescents born preterm. NeuroImage 2014;87:209–19.
[207] Nosarti C, Giouroukou E, Healy E, et al. Grey and white matter distribution in very preterm adolescents mediates neurodevelopmental outcome. Brain 2008;131(Pt 1):205–17.
[208] Ment LR, Kesler S, Vohr B, et al. Longitudinal brain volume changes in preterm and term control subjects during late childhood and adolescence. Pediatrics 2009;123(2):503–11.
[209] Melbourne A, Kendall GS, Cardoso MJ, et al. Preterm birth affects the developmental synergy between cortical folding and cortical connectivity observed on multimodal MRI. NeuroImage 2014;89:23–34.
[210] Hill J, Inder T, Neil J, Dierker D, Harwell J, Van Essen D. Similar patterns of cortical expansion during human development and evolution. Proc Natl Acad Sci USA 2010;107(29):13135–40.
[211] Kim D, Davis EP, Sandman CA, et al. Longer gestation is associated with more efficient brain networks in preadolescent children. NeuroImage 2014;100:619–27.
[212] Martinussen M, Flanders DW, Fischl B, et al. Segmental brain volumes and cognitive and perceptual correlates in 15-year-old adolescents with low birth weight. J Pediatr 2009;155(6):848–53 e1.
[213] Eikenes L, Martinussen MP, Lund LK, et al. Being born small for gestational age reduces white matter integrity in adulthood: a prospective cohort study. Pediatr Res 2012;72(6):649–54.
[214] Muller M, Sigurdsson S, Kjartansson O, et al. Birth size and brain function 75 years later. Pediatrics 2014;134(4):761–70.
[215] Charil A, Laplante DP, Vaillancourt C, King S. Prenatal stress and brain development. Brain Res Rev 2010;65(1):56–79.
[216] Cratty MS, Ward HE, Johnson EA, Azzaro AJ, Birkle DL. Prenatal stress increases corticotropin-releasing factor (CRF) content and release in rat amygdala minces. Brain Res 1995;675(1–2):297–302.
[217] Salm AK, Pavelko M, Krouse EM, Webster W, Kraszpulski M, Birkle DL. Lateral amygdaloid nucleus expansion in adult rats is associated with exposure to prenatal stress. Brain Res Dev Brain Res 2004;148(2):159–67.
[218] Kolb B, Mychasiuk R, Muhammad A, Li Y, Frost DO, Gibb R. Experience and the developing prefrontal cortex. Proc Natl Acad Sci USA 2012;109(Suppl. 2):17186–93.
[219] Garbett KA, Hsiao EY, Kalman S, Patterson PH, Mirnics K. Effects of maternal immune activation on gene expression patterns in the fetal brain. Transl Psychiatry 2012;2:e98.
[220] Lo CL, Zhou FC. Environmental alterations of epigenetics prior to the birth. Int Rev Neurobiol 2014;115:1–49.
[221] Treit S, Lebel C, Baugh L, Rasmussen C, Andrew G, Beaulieu C. Longitudinal MRI reveals altered trajectory of brain development during childhood and adolescence in fetal alcohol spectrum disorders. J Neurosci 2013;33(24):10098–109.
[222] Schuurmans C, Kurrasch DM. Neurodevelopmental consequences of maternal distress: what do we really know? Clin Genet 2013;83(2):108–17.
[223] Christian LM. Psychoneuroimmunology in pregnancy: Immune pathways linking stress with maternal health, adverse birth outcomes, and fetal development. Neurosci Biobehav Rev 2012;36(1):350–61.
[224] Cohen MM, Jing DQ, Yang RR, Tottenham N, Lee FS, Casey BJ. Early-life stress has persistent effects on amygdala function and development in mice and humans. Proc Natl Acad Sci USA 2013;110(45):18274–8.
[225] Burghy CA, Stodola DE, Ruttle PL, et al. Developmental pathways to amygdala-prefrontal function and internalizing symptoms in adolescence. Nat Neurosci 2012;15(12):1736–41.
[226] Sarkar S, Craig MC, Dell'Acqua F, et al. Prenatal stress and limbic-prefrontal white matter microstructure in children aged 6-9 years: a preliminary diffusion tensor imaging study. World J Biol Psychia 2014;15(4):346–52.
[227] Philip NS, Sweet LH, Tyrka AR, et al. Early-life stress is associated with greater default network deactivation during working memory in healthy controls: a preliminary report. Brain Imaging Behav 2013;7(2):204–12.
[228] Kane HS, Dunkel Schetter C, Glynn LM, Hobel CJ, Sandman CA. Pregnancy anxiety and prenatal cortisol trajectories. Biol Psychol 2014;100:13–9.
[229] Qiu A, Rifkin-Graboi A, Chen H, et al. Maternal anxiety and infants' hippocampal development: timing matters. Transl Psychiatry 2013;3:e306.
[230] Barker ED, Jaffee SR, Uher R, Maughan B. The contribution of prenatal and postnatal maternal anxiety and depression to child maladjustment. Depress Anxiety 2011;28(8):696–702.
[231] Evans J, Melotti R, Heron J, et al. The timing of maternal depressive symptoms and child cognitive development: a longitudinal study. J Child Psychol Psychiatry Allied Discip 2012;53(6):632–40.

[232] Korhonen M, Luoma I, Salmelin R, Tamminen T. A longitudinal study of maternal prenatal, postnatal and concurrent depressive symptoms and adolescent well-being. J Affect Disord 2012;136(3):680–92.
[233] Rifkin-Graboi A, Bai J, Chen H, et al. Prenatal maternal depression associates with microstructure of right amygdala in neonates at birth. Biol Psychiatry 2013;74(11):837–44.
[234] Sandman CA, Buss C, Head K, Davis EP. Fetal exposure to maternal depressive symptoms is associated with cortical thickness in late childhood. Biol Psychiatry 2015;77(4):324–34.
[235] Yim IS, Glynn LM, Dunkel-Schetter C, Hobel CJ, Chicz-DeMet A, Sandman CA. Risk of postpartum depressive symptoms with elevated corticotropin-releasing hormone in human pregnancy. Arch Gen Psychiatry 2009;66(2):162–9.
[236] Fallucca E, MacMaster FP, Haddad J, et al. Distinguishing between major depressive disorder and obsessive-compulsive disorder in children by measuring regional cortical thickness. Arch Gen Psychiatry 2011;68(5):527–33.
[237] Peterson BS, Weissman MM. A brain-based endophenotype for major depressive disorder. Annu Rev Med 2011;62:461–74.
[238] Tu PC, Chen LF, Hsieh JC, Bai YM, Li CT, Su TP. Regional cortical thinning in patients with major depressive disorder: a surface-based morphometry study. Psychiatry Res 2012;202(3):206–13.
[239] Peterson BS, Warner V, Bansal R, et al. Cortical thinning in persons at increased familial risk for major depression. Proc Natl Acad Sci USA 2009;106(15):6273–8.
[240] Lange C, Irle E. Enlarged amygdala volume and reduced hippocampal volume in young women with major depression. Psychol Med 2004;34(6):1059–64.
[241] Lupien SJ, Parent S, Evans AC, et al. Larger amygdala but no change in hippocampal volume in 10-year-old children exposed to maternal depressive symptomatology since birth. Proc Natl Acad Sci USA 2011;108(34):14324–9.
[242] Mehta MA, Golembo NI, Nosarti C, et al. Amygdala, hippocampal and corpus callosum size following severe early institutional deprivation: the English and Romanian Adoptees study pilot. J Child Psychol Psychiatry Allied Discip 2009;50(8):943–51.
[243] Tottenham N, Hare TA, Quinn BT, et al. Prolonged institutional rearing is associated with atypically large amygdala volume and difficulties in emotion regulation. Dev Sci 2010;13(1):46–61.
[244] Herman JP, Cullinan WE. Neurocircuitry of stress: central control of the hypothalamo-pituitary-adrenocortical axis. Trends Neurosci 1997;20(2):78–84.
[245] Joels M, Baram TZ. The neuro-symphony of stress. Nat Rev Neurosci 2009;10(6):459–66.
[246] Davis EP, Sandman CA, Buss C, Wing DA, Head K. Fetal glucocorticoid exposure is associated with preadolescent brain development. Biol Psychiatry 2013;74(9):647–55.
[247] Boes AD, McCormick LM, Coryell WH, Nopoulos P. Rostral anterior cingulate cortex volume correlates with depressed mood in normal healthy children. Biol Psychiatry 2008;63(4):391–7.
[248] Cooperstock M, Campbell J. Excess males in preterm birth: interactions with gestational age, race, and multiple birth. Obstetrics Gynecol 1996;88(2):189–93.
[249] Mathews TJ, Hamilton BE. Trend analysis of the sex ratio at birth in the United States. 2005;53(20):1–17.
[250] Hunt PA, Hassold TJ. Sex matters in meiosis. Science 2002;296(5576):2181–3.
[251] Clifton VL. Review: sex and the human placenta: mediating differential strategies of fetal growth and survival. Placenta 2010;31(Suppl.):S33–9.
[252] Peacock JL, Marston L, Marlow N, Calvert SA, Greenough A. Neonatal and infant outcome in boys and girls born very prematurely. Pediatr Res 2012;71(3):305–10.
[253] Wells JC. Natural selection and sex differences in morbidity and mortality in early-life. J Theor Biol 2000;202(1):65–76.
[254] Buss C, Davis EP, Hobel CJ, Sandman CA. Maternal pregnancy-specific anxiety is associated with child executive function at 6-9 years age. Stress 2011;14(6):665–76.
[255] Lyons DM, Parker KJ. Stress inoculation-induced indications of resilience in monkeys. J Trauma Stress 2007;20(4):423–33.
[256] Ellis BJ, Boyce WT, Belsky J, Bakermans-Kranenburg MJ, van Ijzendoorn MH. Differential susceptibility to the environment: an evolutionary–neurodevelopmental theory. Dev Psychopathol 2011;23(1):7–28.
[257] Gluckman PD, Hanson MA, Spencer HG. Predictive adaptive responses and human evolution. Trends Ecol Evol 2005;20(10):527–33.
[258] Gluckman PD, Hanson MA. Developmental origins of disease paradigm: a mechanistic and evolutionary perspective. Pediatr Res 2004;56(3):311–7.
[259] Kubota Y, Karube F, Nomura M, et al. Conserved properties of dendritic trees in four cortical interneuron subtypes. Sci Rep 2011;1:89.
[260] Hickmott PW, Ethell IM. Dendritic plasticity in the adult neocortex. Neurosci 2006;12(1):16–28.
[261] Rosario M, Schuster S, Juttner R, Parthasarathy S, Tarabykin V, Birchmeier W. Neocortical dendritic complexity is controlled during development by NOMA-GAP-dependent inhibition of Cdc42 and activation of cofilin. Genes Dev 2012;26(15):1743–57.
[262] Maynard KR, Stein E. DSCAM contributes to dendrite arborization and spine formation in the developing cerebral cortex. J Neurosci 2012;32(47):16637–50.

[263] Murmu MS, Salomon S, Biala Y, Weinstock M, Braun K, Bock J. Changes of spine density and dendritic complexity in the prefrontal cortex in offspring of mothers exposed to stress during pregnancy. Eur J Neurosci 2006;24(5):1477–87.

[264] Lupien SJ, McEwen BS, Gunnar MR, Heim C. Effects of stress throughout the lifespan on the brain, behaviour and cognition. Nat Rev Neurosci 2009;10(6):434–45.

[265] Bock J, Rether K, Groger N, Xie L, Braun K. Perinatal programming of emotional brain circuits: an integrative view from systems to molecules. Front Neurosci 2014;8:11.

[266] Weinstock M. Sex-dependent changes induced by prenatal stress in cortical and hippocampal morphology and behaviour in rats: an update. Stress 2011;14(6):604–13.

[267] Maras PM, Baram TZ. Sculpting the hippocampus from within: stress, spines, and CRH. Trends Neurosci 2012;35(5):315–24.

[268] Chen YC, Rex CS, Rice CJ, et al. Correlated memory defects and hippocampal dendritic spine loss after acute stress involve corticotropin-releasing hormone signaling. Proc Natl Acad Sci USA 2010;107(29):13123–8.

[269] Avishai-Eliner S, Brunson KL, Sandman CA, Baram TZ. Stressed-out, or in (utero)? Trends Neurosci 2002;25(10):518–24.

CHAPTER

14

Metabolic Disorders and Developmental Origins of Health and Disease

Sara E. Pinney

Department of Pediatrics, Perelman School of Medicine at the University of Pennsylvania, Division of Endocrinology and Diabetes, The Children's Hospital of Philadelphia, Philadelphia, PA, USA

OUTLINE

The Epigenome and Developmental Origins of Health and Disease
http://dx.doi.org/10.1016/B978-0-12-801383-0.00014-1

Copyright © 2016 Elsevier Inc. All rights reserved.

INTRODUCTION

Type 2 diabetes and obesity are metabolic disorders of complex genetics involving interactions between susceptible genetic loci and environmental stimuli. In recent decades the incidence of type 2 diabetes and obesity has reached epidemic proportions creating a serious public health crisis as these conditions are linked with the leading causes of death in the United States (US) and worldwide including heart disease, stroke, and some forms of cancer. From 2012 to 2014, the Centers for Disease Control and Prevention reported that 29.1 million people in the US have diabetes (9.3% of the population) and over one-third of the population is obese [1,2]. The cost of caring for patients with diabetes and obesity in the US reached $390 billion annually [1]. Genome-wide association studies have identified more than 70 single nucleotide polymorphisms (SNPs) associated with type 2 diabetes, but none have resulted in the identification of specific disease-causing mutations [3,4]. These results support the idea that type 2 diabetes and obesity are complex disorders influenced by multiple susceptible genetic loci and a variety of environmental perturbations.

An altered intrauterine milieu is one such environmental insult that can contribute to the development of diabetes and obesity later in life, a relationship initially described by David Barker and Nick Hales [5]. This chapter will focus on altered intrauterine environments that are characterized by intrauterine growth restriction (IUGR), exposure to maternal diabetes, and/or exposure to maternal obesity. These conditions can lead to the development of diabetes and obesity in the offspring by permanently altering gene expression. Epigenetic modifications are defined as mitotically heritable alterations in gene function that are not related to changes in DNA sequence [6]. Environmental perturbations that alter the intrauterine milieu can affect fetal development by modifying gene function of both pluripotent cells that are rapidly replicating and by affecting terminally differentiated cells that are replicating more slowly. During fetal development, the rapidly replicating pancreatic beta cells, hepatocytes, myocytes, and adipocytes are particularly susceptible to exposure to a perturbed intrauterine environment, and their altered function contributes to the development of adult-onset metabolic disorders. We will describe examples where exposure to conditions such as IUGR, maternal diabetes, or maternal obesity affect the function of many different cell types through both site-specific and genome-wide epigenetic modifications, ultimately leading to the development of diabetes and obesity in the offspring.

INTRAUTERINE GROWTH RESTRICTION

The abnormal intrauterine milieu associated with IUGR limits the supply of critical substrates and hormones to the fetus and affects its development by permanently modifying gene expression and function of susceptible cells. Permanent changes in the phenotype of the offspring suggest that IUGR is associated with stable changes in gene expression potentially as a result of epigenetic modifications. Here we will discuss three animal models of IUGR including their effects on metabolic gene expression and associated epigenetic modifications that contribute to functional changes of these genes. We will highlight recent evidence that epigenetic mechanisms contribute directly to the malprogramming of gene expression during the critical periods in fetal and neonatal development, and these time points may represent a critical window that could be utilized to prevent the development of metabolic disease in the future.

IUGR Induced by Maternal Dietary Protein Restriction

Initially established by Snoeck et al., rats born to protein-restricted dams have lower birth weights and develop age-dependent glucose intolerance that progresses to overt diabetes [7]. In this model, dams are fed a diet containing 8% protein throughout gestation and lactation (LP=low protein), compared to offspring of a control dam fed an isocaloric diet containing 20% protein. There are no effects on conception rates or litter size, but placental and offspring birth weights are consistently reduced in this model. Pups of mothers on the LP diet have 5.5% lower birth weights than control [7,8]. Neonates of LP dams have impaired pancreatic development and islet vascularization. Beta cell mass, islet size, and insulin content are reduced due to reduced beta cell proliferation and increased apoptosis.

In general, male offspring have altered insulin secretory capacity and reduced beta cell mass through a reduction in beta cell proliferation rates and an increase in apoptosis [7–10]. The 15-month-old male rat offspring of mothers fed an LP diet have hyperinsulinemia [11]. At 17 months of age, male offspring have fasting hyperglycemia and postprandial hyperinsulinemia. Female offspring at 21 months of age have elevated fasting insulin levels compared to controls but no fasting hyperglycemia [11]. Glucose tolerance measured with intravenous glucose tolerance testing (IVGTT) is comparable in the female offspring and controls, but plasma insulin area under the curve is 1.9 times higher in the female offspring of LP-fed dams [11].

The LP model has been associated with the reduction of both pancreatic duodenal homeobox-1 (Pdx-1) and insulin-like growth factor-II (IGF-II) in the pancreatic islet [12,13]. Pdx-1 is a homeodomain-containing transcription factor that regulates early development of both endocrine and exocrine pancreas and later the differentiation into beta cells [7,14]. Targeted homozygous disruption of *Pdx-1* in mice results in pancreatic agenesis, and homozygous mutations yield a similar phenotype in humans [15]. The *Pdx-1*(+/−) mouse has normal beta cell mass but a defect in glucose-stimulated insulin secretion [15]. There is decreased expression of insulin-like growth factor-II (IGF-II) in fetal and neonatal islets of pups born to pregnant rats on the LP diet [12,13]. IGF-II is thought to act as a beta cell mitogen and prevents apoptosis [12]. Taurine supplementation of the LP diet given to the pregnant rats partially prevents beta cell mass reduction by restoring IGF-II expression and thus normalizes the proliferation and apoptosis rates [12].

Maternal protein restriction has been shown to have profound long-term effects on the liver of the offspring. The ability of glucagon to stimulate hepatic glucose output is impaired in ex vivo experiments at 3 months of age and is related to a reduction in the expression of glucagon receptors [16] and associated with an increase in the activity of the gluconeogenic enzyme phosphoenolpyruvate carboxykinase (PEPCK) and reduced levels of glycolytic glucokinase. These changes persist at 11 months of age despite the fact that the pups are fed standard diet [17]. Ex vivo livers from three-month-old male offspring of LP-fed pregnant rats displayed evidence of hepatic insulin resistance characterized by an increase in expression of insulin receptors [16].

In the 15-month-old male offspring of LP-fed dams, insulin-stimulated glucose uptake is reduced in both muscle and adipose tissue [18,19]. Maternal protein restriction in the LP rat model results in reduced skeletal muscle mass, but young adult LP offspring have improved insulin sensitivity at the level of the skeletal muscle [17].

Decreased expression of specific insulin-signaling proteins in the skeletal muscle in both low birth weight (LBW) rats born to LP dams and humans born with LBW was shown by Ozanne et al. [18,19]. Low birth weight human males and male rat offspring of dams fed a low protein diet had decreased expression of protein kinase c zeta (PKCζ), the phosphoinositide-3-kinase (PI3K) regulatory subunit p85 alpha, and glucose transporter, type 4 (GLUT4) [18]. The similarity of the protein expression profile of the men born with LBW and the in utero LP diet exposed rats suggest that the LP model of IUGR is an appropriate model to study the development of peripheral insulin signaling defects that may lead to insulin resistance in IUGR [18].

Exposure to the LP diet in utero leads to changes in DNA methylation and histone modifications affecting expression of hepatocyte nuclear factor 4, alpha (Hnf4α) in the beta cell [20]. In the LP model of IUGR, Sandovici et al. found that the pancreatic transcription factor Hnf4α is epigenetically regulated by maternal diet and aging in rat islets from offspring of protein-restricted dams [20]. Hnf4α has been linked to both monogenic diabetes and type 2 diabetes. In the normal rat islets, transcriptional activity of *Hnf4α* is controlled by the distal P2 promoter and an additional downstream enhancer region. At baseline, the P2 promoter region of *Hnf4α* in rat islets is associated with minimal DNA methylation as well as enrichment of histone activating marks such as histone 3 acetylation, and histone 3 lysine 4 trimethylation [20]. Through *Hnf4α* promoter-specific quantitative PCR experiments, it was revealed that the decreases in DNA methylation and the enrichment of activating histone modifications corresponded to increased P2 transcript [20]. Using two different established insulin-secreting cell lines INS-1 (which expresses the P2 promoter of *Hnf4α*) and BRIN (which does not express the P2 promoter of *Hnf4α*), the authors demonstrated that changes in DNA methylation and histone modifications could be responsible for regulating *Hnf4α* transcription by inducing demethylation with 5-Aza-C and histone acetylation with trichostatin A [20]. Finally, Sandovici et al. showed that islets isolated from the LP-exposed offspring had significantly decreased expression of the P2 promoter at 3 months of age, which correlated with a modest but statistically significant increase in DNA methylation at the P2 promoter of *Hnf4α* at both 3 and 15 months of age (before the offspring have developed diabetes) [20]. In addition, the offspring of dams exposed to the LP diet had reductions of activating histone modifications and increases in silencing histone modifications at the *Hnf4α* enhancer in three-month-old islets [20]. There were no diet-associated changes in histone modifications at the P1 promoter of *Hnf4α* at either 3 or 15 months of age [20]. The authors conclude that exposure to a suboptimal nutritional environment during early development leads to epigenetic silencing at the enhancer region of *Hnf4α*, which weakens the P2 promoter–enhancer interaction and results in a permanent reduction in *Hnf4α* expression [20].

Uteroplacental Insufficiency Model of IUGR

Uteroplacental insufficiency (UPI), caused by disorders such as preeclampsia, maternal smoking, and abnormalities of uteroplacental development, is one of the most common causes of fetal growth retardation worldwide [21]. The resultant abnormal intrauterine milieu restricts the supply of crucial nutrients to the fetus, thereby limiting fetal growth. In one model of UPI, pregnant Sprague Dawley dams undergo bilateral uterine artery ligation surgery on gestational day 18 (term is day 22) and deliver spontaneously. UPI fetal rats have critical features of a metabolic profile characteristic of growth-retarded human fetuses: decreased levels of glucose, insulin, IGF-1, amino acids, and oxygen [22,23]. Birth weights of UPI offspring are significantly lower than those of controls until approximately 7 weeks of age, when weights of UPI offspring catch up. Between 7 and 10 weeks of age, the growth rate of UPI offspring accelerates and UPI weights surpass those of controls, and by 26 weeks UPI rats are obese. Fetal insulin and leptin levels are significantly decreased compared with controls. There are no significant differences in blood glucose and insulin levels at 1 week of age, but by 2 weeks of age there were significant increases in circulating insulin and leptin levels [24]. By 7–10 weeks of age, IUGR rats develop mild fasting hyperglycemia and hyperinsulinemia. UPI male offspring were glucose intolerant and insulin resistant by 15 weeks of age. First-phase insulin secretion in response to glucose is impaired by 7 weeks of age in UPI males before the onset of hyperglycemia. There are no significant differences in beta cell mass, islet size, or pancreatic mass between 1 and 7 weeks of age. In 15-week-old male offspring from UPI affected dams, relative beta cell mass is 50% of controls, and by 26 weeks of age, beta cell mass is 33% of controls. The loss of beta cell mass is accompanied by a reduction in *Pdx-1* expression [25]. By 6 months of age, UPI offspring develop diabetes, which is similar to type 2 diabetes seen in humans with progressive dysfunction of insulin secretion and insulin action.

Multiple studies have shown that IUGR is associated with increased oxidative stress in the human fetus [25,26]. A key adaption enabling the fetus to survive in a limited energy environment may be the reprogramming of mitochondrial function, but this can have deleterious effects in cells with high energy requirements. Reactive oxygen species production and oxidative stress gradually increase in islets from offspring exposed to UPI [27]. ATP production is impaired and activities of complex I and II of the electron transport chain progressively decline in UPI islets [27]. There is decreased expression of mitochondrial genes in UPI islets, and mitochondrial dysfunction results in impaired insulin secretion [27].

Studies with the UPI model suggest that the aberrant intrauterine milieu permanently impairs insulin signaling in the liver resulting in an augmentation of gluconeogenesis [28]. At age 7–9 weeks, male offspring exposed to UPI have increased hepatic glucose production (HGP) at baseline and insulin suppression of HGP is impaired. Insulin receptor substrate 2 (IRS2) and v-akt murine thymoma viral oncogene homolog 2 (Akt-2) phosphorylation are significantly blunted in the UPI offspring, and PEPCK and glucose-6-phosphate (G-6-P) mRNA levels are increased threefold [28]. These processes occur early in the UPI offspring's adult life, before the onset of hyperglycemia, indicating that UPI causes a primary defect in hepatic metabolism that contributes to the eventual development of overt hyperglycemia along with the beta cell defects described above.

UPI leads to a permanent decrease in the expression of *Pdx-1* in the beta cell that is associated with silencing histone modifications at the proximal promoter of *Pdx-1* in the fetal and early postnatal period followed by an increase in DNA methylation in the same region by 6–10 months of age [29,30]. As early as 24 h after uterine artery ligation, *Pdx-1* mRNA levels are reduced by more than 50% in UPI fetal rats. Suppression of *Pdx-1* expression persists after birth and progressively declines in the UPI-exposed offspring, implicating an epigenetic mechanism. A change in histone acetylation is the first epigenetic modification found in beta cells of UPI rats. Islets isolated from UPI fetuses show a significant decrease in H3 and H4 acetylation and with a loss of binding of upstream transcription factor 1 (USF-1) to the proximal promoter of *Pdx-1* [29]. USF-1 is a critical activator of *Pdx-1* transcription and its decreased binding markedly decreases *Pdx-1* transcription [31]. After birth, histone deacetylation progresses and is followed by a marked decrease in H3K4 trimethylation (H4K4me3) and a significant increase in dimethylation of H3K9 (H3K4me2) in UPI islets [29]. H3K4me3 is usually associated with active gene transcription, whereas H3K9me2 is usually a repressive chromatin mark. Progression of these histone modifications parallels the progressive decrease in *Pdx-1* expression that manifests as defective glucose homeostasis and increased oxidative stress in aging UPI animals [29]. In offspring exposed to UPI, *Pdx-1* is first silenced as a result of recruitment of corepressors, including histone deacetylase 1 (HDAC1) and mSin3A [29]. Binding of these deacetylases facilitates loss of H3K4me3, further repressing *Pdx-1* expression [29]. In fact, inhibiting HDAC activity by trichostatin A normalizes H3K4me3 levels on *Pdx-1* in IUGR islets [29]. These data suggest that the association of HDAC1 with *Pdx-1* in IUGR islets probably serves as a platform for recruiting a demethylase to catalyze H3K4me2. In the UPI model, IUGR induces a self-propagating epigenetic cycle in which the mSin3A/HDAC complex is first recruited to the *Pdx-1* promoter; histone tails are subjected to deacetylation and *Pdx-1* transcription is repressed.

At the neonatal stage, this epigenetic process is reversible through treatment with the diabetes drug Exendin-4, a long-acting glucagon-like peptide 1 (GLP-1) agonist, by increasing histone acetylase activity. The preventative treatment with Exendin-4 during the first 6 days of life might define an important developmental window for therapeutic approaches to developmental programming [30]; however, as H3K9me2 accumulates, dnmt3a is recruited to the promoter and initiates de novo DNA methylation, which locks in the silenced state of *Pdx-1* in the UPI adult pancreas, leading to permanently suppressed expression of *Pdx-1* [30] (Figure 1).

Given the role that epigenetic modifications play in regulating the expression of *Pdx-1* in UPI islets, Thompson et al. examined the role of DNA methylation in regulating gene expression in islets exposed to UPI on a genome-wide scale [32]. Cytosine methylation was altered in 1400 loci in islets from male UPI rats at 7 weeks of age, preceding the development of diabetes. CpG hypermethylation of an intergenic region of GTP cyclohydrolase 1 (*CGH-1*) was associated with a threefold decrease in mRNA expression while CpG hypomethylation at an intergenic region of fibroblast growth factor receptor 1 (*FGFR1*) was associated with increased mRNA expression. CpG hypermethylation was also noted at the transcriptional start site of proprotein convertase subtilisin/ketin type 5 (*PCSK5*) and was associated with significantly decreased mRNA levels. The changes in methylation and concordant changes in mRNA levels were found to affect genes regulating processes known to be altered in IUGR including beta cell proliferation, vascularization, insulin secretion, and cell death [32].

IUGR Induced by Maternal Total Calorie Restriction

The total calorie restriction model of IUGR approximates a generalized poor nutritional state during pregnancy and the effects on the offspring. In this model of IUGR, rats have food intake limited to 50% of ad lib during pregnancy and lactation, which results in offspring with significantly lower birth weights who continue to be small even in adulthood [33]. There are no differences in litter size or in the male-to-female ratio between control and offspring exposed to dams with total calorie restriction. When calorie restriction is limited to 50% during gestation only, the offspring are born with reduced birth weights, but their body weights increase and are greater than control offspring by postnatal day 20 [33]. When the time period during which the maternal nutritional deprivation (50% restriction) is limited to the last week to 10 days of gestation, the pups are born small; some models demonstrate impaired beta cell development, and reduced plasma glucose and insulin concentrations during the neonatal period [34], while other models show no statistical difference in pancreas weight, beta cell fractional area, beta cell mass, beta cell replication, or apoptosis in the neonate [35]. In the neonate, *Pdx-1* expression is not reduced in islets of pups born under conditions of maternal total calorie restriction [35]. At 21 days, the male pups have a 15% reduction in body weight and a 30% reduction in pancreatic weight, while there is no difference in female pups at this age [35]. There is a 50% reduction in absolute beta cell mass and a 30% reduction in beta cell replication but no change in rates of apoptosis in offspring at 21 days. Total calorie-restricted offspring have an 80% reduction *Pdx-1* ductal immunoreactivity compared to control at 21 days [35]. At 3–4 months, the offspring continue to have reduced beta cell mass along with altered insulin response and sensitivity at 3–4 months of age [34,36]. At 8 months of age, the reduction of beta cell mass in this model is still apparent and the beta cells had a 40% lower insulin content compared to controls [34]. The offspring are not overtly hyperglycemic but do have decreased fasting insulin levels [34].

In the total calorie restriction model of IUGR, epigenetic marks associated with gene silencing

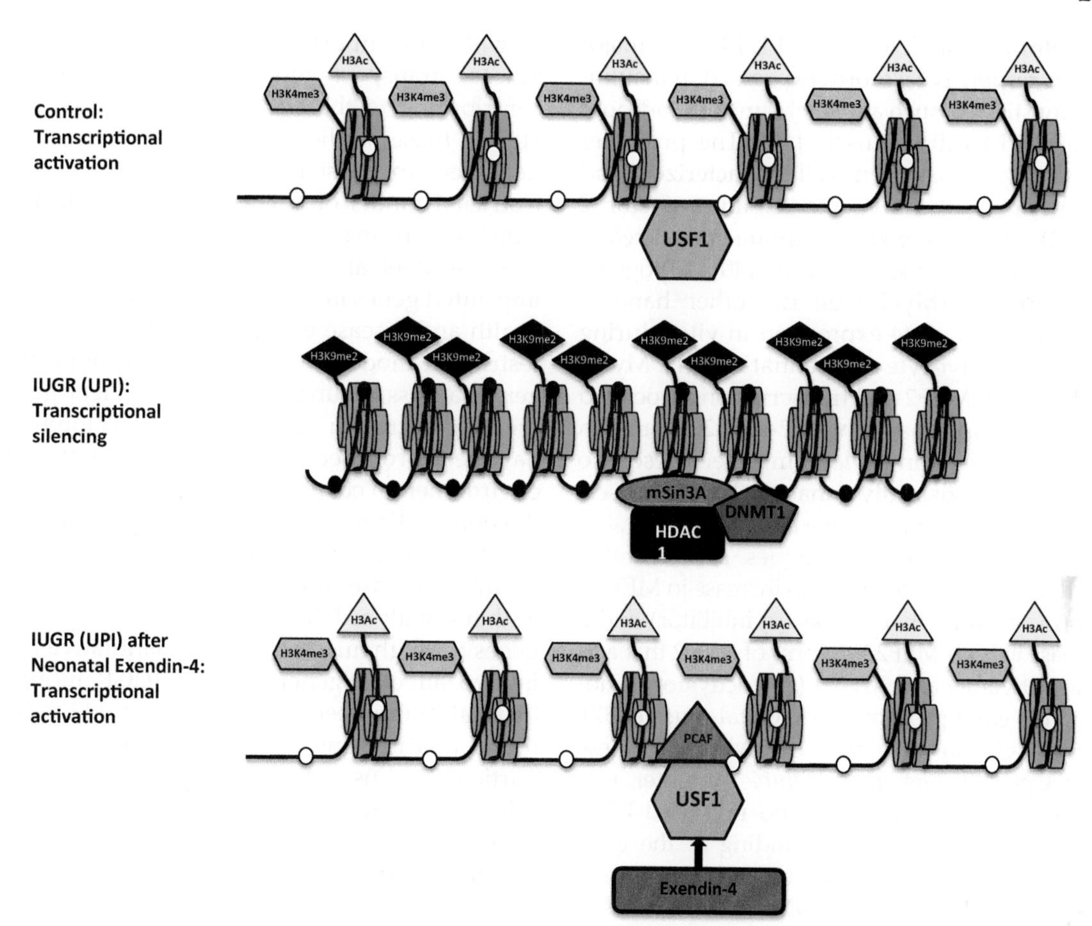

FIGURE 1 Summary of epigenetic changes at the *Pdx-1* promoter in UPI islets [29,30]. In control islets, the *Pdx1* promoter is normally found in an unmethylated (open circles) open chromatin state allowing access to transcription factors like USF-1 (green hexagon) and associated with nucleosomes characterized by acetylated histone H3 (yellow triangles) and with the H3K4me3 (lavender hexagons). In islets collected from offspring exposed to UPI, histone acetylation is lost through association with an mSin3A-HDAC1-DNMT1 repressor complex. The activating histone modification, H3K4me3, disappears while the silencing histone modification, H3K9me2 (black diamonds), accumulates. The chromatin is more compressed and extensive DNA methylation (black circles) locks in the transcriptionally silent state of *Pdx-1* in the UPI islets. A short course of treatment with Exendin-4 during the newborn period leads to increased USF-1 binding at the proximal *Pdx-1* promoter leading to increased histone acetyl transferase activity and increased PCAF binding (orange triangle). Binding of these factors results in increased histone H3 acetylation and increased methylation of H3K4 in addition to the prevention of DNA methylation. The epigenetic modifications result in increased transcription of *Pdx-1* and ultimately prevention of the development of diabetes in the IUGR model.

have been described at the promoter of the glucose transporter 4 (*glut4*) in skeletal muscle [37]. A reduction in glucose transport in muscle is a central basis for insulin resistance in the IUGR offspring [18,38]. Glucose transport, a rate-limiting step in glucose utilization under normal physiological circumstances, occurs by facilitated diffusion [39]. This process is mediated by a family of structurally related membrane-spanning glycoproteins, termed the facilitative glucose transporters (GLUT; Slc2 family of transport proteins; reviewed in Ref. [40]). Of the isoforms cloned to date, GLUT4 is the major insulin-responsive isoform expressed in insulin-sensitive tissues such as skeletal muscle, adipose tissue, and cardiac muscle [40]. The promoter region of *glut4* has been well characterized, and disruption of the myocyte enhancer factor 2 (MEF2)-binding site ablates tissue-specific *glut4* expression in transgenic mice [40]. Myogenic differentiation (MyoD) on the other hand is responsible for *glut4* expression in vitro during myoblast to myocyte differentiation [41]. MyoD binding with MEF2 and transcription associated protein 1 (Tra1) spans the 502- to 420-bl region of the *glut4* gene in skeletal muscle. These two proteins synergistically enhance skeletal muscle *glut4* transcription and gene expression [41].

Raychaudhuri and colleagues demonstrated that IUGR is associated with an increase in MEF2D (a form of MEF2 that acts as an inhibitor) and a decrease in both MEF2A (a form of MEF2 that acts as an activator) and MyoD (a coactivator) binding to the *glut4* promoter in skeletal muscle [37]. Interestingly, no differential methylation of these three CpG clusters in the *glut4* promoter was found. Furthermore, they found increased DNA methyltransferase (DNMT) binding at the *glut4* gene at different ages: DNA methyltransferase 1 (DNMT1) postnatally, and DNMT3a and DNMT3b in adults [37]. The increase in DNMT binding was associated with exposure to increased methyl CpG-binding protein 2 (MeCP2) concentrations. Covalent modifications of the histone code consisted of histone 3 lysine 14 (H3K14) deacetylation mediated by recruitment of HDAC1 and enhanced association of histone deacetylase 4 (HDAC4) enzymes. This set the stage for Suv39H1 methylase-mediated dimethylation of H3K9 and increased recruitment of heterochromatin protein 1, which partially inactivates postnatal and adult IUGR *glut4* gene transcription. These studies demonstrate that perinatal nutrient restriction resulting in IUGR leads to silencing histone modifications in skeletal muscle, which in turn directly decrease *glut4* gene expression. These events effectively create a metabolic knockdown of *glut4*, an important regulator of peripheral glucose transport and insulin resistance and contributing to the adult T2D phenotype [37]. Hence, these studies show that histone modifications can be stably inherited in a calorie-restricted model of IUGR, mimicking the 1944 Dutch famine experience (Table 1).

Radford et al. sought to assess the role of imprinted genes in the developmental origins of health and disease using the IUGR total calorie restriction model in the mouse [42]. Imprinted genes, a class of functionally mono-allelic genes critical for growth and metabolic development, have been proposed to be uniquely susceptible to environmental change and that perturbations of the epigenetic regulation of the imprinting control regions may play a role in the development of adult disease after an early life insult. The authors analyzed the expression of imprinted genes through microarray and quantitative PCR in two affected generations of IUGR induced by total calorie restriction of the dam [42]. They found that imprinted genes as a class were not particularly susceptible to expression changes following in utero under nutrition. Imprinted genes in the developing germ line that were not affected were largely stable in the second generation [42].

Brøns et al. studied men born with LBW who were exposed to 5 days of a high fat diet to investigate whether the presumed exposure to an altered intrauterine milieu (based on their LBW) programmed the ability to adapt to the effect of a high fat diet later in life [43]. The authors showed that DNA methylation and gene expression of peroxisome proliferator-activated receptor gamma, coactivator 1 alpha (*PPARGC1A*) in human muscle are influenced by both exposure to a high fat diet and birth weight [43]. They studied 20 healthy young men with a history of LBW and 26 matched control men with normal birth weights after 5 days of high fat, high calorie diet (50% extra calories and 60% fat). After a high fat overfeeding diet, LBW men had peripheral insulin resistance and

TABLE 1 Rodent Models of IUGR: Effects on Beta Cell Phenotype, Glucose Homeostasis, and Epigenetic Modifications

	Low Protein	Total Calorie Restriction	Uteroplacental Insufficiency
Metabolic changes: neonate	• Decreased beta cell mass, proliferation, islet size and insulin content of beta cells • Increased beta cell apoptosis • Decreased expression *Pdx-1* and *IGF-II*	• Mixed reports: some models with impaired beta cell development, low plasma glucose and insulin; others with no change in pancreas weight, beta cell mass, beta cell replication or apoptosis • *Pdx-1* expression is not decreased	• No difference in beta cell mass, islet size, pancreatic weight • Decreased plasma glucose and insulin concentrations • First-phase insulin secretion and insulin sensitivity impaired • 14 days: reduced beta cell proliferation; *Pdx-1* expression decreased by 60%
Metabolic changes: adult	• 15-month males have hyperinsulinemia • 17-month males have fasting hyperglycemia	• 3–4 months: decreased beta cell mass; decreased insulin sensitivity • 8 months: decreased beta cell mass; 40% decreased insulin content; elevated fasting insulin but no overt hyperglycemia	• 7–10 weeks: mild fasting hyperglycemia and hyperinsulinemia • 15 weeks: beta cell mass reduced 50% • 26 weeks: fasting insulin levels decline and first phase insulin secretion absent
Epigenetic modifications affecting gene expression	• *Hnf4α*: pancreatic transcription factor important for beta cell development and insulin secretion • Epigenetic change: increased DNA methylation at P2 promoter	• *Glut4*: insulin responsive glucose transporter in skeletal muscle • Epigenetic change: decreased H3K14 acetylation; increased H3K9 methylation	• *Pdx-1*: pancreatic transcription factor critical for beta cell development and function • Decreased H3 and H4 acetylation; decreased H3K4me3 methylation; increased H3K9 methylation; increased DNA methylation at proximal promoter

reduced *PPARGC1A* and co-regulated oxidative phosphorylation (OXPHOS) gene expression in muscle. *PPARGC1A* promoter methylation was significantly higher in LBW subjects during the control diet. After the 5 days of high fat feeding, the control subjects saw significant increase in *PPARGC1A* promoter methylation compared to baseline, but the LBW subjects did not show any difference [43]. The increase in *PPARGC1A* promoter methylation after high fat feeding in the control men was reversible and did not correlate with mRNA expression. Although this study was the first to provide experimental support in humans that DNA methylation induced by overfeeding is reversible, the extent to which the persistent increased *PPARGC1A* promoter methylation in LBW subjects contributes to the decreases in mRNA expression of *PPARGC1A* and *OXPHOS* genes still needs to be defined.

GESTATIONAL DIABETES IN PREGNANCY AND METABOLIC PROGRAMMING OF THE OFFSPRING

Diabetic pregnancy induces marked abnormalities in glucose homeostasis and insulin secretion in the fetus that result in abnormal fetal growth [44]. Population-based studies demonstrate that the offspring of diabetic mothers have an increased risk for obesity, glucose intolerance, and type 2 diabetes [45–48]. Individuals whose cord blood or amniotic fluid insulin levels were elevated have a three- to fourfold risk for developing glucose intolerance, obesity, and type 2 diabetes in late childhood [48]. It has been proposed that early exposure to elevated insulin levels may lead to a malprogramming of critical functions related to the development of diabetes and obesity later in life [49].

Exposure to an adverse intrauterine milieu, including the environment of hyperglycemia and hyperinsulinemia that characterized the intrauterine environment of women with diabetes in pregnancy, may disturb epigenetic, structural, and functional adaptive responses responsible for developmental programming. In this section, we will highlight recent publications exploring the role of gestational diabetes (GDM) in the programming of disease in the offspring. We will also review literature aiming to identify mechanisms responsible for the "programming effect" that result from exposure to diabetes mellitus in utero.

Population-Based Studies Examining the Effect of Diabetes in Pregnancy on the Offspring

Two recent studies examined the relationship between maternal glucose concentrations at 28 weeks gestation and measurements of early childhood obesity. Pettitt et al. evaluated the relationship of glycemic levels during pregnancy with anthropometry in offspring of diabetic and nondiabetic pregnant women from the United Kingdom (UK) as part of the multinational hyperglycemia and adverse pregnancy outcome (HAPO) study [50]. The authors found that there was a significant association with maternal 1-h blood sugars during gestational diabetes testing and offspring body mass index (BMI) Z score greater than the 85th percentile at 2 years of age, but other relationships between the offspring's measurements of obesity and maternal glucose levels were indistinguishable from control offspring. Deierlein et al. examined the relationship between maternal glucose concentrations among women without preexisting diabetes or GDM and BMI of their offspring at age 3 years [51]. In their adjusted model, a maternal glucose concentration ≥130 mg/dL was associated with approximately a twofold greater risk of childhood overweight/obesity (adjusted risk ratio of 2.34), after adjusting for maternal prepregnancy BMI. The authors conclude that fetal exposure to high maternal glucose concentrations in the absence of preexisting diabetes may contribute to the development of overweight/obesity in the offspring at 3 years of age.

Tsadok et al. examined the outcomes of obesity and hypertension in 17-year-old offspring born to mothers with GDM in pregnancy [52]. This study utilized the Jerusalem Perinatal Study birth cohort containing 92,408 birth records collected from 1964 to 1976. Follow-up data on offspring BMI and blood pressure were obtained from Israeli military records. The cohort contained 293 women with GDM and 59,499 control women without recorded GDM. After adjusting for birth weight, GDM remained significantly associated with offspring BMI and diastolic blood pressure at age 17. However, when a subcohort was adjusted with self-reported, prepregnancy BMI, the association between GDM and offspring BMI was attenuated. The authors conclude that maternal characteristics have long-term effects on cardiometabolic outcomes of their offspring at age 17 years. Some limitations of this study were the lack of routine GDM screening in Israel at the time of data collection and that the prevalence of GDM in the cohort (0.5%) was lower than the current prevalence of 3–5%. In addition, the authors did not have any data about the severity of the GDM, and the outcome data were only available for 65% of the offspring (38% of the female and 82% of the males).

Crume et al. examined the association between exposure to maternal diabetes in utero and BMI growth trajectories from birth through age 13 years [53]. The authors used a mixed linear effects model to assess difference in BMI and BMI growth velocity from birth through 13 years of age for 95 subjects exposed to diabetes in utero and 409 unexposed subjects enrolled in a retrospective cohort study. They found that the overall BMI growth trajectory (adjusted for sex and race/ethnicity) was not significantly different for exposed and unexposed subjects from birth through 26 months or age ($p = 0.48$). However, the overall BMI growth trajectory from 27 months through 13 years was significantly greater in the offspring exposed to diabetes in utero after adjustment for sex and race/ethnicity. This difference was primarily due to a significantly higher BMI growth velocity among exposed youth between 10 and 13 years, increasing by 4.56 kg/m^2 in the exposed group compared to an increase of 3.51 kg/m^2 in the unexposed group. A limitation of this study is that the authors did not include pubertal status and did not assess whether advanced pubertal stage may have accounted for the increased BMI velocity.

Brandt et al. aimed to identify parental and childhood factors that influence fasting plasma insulin concentrations in children [54]. In the Ulm Birth Cohort containing 422 patients enrolled at birth at the University of Ulm, Germany, from 2000 to 2001, the authors found that fasting plasma insulin levels in eight-year-old children significantly correlated with maternal prepregnancy BMI ($r = 0.16$) as well as maternal fasting insulin concentrations ($r = 0.26$) but not with paternal fasting insulin concentrations. Children with high plasma fasting insulin concentrations at age 8 years had an altered BMI trajectory in childhood calculated after age 1 year.

In summary, the epidemiological studies from 2010 to 2014 support previous studies finding an association between exposure to diabetes in pregnancy and adverse outcomes in the offspring, indicating that the in utero exposure effectively programs the individual to develop various diseases, including obesity and hypertension, later in life.

Oxidative Stress: A Possible Link Between Diabetes in Pregnancy and Chromatin Remodeling in the Offspring

The changes in gene expression in the diabetes-exposed embryo appear to be permanent and indicate that these changes could be induced by epigenetic modifications, including histone modifications and/or DNA methylation. Exposure to oxidative stress can directly mediate both DNA methylation and chromatin remodeling in multiple disease models, and this could be a mechanism by which aberrant epigenetic programming leads to type 2 diabetes and obesity in the offspring [55–59]. In addition to targeted DNA methylation changes in response to external stimuli, random DNA methylation changes also occur during aging in several tissue types and are associated with increased oxidative stress [55,59]. Such changes in DNA methylation patterns affect the expression of multiple genes [57]. Conditions of oxidative stress that can lead to the replacement of guanine with the oxidative radical 8-hydroxy guanine can profoundly affect the methylation of adjacent cytosines [57]. In addition, histones are also susceptible to conditions of oxidative stress due to their abundant lysine residues [60,61]. Members of the jumonji C family, including lysine (K) specific demethylase 1 (KDM1), lysine specific demethylase 3A (KDM3A), ubiquitously transcribed tetratricopeptide repeat, X chromosome (UTX) and lysine specific demethylase 6B (JMJD3), have recently been identified as histone demethylases. These compounds require oxygen as a cofactor and have specificity based on the lysine residue. KDM1, KDM3A, UTX, and JMJD3 directly link epigenetic modifications to oxygen gradients during development, a concept that is particularly relevant to the development of type 2 diabetes given the recent evidence that oxidative stress plays a significant role in beta cell deterioration [62–64].

Evidence of Epigenetic Programming of Gene Expression in Humans Exposed to GDM

Maternal hyperglycemia has been found to be associated with changes in placental DNA methylation of leptin (*lep*), adiponectin (*adipoq*), and mesoderm specific transcript (*MEST*). Bouchard et al. examined placental tissue and cord blood samples from 48 women at term including 23 women diagnosed with impaired glucose tolerance (IGT) at 24–28 weeks of age [65]. Mean placental *lep* gene DNA methylation levels did not differ between women with IGT and normal glucose tolerance (NGT) testing results, nor was there any difference in mean *lep* mRNA levels. Within the group of women with IGT during pregnancy, placental leptin gene DNA methylation levels correlated with glucose levels and with placental *lep* gene mRNA expression. In a second study, Bouchard et al. found that DNA methylation of adiponectin isolated from placenta correlated with maternal 2-h glucose levels from oral glucose tolerance test, but no measurements of *adipoq* mRNA expression isolated from placenta are reported. Decreased DNA methylation of adiponectin isolated from placenta was associated with higher maternal plasma adiponectin levels throughout pregnancy, and the authors indicated that DNA methylation of the *adipoq* gene has the potential to induce sustained glucose metabolism changes in the both the mother and offspring later in life [66]. Haghiac et al. found that maternal adipose tissue is the primary source of circulating adiponectin during pregnancy, and they were unable to detect *adipoq* RNA or protein in placenta. They report that maternal adipose tissue from obese women had significantly decreased *adipoq* mRNA associated with significant increases in DNA methylation compared to lean pregnant women. Haghiac et al. conclude that decreased adiponectin in obese pregnant women may have functional consequences in downregulating signaling through the adiponectin receptor in the placenta [67].

Another study, which aimed to examine the effect of GDM on imprinted genes in the placenta, found that DNA methylation levels of the maternally imprinted gene *MEST* were decreased 4–7% in GDM-exposed placenta and cord blood compared to control, but no correlations to *MEST* mRNA levels were reported. These results may indicate that alterations in maternal glucose metabolism affect genes involved in energy and glucose metabolism, fetal and placental growth, somatic differentiation, and neurodevelopmental and behavioral functions [68].

Ruchat et al. aimed to examine the DNA methylation profile differences in placenta and cord blood samples exposed to GDM ($n=30$) and control ($n=14$), with the hopes of identifying additional genes and their biochemical pathways regulated by exposure to an adverse intrauterine milieu [69]. The authors measured genome-wide changes in placental and cord blood DNA methylation with the Infinium Human 450K Methylation array and found the mean DNA methylation difference between GDM and control placenta to be $5.7 \pm 3.2\%$ and between cord blood samples $3.4 \pm 1.1\%$. Genes with the largest changes in DNA methylation were dipeptidyl-peptidase 6 (*DPP6*) ($\Delta\beta$ –value$=13.0\%$, $p=1.0\times10^{-6}$) in the placenta and stanniocalcin 2 (*STC2*) ($\Delta\beta$ –value$=5.6\%$, $p=7.4\times10^{-5}$). These changes in DNA methylation were not correlated with mRNA levels in placenta or cord blood. The authors identified 781 differentially methylated genes in the placenta, 758 differentially methylated in the cord blood using a $p<0.01$. Using Ingenuity Pathways Analysis, the top pathways emerging from the placental genes were cardiovascular diseases and metabolic diseases and from the cord blood genes were gastro-intestinal diseases, metabolic diseases, and endocrine system disorders.

Quilter et al. used the Infinium HumanMethylation 27K array to assess genome-wide DNA methylation in cord blood samples in infants of women with GDM, infants with prenatal growth restraint, and infants with normal prenatal growth [70]. The authors limited their results to a $\Delta\beta$ greater than $\pm10\%$ ($p<0.001$) and found 75 differentially methylated loci representing 72 genes comparing the different cohorts. The results of the Infinium HumanMethylation 27K were confirmed on Sequenom Massy Array and demonstrated appropriate correlation in

DNA methylation measurements. No correlation between gene-specific DNA methylation levels and mRNA expression was reported. The authors found that eight genes associated with differentially methylated CpGs were common to both exposure to GDM and growth restriction in cord blood of male offspring, including acylphosphatase 2, muscle type (*ACYP2*), associated with coronary artery disease, chromosome 3 open reading frame 31 (*C3orf31*), a mitochondrial translocator assembly and maintenance protein, and vestigial-like family member 4 (*VGLL4*), previously shown to be associated with insulin resistance. When cord blood samples were studied, four loci had significant changes in DNA methylation between the infants exposed to GDM (male and female) and controls, one of which was also identified in the analysis of growth-restricted male cord blood. These results support the hypothesis that common gene pathways are epigenetically altered by the in utero exposure to GDM and growth restriction, but no direct relationship to gene expression was established.

In the above studies describing women with diabetes in pregnancy and the effect on their offspring, it is difficult to understand the significance of the reported changes in DNA methylation in placenta and cord blood since frequently no corresponding changes in gene expression were reported. It is unclear if the gene expression studies are still under investigation or if the power of these studies was not sufficient to detect statistically significant changes in expression in the human cohorts.

OBESITY IN PREGNANCY AND METABOLIC PROGRAMMING OF THE OFFSPRING

Obesity is a growing threat worldwide, and its prevalence has risen dramatically over the last two decades with now more than one-third of the adult population of the US characterized as obese [2]. There are a number of critical periods during childhood that appear to influence the later development of obesity, including early infancy, five to 7 years of age (known as the adiposity rebound period), and puberty [71]. It is becoming increasingly evident that the prenatal stage also represents a window of susceptibility for early-life exposures (reviewed in Refs [72,73]). The period from conception to birth is a time of rapid growth, cellular replication and differentiation, and functional maturation of organ systems. These processes are very sensitive to nutritional alterations, and the abnormal intrauterine metabolic milieu associated with obesity in pregnancy can have long-lasting effects on the development of obesity and diabetes in offspring [74–77]. This section highlights the current evidence that chromatin remodeling and histone modifications play key roles in adipogenesis and the development of obesity.

Stages of the Developing Adipocyte: Determination and Differentiation

Adipocytes, like muscle and bone cells, are generally described as arising from the mesoderm. However, precise lineage tracing studies to define the origin of mesenchymal stem cell (MSC) adipocytes have been difficult to complete [78]. In a review of the developmental origin of adipocytes, Billion et al. noted that in head, facial bones, jaw and associated connective tissue are derived from neural crest cells, a cell population that arises from neuroectoderm. In the head and neck, the neural crest cells yield mesenchymal precursors, which have been shown to differentiate into connective tissue cells, vascular smooth muscle cells, dermis, odontoblasts, cartilage, and bone. The authors propose that these cells, being of mesenchymal origin, could also differentiate into adipose tissue. Various pools of MSCs could be derived from many sources including neural crest cells, hematopoietic pluripotent cells, and mesoderm [78].

Adipogenesis is generally described as a two-step process. The first step is generally described as determination, the process by which pluripotent MSCs are committed to the adipocyte lineage. Determination results in the conversion of a stem cell into a preadipocyte, which cannot be distinguished morphologically from a pluripotent MSC, but it has lost potential to differentiate into other cell types (reviewed in Ref. [79]). In the second stage, called terminal differentiation, the preadipocyte acquires the characteristics of a mature adipocyte, obtaining the cellular machinery necessary for the functions of lipid synthesis and transport, insulin sensitivity, and the ability to secrete adipocyte-specific proteins, also known as adipokines. The molecular processes necessary for terminal differentiation have been more extensively studied and described due to the availability of experimental systems of preadipocyte cultured cell lines such as 3T3-L1 and 3T3-F442A, which are described extensively below.

Experimental Systems for Studying Adipogenesis

Much of our understanding of the complex network of transcription factor activation and cell signaling processes, as well the epigenetic regulation of adipogenesis, comes from the use of various experimental systems that were established for the study of the processes by which preadipocytes are differentiated into mature adipocytes. Most of the research in the field of adipogenesis has been performed with the experimental systems that are described below.

The preadipocyte cell lines, 3T3-L1 and 3T3-F422A, were originally established by Green and associates in 1975–1976 [80]. Although these cell lines were already committed to the adipocyte lineage, they provide a basic model to study the processes involved in terminal differentiation of the adipocyte. Confluent 3T3-L1 cells differentiate upon exposure to a cocktail of adipogenic inducers including fetal bovine serum (FBS), dexamethasone, isobutylmethylxanthine (a cyclic AMP inducer), and insulin. This combination of hormonal inducers activates the adipogenic program, which occurs in two well-defined stages. The stimulated cells immediately reenter the cell cycle and progress through at least two cell-cycle divisions, a phase referred to as clonal expansion while expressing adipogenic transcription factors as well as cell cycle regulators. Following the clonal expansion, the cells undergo terminal differentiation defined by the production of lipid droplets and the expression of multiple metabolic programs characteristic of mature fat cells. Mouse (C3H10T1/2) and human preadipocyte cell lines do not undergo the clonal expansion step and are able to differentiate without postconfluence mitosis. C3H10T1/2 is an adipogenic cell line that was derived from murine bone marrow that can differentiate in vitro into adipocytes, chondrocytes, and myotubes after being treated with a general inhibitor of mammalian methyltransferases, 5-azacytidine and then stimulated with the appropriate adipogenic, chondrogenic, or myogenic signals [79].

Embryonic stem (ES) cells can be differentiated directly into adipocytes using a combination of retinoic acid and proadipogenic hormones. Mouse embryonic fibroblasts (MEFs) isolated after disaggregation of embryos at embryonic day 12–14 can be differentiated into adipocytes or can be immortalized by serial passaging, the introduction of SV40 large T antigen or chemical treatment prior to differentiation. Primary MEFs differentiate with variable efficiency (usually 10–70%) whereas most immortalized MEF lines do not differentiate unless a proadipogenic transcription factor cocktail is introduced. Multipotent precursor cells isolated from adult tissues including adipose tissue, skeletal muscle, and bone marrow provide another source of cells, which can be useful for mesenchymal-cell-fate studies.

Adipogenesis

Here we briefly review the hormonal and transcriptional regulation of adipocyte differentiation that will serve as a platform by which we can further describe the epigenetic regulation of adipogenesis in the following section. We will focus on the transcription factor cascade involving CAAT/enhancer binding proteins (C/EBPs) and peroxisome proliferator-activated receptor gamma (PPARγ). Bone morphogenetic protein-4 (BMP-4) is a transcription factor important to the process of differentiating from a pluripotent cell to an adipocyte and induces white adipose tissue development in C3H10T1/2 cells [81,82], while treatment with bone morphogenetic protein-7 (BMP-7) triggers commitment into the brown adipose lineage [83].

CAAT/Enhancer Binding Proteins

One of the first steps in terminal differentiation is the increased expression and protein accumulation of C/EBPs, specifically C/EBPβ and C/EBPδ, stimulated in vitro by isobutylmethylxanthine and dexamethasone, respectively [84]. The C/EBP family is a group of basic leucine zipper transcription factors, three of which play crucial roles in adipogenesis (α,β,δ; reviewed in Ref. [79]). Early induction of C/EBPβ and C/EBPδ leads to induction of C/EBPα. C/EBPβ and C/EBPδ begin to accumulate within 4h of adipocyte induction but are initially inactive [82]. C/EBPβ-deficient mice have reduced adiposity, but this effect has not been shown to be due to reduced adipogenesis specifically. C/EBPβ and C/EBPδ function in part by inducing the transcription of C/EBPα and PPARγ. After induction of C/EBPβ and C/EBPδ, the cells reenter the cell cycle and undergo mitotic expansion, a step that requires C/EBPβ [85]. C/EBPβ is hypophosphorylated and thus activated by MAPK and GSK3B and goes on to induce the transcription of C/EBPα and PPARγ. By day 2 of the differentiation process, phosphorylated C/EBPα exerts an inhibitory effect on the growth of the cells, which can then exit the cell cycle and begin the process of final differentiation [86,87].

C/EBPα induces many adipogenic genes directly. *C/ebpα*$^{-/-}$ mice are devoid of white adipose tissue except for the mammary gland. In the *c/ebpα*$^{-/-}$ model, the development of brown adipose tissue is delayed but eventually results in relatively normal amounts of brown adipose tissue in the adult. Other isoforms of C/EBP, including C/EBPγ and CHOP, appear to suppress adipogenesis, perhaps through heterodimerization and inactivation of C/EBPβ [88].

PPARγ – A Master Regulator of Adipogenesis

PPARγ is a member of the nuclear receptor superfamily and is necessary for adipogenesis (reviewed in Ref. [79]). Forced expression of PPARγ is sufficient to induce adipocyte differentiation from fibroblasts, and no additional factors have been identified to promote adipogenesis in the absence of PPARγ (reviewed in Ref. [79]). Two isoforms of PPARγ are generated by alternative splicing and promoter usage; both are induced during adipogenesis (reviewed in Ref. [79]). PPARγ1 is found in other cell types besides adipocytes, including colonic epithelium and macrophages. PPARγ2's role in promoting adipogenesis is still being studied. However, evidence supports the idea that PPARγ is not only required for adipogenesis but also is necessary for maintenance of the differentiated state. Adenoviral introduction of a dominant negative PPARγ into mature 3T3-L1 adipocytes caused dedifferentiation with loss of lipid accumulation and decreased expression of adipocyte markers (reviewed in Ref. [79])

Epigenetic Regulation of Adipogenesis

PPARγ and the C/EBP are major transcription factors that play key roles in adipogenesis. Epigenetic events regulating these transcription factors have been described in preadipocyte and adipocyte determination.

Epigenetic Regulation of Preadipocyte Determination

Differentiation of pluripotent cells requires selective silencing and activation of subsets of genes at appropriate time points. Gene activity is determined by chromatin structure and the intervention of chromatin-binding proteins. Studies on embryonic stem cells have shown that the binding of several architectural proteins to chromatin is not as tight as in more developed cells leading to what has been described as "hyperdynamic chromatin" (reviewed in Ref. [89]). Hyperdynamic chromatin has been found in the mesenchymal pluripotent cell line C3H10T1/2, which maintains the ability to give rise to chondroblasts, myoblasts, and preadipocytes. There is an absence of hyperdynamic heterochromatin in the undifferentiated but already committed C2C12 myoblast cell line. Increased chromatin plasticity is a hallmark of pluripotent cells, and these cells maintain a large number of genes that although currently silenced, remain in the poised position, available for transcription [89] (Figure 2).

Pluripotent stem cells are characterized by the presence of a bivalent histone mark. Bivalent histone marks contain both the activating mark of H3 and H4 acetylation and the repressing mark of H3K27 hypermethylation where both histone marks are poised, but neither is dominant in the pluripotent state [89]. Bivalent histone classification has been described in mouse embryonic stem cells and human embryonic fibroblasts. The repressive mark of H3K27 hypermethylation is particularly enriched at the promoters of developmentally important genes such as the adipogenic genes *adipoq*, *lep*, uncoupling protein 1 (*ucp1*), and delta-like homologue 1 (*dlk1*) [89], ensuring that those genes are silenced in the pluripotent state but poised for activation with appropriate differentiation signals.

C3H10T1/2 cells need to be treated with 5-azacytidine (a generalized DNA demethylating agent) in order to differentiate into adipocytes, suggesting that DNA methylation is involved with preadipocyte determination [90]. Several studies using 3T3-L1 cells demonstrate that the promoters of late adipogenic genes, such as *lep* or *glut4*, are methylated in preadipocytes but become demethylated during the process of adipogenesis [89]. Similarly, the *myod1* promoter

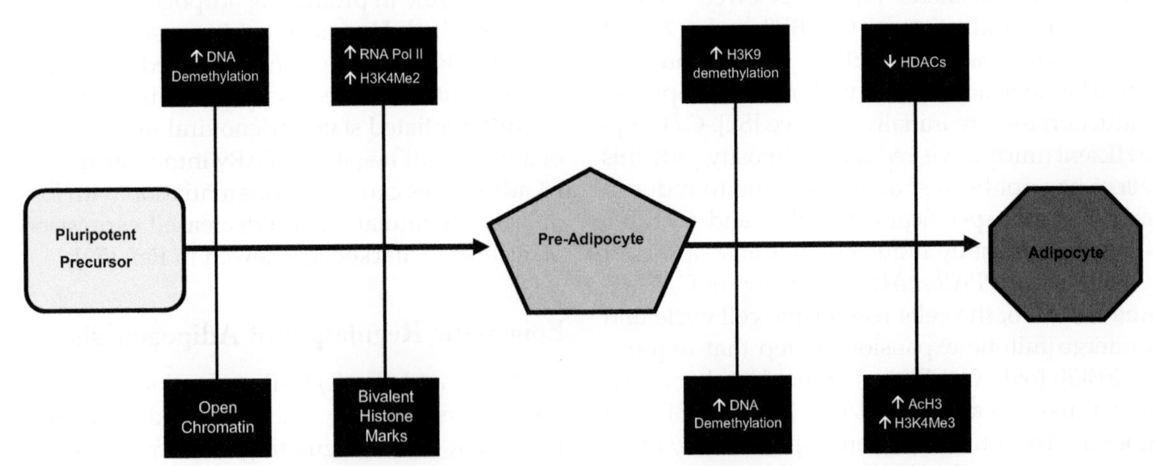

FIGURE 2 **Epigenetic modifications contributing to adipocyte development.** Pluripotent adipocyte precursors have been shown to associate with loosely related chromatin proteins. During determination, DNA demethylation of key genes such as *BMP-4* takes place, directing cells toward adipogenesis [82]. RNA Pol II and the active transcription mark H3K4me2 can be detected at the promoters of genes that are transcriptionally silent at this stage including *glut4* and *lep*. This signal is not found in pluripotent cells and is therefore a sign of cells committed to the adipocyte lineage [81]. During differentiation of the committed adipocyte precursors, further DNA demethylation takes place [81]. H3K9 is demethylated at the promoter of late adipogenic genes, and H3 acetylation and H3K4 trimethylation are increased. In addition, there is a global decrease in histone deacetylase enzymes at this stage [96,97].

has been shown to be methylated in C3H10T1/2 cells but after 5-azacytidine treatment, becomes demethylated and the cells are committed to the myogenic lineage. Another study found that C3H109T1/2 clonal line spontaneously undergoes adipogenesis after treatment with 5-azacytidine, correlating with demethylation of the *bmp4* gene and increased expression [81]. These findings indicate that DNA demethylation plays an important role in the adipocyte determination process.

Chromatin Remodeling and C/EBP Transcription Factors in Adipogenesis

Although adipocyte-specific genes are not expressed prior to differentiation, gene expression is prevented in part by the incorporation of a repressive chromatin structure. C/EBPβ binds to the *cebpα* promoter as early as 4h after induction of the terminal differentiation of the adipocyte. However, its activation can be blocked by the presence of the repressive complex composed of histone deacetylases mSin3A (mSin3A/HDAC1) and histone deacetylase 1 (HDAC1) [89]. C/EBPβ is associated with HDAC1, preventing it from binding to the *cepbα* promoter. Treatment with glucocorticoids activates GCN5, a histone acetyltransferase, resulting in the dissociation of HDAC1 and C/EBPβ, and the activation of the *cebpα* promoter by C/EBPβ [89].

In addition to regulating the expression of adipogenic genes, one of the main functions of C/EBPα is to promote the cells to exit the cell cycle, a crucial event for any model of differentiation [89]. Studies with mutated constructs of NIH-3T3 cells, a mouse embryonic fibroblast cell line, have shown that C/EBPα requires interaction with a functional and active SWI/SNF complex in order to exert its well-established proliferation-arrest effect [89].

Epigenetic Regulation of PPARγ: The Role of Histone Acetylation

The epigenetic regulation of PPARγ has been studied in detail and involves many co-activators and co-repressors. The protein Rb, known to exert a repressive effect on the initial phases of adipogenesis, has been found to associate with histone deacetylase 3 (HDAC3) in 3T3-L1 fibroblasts [89]. While Rb is active and dephosphorylated, the complex Rb-HDAC3 interacts with PPARγ, resulting in the recruitment of deacetylase activity to the target promoters of PPARγ and resulting in the repression of gene transcription. When Rb becomes phosphorylated, the Rb-PPARγ interaction is disrupted. PPARγ is then free to associate with histone acetyltransferases such as CREB binding protein (CBP) and p300 and results in increased gene transcription. CBP and p300 establish an important interaction with PPARγ, and the downregulation of CBP and p300 results in decreased adipogenesis in 3T3-L1 cells [91,92].

PPARγ recruits co-repressors to the promoters of adipogenic genes, including nuclear receptor co-repressor 1(NCoR1) and nuclear receptor co-repressor 2 (SMRT), resulting in the downregulation of transcriptional activity. In 3T3-L1 cells, the addition of pioglitazone, a PPARγ ligand, breaks the PPARγ–NCoR association and results in increased PPARγ transcriptional activity [93]. At the same time, when PPARγ binds to a ligand, this also increases the interaction between PPARγ with the histone acetyltransferases CBP and p300, which results in increased transcriptional activity [94].

As seen from the examples described above, histone acetylation plays a central role in the regulation of adipogenesis and the activity of key adipogenic transcription factors. Most HDAC inhibitors cause cell cycle arrest and increase differentiation, and for this reason many histone deacetylases are being targeted as anticancer therapies [95]. Histone acetylation increases throughout the process of adipogenesis at the promoter regions of adipogenic genes, and decreased HDAC expression occurs during the terminal differentiation process [89,96]. Inhibition of HDAC activity increases adipogenesis, and the overexpression of HDAC1 blocks adipocyte development [96,97].

Histone Methylation in Adipogenesis

Methylation of H3K4 is related to active gene transcription in a similar manner as histone acetylation. As mentioned earlier, the promoters of adipogenic genes have increased dimethyl H3K4 in the preadipocyte state, marking those genes as poised for transcription even though they are silent as preadipocytes [89]. Although increased dimethyl H3K4 is seen at the promoters of the preadipogenic genes, during the differentiation phase, H3K4 dimethylation is increased both at the promoters and the coding regions of those genes, coinciding with the initiation of transcription [89]. Increased H3K4 trimethylation is seen at the promoters of adaptor-related protein complex 1 mu 1 (*apm1*), *lep*, and *glut4* only after the start of transcription, whereas the increased H3K4 trimethyl mark is only seen in the coding region of these genes in fully differentiated adipocytes [89].

Linking In Utero Exposures to the Development of Obesity in the Offspring

Studies of the survivors of the Dutch famine at the end of World War II showed that in utero or early postnatal exposure to famine increased the risk of diabetes and obesity in adulthood [98]. In addition, more recent studies examining the incidence of obesity in the offspring born to women before and after surgical weight loss surgery indicate that maternal obesity before and during pregnancy is a contributing factor in the development of obesity in the offspring [99].

Alterations in gene expression induced by epigenetic modifications are the leading mechanism for explaining how exposure to maternal obesity or a nutrient-rich perinatal environment contributes to the developmental programming of obesity in the offspring [100]. Studies examining exposure to perinatal overnutrition through the rodent (rat) small litter model have demonstrated that insulin receptor substrate 1 (*Irs1*) mRNA and protein levels were lower in female offspring skeletal muscle at postnatal day 21 (immediately after weaning) and postnatal day 140. Analysis of DNA methylation at CpG sites within the promoter region of *Irs1* showed very little CpG methylation in general (ranging from 1% to 6%) and no site-specific changes in *Irs1* promoter DNA methylation at postnatal day 21 and minimal (~2% or less) statistically significant changes in *Irs1*-promoter DNA methylation at three CpG regions at postnatal day 140. No analysis of histone modifications or nonpromoter DNA methylation was performed. These results indicate that there are likely additional mechanisms contributing to the changes in mRNA and protein expression of *Irs1* at these ages studied in this model [100].

Using the same small litter perinatal overnutrition rodent model, Li et al. showed that female adult mice had increased adiposity and body weight along with decreased physical activity and energy expenditure but no change in food intake [101]. Genome-wide DNA methylation of the hypothalamus of overnutrition exposed and control female mice was measured via MCAM (a restriction enzyme–based method combined with hybridization to promoter arrays to measure changes in DNA methylation between groups), and no differences were detected between groups [101]. Subsequent studies measuring DNA methylation of candidate genes revealed small changes in DNA methylation at CpG sites between the hypothalamus of overnutrition exposed and control offspring at four candidate genes measured at postnatal days 25 and 180 (p values ranging from 0.04 to 0.09) [101]. However, the authors note that the hypothalamus is composed of mixed cell types, so the significance of this finding is not clear [101].

Finally, Yang et al. used a model of maternal obesity during pregnancy to study the adipogenic potential in the offspring [102]. In fetal tissue of mice exposed to maternal obesity during pregnancy, mRNA levels of the adipogenic transcription factor *Zfp423* were reduced. Mouse embryonic fibroblasts (MEFs) isolated from maternal obesity exposed offspring had increased

Zfp423 mRNA levels, and since *Zfp423* is a critical transcription factor initiating adipogenic commitment, a larger fraction of MEFs exposed to maternal obesity differentiated into adipocytes [102]. Chromatin immunoprecipitation sequencing studies demonstrated that the *Zfp423* promoter region is characterized as bivalent with enrichment of both H3K27me3 and H3K4me3. Bisulfite sequencing of the *Zfp423* promoter near transcriptional start site (TSS) revealed very little DNA methylation in both maternal obesity exposed and control MEFs, but there was approximately a 15% decrease in DNA methylation in the maternal obesity exposed MEFs at CpG sites approximately 1400 bps upstream from the TSS overlapping the bivalent regions occupied by both H3K27me3 and H3K4me3. These results suggest that exposure to maternal obesity may influence adipogenesis through effects on *Zfp423* DNA methylation as well as expression.

CONCLUSION

Epigenetic regulation of gene expression is an important mechanism contributing to the developing beta cell, hepatocyte, skeletal myocyte, and adipocyte. There is emerging evidence that epigenetic regulation of gene expression is important to the programming of diabetes and obesity through in utero exposures. The effects that histone acetylation, histone methylation, and DNA methylation have on developing cells in vivo are still being evaluated. Pharmaceutical treatments are being developed for treating and restoring aberrant epigenetic marks that contribute to disease, and these drugs are current therapies for various forms of cancer. Histone acetyltransferases (HATs) and DNA methyltransferases (DNMTs) have been linked to the pathology of cancer, asthma, and viral infections, and the development of HAT inhibitors and DNMT inhibitors is an active area of research [103]. Once these agents are established as safe and effective treatments with well-defined and specific targets, they may be useful as preventative therapies for populations at risk for developing diabetes and obesity due to exposure to an altered intrauterine milieu. However, much remains to be learned about how the epigenome is affected by intrauterine exposures, and this includes defining the developmental window(s) for both the aberrant exposures that lead to disease and the time frame during which a preventative treatment would be effective. These concerns must be addressed before we can consider manipulating the human epigenome in a developing infant or child for a therapeutic or preventative purpose.

References

[1] Center for Disease Control and Prevention. National diabetes statistics report: estimates of diabetes and its Burden in the United States, 2014. In: Services UDoHaH. 2014. Atlanta, GA.

[2] Ogden CL, Carroll MD, Kit BK, Flegal KM. Prevalence of childhood and adult obesity in the United States, 2011–2012. JAMA 2014;311(8):806–14.

[3] McCarthy MI. Genomics, type 2 diabetes, and obesity. N Engl J Med 2010;363(24):2339–50.

[4] Morris AP, Voight BF, Teslovich TM, Ferreira T, Segre AV, Steinthorsdottir V, et al. Large-scale association analysis provides insights into the genetic architecture and pathophysiology of type 2 diabetes. Nat Genet 2012;44(9):981–90.

[5] Barker DJ, Hales CN, Fall CH, Osmond C, Phipps K, Clark PM. Type 2 (non-insulin-dependent) diabetes mellitus, hypertension and hyperlipidaemia (syndrome X): relation to reduced fetal growth. Diabetologia 1993;36(1):62–7.

[6] Russo VEA, Martienssen RA. Epigenetic mechanisms of gene regulation. Cold Spring Harbor, NY: Cold Spring Harbor Laboratory Press; 1996.

[7] Snoeck A, Remacle C, Reusens B, Hoet JJ. Effect of a low protein diet during pregnancy on the fetal rat endocrine pancreas. Biol Neonate 1990;57(2):107–18.

[8] Dahri S, Snoeck A, Reusens-Billen B, Remacle C, Hoet JJ. Islet function in offspring of mothers on low-protein diet during gestation. Diabetes 1991;40(Suppl. 2): 115–20.

[9] Dahri S, Reusens B, Remacle C, Hoet JJ. Nutritional influences on pancreatic development and potential links with non-insulin-dependent diabetes. Proc Nutr Soc 1995;54(2):345–56.

[10] Wilson MR, Hughes SJ. The effect of maternal protein deficiency during pregnancy and lactation on glucose tolerance and pancreatic islet function in adult rat offspring. J Endocrinol 1997;154(1):177–85.
[11] Hales CN, Desai M, Ozanne SE, Crowther NJ. Fishing in the stream of diabetes: from measuring insulin to the control of fetal organogenesis. Biochem Soc Trans 1996;24(2):341–50.
[12] Petrik J, Arany E, McDonald TJ, Hill DJ. Apoptosis in the pancreatic islet cells of the neonatal rat is associated with a reduced expression of insulin-like growth factor II that may act as a survival factor. Endocrinology 1998;139(6):2994–3004.
[13] Petrik J, Reusens B, Arany E, Remacle C, Coelho C, Hoet JJ, et al. A low protein diet alters the balance of islet cell replication and apoptosis in the fetal and neonatal rat and is associated with a reduced pancreatic expression of insulin-like growth factor-II. Endocrinology 1999;140(10):4861–73.
[14] Arantes VC, Teixeira VP, Reis MA, Latorraca MQ, Leite AR, Carneiro EM, et al. Expression of PDX-1 is reduced in pancreatic islets from pups of rat dams fed a low protein diet during gestation and lactation. J Nutr 2002;132(10):3030–5.
[15] Bernardo AS, Hay CW, Docherty K. Pancreatic transcription factors and their role in the birth, life and survival of the pancreatic beta cell. Mol Cell Endocrinol 2008;294(1–2):1–9.
[16] Ozanne SE, Smith GD, Tikerpae J, Hales CN. Altered regulation of hepatic glucose output in the male offspring of protein-malnourished rat dams. Am J Physiol 1996;270(4 Pt 1):E559–64.
[17] Desai M, Crowther NJ, Lucas A, Hales CN. Organ-selective growth in the offspring of protein-restricted mothers. Br J Nutr 1996;76(4):591–603.
[18] Ozanne SE, Jensen CB, Tingey KJ, Storgaard H, Madsbad S, Vaag AA. Low birthweight is associated with specific changes in muscle insulin-signalling protein expression. Diabetologia 2005;48(3):547–52.
[19] Ozanne SE, Olsen GS, Hansen LL, Tingey KJ, Nave BT, Wang CL, et al. Early growth restriction leads to down regulation of protein kinase C zeta and insulin resistance in skeletal muscle. J Endocrinol 2003;177(2):235–41.
[20] Sandovici I, Smith NH, Nitert MD, Ackers-Johnson M, Uribe-Lewis S, Ito Y, et al. Maternal diet and aging alter the epigenetic control of a promoter-enhancer interaction at the Hnf4a gene in rat pancreatic islets. Proc Natl Acad Sci USA 2011;108(13):5449–54.
[21] Kingdom J, Huppertz B, Seaward G, Kaufmann P. Development of the placental villous tree and its consequences for fetal growth. Eur J Obstet Gynecol Reprod Biol 2000;92(1):35–43.
[22] Simmons RA, Gounis AS, Bangalore SA, Ogata ES. Intrauterine growth retardation: fetal glucose transport is diminished in lung but spared in brain. Pediatr Res 1992;31(1):59–63.
[23] Unterman T, Lascon R, Gotway MB, Oehler D, Gounis A, Simmons RA, et al. Circulating levels of insulin-like growth factor binding protein-1 (IGFBP-1) and hepatic mRNA are increased in the small for gestational age (SGA) fetal rat. Endocrinology 1990;127(4):2035–7.
[24] Jaeckle Santos LJ, Li C, Doulias PT, Ischiropoulos H, Worthen GS, Simmons RA. Neutralizing Th2 inflammation in neonatal islets prevents beta cell failure in adult IUGR rats. Diabetes 2014;63(5):1672–84.
[25] Stoffers DA, Desai BM, DeLeon DD, Simmons RA. Neonatal exendin-4 prevents the development of diabetes in the intrauterine growth retarded rat. Diabetes 2003;52(3):734–40.
[26] Myatt L, Eis AL, Brockman DE, Kossenjans W, Greer IA, Lyall F. Differential localization of superoxide dismutase isoforms in placental villous tissue of normotensive, pre-eclamptic, and intrauterine growth-restricted pregnancies. J Histochem Cytochem 1997;45(10):1433–8.
[27] Simmons RA, Suponitsky-Kroyter I, Selak MA. Progressive accumulation of mitochondrial DNA mutations and decline in mitochondrial function lead to beta cell failure. J Biol Chem 2005;280(31):28785–91.
[28] Vuguin P, Raab E, Liu B, Barzilai N, Simmons R. Hepatic insulin resistance precedes the development of diabetes in a model of intrauterine growth retardation. Diabetes 2004;53(10):2617–22.
[29] Park JH, Stoffers DA, Nicholls RD, Simmons RA. Development of type 2 diabetes following intrauterine growth retardation in rats is associated with progressive epigenetic silencing of Pdx1. J Clin Invest 2008;118(6):2316–24.
[30] Pinney SE, Jaeckle Santos LJ, Han Y, Stoffers DA, Simmons RA. Exendin-4 increases histone acetylase activity and reverses epigenetic modifications that silence Pdx1 in the intrauterine growth retarded rat. Diabetologia 2011;54(10):2606–14.
[31] Qian J, Kaytor EN, Towle HC, Olson LK. Upstream stimulatory factor regulates Pdx-1 gene expression in differentiated pancreatic beta cells. Biochem J 1999; 341(Pt 2):315–22.
[32] Thompson RF, Fazzari MJ, Niu H, Barzilai N, Simmons RA, Greally JM. Experimental intrauterine growth restriction induces alterations in DNA methylation and gene expression in pancreatic islets of rats. J Biol Chem 2010;285(20):15111–8.
[33] Holemans K, Verhaeghe J, Dequeker J, Van Assche FA. Insulin sensitivity in adult female rats subjected to malnutrition during the perinatal period. J Soc Gynecol Invest 1996;3(2):71–7.

[34] Garofano A, Czernichow P, Breant B. Beta cell mass and proliferation following late fetal and early postnatal malnutrition in the rat. Diabetologia 1998;41(9):1114–20.

[35] Matveyenko AV, Singh I, Shin BC, Georgia S, Devaskar SU. Differential effects of prenatal and postnatal nutritional environment on ss-cell mass development and turnover in male and female rats. Endocrinology 2010;151(12):5647–56.

[36] Blondeau B, Garofano A, Czernichow P, Breant B. Age-dependent inability of the endocrine pancreas to adapt to pregnancy: a long-term consequence of perinatal malnutrition in the rat. Endocrinology 1999;140(9):4208–13.

[37] Raychaudhuri N, Raychaudhuri S, Thamotharan M, Devaskar SU. Histone code modifications repress glucose transporter 4 expression in the intrauterine growth-restricted offspring. J Biol Chem 2008;283(20):13611–26.

[38] Thamotharan M, Shin BC, Suddirikku DT, Thamotharan S, Garg M, Devaskar SU. GLUT4 expression and subcellular localization in the intrauterine growth-restricted adult rat female offspring. Am J Physiol Endocrinol Metab 2005;288(5):E935–47.

[39] Fueger PT, Shearer J, Bracy DP, Posey KA, Pencek RR, McGuinness OP, et al. Control of muscle glucose uptake: test of the rate-limiting step paradigm in conscious, unrestrained mice. J Physiol 2005;562(Pt 3):925–35.

[40] Karnieli E, Armoni M. Transcriptional regulation of the insulin-responsive glucose transporter GLUT4 gene: from physiology to pathology. Am J Physiol Endocrinol Metab 2008;295(1):E38–45.

[41] Moreno H, Serrano AL, Santalucia T, Guma A, Canto C, Brand NJ, et al. Differential regulation of the muscle-specific GLUT4 enhancer in regenerating and adult skeletal muscle. J Biol Chem 2003;278(42):40557–64.

[42] Radford EJ, Isganaitis E, Jimenez-Chillaron J, Schroeder J, Molla M, Andrews S, et al. An unbiased assessment of the role of imprinted genes in an intergenerational model of developmental programming. PLoS Genet 2012;8(4):e1002605.

[43] Brons C, Jacobsen S, Nilsson E, Ronn T, Jensen CB, Storgaard H, et al. Deoxyribonucleic acid methylation and gene expression of PPARGC1A in human muscle is influenced by high-fat overfeeding in a birth-weight-dependent manner. J Clin Endocrinol Metab 2010;95(6):3048–56.

[44] Pettitt DJ, Baird HR, Aleck KA, Bennett PH, Knowler WC. Excessive obesity in offspring of Pima Indian women with diabetes during pregnancy. N Engl J Med 1983;308(5):242–5.

[45] Pettitt DJ, Aleck KA, Baird HR, Carraher MJ, Bennett PH, Knowler WC. Congenital susceptibility to NIDDM. Role of intrauterine environment. Diabetes 1988;37(5):622–8.

[46] Martin AO, Simpson JL, Ober C, Freinkel N. Frequency of diabetes mellitus in mothers of probands with gestational diabetes: possible maternal influence on the predisposition to gestational diabetes. Am J Obstet Gynecol 1985;151(4):471–5.

[47] Silverman BL, Metzger BE, Cho NH, Loeb CA. Impaired glucose tolerance in adolescent offspring of diabetic mothers. Relationship to fetal hyperinsulinism. Diabetes Care 1995;18(5):611–7.

[48] Atkins V, Flozak AS, Ogata ES, Simmons RA. The effects of severe maternal diabetes on glucose transport in the fetal rat. Endocrinology 1994;135(1):409–15.

[49] Plagemann A. Perinatal programming and functional teratogenesis: impact on body weight regulation and obesity. Physiol Behav 2005;86(5):661–8.

[50] Pettitt DJ, McKenna S, McLaughlin C, Patterson CC, Hadden DR, McCance DR. Maternal glucose at 28 weeks of gestation is not associated with obesity in 2-year-old offspring: the Belfast Hyperglycemia and Adverse Pregnancy Outcome (HAPO) family study. Diabetes Care 2010;33(6):1219–23.

[51] Deierlein AL, Siega-Riz AM, Chantala K, Herring AH. The association between maternal glucose concentration and child BMI at age 3 years. Diabetes Care 2011;34(2):480–4.

[52] Tsadok MA, Friedlander Y, Paltiel O, Manor O, Meiner V, Hochner H, et al. Obesity and blood pressure in 17-year-old offspring of mothers with gestational diabetes: insights from the Jerusalem Perinatal Study. Exp Diabetes Res 2011;2011:906154.

[53] Crume TL, Ogden L, Daniels S, Hamman RF, Norris JM, Dabelea D. The impact of in utero exposure to diabetes on childhood body mass index growth trajectories: the EPOCH study. J Pediatr 2011;158(6):941–6.

[54] Brandt S, Moss A, Lennerz B, Koenig W, Weyermann M, Rothenbacher D, et al. Plasma insulin levels in childhood are related to maternal factors - results of the Ulm Birth Cohort Study. Pediatr Diabetes 2014;15(6):453–63.

[55] Bollati V, Schwartz J, Wright R, Litonjua A, Tarantini L, Suh H, et al. Decline in genomic DNA methylation through aging in a cohort of elderly subjects. Mech Ageing Dev 2009;130(4):234–9.

[56] Cooney CA, Dave AA, Wolff GL. Maternal methyl supplements in mice affect epigenetic variation and DNA methylation of offspring. J Nutr 2002;132(8 Suppl.):2393S–400S.

[57] Franco R, Schoneveld O, Georgakilas AG, Panayiotidis MI. Oxidative stress, DNA methylation and carcinogenesis. Cancer Lett 2008;266(1):6–11.

[58] Martin DI, Cropley JE, Suter CM. Environmental influence on epigenetic inheritance at the Avy allele. Nutr Rev 2008;66(Suppl. 1):S12–4.

[59] So K, Tamura G, Honda T, Homma N, Waki T, Togawa N, et al. Multiple tumor suppressor genes are

increasingly methylated with age in non-neoplastic gastric epithelia. Cancer Sci 2006;97(11):1155–8.

[60] Ruchko MV, Gorodnya OM, Pastukh VM, Swiger BM, Middleton NS, Wilson GL, et al. Hypoxia-induced oxidative base modifications in the VEGF hypoxia-response element are associated with transcriptionally active nucleosomes. Free Radic Biol Med 2009;46(3):352–9.

[61] Tikoo K, Meena RL, Kabra DG, Gaikwad AB. Change in post-translational modifications of histone H3, heat-shock protein-27 and MAP kinase p38 expression by curcumin in streptozotocin-induced type I diabetic nephropathy. Br J Pharmacol 2008;153(6):1225–31.

[62] Ihara Y, Toyokuni S, Uchida K, Odaka H, Tanaka T, Ikeda H, et al. Hyperglycemia causes oxidative stress in pancreatic beta cells of GK rats, a model of type 2 diabetes. Diabetes 1999;48(4):927–32.

[63] Sakai K, Matsumoto K, Nishikawa T, Suefuji M, Nakamaru K, Hirashima Y, et al. Mitochondrial reactive oxygen species reduce insulin secretion by pancreatic beta cells. Biochem Biophys Res Commun 2003;300(1):216–22.

[64] Silva JP, Kohler M, Graff C, Oldfors A, Magnuson MA, Berggren PO, et al. Impaired insulin secretion and beta cell loss in tissue-specific knockout mice with mitochondrial diabetes. Nat Genet 2000;26(3):336–40.

[65] Bouchard L, Thibault S, Guay SP, Santure M, Monpetit A, St-Pierre J, et al. Leptin gene epigenetic adaptation to impaired glucose metabolism during pregnancy. Diabetes Care 2010;33(11):2436–41.

[66] Bouchard L, Hivert MF, Guay SP, St-Pierre J, Perron P, Brisson D. Placental adiponectin gene DNA methylation levels are associated with mothers' blood glucose concentration. Diabetes 2012;61(5):1272–80.

[67] Haghiac M, Basu S, Presley L, Serre D, Catalano PM, Hauguel-de Mouzon S. Patterns of adiponectin expression in term pregnancy: impact of obesity. J Clin Endocrinol Metab 2014;99(9):3427–34.

[68] El Hajj N, Pliushch G, Schneider E, Dittrich M, Muller T, Korenkov M, et al. Metabolic programming of MEST DNA methylation by intrauterine exposure to gestational diabetes mellitus. Diabetes 2013;62(4):1320–8.

[69] Ruchat SM, Houde AA, Voisin G, St-Pierre J, Perron P, Baillargeon JP, et al. Gestational diabetes mellitus epigenetically affects genes predominantly involved in metabolic diseases. Epigenetics 2013;8(9):935–43.

[70] Quilter CR, Cooper WN, Cliffe KM, Skinner BM, Prentice PM, Nelson L, et al. Impact on offspring methylation patterns of maternal gestational diabetes mellitus and intrauterine growth restraint suggest common genes and pathways linked to subsequent type 2 diabetes risk. FASEB J 2014;28.

[71] Dietz WH. Overweight in childhood and adolescence. N Engl J Med 2004;350(9):855–7.

[72] Simmons R. Perinatal programming of obesity. Semin Perinatol 2008;32(5):371–4.

[73] Taylor PD, Poston L. Developmental programming of obesity in mammals. Exp Physiol 2007;92(2):287–98.

[74] Bayol SA, Simbi BH, Stickland NC. A maternal cafeteria diet during gestation and lactation promotes adiposity and impairs skeletal muscle development and metabolism in rat offspring at weaning. J Physiol 2005;567(Pt 3):951–61.

[75] Catalano PM. Obesity and pregnancy–the propagation of a viscous cycle? J Clin Endocrinol Metab 2003;88(8):3505–6.

[76] Levin BE, Govek E. Gestational obesity accentuates obesity in obesity-prone progeny. Am J Physiol 1998; 275(4 Pt 2):R1374–9.

[77] Reifsnyder PC, Churchill G, Leiter EH. Maternal environment and genotype interact to establish diabesity in mice. Genome Res 2000;10(10):1568–78.

[78] Billon N, Monteiro MC, Dani C. Developmental origin of adipocytes: new insights into a pending question. Biol Cell 2008;100(10):563–75.

[79] Rosen ED, MacDougald OA. Adipocyte differentiation from the inside out. Nat Rev Mol Cell Biol 2006; 7(12):885–96.

[80] Green H, Kehinde O. Spontaneous heritable changes leading to increased adipose conversion in 3T3 cells. Cell 1976;7(1):105–13.

[81] Bowers RR, Kim JW, Otto TC, Lane MD. Stable stem cell commitment to the adipocyte lineage by inhibition of DNA methylation: role of the BMP-4 gene. Proc Natl Acad Sci USA 2006;103(35):13022–7.

[82] Tang QQ, Lane MD. Activation and centromeric localization of CCAAT/enhancer-binding proteins during the mitotic clonal expansion of adipocyte differentiation. Genes Dev 1999;13(17):2231–41.

[83] Tseng YH, Kokkotou E, Schulz TJ, Huang TL, Winnay JN, Taniguchi CM, et al. New role of bone morphogenetic protein 7 in brown adipogenesis and energy expenditure. Nature 2008;454(7207):1000–4.

[84] Yeh WC, Cao Z, Classon M, McKnight SL. Cascade regulation of terminal adipocyte differentiation by three members of the C/EBP family of leucine zipper proteins. Genes Dev 1995;9(2):168–81.

[85] Tang QQ, Otto TC, Lane MD. Mitotic clonal expansion: a synchronous process required for adipogenesis. Proc Natl Acad Sci USA 2003;100(1):44–9.

[86] Reichert M, Eick D. Analysis of cell cycle arrest in adipocyte differentiation. Oncogene 1999;18(2):459–66.

[87] Wang GL, Shi X, Salisbury E, Sun Y, Albrecht JH, Smith RG, et al. Cyclin D3 maintains growth-inhibitory activity of C/EBPalpha by stabilizing C/EBPalpha-cdk2 and C/EBPalpha-Brm complexes. Mol Cell Biol 2006;26(7):2570–82.

[88] Linhart HG, Ishimura-Oka K, DeMayo F, Kibe T, Repka D, Poindexter B, et al. C/EBPalpha is required for differentiation of white, but not brown, adipose tissue. Proc Natl Acad Sci USA 2001;98(22):12532–7.

[89] Musri MM, Gomis R, Parrizas M. Chromatin and chromatin-modifying proteins in adipogenesis. Biochem Cell Biol 2007;85(4):397–410.

[90] Taylor SM, Jones PA. Multiple new phenotypes induced in 10T1/2 and 3T3 cells treated with 5-azacytidine. Cell 1979;17(4):771–9.

[91] Chen S, Johnson BA, Li Y, Aster S, McKeever B, Mosley R, et al. Both coactivator LXXLL motif-dependent and -independent interactions are required for peroxisome proliferator-activated receptor gamma (PPARgamma) function. J Biol Chem 2000;275(6):3733–6.

[92] Takahashi N, Kawada T, Yamamoto T, Goto T, Taimatsu A, Aoki N, et al. Overexpression and ribozyme-mediated targeting of transcriptional coactivators CREB-binding protein and p300 revealed their indispensable roles in adipocyte differentiation through the regulation of peroxisome proliferator-activated receptor gamma. J Biol Chem 2002;277(19):16906–12.

[93] Yu C, Markan K, Temple KA, Deplewski D, Brady MJ, Cohen RN. The nuclear receptor corepressors NCoR and SMRT decrease peroxisome proliferator-activated receptor gamma transcriptional activity and repress 3T3-L1 adipogenesis. J Biol Chem 2005;280(14):13600–5.

[94] Gelman L, Zhou G, Fajas L, Raspe E, Fruchart JC, Auwerx J. p300 interacts with the N- and C-terminal part of PPARgamma2 in a ligand-independent and -dependent manner, respectively. J Biol Chem 1999; 274(12):7681–8.

[95] Huang L. Targeting histone deacetylases for the treatment of cancer and inflammatory diseases. J Cell Physiol 2006;209(3):611–6.

[96] Yoo EJ, Chung JJ, Choe SS, Kim KH, Kim JB. Down-regulation of histone deacetylases stimulates adipocyte differentiation. J Biol Chem 2006;281(10):6608–15.

[97] Wiper-Bergeron N, Salem HA, Tomlinson JJ, Wu D, Hache RJ. Glucocorticoid-stimulated preadipocyte differentiation is mediated through acetylation of C/EBPbeta by GCN5. Proc Natl Acad Sci USA 2007;104(8):2703–8.

[98] Ravelli AC, van Der Meulen JH, Osmond C, Barker DJ, Bleker OP. Obesity at the age of 50 y in men and women exposed to famine prenatally. Am J Clin Nutr 1999;70(5):811–6.

[99] Smith J, Cianflone K, Biron S, Hould FS, Lebel S, Marceau S, et al. Effects of maternal surgical weight loss in mothers on intergenerational transmission of obesity. J Clin Endocrinol Metab 2009;94(11): 4275–83.

[100] Liu HW, Mahmood S, Srinivasan M, Smiraglia DJ, Patel MS. Developmental programming in skeletal muscle in response to overnourishment in the immediate postnatal life in rats. J Nutr Biochem 2013;24(11): 1859–69.

[101] Li G, Kohorst JJ, Zhang W, Laritsky E, Kunde-Ramamoorthy G, Baker MS, et al. Early postnatal nutrition determines adult physical activity and energy expenditure in female mice. Diabetes 2013;62(8): 2773–83.

[102] Yang QY, Liang JF, Rogers CJ, Zhao JX, Zhu MJ, Du M. Maternal obesity induces epigenetic modifications to facilitate Zfp423 expression and enhance adipogenic differentiation in fetal mice. Diabetes 2013;62(11): 3727–35.

[103] Foulks JM, Parnell KM, Nix RN, Chau S, Swierczek K, Saunders M, et al. Epigenetic drug discovery: targeting DNA methyltransferases. J Biomol Screen 2012; 17(1):2–17.

CHAPTER

15

The Developmental Origins of Renal Dysfunction

James S.M. Cuffe, Sarah L. Walton, Karen M. Moritz

School of Biomedical Science, The University of Queensland, St Lucia, QLD, Australia

OUTLINE

The Epigenome and Developmental Origins of Health and Disease
http://dx.doi.org/10.1016/B978-0-12-801383-0.00015-3

Copyright © 2016 Elsevier Inc. All rights reserved.

INTRODUCTION

Chronic kidney disease (CKD) is now recognized as a major cause of global mortality with prevalence rates of between 8% and 16% [1]. It is also amongst the fastest growing cause of death worldwide, with deaths attributable to CKD increasing more than 80% over the last 20 years [2]. Risk factors for CKD are similar to those for cardiovascular disease (CVD) and diabetes and include poor diet, physical inactivity, and obesity. Of particular concern, many young people without comorbidities such as hypertension or diabetes are developing CKD, and this has enormous social and economic costs to society. As detailed elsewhere in this book, both CVD and metabolic disorders are strongly associated with exposure to a suboptimal environment during early life. The "Developmental Origins of Health and Disease" (DOHaD) hypothesis was largely based on findings linking a low birth weight, a surrogate marker of a poor intrauterine environment, to increased risk of developing these diseases in adulthood. Less focus has been placed on the "programming" of chronic renal disease and renal dysfunction. This is likely due to the fact that renal dysfunction is often not diagnosed until later in life when renal function has been reduced by up to 80–90%. It is also suggested that lack of awareness about CKD by both patients and doctors can delay diagnosis [1]. The limited epidemiological data available suggest that like many other adult onset diseases, CKD can be linked to early life events. Animal models have provided compelling evidence that alterations in normal renal developmental processes occur as a result of maternal perturbations and contribute to the formation of fewer nephrons in the kidney. As no new nephrons can be formed following cessation of nephrogenesis, this permanent reduction in nephron endowment places an individual at risk of subsequent disease. In this chapter, we shall focus on causes and mechanisms contributing to the programming of impaired renal development and kidney dysfunction. As impaired kidney development is also intrinsically linked to hypertension, salient discussion will also be given to the programming of high blood pressure.

IMPACT OF THE PRENATAL ENVIRONMENT ON RENAL DYSFUNCTION AND CHRONIC KIDNEY DISEASE

Recent studies have identified CKD as making an important contribution to the global burden of disease [2]. CKD is classified based on estimated glomerular filtration rate (eGFR) using a five-stage system. Stage 1 is classified as normal function with an estimated eGFR of >90 ml/min, whilst stage 5 is considered end-stage renal disease (ESRD) with an eGFR <15 ml/min at which point an individual usually requires renal replacement therapy such as dialysis. Diagnosis of CKD usually only occurs in the later stages (stages 3–5) when there are major increases in serum creatinine (a doubling from baseline concentrations) and decreases in eGFR, often to less than 30% of that in healthy individuals. Early signs of declining renal function include increased serum creatinine as well as urinary excretion of protein and/or albumin (proteinuria and albuminuria), which can be readily measured in a urine sample. Early recognition of declining renal function is important as lifestyle interventions and treatment of comorbid conditions (hypertension and/or diabetes) can dramatically slow the progression to renal failure.

A number of epidemiological studies have now linked birth weight to ESRD with babies at both end of the birth weight spectrum shown to be at risk. One of the first studies in this area examined risk factors for ESRD in the southwestern USA where rates of renal failure are very high [3]. The study found a strong association between ESRD and low birth weight, particularly if the renal disease was due to underlying hypertension. However, if the renal disease was primarily due to diabetes (diabetic nephropathy), babies born of a high birth weight were most at risk. Other studies have also demonstrated an association between low birth weight and ESRD [4], although in one study, this was only found to be significant in males [5]. A meta-analysis demonstrated a reduction in eGFR, and long-term risk of CKD was associated with a low birth weight although it suggested that further large-scale population-based studies were necessary to validate these findings [6]. Earlier signs of renal disease such as proteinuria have been shown to be increased in people born of a low birth weight [6], whilst high birth weight babies, most often those born to mothers with gestational diabetes, also have an increased risk of proteinuria [7].

A number of indigenous communities appear to be particularly susceptible to the programming of renal dysfunction. The Australian Aboriginal people have amongst the highest rates of CKD worldwide, and this is associated with increased rates of low birth weight babies [8] and increased proteinuria in young children. This population also has high prevalence of early life infections resulting in poststreptococcal glomerulonephritis [9]. This highlights the important interactions between the pre- and postnatal environments; a poor intrauterine environment may contribute to increased disease risk due to the low birth weight, with the postnatal environment providing a "second hit" that further increases risk of disease. Childhood obesity has also been shown to exacerbate poor renal outcomes in aboriginal children of a low birth weight [9].

KIDNEY DEVELOPMENT AND NEPHRON ENDOWMENT

A programmed susceptibility to renal dysfunction and disease is likely to be due in part to altered renal development. The filtering units of the kidney, the nephrons, are complex structures consisting of a filtering component (the glomerulus) and a series of tubules where reabsorption (and some secretion) of electrolytes and water occurs. The development of the mammalian kidneys is marked by the development of a series of three excretory organs, each becoming more advanced: the pronephros, mesonephros, and metanephros (for review, see [10]). In the human, the pro- and mesonephros are transitory organs and the definitive kidneys, the metanephroi, commence development during the fifth week of gestation. Nephrons arise from an embryonic nephron progenitor population within the metanephric cap mesenchyme through a cyclical process termed *nephrogenesis*. The cap mesenchyme signals to the Wolffian duct to form the ureteric bud, which invades the metanephric mesenchyme. Reciprocal signaling between the cap mesenchyme and the ureteric epithelium occurs, leading to the proliferation, migration, and branching of the ureteric tips. This process of branching morphogenesis is a major determinant of nephron number as new nephrons can only form at the ureteric tip. Ultimately, condensation and aggregation with the metanephric mesenchyme results in formation of a pretubular aggregate and mesenchymal-to-epithelial transition (MET) to form an epithelial renal vesicle. Gene expression within the renal vesicle is polarized, allowing the formation of distinct tubular segments (tubulogenesis). A process of elongation, patterning, and segmentation produces a nephron consisting of the glomerulus, proximal tubules, the loop of Henle, and distal tubules. The immature nephron then connects with the collecting ducts, which are formed from the Wolffian duct and subsequently undergoes a process of maturation.

In the human, nephron formation is complete by ~36 weeks gestation, and thus babies born at term have their full complement of nephrons. Whilst the average human kidney contains ~1 million nephrons, it is now appreciated that there is a 10-fold range in normal nephron number from approximately 200,000 to 2 million [11]. The reasons for this large variation are unknown, but the timing of nephrogenesis in the human highlights that optimal in utero conditions are necessary to ensure adequate nephron endowment. This concept is supported by the fact that nephron endowment is strongly linked to fetal growth with an estimated increase of almost 250,000–350,000 nephrons for every kilogram increase in birth weight [12]. In the human, other factors known to influence kidney size at birth, and thus potentially nephron number, include prematurity and vitamin A deficiency [11].

THE IMPORTANCE OF NEPHRON ENDOWMENT: LINKS TO BLOOD PRESSURE AND RENAL DISEASE

The importance of nephron number, in the pathogenesis of essential hypertension, was first proposed by Brenner who postulated that a low number of nephrons in the kidney is correlated with low birth weight and can increase the risk of hypertension and renal disease [13,14]. A congenital nephron deficit results in a diminished filtration surface area and hence a reduced capacity of the kidney to excrete normal sodium load. Sodium retention leads to the expansion of extracellular fluid volume and plasma volume, which acts to increase cardiac output, total peripheral resistance, and mean arterial pressure. This can contribute to glomerular hypertension and scarring and causes a positive feedback loop, where hypertension is worsened by the progressive glomerular damage (Figure 1). A landmark study found that the number of glomeruli (nephrons) was lower in patients with hypertension than matched normotensive controls [15]. This was also shown in studies of children with unilateral renal agenesis (one kidney) of which 50% have reduced GFR and develop hypertension and albuminuria by the age of 18. Ethnicities with a predisposition to hypertension have on average fewer, but larger, glomeruli [16–18]. Females tend to have greater age-related increases in blood pressure following menopause [19], and it has been suggested

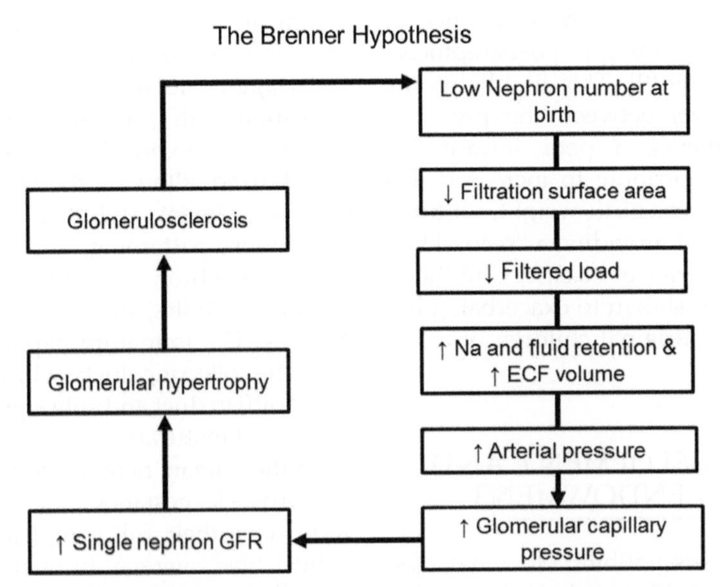

FIGURE 1 The Brenner hypothesis depicting the mechanisms through which a congenital nephron deficit may contribute to hypertension and renal disease in adulthood.

that this may be due to the fact that females tend to have smaller kidneys than males, with fewer glomeruli, and thus increased susceptibility to hypertension [20]. There is also an age-related decline in nephrons that correlates with an age-related increase in the incidence of hypertension [21]. Although a nephron deficit does not universally result in hypertension, these data suggest that people with fewer nephrons are more susceptible to elevations in blood pressure and subsequent glomerular damage.

EXPLORING THE MECHANISMS OF RENAL PROGRAMMING OF LOW NEPHRON ENDOWMENT: EVIDENCE FROM ANIMAL MODELS

Animal models designed to elucidate mechanisms underpinning the developmental origins of disease have predominantly used perturbations that result in growth restriction including maternal dietary restrictions, hypoxia, and uteroplacental insufficiency. These models have been designed to mimic common maternal and/or fetal suboptimal conditions and examine the physiological consequences in offspring including blood pressure, glucose regulation, and to a lesser extent, renal function. Early descriptive studies linking a perturbation to a change in offspring physiology relied heavily on large animal species such as sheep due to the relative ease of physiological experimentation. The sheep model also allows fetal physiology to be studied in utero as cannulation of the ovine fetus is widely utilized. The vast majority of studies, however, have utilized the rat, which still allows robust physiological measures but has a shorter gestation and greater availability of molecular tools. More recently, some researchers have moved to the mouse for investigation into the molecular mechanisms responsible for impairing fetal development and regulating disease outcomes. Additional animals such as nonhuman primates, spiny mice, guinea pigs, and rabbits have all been used experimentally with each providing specific benefits unique to the species of interest. A key factor that must be considered in each of these animal models is the timing of nephrogenesis in each species [22]. In species such as the sheep and spiny mouse, nephrogenesis is a relatively long process and is complete prior to birth as is the case in humans. In contrast, nephrogenesis begins considerably later in mouse and rat gestation and is not complete until several days after birth (see Figure 2). It is therefore important

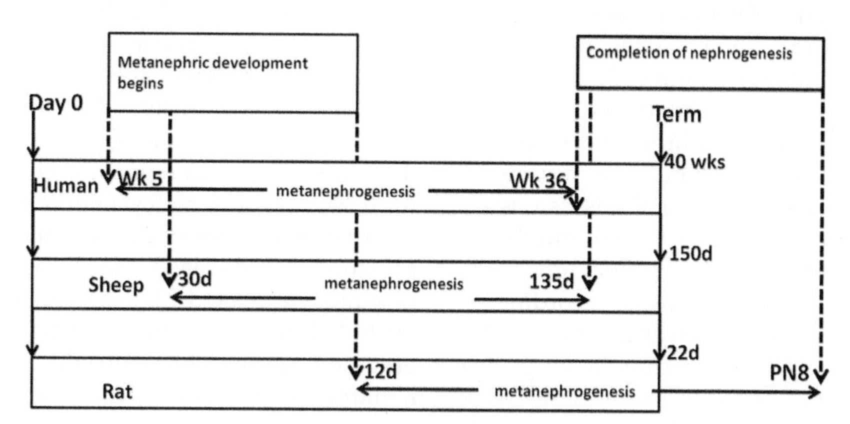

FIGURE 2 Comparison of the timing of nephrogenesis between three species: human, sheep, and rat. Note that nephrogenesis is complete prior to birth in some species (human and sheep) but continues into the postnatal period in other species such as the rat. *Reproduced (with permission) from Singh et al. [22].*

to not only consider the timing of gestation at which a perturbation occurs, but also to take into account the relative stage of renal development.

In what follows, we shall discuss in detail a number of commonly used animal models with a focus on the effects on renal development (as indicated by nephron number) and blood pressure/renal function outcomes. An important point to note when comparing studies is the methodology used to determine the number of nephrons and blood pressure. Assessing final nephron endowment in both animal models and human studies of developmental programming is highly valuable, as it provides insights into the efficiency and success of nephrogenesis in utero. However, there are many technical limitations in assessing nephron number and structure in kidney tissue. Clinically, kidney volume estimated by ultrasound is used as an estimate of the final complement of nephrons; however, the compensatory capacity of the renal tubules means that this estimate is only likely to be useful in very young children. Currently, an unbiased stereological method known as the physical dissector/fractionator method is the gold standard for determining glomeruli (and thus nephron) number estimation in both human and animal studies [23]. However, this process is time-consuming and requires excision of the kidney at postmortem and therefore cannot be used to assess nephron number in a live person/animal. Acid maceration is commonly used, however, reproducibility is limited and may lead to underestimation of nephron endowment if glomeruli are prone to degradation during the process. Other histological methods can assess glomerular density, often in a small number of sections, but this may not accurately reflect glomerular number. Similarly, whilst recordings of blood pressure performed over 2–3 days (either by indwelling catheter or radiotelemetry) provide the most accurate measure of basal blood pressure, many studies utilize tail-cuff methods, which may induce a cardiovascular stress response. The advantage of the tail-cuff method is that it enables multiple measures of blood pressure over the life course, but caution must be taken to train animals to the procedure to minimize the effects of stress. In Table 1, we provide a brief summary of the outcomes recorded in a range of animal models (described in detail below). It should be recognized that this table is not designed to be a comprehensive list of all studies in the area but rather a selection to highlight the wide range of perturbations that affect kidney development. Where possible, our discussion includes studies that use unbiased stereology for nephron number determination and either in-dwelling catheter or radiotelemetry for blood pressure measurement.

Maternal Nutrition

Many animal models of programmed disease have utilized global calorie restriction, protein deficient diets, or diets lacking particular nutrients to demonstrate that poor maternal nutrition can result in impaired nephrogenesis and hypertension in offspring. Such studies have been extensively discussed in earlier reviews [10,24]. In summary, studies in the rat have demonstrated that protein restriction (9% casein diet compared to 18%) can regulate nephron number depending on the timing and severity of the exposure as well as the sex of the animal. When rats were protein restricted in just the first week of pregnancy, no effect on nephron development was observed at any age. In contrast, when rats were exposed to a low-protein diet through gestation or through the second or third week of pregnancy, nephron number was reduced at 4 weeks of age at which time blood pressure was increased and glomerular filtration rate unaffected [25]. A similar study demonstrated that protein restriction throughout gestation reduced birth weight, reduced adult nephron number and glomerular filtration rate, and increased blood pressure [26]. These outcomes were at least in part dependent upon the sex of offspring. In female offspring, a more

TABLE 1 Effects on Fetal Growth and Renal and Blood Pressure Outcomes in Animal Models Designed to Test the DOHaD Hypothesis

Maternal Perturbation	Species	Sexes Studied	Birth Weight/ Fetal Weight	Reduction in Nephron Endowment	Renal Outcomes	Change in Arterial Pressure
UNDERNUTRITION						
9% Casein diet [25]	Rat	Pooled	↔	↓13%	↔GFR	↑13 mmHg
8.5% Low protein diet [26,27]	Rat	Males Females	↓	↓25% males ↔females	↓GFR males ↔females	↑10 mmHg ↔females
50% Nutrient restriction [29]	Sheep	Males	↔	↓11%	Impaired sodium handling	↑17 mmHg
OVERNUTRITION						
High fat/high fructose [40]	Rat	Pooled	↑	–	Albuminuria	↑17 mmHg
MICRONUTRIENT DEFICIENCIES						
Iron deficiency [41]	Rat	Males Females	↓	↓	–	↑18 mmHg
Vitamin A deficiency [43]	Rat	Pooled	↔	↓20%	–	–
Zinc deficiency [42]	Rat	Males	↓	↓	↓GFR Proteinuria	↑SBP
Vitamin D deficiency [45]	Rat	-	↔	↑20%	–	–
HYPOXIA						
Maternal hypoxia [54]	Rat	Males Females	↓	↓26% males ↓52% females	–	–
Cigarette smoke exposure [56]	Rat	Males	↔	↓	↑urinary albumin: creatinine	–
Uteroplacental insufficiency [58,59]	Rat	Males females	↓	↓	↑plasma creatinine in aged females	↑SBP in males
GLUCOCORTICOID EXPOSURE						
Dexamethasone [62]	Sheep	Females	↔	↓38%	–	↑10 mmHg
Betamethasone [63]	Sheep	Males Females	↔	↓26% males and females	↓GFR in males	↑7–9 mmHg
Corticosterone [69]	Rats	Males Females	↔	↓21% males ↓19% females	–	↑10 mmHg
Dexamethasone [65]	Spiny Mouse	Males Females	↔	↓13% males ↓17% females	–	↔
ETHANOL CONSUMPTION						
Intravenous ethanol infusion [71]	Sheep	Pooled	↔	↓11%	–	–
Ethanol (gavage) [72]	Rat	Males Females	↓	↓15% males ↓10% females	↑GFR males ↓GFR females Proteinuria	↑10%

Further details on the models are described in the text.

severe protein restriction (5% casein diet) was required to program hypertension [27]. In addition, the postnatal diet influenced long-term outcomes as male rats exposed to a low-protein diet throughout pregnancy and then maintained on a low-protein diet throughout life did not develop hypertension despite a nephron deficit and reduced GFR [28]. A number of studies have shown a maternal low-protein diet throughout pregnancy results in offspring with altered electrolyte handling. This includes increased sodium excretion, which was associated with a decrease in expression and activity of the Na^+,K^+-ATPase pump, in the collecting ducts [29]. Whilst an increase in sodium excretion contradicts the Brenner hypothesis (Figure 1), the authors found that the renal sodium loss resulted in increased sodium appetite, suggesting that the increased salt (and thus food intake) drove accelerated growth and increased blood pressure [30].

Overnutrition and/or maternal obesity have been shown to program for disease outcomes in multiple animal models. Worldwide, 1.4 billion people are overweight or obese [31], and the prevalence rates during pregnancy are thought to be approximately 40% in Western nations [32]. While maternal obesity has long been associated with increased birth weight, low birth weight is an equally common outcome [33]. This U-shaped association suggests that the programming of disease outcomes by maternal obesity is complex. With respect to the programming of CKD, there are clear links between obesity, renal function, and cardiovascular disease in adults, with obesity being the greatest risk factor for CKD [34]. Models of maternal overnutrition and/or obesity usually include feeding rodents a high fat or a high fat/high carbohydrate diet prior to pregnancy to induce an obesogenic state, although many of these may not mimic aspects of human obesity. Relatively short-term exposure of rats to a high-fat diet (for 10 days prior to pregnancy until the end of lactation) has been shown to result in female-specific hypertension at 12 months in association with normal nephron endowment but reduced renal renin and Na^+,K^+-ATPase activity [35]. Several rodent studies have taken a longer-term approach to determine the effects of overnutrition prior to, during gestation, and during lactation on long-term kidney health. Feeding rats a high-fat/high-fructose diet from 6 weeks prior to mating, throughout pregnancy and subsequent lactation, resulted in offspring that developed albuminuria, which was exacerbated by postnatal overnutrition causing tubulointerstitial fibrosis and increased TGF-β expression [36]. In a separate study, inducing obesity in rats 5 weeks prior to mating results in the programming of increased adiposity in offspring, hyperleptinemia, and hypertension in offspring in association with increased renal norepinephrine and increased renal renin expression [37].

In many such studies, it is unclear as to whether nutrient excess or maternal obesity itself is regulating kidney development. An interesting observation is that excess glucose levels, either caused by simple glucose infusion during nephrogenesis or beta cell destruction using streptozotocin, reduces nephron number in rats [38]. Similarly, streptozotocin-induced maternal diabetes in mice results in offspring hypertension in association with glomerular hypertrophy, increased extracellular matrix accumulation, as well as increased expression of components of the renin-angiotensin system and fibrosis genes [39]. Additional studies have suggested that changes in the hormonal profile associated with increased food intake can lead to programming of disease outcomes. Rabbits exposed to a high-fat diet during pregnancy deliver offspring who later develop stress-induced hypertension, increased heart rate, and increased renal sympathetic nerve activity in association with ghrelin and leptin resistance [40]. While it is clear that calorie excess results in disturbed fetal development and long-term disease outcomes, it is interesting to note that foods rich in calories are often poor in micronutrient levels.

In addition to global undernutrition, an estimated 2 billion people in the world suffer from at least one type of micronutrient deficiency. These deficiencies can occur in well-fed women and may include, but are not limited to, iron, zinc, vitamin A, and vitamin D deficiencies. Outcomes range from poor pregnancy outcomes, increased risk of postnatal morbidity and mortality, and impaired physical and cognitive abilities. Iron deficiency is the most common, and estimates suggest that nearly one in two women in developing countries is iron-deficient both before and during pregnancy, highlighting the importance of considering both the pre-/periconceptional period and the prenatal period in the context of developmental programming. Rodent models of maternal iron and vitamin A deficiencies in utero have shown a reduction in nephron endowment [41–43], and maternal anemia has been shown to result in increased blood pressure in adult offspring [44]. A comprehensive study of renal function in male rat offspring born to zinc-restricted mothers demonstrated reduced GFR and proteinuria, which was associated with increased systolic blood pressure [42]. An emerging area of great interest is the long-term effects of maternal vitamin D deficiency. A model of vitamin D deficiency during pregnancy in rats has reported a 20% increase in nephron number in offspring kidneys, with an overall decrease in size of the renal corpuscle [45]. Whether increased nephron endowment confers an advantage or disadvantage is unknown, as the necessary functional studies have not yet been performed.

Maternal Hypoxia and Uteroplacental Insufficiency

Reduced oxygen supply to the fetus, or hypoxia, is the most common complication seen in pregnancy and has a wide etiology including high-altitude living [46], maternal anemia, maternal asthma, maternal smoking [47], sleep apnea [48], and poor placentation leading to reduced fetoplacental perfusion [49]. Severe hypoxic episodes are associated with reduced fetal survival and severe perinatal injury. Milder forms of hypoxia are perhaps more common and initiate a range of protective mechanisms in the fetus in order to maximize survival and growth of critical organs. However, intrauterine growth restriction and subsequent low birth weights are common features of pregnancies complicated by reduced oxygenation, and this may have an impact on the overall health of the offspring in later life. The centralization of blood flow in the chronically hypoxic fetus means that growth restriction is frequently asymmetric, with brain, heart, and adrenal gland size preserved at the expense of peripheral organs such as the kidneys. Although the preservation of critical organs is necessary for immediate fetal survival, the structural and functional consequences of impaired peripheral organ development can be immense in later life. There is a dearth of information surrounding the impact of hypoxia on long-term renal outcomes, although short-term consequences have been well described. A study in fetal lambs during acute bouts of maternal hypoxemia showed an increase in renal vascular resistance and a reduction in renal blood flow by 20% [50–52]. Glomerular filtration rate was maintained, suggesting renal vasoconstriction at the efferent arteriolar level [51]. The reduction in renal blood flow increased plasma renin activity and plasma concentrations of vasopressin, leading to increased water reabsorption. Furthermore, the limited renal blood flow reduced renal oxygen delivery and ultimately renal oxygen consumption. In addition, the renal metabolism of carbohydrates can be altered by changes in fetal oxygenation [53]. A recent study in rats has shown decreased nephron number in male and female offspring in response to gestational hypoxia. Contrary to current literature that states nephrogenesis is complete in early postnatal life, nephron number was shown to increase almost twofold from P7 to 3 months of age, with a nephron deficit remaining in the hypoxia-exposed offspring [54].

However, the acid maceration technique was used to quantity nephron number and is not as reliable as the physical-disector fractionator method, the current gold standard of glomerular number estimation [23]. A very recent study has used two different hypoxic insults and characterized impacts on renal development. The study demonstrated that severe, short-term gestational hypoxia (5.5–7.5% maternal O_2 for 10h from embryonic day 9.5–10.5) resulted in congenital abnormalities of the kidney and urinary tract including duplex kidney, whilst a more modest hypoxia in mid-gestation pregnancy (12% from E12.5–14.5) resulted in fetal growth restriction and a decrease in nephron endowment [55]. In a mouse model of cigarette smoke exposure, renal development has been shown to be delayed and male offspring have fewer nephrons and an increased albumin/creatinine ratio [56]. Further, in-depth study is warranted in order to elucidate the impact of gestational hypoxia on long-term kidney health.

Inadequate blood supply to the fetus, as occurs with uteroplacental insufficiency, is the most common cause of fetal growth restriction in the Western world and results in insufficient nutrient and oxygen supply to the developing fetus. This has been modeled in a number of experimental studies. In the rat, uteroplacental insufficiency can be induced by ligation of the uterine arteries and/or veins. If the ligation occurs around day 14 in the rat, the dam develops characteristics of preeclampsia including late-gestation hypertension and proteinuria [57]. If performed at day 18 of pregnancy, the model results in late-gestation growth restriction and programming of a number of sex-specific deficits including a reduction in nephron endowment (both males and females), increases in blood pressure (in male offspring) [58], and increases in serum creatinine (in aged females) [59]. Interestingly, the nephron deficit and hypertension in male offspring could be ameliorated by cross-fostering growth-restricted pups onto a control dam on day 1 of life [60], suggesting improved nutrition can stimulate the late stages of nephrogenesis. Whilst this would not be an effective strategy to improve nephron endowment in a growth-restricted term infant where nephrogenesis is likely to be complete, it is relevant to the treatment and care of premature babies where nephrogenesis is ongoing. Other models of uteroplacental insufficiency have been developed in the sheep where removal of placentation sites or injection of microspheres limits blood supply to the fetus. This has been shown to result in fetal growth restriction and a decrease in nephron number in late gestation [61].

Maternal Glucocorticoid Exposure

A common maternal perturbation known to occur during pregnancy is that of increased glucocorticoid exposure. Maternal glucocorticoid administration during pregnancy is standard clinical practice for women at risk of preterm delivery. Additionally, naturally synthesized glucocorticoids (predominantly cortisol in the human and sheep and corticosterone in the rodent) are elevated in times of stress. Many of the animal models used to explore the mechanisms of developmental programming are likely to include a maternal stress component. Whilst the placenta is thought to inactivate a significant proportion of maternal glucocorticoids via the actions of the enzyme 11 beta hydroxysteroid dehydrogenase type II, it is now recognized that this system has distinct limitations and in times of maternal stress, the fetus is also likely to be exposed to elevated glucocorticoid concentrations. There is considerable literature to support a role for elevated glucocorticoids in altering renal development prenatally as well as inducing long-term renal pathologies in offspring. The majority of this research has been performed using synthetic glucocorticoids such as betamethasone or dexamethasone in sheep and rodent models. Studies in sheep (term = 150 days) demonstrated that maternal administration of dexamethasone at a dose of 0.48 mg/h for 48 h during early pregnancy (day 27/28) resulted in reduced nephron endowment, increased glomerular volume, enlarged and dilated proximal tubules, and tubulointerstitial and vascular periadventitial fibrosis at 7 years of age [62]. These renal pathologies are likely to have contributed toward the observed hypertension in these animals. More recent studies have shown that betamethasone administration in mid-pregnancy (0.17 mg/kg/day on days 80/81) induces a 26% nephron deficit in males and females and reduced GFR in males. These sheep also developed hypertension, but there was no correlation between nephron number and blood pressure on an individual animal level suggesting that while nephron number may have contributed to the observed hypertension, additional factors may have been involved [63]. When betamethasone was administered in late pregnancy in the sheep (0.5 mg/kg on days 104, 111, and 118), kidney weight was temporarily reduced at day 116 and 122 but was of normal size thereafter [64]. Dexamethasone administration during mid-gestation in the spiny mouse also reduced nephron number, and this was associated with changes in mRNA expression of genes that regulate kidney development. Although offspring did not have alterations in blood pressure at 20 weeks of age, they had elevations in heart rate after surgical stress suggesting programming of cardiovascular responses to stress [65]. Other studies have demonstrated that dexamethasone can also reduce nephron endowment and lead to hypertension in rats [66] and mice [67], although outcomes were dependent upon the sex of the offspring, the timing of exposure, and the dose used.

While less well characterized, sheep and rodent studies have also demonstrated the effects of endogenous glucocorticoids on the developing kidney. Administration of cortisol (5 mg/h) for 48 h from day 26 of gestation in the sheep induced a greater nephron deficit (40%) than did maternal dexamethasone exposure (0.48 mg/h, 25%) when assessed at day 140 of gestation. Offspring exposed to either glucocorticoid were programmed to have altered mRNA levels of genes that regulate sodium transport. Furthermore, mean glomerular volume was increased by maternal cortisol exposure but was not affected by maternal dexamethasone exposure, and blood pressure was elevated in 4- to 5-year-old female sheep in cortisol-exposed offspring [68]. More recently, corticosterone administered to rats has been shown to increase mean arterial pressure, reduce nephron number, and alter kidney mRNA expression including components of the renin-angiotensin system (discussed in detail below) [69]. These studies demonstrate that both natural and synthetic glucocorticoids can severely impair renal development and contribute to programming of hypertension and renal dysfunction.

Maternal Substance Exposure during Pregnancy

Excessive alcohol (ethanol) consumption during pregnancy is known to cause behavioral and developmental abnormalities, which are encompassed under the broad term of *fetal alcohol spectrum disorder*. The most severe form is fetal alcohol syndrome, which is associated with significant renal malformations [70]. However, milder forms of ethanol exposure in utero have not been explored in great detail in the human population but have been examined in both sheep and rodent models. In an ovine model of daily intravenous infusion of ethanol during the latter half of pregnancy, offspring were born with normal kidney size and morphology but an 11% reduction in nephron endowment [71]. A similar model of a short-term late-gestational "binge" alcohol exposure in the rat explored the impact of prenatal ethanol exposure on renal and cardiovascular outcomes in adulthood [72]. Male offspring had a 15% nephron deficit, slightly more severe than the 10% nephron deficit seen in females. These structural deficits were accompanied by renal dysfunction, proteinuria, and hypertension in adult male and female offspring.

Renal tubular dysfunction in the absence of gross morphological differences has also been reported in a similar rodent model of prenatal ethanol exposure [73]. These data suggest that the developing fetal kidney is also susceptible to the teratogenicity of even low-dose ethanol consumption, which may have long-lasting implications for adult health.

Programmed renal outcomes have also been reported both in studies of maternal tobacco smoking and illicit drug use. As mentioned briefly above, maternal smoking has been shown to induce growth restriction and result in adverse renal outcomes in both human studies and animal models. A 2011 prospective study investigating over 1000 human pregnancies demonstrated that cigarette smoking during pregnancy altered kidney volume in a dose-dependent manner [74]. These outcomes are largely attributed to the hypoxic environment that is induced by the act of cigarette smoking, however studies have shown that nicotine exposure itself can be detrimental to fetal development. To isolate the effects of nicotine exposure on offspring outcomes, nicotine was administered to pregnant rats via subcutaneous osmotic minipumps throughout gestation. Nicotine administration itself was found to increase the renal AT1R/AT2R ratio and decrease kidney weight in both sexes [75]. In a similar study, intrauterine nicotine exposure impaired cardiovascular outcomes, although this was dependent upon the genetic background of the rats used [76]. While these impaired cardiovascular outcomes may be associated with impaired renal development, to date a comprehensive analysis of renal structure and function has not been conducted in offspring of nicotine-exposed animals. Studies of in utero drug exposure have focused primarily on the neurological outcomes of the offspring, although one epidemiological review of renal tract abnormalities highlighted an increased incidence in infants exposed to cocaine [77]. This suggests that an in-depth examination of the long-term renal outcomes in response to in utero drug exposure is warranted.

WHY IS THE KIDNEY SUSCEPTIBLE TO PROGRAMMING?

Other chapters of this book have discussed the ability of maternal perturbations to affect a wide range of organs and systems, however the kidney appears to be particularly susceptible with reduced kidney size and a reduction in the final complement of nephrons commonly reported. The reasons for the sensitivity of the kidney to developmental insults are essentially unknown but may be due in part to the hemodynamics of the fetal renal system, which are quite different in utero compared to ex utero life. The fetal kidney produces hypotonic urine at a high flow rate, which is excreted into the amniotic fluid ensuring healthy fetal growth and development. However, only 3–5% of total cardiac output is distributed to both fetal kidneys [78] compared to 20–25% of total cardiac output in the adult. In addition, fetal GFR is only half of what is seen in the adult. The relatively low functional capacity of the fetal kidney due to low renal perfusion and GFR requires the placenta to perform a significant proportion of fetal fluid balance maintenance. Upon birth, the neonatal kidney abruptly loses placental support and requires the already-formed nephrons to undergo profound structural and functional maturation to meet the physiological demands of ex utero life. This includes increased glomerular filtration rate (GFR), renal blood flow, and tubular reabsorption of sodium [79]. As testament to the fact the kidneys are not essential for prenatal development, infants with renal agenesis (Potter sequence) are able to survive in utero, albeit with morphogenetic malformations caused by insufficient amniotic fluid volume (oligohydramnios). These infants are unable to survive after birth, however, due to renal failure and pulmonary insufficiency. It has been suggested that in times of fetal stress such as underoxygenation and impaired nutrient delivery, the fetus redistributes cardiac output toward immediately vital organs such as the brain and heart at the expense of "nonessential" peripheral organs such as the kidney. Reduced perfusion of the kidneys may lead to impaired nephrogenesis, resulting in decreased nephron number. Together, the high susceptibility of the kidney to in utero insults is likely due to the combined effects of a low fetal renal blood flow, the fact that the kidneys are not essential for prenatal survival, and the inability to form new nephrons after birth.

MECHANISMS LEADING TO IMPAIRED RENAL STRUCTURE AND FUNCTION

As the Brenner hypothesis suggests, the nephron deficit caused by a prenatal perturbation, combined with alterations in renal function, particularly sodium excretion, are likely to contribute to the adult disease outcomes. This leads to two important questions: firstly, what are the molecular pathways through which a prenatal perturbation can inhibit nephrogenesis; and secondly, what are the long-term "programmed" renal changes that result in altered function in the adult offspring? Below we shall focus our discussion on these questions with emphasis on the important role of factors controlling sodium homeostasis, that is, the renal sodium channels and the renal renin-angiotensin system. We shall also consider the evidence that epigenetic changes play a role in the programming of altered kidney development and impaired renal function in offspring. Figure 3 outlines the links between the disturbances in kidney development with impairments in offspring renal function that may ultimately contribute to chronic renal disease.

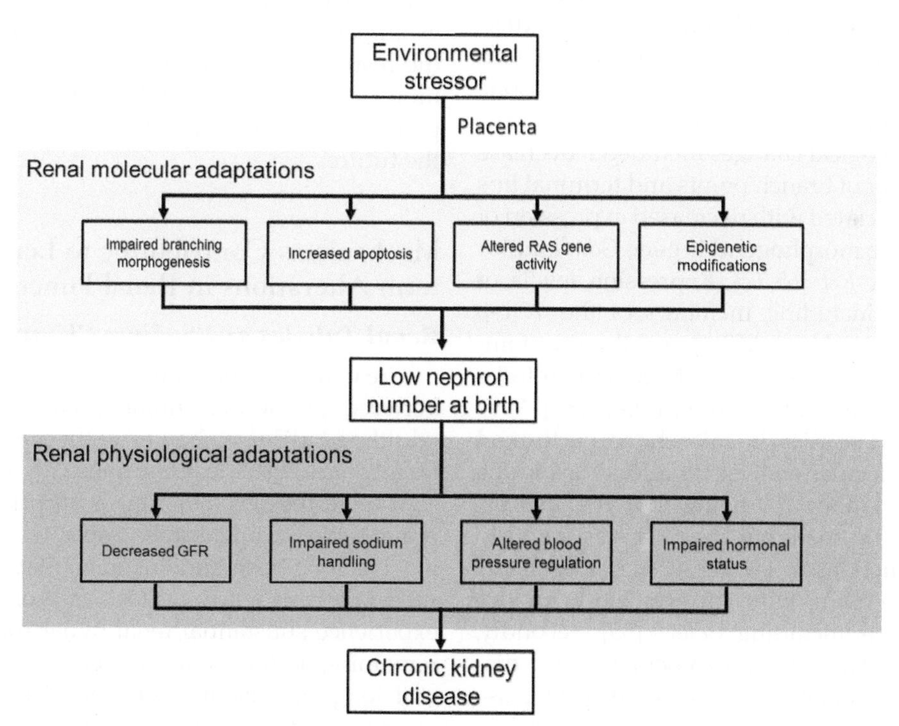

FIGURE 3 Potential pre- and postnatal pathways through which a suboptimal in utero environment may contribute to the programming of chronic kidney disease.

Mechanisms Contributing to Altered Renal Development

Despite the focus on nephron endowment in the offspring, there is a relative paucity of information concerning the molecular disruptions to the developing kidney by maternal insults. Mechanisms that restrict nephron formation could occur at any stage of kidney development: branching morphogenesis, condensation of the mesenchyme around the epithelium, mesenchymal-to-epithelial transition, or nephrogenesis cessation. Thus, maternal perturbations during particular "critical windows" of renal development may impact different stages of nephrogenesis, and of relevance to animal studies, this may occur at different stages of gestation (Figure 1).

Impairment of branching morphogenesis, a key regulator of nephron endowment in the developing kidney, has been reported in a range of developmental programming studies. In vitro, metanephric organ cultures have been used to show a direct effect of dexamethasone [80] or alcohol [72] on branching morphogenesis. The morphological changes included a decrease in the number of branch points and terminal tips and was associated with decreased expression of the branching morphogenesis gene *Gdnf* and/or its receptor, *c-Ret* [65,72]. Expression levels of other genes including members of the TGF-β superfamily (such as BMP-4) and the Wnt family (Wnt-4 and Wnt-11) have been shown to be altered by maternal low protein diets [81]. Very recently, two studies in mice have performed detailed molecular analysis of alterations in the developing kidney following maternal perturbations. Firstly, maternal hypoxia was found to induce reductions in β-catenin signaling in the ureteric tree along with suppression of downstream targets including *Ccnd1* [55]. Secondly, a mouse model of streptozotocin-induced diabetes during pregnancy was found to increase expression of hedgehog interacting protein (HHIP) resulting in alterations in TGF-β1 [82].

Activation of apoptotic pathways in the developing kidney has also been proposed as a mechanism contributing to a low nephron endowment. Increased apoptosis has been observed in a variety of rat and sheep models of maternal insults including maternal low protein [83] and hypoxia [84]. These data suggest that increased levels of apoptosis in the fetal kidney may contribute to decreased nephron endowment, however, whether apoptosis is localized to a specific renal cell type during development has not be explored definitively.

Collectively, these studies suggest that a nephron deficit may be due to suppression of key genes involved in the nephrogenic pathway. It is important to note that most studies focus on mRNA expression rather than protein, and several vital components of the nephrogenic process such as condensation and epithelialization have not been examined in the context of reduced nephron endowment. Thus, a more thorough examination into the mechanisms involved in the restriction of nephrogenesis, particularly in mouse models where molecular developmental pathways are well characterized, is warranted in the future.

Mechanisms Contributing to Long-term Alterations in Renal Function

Renal Tubules and Sodium Channels

Great focus has been placed upon the link between glomerular number and the risk of adulthood disease, however the literature has largely ignored the importance of the associated renal tubules [85]. This is surprising as the remarkable compensatory capacity of the kidney has long been known, given that nephrectomized patients who have lost ~50% of renal mass experience substantial renal hypertrophy of the remaining kidney and can persist with minimal long-term health outcomes. This compensatory growth has been attributed to primarily the proximal tubule, as the major site of sodium reabsorption in the nephron, and secondly to the distal tubular segments. An early study reported hypertrophy and hyperplasia of proximal and distal tubular segments following uninephrectomy in the adult rat [86], and a model of diabetic nephropathy in the rat also showed hypertrophy of the proximal and distal tubular segments [87]. The outcomes of tubular hypertrophy are likely to maintain sodium homeostasis in response to the increase in single nephron GFR and increased tubular fluid delivery [88]. Consequently, an examination of tubular structure in a model of maternal programming is needed to determine if or indeed how the kidney adapts to a significant reduction in nephron number.

It is important to note that postnatal maturation of renal tubules from fetal to adult life is associated with marked increases in the renal expression of sodium channels and transporters. This accounts for the increased capacity of the adult kidney to reabsorb sodium compared to the fetal kidney, which produces hypotonic urine in the unstressed mammalian fetus [10]. The predominant sodium transporter responsible for sodium balance is the sodium-hydrogen exchanger 3 (NHE3) expressed in the apical membrane and brush border of the proximal tubule as well as the thick ascending limb. NHE3 absorbs large amounts of sodium at high capacity, as well as the majority of HCO3- absorption. Other major apical sodium transporters that contribute to sodium reabsorption are the electroneural sodium-chloride cotransporter (NCC) expressed along the entire distal tubule, the amiloride-sensitive epithelial sodium channel (ENaC) expressed in the collecting duct and increasingly along the distal tubule, the antiporter enzyme Na^+,K^+-ATPase located in the plasma membrane of tubular cells and the type-2 Na-K-2Cl cotransporter in the thick ascending limb of Henle (NKCC2). Importantly, studies have reported higher expression levels of ENaCs and their subunits in females rather than males, highlighting the importance of studying the sexes separately in programming models [89].

The importance of each sodium channel and transporter has been demonstrated in both knockout mice studies and congenital defects in the human. Bartter syndrome in the human population is caused by a congenital mutation in the NKCC2 transporter and causes severe volume depletion and excess polyhydramnios (excess amniotic fluid), which frequently leads to premature birth [90]. Similarly, a mutation in the NCC transporter leads to a renal tubular disorder in humans known as Gitelman syndrome, which involves abnormal fluid and electrolyte homeostatic mechanisms [91]. Partial inactivity of the ENaC $\alpha1$ subunit in the mouse leads to a salt-wasting phenotype [92], and the NHE3 knockout mouse have hypotension and a marked inability to reabsorb bicarbonate and fluid in the proximal tubules [93].

As noted above, alterations to renal handling of sodium are increasingly reported in developmental programming studies, particularly in protein-deprivation models in rats. Offspring born to mothers fed a low-protein (9%, w/w, protein) diet had normal GFR, although urinary sodium excretion was increased suggesting impaired tubular reabsorption of sodium. Gene expression and protein analyses revealed that NKCC2 levels were elevated in the kidney, with no changes to NCC. This combined with immunohistochemical evidence that the Na^+,K^+-ATPase $\alpha1$ subunit was lost from the inner medulla suggests that the programming of hypertension and impaired sodium handling is intricately linked to the expression of sodium channels. These observations were also observed in a similar model of low-protein (6%, w/w, protein) diet in rats during the latter half of pregnancy, with offspring being born with low nephron numbers [94]. With age, offspring exhibit hypertension, slightly lower GFR, and significantly increased expression of NKCC2 (or bumetanide-sensitive cotransporter BSC1) and cotransporter thiazide-sensitive Na-Cl cotransporter (TSC) in the kidneys suggesting renal sodium handling and reabsorption may be programmed by maternal insult. Singh and colleagues [95] reported that sheep uninephrectomized in fetal life develop hypertension and low GFR. Deficits in renal sodium handling were observed in response to a salt load, and apical NHE3, ENaC β and γ subunits, and basolateral Na^+/K^+-ATPase β and γ subunits were significantly elevated in uninephrectomized animals, while ENaC α subunit expression was reduced. Similar increases in sodium channel and transporter expression were observed in a model of dexamethasone treatment for 2 days mid-gestation in the rat, with increased protein levels of NHE3 exchanger at the proximal tubules as well as a 50% increase in activity [96]. Together these data suggest that maternal perturbations can program changes in renal sodium handling as well as a reduction in nephron number, of which both may be implicated in the progression of hypertension in adult life.

Renin-Angiotensin System

The renin-angiotensin system (RAS) is a complex systemic cascade of interacting peptides and enzymes known to play a key role in the regulation of adult fluid and electrolyte homeostasis. Angiotensinogen is produced primarily in the liver and is secreted into the systemic circulation where it is converted into angiotensin I by the enzyme renin, which is secreted from the juxtaglomerular apparatus of the kidney. While other tissues express the gene responsible for producing the active renin enzyme, it is only the kidney that secretes the active renin, which is fully capable of enzymatic cleavage of angiotensinogen. Angiotensin I is converted into angiotensin II by the angiotensin-converting enzyme (ACE) localized to endothelial cells throughout the body as well as the brush border of the renal proximal tubules. Angiotensin II is the most biologically active effector peptide of the RAS and binds to either the angiotensin II type 1 (Agtr1) receptor located within the renal tubule to increase sodium reabsorption or the Agtr2 receptor to decrease sodium reabsorption. While in species such as the human and sheep, a single Agtr1 isoform exists, in the rodent, two genes collectively perform these functions, namely Agtr1a and Agtr1b. In addition to the direct role of Agtr1 and Agtr2 in the kidney itself, signaling through these receptors in other tissues can indirectly influence renal function via adrenal production of aldosterone or through vascular effects. In addition to this predominant arm of the RAS, a number of additional peptides, enzymes, and receptors have all been shown to play a role in renal function and adult disease. Interestingly, in many organs, including the kidney, all components of the RAS are thought to coexist suggesting a role for a complete local RAS in such tissues. Programmed alterations to this system can have profound effects on both renal function and cardiovascular outcomes.

In addition to its role in adult physiology, the RAS is known to be involved in the development of many fetal organs and plays a key role in renal organogenesis. This is supported by the fact that when genes for renin, ACE, or Agtr1 are knocked out of the genome, mice develop a number of congenital abnormalities of the kidney and have decreased survival rates (reviewed by Kobori [97]). During development, Agtr1 signaling is generally accepted to have proliferative effects, while the activation of Agtr2 is thought to lead to antiproliferative, apoptotic phenotypes. Thus, while decreased activation of Agtr1 in adult life may reduce sodium reabsorption and reduce blood pressure, in fetal life, this may result in impairment of renal formation. Given the dual roles of the RAS in renal development and adult renal function, it is not surprising that the RAS is one of the best characterized systems shown to be programmed by in utero insults (for review, see [98]). It is also interesting to note that there are sex differences in the basal expression of many components of the renal RAS, however, the sex of the fetus has not been taken into account when investigating the renal RAS in programming models [98].

Renal expression of renin, ACE, and angiotensin II receptors have all been shown to be programmed in offspring prenatally exposed to a number of maternal perturbations (summarized in Table 2). Maternal exposure to a low-protein diet throughout pregnancy in the rat has been shown to reduce nephron number and increase blood pressure in offspring in association with a decrease in offspring Agtr2 mRNA levels [99]. A similar study in the rat demonstrated that protein restriction reduces renal renin mRNA levels, renal renin concentrations, and renal tissue angiotensin II levels [26]. In addition, in the previously mentioned rat model of high-fat intake during pregnancy and lactation, offspring developed increased blood pressure despite a reduction in renal renin activity [35].

In addition to these postnatal programmed changes to the renal RAS, dysregulation of the

TABLE 2 Effects of a Variety of Maternal Perturbations on the Renal Renin-Angiotensin System (RAS)

Perturbation	Species	Sexes Studied	Changes to Renal RAS during Nephrogenesis	Changes to Renal RAS after Completion of Nephrogenesis
IMPAIRED NUTRITION				
6% Low-protein diet [101]	Rat	Pooled	↓Agtr1b, ↑Agtr2	↑Agtr1a
90 g/kg Low-protein diet [99]	Rat	Pooled for RAS studies	NA	↓Agtr2 ↑MAP in response to angiotensin II
8.5% Low-protein diet [26]	Rat	Pooled for RAS studies	NA	↓ Renin, ↓ angiotensin II
50% Nutrient restriction [111]	Sheep	Males and females	↓Agtr1 protein	↑ Renin
20% High-fat diet [35]	Rat	Males and females	NA	↓ Renin activity in both sexes
8% High-salt diet [112]	Sheep	Not specified	NA	↑Agtr1a, ↑AGT
HYPOXIA				
Maternal hypoxia [54]	Rat	Males and females	↔Agtr1a, ↔Agtr1b, ↔Agtr2 in both sexes	↓Agtr1a, ↓Agtr1b ↔Agtr2 in both sexes
Uteroplacental insufficiency [102]	Rat	Males only	↓ Renin, ↓Angiotensinogen	↑ Renin, ↑Angiotensinogen
Uteroplacental insufficiency [59,60]	Rat	Males Females	NA	↑Agtr1a males ↔Agtr1a females
Maternal nicotine exposure [75]	Rat	Males and females	NA	↓Agtr2, ↑Agtr1/2 ratio
GLUCOCORTICOID EXPOSURE				
Dexamethasone [100]	Sheep	Pooled	NA	↑ Angiotensinogen, ↑Agtr1, ↑Agtr2
Cortisol [100]	Sheep	Pooled	NA	↑Agtr1
Corticosterone [69]	Rat	Males and females studied separately in offspring studies	↑Agtr1a, ↓Agtr1b, ↑ Agtr2	↑Agtr1a, ↑Agtr1b, ↑ Agtr2 in males ↔Agtr1a, ↑Agtr1b, ↑ Agtr2 in females

NA, no data available.

fetal renal RAS has been shown to be a common finding in both rodent and sheep models of maternal insult (summarized in Table 2). A notable study demonstrated that dexamethasone exposure for 48h from D27 of pregnancy in the sheep induces high blood pressure in offspring in association with increased mRNA levels of renal angiotensinogen, Agtr1, and Agrt2 at 130 days of gestation [100]. In a number of programming studies, fetal adaptations to the renal RAS and programmed changes to the RAS in the kidneys of offspring occur in response to the same maternal challenge suggesting that adaptive or maladaptive changes to the renal RAS that occur in utero may persist into adult life where they may have opposing effects on adult physiology. An example of this age-dependent dysregulation of the RAS is demonstrated in a rat model of corticosterone exposure in which renal Agtr1a and Agtr2 were increased and Agtr1b was decreased during nephrogenesis and both Agtr1b and Agtr2 were increased in adolescent offspring, at which time Agtr1 was unaffected [69]. Similarly, in a rat model of protein restriction, Agtr1b was decreased during kidney development and Agtr1a was increased in postnatal life [101]. This biphasic fetal and postnatal dysregulation of the RAS has also been demonstrated in a model of intrauterine growth restriction induced through uteroplacental insufficiency that is known to induce hypertension in offspring. In this model, renal renin and angiotensinogen mRNA levels were decreased at birth but were increased in adult offspring. This was additionally associated with an increase in offspring ACE activity [102]. Finally, in a model of fetal hypoxia known to impair nephron endowment, while AGTR1 protein levels were unchanged during nephrogenesis, AGTR1 protein levels as well as Agtr1a and Agtr1b mRNA levels were decreased in adult offspring [54]. It is interesting to note that in some of the models described above, alterations to the fetal renal RAS may be protective in the short term but in the longer term are detrimental to offspring health.

Epigenetic and Transgenerational Regulation of Renal Function

Many of the studies detailed above demonstrate key changes to gene expression that persist into adult life. While in many cases the postnatal dysregulation of gene expression may be due to programmed adaptations that occur in adulthood in response to structural or physiological impairments, recent evidence has strongly suggested that epigenetic regulation of the genome may set up postnatal regulation of gene expression during the prenatal period. This book has discussed in depth the role of epigenetics in programming outcomes with similar mechanisms likely to be involved in all organs and systems, nevertheless it is important to discuss here what is known about the role of epigenetics in the lifelong programming of impaired renal function. It is important to note that while epigenetic regulation of programming outcomes in other tissues is relatively well understood, until recently, few studies had mechanistically investigated the role of epigenetics in the dysregulation of kidney development and lifelong kidney function. Initially studies demonstrated clear links between kidney dysfunction in adulthood and altered epigenetic status. These studies did not link disease outcomes with in utero events but rather demonstrated the importance of epigenetics in the progression of kidney disease. An important study demonstrated that renal fibrosis could be minimized by preventing DNA methylation. Folic acid-induced kidney fibrosis was ameliorated by administration of a demethylating agent (5′-azacytidine), which also reduced the accumulation of activated fibroblasts and prevented kidney failure [103]. Histone modification has also been shown to be an important regulator of renal disease as disrupted histone acetylation, demethylation, and phosphorylation were all detected in mouse models of advanced diabetic nephropathy [104].

Human studies have found links between renal epigenetic modifications and kidney disease in adulthood with some suggestions that these modifications may have been established in utero. Genome-wide cytosine methylation profiling of kidney tubular epithelial cells from healthy patients and those with chronic kidney disease identified more than 4000 differentially methylated regions between groups. While many of the identified methylated regions were enriched for genes related to fibrosis and kidney disease itself, a large number of differentially methylated regions contained binding sites for key kidney developmental genes that are not expressed in adulthood. The authors of this study suggest that many of the epigenetic differences between these control and diseased patients are likely to have been established in utero and may have programmed disease outcomes in these individuals [105]. While studies such as these suggest that adult disease may be regulated by epigenetic mechanisms, several recent animal models have been used to demonstrate epigenetic dysregulation after direct manipulation of the maternal/fetal environment. In a baboon study that investigated the effects of maternal undernutrition on global methylation patterns, nutrient restriction was found to increase global DNA methylation in the kidney toward the end of gestation [106]. While the effects of this global methylation on fetal outcomes were not fully described, it is likely that methylation of individual genes may play a key role in the long-term dysregulation of renal gene expression. More specific analyses of the methylation status of individual genes were performed in a model of uteroplacental insufficiency in the rat. This study demonstrated renal cellular apoptosis along with decreased CpG methylation of the p53 promoter and the associated increases in p53 gene expression [107]. In addition, class one histone deacetylase mRNA levels have been reported to be increased in the kidneys of offspring programmed to develop hypertension through prenatal exposure to glucocorticoids [108].

The role of epigenetic regulation of programmed disease has become a central focus of investigation particularly in models where the period of exposure occurs prior to pregnancy or organ development as well as when investigating intergenerational programming outcomes. A recent study has investigated the effects of pesticide exposure during fetal gonad development on offspring outcomes. This model not only investigated the direct progeny, but also subsequent generations who originated from gametes that were formed during the pesticide exposure. The F1 generation demonstrated kidney disease in response to prenatal pesticide exposure in both males and females. Interestingly, F3 offspring similarly demonstrated kidney disease in response to being ancestrally exposed to pesticides during development. This study revealed that sperm of the F3 offspring ancestrally exposed to pesticide contained more than 300 epimutations and suggested that these epigenetic changes are likely to be responsible for the intergenerational programming of kidney disease reported in this study [109]. A model of uteroplacental insufficiency has clearly demonstrated that regulators of impaired nephron formation can be passed on to not only the first generation but also subsequent generations, suggesting intergenerational programming, possibly through epigenetic dysregulation [110]. Collectively, while accumulating evidence suggests that epigenetics plays a central role in the regulation of programmed kidney dysfunction, future studies will help to further define the mechanisms through which this occurs.

CONCLUSIONS

Human epidemiological studies combined with experimental data from animal models provide strong evidence for a key role of impaired renal development contributing to the onset of chronic kidney disease in adulthood.

The congenital nephron deficit that occurs following a range of maternal perturbations likely also contributes to the programming of high blood pressure, a major risk factor for cardiovascular disease and diabetes. The current challenge for clinicians is the ability to diagnose an infant with a low nephron endowment, and thus at risk of later disease. This is crucial to allow preventative measures and/or early intervention but will require advances in medical imaging to allow accurate determination of nephron number. Furthermore, there is an urgent need to gain insights into the mechanisms that result in impairments in renal development including the role of epigenetics. Researchers in the DOHaD field should be considering expanding the use of mouse models of developmental programming and exploiting the knowledge of the mouse genome and molecular tools available.

References

[1] Jha V, Garcia-Garcia G, Iseki K, et al. Chronic kidney disease: global dimension and perspectives. Lancet 2013;382(9888):260–72.

[2] Radhakrishnan J, Remuzzi G, Saran R, et al. Taming the chronic kidney disease epidemic: a global view of surveillance efforts. Kidney Int 2014;86(2):246–50.

[3] Lackland DT, Bendall HE, Osmond C, Egan BM, Barker DJ. Low birth weights contribute to high rates of early-onset chronic renal failure in the Southeastern United States. Arch Intern Med 2000;160(10):1472–6.

[4] Vikse BE, Irgens LM, Leivestad T, Hallan S, Iversen BM. Low birth weight increases risk for end-stage renal disease. J Am Soc Nephrol 2008;19(1):151–7.

[5] Li S, Chen SC, Shlipak M, et al. Low birth weight is associated with chronic kidney disease only in men. Kidney Int 2008;73(5):637–42.

[6] White SL, Perkovic V, Cass A, et al. Is low birth weight an antecedent of CKD in later life? A systematic review of observational studies. Am J Kidney Dis 2009;54(2):248–61.

[7] Nelson RG, Morgenstern H, Bennett PH. An epidemic of proteinuria in Pima Indians with type 2 diabetes mellitus. Kidney Int 1998;54(6):2081–8.

[8] Hoy WE, Rees M, Kile E, Mathews JD, Wang Z. A new dimension to the Barker hypothesis: low birthweight and susceptibility to renal disease. Kidney Int 1999;56(3):1072–7.

[9] Hoy WE, White AV, Tipiloura B, et al. The influence of birthweight, past poststreptococcal glomerulonephritis and current body mass index on levels of albuminuria in young adults: the multideterminant model of renal disease in a remote Australian Aboriginal population with high rates of renal disease and renal failure. Nephrol Dial Transpl 2014;24.

[10] Moritz KM, Wintour EM, Black MJ, Bertram JF, Caruana G. Factors influencing mammalian kidney development: implications for health in adult life. Adv Anat Embryol Cell Biol 2008;196:1–78.

[11] Luyckx VA, Bertram JF, Brenner BM, et al. Effect of fetal and child health on kidney development and long-term risk of hypertension and kidney disease. Lancet 2013;382(9888):273–83.

[12] Hughson M, Farris III AB, Douglas-Denton R, Hoy WE, Bertram JF. Glomerular number and size in autopsy kidneys: the relationship to birth weight. Kidney Int 2003;63(6):2113–22.

[13] Brenner BM, Chertow GM. Congenital oligonephropathy and the etiology of adult hypertension and progressive renal injury. Am J Kidney Dis 1994;23(2):171–5.

[14] Brenner BM, Garcia DL, Anderson S. Glomeruli and blood pressure. Less of one, more the other? Am J Hypertens 1988;1(4 Pt 1):335–47.

[15] Keller G, Zimmer G, Mall G, Ritz E, Amann K. Nephron number in patients with primary hypertension. N Engl J Med 2003;348(2):101–8.

[16] Hughson M, Farris III AB, Douglas-Denton R, Hoy WE, Bertram JF. Glomerular number and size in autopsy kidneys: the relationship to birth weight. Kidney Int 2003;63(6):2113–22.

[17] Hughson MD, Douglas-Denton R, Bertram JF, Hoy WE. Hypertension, glomerular number, and birth weight in African Americans and white subjects in the southeastern United States. Kidney Int 2006;69(4):671–8.

[18] Schmidt K, Pesce C, Liu Q, et al. Large glomerular size in Pima Indians: lack of change with diabetic nephropathy. J Am Soc Nephrol 1992;3(2):229–35.

[19] Parraguez VH, Atlagich M, Araneda O, et al. Effects of antioxidant vitamins on newborn and placental traits in gestations at high altitude: comparative study in high and low altitude native sheep. Reprod Fert Dev 2011;23(2):285–96.

[20] Halliwell B, Gutteridge JM. Free radicals in biology and medicine. Oxford: Oxford University Press; 1999.

[21] Zamudio S, Kovalenko O, Vanderlelie J, et al. Chronic hypoxia in vivo reduces placental oxidative stress. Placenta 2007;28(8–9):846–53.

[22] Singh RR, Cuffe JS, Moritz KM. Short- and long-term effects of exposure to natural and synthetic glucocorticoids during development. Clin Exp Pharmacol Physiol 2012;39(11):979–89.

[23] Cullen-McEwen LA, Douglas-Denton RN, Bertram JF. Estimating total nephron number in the adult kidney using the physical disector/fractionator combination. Methods Mol Biol 2012;886:333–50.
[24] Bagby SP. Maternal nutrition, low nephron number, and hypertension in later life: pathways of nutritional programming. J Nutr 2007;137(4):1066–72.
[25] Langley-Evans SC, Welham SJ, Jackson AA. Fetal exposure to a maternal low protein diet impairs nephrogenesis and promotes hypertension in the rat. Life Sci 1999;64(11):965–74.
[26] Woods LL, Ingelfinger JR, Nyengaard JR, Rasch R. Maternal protein restriction suppresses the newborn renin-angiotensin system and programs adult hypertension in rats. Pediatr Res 2001;49(4):460–7.
[27] Woods LL, Weeks DA, Rasch R. Programming of adult blood pressure by maternal protein restriction: role of nephrogenesis. Kidney Int 2004;65(4):1339–48.
[28] Hoppe CC, Evans RG, Moritz KM, et al. Combined prenatal and postnatal protein restriction influences adult kidney structure, function, and arterial pressure. Am J Physiol Regul Integr Comp Physiol 2007;292(1):R462–9.
[29] Alwasel SH, Ashton N. Prenatal programming of renal sodium handling in the rat. Clin Sci 2009;117(2): 75–84.
[30] Alwasel SH, Barker DJ, Ashton N. Prenatal programming of renal salt wasting resets postnatal salt appetite, which drives food intake in the rat. Clin Sci 2012;122(6):281–8.
[31] World Health Organisation. Obesity and overweight: fact sheet. 2012. N311.
[32] Athukorala C, Rumbold AR, Willson KJ, Crowther CA. The risk of adverse pregnancy outcomes in women who are overweight or obese. BMC Pregnancy Childbirth 2010;10(56):1471–2393.
[33] Blackmore HL, Ozanne SE. Maternal diet-induced obesity and offspring cardiovascular health. J Dev Orig Health Dis 2013;4(5):338–47.
[34] Wickman C, Kramer H. Obesity and kidney disease: potential mechanisms. Semin Nephrol 2013; 33(1):14–22.
[35] Armitage JA, Lakasing L, Taylor PD, et al. Developmental programming of aortic and renal structure in offspring of rats fed fat-rich diets in pregnancy. J Physiol 2005;565(Pt 1):171–84.
[36] Jackson CM, Alexander BT, Roach L, et al. Exposure to maternal overnutrition and a high-fat diet during early postnatal development increases susceptibility to renal and metabolic injury later in life. Am J Physiol Ren Physiol 2012;302(6):7.
[37] Samuelsson AM, Morris A, Igosheva N, et al. Evidence for sympathetic origins of hypertension in juvenile offspring of obese rats. Hypertension 2010;55(1):76–82.
[38] Amri K, Freund N, Vilar J, Merlet-Benichou C, Lelievre-Pegorier M. Adverse effects of hyperglycemia on kidney development in rats: in vivo and in vitro studies. Diabetes 1999;48(11):2240–5.
[39] Chen YW, Chenier I, Tran S, Scotcher M, Chang SY, Zhang SL. Maternal diabetes programs hypertension and kidney injury in offspring. Pediatr Nephrol 2010;25(7):1319–29.
[40] Prior LJ, Davern PJ, Burke SL, Lim K, Armitage JA, Head GA. Exposure to a high-fat diet during development alters leptin and ghrelin sensitivity and elevates renal sympathetic nerve activity and arterial pressure in rabbits. Hypertension 2014;63(2):338–45.
[41] Lewis RM, Forhead AJ, Petry CJ, Ozanne SE, Hales CN. Long-term programming of blood pressure by maternal dietary iron restriction in the rat. Brit J Nutr 2002; 88(3):283–90.
[42] Tomat AL, Inserra F, Veiras L, et al. Moderate zinc restriction during fetal and postnatal growth of rats: effects on adult arterial blood pressure and kidney. Am J Physiol Regul Integr Comp Physiol 2008;295(2):R543–9.
[43] Lelièvre-Pégorier M, Vilar J, Ferrier M-L, et al. Mild vitamin A deficiency leads to inborn nephron deficit in the rat. Kidney Int 1998;54(5):1455–62.
[44] Crowe C, Dandekar P, Fox M, Dhingra K, Bennet L, Hanson MA. The effects of anaemia on heart, placenta and body weight, and blood pressure in fetal and neonatal rats. J Physiol 1995;488(Pt 2):515–9.
[45] Maka N, Makrakis J, Parkington HC, Tare M, Morley R, Black MJ. Vitamin D deficiency during pregnancy and lactation stimulates nephrogenesis in rat offspring. Pediatr Nephrol 2008;23(1):55–61.
[46] Giussani DA, Phillips PS, Anstee S, Barker DJ. Effects of altitude versus economic status on birth weight and body shape at birth. Pediatr Res 2001;49(4):490–4.
[47] Bulterys MG, Greenland S, Kraus JF. Chronic fetal hypoxia and sudden infant death syndrome: interaction between maternal smoking and low hematocrit during pregnancy. Pediatrics 1990;86(4):535–40.
[48] Sahin FK, Koken G, Cosar E, et al. Obstructive sleep apnea in pregnancy and fetal outcome. Int J Gynaecol Obstet 2008;100(2):141–6.
[49] Krebs C, Macara LM, Leiser R, Bowman AW, Greer IA, Kingdom JCP. Intrauterine growth restriction with absent end-diastolic flow velocity in the umbilical artery is associated with maldevelopment of the placental terminal villous tree. Am J Obstet Gynecol 1996;175(6):1534–42.
[50] Cohn HE, Sacks EJ, Heymann MA, Rudolph AM. Cardiovascular responses to hypoxemia and acidemia in fetal lambs. Am J Obstet Gynecol 1974;120(6):817–24.
[51] Robillard JE, Weitzman RE, Burmeister L, Smith FG. Developmental aspects of the renal response to hypoxemia in the lamb fetus. Circ Res 1981;48(1):128–38.

[52] Weismann DN, Robillard JE. Renal hemodynamic responses to hypoxemia during development: relationships to circulating vasoactive substances. Pediatr Res 1988;23(2):155–62.

[53] Iwamoto HS, Rudolph AM. Metabolic responses of the kidney in fetal sheep: effect of acute and spontaneous hypoxemia. Am J Physiol Ren Physiol 1985;249(6):F836–41.

[54] Gonzalez-Rodriguez Jr P, Tong W, Xue Q, Li Y, Hu S, Zhang L. Fetal hypoxia results in programming of aberrant angiotensin II receptor expression patterns and kidney development. Int J Med Sci 2013;10(5):532–8.

[55] Wilkinson L, Neal CS, Singh RR, et al. Renal development defects resulting from in utero hypoxia are associated with suppression of ureteric beta-catenin signalling. Kidney Int October 2014. Accepted.

[56] Al-Odat I, Chen H, Chan YL, et al. The impact of maternal cigarette smoke exposure in a rodent model on renal development in the offspring. PLoS One 2014;9(7):e103443.

[57] Li J, LaMarca B, Reckelhoff JF. A model of preeclampsia in rats: the reduced uterine perfusion pressure (RUPP) model. Am J Physiol Heart Circ Physiol 2012; 303(1):20.

[58] Wlodek ME, Westcott K, Siebel AL, Owens JA, Moritz KM. Growth restriction before or after birth reduces nephron number and increases blood pressure in male rats. Kidney Int 2008;74(2):187–95.

[59] Moritz KM, Mazzuca MQ, Siebel AL, et al. Uteroplacental insufficiency causes a nephron deficit, modest renal insufficiency but no hypertension with ageing in female rats. J Physiol 2009;587(Pt 11):2635–46.

[60] Wlodek ME, Mibus A, Tan A, Siebel AL, Owens JA, Moritz KM. Normal lactational environment restores nephron endowment and prevents hypertension after placental restriction in the rat. J Am Soc Nephrol 2007;18(6):1688–96.

[61] Zohdi V, Moritz KM, Bubb KJ, et al. Nephrogenesis and the renal renin-angiotensin system in fetal sheep: effects of intrauterine growth restriction during late gestation. Am J Physiol Regul Integr Comp Physiol 2007;293(3):R1267–73.

[62] Wintour EM, Moritz KM, Johnson K, Ricardo S, Samuel CS, Dodic M. Reduced nephron number in adult sheep, hypertensive as a result of prenatal glucocorticoid treatment. J Physiol 2003;549(Pt 3):929–35.

[63] Zhang J, Massmann GA, Rose JC, Figueroa JP. Differential effects of clinical doses of antenatal betamethasone on nephron endowment and glomerular filtration rate in adult sheep. Reprod Sci 2010;17(2):186–95.

[64] Li S, Sloboda DM, Moss TJ, et al. Effects of glucocorticoid treatment given in early or late gestation on growth and development in sheep. J Dev Orig Health Dis 2013;4(2):146–56.

[65] Dickinson H, Walker DW, Wintour EM, Moritz K. Maternal dexamethasone treatment at midgestation reduces nephron number and alters renal gene expression in the fetal spiny mouse. Am J Physiol Regul Integr Comp Physiol 2007;292(1):R453–61.

[66] Ortiz LA, Quan A, Zarzar F, Weinberg A, Baum M. Prenatal dexamethasone programs hypertension and renal injury in the rat. Hypertension 2003;41(2):328–34.

[67] O'Sullivan L, Cuffe JS, Paravicini TM, et al. Prenatal exposure to dexamethasone in the mouse alters cardiac growth patterns and increases pulse pressure in aged male offspring. PLoS One 2013;8(7):e69149.

[68] Moritz KM, De Matteo R, Dodic M, et al. Prenatal glucocorticoid exposure in the sheep alters renal development in utero: implications for adult renal function and blood pressure control. Am J Physiol Regul Integr Comp Physiol 2011;301(2):18.

[69] Singh RR, Cullen-McEwen LA, Kett MM, et al. Prenatal corticosterone exposure results in altered AT1/AT2, nephron deficit and hypertension in the rat offspring. J Physiol 2007;579(Pt 2):503–13.

[70] Qazi Q, Masakawa A, Milman D, McGann B, Chua A, Haller J. Renal anomalies in fetal alcohol syndrome. Pediatrics 1979;63(6):886–9.

[71] Gray SP, Kenna K, Bertram JF, et al. Repeated ethanol exposure during late gestation decreases nephron endowment in fetal sheep. 2008.

[72] Gray SP, Denton KM, Cullen-McEwen L, Bertram JF, Moritz KM. Prenatal exposure to alcohol reduces nephron number and raises blood pressure in progeny. J Am Soc Nephrol 2010;21(11):1891–902.

[73] Assadi FK, Manaligod JR, Fleischmann LE, Zajac CS. Effects of prenatal ethanol exposure on postnatal renal function and structure in the rat. Alcohol 1991;8(4):259–63.

[74] Taal HR, Geelhoed JJ, Steegers EA, et al. Maternal smoking during pregnancy and kidney volume in the offspring: the Generation R Study. Pediatr Nephrol 2011;26(8):1275–83.

[75] Mao C, Wu J, Xiao D, et al. The effect of fetal and neonatal nicotine exposure on renal development of AT(1) and AT(2) receptors. Reprod Toxicol 2009;27(2): 149–54.

[76] Pausova Z, Paus T, Sedova L, Berube J. Prenatal exposure to nicotine modifies kidney weight and blood pressure in genetically susceptible rats: a case of gene-environment interaction. Kidney Int 2003;64(3):829–35.

[77] Battin M, Albersheim S, Newman D. Congenital genitourinary tract abnormalities following cocaine exposure in utero. Am J Perinatol 1995;12(06):425–8.

[78] Behrman RE, Lees MH, Peterson EN, De Lannoy CW, Seeds AE. Distribution of the circulation in the normal and asphyxiated fetal primate. Am J Obstet Gynecol 1970;108(6):956–69.

[79] Aperia A, Broberger O, Elinder G, Herin P, Zetterstraum R. Postnatal development of renal function in pre-term and full-term infants. Acta Paediatr 1981;70(2):183–7.
[80] Singh RR, Moritz KM, Bertram JF, Cullen-McEwen LA. Effects of dexamethasone exposure on rat metanephric development: in vitro and in vivo studies. Am J Physiol Ren Physiol 2007;293(2):F548–54.
[81] Abdel-Hakeem AK, Henry TQ, Magee TR, et al. Mechanisms of impaired nephrogenesis with fetal growth restriction: altered renal transcription and growth factor expression. Am J Obstet Gynecol 2008;199(3):17.
[82] Zhao XP, Liao MC, Chang SY, et al. Maternal diabetes modulates kidney formation in murine progeny: the role of hedgehog interacting protein (HHIP). Diabetologia 2014;57(9):1986–96.
[83] Welham SJ, Wade A, Woolf AS. Protein restriction in pregnancy is associated with increased apoptosis of mesenchymal cells at the start of rat metanephrogenesis. Kidney Int 2002;61(4):1231–42.
[84] Xia S, Lv J, Gao Q, et al. Prenatal exposure to hypoxia induced Beclin 1 signaling-Mediated renal Autophagy and altered renal development in rat fetuses. Reprod Sci 2014;28. http://dx.doi.org/10.1177/1933719114536474.
[85] Orlov SN, Mongin AA. Salt-sensing mechanisms in blood pressure regulation and hypertension. Am J Physiol Heart Circ Physiol 2007;293(4):H2039–53.
[86] Hayslett JP, Kashgarian M, Epstein FH. Functional correlates of compensatory renal hypertrophy. J Clin Invest 1968;47(4):774.
[87] Nyengaard JR, Rasch R. The impact of experimental diabetes mellitus in rats on glomerular capillary number and sizes. Diabetologia 1993;36(3):189–94.
[88] Baum M, Quigley R, Satlin L. Maturational changes in renal tubular transport. Curr Opin Nephrol Hypertens 2003;12(5):521–6.
[89] Gambling L, Dunford S, Wilson CA, McArdle HJ, Baines DL. Estrogen and progesterone regulate alpha, beta, and gammaENaC subunit mRNA levels in female rat kidney. Kidney Int 2004;65(5):1774–81.
[90] Simon DB, Karet FE, Hamdan JM, Di Pietro A, Sanjad SA, Lifton RP. Bartter's syndrome, hypokalaemic alkalosis with hypercalciuria, is caused by mutations in the Na–K–2Cl cotransporter NKCC2. Nat Genet 1996;13(2):183–8.
[91] de Jong JC, van der Vliet WA, van den Heuvel LP, Willems PH, Knoers NV, Bindels RJ. Functional expression of mutations in the human NaCl cotransporter: evidence for impaired routing mechanisms in Gitelman's syndrome. J Am Soc Nephrol 2002;13(6):1442–8.
[92] Wang Q, Hummler E, Maillard M, et al. Compensatory up-regulation of angiotensin II subtype 1 receptors in [agr]ENaC knockout heterozygous mice. Kidney Int 2001;59(6):2216–21.
[93] Schultheis PJ, Clarke LL, Meneton P, et al. Renal and intestinal absorptive defects in mice lacking the NHE3 Na^+/H^+ exchanger. Nat Genet 1998;19(3):282–5.
[94] Manning J, Beutler K, Knepper MA, Vehaskari VM. Upregulation of renal BSC1 and TSC in prenatally programmed hypertension. Am J Physiol Ren Physiol 2002;283(1):F202–6.
[95] Singh RR, Denton KM, Bertram JF, Jefferies AJ, Moritz KM. Reduced nephron endowment due to fetal uninephrectomy impairs renal sodium handling in male sheep. Clin Sci (Lond) 2010;118(11):669–80.
[96] Dagan A, Gattineni J, Cook V, Baum M. Prenatal programming of rat proximal tubule Na^+/H^+ exchanger by dexamethasone. Am J Physiol Regul Integr Comp Physiol 2007;292(3):R1230–5.
[97] Kobori H, Nangaku M, Navar LG, Nishiyama A. The intrarenal renin-angiotensin system: from physiology to the pathobiology of hypertension and kidney disease. Pharmacol Rev 2007;59(3):251–87.
[98] Moritz KM, Cuffe JS, Wilson LB, et al. Review: sex specific programming: a critical role for the renal renin-angiotensin system. Placenta 2010;31(Suppl.):S40–6.
[99] McMullen S, Gardner DS, Langley-Evans SC. Prenatal programming of angiotensin II type 2 receptor expression in the rat. Br J Nutr 2004;91(1):133–40.
[100] Moritz KM, Johnson K, Douglas-Denton R, Wintour EM, Dodic M. Maternal glucocorticoid treatment programs alterations in the renin-angiotensin system of the ovine fetal kidney. Endocrinology 2002;143(11):4455–63.
[101] Vehaskari VM, Stewart T, Lafont D, Soyez C, Seth D, Manning J. Kidney angiotensin and angiotensin receptor expression in prenatally programmed hypertension. Am J Physiol Ren Physiol 2004;287(2):20.
[102] Grigore D, Ojeda NB, Robertson EB, et al. Placental insufficiency results in temporal alterations in the renin angiotensin system in male hypertensive growth restricted offspring. Am J Physiol Regul Integr Comp Physiol 2007;293(2):30.
[103] Bechtel W, McGoohan S, Zeisberg EM, et al. Methylation determines fibroblast activation and fibrogenesis in the kidney. Nat Med 2010;16(5):544–50.
[104] Sayyed SG, Gaikwad AB, Lichtnekert J, et al. Progressive glomerulosclerosis in type 2 diabetes is associated with renal histone H3K9 and H3K23 acetylation, H3K4 dimethylation and phosphorylation at serine 10. Nephrol Dial Transpl 2010;25(6):1811–7.
[105] Ko YA, Mohtat D, Suzuki M, et al. Cytosine methylation changes in enhancer regions of core pro-fibrotic genes characterize kidney fibrosis development. Genome Biol 2013;14(10):R108.
[106] Unterberger A, Szyf M, Nathanielsz PW, Cox LA. Organ and gestational age effects of maternal nutrient restriction on global methylation in fetal baboons. J Med Primatol 2009;38(4):219–27.

[107] Pham TD, MacLennan NK, Chiu CT, Laksana GS, Hsu JL, Lane RH. Uteroplacental insufficiency increases apoptosis and alters p53 gene methylation in the full-term IUGR rat kidney. Am J Physiol Regul Integr Comp Physiol 2003;285(5):17.

[108] Tain YL, Sheen JM, Chen CC, et al. Maternal citrulline supplementation prevents prenatal dexamethasone-induced programmed hypertension. Free Radic Res 2014;48(5):580–6.

[109] Manikkam M, Haque MM, Guerrero-Bosagna C, Nilsson EE, Skinner MK. Pesticide methoxychlor promotes the epigenetic transgenerational inheritance of adult-onset disease through the female germline. PLoS One 2014;9(7).

[110] Gallo LA, Tran M, Cullen-McEwen LA, et al. Transgenerational programming of fetal nephron deficits and sex-specific adult hypertension in rats. Reprod Fertil Dev 2014;26(7):1032–43.

[111] Gilbert JS, Ford SP, Lang AL, et al. Nutrient restriction impairs nephrogenesis in a gender-specific manner in the ovine fetus. Pediatr Res 2007;61(1):42–7.

[112] Mao C, Liu R, Bo L, et al. High-salt diets during pregnancy affected fetal and offspring renal renin-angiotensin system. J Endocrinol 2013;218(1):61–73.

CHAPTER

16

Cancer and Developmental Origins of Health and Disease—Epigenetic Reprogramming as a Mediator

Shuk-Mei Ho[1,2,3,4], *Ana Cheong*[1,2], *Sarah To*[1], *Vinothini Janakiram*[1,2], *Pheruza Tarapore*[1,2,3], *Yuet-Kin Leung*[1,2,3]

[1]Department of Environmental Health, University of Cincinnati College of Medicine, Cincinnati, OH, USA; [2]Center for Environmental Genetics, University of Cincinnati Medical Center, Cincinnati, OH, USA; [3]Cincinnati Cancer Center, Cincinnati, OH, USA; [4]Cincinnati Veteran Affairs Medical Center, Cincinnati, OH, USA

OUTLINE

The Epigenome and Developmental Origins of Health and Disease
http://dx.doi.org/10.1016/B978-0-12-801383-0.00016-5

Copyright © 2016 Elsevier Inc. All rights reserved.

INTRODUCTION

Cancer cells are characterized as having abnormal cellular behaviors induced by aberrant gene expression due to progressive accumulation of loss- or gain-of-function mutations and epimutations, gene amplification or fusion, change in chromosome copy numbers, and various other genetic and epigenetic mechanisms altering higher-order chromatin structures. During malignant transformation, normal somatic cells, which are customarily quiescent, actively proliferate, grow in nonadherent culturing conditions, migrate, and invade into extracellular matrices. As cancer advances, cells lose contact inhibition, become apoptotic resistant, gain replication immortality, switch to aerobic glycolysis, escape immune surveillance, and metastasize to neighboring and distant organs [1]. Functional failure of these vital organs finally takes the life of patients. However, what induces malignant transformation and how early does this process begin are not well understood. Importantly, how much of the process is attributable to genetics and how much is caused by environmental factors varies greatly among cancers and often remains poorly defined. This chapter focuses on addressing the hypotheses that certain cancers have an early-life origin and begin at various susceptible developmental windows and that alteration in the epigenome is an important mechanism mediating this phenomenon. Since cancer and its progression are frequently associated with exposures to less-than-optimal levels of endogenous signaling molecules or exogenous factors present in our environment, we also categorize the common cancer-initiating/promoting agents and the epigenetic modifications induced by them in their target genes. One data gap resides in identifying epigenetic changes that are the "drivers" responsible for developmental origin of cancer versus the "passengers," which are the changes that occur as a consequence of neoplastic transformation. Success in filling this void may present a huge opportunity for cancer prevention because it is generally believed that epigenetic changes may be reversible upon removal of the inducers and/or their mediators.

EPIGENETIC REGULATION IN CANCER

Aberrant transcriptomic changes are commonly observed in cancer cells. Traditionally, these changes are attributable to accumulation of genetic alterations such as mutations, chromosomal amplifications and deletions, gene fusions, and other deleterious changes that activate/inactivate oncogenes/tumor suppressors. A more modern viewpoint now argues that many of these transcriptome changes are results of epigenetic changes that occur during malignant transformation. Epigenetic modifications are inheritable changes in gene expression, which do not involve alterations in the primary sequences in the DNA of the affected cells. These changes may involve DNA methylation, histone modifications, and aberrant expression of noncoding RNA (ncRNA), occurring singularly or co-jointly, in various types of cancer and/or in their parent cell lineages prior to malignant transformation. Such changes have been shown to cause cell cycle arrest, dysregulation of transcription factors, disruption of cell–cell and cell–substratum interaction, and promotion of cell migration and invasion, which are key characteristics of cancer cells.

DNA methylation is one of the most and earliest studied epigenetic changes in cancer. It refers to the addition of a methyl group to the 5-position cytosine mainly within the CpG dinucleotide. Although CpGs are underrepresented in the genome, over 70% of them are clustered as CpG islands in gene promoter regions. DNA cytosine-5 methylation is mediated by the action of maintenance (DNA methylation transferase (DNMT1)) or the de novo (DNMT3A, DNMT3B, DNMT3L) DNA methyltransferase enzymes [2].

Biallelic methylated CpGs recruit methyl-CpG-binding domain (MBD) proteins, such as methyl CpG binding protein 2 (MeCP2) and the MBDs, that further recruit histone deacetylases and other chromatin-remodeling molecules to the region resulting in compaction of the promoter region limiting access to transcription factors and silencing of gene expression [3]. DNMTs, especially the de novo enzyme genes, are upregulated in most cancer types and associated with poor prognosis and lower patient survival rates, leading to the hypothesis that they promote silencing of tumor suppressor genes [4].

Emerging evidence suggests a role for ten-eleven translocation (TET) family of α-ketoglutarate (α-KG)-dependent dioxygenase during cancer initiation and progression [5]. Mammalian TET, comprising TET1, TET2, and TET3, catalyze sequential oxidation of 5′-methylcytosine (5MC) to 5′-hydroxymethylcytosine (5hMC), which is the first step in the overall demethylation process [5]. Clinical studies have shown that the lower 5hMC level in various cancers is associated with low TET expression levels (TET1 < TET2 < TET3) and tumor development [6]; and a large number of mutations are detected in *TET2* in myeloid malignancies [7]. However, the precise role of TET during cancer progression is still unknown.

DNA methylation is closely associated with histone modifications. Histones (H1, H2, H3, and H4) are structural proteins involved in chromatin formation. During DNA condensation, two H3-H4 dimers are bridged to form the histone octamer, which is wrapped 1.6 times by a 147-base pair DNA sequence. As a result, the DNA is blocked from the binding of transcription factors and thus gene expression is silenced [8]. Histone activities are modulated by modifications to amino acid residues on its tail including acetylation, methylation, sumoylation, ADP-ribosylation, or ubiquination at specific residues [9]. Different modifications in the same residue can result in different transcription activities. For example, trimethylation at lysine 27 of histone 3 (H3K27me3) represses gene expression [10] whereas acetylation at lysine 9 of histone 3 (H3K9Ac) activates it [11].

There are two major proteins responsible for histone methylation. Polycomb group protein enhancer of zeste homolog 2 (EZH2) catalyzes trimethylation of H3K27 and is associated with transcriptional repression [12]. High EZH2 expression has been shown to be associated with metastatic prostate cancer (PCa), invasive breast cancer, and other cancers including bladder, gastric, lung, and hepatocellular carcinoma [13]. Overexpressing EZH2 in immortalized breast epithelial cells promotes anchorage independent colony growth and invasion, which are features during neoplastic transformation. Moreover, suppressing EZH2 expression and H3K27me3 by 3-deazaneplanocin (DZNep) reduces *in vitro* breast and prostate cancer cell proliferation [12]. This suggests high EZH2 expression in association with methylation of H3K27 promotes cancer progression.

Heterochromatin protein 1 (HP1) is another histone methylation-associated protein. In mammals, there are three HP1 isoforms: HP1α, HP1β, and HP1γ [14]. HP1 functions by recruiting histone methyltransferases (HMTs) to mediate methylation at lysine 9 of histone 3. JmjC domain family of demethylases is an HMT that catalyzes the removal of mono- and dimethylation from H3K9 (lysine (K)-specific demethylase 3A (JHDM2A); [15]), trimethylation of H3K9 (jumonji domain containing 2c (JMJD2C] [16]; and lysine (K)-specific demethylase 4B (JMJD2B) [17]). Further cell-based study demonstrated that suppressing *JMJD2B* inhibits estradiol (E2)-induced G1/S cell cycle transition and tumorigenesis [18] whereas inducing another HMT, *G2A*, promotes migration and invasion of poorly invasive lung cancer [19]. Together, this suggests a role for EZH2 and HP1-mediated histone modification during cancer progression.

Noncoding RNA (ncRNA) is another form of cancer-related epigenetic regulation. MicroRNA (miR) is a class of ncRNA that is approximately

22 nucleotides in length and specifically silences genes to regulate cellular activities including cancer progression [20]. Genome-wide studies show that cancer cells display differential microRNA profiles when compared with the noncancerous cells [21]. Cell-based analysis further illustrates that tumor-suppressing microRNAs including miR-let-7 silence oncogenes high mobility group A hook 2 (*HMGA2*) and *K-RAS* and inhibit cell cycle progression and tumor formation in multiple cancers [22]. Furthermore, another member of this gene family, high mobility group A hook 1 (*HMGA1*), serves as a downstream target of a fulvestrant-induced miR in inhibiting growth of prostate cancer cells (miR-765 [23]). In contrast, oncogenic microRNAs like miR-21 induce cancer cell invasion [24,25] and chemoresistance [26–28].

DEVELOPMENTAL ORIGIN OF ADULT DISEASE

Ongoing research has established that cancer, and its associated epigenetic alterations, may trace its origins back as far as fetal development [29]. This view originates from the theory of developmental origins of adult health and disease [30]. It posits that the developing fetus takes clues from suboptimal maternal physiological states such as malnutrition and reprograms its organs with adaptive traits to overcome hardships after birth. As the theory evolves, the windows of reprogramming for some tissues have extended beyond the *in utero* environment to include preconception, perinatal, peripubertal, pubertal, pregnancy in women, and aging [31]. Changes in epigenetic markers occur at these important windows of susceptibility [32–34]. For example, maternal supplementation with folic acid (*in utero* exposure) significantly reduced global DNA methylation, whereas post weaning folic acid supplementation significantly decreased DNA methyltransferase activity in the mammary glands of offspring [35]. Moreover, the signals/agents capable of reprogramming have greatly expanded to encompass various kinds of environmental exposures [36]. Finally, epigenetics has emerged as a key mechanism for mediating developmental origin of adult disease (DOHaD) reprogramming events [29,31,36].

The prevailing hypothesis states that rapid proliferation of cells and massive tissue reorganization that occurs during developmental stages leaves their epigenomes in a particularly fluid state susceptible to reprogramming. Even slight alteration in epigenomic patterns caused by endogenous or exogenous factors during a development window may have profound effects on regulatory networks and cellular/tissue behaviors, predisposing individuals to later-life carcinogenic events and altering the course of tumor progression on the cancer forms [37].

DEVELOPMENTAL ORIGIN OF ADULT DISEASE AND BREAST CANCER

Though the etiology of breast cancer is multifactorial, exposure to estrogen *in utero* is a major risk factor for development of breast cancer in adulthood. Increasingly, there is evidence that exposure to hormonally active agents including endocrine disruptors could alter mammary gland development and predispose to breast cancer [38–45]. The positive correlation between increased intrauterine levels of estrogen from twin studies and breast cancer in daughters born from such pregnancies further supports this link [41]. Moreover, women who were prenatally exposed to diethylstilbestrol (DES) exhibit an increased risk of breast cancer and clear cell carcinoma of the vagina [46,47]. Also, mice exposed in utero to DES or bisphenol A (BPA) show increase in *Ezh2* expression and in histone H3 trimethylation [39] in the mammary gland.

Recently, DOHaD studies on mammary gland development and tumorigenesis in animal studies had placed strong emphasis on specific gene expression changes, possibly reprogrammed by epigenetics [40,48–52]. In most studies of DOHaD and mammary carcinogenesis, the prenatal reprogramming is the main focus, although the end points were examined at various life stages. The following examples are arranged according to the chronological age at which the effects were studied. Exposure to BPA during gestation (gestational day (GD) 8–18) was found to induce morphological alterations in the stroma and epithelium of the GD18 fetal mammary gland [53]. Transcriptome analysis of the primary periductal stroma showed increased expression of adipogenesis pathway genes (peroxisome proliferation-activated receptor gamma (*Ppar*γ), low-density lipoprotein receptor (*Ldlr*), G protein-coupled receptor 81 (*Gpr81*) and fatty acid binding protein 4 (*Fabp4*)) and a decreased expression of extracellular matrix (ECM) components, such as tenascin-C (TnC) [54]. Moreover, BPA exposure resulted in higher levels of proactivation histone H3K4me3 at the transcriptional initiation site of the α-lactalbumin gene at postnatal day (PND) 4, concomitantly enhancing mRNA expression of this gene [55]. Additionally, in mice exposed pre- and postnatally to BPA, a dose-dependent increase in terminal end buds (TEBs), and in the mRNA expression levels of estrogen-regulated genes such as amphiregulin [56] and secretory leukoprotease inhibitor [44] were observed. In rats, prenatal exposure to BPA also resulted in an increase in TEBs at PND21, accompanied by upregulation of proteins involved in the epidermal growth factor receptor (EGFR), the fibroblast growth factor receptor (FGFR), phosphoinositide 3-kinase (PI3K), and p53 pathways [57]. When the effect of prenatal BPA exposure on protein expression on PND50 was compared to PND100, differentially regulated proteins were identified in the mammary gland including estrogen receptor-alpha, progesterone receptor-A, B-cell lymphoma-2 (Bcl-2), steroid receptor coactivators, EGFR, phospho-insulin-like growth factor 1 receptor, and phospho-Raf [48]. In a separate study, *in utero* exposure to genistein (GEN) increased the incidence of carcinogen-induced mammary tumors in rats, when compared with controls, and at PND60, the number of estrogen receptor binding sites was significantly elevated whilst the protein kinase C (PKC) activity was significantly reduced in the mammary glands [58]. Finally, fetal exposure to BPA resulted in development of carcinomas *in situ* in the mammary glands of 33% of rats exposed to BPA [59]. Neoplasia was observed in young adult rats (PND 50 and 95) as well as an increase in the number of intraductal hyperplasias [60].

Alteration in maternal nutrition is another factor that could affect the intrauterine milieu predisposing offspring to greater disease susceptibility, according to the DOHaD theory. Although initial studies focused predominantly on the effects of maternal undernutrition [61–63], recent studies have sought to investigate the effects of maternal overnutrition, especially high fat diet (HFD). Maternal HFD can alter the development of various organs [64,65] such as the mammary glands [49–52], and thus the offspring could be more susceptible to breast cancer in later life. In this regard, it has been shown that exposure to lard-based high fat diet during fetal and lactation periods decreases later-life susceptibility to breast cancer, accompanied by lower expression of nuclear factor kappa B (*Nfκb*) and *p65*, and higher cyclin-dependent kinase inhibitor 1A (*Cdkn1A; p21*) expression and global levels of trimethylation at lysine 9 of histone H3 (H3K9me3) in the mammary glands of rats [66]. We also reported maternal high butterfat diet being protective for 7,12-dimethylbenz[a]anthracene (DMBA)-induced mammary tumors in offspring [67].

The idea that epigenetic changes occurring upon environmental exposures during the developmental susceptibility windows can be transmitted across generations is another intriguing possibility now being explored. In support, mammary tumorigenesis was shown to be higher in daughters and granddaughters of dams fed high fat or ethinyl-estradiol (EE2), a common estrogen in oral contraceptive, in their diet when pregnant [68]. Both high fat and EE2 *in utero* exposures increased the transcript levels of the gene *Dnmt1*, which encodes the maintenance DNA methylation transferase, in filial 1 (F_1) generation mammary glands. This increase continued to F_2 and F_3 generation in the offspring of EE2-treated dams but not in those born to high-fat-diet-fed females. Similarly, mammary mRNA levels of genes encoding de novo DNA methyltransferases, *Dnmt3a* and *Dnmt3b*, were increased in the F_2 and F_3 generations of the EE2 offspring, respectively, but were not in the high fat offspring [68]. Furthermore, 375 gene-promoter regions were identified as differentially methylated between the control and EE2 offspring in all three (F_1, F_2, and F_3) generations. Further analysis revealed that certain hypermethylated promoter regions were associated with polycomb group target genes paired box-6 (*Pax6*), runt-related transcription factor 3 (*Runx3*), forkhead box E3 (*Foxe3*), GATA binding protein 4 (*Gata4*), and nerve growth factor (*Vgf*). Since polycomb group target genes are frequently dysregulated in cancer cells, this suggests a mechanism linking epigenetic alteration during the development susceptibility window, with adult cancer development in a rodent model.

Thus far, almost all DOHaD and mammary cancer animal studies have focused on fetal exposure as a susceptible window for reprogramming, although conceptually puberty and pregnancy should clearly be investigated in the future.

DEVELOPMENTAL ORIGIN OF ADULT DISEASE AND PROSTATE CANCER

Epigenetic alterations are a common feature of prostate cancer, and in addition to genetic alterations, they have a critical role in cancer initiation and progression. The most common epigenetic aberrations identified to date in prostate cancer with respect to the biomarker potential have recently been reviewed [69,70], as have the epigenetic alterations induced by diet [71]. Growing evidence indicates that lifestyle and environmental exposures may disrupt the epigenetic balance and compromise the stability of the epigenome in normal cells leading to the development of PCa. Epidemiologic evidence links specific pesticides, PCBs, and inorganic arsenic (iAs) exposures to elevated PCa risk [72]. Studies in animal models also indicate that cadmium, ultraviolet (UV) filters, and BPA can augment prostate cancer risk [72]. Importantly, there appears to be heightened sensitivity of the prostate to environmental exposures during critical developmental windows, including *in utero* and neonatal time points as well as during puberty.

Data from several recent studies indicate that early-life exposure to endocrine disruptors such as BPA and DES may influence epigenetic programming of endocrine signaling pathways. Transient neonatal exposure of rats to low-dose BPA results in increased prostate gland susceptibility to adult-onset precancerous lesions and hormonal carcinogenesis [73]. These adverse adult outcomes are accompanied by alterations in the DNA methylation patterns of multiple signal transduction genes: Na-K-Cl cotransporter solute carrier family 12, member 2 (*Slc12a2*), mitogen-activated protein kinase/extracellular signal-regulated kinase (MAPK/ERK) pathway (G-protein coupled receptor 14 (*Gpcr14*) and platelet-derived growth factor alpha (*Pdgfrα*)), phosphokinase C pathway (phosphatidylinositol-4,5-bisphosphate phosphodiesterase beta 3 (*Plcβ3*)), cAMP pathways (phosphodiesterase type 4 variant 4 (*Pde4d4*) and hippocalcin-like protein 1 (*Hpcal1*)), and neural or cardiac development (carbonic anhydrase-related xI protein (*CarxI*) and TNNI3 interacting kinase (*Cark*)). Specifically, for *Pde4d4*, a specific methylation cluster was identified in the 5′-flanking CpG island that was gradually hypermethylated with aging in normal prostates, resulting in loss of gene expression. However, in prostates exposed to neonatal estradiol or BPA, this region became hypomethylated with aging, resulting in persistent and elevated *Pde4d4* expression [73,74]. Additionally, neonatal exposure to estradiol/BPA altered promoter methylation and expression of nucleosomal binding protein 1 (*Nsbp1*; hypomethylation) and *Hpcal1* (hypermethylation) genes, and the transcriptional programs of genes involved in DNA methylation/demethylation such as *Dnmt3a/b* and *Mbd2/4* in the rat prostate gland throughout life [75]. We speculate that the distinctly different fate of early-life epigenetic marks during adulthood reflect the complex nature of lifelong editing of early-life epigenetic reprogramming by BPA. Recently, the secretoglobin family 2A (*Scgb2a1*) was found to be as significantly upregulated in the prostate of adult rats exposed neonatally to BPA, due to hypomethylation of the CpG island at the promoter region upstream of the transcription start site [76]. In addition to BPA, maternal exposure to DES during pregnancy was found to result in extensive hypertrophy and squamous metaplasia in human male offspring [77]. Moreover, studies with DES and estrogen in rodent models predict marked abnormalities in the adult prostate including increased susceptibility to adult-onset carcinogenesis following early DES exposures [78–81]. Mixtures of endocrine disrupters relevant for human exposure were also found to elicit persistent effects on the rat prostate following perinatal exposure, suggesting that human perinatal exposure to environmental chemicals may increase the risk of PCa later in life [82].

Vinclozolin is a well-characterized antiandrogenic pesticide. In rodent models, prenatal exposure to this chemical increased the risk of prostate disease in male offspring and was transmitted for four generations [83]. Characterization of regulated genes demonstrated involvement of the calcium and wingless (Wnt) pathways. Moreover, a number of genes were identified that are also associated with prostate disease and cancer, including β-microseminoprotein (muscle specific protein; *Msp*) and tumor necrosis factor receptor superfamily 6 (Fas-associated protein with death domain; *Fadd*).

Fetal alcohol exposure increases susceptibility to carcinogenesis and promotes tumor progression in prostate gland [84]. Prenatal ethanol exposures induced histophysiological changes in the prostate as well as increases in the susceptibility of the prostate to develop neoplasia during adulthood [85] in response to *N*-methyl-*N*-nitosourea (NMU) and testosterone.

Hydrocarbon mixtures involving jet fuel (jet propellant-8; JP-8) increase susceptibility to prostate disease and a significant increase in a delayed pubertal onset in male pups exposed in utero [86].

In addition to environmental factors, maternal protein malnutrition in Wistar rat dams was found to delay prostate development, growth, and maturation until adulthood, probably as a result of low testosterone stimuli. The higher incidence of cellular dysplasia and prostatitis suggested that maternal protein malnutrition increases prostate susceptibility to diseases with aging [87].

OTHER CANCERS

Other than the most commonly studied breast and prostate cancers, epidemiologic studies have also established a link between early-life exposure and the risk or incidence of other cancers in later stages of life. The most studied is the effect of *in utero* and early childhood arsenic exposure through drinking water, which has been shown increasing the incidence of lung [88] and bladder cancer during adulthood [88,89]. Exposure to pesticides including dichlorodiphenyltrichloroethane (DDT) during this critical window is also correlated with the increased rates of breast cancer [90] and childhood leukemia [91]. Other studies also showed that light exposure at early stages of life increases cancer risk. Specifically, childhood exposure to light is associated with risk of developing melanoma [92]. Diagnostic radiology during early pregnancy also introduces prenatal and postnatal exposure to X-ray, which is positively correlated to leukemia and non-Hodgkin's lymphoma in adult life [93].

Furthermore, *in utero* exposure to the endocrine disruptor DES raises the concern of developing ovarian cancer in adulthood [94]. Yet, the mechanism is not fully understood. Using rodent models, neonatal exposure to DES causes uterine epithelial cancer in most of the mice at age 18 months by inducing hypomethylation of *c-fos* at exon 4 of the uterine epithelial cells [95,96]. Neonatal exposure of CD-1 mice to DES or GEN was also found to induce uterine adenocarcinoma in aging animals. In uteri of immature control mice, *Nsbp1* promoter was hypomethylated but became hypermethylated during puberty, with decrease in gene expression. In contrast, in neonatal DES/GEN-treated mice, the *Nsbp1* promoter stayed hypomethylated, and the gene exhibited persistent overexpression throughout life. Thus, the early-life reprogramming of uterine *Nsbp1* expression by neonatal DES/GEN exposure appears to be mediated by an epigenetic mechanism that interacts with ovarian hormones in adulthood [97].

Inorganic arsenic (iAs) is a known transplacental carcinogen in liver [98]. Gestational exposure to iAs induced alternations in DNA methylation and a complex set of aberrant gene expression in the newborn liver [99]. Combined prenatal exposure to iAs and folate showed that *Dnmt3a* was significantly downregulated in mice fetal liver, suggesting that methylated iAs may adversely affect DNA methylation globally [100].

Early development of colon has been shown to be reprogrammed by GEN and soy protein isolate [101]. Zhang et al. (2013) revealed that induction of secreted frizzled-related protein 2 (*SFRP2*), *SFRP5*, and wingless type MMTV integration site family, member 5A (*WNT5A*) by azoxymethane, an inducer for colon cancer, were epigenetically suppressed by both phytoestrogen-based diets through augmentation of promoter methylation, reduction in acetylation of histone H3 and H3K9me3 [101]. In another colon cancer mice model, similar *Wnt* signaling pathway (adenomatous polyposis coli (*Apc*)), secreted frizzled-related protein 1 (*Sfrp1*), Wnt inhibitory factor 1 (*Wif1*), and *Wnt5a* were also found disrupted and epigenetically regulated in the offspring by depleting vitamin B in dams' diet during gestation until weaning period [102]. Interestingly, supplementing folate in maternal diet, but not postweaning diet, showed significant reduction in incidence of azoxymethane-induced colon cancer [103]. This consolidates that early life is a critical window of susceptibility for the onset of cancer in adulthood and further suggests epigenetics may be a key to disease predisposition. However, evidence that directly links early-life exposure, epigenetics, and carcinogenicity of these environmental agents is still scarce.

EPIGENETICALLY ACTIVE AGENTS INVOLVED IN CARCINOGENESIS

All these studies thus point to the ability of fetal environmental exposures to trigger changes in the postnatal and adult epigenome, altering gene expression patterns in various organs. These events may later contribute toward development of preneoplastic and neoplastic lesions that manifest during adulthood. The epigenetic alterations underlying changes in expression of genes following exposures, and their deleterious implications for cancers and other disease development and progression later in adult life, cannot be ignored. However, as is evident from the above studies, there is a huge data gap linking the epigenetic markers altered in response to environmental insults to their contributions in reprogramming the functions of the mature organs that ultimately determine the cancer susceptibility or course of cancer progression in later life.

Much of the current knowledge is correlative in nature and lacks the power to provide concrete evidence for causal effects. Although it is apparent that different environmental insults lead to epigenetic reprogramming of different sets of genes, the depth and breadth of these data are insufficient to address the question of whether epigenetic alternation of gene expression is a root cause of developmental origin of cancer. Hence, in an attempt to provide more insights to fill this data gap, it is prudent to examine the direct effects of exogenous and endogenous factors on epigenetic regulation resulting in reprogramming genes involved in adult disease.

EXOGENOUS FACTORS

Heavy Metals

Environmental or occupational exposure to heavy metal is associated with the prevalence of cancer. Accumulation of chromium, cadmium, lead, and selenium has shown increasing risks of pancreatic, renal, pulmonary, kidney, prostate, and breast cancer through epigenetic regulation of cancer-related genes. The mechanisms involved yet vary among different metals. Several population studies demonstrate a strong association between chromium exposure and higher risk of chromate-related lung cancer [104,105] through hypermethylation of tumor-suppressing genes cyclin-dependent kinase 2A (*CDK2A*) [104], mutL homolog 1 (*MLH1*), and *APC* [105]. Chromium exposure also upregulates *G9A*, an HMT, to increase dimethylation at lysine 9 of histone 3 (H3K9me2) and decrease trimethylation at lysine 4 of histone 3 (H3K4), which are associated with *MLH1* suppression in lung carcinoma cells.

Cadmium is related to malignant transformation and cancer cell aggressiveness and is associated with global hypermethylation by inducing DNMT activities, which is commonly observed in lung [106] and liver cancer [107]. It has been shown in normal human breast epithelial cell MCF-10a that chronic chromium exposure induces colony formation, invasion, and growth of tumors that possess basal-like breast cancer cell properties including the loss of tumor suppressor gene breast cancer gene 1 (*BRCA1*), estrogen receptor-α (*ESR1*), and epidermal growth factor receptor (*EGFR*; *HER2*) and gain of oncogenic proteins C-MYC, K-RAS, and breast stem cell markers cytokeratin-5 and p63 [108].

Lead has strong association with lung, stomach, breast, and to a lesser extent, kidney and brain cancer [109,110] by hypermethylation of tumor-suppressor genes including *CDK2A*. It may also act through microRNA to induce cancer progression. For example, lead exposure induces oncogenic miR-222 and decreases tumor-suppressing miR-146a expression in peripheral blood leukocytes. Yet, the effect of lead on cancer progression is understudied and warrants further investigation.

Selenium is an essential element, albeit with narrow safety margins. Unlike other heavy metals, selenium disrupts methionine-homocytosine metabolic pathway and causes global DNA hypermethylation [111]. It downregulates *DNMT1* and *DNMT3A* to suppress histone methylation, and increases histone acetylation in multiple cancers [112,113]. In prostate cells, selenium restores expression of silenced tumor suppressor genes by hypomethylation [112]. However, the same exposure in liver cells epigenetically silenced oncogenic *tumor protein p53* (*TP53*) [114]. It is thus possible that selenium may induce or protect against tumor development in a cell-type specific manner.

Endocrine-Disrupting Chemicals

The potential link between xenoestrogens and cancer is of concern, due to the widespread nature of xenoestrogen exposure in the community. Three of the most commonly occurring xenoestrogens are BPA, DES, and polycyclic aromatic hydrocarbons (PAHs). The commercial chemical BPA has widespread use in the manufacture of plastics and metal cans, and up to 95% of the population has detectible levels of BPA in their bodies, highlighting its vast distribution. Developing concerns, however, have been raised about the effect that BPA has on the epigenome and how this contributes to the development of hormone-responsive cancers. BPA-induced changes to the epigenome in the breast are the most well studied, with *in vitro* studies suggesting distinct ncRNA profiles are induced by BPA, which differs from estradiol [115]. Distinct DNA methylation profiles are also detected in cell lines, affecting key antiapoptotic and tumor-developmental genes such as *BRCA1*, cyclin A1 (*CCNA1*), cyclin-dependent kinase inhibitor 2A (*CDKN2A*), thrombospondin 1 (*THBS1*), and tumor necrosis factor receptor superfamily membrane 10C (*TNFRSF10C*) [116,117]. Animal models exposed to BPA show accelerated mammary gland development and increased terminal end bud formation, leading to a greater susceptibility to carcinogen-induced tumorigenesis later in life [44,56,118,119]. However, it is not clear whether epigenetics is the cause of this. The synthetic nonsteroidal estrogen DES was commonly prescribed to women up until the 1970s, and was also used to accelerate growth of livestock. However, its widespread use was banned once adverse health effects became apparent. Due to the nature of its design, DES is linked to hormonal cancers such as breast and uterine disease [120,121]. The role of epigenetics in this process is being investigated, and is thought to involve DNA methylation and histone modifications affecting key genes homeobox A10 (*HOXA10*), *C-FOS*, and the p53-related proapoptotic pathway [95,122,123]. PAH exposure is also thought to contribute to cancers of the breast, bladder, and lung [124–126]. This is of particular concern to those working in coke ovens, firefighters, and smokers, all of whom are exposed to high levels of PAHs [127,128]. PAHs are known to form DNA adducts, which may occur close to sites of CpG methylation [129].

Xenoestrogens

Distinct from xenoestrogens, phytoestrogens naturally occur in plant material and are ingested in the diet. Soy phytoestrogens, namely genistein and equol, are widely consumed and due to their structural similarity to estrogen, are known to cause both estrogenic and anti-estrogenic effects [130]. There is contention, however, as to whether these phytoestrogens have a protective or negative effect on breast cancer risk. Some animal studies and human population cohorts have suggested that they are protective and that a diet high in soy products is beneficial [131,132]. Conversely, the proliferative effect of phytoestrogen was shown in breast cell models [133]. Other cancers that have been associated with dietary phytoestrogens include prostate [134], pancreatic [135], lung [136], and colorectal [137]. Studies in breast cancer cells MCF-7 and MCF-10A revealed the capacity of phytoestrogens to inhibit the expression and activity of the three DNA methyltransferase enzymes, mechanistically explaining changes in global methylation patterns observed in breast cells exposed to phytoestrogens [138]. miRNA targeting critical cancer pathway genes Ras-related C3 botulinum toxin substrate 1 (*RAC1*), *EGFR*, and E1A binding protein P300 (*EP300*) were upregulated in prostate cancer [139], and histone modifications due to the effects of phytoestrogens have also been shown [140]. These demonstrate a diverse range of effects that may be exerted by phytoestrogens.

Pesticides and Insecticides

Agricultural exposure to pesticides and insecticides raises the concern of cancer risk. Though these widely used chemicals may not be classified by the Environmental Protection Agency (EPA) as carcinogenic, their epigenetic regulation is associated with increased risk of prostate cancer, adult leukemia, non-Hodgkin's lymphoma, and multiple myeloma [141].

Gestational exposure of Sprague–Dawley rats to pesticides, such as PCB and methylmercury (MeHg) but not organochlorinated pesticides, suppresses hepatic *Dnmt1* and *Dnmt3b* expression to induce oncogene expression through hypomethylation [142]. Similar hypomethylation was also observed in mice exposed to pesticides (dichloroacetic and tricholoracetic acids), which upregulate oncogene *Myc* to increase kidney and liver tumor formation [143]. Moreover, when erythroblastic leukemia-derived K562 cells were exposed to an organophosphate insecticide diazinon, genome-wide methylation analysis identified hypermethylation of tumor-suppressing genes [141]. These studies reveal hypermethylation of onco- and tumor-initiating genes is a mechanism of pesticides and insecticides to promote carcinogenesis. However, whether these chemicals act through histone modification and/or noncoding RNA is not known. A study using dopaminergic neuronal N27 cells demonstrates that dieldrin, an insecticide and by-product of the pesticide Aldrin (EPA), induces hyperacetylation of histones H3 and H4, which is associated with the accumulation of cAMP response element binding protein (CREB) and increases apoptotic cell death [144]. This postulates that pesticides and insecticides may act through similar mechanism to initiate tumor formation and promote cancer progression.

Agent Orange

The link between Agent Orange (AO) and cancer has always been a concern to military veterans. The phenoxyherbicide AO is a defoliant chemical widely used by the US military during the Vietnam War [145]. Several epidemiology studies have shown that AO exposure is associated with soft tissue sarcoma, non-Hodgkin's lymphoma [145], and prostate [147–149] and testicular cancer [150]. Specifically in prostate cancer, AO exposure is correlated with risk of aggressive disease and a rapid shorter prostate-specific antigen (PSA) doubling time after recurrence [149,151]. However, AO is not carcinogenic. On the contrary, its contaminant dioxin, also known as TCDD (2,3,7,8-trichlorodibenzo[p]dioxin), is classified as a "Class I" human carcinogen by the Environmental Protection Agency. Prenatal TCDD-exposed rats have been shown possessing more premature terminal end buds and fewer mature lobules at day 50 postpartum [152]. Offspring of these exposed females also show higher incidence of breast cancer, revealing the predisposal of breast cancer risk by TCDD [153]. Inheritance of cancer is often an epigenetic effect. Studies using nontumorigenic, immortalized human epithelial cells demonstrate that TCDD induces abnormal growth rate, aryl hydrocarbon hydrolase activity, cell morphological changes, and colony formation in soft agar of the normal cells [154–157]. These malignant transformed cells are further characterized with the loss of cytochrome aromatase P450 1A1 (*CYP1A1*), and tumor suppressor proteins CDK2A, CDKN1A, and TP53 by hypermethylation and recruitment of acetylated H3 and H4 at their promoter regions.

Light

High exposure to light risks cancer. Epidemiology studies show occupational exposure to sunlight from outdoor activities among construction workers increases risk of hematologic malignancy such as myeloid leukemia, lymphocytic leukemia, non-Hodgkin's lymphoma, malignant melanoma in the eye, and tumors in head, face, and neck [158], which are correlated to epigenetic regulation. UV and solar-stimulated light has been shown inducing C→T transition mutation at the dipyrimidine sequences [159,160]. Hypermethylation of the resulted 5′-methylcytosine residues at the promoter of *T53* suppresses tumorigenesis. Sun exposure also upregulates keratin 5 (*KRT5*), which encodes a keratin protein normally absent in epidermis through hypomethylation of *KRT5* promoter. The role of *KRT5* on intermediate filament architecture of epithelial cells [161] suggests that light exposure may epigenetically induce malignant transformation of epidermal cells. In the brain, light exposure is associated with increased H3 phosphorylation and expression of the proto-oncogene and transcription factor *FOS* [162]. On the contrary, light exposure may act through melatonin production in pineal gland to induce cancer progression. Melatonin, the hormone that regulates sleep cycle and circadian rhythm, is downregulated by light [163]. It is used in various cancer therapeutic regimens owing to its promising effect on improving patient survival rate. In triple-negative breast cancer cells MDA-MB231, melatonin inhibits cell proliferation and tumor growth by upregulating *EGFR* and insulin-growth factor receptor (IGFR), and downregulating vascular epithelial growth factor receptor (*VEGFR*) and cell proliferation marker *KI67* [164]. Melatonin is further shown associated with increased H3 acetylation and upregulation of nestin (*NES*), tubulin beta 3 class III (*TUBB3*), and nuclear receptor subfamily 4, group A, member 2 (*NR4A2*) to induce neural stem cell differentiation [165]. The epigenetic mechanism of melatonin on regulating cancer progression is yet to be discovered.

ENDOGENOUS FACTORS

Oxidants

Oxidants are prevalent carcinogens. Environmental exposure, metabolic reactions, and inflammation often increase levels of reactive oxygen species, superoxide anions, and hydrogen peroxide, which have been shown inducing DNA adducts and mutations to promote carcinogenesis. Oxidative stress resulting from aberrant levels of enzymatic (superoxide dismutase, catalase, glutathione peroxidase) and nonenzymatic (vitamin C, vitamin E, carotenoids, thiol antioxidant, flavonoids, selenium) antioxidants [166] has been shown associated with prostate [167], breast [168], and lung [169] cancers. However, the epigenetic effect of oxidants during carcinogenesis is mostly studied in melanocytes. As mouse nontumorigenic melanocyte melan-a cells undergo malignant transformation, the increase in intracellular-free oxygen radicals increased glutathione expression and induced global gene hypermethylation through upregulating *Dnmt3b* but not *Dnmt1* [170]. The activation of Rac1, a small GTPase that functions as an nicotinamide adenine dinucleotide phosphate (NADPH) oxidase, has been shown elevating superoxide anion levels, which inhibits ERK signaling and increases Dnmt1 expression to cause global hypermethylation of genes involved in anoikis, cell proliferation, clonogenicity, and latency for malignant transformation [171]. Progression of melanocyte malignant transformation has also been shown associated with the recruitment of trimethylation at lysine 4 of histone 4 (H4K4me3) and dissociation of trimethylation at lysine 4 of histone 3 (H3K4me3), which upregulate EZH2 and sirtuin-1 (SIRT1) to induce metastasis-related gene expression [172]. The association between histone modification and cancer progression thus warrants further investigation.

Vitamins

Vitamins B12 and D are potential epigenetic regulators for cancer. Epidemiologic studies show high plasma vitamin B12 is associated with increased risk of prostate [173] and colon [174] cancers. Vitamin B12 and folic acid are methyl donors that enhance S-adenosylmethionine (SAM)-dependent methylation. The disrupted methionine cycle may lead to hypomethylation of long interspersed nucleotide element 1 (*LINE-1*) and hypermethylation of Ras-associated domain family 1 isoform A (*RASSF1A*) and methylene tetrahydrofolate reductase (*MTHFR*), which are associated with colon [174] and lung [175] cancer, respectively. Vitamin D, on the contrary, has a tumor suppressive effect. High plasma vitamin D level is associated with lower risk of breast cancer [176]. It has been shown suppressing melanoma cell proliferation by downregulating oncogenic miR-125b [177]. Further analysis using calcitriol, the active form of vitamin D, demonstrates hypermethylation of cytochrome 24 A1 (*CYP24A1*) and recruitment of H3K9Ac, and H3K9me2 at the vitamin D response element of *CDKN1A* to suppress cell proliferation [178]. In colon cancer HCT116 cells, calcitriol induces miR-627 expression to downregulate histone methylation-related iron- and 2-oxoglutarate-dependent dioxygenase (*JMJD1A*) and its related cell-cycle genes *HOXA1* and cyclin D1 (*CCND1*) [179]. Simultaneously, it recruits H3K27me2/3 and dissociates H3K4me3 to upregulate growth differentiation factor 15 (*GDF15*) and suppresses cancer progression. The multi cancer-suppressing effect of vitamin D thus draws attention to its potential therapeutic use in various cancers.

Endogenous Hormones

Estrogen, progesterone, and androgens are all key contributors to both normal physiologic processes and cancer. Their role in modulating the epigenome and how this leads to carcinogenesis is becoming increasingly apparent. Estrogen is the most well understood in this regard, with several studies highlighting how it may modify DNA methylation, histone modifications, and ncRNAs in hormone-responsive cancers such as breast, endometrial, and ovarian. This is particularly highlighted in breast cancer, where distinct global methylation patterns are observed between ER^+ and ER^- breast cancer [180]. The lncRNA HOTAIR is most commonly associated with estrogen induction in the breast [181], showing that estrogen affects diverse epigenetic processes. The actions of progesterone and androgens on epigenetics in cancer are less well understood, however, progesterone action has been linked to miRNA regulation in ovarian cancer [182]. Androgens have been associated with lower histone H3 trimethylation levels in prostate cancer [183], an effect that increases protein transactivation including that of the androgen receptor (AR) itself.

Fatty Acids and Glucose

Endogenous fatty acids appear to affect the methylation status of genes important to cancer. Hypomethylation of apoptotic-related genes cell death-inducing DFFA-like effector b (*CIDEB*), death-associated protein kinase 1 (*DAPK1*), and tumor necrosis factor receptor superfamily membrane 25 (*TNFRSF25*) is observed in colon cancer cells upon treatment with the docosahexaenoic fatty acid [184]. Differential methylation of a single CpG site rather than the overall methylation levels can also affect gene expression through fatty acid-induced changes at the CCAAT/enhancer-binding protein beta (*CEBP-β*) site within the interleukin-8 (*IL-8*) promoter. This affected the binding capacity of transcription factors and led to an overall decrease in IL-8 expression in astrocytoma cells [185]. Although glucose has been implicated in epigenetic effects induced in diabetes [186], limited evidence exists for its effect in cancer. One study using triple-negative breast cancer cells has identified that glucose induces changes to promoter methylation in key cancer-related genes protein tyrosine phosphate nonreceptor type 12 (*PTPN12*), vimentin (*VIM*), and snail family zinc finger 1 (*SNAI1*) [187]. In contrast, glucose restriction in human lung fibroblasts has been shown upregulating human telomerase reverse transcriptase (*hTERT*), a key enzymatic component of telomerase highly expressed in tumors through histone modifications, which couples with hypermethylation to downregulate *CDKN2A* and delay senescence to promote malignant transformation [188]. In prostate cancer, glucose starvation upregulates Kruppel-like factor 6 (*KLF6*), a tumor suppressor involved in cell growth, differentiation, adhesion, and endothelial motility, through histone modification to induce apoptotic cell death [189]. This implies glucose may be tumor suppressing or oncogenic in a cancer-dependent manner.

CONCLUSIONS

Although much circumstantial evidence from population studies hints to the existence of developmental origins of cancer, definitive proof that it exists still awaits strong supportive data. Except for childhood cancers, the study of DOHaD, epigenetics, and cancer in humans is difficult as the time line (in decades) is so long that well-annotated, longitudinal epidemiological studies are needed to establish clear patterns of correlations. Furthermore, many such studies lack storage of the tissues of origin of the cancers and surrogate cell types such as blood lymphocytes are being used to examine epigenetic changes. Associations are often made between the epigenomes of the surrogate cells and the full-blown cancers, often creating unnecessary conundrum in data interpretation. Therefore, the development of animal models to study windows of susceptibility in developmental origins of cancer is critical to the advancement of the field. Models must be established that allow for controlled developmental exposure to signaling molecules, toxicants, or lifestyle factors and examination of the state of the epigenomes of the target organs.

Human studies are necessary early for the identification of environmental toxicants/agents with adverse effects on human health. However, human cohort studies usually have underlying problems of sample size, genetic variability, multiple-agent exposures, and degree of sensitivity to whole body exposures. These factors make the statistical analyses and the interpretation of data extremely complicated, thus arguing for validation in carefully designed animal studies. However, the differences in metabolizing environmental toxicants, and the presence of gene orthologues between human and animals can complicate the interpretation of findings of animal-based studies. In this context, establishing *in vivo* model systems using engrafted human materials, for example, prostaspheres [190,191] or mammospheres [192], might be useful in delineating the action of these environmental toxicants on genetic reprogramming.

Thus far, animal studies have mainly focused on fetal or perinatal life exposure; the vulnerability of other developmental windows has not been studied in any depth or breadth. Well-designed animal studies could open an avenue for the scientific community to think about the most critical "exposure windows" for each disease including cancer. Tomasetti and Vogelstein [193] recently raised an interesting debatable topic that only one-third of variance in cancer risk is attributable to environmental influences or genetics and the majority is due to random mutational events left unrepaired during proliferation of normal stem cells. This provocative estimate needs to be debated and validated by further experimental and human studies. However, the authors did not account for the fact that the stem cell populations in each organ of an individual could be a susceptible target for developmental reprogramming making them more or less vulnerable to DNA damaging agents, or having higher or lower fidelity for repair.

The concept that DOHaD can be compared to "repressed memories" that travel along the entire life of an individual or even across generations awaiting retrieval by special cues specific to the disease is worthy of exploring using existing and futuristic technologies. At present, we only have anecdotal evidence in support of the concept of a developmental origin of some cancers and the epigenetics is at work. In spite of data accumulated from several well-designed animal studies [52,194–196] and a handful of cohort studies, the controversy abounds on the strength and interpretation of these data. Challenges include the long time line between the reprogramming events and the manifestation of cancer, as well as the likely requirement of the retrieval cues or the second hits, which in many cases could be multiple events and of a complex nature. Additionally, the exposures responsible for alteration of the epigenetic events that govern the behavior of the cancer stem cells tend to be complex and not restricting to a single agent, but a mixture. Adding to the complexity is the possibility that multiple developmental windows are involved. To sort out the weight and the temporal order required for developmental influences on cancer, a multidimensional model taking into consideration multiple exposures, multiple windows, transversal through life stages, the retrieval cues, the genetic background, and more, is needed.

Finally, development and the advent of minimally invasive tools for measurement of changes, high-sensitivity detection methods, high-throughput techniques for data collection and analyses, longitudinal nature of studies, single-cell biology, real-time and 24-7 monitoring methodologies, and other novel new tools for both animal and human studies are needed to provide solid evidence to prove this tenderizing concept in the future.

Glossary Terms, Acronyms, and Abbreviations

5hMC 5′-Hydroxymethylcytosine
5MC 5′-methylcytosine
AO Agent Orange
APC Adenomatous polyposis coli
BPA Bisphenol-A
CDK2A Cyclin dependent kinase inhibitor 2A
CpG -C-phosphate-G-
CYP1A1 Cytochrome aromatase P450 1A1
DDT Dichlorodiphenyltrichloroethane
DES Diethylstilbestrol
DNMT DNA methyltransferase enzyme
DOHaD Developmental origin of health and disease
DZNep 3-Deazaneplanocin
EE2 Ethinyl-estradiol
EGFR Epidermal growth factor receptor
EPA Aldrin
EZH2 Enhancer of zeste homolog 2
Fabp4 Fatty acid binding protein 4
Fadd Tumor necrosis factor receptor superfamily 6
FGF Fibroblast growth factor
GD Gestation day
GEN Genistein
GPR81 G-protein coupled receptor
HFD High fat diet
HMGA1 High-mobility group A hook 1
HMGA2 High-mobility group A hook 2
HMT Histone methyltransferases
HPCal1 Hippocalcin-like protein 1
HP1 Heterochromatin protein 1
H1, H2, H3, and H4 Histones
H3K9Ac Acetylation at lysine 9 of histone 3
H3K9me3 Trimethylation at lysine 27 of histone 3
IGFR Insulin growth factor receptor
JHDM2A JmjC domain family of demethylase
Ldlr Low-density lipoprotein receptor
MBD Methyl-CpG-binding domain
MeHg Methylmercury
MLH1 mutL homolog 1
miR microRNA
Msp Microseminoprotein
MTHFR Methylene tetrahydrofolate reductase
ncRNA Noncoding RNA
Nsbp1 Nucleosomal binding protein 1
PAH Polycyclic aromatic hydrocarbons
PCa Prostate cancer
PCB Polychlorinated biphenyls
Pde4d4 Phosphodiesterase 4 variant 4
PND Postnatal day
PPARδ Peroxisome proliferator-activated receptor gamma
SAM S-adenosylmethionine
TCDD 2,3,7,8-Trichlorodibenzo[p]dioxin
TEB Terminal end bud
TET Ten-eleven translocation family
UV Ultraviolet
VEGFR Vascular epithelial growth factor receptor

Acknowledgments

Research was supported in part by grants from the National Institute of Environmental Health Sciences: P30ES006096, U01ES019480, U01ES020988, R01ES022071; a Congressionally Directed Medical Research Program Department of Defense Award PC094619; and a National Health and Medical Research Council of Australia GNT1070112.

References

[1] Hanahan D, Weinberg RA. Hallmarks of cancer: the next generation. Cell 2011;144(5):646–74.

[2] Robertson KD. DNA methylation, methyltransferases, and cancer. Oncogene 2001;20(24):3139–55.

[3] Jones PA, Baylin SB. The fundamental role of epigenetic events in cancer. Nat Rev Genet 2002;3(6):415–28.

[4] Subramaniam D, Thombre R, Dhar A, Anant S. DNA methyltransferases: a novel target for prevention and therapy. Front Oncol 2014;4:80.

[5] Pfeifer GP, Kadam S, Jin SG. 5-Hydroxymethylcytosine and its potential roles in development and cancer. Epigenetics Chromatin 2013;6(1):10.

[6] Yang H, Liu Y, Bai F, et al. Tumor development is associated with decrease of TET gene expression and 5-methylcytosine hydroxylation. Oncogene 2013;32(5):663–9.

[7] Delhommeau F, Dupont S, Della Valle V, et al. Mutation in TET2 in myeloid cancers. N Engl J Med 2009;360(22):2289–301.

[8] Sawan C, Herceg Z. Histone modifications and cancer. Adv Genet 2010;70:57–85.

[9] Kouzarides T. Chromatin modifications and their function. Cell 2007;128(4):693–705.

[10] Agger K, Cloos PA, Christensen J, et al. UTX and JMJD3 are histone H3K27 demethylases involved in HOX gene regulation and development. Nature 2007;449(7163):731–4.

[11] Fernandez-Sanchez A, Baragano Raneros A, Carvajal Palao R, et al. DNA demethylation and histone H3K9 acetylation determine the active transcription of the NKG2D gene in human $CD8^+$ T and NK cells. Epigenetics 2013;8(1):66–78.

[12] Chase A, Cross NC. Aberrations of EZH2 in cancer. Clin Cancer Res 2011;17(9):2613–8.

[13] Varambally S, Dhanasekaran SM, Zhou M, et al. The polycomb group protein EZH2 is involved in progression of prostate cancer. Nature 2002;419(6907):624–9.

[14] Singh PB, Miller JR, Pearce J, et al. A sequence motif found in a Drosophila heterochromatin protein is conserved in animals and plants. Nucleic Acids Res 1991;19(4):789–94.

[15] Yamane K, Toumazou C, Tsukada Y, et al. JHDM2A, a JmjC-containing H3K9 demethylase, facilitates transcription activation by androgen receptor. Cell 2006;125(3):483–95.

[16] Wissmann M, Yin N, Muller JM, et al. Cooperative demethylation by JMJD2C and LSD1 promotes androgen receptor-dependent gene expression. Nat Cell Biol 2007;9(3):347–53.

[17] Fodor BD, Kubicek S, Yonezawa M, et al. Jmjd2b antagonizes H3K9 trimethylation at pericentric heterochromatin in mammalian cells. Genes Dev 2006;20(12):1557–62.

[18] Shi L, Sun L, Li Q, et al. Histone demethylase JMJD2B coordinates H3K4/H3K9 methylation and promotes hormonally responsive breast carcinogenesis. Proc Natl Acad Sci USA 2011;108(18):7541–6.

[19] Chen MW, Hua KT, Kao HJ, et al. H3K9 histone methyltransferase G9a promotes lung cancer invasion and metastasis by silencing the cell adhesion molecule Ep-CAM. Cancer Res 2010;70(20):7830–40.

[20] Bushati N, Cohen SM. microRNA functions. Annu Rev Cell Dev Biol 2007;23:175–205.

[21] Lu J, Getz G, Miska EA, et al. MicroRNA expression profiles classify human cancers. Nature 2005;435(7043):834–8.

[22] Kumar MS, Erkeland SJ, Pester RE, et al. Suppression of non-small cell lung tumor development by the let-7 microRNA family. Proc Natl Acad Sci USA 2008;105(10):3903–8.

[23] Leung YK, Chan QK, Ng CF, et al. Hsa-miRNA-765 as a key mediator for inhibiting growth, migration and invasion in fulvestrant-treated prostate cancer. PLoS One 2014;9(5):e98037.

[24] Dey N, Das F, Ghosh-Choudhury N, et al. microRNA-21 governs TORC1 activation in renal cancer cell proliferation and invasion. PLoS One 2012;7(6):e37366.

[25] Han M, Liu M, Wang Y, et al. Re-expression of miR-21 contributes to migration and invasion by inducing epithelial-mesenchymal transition consistent with cancer stem cell characteristics in MCF-7 cells. Mol Cell Biochem 2012;363(1–2):427–36.

[26] Echevarria-Vargas IM, Valiyeva F, Vivas-Mejia PE. Upregulation of miR-21 in cisplatin resistant ovarian cancer via JNK-1/c-Jun pathway. PLoS One 2014;9(5):e97094.

[27] Eto K, Iwatsuki M, Watanabe M, et al. The microRNA-21/PTEN pathway regulates the sensitivity of HER2-positive gastric cancer cells to trastuzumab. Ann Surg Oncol 2014;21(1):343–50.

[28] Lei BX, Liu ZH, Li ZJ, Li C, Deng YF. miR-21 induces cell proliferation and suppresses the chemosensitivity in glioblastoma cells via downregulation of FOXO1. Int J Clin Exp Med 2014;7(8):2060–6.

[29] Walker CL, Ho SM. Developmental reprogramming of cancer susceptibility. Nat Rev Cancer 2012;12(7):479–86.

[30] Barker DJ. The origins of the developmental origins theory. J Intern Med 2007;261(5):412–7.

[31] Guintivano J, Kaminsky ZA. Role of epigenetic factors in the development of mental illness throughout life. Neurosci Res 2014.

[32] Fudvoye J, Bourguignon JP, Parent AS. Endocrine-disrupting chemicals and human growth and maturation: a focus on early critical windows of exposure. Vitamins Hormones 2014;94:1–25.

[33] Rzeczkowska PA, Hou H, Wilson MD, Palmert MR. Epigenetics: a new player in the regulation of mammalian puberty. Neuroendocrinology 2014;99(3–4):139–55.
[34] Kanherkar RR, Bhatia-Dey N, Csoka AB. Epigenetics across the human lifespan. Front Cell Dev Biol 2014; 2:49.
[35] Ly A, Lee H, Chen J, et al. Effect of maternal and post-weaning folic acid supplementation on mammary tumor risk in the offspring. Cancer Res 2011;71(3):988–97.
[36] Ho SM, Johnson A, Tarapore P, Janakiram V, Zhang X, Leung YK. Environmental epigenetics and its implication on disease risk and health outcomes. ILAR J 2012;53(3–4):289–305.
[37] Wong RL, Walker CL. Molecular pathways: environmental estrogens activate nongenomic signaling to developmentally reprogram the epigenome. Clin Cancer Res 2013;19(14):3732–7.
[38] Braun MM, Ahlbom A, Floderus B, Brinton LA, Hoover RN. Effect of twinship on incidence of cancer of the testis, breast, and other sites (Sweden). Cancer Causes Control 1995;6(6):519–24.
[39] Doherty LF, Bromer JG, Zhou Y, Aldad TS, Taylor HS. In utero exposure to diethylstilbestrol (DES) or bisphenol-A (BPA) increases EZH2 expression in the mammary gland: an epigenetic mechanism linking endocrine disruptors to breast cancer. Hormones Cancer 2010;1(3):146–55.
[40] Durando M, Kass L, Piva J, et al. Prenatal bisphenol A exposure induces preneoplastic lesions in the mammary gland in Wistar rats. Environ Health Perspect 2007;115(1):80–6.
[41] Ekbom A, Trichopoulos D, Adami HO, Hsieh CC, Lan SJ. Evidence of prenatal influences on breast cancer risk. Lancet 1992;340(8826):1015–8.
[42] Palmer JR, Hatch EE, Rosenberg CL, et al. Risk of breast cancer in women exposed to diethylstilbestrol in utero: preliminary results (United States). Cancer Causes Control 2002;13(8):753–8.
[43] Potischman N, Troisi R. In-utero and early life exposures in relation to risk of breast cancer. Cancer Causes Control 1999;10(6):561–73.
[44] Soto AM, Brisken C, Schaeberle C, Sonnenschein C. Does cancer start in the womb? altered mammary gland development and predisposition to breast cancer due to in utero exposure to endocrine disruptors. J Mammary Gland Biol Neoplasia 2013;18(2): 199–208.
[45] Watkins DJ, Tellez-Rojo MM, Ferguson KK, et al. In utero and peripubertal exposure to phthalates and BPA in relation to female sexual maturation. Environ Res 2014;134:233–41.
[46] Palmer JR, Wise LA, Hatch EE, et al. Prenatal diethylstilbestrol exposure and risk of breast cancer. Cancer Epidemiol Biomarkers Prev 2006;15(8):1509–14.
[47] Williams RR, Schweitzer RJ. Clear-cell adenocarcinoma of the vagina in a girl whose mother had taken diethylstilbestrol. Calif Med 1973;118(6):53–5.
[48] Betancourt AM, Eltoum IA, Desmond RA, Russo J, Lamartiniere CA. In utero exposure to bisphenol A shifts the window of susceptibility for mammary carcinogenesis in the rat. Environ Health Perspect 2010;118(11):1614–9.
[49] Hilakivi-Clarke L, Cho E, Cabanes A, et al. Dietary modulation of pregnancy estrogen levels and breast cancer risk among female rat offspring. Clin Cancer Res 2002;8(11):3601–10.
[50] Hilakivi-Clarke L, Clarke R, Lippman M. The influence of maternal diet on breast cancer risk among female offspring. Nutrition 1999;15(5):392–401.
[51] Hilakivi-Clarke L, Clarke R, Onojafe I, Raygada M, Cho E, Lippman M. A maternal diet high in n – 6 polyunsaturated fats alters mammary gland development, puberty onset, and breast cancer risk among female rat offspring. Proc Natl Acad Sci USA 1997;94(17):9372–7.
[52] Lo CY, Hsieh PH, Chen HF, Su HM. A maternal high-fat diet during pregnancy in rats results in a greater risk of carcinogen-induced mammary tumors in the female offspring than exposure to a high-fat diet in postnatal life. Int J Cancer 2009;125(4):767–73.
[53] Paulose T, Speroni L, Sonnenschein C, Soto AM. Estrogens in the wrong place at the wrong time: fetal BPA exposure and mammary cancer. Reprod Toxicol 2014;54:58–65.
[54] Wadia PR, Cabaton NJ, Borrero MD, et al. Low-dose BPA exposure alters the mesenchymal and epithelial transcriptomes of the mouse fetal mammary gland. PLoS One 2013;8(5):e63902.
[55] Dhimolea E, Wadia PR, Murray TJ, et al. Prenatal exposure to BPA alters the epigenome of the rat mammary gland and increases the propensity to neoplastic development. PLoS One 2014;9(7):e99800.
[56] Ayyanan A, Laribi O, Schuepbach-Mallepell S, et al. Perinatal exposure to bisphenol a increases adult mammary gland progesterone response and cell number. Mol Endocrinol 2011;25(11):1915–23.
[57] Betancourt AM, Mobley JA, Russo J, Lamartiniere CA. Proteomic analysis in mammary glands of rat offspring exposed in utero to bisphenol A. J Proteomics 2010;73(6):1241–53.
[58] Hilakivi-Clarke L, Cho E, Onojafe I, Raygada M, Clarke R. Maternal exposure to genistein during pregnancy increases carcinogen-induced mammary tumorigenesis in female rat offspring. Oncol Rep 1999;6(5):1089–95.
[59] Murray TJ, Maffini MV, Ucci AA, Sonnenschein C, Soto AM. Induction of mammary gland ductal hyperplasias and carcinoma in situ following fetal bisphenol A exposure. Reprod Toxicol 2007;23(3):383–90.

[60] Medina D. The preneoplastic phenotype in murine mammary tumorigenesis. J Mammary Gland Biol Neoplasia 2000;5(4):393–407.

[61] Barker DJ. The developmental origins of adult disease. J Am Coll Nutr 2004;23(6 Suppl):588S–95S.

[62] Langley-Evans SC, McMullen S. Developmental origins of adult disease. Med Princ Pract 2010;19(2):87–98.

[63] Taylor PD, Poston L. Developmental programming of obesity in mammals. Exp Physiol 2007;92(2):287–98.

[64] Nagaoka T, Onodera H, Hayashi Y, Maekawa A. Influence of high-fat diets on the occurrence of spontaneous uterine endometrial adenocarcinomas in rats. Teratogenesis Carcinog Mutagen 1995;15(4):167–77.

[65] Plata Mdel M, Williams L, Seki Y, et al. Critical periods of increased fetal vulnerability to a maternal high fat diet. Reproductive Biol Endocrinol 2014;12:80.

[66] de Oliveira Andrade F, Fontelles CC, Rosim MP, et al. Exposure to lard-based high-fat diet during fetal and lactation periods modifies breast cancer susceptibility in adulthood in rats. J Nutr Biochem 2014;25(6):613–22.

[67] Janakiram V, Leung Y-K, Ho S-M. In utero exposure of non-monotonic dose of bisphenol a and high fat diets: high fat butter; high fat olive oil; high fat safflower oil on breast cancer risk. Impacts Endocr Disrupting Chem Physiol Funct 2014. SUN-0374-SUN-.

[68] de Assis S, Warri A, Cruz MI, et al. High-fat or ethinyl-oestradiol intake during pregnancy increases mammary cancer risk in several generations of offspring. Nat Commun 2012;3:1053.

[69] Blute Jr ML, Damaschke NA, Jarrard DF. The epigenetics of prostate cancer diagnosis and prognosis: update on clinical applications. Curr Opin Urol 2015;25(1):83–8.

[70] Valdes-Mora F, Clark SJ. Prostate cancer epigenetic biomarkers: next-generation technologies. Oncogene 2014;34(13):1609–18.

[71] Labbe DP, Zadra G, Ebot EM, et al. Role of diet in prostate cancer: the epigenetic link. Oncogene 2014.

[72] Prins GS. Endocrine disruptors and prostate cancer risk. Endocr Relat Cancer 2008;15(3):649–56.

[73] Ho SM, Tang WY, Belmonte de Frausto J, Prins GS. Developmental exposure to estradiol and bisphenol A increases susceptibility to prostate carcinogenesis and epigenetically regulates phosphodiesterase type 4 variant 4. Cancer Res 2006;66(11):5624–32.

[74] Prins GS, Tang WY, Belmonte J, Ho SM. Perinatal exposure to oestradiol and bisphenol A alters the prostate epigenome and increases susceptibility to carcinogenesis. Basic Clin Pharmacol Toxicol 2008;102(2):134–8.

[75] Tang WY, Morey LM, Cheung YY, Birch L, Prins GS, Ho SM. Neonatal exposure to estradiol/bisphenol A alters promoter methylation and expression of Nsbp1 and Hpcal1 genes and transcriptional programs of Dnmt3a/b and Mbd2/4 in the rat prostate gland throughout life. Endocrinology 2012;153(1):42–55.

[76] Wong RLYW Q, Treviño LS, Bosland MC, Chen J, Medevedoic M, Prins GS, et al. Identification of secretaglobin Scgb2a1 as a target for developmental reprogramming by BPA in the rat prostate. Epigenetics 2015;10(2):127–34.

[77] Driscoll SG, Taylor SH. Effects of prenatal maternal estrogen on the male urogenital system. Obstetrics Gynecol 1980;56(5):537–42.

[78] Rajfer J, Coffey DS. Sex steroid imprinting of the immature prostate. Long-term effects. Investig Urol 1978;16(3):186–90.

[79] Arai Y, Mori T, Suzuki Y, Bern HA. Long-term effects of perinatal exposure to sex steroids and diethylstilbestrol on the reproductive system of male mammals. Int Rev Cytol 1983;84:235–68.

[80] Prins GS, Birch L, Habermann H, et al. Influence of neonatal estrogens on rat prostate development. Reprod Fertil Dev 2001;13(4):241–52.

[81] Huang L, Pu Y, Alam S, Birch L, Prins GS. Estrogenic regulation of signaling pathways and homeobox genes during rat prostate development. J Androl 2004;25(3):330–7.

[82] Boberg J, Johansson HK, Hadrup N, et al. Perinatal exposure to mixtures of anti-androgenic chemicals causes proliferative lesions in rat prostate. Prostate 2015;75(2):126–40.

[83] Anway MD, Skinner MK. Transgenerational effects of the endocrine disruptor vinclozolin on the prostate transcriptome and adult onset disease. Prostate 2008;68(5):517–29.

[84] Sarkar DK. Fetal alcohol exposure increases susceptibility to carcinogenesis and promotes tumor progression in prostate gland. Adv Exp Med Biol 2015;815:389–402.

[85] Murugan S, Zhang C, Mojtahedzadeh S, Sarkar DK. Alcohol exposure in utero increases susceptibility to prostate tumorigenesis in rat offspring. Alcohol Clin Exp Res 2013;37(11):1901–9.

[86] Tracey R, Manikkam M, Guerrero-Bosagna C, Skinner MK. Hydrocarbons (jet fuel JP-8) induce epigenetic transgenerational inheritance of obesity, reproductive disease and sperm epimutations. Reprod Toxicol 2013;36:104–16.

[87] Rinaldi JC, Justulin Jr LA, Lacorte LM, et al. Implications of intrauterine protein malnutrition on prostate growth, maturation and aging. Life Sci 2013;92(13):763–74.

[88] Smith AH, Marshall G, Yuan Y, et al. Increased mortality from lung cancer and bronchiectasis in young adults after exposure to arsenic in utero and in early childhood. Environ Health Perspect 2006;114(8):1293–6.

[89] Steinmaus C, Ferreccio C, Acevedo J, et al. Increased lung and bladder cancer incidence in adults after in utero and early-life arsenic exposure. Cancer Epidemiol Biomarkers Prev 2014;23(8):1529–38.

[90] Cohn BA, Wolff MS, Cirillo PM, Sholtz RI. DDT and breast cancer in young women: new data on the significance of age at exposure. Environ Health Perspect 2007;115(10):1406–14.
[91] Taub JW, Ge Y. The prenatal origin of childhood acute lymphoblastic leukemia. Leukemia lymphoma 2004;45(1):19–25.
[92] Whiteman DC, Whiteman CA, Green AC. Childhood sun exposure as a risk factor for melanoma: a systematic review of epidemiologic studies. Cancer Causes Control 2001;12(1):69–82.
[93] Schulze-Rath R, Hammer GP, Blettner M. Are pre- or postnatal diagnostic X-rays a risk factor for childhood cancer? A systematic review. Radiat Environ Biophys 2008;47(3):301–12.
[94] Titus-Ernstoff L, Troisi R, Hatch EE, et al. Offspring of women exposed in utero to diethylstilbestrol (DES): a preliminary report of benign and malignant pathology in the third generation. Epidemiology 2008;19(2):251–7.
[95] Li S, Hansman R, Newbold R, Davis B, McLachlan JA, Barrett JC. Neonatal diethylstilbestrol exposure induces persistent elevation of c-fos expression and hypomethylation in its exon-4 in mouse uterus. Mol Carcinog 2003;38(2):78–84.
[96] Newbold RR, Bullock BC, McLachlan JA. Uterine adenocarcinoma in mice following developmental treatment with estrogens: a model for hormonal carcinogenesis. Cancer Res 1990;50(23):7677–81.
[97] Tang WY, Newbold R, Mardilovich K, et al. Persistent hypomethylation in the promoter of nucleosomal binding protein 1 (Nsbp1) correlates with overexpression of Nsbp1 in mouse uteri neonatally exposed to diethylstilbestrol or genistein. Endocrinology 2008;149(12):5922–31.
[98] Liu J, Waalkes MP. Liver is a target of arsenic carcinogenesis. Toxicol Sci 2008;105(1):24–32.
[99] Xie Y, Liu J, Benbrahim-Tallaa L, et al. Aberrant DNA methylation and gene expression in livers of newborn mice transplacentally exposed to a hepatocarcinogenic dose of inorganic arsenic. Toxicology 2007;236(1–2):7–15.
[100] Tsang V, Fry RC, Niculescu MD, et al. The epigenetic effects of a high prenatal folate intake in male mouse fetuses exposed in utero to arsenic. Toxicol Appl Pharmacol 2012;264(3):439–50.
[101] Zhang Y, Li Q, Chen H. DNA methylation and histone modifications of Wnt genes by genistein during colon cancer development. Carcinogenesis 2013;34(8):1756–63.
[102] Ciappio ED, Liu Z, Brooks RS, Mason JB, Bronson RT, Crott JW. Maternal B vitamin supplementation from preconception through weaning suppresses intestinal tumorigenesis in Apc1638N mouse offspring. Gut 2011;60(12):1695–702.
[103] Sie KK, Medline A, van Weel J, et al. Effect of maternal and postweaning folic acid supplementation on colorectal cancer risk in the offspring. Gut 2011;60(12):1687–94.
[104] Ali AH, Kondo K, Namura T, et al. Aberrant DNA methylation of some tumor suppressor genes in lung cancers from workers with chromate exposure. Mol Carcinog 2011;50(2):89–99.
[105] Kondo K, Takahashi Y, Hirose Y, et al. The reduced expression and aberrant methylation of p16(INK4a) in chromate workers with lung cancer. Lung Cancer 2006;53(3):295–302.
[106] Zhou ZH, Lei YX, Wang CX. Analysis of aberrant methylation in DNA repair genes during malignant transformation of human bronchial epithelial cells induced by cadmium. Toxicol Sci 2012;125(2):412–7.
[107] Takiguchi M, Achanzar WE, Qu W, Li G, Waalkes MP. Effects of cadmium on DNA-(Cytosine-5) methyltransferase activity and DNA methylation status during cadmium-induced cellular transformation. Exp Cell Res 2003;286(2):355–65.
[108] Benbrahim-Tallaa L, Tokar EJ, Diwan BA, Dill AL, Coppin JF, Waalkes MP. Cadmium malignantly transforms normal human breast epithelial cells into a basal-like phenotype. Environ Health Perspect 2009;117(12):1847–52.
[109] Ionescu JG, Novotny J, Stejskal V, Latsch A, Blaurock-Busch E, Eisenmann-Klein M. Increased levels of transition metals in breast cancer tissue. Neuro Endocrinol Lett 2006;27(Suppl. 1):36–9.
[110] Steenland K, Boffetta P. Lead and cancer in humans: where are we now? Am J Ind Med 2000;38(3):295–9.
[111] Fong M, Henson DE, Devesa SS, Anderson WF. Inter- and intra-ethnic differences for female breast carcinoma incidence in the continental United States and in the state of Hawaii. Breast Cancer Res Treat 2006;97(1):57–65.
[112] Xiang N, Zhao R, Song G, Zhong W. Selenite reactivates silenced genes by modifying DNA methylation and histones in prostate cancer cells. Carcinogenesis 2008;29(11):2175–81.
[113] de Miranda JX, Andrade FD, Conti AD, Dagli ML, Moreno FS, Ong TP. Effects of selenium compounds on proliferation and epigenetic marks of breast cancer cells. J Trace Elem Med Biol 2014;28(4):486–91.
[114] Zeng H, Yan L, Cheng WH, Uthus EO. Dietary selenomethionine increases exon-specific DNA methylation of the p53 gene in rat liver and colon mucosa. J Nutr 2011;141(8):1464–8.
[115] Tilghman SL, Bratton MR, Segar HC, et al. Endocrine disruptor regulation of microRNA expression in breast carcinoma cells. PLoS One 2012;7(3):e32754.
[116] Fernandez SV, Huang Y, Snider KE, Zhou Y, Pogash TJ, Russo J. Expression and DNA methylation changes in human breast epithelial cells after bisphenol A exposure. Int J Oncol 2012;41(1):369–77.

[117] Qin XY, Fukuda T, Yang L, et al. Effects of bisphenol A exposure on the proliferation and senescence of normal human mammary epithelial cells. Cancer Biol Ther 2012;13(5):296–306.

[118] Hermoni-Levine M, Rahamimoff H. Role of the phospholipid environment in modulating the activity of the rat brain synaptic plasma membrane Ca2(+)-ATPase. Biochemistry 1990;29(20):4940–50.

[119] Lamartiniere CA, Jenkins S, Betancourt AM, Wang J, Russo J. Exposure to the endocrine disruptor bisphenol a alters susceptibility for mammary Cancer. Horm Mol Biol Clin Investig 2011;5(2):45–52.

[120] Davis VL, Newbold RR, Couse JF, et al. Expression of a dominant negative estrogen receptor alpha variant in transgenic mice accelerates uterine cancer induced by the potent estrogen diethylstilbestrol. Reprod Toxicol 2012;34(4):512–21.

[121] Rothschild TC, Boylan ES, Calhoon RE, Vonderhaar BK. Transplacental effects of diethylstilbestrol on mammary development and tumorigenesis in female ACI rats. Cancer Res 1987;47(16):4508–16.

[122] Hsu PY, Deatherage DE, Rodriguez BA, et al. Xenoestrogen-induced epigenetic repression of microRNA-9-3 in breast epithelial cells. Cancer Res 2009;69(14):5936–45.

[123] Pistek VL, Furst RW, Kliem H, Bauersachs S, Meyer HH, Ulbrich SE. HOXA10 mRNA expression and promoter DNA methylation in female pig offspring after in utero estradiol-17β exposure. J Steroid Biochem Mol Biol 2013;138:435–44.

[124] Boffetta P, Jourenkova N, Gustavsson P. Cancer risk from occupational and environmental exposure to polycyclic aromatic hydrocarbons. Cancer Causes Control 1997;8(3):444–72.

[125] Clavel J, Mandereau L, Limasset JC, Hemon D, Cordier S. Occupational exposure to polycyclic aromatic hydrocarbons and the risk of bladder cancer: a French case-control study. Int J Epidemiol 1994;23(6):1145–53.

[126] Gammon MD, Santella RM. PAH, genetic susceptibility and breast cancer risk: an update from the Long Island Breast Cancer Study Project. Eur J Cancer 2008;44(5):636–40.

[127] Costantino JP, Redmond CK, Bearden A. Occupationally related cancer risk among coke oven workers: 30 years of follow-up. J Occup Environ Med 1995;37(5):597–604.

[128] Nadon L, Siemiatycki J, Dewar R, Krewski D, Gerin M. Cancer risk due to occupational exposure to polycyclic aromatic hydrocarbons. Am J Ind Med 1995;28(3):303–24.

[129] Tretyakova N, Matter B, Jones R, Shallop A. Formation of benzo[a]pyrene diol epoxide-DNA adducts at specific guanines within K-ras and p53 gene sequences: stable isotope-labeling mass spectrometry approach. Biochemistry 2002;41(30):9535–44.

[130] Cos P, De Bruyne T, Apers S, Vanden Berghe D, Pieters L, Vlietinck AJ. Phytoestrogens: recent developments. Planta Med 2003;69(7):589–99.

[131] Su Y, Eason RR, Geng Y, Till SR, Badger TM, Simmen RC. In utero exposure to maternal diets containing soy protein isolate, but not genistein alone, protects young adult rat offspring from NMU-induced mammary tumorigenesis. Carcinogenesis 2007;28(5):1046–51.

[132] Trock BJ, Hilakivi-Clarke L, Clarke R. Meta-analysis of soy intake and breast cancer risk. J Natl Cancer Inst 2006;98(7):459–71.

[133] Kao YC, Zhou C, Sherman M, Laughton CA, Chen S. Molecular basis of the inhibition of human aromatase (estrogen synthetase) by flavone and isoflavone phytoestrogens: a site-directed mutagenesis study. Environ Health Perspect 1998;106(2):85–92.

[134] Ozasa K, Nakao M, Watanabe Y, et al. Serum phytoestrogens and prostate cancer risk in a nested case-control study among Japanese men. Cancer Sci 2004;95(1):65–71.

[135] Nothlings U, Murphy SP, Wilkens LR, Henderson BE, Kolonel LN. Flavonols and pancreatic cancer risk: the multiethnic cohort study. Am J Epidemiol 2007;166(8):924–31.

[136] Shimazu T, Inoue M, Sasazuki S, et al. Isoflavone intake and risk of lung cancer: a prospective cohort study in Japan. Am J Clin Nutr 2010;91(3):722–8.

[137] Yuan JM, Gao YT, Yang CS, Yu MC. Urinary biomarkers of tea polyphenols and risk of colorectal cancer in the Shanghai Cohort Study. Int J Cancer 2007;120(6):1344–50.

[138] Li Y, Liu L, Andrews LG, Tollefsbol TO. Genistein depletes telomerase activity through cross-talk between genetic and epigenetic mechanisms. Int J Cancer 2009;125(2):286–96.

[139] Motojima S, Makino S. Chronic eosinophilic pneumonia. Nihon Rinsho 1990;48(3):602–7.

[140] Dagdemir A, Durif J, Ngollo M, Bignon YJ, Bernard-Gallon D. Histone lysine trimethylation or acetylation can be modulated by phytoestrogen, estrogen or anti-HDAC in breast cancer cell lines. Epigenomics 2013;5(1):51–63.

[141] Zhang X, Wallace AD, Du P, et al. Genome-wide study of DNA methylation alterations in response to diazinon exposure in vitro. Environ Toxicol Pharmacol 2012;34(3):959–68.

[142] Desaulniers D, Xiao GH, Lian H, et al. Effects of mixtures of polychlorinated biphenyls, methylmercury, and organochlorine pesticides on hepatic DNA methylation in prepubertal female Sprague-Dawley rats. Int J Toxicol 2009;28(4):294–307.

[143] Luo J, Li YN, Wang F, Zhang WM, Geng X. S-adenosylmethionine inhibits the growth of cancer cells by reversing the hypomethylation status of c-myc and H-ras in human gastric cancer and colon cancer. Int J Biol Sci 2010;6(7):784–95.

[144] Song C, Kanthasamy A, Anantharam V, Sun F, Kanthasamy AG. Environmental neurotoxic pesticide increases histone acetylation to promote apoptosis in dopaminergic neuronal cells: relevance to epigenetic mechanisms of neurodegeneration. Mol Pharmacol 2010;77(4):621–32.

[145] Atallah E, Schiffer CA. Agent Orange, prostate cancer irradiation and acute promyelocyic leukemia (APL): is there a link? Leuk Res 2007;31(5):720–1.

[146] Schecter A, Needham L, Pavuk M, et al. Agent Orange exposure, Vietnam war veterans, and the risk of prostate cancer. Cancer 2009;115(14):3369–71.

[147] Ansbaugh N, Shannon J, Mori M, Farris PE, Garzotto M. Agent Orange as a risk factor for high-grade prostate cancer. Cancer 2013;119(13):2399–404.

[148] Giri VN, Cassidy AE, Beebe-Dimmer J, et al. Association between Agent Orange and prostate cancer: a pilot case-control study. Urology 2004;63(4):757–60. Discussion 60–1.

[149] Shah SR, Freedland SJ, Aronson WJ, et al. Exposure to Agent Orange is a significant predictor of prostate-specific antigen (PSA)-based recurrence and a rapid PSA doubling time after radical prostatectomy. BJU Int 2009;103(9):1168–72.

[150] Bullman TA, Watanabe KK, Kang HK. Risk of testicular cancer associated with surrogate measures of Agent Orange exposure among Vietnam veterans on the Agent Orange Registry. Ann Epidemiol 1994;4(1):11–6.

[151] Young AL, Giesy JP, Jones PD, Newton M. Environmental fate and bioavailability of Agent Orange and its associated dioxin during the Vietnam War. Environ Sci Pollut Res Int 2004;11(6):359–70.

[152] Jenkins S, Rowell C, Wang J, Lamartiniere CA. Prenatal TCDD exposure predisposes for mammary cancer in rats. Reprod Toxicol 2007;23(3):391–6.

[153] Brown NM, Manzolillo PA, Zhang JX, Wang J, Lamartiniere CA. Prenatal TCDD and predisposition to mammary cancer in the rat. Carcinogenesis 1998;19(9):1623–9.

[154] Yang JH, Thraves P, Dritschilo A, Rhim JS. Neoplastic transformation of immortalized human keratinocytes by 2,3,7,8-tetrachlorodibenzo-p-dioxin. Cancer Res 1992;52(12):3478–82.

[155] Okino ST, Pookot D, Li LC, et al. Epigenetic inactivation of the dioxin-responsive cytochrome P4501A1 gene in human prostate cancer. Cancer Res 2006;66(15):7420–8.

[156] Ray SS, Swanson HI. Alteration of keratinocyte differentiation and senescence by the tumor promoter dioxin. Toxicol Appl Pharmacol 2003;192(2):131–45.

[157] Ray SS, Swanson HI. Dioxin-induced immortalization of normal human keratinocytes and silencing of p53 and p16INK4a. J Biol Chem 2004;279(26): 27187–93.

[158] Hakansson N, Floderus B, Gustavsson P, Feychting M, Hallin N. Occupational sunlight exposure and cancer incidence among Swedish construction workers. Epidemiology 2001;12(5):552–7.

[159] You YH, Li C, Pfeifer GP. Involvement of 5-methylcytosine in sunlight-induced mutagenesis. J Mol Biol 1999;293(3):493–503.

[160] You YH, Pfeifer GP. Similarities in sunlight-induced mutational spectra of CpG-methylated transgenes and the p53 gene in skin cancer point to an important role of 5-methylcytosine residues in solar UV mutagenesis. J Mol Biol 2001;305(3):389–99.

[161] Gronniger E, Weber B, Heil O, et al. Aging and chronic sun exposure cause distinct epigenetic changes in human skin. PLoS Genet 2010;6(5):e1000971.

[162] Crosio C, Cermakian N, Allis CD, Sassone-Corsi P. Light induces chromatin modification in cells of the mammalian circadian clock. Nat Neurosci 2000;3(12):1241–7.

[163] Hong Y, Won J, Lee Y, et al. Melatonin treatment induces interplay of apoptosis, autophagy, and senescence in human colorectal cancer cells. J Pineal Res 2014;56(3):264–74.

[164] Jardim-Perassi BV, Arbab AS, Ferreira LC, et al. Effect of melatonin on tumor growth and angiogenesis in xenograft model of breast cancer. PLoS One 2014;9(1):e85311.

[165] Sharma R, Ottenhof T, Rzeczkowska PA, Niles LP. Epigenetic targets for melatonin: induction of histone H3 hyperacetylation and gene expression in C17.2 neural stem cells. J Pineal Res 2008;45(3):277–84.

[166] Valko M, Rhodes CJ, Moncol J, Izakovic M, Mazur M. Free radicals, metals and antioxidants in oxidative stress-induced cancer. Chem Biol Interact 2006;160(1):1–40.

[167] Lakkur S, Goodman M, Bostick RM, et al. Oxidative balance score and risk for incident prostate cancer in a prospective U.S. cohort study. Ann Epidemiol 2014;24(6):475–8. e4.

[168] Payne SL, Fogelgren B, Hess AR, et al. Lysyl oxidase regulates breast cancer cell migration and adhesion through a hydrogen peroxide-mediated mechanism. Cancer Res 2005;65(24):11429–36.

[169] Muscarella LA, Parrella P, D'Alessandro V, et al. Frequent epigenetics inactivation of KEAP1 gene in non-small cell lung cancer. Epigenetics 2011;6(6):710–9.

[170] Campos AC, Molognoni F, Melo FH, et al. Oxidative stress modulates DNA methylation during melanocyte anchorage blockade associated with malignant transformation. Neoplasia 2007;9(12):1111–21.

[171] Molognoni F, de Melo FH, da Silva CT, Jasiulionis MG. Ras and Rac1, frequently mutated in melanomas, are activated by superoxide anion, modulate Dnmt1 level and are causally related to melanocyte malignant transformation. PLoS One 2013;8(12):e81937.

[172] Molognoni F, Cruz AT, Meliso FM, et al. Epigenetic reprogramming as a key contributor to melanocyte malignant transformation. Epigenetics 2011;6(4):450–64.
[173] Collin SM, Metcalfe C, Refsum H, et al. Circulating folate, vitamin B12, homocysteine, vitamin B12 transport proteins, and risk of prostate cancer: a case-control study, systematic review, and meta-analysis. Cancer Epidemiol Biomarkers Prev 2010;19(6):1632–42.
[174] Schernhammer ES, Giovannucci E, Kawasaki T, Rosner B, Fuchs CS, Ogino S. Dietary folate, alcohol and B vitamins in relation to LINE-1 hypomethylation in colon cancer. Gut 2010;59(6):794–9.
[175] Vineis P, Chuang SC, Vaissiere T, et al. DNA methylation changes associated with cancer risk factors and blood levels of vitamin metabolites in a prospective study. Epigenetics 2011;6(2):195–201.
[176] Garland CF, Gorham ED, Mohr SB, et al. Vitamin D and prevention of breast cancer: pooled analysis. J Steroid Biochem Mol Biol 2007;103(3–5):708–11.
[177] Essa S, Denzer N, Mahlknecht U, et al. VDR microRNA expression and epigenetic silencing of vitamin D signaling in melanoma cells. J Steroid Biochem Mol Biol 2010;121(1–2):110–3.
[178] Thorne JL, Maguire O, Doig CL, et al. Epigenetic control of a VDR-governed feed-forward loop that regulates p21(waf1/cip1) expression and function in non-malignant prostate cells. Nucleic Acids Res 2011;39(6):2045–56.
[179] Padi SK, Zhang Q, Rustum YM, Morrison C, Guo B. MicroRNA-627 mediates the epigenetic mechanisms of vitamin D to suppress proliferation of human colorectal cancer cells and growth of xenograft tumors in mice. Gastroenterology 2013;145(2):437–46.
[180] Ung M, Ma X, Johnson KC, Christensen BC, Cheng C. Effect of estrogen receptor alpha binding on functional DNA methylation in breast cancer. Epigenetics 2014;9(4):523–32.
[181] Bhan A, Hussain I, Ansari KI, Kasiri S, Bashyal A, Mandal SS. Antisense transcript long noncoding RNA (lncRNA) HOTAIR is transcriptionally induced by estradiol. J Mol Biol 2013;425(19):3707–22.
[182] Xie YL, Yang YJ, Tang C, et al. Estrogen combined with progesterone decreases cell proliferation and inhibits the expression of Bcl-2 via microRNA let-7a and miR-34b in ovarian cancer cells. Clin Transl Oncol 2014;16(10):898–905.
[183] Casati L, Sendra R, Poletti A, Negri-Cesi P, Celotti F. Androgen receptor activation by polychlorinated biphenyls: epigenetic effects mediated by the histone demethylase Jarid1b. Epigenetics 2013;8(10):1061–8.
[184] Cho Y, Turner ND, Davidson LA, Chapkin RS, Carroll RJ, Lupton JR. Colon cancer cell apoptosis is induced by combined exposure to the n-3 fatty acid docosahexaenoic acid and butyrate through promoter methylation. Exp Biol Med 2014;239(3):302–10.
[185] Venza I, Visalli M, Fortunato C, et al. PGE2 induces interleukin-8 derepression in human astrocytoma through coordinated DNA demethylation and histone hyperacetylation. Epigenetics 2012;7(11):1315–30.
[186] El-Osta A, Brasacchio D, Yao D, et al. Transient high glucose causes persistent epigenetic changes and altered gene expression during subsequent normoglycemia. J Exp Med 2008;205(10):2409–17.
[187] Gupta C, Kaur J, Tikoo K. Regulation of MDA-MB-231 cell proliferation by GSK-3β involves epigenetic modifications under high glucose conditions. Exp Cell Res 2014;324(1):75–83.
[188] Li Y, Liu L, Tollefsbol TO. Glucose restriction can extend normal cell lifespan and impair precancerous cell growth through epigenetic control of hTERT and p16 expression. FASEB J 2010;24(5):1442–53.
[189] Ujhazy E, Navarova J, Dubovicky M, Bezek S, Mach M. Important issues in developmental toxicity testing. Interdiscip Toxicol 2008;1(1):27–8.
[190] Hu WY, Shi GB, Lam HM, et al. Estrogen-initiated transformation of prostate epithelium derived from normal human prostate stem-progenitor cells. Endocrinology 2011;152(6):2150–63.
[191] Prins GS, Hu WY, Shi GB, et al. Bisphenol A promotes human prostate stem-progenitor cell self-renewal and increases in vivo carcinogenesis in human prostate epithelium. Endocrinology 2014;155(3):805–17.
[192] Grimshaw MJ, Cooper L, Papazisis K, et al. Mammosphere culture of metastatic breast cancer cells enriches for tumorigenic breast cancer cells. Breast Cancer Res 2008;10(3):R52.
[193] Tomasetti C, Vogelstein B. Cancer etiology. Variation in cancer risk among tissues can be explained by the number of stem cell divisions. Science 2015;347(6217):78–81.
[194] Miyahara M, Miyahara M, Nakadate M, Suzuki I, Odashima S. Sensitivity difference of rat ascites hepatoma AH-13 and mouse leukemia L-1210 to nitrosourea derivatives. Gan 1978;69(2):187–93.
[195] Bachman AN, Curtin GM, Doolittle DJ, Goodman JI. Altered methylation in gene-specific and GC-rich regions of DNA is progressive and nonrandom during promotion of skin tumorigenesis. Toxicol Sci 2006;91(2):406–18.
[196] Dunn KL, Espino PS, Drobic B, He S, Davie JR. The Ras-MAPK signal transduction pathway, cancer and chromatin remodeling. Biochem Cell Biol 2005;83(1):1–14.

CHAPTER

17

Epigenetic Regulation of Gastrointestinal Epithelial Barrier and Developmental Origins of Health and Disease

J.P. Lallès[1,2], C. Michel[2,3,4], V. Theodorou[5], J.P. Segain[2,3,4]

[1]Institut National de la Recherche Agronomique, UR1341 ADNC, Saint Gilles, France; [2]Centre de Recherche en Nutrition Humaine-Ouest, Nantes, France; [3]Institut National de la Recherche Agronomique/Université de Nantes, UMR1280, Nantes, France; [4]Institut des Maladies de l'Appareil Digestif, Nantes, France; [5]Institut National de la Recherche Agronomique, UMR Toxalim, Toulouse, France

OUTLINE

The Epigenome and Developmental Origins of Health and Disease
http://dx.doi.org/10.1016/B978-0-12-801383-0.00017-7

Copyright © 2016 Elsevier Inc. All rights reserved.

INTRODUCTION

The primary function of the gastrointestinal (GI) tract is to digest food and absorb nutrients and water. The epithelial surface of the GI tract also has a crucial barrier function to prevent entry of dietary antigens and microbes into the body. The intestinal epithelial barrier mainly involves a monolayer of intestinal epithelial cells (IECs) that are interconnected by tight junctions [1]. Regulation of intestinal barrier function is a complex task involving integrated actions of epithelial, immune, and nervous cells but also of the gut microbiota [1,2]. Barrier alterations are quite often involved in the etiology and pathogenesis of chronic GI inflammatory diseases [3].

Dysregulation of the cross talk between GI cells and the microbiota is a common feature of several GI pathologies such as inflammatory bowel disease (IBD), irritable bowel syndrome (IBS), and colorectal cancer (CRC) [4–6]. A particular focus has been devoted to them in this chapter because of the environmental impact on disease risk characterizing IBD, IBS, and CRC as well as the public health importance and economic burden of these diseases. Indeed, IBD is a chronic disabling GI disorder impacting every aspect of the affected individual's life and account for substantial costs to health care [7]. In addition, patients with long-standing IBD have an increased risk of colorectal cancer [8]. IBS, a common functional disorder, despite lower morbidity than IBD, is highly prevalent affecting up to 20% of the population in the Western world; it is estimated that IBS constitutes an important part of the gastroenterology practice generating high direct and indirect health costs [9]. CRC is the third most common cancer and the fourth most common cancer cause of death globally, accounting for roughly 1.2 million new cases and 600,000 deaths per year [10].

It is now becoming clear that beside genetic susceptibility, environmental factors, including food and dietary habits as well as other lifestyle features, influence individual variation in disease risk for these three important pathologies [11–13]. Environmental factors modulate gene expression in most cell types through epigenetic mechanisms, including DNA methylation, posttranslational histone modifications, and noncoding RNAs, such as microRNAs [14,15]. Nutrients and various food compounds can, for instance, inhibit or activate (as substrate) epigenetic-regulating enzymes including DNA and histone methyltransferases (DNMT and HMT), histone deacetylases (HDACs), or histone acetyltransferase (HAT). Epigenetically silenced genes can be reactivated using, for example, HDAC inhibitors in cancer cells [16]. Some studies suggest that the gut microbiota may contribute to programming processes [17–20], particularly in the context of metabolic and inflammatory diseases [21–25]. This hypothesis, although still to be proven, is supported by both the putative ability of the gut microbiota to modulate the epigenome [26–29] and/or its propensity to be durably shaped in neonatal life [30,31]; this would, therefore, condition its future interactions with the physiology of the adult host.

IECs are particularly exposed to nutrient-induced epigenetic modifications, such as folate and butyrate, that may in turn positively or negatively modulate gene programs involved in cell proliferation, differentiation, and apoptosis [32–34]. Therefore, an important question is also how our diet participates, directly or indirectly (e.g., through the microbiota), in modulation of the GI epithelial epigenome. Numerous bioactive food components and nutrients (e.g., methyl donors) are known for regulating cell epigenetic machinery, and this provides a link between cell metabolism and control of gene expression [35,36]. As expected, due to the involvement of methyl groups in epigenetic modifications of genomic DNA and of histones, modification of dietary supply of methyl donors (e.g., methionine, pyridoxine (vitamin B6), folate (vitamin B9), betaine, choline, vitamin B12) directly impacts methylation processes at the GI epithelial level [37,38]. Importantly, other pathways related to methylation/demethylation, and food components other than methyl donors, have been shown in recent years to modulate the GI epithelial epigenome. Much of the available data come from investigations with IEC in vitro; in vivo data supporting cellular and molecular results are still scarce [17,39].

Early studies reported that nutrition during critical periods of development induces epigenetic regulation of transposable elements and imprinted genes [40]. A pioneering study was carried out in viable yellow agouti mice, demonstrating that supplementation of maternal diet with methyl donors modified the coat color of the offspring from yellow to brown proportionally to the degree of methylation of a transposable element controlling agouti gene expression [40]. Furthermore, it was found that individuals prenatally exposed to famine during the "Dutch Hunger Winter" of 1944–1945 had less DNA methylation of the imprinted insulin-like growth factor *(IGF) 2* gene compared to sex-matched siblings unexposed to the famine [41]. Thus, these examples support the hypothesis that early environmental factors influence developmental programming of the epigenome. There is now mounting evidence in human and animal studies that perinatal environment can induce epigenetic modifications in several important metabolic organs such as liver, skeletal muscle, pancreas, or adipose tissue, affecting their structure and function later in life (for review, see this book and Ref. [42]). However, data on the GI system are still limited.

The aim of this chapter is to analyze the evidence for early influences on epigenetic modifications of the GI epithelium in connection with the "developmental origin of health and disease" (DOHaD) theory, the possible role played by the gut microbiota, the links between early epigenetic modifications and GI diseases in later life, and, finally, dietary modulation of epigenetic changes at the GI epithelial level.

EPIGENETICS AND GI TRACT DEVELOPMENT AND MATURATION

Epigenetics includes all the epigenetic modifications that affect gene expression without changing the nucleotide sequence of the genome. The epigenome describes the epigenetic code that is constituted by cell- and tissue-specific patterns of DNA methylation and posttranslational modifications of histones (methylation, acetylation, phosphorylation, biotinylation, ubiquitinylation, and sumoylation) [14,15,35]. Epigenetic marks are installed, read, and erased by DNA- and histone-modifying enzymes acting synergistically or antagonistically, including DNMT, HMT, HDMT, HDAC, or HAT. DNA methylation status in combination with distinct histone marks will determine the "active" or "inactive" chromatin state. Establishment and maintenance of epigenetic patterns are essential steps during development for the differentiation of stem cells into specific cell types that form organs and tissue expressing distinct gene expression programs [43]. The two important features of epigenetics that are to be considered in the DOHaD hypothesis are the plasticity of the epigenetic code and its inheritance through mitosis and/or meiosis. Because of the high degree of plasticity of the epigenome during the developmental process, it seems obvious that any epigenetic modification induced by intrinsic or extrinsic factors will alter cell differentiation and organ and tissue formation [14,43,44]. Developmental modification of the epigenome in somatic (adult) stem cells, such as those found in gut epithelium, can have lifelong effects altering the self-renewal process of intestinal epithelium.

The gut epithelium is constantly regenerated in order to maintain normal GI function. Regeneration of the epithelium is performed by multipotent stem cells that have the capacity to self-renew rapidly during the whole life [43,45]. Intestinal stem cells proliferate, migrate, and differentiate in the "transit amplifying" zone into all mature cell types of the intestinal epithelium including enterocytes and goblet, enteroendocrine, and Paneth cells. Self-renewal of the intestinal epithelium follows a well-orchestrated program involving the coordinated action of signal transduction and transcription pathways for proper cell fate [45]. It has become clear over the past few years that epigenetic mechanisms play an important role in the self-renewal genetic program of intestinal epithelium [43]. However, how epigenetic mechanisms regulate in time and space gene transcription program of gut epithelium renewal still needs to be better characterized.

Stem cells contain a less compact chromatin suggesting that it may be more permissive for transcription of inactive genes [44]. Recent work in zebra fish showed that DNA demethylase activity is important for maintaining stem cells in an undifferentiated state [46,47]. Knockdown of DNA demethylase components restored DNA methylation and induced intestinal cell differentiation [47]. Interestingly, these authors reported that the adenomatous polyposis coli tumor suppressor downregulated DNMT expression, resulting in DNA methylation and cell differentiation [46,47]. From these findings, it was suggested that early adenomatous polyposis coli mutation in CRC could initiate DNA demethylation, thus contributing to cell proliferation, in association with activation of the Wnt/β-catenin signaling pathway [48]. Another characteristic of stem cells is that many nontranscribed genes contain active histone marks, such as di- and trimethylated H3K4 and acetylated H3K9, but also repressive trimethylated histone H3K27 in their promoter region [49,50]. This epigenetic mechanism "poised" for expression genes is involved in cell differentiation that are ready for rapid transcriptional activation upon differentiation [44,45]. During differentiation, bivalent chromatin profiles are modified. Genes specific for a particular cell fate only maintain active histone marks, whereas genes specific for other cell types keep the repressive trimethylated H3K27 [49,50]. Methylation of H3K27 is catalyzed by the Polycomb group of protein repressors including Polycomb repressor complex 1 and 2. Interestingly, Polycomb group proteins can recruit DNMTs that in turn cooperate in the induction of a repressive chromatin through DNA methylation [44]. Recent in vitro and in vivo data suggest that, in addition to maintaining cell stemness, Polycomb repressor complex 2 could control cell fate decision in the "transit amplifying" zone by repressing the expression of transcription factors Cdx2 and hepatic nuclear factor 1a, which are involved in terminal enterocyte differentiation [51]. On the other hand, HDAC1, 2, and 3 are also important regulators of gut epithelial cell maturation. HDAC1, 2, and 3 are downregulated along small intestine crypt-villus and colonic crypt axes [52]. Using an ex vivo model of fetal mouse intestine development, Tou et al. showed that HDAC1 and 2 inhibit cell differentiation and that their expression levels decline concomitantly with the activation of differentiation genes [53]; accordingly, HDAC2 was also shown to bind to differentiation gene promoters [53]. HDAC3 inhibited IEC maturation while HDAC3 silencing induced expression of the cell cycle inhibitor p21 and induced cell differentiation [52]. Moreover, the role of HDAC in epithelial differentiation has been largely demonstrated by the use of HDAC inhibitors (e.g., butyrate), showing that HDAC inhibition induced expression of terminal differentiation markers [52].

These studies illustrate the importance of epigenetic reprogramming that occurs during normal self-renewing of gut epithelium. However, modification of the epigenetic program is also a common feature of GI disease, as well as of aging [54]. Importantly, hypermethylation of gene promoters and genome-wide DNA methylation observed in CRC also occur in normal aged colonic tissue and are often common with Polycomb group target genes in embryonic stem cells [48,55]. Accordingly, a recent study showed that transgenic expression of DNMT3b in mouse colon induces de novo DNA methylation of genes that are also methylated in CRC [56]. Furthermore, DNMT3b-induced methylation was maintained even after suppressing DNMT3b expression. These data suggest that methylation in CRC or aged colon is not a stochastic process and that transient induction of DNA methylation, most likely in intestinal stem cells, permanently may alter cell epigenome.

Many lines of evidence, including data from epidemiological, clinical, and experimental studies, now indicate that environmental factors during critical periods of development influence epigenome shaping in different organs, thus affecting their structure and function later in life (for review, see this book and Ref. [42]). Although data on the GI epithelium are limited, recent studies have underlined the influence of dietary methyl donors during the prenatal and postnatal periods on DNA methylation in mouse colon [57–60].

THE GUT MICROBIOTA AND PROGRAMMING

The Gut Microbiota and Associated Metabolic Activities

The gut microbiota is now recognized as a true organ that constitutes the largest reservoir of microorganisms in our body [61,62]. Indeed, the human GI tract hosts more than 100 trillion bacteria and archaea, with maximal bacterial densities occurring in the colon [62,63]. Approximately 1000 different bacterial species have been detected as potential components of the gut microbiota. However, the microbiota of each individual is composed of some 150 to 400 species, with fewer than 100 being shared by most individuals [62,64,65]. Sequencing of the combined genomes of all the dominant microorganisms included in the microbiota reveals approximately 10 million different bacterial genes [66], that is, approximately 1000-fold more than the human genome, of which 500,000 would be present in each microbiota [67]. This would correspond to an overall potential of 19,000 metabolic functions, with at least 6000 being common to all individuals [67]. Half of these functions are still unknown but among the identified ones (see Clusters of Orthologous Groups definitions at: http://www.ncbi.nlm.nih.gov/COG/), "carbohydrate metabolism and transport" and "amino acid metabolism and transport" are the most represented [68,69]. This picture is globally coherent with the considerable metabolic capability of the gut microbiota that may be even higher than that of the liver [61,70] and that largely contribute to many of the metabolic processes essential for human body homeostasis [64,71]. Actually, the microbiota appear to be able to metabolically process most of the components reaching the large intestine, that is, all indigestible dietary, endogenous, or xenobiotic compounds [70]. Besides the well-described metabolism of indigestible carbohydrates and proteins [72], the gut microbiota are also involved in numerous other key metabolisms, among which, albeit not exhaustively, are: bile acid metabolism [73]; phosphatidylcholine and choline conversion [74]; polyphenol metabolism [70]; cholesterol conversion [75]; vitamin synthesis [76]; and cruciferous glucosinolate degradation [77]. As pointed out below, some of these metabolisms can generate compounds involved in epigenetic processes, the activity of which may concern the whole body as evidenced by the changes in plasma metabolites observed between germ-free and conventional animals [78], or in animals colonized with microbiota of different complexities [79].

The Gut Microbiota and Epigenetic Processes

The actual contribution of the microbiota to the epigenetic (re)programming has not been empirically established yet but appears likely.

First, numerous bacterial pathogens have been shown to influence the host epigenome through activation of signaling cascades [29,80,81]. As an example, *Helicobacter pylori* induces, via the infection-associated inflammatory response it triggers, aberrant DNA methylation, which is only partly reversed after eradication of the pathogen [82]. Whether procaryotes could be directly active in epigenetic modifications of eucaryotic cells is still debated [26,29].

Second, gut bacteria can interfere with the bioavailable pool of numerous compounds known to be involved in the induction of posttranslational modifications of histones or in DNA methylation. This capability has led Shenderov to state that low molecular weight molecules of indigenous microbiota may be key determinants of host epigenetic modulation [26]. Among those microbial compounds, short-chain fatty acids, mainly acetate, propionate, and especially butyrate (discussed in more detail below) are the best-known end products of gut microbiota metabolism [72]. They can influence histone acetylation, thereby affecting DNA accessibility for transcription. Indeed, acetate constitutes a donor for acetyl groups and butyrate is an HDAC inhibitor [83]. Some of the vitamins produced by gut bacteria [76], particularly those not synthetized by mammals, are also of tremendous interest with respect to epigenetic processes. In particular, members of the *Bifidobacterium* and *Lactobacillus* genera produce folate [84], which is central to the availability of methyl groups for DNA methylation [27]. Some other gut bacteria produce biotin [84], which serves as substrate for histone biotinylation [26]. Bacterial end products from the metabolism of cruciferous vegetables or garlic, such as isothiocyanate, which is produced from glucosinolate, can inhibit HDAC enzyme activity [85]. Conversely, gut microbiota metabolism can also decrease the pool of bioavailable compounds necessary for epigenetic reprogramming. As an example, the gut microbiota degrades choline [74] and, therefore, decreases the available pool of methyl donors [26]. Some gut bacteria can accumulate high intracellular amounts of minerals, such as selenium [86], and possibly compete with the host for these compounds when their availability becomes limiting [87]. This may affect DNA methylation since selenium can interfere with DNMT enzyme [88].

Finally, few findings suggest that this potential actually translates into epigenetic reprogramming by gut bacteria at least in the GI tract. Indeed, epigenomic machinery, particularly HDAC3, has been shown as crucial for establishing the normal host–microbiota dialogue [89], and differences in the methylation level of some genes (e.g., *Toll-like receptor 4, chemokine (C-X-C motif) ligand 16*) have been reported in colonocytes of germ-free versus conventional mice [90,91]. Of particular interest with regard to programming, the decrease in the hypermethylation of *chemokine (C-X-C motif) ligand 16* gene, which is induced by neonatal gut colonization, is not observed when colonization is performed at an adult age [91].

The Gut Microbiota as a Long-term Memory of Perinatal Life

Besides its possible direct contribution to epigenetic changes, the gut microbiota may also contribute to programming processes by acting as a long-term memory of perinatal life. Such a role is theoretically possible in view of: (1) the ability of the gut microbiota to interact with host physiology far beyond the GI tract [92]; (2) the fact that perinatal environment can affect neonatal microbiota setup [93]; and (3) the assumption that the gut microbiota would be durably shaped during neonatal period [31]. In this scenario, the perinatal environment would program adult host–microbiota interactions through sustainable shaping of the composition of the neonatal microbiota. Whether early alterations in gut microbiota persist until adulthood remains to be ascertained. Indeed, prebiotics given before weaning, which induces by itself immediate changes in the gut microbiota, alter adult microbiota composition only under specific conditions of age and diet in mice [94]. Moreover, the high variability of microbiota composition in infants between 2 and 6 years of age [95,96], together with changes in gut microbiota composition observed after weaning in rats [97,98], somewhat contradict the sustainability of early microbiota changes in later life. Few studies suggest that early intervention may have a short-term programming effect on gut microbiota. For instance, changes in gut microbiota were detected at 1 to 2 years of age in children who had been supplemented with probiotics or prebiotics during the first 6 months of life [99,100]. Similarly, children born by caesarean section have an altered microbiota [101] still detectable at 7 years of age [102]. Our recent work performed in rats has shown that programming of gut microbiota composition can occur, but this depends on the type of oligosaccharides used to modulate the neonatal microbiota [20]. This more likely suggests that microbiota programming depends on the specificity of the neonatal impact on microbiota. This would support the idea that the first colonizers play an important role in the development of the gut microbiota and may impact its long-term composition and activity [103,104]. Whether such a microbiota programming actually programs host physiology has still to be established.

The Particular Role of Butyrate in Epigenetic Modulation of GI Epithelium

Butyrate is a short-chain fatty acid produced by bacterial fermentation of dietary fiber in the colon. It is a well-known HDAC inhibitor and the principal energy source for colonocytes [83,105]. Butyrate-induced acetylation of histones in colonocytes modulates gene expression programs involved in cell proliferation, differentiation, and apoptosis [33,106]. Butyrate oxidation also provides acetyl-CoA for HAT activity and histone acetylation, thus demonstrating the interplay between metabolism and epigenetics in the regulation of gene expression [34]. In germ-free mice, butyrate deficiency leads to metabolic starvation and gene expression alteration in the colon [107]. As an example, butyrate has also been shown to modulate intestinal alkaline phosphatase (IAP) gene expression epigenetically through histone acetylation [108–110]. Conversely, gut microbiota composition and inflammatory potential are under IAP control [111–113]. IAP is a major host anti-inflammatory enzyme detoxifying proinflammatory microbe-associated molecular patterns (e.g., lipopolysaccharide) [111,112,114]. IAP is programmed in a pig model of early gut microbiota disturbance [115], but underlying epigenetic changes have not been investigated.

DEVELOPMENTAL ORIGINS OF GI DISEASES

In 2006, Waterland raised the hypothesis that nutrition during fetal and postnatal periods could induce epigenetic modifications in the developmental gene program of the GI tract [116].

These adaptive responses to early nutrition, referred to as "metabolic imprinting," would permanently alter GI structure and function, causing an increased risk of GI diseases in adulthood. Although metabolic imprinting and associated epigenetic modifications have been described in several organs, including liver, pancreas, or adipose tissue (for review, see this book and Ref. [42]), little is known about GI metabolic programming.

There is mounting evidence that IBD and CRC are also characterized by both metabolic and epigenetic alterations [117–118]. A well-described epigenetic instability in CRC is the CpG island methylator phenotype, which is characterized by numerous hypermethylated promoters of tumor suppressor and DNA repair genes [119]. Thus, it is now generally accepted that dietary factors may play a role in initiating or promoting GI diseases. This is particularly well established for IBD, IBS, and CRC [11–13]. However, the role of environmental factors during critical periods of development (i.e., gestational and neonatal) in inducing stable epigenetic changes that influence the risk of developing GI diseases later in life has been poorly investigated.

Inflammatory Bowel Disease

A developmental origin for epigenetic modifications in IBD has been hypothesized in a recent review [25]. However, there are very few epidemiological and experimental studies exploring the impact of perinatal nutritional environment on the risk of GI diseases and overall on the underlying epigenetic mechanisms. Most investigations have used birth body weight and/or length as surrogates of poor in utero nutrition and thereby intrauterine growth restriction.

Epidemiology

IBD includes ulcerative colitis and Crohn's disease. It is now generally accepted that these chronic inflammatory diseases are characterized by a dysregulated immune response to the intestinal microbiota in genetically predisposed patients [120]. Limited epidemiological studies have investigated the influence of early factors, including birth weight, preterm birth, and breast-feeding on the risk of IBD in adulthood [121–123]. However, many of them reported inconclusive results because of low patient numbers and lack of information concerning duration of breast-feeding and the diagnosis of the onset of IBD. A case-control study reported an association between preterm birth, but not breast-feeding, and the risk of IBD development later in life, particularly in the first year [121]. On the contrary, a meta-analysis found that breast-feeding could lower the risk of IBD in adulthood [124]. Finally, a recent prospective study in a cohort of women enrolled in the Nurses Health Studies did not find any association between breast-feeding, birth weight, and IBD risk in adulthood [125].

Animal Studies

Few experimental data exist demonstrating that perinatal environment could induce stable epigenetic modifications altering GI function later in life, thereby increasing GI disease risk [17]. For instance, intrauterine growth-restricted rats born to mothers with protein restriction during gestation and/or lactation displayed colonic barrier alterations, including decreased mucin expression and increased permeability in adulthood [126,127]. They also had blunted mucosal and luminal IAP activities and mucosal *IAP* gene expression in response to high fat diet in adulthood [128], but data on epigenetic changes are lacking.

Two recent studies by Kellermayer's group have demonstrated in mice that nutritional environment during gestation may reprogram the fetal epigenome, resulting in increased susceptibility of offspring to dextran sulfate sodium-induced colitis in adulthood [59,60]. The authors showed that maternal dietary supplementation with methyl donors induced prolonged DNA methylation in association with gene expression changes in colonic mucosa of adult offspring. Interestingly, this adaptive response of colonic mucosa to fetal nutrition also conditioned the stable colonization of the gut by a proinflammatory (colitogenic) microbiota. However, at what stage and how methyl donors modulate the epigenome during fetal development remain to be elucidated.

Irritable Bowel Syndrome

IBS is a common and highly prevalent disorder afflicting 5–20% of the global population, according to diagnosis based on Rome III criteria [9]. Despite low morbidity, this illness strongly impairs patients' quality of life, generates important heath care costs, and is associated with psychiatric comorbidities such as anxiety and depression [129]. Among features of IBS, abdominal pain, bloating, and altered bowel habits are the most representative. The precise IBS pathophysiology remains poorly understood probably due to the heterogeneity of IBS populations and the multifactorial etiology of this disorder. Furthermore, IBS symptoms can be exacerbated by stressful life events and inappropriate central coping mechanisms [130], underlying the involvement of the brain gut axis in the symptoms regulation. This context suggests that either genetic [131] or environmental [132,133] factors may alter central and peripheral pathophysiology of IBS, but the debate is still open since the role of genetics remains controversial and few environmental factors have been linked to IBS.

Epidemiology

The genetic contribution to the development of the IBS phenotype is emerging from population-based studies, and several studies have depicted family associations in IBS symptomatology. Locke and coworkers evaluated the association between functional GI symptoms and a family history of abdominal pain or bowel problems [134]. Based on a family cluster study of 643 subjects, they reported that subjects having first-degree relatives with GI disorders were more likely to experience IBS, whereas those who reported having a spouse with bowel problems (controls) were not [134]. The same group published three years later another prospective study aimed at evaluating the occurrence of familial aggregation in IBS [135]. In this study, IBS prevalence was 17% in patient's relatives included in the study versus 7% in spouse's relatives (controls). However, the associations observed in both studies may result from both genetic and environmental factors since the impact of a common intrafamily context cannot be excluded.

Another interesting approach consists in evaluating intergenerational transmission of IBS features. Levy et al. did so for GI illness behavior [136]. Interestingly, children from parents with IBS had more ambulatory care visits for GI symptoms or other causes versus controls. Thus, specific illness behavior may be learned, and the learning process from the environment has equal influence as heredity in IBS. Interesting data arise also from twin studies. Concordance rate of IBS or functional bowel disorders was greater in monozygotic than in dizygotic twins (17.2% vs 8.4%) [137,138]. These observations support a genetic contribution in IBS. However, a large-scale study with 1870 twin pairs concluded that genetic factors are of little or no influence on IBS where the predominant influences appear to be environmental [139]. Therefore, twin studies indicate that, despite the presence of a genetic component, environmental factors are strong contributors, even stronger than heredity itself in IBS.

Environmental Factors and Epigenetics in IBS

Chronic stress appears as a risk factor to be shared by the majority of IBS patients contributing to exacerbation of symptoms. In addition, several studies depict that psychologically stressful experience in early life such as physical, sexual, or emotional abuse, adverse family interactions, and even wartime increases the risk of IBS occurrence in adulthood [140,141]. Therefore, a systematic review concluded that research examining the effect of affluent childhood socioeconomic status and early childhood stress in the evolution of functional GI disorders may help identify causative factors of IBS [132]. Taken together, these data strongly support the hypothesis that all sorts of stress occurring during the early postnatal periods might contribute to the development of IBS in adulthood.

These observations, combined with the lack of conclusive results regarding IBS genotype characterization, encouraged investigations in the field of epigenetics. Further support to the epigenetic modifications possibly leading to an IBS phenotype comes from the important role of environmental factors in the twin studies referred to above. They suggested that clustering into families or even transfer across generations can occur independently of changes in DNA sequence. If there is a consensus on the early-life adverse events as predisposing factor for IBS development later in life, generational transfer of IBS hallmarks related to those adverse events has not yet been investigated directly in IBS patients.

Animal Studies

A relevant animal model for understanding the physiopathology of early-life stress leading to long-lasting GI and central effects is the neonatal maternal deprivation (NMD) model in rodents. Interestingly, the hypothalamo-pituitary-adrenal (HPA) axis in early postnatal life is found less responsive to environmental factors compared to the adult, except for maternal separation [142]. This HPA axis hyperresponsiveness induced by maternal separation is a long-lasting event [142].

GI disturbances have also been observed beside long-lasting behavioral effects and altered stress reactivity induced by NMD. Accordingly, NMD modified gut immunity in adulthood as reflected by increased density of mast cells in colonic mucosa and increased numbers of neutrophils both in colonic and jejunal mucosa in 12-week-old rats [143]. NMD also induced visceral hypersensitivity, increased gut permeability, and bacterial translocation under baseline and stress conditions applied in adulthood [143,144]. Interestingly, these data obtained from animal studies are in line with results from clinical studies in IBS. Indeed, HPA axis hyperresponsiveness to a visceral stressor related to early adverse life events has been observed in IBS patients [145]. Increased gut permeability and visceral hypersensitivity are common features described in all IBS subtypes [146,147]. Furthermore, low-grade mucosal inflammation and increased mast cell density in the intestinal mucosa of IBS patients have been reported [148].

Generational Transfer and Epigenetics

The strong imprinting in adulthood of early-life stressful experience has led to considering the possibility for possible generational transfer in IBS. Indeed, the pioneering work by Meaney's group reported nongenomic transmission across generations of maternal behavior and stress responses in rats [149]. Female offspring reared by mothers exhibiting high nursing behavior (pup licking/grooming and arched-back nursing) showed increased maternal care toward their pups, suggesting that individual differences in maternal behavior are transferred across generations. In addition, offspring reared by mothers exhibiting high nursing behavior had higher levels of glucocorticoid receptors in the hippocampus and less fearful behavior compared with those reared by mothers with low nursing behavior. Importantly, pups raised by low nursing mothers displayed higher levels of methylation in the promoter region of the glucocorticoid receptor gene and reduced activation of the nerve growth factor–inducible protein A transcription factor and acetylation to the promoter region of glucocorticoid receptor gene [150]. However, in an NMD model, despite clear HPA axis dysfunction, no alteration in the methylation status of exon 1(7) glucocorticoid receptor promoter region in the hippocampus was observed. This suggests that in this model, NMD-induced abnormal behavior and HPA axis responsiveness may occur independently of epigenetic changes to exon 1(7) glucocorticoid receptor promoter [151].

Epigenetic programming of glucocorticoid receptor expression has also been studied in humans. McGowan and coworkers have found decreased levels of glucocorticoid receptor mRNA and increased cytosine methylation of a neuron-specific glucocorticoid receptor (NR3C1) promoter in the hippocampus of suicide victims with a history of childhood abuse [152–154].

The GI incidences of early-life stress can also be transmitted across generations. Indeed, visceral hypersensitivity induced by NMD in adult rats can be transferred to the next generation without any additional exposure of F2 individuals to maternal separation [155]. Furthermore, using a cross-fostering approach of the F2 offspring, maternal care was shown to be crucial to the observed transfer [155]. However, no epigenetic programming of the glucocorticoid receptor has been observed in this study.

Colorectal Cancer

Epidemiological Studies

Birth body weight is commonly used as a proxy for fetal growth and may be used as an indirect marker for fetal exposure to growth-stimulating factors, such as IGFs. Studies investigating the relationship between birth weight and CRC risk are inconsistent. In a cohort of men and women, the analysis of self-reported birth weight showed an association between low weight and CRC [156]. Another prospective population-based study in a Norwegian cohort suggested that men, but not women, born with a relative small size had an increased risk of CRC [157]. But in a cohort study of Swedish twins, there was no evidence for an influence of birth body weight or length on CRC risk [158]. Noteworthy, none of the mentioned studies looked at a role for epigenetic mechanisms. Hughes and coworkers analyzed the CpG island methylator phenotype status of individuals exposed to the Dutch famine using DNA isolated from paraffin-embedded CRC tissue [159]. Interestingly, they found that famine-exposed individuals had a decreased risk of developing a CpG island methylator phenotype tumor in comparison to nonexposed subjects. This is the first study indicating that transient exposure to a deleterious environment during adolescence or young adulthood may induce persistent epigenetic modifications that influence CRC development later in life.

Animal Studies

There is actually no study demonstrating the influence of early environment on the risk of CRC development later in life. However, an in vivo study in $Apc^{Min/+}$ mice showed that the effect of aspirin in the suppression of intestinal tumorigenesis becomes apparent when aspirin treatment covers the period between conception and weaning but not when administered in adult mice [160]. Although the mechanism of aspirin action was not investigated, this data suggests that loss of adenomatous polyposis coli function could early program tumorigenesis, opening the opportunity for perinatal prevention strategies of CRC development.

FOOD AND EPIGENETICS OF THE GI EPITHELIUM

Food components that are shown to be active as epigenetic modulators can act through various mechanisms: as methyl donors and associated precursors, as epigenetic enzyme modulators/inhibitors, and finally, indirectly by modulating cellular redox status. Part of the observed effect may be through the gut microbiota (see the Section The Gut Microbiota and Programming).

Of note, most available data deal with GI epithelial cell cultures, and information on long-term influences of food components on GI epigenome are virtually absent. Available in vitro and in vivo data are gathered in Tables 1 and 2, respectively.

Methyl Donors and Precursors

As already mentioned earlier in this chapter, the effect of methyl donor deficiency or supply on the epigenome and related phenotypes is established [25,115]. Regarding the GI epigenome, aberrant DNA methylation has been observed with supraphysiological folic acid concentrations in colonic epithelial cells [161]. In vivo, folate depletion during gestation and lactation decreased DNA methylation in the small intestine of adult mouse offspring [57,58]. Importantly, folate supply after weaning did not change intestinal epigenetic marking in this study, highlighting the critical window of action of methyl donors. Maternal methyl donor supply was also shown to increase mouse offspring susceptibility to colitis, especially in males [59,60]. In one study, 18 genes out of 155 (12%) with both methylation and expression changes were identified; 10 genes displayed reduced methylation associated with increased expression in six cases; and increased methylation observed in eight genes was associated with reduced gene expression in six cases [45]. The second study supports the notion that prenatal IBD may have prenatal nutritional origins, leading to a "colitogenic" microbiome in offspring [45]. Relationships between susceptibility to colitis, modulation of gene methylation and expression, and the influence of the GI microbiota appeared complex. Finally, methyl donor deficiency led also to alterations in small intestinal differentiation and barrier function in rats [162], but underlying epigenetic mechanisms were not investigated.

Cellular transmethylation reactions are strongly influenced by S-adenosyl methionine to S-adenosyl homocysteine ratio, with low ratios reducing methyl transfers [39]. Cysteine modulates cellular antioxidant glutathione and methyl donor S-adenosyl methionine concentrations [163], and this is of major relevance to the GI tract. Cysteine is intestinally absorbed through excitatory amino acid transporter (EAAT3/EAAC1/SLC1A1) at the apical membrane of enterocytes [164]. Recently, an original mechanism linking food-derived opioid peptides, intestinal cysteine absorption, and changes in DNA methylation has been disclosed [165]. In brief, opioid peptides from cow's milk (bovine β-casomorphin-7) and wheat α-gliadin (gliadinomorphin-7), acting on μ-opioid receptor actually reduced cysteine absorption in Caco2 cells. This in turn affected cell redox and methylation status, with an effect on genomic DNA methylation and gene transcription. Importantly, neuronal and immune cells highly represented in the GI tract also express μ-opioid receptors [165,166]. These cells may respond to dietary opioid peptides by epigenetic changes that may affect different regulations (e.g., immune, neuronal) at the GI level.

Although milk and wheat gluten are the main sources of opioid peptides, other proteins from vegetables (soybean β-conglycinin, rice albumin, spinach) and meat (hemoglobin, serum albumin) also display opioid receptor agonist activities [167,168] and, therefore, may contribute to modulate DNA methylation in the GI tract.

Dietary Modulators/Inhibitors of Epigenetic Enzymes

HDAC Inhibitors from Cruciferous Vegetables

Cruciferous vegetables contain sulfur-containing compounds like sulforaphane in broccoli and phenethyl isothiocyanate in watercress that were able to prevent the development of aberrant colonic crypt foci in rats [16,169]. As already mentioned, part of these compounds may result from bacterial fermentation [77]. Later,

TABLE 1 Dietary Compounds and Epigenetic Modifications of Colon Cancer Epithelial Cells In Vitro

Dietary Compound	Dose/Range	Colon Cancer Cell Line	Epigenetic Modifications	References
Folate (various forms)	Physiological (4.5–68 nM) Supraphysiological (2.3–9.1 μM)	FHC	Increased CpG methylation of *ESR1* and *p16* gene promoter regions and reduced global DNA methylation of *LINE-1* with the supraphysiological doses (no effects with physiological doses)	[161]
Sulforaphane and its cysteine and methylcysteine metabolites	3–15 μM	HCT116	Both global and localized (e.g., *p21* promoter region) histone methylation increased in a sulforaphane dose-dependent manner	[170]
Sulforaphane and related isothiocyanates	15 μM	HCT116, HT29, SW48, SW480	Enhanced acetylation of *CtIP* gene in HCT116 cells only	[172]
Diallyl disulfide; Allyl mercaptan	200 μM	Caco2; HT29	Increased histone H3 acetylation in both cell lines, and of histone H4 in Caco2 cells only	[175,176]
Allyl mercaptan	2–200 μM	HT29	Hyperacetylation of histone H3 on the *p21WAF1* gene promoter region. Also hyperacetylation of histone H4	[178]
S-allyl mercaptocysteine (SAMC)	25–500 μM	Caco2	Acetylation of histone H4-0 was decreased at all SAMC concentrations tested, while those of histones H4-2, H4-3, and H4-4 were increased at the highest SAMC concentration	[179]
(-)-Epigallocatechin-3-gallate	5–20 μM	HT29	CpG island demethylation of *P16INK4α* gene promoter region	[182]
Anthocyanins (from black raspberry)	0.5–25 μg/ml	Caco2, HCT116, SW480	Increased histone H3 acetylation of the *DKK1* gene promoter region in SW480 and HCT15 cells only	[185]
Quercetin	5–20 μM	RKO	CpG island demethylation of *P16INK4α* gene promoter region	[187]
Genistein	75 μM	DLD-1, SW480, SW1116	Demethylation of *WNT5a* promoter CpG islands in SW116 cell line only	[188]
Genistein	50–75 μM	DLD-1, HCT15, HT29, RKO, SW48, SW480	Histone H3 acetylation of the *DKK1* promoter region in HCT15 and SW480 cell lines only	[189]

TABLE 2 Dietary Compounds and Epigenetic Modifications of Colonic Epithelial Cells In Vivo

Dietary Compound	Dose/Range	Species/Model	Epigenetic Modifications	References
STUDIES IN ANIMALS				
Folate	Maternal folate deficiency; offspring folate deficiency postweaning	C57Bl/6J or Apc^{Min} mice	Decreased *p53* DNA methylation in the small intestine of adult (96 days of age) offspring born to wild-type or Apc^{Min} mice. *Apc* DNA methylation increased in Apc^{Min} mice only	[58]
Methyl donors	Supplementation: 5 mg/kg folic acid; 0.5 g/kg vitamin B12; 5 g/kg betaine; 5.76 g/kg choline	C57Bl/6J mice (−/+ DSS 3%)	Decreased methylation of *Ptpn22*- and *Ppara*-associated *Sma*I/*Xma*I in adult (90 days of age) offspring colonic mucosa	[59]
Diallyl disulfide	200 mg/kg body weight	Wistar male rats	Increased acetylation of histones H3 and H4 in colonocytes	[177]
Black raspberry (rich in anthocyanins)	50 g/kg diet	C57Bl/6 mice (−/+ DSS 1%)	Demethylation of *dKK2*, *dKK3*, and *Apc* gene promoter regions in the colon	[185]
Genistein	140 mg/kg diet	Sprague–Dawley rats	Decreased histone H3 acetylation of the promoter region of *Sfrp2*, *Sfrp5*, and *Wnt5a* genes	[190]
STUDIES IN HUMANS				
Black raspberry (rich in anthocyanins)	60 g/day	Adult patients with colorectal cancer	Demethylation of promoter regions of the tumor suppressor genes *SFRP2*, *SFRP5*, and *WIF1*	[186]

sulforaphane was shown to inhibit HDAC activity and to increase histone H3 and H4 acetylation on the promoter region of *P21* gene in human CRC cell lines [170]. HDAC activity and increased HDAC protein turnover were recently shown to be proportional to sulforaphane alkyl chain length (sulforaphane < 6-sulforaphane < 9-sulforaphane) in various human colon cancer cell lines [171]. Importantly, sulforaphane given for 70 days to Apc^{Min} mice (that develop multiple intestinal neoplasia) displayed decreased HDAC activity leading to concomitant increase in histone H3 and H4 acetylation on *p21* gene promoter region, and this was associated with a reduction of spontaneous gut polyps [172]. Sulforaphane-specific protective responses against intestinal polyposis in Apc^{Min} mice appeared to be dose-dependent [173]. In humans, single intake of broccoli sprouts led to a reduction in HDAC activity in peripheral blood mononuclear cells [174]. Despite these encouraging results, chronic studies in humans have not been reported.

Allyl Compounds from Garlic

Diallyl disulfide from garlic, and its metabolite allyl mercaptan, were shown to increase histone H3 acetylation in Caco2 and HT-29 cell lines [175]. Histone H4 hyperacetylation was not consistently reported in these studies [173,176]. Epigenetic modification of histone H3 was associated with *P21* gene expression and inhibition of cell proliferation, with a blockade of cells in G2 phase [175]. Single treatment with diallyl disulfide led to transient effect on cells while repeated treatment induced more prolonged histone H3 hyperacetylation [176]. Importantly, diallyl disulfide-induced hyperacetylation of histones H3 and H4 was also demonstrated in colonocytes of rats [177].

Allyl mercaptan appears to be much more potent than diallyl disulfide in decreasing HDAC activity in IEC in vitro [175]. It was shown to strongly inhibit HDAC activity and stimulate histone H3 acetylation in HT-29 cells [178]. S-allyl-mercaptocysteine, another garlic sulfur compound, was also shown to induce histone hyperacetylation in Caco2 cells [179]. Other cruciferous compounds like 3,3′-diindolyl-methane, a breakdown product of glucobrassicin, were shown to downregulate HDAC activity by inducing proteosomal degradation of class-1 HDAC enzymes specifically [180].

Dietary Polyphenols

Despite abundant literature on plant polyphenols as possible substances against various types of cancers, studies dedicated to CRC are limited [181]. Tea polyphenol (-)-epigallocatechin-3-gallate proved to inhibit DNMT and reactivate methylation-silenced genes in HT-29 cells [182]. *p*-Coumaric acid, 3-(4-OH-phenyl)-propionate, and caffeic acid were the most potent HDAC inhibitors among various polyphenols tested on HT-29 cells [183]. However, the authors concluded that the role played by these compounds in the colon might be limited compared to butyrate produced by butyrogenic bacteria. Anthocyanins from black raspberries, acting by gene promoter region demethylation, may be responsible for the activation of tumor suppression genes in different CRC cell lines [184]. In vivo, black raspberry was shown to suppress chemical-induced ulcerative colitis by demethylating such suppressor genes in mice [185]. Importantly, this was recently confirmed in humans [186]. Finally, the flavonoid quercetin was shown to demethylate the promoter region of a cancer-related gene *p16INK4a* in human CRC cells [187]. Whether quercetin acts on IEC epigenetics in vivo has not been reported.

Dietary Isoflavones

Recent data report that genistein, an isoflavone present in soybean, was able to increase *Wnt* gene expression by demethylation of its promoter region in CRC cells [188]. Later, these authors demonstrated that genistein (and soy protein isolate) controlled *Wnt* gene responses by modulating methylations on both histone H3 and genomic DNA in various CRC cell lines [189,190].

Other Food Compounds and Interactions

As already mentioned, selenium may affect DNA methylation through DNMT modulation [88]. Dietary iron (Fe) salts and ascorbic (Asc) acid can favor the formation of oxygen-free radicals inducing altered cell redox capacity and increase inflammation. Yara and coworkers showed in Caco2/5 cell lines that Fe/ascorbic acid raised superoxide dismutase 2 and reduced glutathione peroxidase activities [191]. This was associated with a modulation of promoter methylation of corresponding genes (decreased for superoxide dismutase 2 and enhanced for glutathione peroxidase). Importantly, variations in superoxide dismutase 2 gene methylation and in histone methylation and acetylation are involved in cancer, and this enzyme might act as an important modulator of epigenome stability [192]. However, Chen and coworkers did not find any association between superoxide dismutase 2 gene variants and CRC in humans [193].

Interactions between bioactive compounds with epigenetic modulatory activities are important to investigate because those with protective

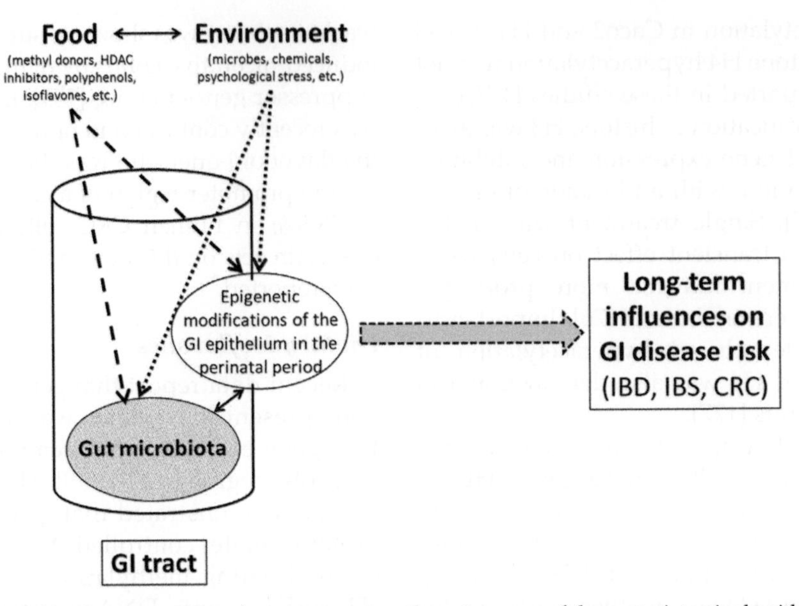

FIGURE 1 Effects of food components and environment on the epigenome of the gastrointestinal epithelium in the perinatal period and long-term influences on chronic gastrointestinal disease risk. GI, gastrointestinal; IBD, inflammatory bowel disease; IBS, irritable bowel syndrome; CRC, colorectal cancer.

bioactivity taken separately may have synergistic, no, or even adverse effects when associated, for example, in the diet. There are unfortunately only few published examples of interaction studies and no data on possible underlying interactions at the epigenetic level. For instance, combining HDAC and DNMT inhibitors could inhibit cancer development better [194]. Ten weeks of sulforaphane and the anticancer drug dibenzoylmethane, given orally alone or in combination, to Apc^{Min} mice inhibited the development of intestinal adenomas equally while a synergistic effect between these molecules was observed on colon cancer [195]. Interaction studies between sulforaphane and tea polyphenol epigallocathechin-3-gallate revealed discrepancies. Nair and coworkers [196] reported drastic enhancement of transcriptional activation of *AP-1* reporter gene in HT-29 cells. Cell HDAC activity remained unaltered [196]. Sulforaphane and tea polyphenol together limited intestinal colonic cancer in two independent studies in vivo [169,182]. However, antagonism between them was revealed in other studies. They had no protective effects on gut carcinoma development when given together to Apc^{Min} mice [172,195]. Whether these antagonisms operate at the epigenetic level is not known (Figure 1).

CONCLUSIONS AND FUTURE DIRECTIONS

There is now mounting evidence for an involvement of complex gene–environment interactions in the development of GI pathologies, including IBD, IBS, and CRC (Figure 1). Epigenetic modifications have been identified in these diseases, but their relationship with environmental factors, in particular with diet and gut microbiota, and the timing of these modifications, is not yet established. Although the DOHaD concept provides significant evidence that fetal and postnatal periods are critical windows during which epigenome programming is greatly shaped by environmental factors, the advancement of this research area applied to GI pathologies is still in its infancy. For many noncommunicable diseases, including metabolic syndrome, diabetes, cardiovascular diseases, and so on, human and animal studies have provided robust proof of the involvement of early influences in disease risk, and associated epigenetic marks have been identified in relevant tissues. Yet, epigenetic mechanisms are of particular importance in self-renewal and maturation processes of the GI epithelium. Thus, early modifications of the epigenome of intestinal stem cells would have long-lasting effects on epithelial barrier function and could modify (adapt) the response of the epithelium to luminal factors such as nutrients and microbes.

Investigating the DOHaD hypothesis in GI pathologies is not a simple task because regulation of the epithelial barrier involves many cell types that are as many targets of epigenetic imprinting. One major challenge would be to identify tissue- and cell type-specific epigenetic marks that might serve as biomarkers for specific perinatal environments and to determine the prognostic value of such marks in terms of health or disease risk for the GI tract. Determining how and when these epigenetic marks are being established will certainly need relevant animal models. Also, it would be useful to look at potential correlations between these epigenetic signatures and genetic polymorphisms identified in genome-wide association studies. To reach this goal, it will be necessary to perform well-designed epidemiologic studies in large prospective cohorts.

To conclude, investigating the DOHaD concept in the GI system will need collaborative approaches in the fields of developmental biology, genetics, epigenetics, microbiology, and nutrition, this in concert with a systems biology approach.

ABBREVIATIONS

CpG Cytosine-phosphate-guanidine
CRC Colorectal cancer
DNMT DNA methyltransferase
DOHaD Developmental origin of health and disease
GI Gastrointestinal
HAT Histone acetyltransferase
HDAC Histone deacetylase
HMT Histone methyltransferase
HPA Hypothalamic-pituitary-adrenal (axis)
IAP Intestinal alkaline phosphatase
IBD Inflammatory bowel disease
IBS Irritable bowel syndrome
IEC Intestinal epithelial cell
IGF Insulin-like growth factor
NMD Neonatal maternal deprivation (model)

References

[1] Turner JR. Intestinal mucosal barrier function in health and disease. Nat Rev Immunol 2009;9:799–809.

[2] Huang XZ, Zhu LB, Li ZR, Lin J. Bacterial colonization and intestinal mucosal barrier development. World J Clin Pediatr 2013;2:46–53.

[3] Pastorelli L, De Salvo C, Mercado JR, Vecchi M, Pizarro TT. Central role of the gut epithelial barrier in the pathogenesis of chronic intestinal inflammation: lessons learned from animal models and human genetics. Front Immunol 2013;4:280.

[4] Shanahan F, Quigley EM. Manipulation of the microbiota for treatment of IBS and IBD-challenges and controversies. Gastroenterology 2014;146:1554–63.

[5] Katiraei P, Gilberto Bultron G. Need for a comprehensive medical approach to the neuroimmuno-gastroenterology of irritable bowel syndrome. World J Gastroenterol 2011;17:2791–800.

[6] Viaud S, Daillère R, Boneca IG, Lepage P, Pittet MJ, Ghiringhelli F, et al. Harnessing the intestinal microbiome for optimal therapeutic immunomodulation. Cancer Res 2014;74:4217–21.

[7] Burisch J, Jess T, Martinato M, Lakatos PL, ECCO -EpiCom. The burden of inflammatory bowel disease in Europe. J Crohns Colitis 2013;7(4):322–37.

[8] Herszényi L, Barabás L, Miheller P, Tulassay Z. Colorectal cancer in patients with inflammatory bowel disease: the true impact of the risk. Dig Dis 2015;33(1):52–7.

[9] Hungin AP, Chang L, Locke GR, Dennis EH, Barghout V. Irritable bowel syndrome in the United States: prevalence, symptom patterns and impact. Aliment Pharmacol Ther 2005;21:1365–75.

[10] Brenner H, Kloor M, Pox CP. Colorectal cancer. Lancet 2014;383(9927):1490–502.

[11] Renz H, von Mutius E, Brandtzaeg P, Cookson WO, Autenrieth IB, Haller D. Gene-environment interactions in chronic inflammatory disease. Nat Immunol 2011;12:273–7.

[12] Vaiopoulou A, Karamanolis G, Psaltopoulou T, Karatzias G, Gazouli M. Molecular basis of the irritable bowel syndrome. World J Gastroenterol 2014;20:376–83.

[13] Arasaradnam RP, Commane DM, Bradburn D, Mathers JC. A review of dietary factors and its influence on DNA methylation in colorectal carcinogenesis. Epigenetics 2008;3:193–8.

[14] Jaenisch R, Bird A. Epigenetic regulation of gene expression: how the genome integrates intrinsic and environmental signals. Nat Genet 2003;(33 Suppl.):245–54.

[15] Feil R, Fraga MF. Epigenetics and the environment: emerging patterns and implications. Nat Rev Genet 2012;13:97–109.

[16] Nian H, Delage B, Ho E, Dashwood RH. Modulation of histone deacetylase activity by dietary isothiocyanates and allyl sulfides: studies with sulforaphane and garlic organosulfur compounds. Environ Mol Mutagen 2009;50:213–21.

[17] Lallès JP. Long term effects of pre- and early postnatal nutrition and environment on the gut. J Anim Sci 2012;90(Suppl. 4):421–9.

[18] Wall R, Ross RP, Ryan CA, Hussey S, Murphy B, Fitzgerald GF, et al. Role of gut microbiota in early infant development. Clin Med Pediatr 2009;3:45–54.

[19] Canani RB, Costanzo MD, Leone L, Bedogni G, Brambilla P, Cianfarani S, et al. Epigenetic mechanisms elicited by nutrition in early life. Nutr Res Rev 2011;24:198–205.

[20] Morel FB, Oozeer R, Piloquet H, Moyon T, Pagniez A, Knol J, Darmaun D, et al. Preweaning modulation of intestinal microbiota by oligosaccharides or amoxicillin can contribute to programming of adult microbiota in rats. Nutrition 2015;31:515–22.

[21] Haemer MA, Huang TT, Daniels SR. The effect of neurohormonal factors, epigenetic factors, and gut microbiota on risk of obesity. Prev Chronic Dis 2009;6:A96.

[22] Reinhardt C, Reigstad CS, Backhed F. Intestinal microbiota during infancy and its implications for obesity. J Pediatr Gastroenterol Nutr 2009;48:249–56.

[23] Backhed F. Programming of host metabolism by the gut microbiota. Ann Nutr Metab 2011;58(Suppl. 2):44–52.

[24] Kaplan JL, Walker WA. Early gut colonization and subsequent obesity risk. Curr Opin Clin Nutr Metab Care 2012;15:278–84.

[25] Kellermayer R. Epigenetics and the developmental origins of inflammatory bowel diseases. Can J Gastroenterol 2012;26:909–15.

[26] Shenderov BA. Gut indigenous microbiota and epigenetics. Microb Ecol Health Dis 2012;23.

[27] Mischke M, Plosch T. More than just a gut instinct-the potential interplay between a baby's nutrition, its gut microbiome, and the epigenome. Am J Physiol Regul Integr Comp Physiol 2013;304:R1065–9.

[28] Shenderov BA, Midtvedt T. Epigenomic programing: a future way to health? Microb Ecol Health Dis 2014;25.

[29] Takahashi K. Influence of bacteria on epigenetic gene control. Cell Mol Life Sci 2014;71:1045–54.

[30] Mackie RI, Sghir A, Gaskins HR. Developmental microbial ecology of the neonatal gastrointestinal tract. Am J Clin Nutr 1999;69:1035S–45S.

[31] Yatsunenko T, Rey FE, Manary MJ, Trehan I, Dominguez-Bello MG, Contreras M, et al. Human gut microbiome viewed across age and geography. Nature 2012;486:222–7.

[32] Crott JW, Liu Z, Keyes MK, Choi SW, Jang H, Moyer MP, et al. Moderate folate depletion modulates the expression of selected genes involved in cell cycle, intracellular signaling and folate uptake in human colonic epithelial cell lines. J Nutr Biochem 2008;19:328–35.

[33] Mariadason JM, Corner GA, Augenlicht LH. Genetic reprogramming in pathways of colonic cell maturation induced by short chain fatty acids: comparison with trichostatin A, sulindac, and curcumin and implications for chemoprevention of colon cancer. Cancer Res 2000;60:4561–72.

[34] Donohoe DR, Collins LB, Wali A, Bigler R, Sun W, Bultman SJ. The Warburg effect dictates the mechanism of butyrate-mediated histone acetylation and cell proliferation. Mol Cell 2012;48:612–26.

[35] Choi SW, Friso S. Epigenetics: a new bridge between nutrition and health. Adv Nutr 2010;1:8–16.

[36] Donohoe DR, Bultman SJ. Metaboloepigenetics: interrelationships between energy metabolism and epigenetic control of gene expression. J Cell Physiol 2012;227:3169–77.

[37] McKay JA, Mathers JC. Diet induced epigenetic changes and their implications for health. Acta Physiol (Oxf) 2011;202:103–18.

[38] Anderson OS, Sant KE, Dolinoy DC. Nutrition and epigenetics: an interplay of dietary methyl donors, one-carbon metabolism and DNA methylation. J Nutr Biochem 2012;23:853–9.

[39] Johnson IT, Belshaw NJ. The effect of diet on the intestinal epigenome. Epigenomics 2014;6:239–51.

[40] Waterland RA, Jirtle RL. Transposable elements: targets for early nutritional effects on epigenetic gene regulation. Mol Cell Biol 2003;23:5293–300.

[41] Heijmans BT, Tobi EW, Stein AD, Putter H, Blauw GJ, Susser ES, et al. Persistent epigenetic differences associated with prenatal exposure to famine in humans. Proc Natl Acad Sci USA 2008;105:17046–9.

[42] Joss-Moore LA, Lane RH. The developmental origins of adult disease. Curr Opin Pediatr 2009;21:230–4.

[43] Vincent A, Van Seuningen I. Epigenetics, stem cells and epithelial cell fate. Differentiation 2009;78:99–107.

[44] Spivakov M, Fisher AG. Epigenetic signatures of stem-cell identity. Nat Rev Genet 2007;8:263–71.

[45] van der Flier LG, Clevers H. Stem cells, self-renewal, and differentiation in the intestinal epithelium. Annu Rev Physiol 2009;71:241–60.

[46] Nadauld LD, Phelps R, Moore BC, Eisinger A, Sandoval IT, Chidester S, et al. Adenomatous polyposis coli control of C-terminal binding protein-1 stability regulates expression of intestinal retinol dehydrogenases. J Biol Chem 2006;281:37828–35.

[47] Rai K, Sarkar S, Broadbent TJ, Voas M, Grossmann KF, Nadauld LD, et al. DNA demethylase activity maintains intestinal cells in an undifferentiated state following loss of APC. Cell 2010;142:930–42.

[48] Hammoud SS, Cairns BR, Jones DA. Epigenetic regulation of colon cancer and intestinal stem cells. Curr Opin Cell Biol 2013;25:177–83.

[49] Azuara V, Perry P, Sauer S, Spivakov M, Jørgensen HF, John RM, et al. Chromatin signatures of pluripotent cell lines. Nat Cell Biol 2006;8:532–8.

[50] Bernstein BE, Mikkelsen TS, Xie X, Kamal M, Huebert DJ, Cuff J, et al. A bivalent chromatin structure marks key developmental genes in embryonic stem cells. Cell 2006;125:315–26.

[51] Benoit YD, Lepage MB, Khalfaoui T, Tremblay E, Basora N, Carrier JC, et al. Polycomb repressive complex 2 impedes intestinal cell terminal differentiation. J Cell Sci 2012;125:3454–63.

[52] Wilson AJ, Byun DS, Popova N, Murray LB, L'Italien K, Sowa Y, et al. Histone deacetylase 3 (HDAC3) and other class I HDACs regulate colon cell maturation and p21 expression and are deregulated in human colon cancer. J Biol Chem 2006;281:13548–58.

[53] Tou L, Liu Q, Shivdasani RA. Regulation of mammalian epithelial differentiation and intestine development by class I histone deacetylases. Mol Cell Biol 2004;24:3132–9.

[54] de Rooij SR, Roseboom TJ. The developmental origins of ageing: study protocol for the Dutch famine birth cohort study on ageing. BMJ Open 2013;3:e003167.

[55] Steegenga WT, de Wit NJ, Boekschoten MV, Ijssennagger N, Lute C, Keshtkar S, et al. Structural, functional and molecular analysis of the effects of aging in the small intestine and colon of C57BL/6J mice. BMC Med Genomics 2012;5:38.

[56] Steine EJ, Ehrich M, Bell GW, Raj A, Reddy S, van Oudenaarden A, et al. Genes methylated by DNA methyltransferase 3b are similar in mouse intestine and human colon cancer. J Clin Invest 2011;121:1748–52.

[57] McKay JA, Waltham KJ, Williams EA, Mathers JC. Folate depletion during pregnancy and lactation reduces genomic DNA methylation in murine adult offspring. Genes Nutr 2011;6:189–96.

[58] McKay JA, Williams EA, Mathers JC. Effect of maternal and post-weaning folate supply on gene-specific DNA methylation in the small intestine of weaning and adult apc and wild type mice. Front Genet 2011;2:23.

[59] Schaible TD, Harris RA, Dowd SE, Smith CW, Kellermayer R. Maternal methyl-donor supplementation induces prolonged murine offspring colitis susceptibility in association with mucosal epigenetic and microbiomic changes. Hum Mol Genet 2011;20:1687–96.

[60] Mir SA, Nagy-Szakal D, Dowd SE, Szigeti RG, Smith CW, Kellermayer R. Prenatal methyl-donor supplementation augments colitis in young adult mice. PLoS One 2013;8:e73162.

[61] Blottière HM, de Vos WM, Ehrlich SD, Doré J. Human intestinal metagenomics: state of the art and future. Curr Opin Microbiol 2013;16:232–9.

[62] Lepage P, Leclerc MC, Joossens M, Mondot S, Blottière HM, Raes J, et al. A metagenomic insight into our gut's microbiome. Gut 2013;62:146–58.

[63] Doré J, Corthier G. The human intestinal microbia. Gastroentérol Clin Biol 2010;34:7–16.

[64] Bäckhed F, Ley RE, Sonnenburg JL, Peterson DA, Gordon JI. Host-bacterial mutualism in the human intestine. Science 2005;307:1915–20.

[65] Tap J, Mondot S, Levenez F, Pelletier E, Caron C, Furet JP, et al. Towards the human intestinal microbiota phylogenetic core. Environ Microbiol 2009;11:2574–84.

[66] Li J, Jia H, Cai X, Zhong H, Feng Q, Sunagawa S, MetaHIT Consortium, et al. An integrated catalog of reference genes in the human gut microbiome. Nat Biotechnol 2014;32:834–41.

[67] Qin J, Li R, Raes J, Arumugam M, Arumugam M, Burgdorf KS, MetaHIT Consortium, et al. A human gut microbial gene catalogue established by metagenomic sequencing. Nature 2010;464:59–65.

[68] Kurokawa K, Itoh T, Kuwahara T, Oshima K, Toh H, Toyoda A, et al. Comparative metagenomics revealed commonly enriched gene sets in human gut microbiomes. DNA Res 2007;14:169–81.

[69] Arumugam M, Raes J, Pelletier E, Le Paslier D, Yamada T, Mende DR, MetaHIT Consortium, et al. Enterotypes of the human gut microbiome. Nature 2011;473:174–80.

[70] Possemiers S, Bolca S, Verstraete W, Heyerick A. The intestinal microbiome: a separate organ inside the body with the metabolic potential to influence the bioactivity of botanicals. Fitoterapia 2011;82:53–66.

[71] Sekirov I, Russell SL, Antunes LC, Finlay BB. Gut microbiota in health and disease. Physiol Rev 2010;90:859–904.

[72] Macfarlane GT, Macfarlane S. Bacteria, colonic fermentation, and gastrointestinal health. J AOAC Int 2012;95:50–60.

[73] Sayin SI, Wahlström A, Felin J, Jäntti S, Marschall HU, Bamberg K, et al. Gut microbiota regulates bile acid metabolism by reducing the levels of tauro-beta-muricholic acid, a naturally occurring FXR antagonist. Cell Metab 2013;17:225–35.

[74] Wang Z, Klipfell E, Bennett BJ, Koeth R, Levison BS, Dugar B, et al. Gut flora metabolism of phosphatidylcholine promotes cardiovascular disease. Nature 2011;472:57–63.

[75] Gérard P, Béguet F, Lepercq P, Rigottier-Gois L, Rochet V, Andrieux C, et al. Gnotobiotic rats harboring human intestinal microbiota as a model for studying cholesterol-to-coprostanol conversion. FEMS Microbiol Ecol 2004;47:337–43.

[76] Hill MJ. Intestinal flora and endogenous vitamin synthesis. Eur J Cancer Prev 1997;6(Suppl. 1):S43–5.

[77] Krul C, Humblot C, Philippe C, Vermeulen M, van Nuenen M, Havenaar R, et al. Metabolism of sinigrin (2-propenyl glucosinolate) by the human colonic microflora in a dynamic in vitro large-intestinal model. Carcinogenesis 2002;23:1009–16.

[78] Wikoff WR, Anfora AT, Liu J, Schultz PG, Lesley SA, Peters EC, et al. Metabolomics analysis reveals large effects of gut microflora on mammalian blood metabolites. Proc Natl Acad Sci USA 2009;106:3698–703.

[79] Rezzonico E, Mestdagh R, Delley M, Combremont S, Dumas ME, Holmes E, et al. Bacterial adaptation to the gut environment favors successful colonization: microbial and metabonomic characterization of a simplified microbiota mouse model. Gut Microbes 2011;2:307–18.

[80] Bierne H, Hamon M, Cossart P. Epigenetics and bacterial infections. Cold Spring Harb Perspect Med 2012;2:a010272.

[81] Al Akeel R. Role of epigenetic reprogramming of host genes in bacterial pathogenesis. Saudi J Biol Sci 2013;20:305–9.

[82] Niwa T, Tsukamoto T, Toyoda T, Mori A, Tanaka H, Maekita T, et al. Inflammatory processes triggered by *Helicobacter pylori* infection cause aberrant DNA methylation in gastric epithelial cells. Cancer Res 2010;70:1430–40.

[83] Boffa LC, Vidali G, Mann RS, Allfrey VG. Suppression of histone deacetylation in vivo and in vitro by sodium butyrate. J Biol Chem 1978;253:3364–6.

[84] LeBlanc JG, Milani C, de Giori GS, Sesma F, van Sinderen D, Ventura M. Bacteria as vitamin suppliers to their host: a gut microbiota perspective. Curr Opin Biotechnol 2013;24:160–8.

[85] Kim GW, Gocevski G, Wu CJ, Yang XJ. Dietary, metabolic, and potentially environmental modulation of the lysine acetylation machinery. Int J Cell Biol 2010;2010:632739.

[86] Pessione E. Lactic acid bacteria contribution to gut microbiota complexity: lights and shadows. Front Cell Infect Microbiol 2012;2:86.

[87] Hrdina J, Banning A, Kipp A, Loh G, Blaut M, Brigelius-Flohé R. The gastrointestinal microbiota affects the selenium status and selenoprotein expression in mice. J Nutr Biochem 2009;20:638–48.

[88] Barrera LN, Cassidy A, Johnson IT, Bao Y, Belshaw NJ. Epigenetic and antioxidant effects of dietary isothiocyanates and selenium: potential implications for cancer chemoprevention. Proc Nutr Soc 2012;71:237–45.

[89] Alenghat T, Osborne LC, Saenz SA, Kobuley D, Ziegler CG, Mullican SE, et al. Histone deacetylase 3 coordinates commensal-bacteria-dependent intestinal homeostasis. Nature 2013;504:153–7.

[90] Takahashi K, Sugi Y, Nakano K, Tsuda M, Kurihara K, Hosono A, et al. Epigenetic control of the host gene by commensal bacteria in large intestinal epithelial cells. J Biol Chem 2011;286:35755–62.

[91] Olszak T, An D, Zeissig S, Vera MP, Richter J, Franke A, et al. Microbial exposure during early life has persistent effects on natural killer T cell function. Science 2012;336:489–93.

[92] Burcelin R, Serino M, Chabo C, Blasco-Baque V, Amar J. Gut microbiota and diabetes: from pathogenesis to therapeutic perspective. Acta Diabetol 2011;48:257–73.

[93] Marques TM, Wall R, Ross RP, Fitzgerald GF, Ryan CA, Stanton C. Programming infant gut microbiota: influence of dietary and environmental factors. Curr Opin Biotechnol 2010;21:149–56.

[94] Fujiwara R, Takemura N, Watanabe J, Sonoyama K. Maternal consumption of fructo-oligosaccharide diminishes the severity of skin inflammation in offspring of NC/Nga mice. Br J Nutr 2010;103:530–8.

[95] Ellis-Pegler RB, Crabtree C, Lambert HP. The faecal flora of children in the United Kingdom. J Hyg (Lond) 1975;75:135–42.

[96] Tannock GW. What immunologists should know about bacterial communities of the human bowel. Semin Immunol 2007;19:94–105.

[97] Inoue R, Ushida K. Development of the intestinal microbiota in rats and its possible interactions with the evolution of the luminal IgA in the intestine. FEMS Microbiol Ecol 2003;45:147–53.

[98] Fança-Berthon P, Hoebler C, Mouzet E, David A, Michel C. Intrauterine growth restriction not only modifies the cecocolonic microbiota in neonatal rats but also affects its activity in young adult rats. J Pediatr Gastroenterol Nutr 2010;51:402–13.

[99] Rinne M, Kalliomaki M, Salminen S, Isolauri E. Probiotic intervention in the first months of life: short-term effects on gastrointestinal symptoms and long-term effects on gut microbiota. J Pediatr Gastroenterol Nutr 2006;43:200–5.

[100] Salvini F, Riva E, Salvatici E, Boehm G, Jelinek J, Banderali G, et al. A specific prebiotic mixture added to starting infant formula has long-lasting bifidogenic effects. J Nutr 2011;141:1335–9.
[101] Gronlund MM, Lehtonen OP, Eerola E, Kero P. Fecal microflora in healthy infants born by different methods of delivery: permanent changes in intestinal flora after cesarean delivery. J Pediatr Gastroenterol Nutr 1999;28:19–25.
[102] Salminen S, Gibson GR, McCartney AL, Isolauri E. Influence of mode of delivery on gut microbiota composition in seven year old children. Gut 2004;53:1388–9.
[103] Eggesbø M, Moen B, Peddada S, Baird D, Rugtveit J, Midtvedt T, et al. Development of gut microbiota in infants not exposed to medical interventions. APMIS 2011;119:17–35.
[104] Scholtens PA, Oozeer R, Martin R, Amor KB, Knol J. The early settlers: intestinal microbiology in early life. Annu Rev Food Sci Technol 2012;3:425–47.
[105] Roediger WE. Utilization of nutrients by isolated epithelial cells of the rat colon. Gastroenterology 1982;83:424–9.
[106] Davie JR. Inhibition of histone deacetylase activity by butyrate. J Nutr 2003;133:2485S–93S.
[107] Donohoe DR, Garge N, Zhang X, Sun W, O'Connell TM, Bunger MK, et al. The microbiome and butyrate regulate energy metabolism and autophagy in the mammalian colon. Cell Metab 2011;13:517–26.
[108] Hinnebusch BF, Meng S, Wu JT, Archer SY, Hodin RA. The effects of short-chain fatty acids on human colon cancer cell phenotype are associated with histone hyperacetylation. J Nutr 2002;132:1012–7.
[109] Hinnebusch BF, Henderson JW, Siddique A, Malo MS, Zhang W, Abedrapo MA, et al. Transcriptional activation of the enterocyte differentiation marker intestinal alkaline phosphatase is associated with changes in the acetylation state of histone H3 at a specific site within its promoter region in vitro. J Gastrointest Surg 2003;7:237–44. Discussion 244–5.
[110] Shin J, Carr A, Corner GA, Tögel L, Dávaos-Salas M, Tran H, et al. The intestinal epithelial cell differentiation marker intestinal alkaline phosphatase (ALPi) is selectively induced by histone deacetylase inhibitors (HDACi) in colon cancer cells in a Kruppel-like factor 5 (KLF5)-dependent manner. J Biol Chem 2014;289:25306–16.
[111] Lallès JP. Intestinal alkaline phosphatase: multiple biological roles in maintenance of intestinal homeostasis and modulation by diet. Nutr Rev 2010;68:323–32.
[112] Lallès JP. Intestinal alkaline phosphatase: novel functions and protective effects. Nutr Rev 2014;72:82–94.
[113] Malo MS, Moaven O, Muhammad N, Biswas B, Alam SN, Economopoulos KP, et al. Intestinal alkaline phosphatase promotes gut bacterial growth by reducing the concentration of luminal nucleotide triphosphates. Am J Physiol Gastrointest Liver Physiol 2014;306:G826–38.
[114] Cani PD, Delzenne NM. The role of the gut microbiota in energy metabolism and metabolic disease. Curr Pharm Des 2009;15:1546–58.
[115] Arnal ME, Zhang J, Messori S, Bosi P, Smidt H, Lallès JP. Early changes in microbial colonization selectively modulate intestinal enzymes, but not inducible heat shock proteins in young adult Swine. PLoS One 2014;9:e87967. Erratum in: PLoS One 2014;9:e98730.
[116] Waterland RA. Epigenetic mechanisms and gastrointestinal development. J Pediatr 2006;149:S137–42.
[117] Thibault R, Blachier F, Darcy-Vrillon B, de Coppet P, Bourreille A, Segain JP. Butyrate utilization by the colonic mucosa in inflammatory bowel diseases: a transport deficiency. Inflamm Bowel Dis 2010;16:684–95.
[118] Rodríguez-Paredes M, Esteller M. Cancer epigenetics reaches mainstream oncology. Nat Med 2011;17:330–9.
[119] Toyota M, Ahuja N, Ohe-Toyota M, Herman JG, Baylin SB, Issa JP. CpG island methylator phenotype in colorectal cancer. Proc Natl Acad Sci USA 1999;96:8681–6.
[120] Huttenhower C, Kostic AD, Xavier RJ. Inflammatory bowel disease as a model for translating the microbiome. Immunity 2014;40:843–54.
[121] Sonntag B, Stolze B, Heinecke A, Luegering A, Heidemann J, Lebiedz P, et al. Preterm birth but not mode of delivery is associated with an increased risk of developing inflammatory bowel disease later in life. Inflamm Bowel Dis 2007;13:1385–90.
[122] Corrao G, Tragnone A, Caprilli R, Trallori G, Papi C, Andreoli A, et al. Risk of inflammatory bowel disease attributable to smoking, oral contraception and breastfeeding in Italy: a nationwide case-control study. Cooperative Investigators of the Italian Group for the Study of the Colon and the Rectum (GISC). Int J Epidemiol 1998;27:397–404.
[123] Ekbom A, Adami HO, Helmick CG, Jonzon A, Zack MM. Perinatal risk factors for inflammatory bowel disease: a case-control study. Am J Epidemiol 1990;132:1111–9.
[124] Klement E, Cohen RV, Boxman J, Joseph A, Reif S. Breastfeeding and risk of inflammatory bowel disease: a systematic review with meta-analysis. Am J Clin Nutr 2004;80:1342–52.
[125] Khalili H, Ananthakrishnan AN, Higuchi LM, Richter JM, Fuchs CS, Chan AT. Early life factors and risk of inflammatory bowel disease in adulthood. Inflamm Bowel Dis 2013;19:542–7.
[126] Fança-Berthon P, Michel C, Pagniez A, Rival M, Van Seuningen I, Darmaun D, et al. Intrauterine growth restriction alters postnatal colonic barrier maturation in rats. Pediatr Res 2009;66:47–52.
[127] Le Dréan G, Haure-Mirande V, Ferrier L, Bonnet C, Hulin P, de Coppet P, et al. Visceral adipose tissue and leptin increase colonic epithelial tight junction permeability via a RhoA-ROCK-dependent pathway. FASEB J 2014;28:1059–70.

[128] Lallès JP, Orozco-Solís R, Bolaños-Jiménez F, de Coppet P, Le Dréan G, Segain JP. Perinatal undernutrition alters intestinal alkaline phosphatase and its main transcription factors KLF4 and Cdx1 in adult offspring fed a high-fat diet. J Nutr Biochem 2012;23:1490–7.

[129] Lydiard RB. Irritable bowel syndrome, anxiety, and depression: what are the links? J Clin Psychiatry 2001;62:38–45.

[130] Levy RL, Olden KW, Naliboff BD, Bradley LA, Francisconi C, Drossman DA, et al. Psychosocial aspects of functional gastrointestinal disorders. Gastroenterology 2006;130:1447–58.

[131] Saito YA, Talley NJ. Genetics of irritable bowel syndrome. Am J Gastroenterol 2008;103:2100–4.

[132] Chitkara DK, van Tilburg MA, Blois-Martin N, Whitehead WE. Early life risk factors that contribute to irritable bowel syndrome in adults: a systematic review. Am J Gastroenterol 2008;103:765–74.

[133] Spiller R, Garsed K. Postinfectious irritable bowel syndrome. Gastroenterology 2009;136:1979–88.

[134] Locke 3rd GR, Zinsmeister AR, Talley NJ, Fett SL, Melton 3rd LJ. Familial association in adults with functional gastrointestinal disorders. Mayo Clin Proc 2000;75:907–12.

[135] Kalantar JS, Locke 3rd GR, Zinsmeister AR, Beighley CM, Talley NJ. Familial aggregation of irritable bowel syndrome: a prospective study. Gut 2003;52:1703–7.

[136] Levy RL, Whitehead WE, von Korff MR, Feld AD. Intergenerational transmission of gastrointestinal illness behavior. Am J Gastroenterol 2000;95:451–6.

[137] Levy RL, Jones KR, Whitehead WE, Feld SI, Talley NJ, Corey LA. Irritable bowel syndrome in twins: heredity and social learning both contribute to etiology. Gastroenterology 2001;121:799–804.

[138] Morris-Yates A, Talley NJ, Boyce PM, Nandurkar S, Andrews G. Evidence of a genetic contribution to functional bowel disorders. Am J Gastroenterol 1998;93:1311–7.

[139] Mohammed I, Cherkas LF, Riley SA, Spector TD, Trudgill NJ. Genetic influences in irritable bowel syndrome: a twin study. Am J Gastroenterol 2005;100: 1340–4.

[140] Delvaux M, Denis P, Allemand H. Sexual abuse is more frequently reported by IBS patients than by patients with organic digestive diseases or controls. Results of a multicenter inquiry. French Club of Digestive Motility. Eur J Gastroenterol Hepatol 1997;9:345–52.

[141] Klooker TK, Braak B, Painter RC, de Rooij SR, van Elburg RM, van den Wijngaard RM, et al. Exposure to severe wartime conditions in early life is associated with an increased risk of irritable bowel syndrome: a population-based cohort study. Am J Gastroenterol 2009;104:2250–6.

[142] Ladd CO, Huot RL, Thrivikraman KV, Nemeroff CB, Meaney MJ, Plotsky PM. Long-term behavioral and neuroendocrine adaptations to adverse early experience. Prog Brain Res 2000;122:81–103.

[143] Barreau F, Ferrier L, Fioramonti J, Bueno L. Neonatal maternal deprivation triggers long term alterations in colonic epithelial barrier and mucosal immunity in rats. Gut 2004;53:501–6.

[144] Coutinho SV, Plotsky PM, Sablad M, Miller JC, Zhou H, Bayati AI, et al. Neonatal maternal separation alters stress-induced responses to viscerosomatic nociceptive stimuli in rat. Am J Physiol Gastrointest Liver Physiol 2002;282:G307–16.

[145] Videlock EJ, Adeyemo M, Licudine A, Hirano M, Ohning G, Mayer M, et al. Childhood trauma is associated with hypothalamic-pituitary-adrenal axis responsiveness in irritable bowel syndrome. Gastroenterology 2009;137:1954–62.

[146] Piche T, Barbara G, Aubert P, Bruley des Varannes S, Dainese R, Nano JL, et al. Impaired intestinal barrier integrity in the colon of patients with irritable bowel syndrome: involvement of soluble mediators. Gut 2009;58:196–201.

[147] Camilleri M, Lasch K, Zhou W. Irritable bowel syndrome: methods, mechanisms, and pathophysiology. The confluence of increased permeability, inflammation, and pain in irritable bowel syndrome. Am J Physiol Gastrointest Liver Physiol 2012;303:G775–85.

[148] Barbara G, Stanghellini V, De Giorgio R, Cremon C, Cottrell GS, Santini D, et al. Activated mast cells in proximity to colonic nerves correlate with abdominal pain in irritable bowel syndrome. Gastroenterology 2004;126:693–702.

[149] Francis D, Diorio J, Liu D, Meaney MJ. Nongenomic transmission across generations of maternal behavior and stress responses in the rat. Science 1999;286:1155–8.

[150] Szyf M, McGowan P, Meaney MJ. The social environment and the epigenome. Environ Mol Mutagen 2008;49:46–60.

[151] Daniels WM, Fairbairn LR, van Tilburg G, McEvoy CR, Zigmond MJ, Russell VA, et al. Maternal separation alters nerve growth factor and corticosterone levels but not the DNA methylation status of the exon 1(7) glucocorticoid receptor promoter region. Metab Brain Dis 2009;24:615–27.

[152] McGowan PO, Sasaki A, D'Alessio AC, Dymov S, Labonté B, Szyf M, et al. Epigenetic regulation of the glucocorticoid receptor in human brain associates with childhood abuse. Nat Neurosci 2009;12: 342–8.

[153] McGowan PO, Sasaki A, Huang TC, Unterberger A, Suderman M, Ernst C, et al. Promoter-wide hypermethylation of the ribosomal RNA gene promoter in the suicide brain. PLoS One 2008;3:e2085.

[154] van den Wijngaard RM, Stanisor OI, van Diest SA, Welting O, Wouters MM, Cailotto C, et al. Susceptibility to stress induced visceral hypersensitivity in maternally separated rats is transferred across generations. Neurogastroenterol Motil 2013;25:e780–90.

[155] Murgatroyd C, Patchev AV, Wu Y, Micale V, Bockmühl Y, Fischer D, et al. Dynamic DNA methylation programs persistent adverse effects of early life stress. Nat Neurosci 2009;12:1559–66.

[156] Sandhu MS, Luben R, Day NE, Khaw KT. Self-reported birth weight and subsequent risk of colorectal cancer. Cancer Epidemiol Biomarkers Prev 2002;11:935–8.

[157] Nilsen TI, Romundstad PR, Troisi R, Potischman N, Vatten LJ. Birth size and colorectal cancer risk: a prospective population based study. Gut 2005;54:1728–32.

[158] Cnattingius S, Lundberg F, Iliadou A. Birth characteristics and risk of colorectal cancer: a study among Swedish twins. Br J Cancer 2009;100:803–6.

[159] Hughes LA, van den Brandt PA, de Bruïne AP, Wouters KA, Hulsmans S, Spiertz A, et al. Early life exposure to famine and colorectal cancer risk: a role for epigenetic mechanisms. PLoS One 2009;4:e7951.

[160] Sansom OJ, Stark LA, Dunlop MG, Clarke AR. Suppression of intestinal and mammary neoplasia by lifetime administration of aspirin in Apc(Min/+) and Apc(Min/+), Msh2(−/−) mice. Cancer Res 2001;61:7060–4.

[161] Charles MA, Johnson IT, Belshaw NJ. Supra-physiological folic acid concentrations induce aberrant DNA methylation in normal human cells in vitro. Epigenetics 2012;7:689–94.

[162] Bressenot A, Pooya S, Bossenmeyer-Pourie C, Gauchotte G, Germain A, Chevaux JB, et al. Methyl donor deficiency affects small-intestinal differentiation and barrier function in rats. Br J Nutr 2013;109:667–77.

[163] Bauchart-Thevret C, Stoll B, Burrin DG. Intestinal metabolism of sulfur amino acids. Nutr Res Rev 2009;22:175–87.

[164] Thwaites DT, Anderson CM. H^+-coupled nutrient, micronutrient and drug transporters in the mammalian small intestine. Exp Physiol 2007;92:603–19.

[165] Trivedi MS, Shah JS, Al-Mughairy S, Hodgson NW, Simms B, Trooskens GA, et al. Food-derived opioid peptides inhibit cysteine uptake with redox and epigenetic consequences. J Nutr Biochem 2014;25: 1011–8.

[166] Waly MI, Hornig M, Trivedi M, Hodgson N, Kini R, Ohta A, et al. Prenatal and postnatal epigenetic programming: implications for GI, immune, and neuronal function in autism. Autism Res Treat 2012;2012:190930.

[167] Teschemacher H. Opioid receptor ligands derived from food proteins. Curr Pharm Des 2003;9:1331–44.

[168] Kaneko K, Iwasaki M, Yoshikawa M, Ohinata K. Orally administered soymorphins, soy-derived opioid peptides, suppress feeding and intestinal transit via gut mu(1)-receptor coupled to 5-HT(1A), D(2), and GABA(B) systems. Am J Physiol Gastrointest Liver Physiol 2010;299:G799–805.

[169] Chung FL, Conaway CC, Rao CV, Reddy BS. Chemoprevention of colonic aberrant crypt foci in Fischer rats by sulforaphane and phenethyl isothiocyanate. Carcinogenesis 2000;21:2287–91.

[170] Myzak MC, Karplus PA, Chung FL, Dashwood RH. A novel mechanism of chemoprotection by sulforaphane: inhibition of histone deacetylase. Cancer Res 2004;64:5767–74.

[171] Rajendran P, Kidane AI, Yu TW, Dashwood WM, Bisson WH, Löhr CV, et al. HDAC turnover, CtIP acetylation and dysregulated DNA damage signaling in colon cancer cells treated with sulforaphane and related dietary isothiocyanates. Epigenetics 2013;8:612–23.

[172] Myzak MC, Dashwood WM, Orner GA, Ho E, Dashwood RH. Sulforaphane inhibits histone deacetylase in vivo and suppresses tumorigenesis in Apc-minus mice. FASEB J 2006;20:506–8.

[173] Hu R, Khor TO, Shen G, Jeong WS, Hebbar V, Chen C, et al. Cancer chemoprevention of intestinal polyposis in ApcMin/+ mice by sulforaphane, a natural product derived from cruciferous vegetable. Carcinogenesis 2006;27:2038–46.

[174] Myzak MC, Tong P, Dashwood WM, Dashwood RH, Ho E. Sulforaphane retards the growth of human PC-3 xenografts and inhibits HDAC activity in human subjects. Exp Biol Med (Maywood) 2007;232:227–34.

[175] Druesne N, Pagniez A, Mayeur C, Thomas M, Cherbuy C, Duée PH, et al. Diallyl disulfide (DADS) increases histone acetylation and p21(waf1/cip1) expression in human colon tumor cell lines. Carcinogenesis 2004;25:1227–36.

[176] Druesne N, Pagniez A, Mayeur C, Thomas M, Cherbuy C, Duée PH, et al. Repetitive treatments of colon HT-29 cells with diallyl disulfide induce a prolonged hyperacetylation of histone H3 K14. Ann NY Acad Sci 2004;1030:612–21.

[177] Druesne-Pecollo N, Chaumontet C, Pagniez A, Vaugelade P, Bruneau A, Thomas M, et al. In vivo treatment by diallyl disulfide increases histone acetylation in rat colonocytes. Biochem Biophys Res Commun 2007;354:140–7.

[178] Nian H, Delage B, Pinto JT, Dashwood RH. Allyl mercaptan, a garlic-derived organosulfur compound, inhibits histone deacetylase and enhances Sp3 binding on the P21WAF1 promoter. Carcinogenesis 2008;29:1816–24.

[179] Lea MA, Rasheed M, Randolph VM, Khan F, Shareef A, desBordes C. Induction of histone acetylation and inhibition of growth of mouse erythroleukemia cells by S-allylmercaptocysteine. Nutr Cancer 2002;43:90–102.

[180] Li Y, Li X, Guo B. Chemopreventive agent 3,3′-diindolylmethane selectively induces proteasomal degradation of class I histone deacetylases. Cancer Res 2010;70:646–54.

[181] Joven J, Micol V, Segura-Carretero A, Alonso-Villaverde C, Menéndez JA, Bioactive Food Components Platform. Polyphenols and the modulation of gene expression pathways: can we eat our way out of the danger of chronic disease? Crit Rev Food Sci Nutr 2014;54:985–1001.

[182] Fang MZ, Wang Y, Ai N, Hou Z, Sun Y, Lu H, et al. Tea polyphenol (-)-epigallocatechin-3-gallate inhibits DNA methyltransferase and reactivates methylation-silenced genes in cancer cell lines. Cancer Res 2003;63:7563–70.

[183] Waldecker M, Kautenburger T, Daumann H, Busch C, Schrenk D. Inhibition of histone-deacetylase activity by short-chain fatty acids and some polyphenol metabolites formed in the colon. J Nutr Biochem 2008;19:587–93.

[184] Wang LS, Kuo CT, Cho SJ, Seguin C, Siddiqui J, Stoner K, et al. Black raspberry-derived anthocyanins demethylate tumor suppressor genes through the inhibition of DNMT1 and DNMT3B in colon cancer cells. Nutr Cancer 2013;65:118–25.

[185] Wang LS, Kuo CT, Stoner K, Yearsley M, Oshima K, Yu J, et al. Dietary black raspberries modulate DNA methylation in dextran sodium sulfate (DSS)-induced ulcerative colitis. Carcinogenesis 2013;34:2842–50.

[186] Wang LS, Arnold M, Huang YW, Sardo C, Seguin C, Martin E, et al. Modulation of genetic and epigenetic biomarkers of colorectal cancer in humans by black raspberries: a phase I pilot study. Clin Cancer Res 2011;17:598–610.

[187] Tan S, Wang C, Lu C, Zhao B, Cui Y, Shi X, et al. Quercetin is able to demethylate the p16INK4a gene promoter. Chemotherapy 2009;55:6–10.

[188] Wang Z, Chen H. Genistein increases gene expression by demethylation of WNT5a promoter in colon cancer cell line SW1116. Anticancer Res 2010;30:4537–45.

[189] Wang H, Li Q, Chen H. Genistein affects histone modifications on Dickkopf-related protein 1 (DKK1) gene in SW480 human colon cancer cell line. PLoS One 2012;7:e40955.

[190] Zhang Y, Li Q, Chen H. DNA methylation and histone modifications of Wnt genes by genistein during colon cancer development. Carcinogenesis 2013;34:1756–63.

[191] Yara S, Lavoie JC, Beaulieu JF, Delvin E, Amre D, Marcil V, et al. Iron-ascorbate-mediated lipid peroxidation causes epigenetic changes in the antioxidant defense in intestinal epithelial cells: impact on inflammation. PLoS One 2013;8:e63456.

[192] Cyr AR, Hitchler MJ, Domann FE. Regulation of SOD2 in cancer by histone modifications and CpG methylation: closing the loop between redox biology and epigenetics. Antioxid Redox Signal 2013;18:1946–55.

[193] Chen C, Wang L, Liao Q, Xu L, Huang Y, Zhang C, et al. Association between six genetic polymorphisms and colorectal cancer: a meta-analysis. Genet Test Mol Biomarkers 2014;18:187–95.

[194] Dashwood RH. Frontiers in polyphenols and cancer prevention. J Nutr 2007;137(1 Suppl.):267S–9S.

[195] Shen G, Khor TO, Hu R, Yu S, Nair S, Ho CT, et al. Chemoprevention of familial adenomatous polyposis by natural dietary compounds sulforaphane and dibenzoylmethane alone and in combination in Apc$^{Min/+}$ mouse. Cancer Res 2007;67:9937–44.

[196] Nair S, Hebbar V, Shen G, Gopalakrishnan A, Khor TO, Yu S, et al. Synergistic effects of a combination of dietary factors sulforaphane and (-) epigallocatechin-3-gallate in HT-29 AP-1 human colon carcinoma cells. Pharm Res 2008;25:387–99.

CHAPTER

18

How the Father Might Epigenetically Program the Risk for Developmental Origins of Health and Disease Effects in His Offspring

Kristin E. Murphy, Timothy G. Jenkins, Douglas T. Carrell

Department of Surgery (Urology), University of Utah School of Medicine, Salt Lake City, UT, USA

OUTLINE

INTRODUCTION

The father, in comparison to the mother, at first glance contributes seemingly little to the developing embryo. The mother provides the environment in which the fetus develops, as well as the nutrients for the embryo to survive. Upon fertilization, the egg harbors virtually all of the mRNA transcripts and proteins that support and control early embryonic development prior to zygotic gene transcription [1,2]. In contrast, the sperm has been viewed to supply little more than a chromosomal complement to the offspring. While the genetic information

The Epigenome and Developmental Origins of Health and Disease
http://dx.doi.org/10.1016/B978-0-12-801383-0.00018-9

Copyright © 2016 Elsevier Inc. All rights reserved.

transmitted from the father to the embryo has been known to be essential in determining the offspring health, it has become evident in recent years that epigenetic information is also inherited through the paternal germ line, and may play an important role in the developmental origins of adult health and disease [3,4].

Epigenetics refers to stably heritable changes that influence gene function without altering DNA sequence [5]. These changes include chemical modifications to the DNA or the associated nucleoproteins, such as DNA methylation and histone modifications, and also encompass histone variants and small, nonprotein-coding RNAs [6,7]. Importantly, these factors can influence gene expression by a number of mechanisms, such as promoting or inhibiting transcription factor binding, altering chromatin structure and DNA accessibility, or directing RNA degradation. However, unlike DNA mutations, epigenetic modifications are impermanent, and can be incorporated or removed in a cell-specific manner during one's lifetime in response to environmental exposure, lifestyle, or natural aging processes. Despite their plasticity, epigenetic marks are stable in a sense that they can be maintained through mitotic and meiotic cell divisions, making them heritable from one generation to the next [8,9].

In order for epigenetic information to pass from father to offspring, it must be present within the germ line. Male germ cells are a morphologically and molecularly distinct cell type, which exhibit a unique epigenetic profile. Because fertilization occurs within the female reproductive tract, sperm have evolved a number of specialized mechanisms, such as motility and nuclear compaction, in order to overcome obstacles to fertilization [10]. Epigenetic modifications play a vital role in the spermatocyte differentiation and morphogenesis processes required to achieve fertilization [11]. However, once the sperm and oocyte fuse upon fertilization, the sperm epigenome must be restored back to an undifferentiated state [12].

Beginning in the early zygote, many sperm-specific epigenetic modifications are globally erased and then reprogrammed for embryonic development and totipotent conversion [8,11, 13–15]. Although parental epigenetic marks are largely erased at this time, studies have shown that reprogramming is incomplete. In fact, DNA methylation, histone modifications, and small RNAs are all potential mechanisms for nongenetic inheritance from father to offspring [16] (Figure 1). Because sperm are continually produced from puberty onward, they may incur epigenetic alterations as a result of the environment. Age, diet, smoking, alcohol, and numerous other exposures have all been shown to impact sperm DNA methylation and in some cases are associated with offspring health. Together, these findings have led to a hypothesis that environmentally triggered epigenetic alterations in the father are inherited by offspring [4]. In this way, the father might epigenetically program the risk for developmental origins of health and disease (DOHaD) in his offspring.

This chapter will comprehensively review what is currently known about sperm epigenetics, potential modes of inheritance to offspring, and paternal environmental factors that may contribute to offspring phenotype. Although genome-wide studies have recently elucidated the epigenetics of sperm at a base pair resolution, in many cases, additional research will be necessary to understand the extent to which the father contributes to DOHaD.

SPERM EPIGENETICS

DNA Methylation

The most well-characterized epigenetic mark present in sperm is DNA methylation of a cytosine base, also known as 5-methyl-cytosine (5mC), in which a methyl group is covalently linked to the fifth position on the cytosine carbon ring [17]. Methylation at cytosine-phosphate-guanine

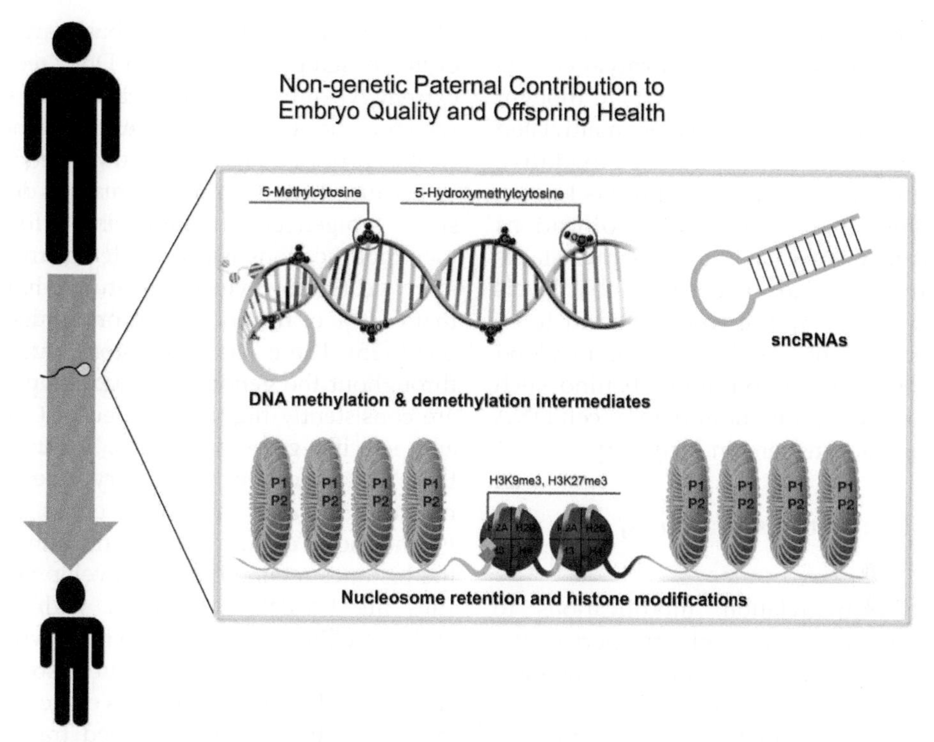

FIGURE 1 Potential epigenetic mechanisms of inheritance from father to offspring include DNA methylation modifications and demethylation intermediate modifications at loci that resist zygotic DNA methylation reprogramming, small noncoding RNAs (sncRNAs), and histone modifications at loci where histones are retained in sperm.

dinucleotides (CpGs) in gene promoters is generally associated with transcriptional silencing. Mechanistically, the 5mC modifications may promote the binding of transcriptional inhibitors or prevent the binding of transcription factors [18].

The importance of DNA methylation for spermatocyte development is highlighted by mouse models that are deficient in DNA methyltransferase (DNMT) enzymes, which catalyze the addition of methyl groups to DNA. DNMT1, the maintenance methyltransferase, functions to conserve methylation marks on newly synthesized DNA upon cell division [19], while de novo methyltransferases DNMT3A, DNMT3B, and catalytically inactive DNMT3L, establish new methyl marks [20]. In studies on male mice lacking DNMT3L, DNA hypomethylation was detected, transposable elements were aberrantly activated, and meiosis I was impaired [21]. Mouse lacking DNMT3A, on the other hand, were found to have defective paternal imprints, which are established in the male germ line [22].

Transcription in mammalian mature sperm is considered inactive; therefore, the methylation patterns observed in sperm are thought to chiefly reflect prior transcriptional regulation during spermatocyte development. This is supported by genome-wide studies that have found hypomethylated promoters at germ line-specific genes in sperm [23]. Interestingly, promoters of specific genes that are critical for early embryonic development are also hypomethylated in sperm, suggesting that the male germ cell harbors DNA methylation patterns in preparation for regulating embryonic transcription following fertilization (Figure 1) [24,25].

In addition to 5mC, 5-hydroxymethyl-cytosine (5hmC), and to a lesser extent 5-formyl-cytosine (5fC) and 5-carboxyl-cytocine (5caC) are thought to represent intermediates between methylated and unmethylated states during oxidative DNA demethylation [26]. While little is known about whether these epigenetic modifications have regulatory roles, a unique 5hmC status was recently identified in sperm. Global levels of 5hmC in human sperm were detected to be only around 30% of the 5hmC levels in blood [27]. The causes and implications behind such differences between somatic and germ cell DNA modification patterns remain a mystery.

Histone Modifications, Variants, and Protamines

Like DNA methylation, modifications to histones are associated with transcriptional changes. In somatic cells, different chemical additions to the core histones that make up nucleosomes, such as methylation, acetylation, phosphorylation, and SUMOylation, are associated with either gene activation or gene repression. By altering chromatin compaction, these modifications impact RNA polymerase accessibility to gene promoters, in turn, regulating transcription [28]. During spermatogenesis, histone modifications associated with heterochromatin formation are crucial for male germ cells to complete meiosis. Mice with germ cells deficient for SUV39H1/2 or G9A, enzymes that are responsible for catalyzing H3K9 methylation, are subfertile and display prophase I defects [16,29,30]. Furthermore, H3K4me3, H3K9ac, and H2AX phosphorylation (γH2AX) are concentrated at DNA double strand break sites that are necessary for meiotic recombination [31].

Following meiosis, a unique process known as protamination occurs during sperm maturation, in which protamine proteins nearly globally replace histones in a stepwise manner [16]. First, transition proteins replace histones, and then protamines replace the transition proteins. The interaction between the basic arginine and cysteine-rich protamines and DNA results in an extremely compact toroidal chromatin structure, which is thought to promote sperm motility, as well as protect DNA against damage [3,32–34].

Because histones are not readily detected in sperm, epigenetic inheritance in the form of histone modifications was considered irrelevant for many years. However, recent studies have shown that 5–10% of histones resist protamine replacement [25]. These retained histones are observed throughout the genome, but unlike protamines, are consistently highly enriched at CG-rich loci and specific gene promoters, including genes that are important for embryo development, microRNAs, and imprinted genes [25,35–37]. Furthermore, in both mouse and human sperm, the histones retained at the developmental gene sites are bivalently enriched with H3K4me3 and H3K27me3 [25,38,39], which when found together at gene promoters are also associated with DNA demethylation. This epigenetic profile is thought to promote a poised transcriptional state that facilitates activation of transcription of the paternal genes within the embryo and maintain pluripotency within embryonic stem (ES) cells [25,37,40]. The importance of bivalent epigenetic modifications to sperm chromatin is highlighted by evolutionary conservation of these modifications. Despite the fact that sperm from zebra fish, a species evolutionarily distinct from human, lack protamines, they possess remarkably similar bivalent epigenetic modifications on developmental regulator loci that correlate with an earlier embryonic timing of gene activation [41]. Taken together, these findings support a model in which histone retention is programmatic for the next generation, aiding the expression of early embryonic development genes.

mRNA, Proteins, and Small Noncoding RNA

The mature sperm is transcriptionally inactive and contains minimal amounts of cytoplasm, so RNA molecules are not abundant. However, mRNA transcripts have been detected in mammalian, including human, sperm [42–44]. These mRNA molecules are thought to accumulate during spermatogenesis [45], and many identified transcripts encode proteins that are critical for spermatocyte function, representing remnants of past transcriptional states. On the other hand, some transcripts found in the sperm code for proteins that are thought to function after fertilization, such as the calcium oscillator *Plc-z* (Phospholipase C-z) transcript, which has shown to be translated within the oocyte [46,47]. A number of paternally derived mRNAs that are not present within the oocyte have been detected in the early zygote to date, suggesting a posttranscriptional mechanism of paternal inheritance to offspring [44,48]. While the amount of mRNA contributed by the sperm pales in comparison to that harbored within the egg, nonetheless, it may be significant for offspring development.

Another class of RNAs found in sperm is small noncoding RNAs (sncRNA), which are single-stranded RNAs of various sizes in the range of 20 nucleotides. Gene-silencing functions have been attributed to sncRNAs due to secondary structure and sequence homology, making them regulatory molecules [49]. The importance of sncRNA during sperm development has been demonstrated using mouse mutants. For example, PIWI interacting RNA (piRNA) pathway proteins, which are also involved in transposable element DNA methylation, are required for male meiosis. Genetic ablation of all three piRNA biogenesis pathway genes, *Miwi*, *Mili*, and *Miwi2*, results in spermatogenesis arrest during meiotic prophase I [50–53]. Increased expression of transposable elements was also detected in *Miwi2* mutants, bringing about the possibility that RNAs are required for transposable element methylation in order for meiosis to proceed through prophase I [52,54]. In addition to their role in spermatogenesis, piRNAs have been discovered in the mature sperm cell [24,55,56]. While a function has not been attributed to piRNAs in the fertilized embryo, transposable element silencing is a reasonable hypothesis.

While a smaller pool of microRNAs (miRNAs) was detected in mature sperm, this class of sncRNAs holds a significant potential for controlling embryo development due to the ability of miRNAs to directly regulate gene expression [24,53,55–57]. In fact, genetic studies have shed light on requirements for specific sperm-derived miRNAs for early embryogenesis. For example, a requirement for sperm-derived mir-34c for the first embryonic cleavage has been demonstrated in mouse [58]. Additional miRNAs present in sperm have been implicated in early embryonic development [57], further supporting a role for paternally inherited sncRNAs in the offspring.

Like mRNA, proteins other than protamines are largely absent from the mature sperm. However, a recent study identified almost 500 proteins within human sperm, with functions ranging from cell metabolism to transcription to structural roles [36]. While it remains unclear whether these proteins are inherited or utilized by the early embryo, this represents yet another possible nongenetic mechanism of inheritance from father to offspring.

ZYGOTIC EPIGENETIC REPROGRAMMING

Although abundant epigenetic information is present within the male germ cell, it is not all transmitted to the embryo. A wave of epigenetic reprogramming occurs just after fertilization, in which parental DNA methylation and histone modifications are erased and then reestablished in order to reset the epigenetic landscape for embryonic pluripotency. However, not all genomic loci undergo epigenetic reprogramming, and those loci that escape epigenetic erasure represent likely candidates for transmittal of paternal epigenetic modifications to offspring.

DNA demethylation in the maternal pronucleus largely adheres to a passive replication-dependent rate over multiple cell divisions by the exclusion of maintenance DNA methyltransferase DNMT1 [59,60], but the paternal pronucleus undergoes rapid and active DNA demethylation following fertilization (Figure 2) [8]. Because much of the paternal demethylation occurs before the first cell division, passive replication-dependent mechanisms cannot be responsible. Ten–eleven translocation methylcytosine dioxygenase-3 (TET3)-mediated oxidation, in which 5mC is converted to 5hmC, 5fC, and 5caC intermediates, has been implicated in this active DNA demethylation process. In support of this, as 5hmC levels rise, 5mC levels fall, and TET3 localizes to the paternal pronucleus [61,62]. Knockdown of *Tet3* in mice also leads to reduced 5hmC and increased 5mC levels in the paternal, but not the maternal, pronucleus [63]. The base excision repair (BER) pathway, in addition to TET3 oxidation, may also contribute to active demethylation through deamination, as inhibition of BER factors in mouse causes reduction of paternal DNA demethylation [61,64].

The functional role of such rapid paternal DNA demethylation in the zygote remains unknown. One hypothesis proposes that zygotic genome activation requires active paternal demethylation [65], but evidence to support this model is lacking. No known functions are attributed to the 5hmC, 5fC, and 5caC

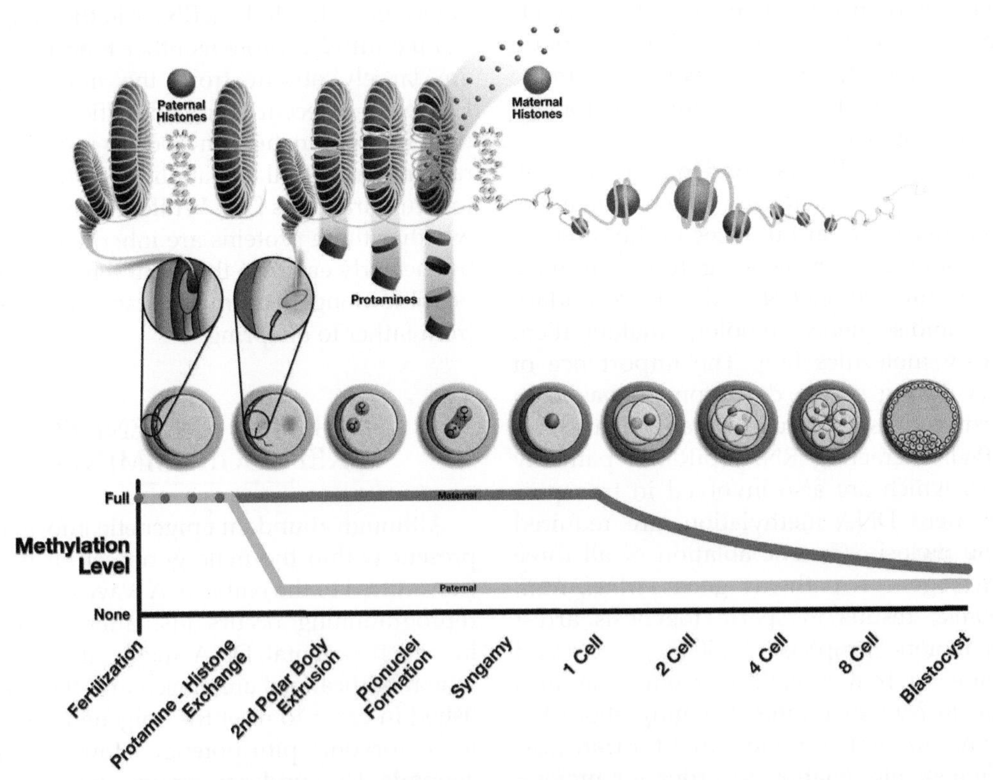

FIGURE 2 **Paternal epigenetic reprogramming following fertilization.** Immediately following fertilization, paternal protamines are replaced with maternal histones (top panel) and the paternal pronucleus becomes decondensed. In addition, paternal DNA methylation is globally, rapidly erased by active mechanisms (graph). *Reproduced from Jenkins TG, Carrell DT. Dynamic alterations in the paternal epigenetic landscape following fertilization. Front Genet 2012;3:143.*

intermediate epigenetic modifications, but they may play important roles in the paternal or zygotic genomes. A recent study characterized genome-wide DNA methylation in mouse zygotes, and showed that (1) the maternal pronucleus, in addition to the paternal pronucleus, undergoes limited active DNA demethylation, and (2) the paternal pronucleus is demethylated by a combination of active and passive mechanisms [66]. These findings are intriguing because specific genes may be regulated based on the rate of DNA demethylation in either the maternal or paternal genomes in the earliest stages of embryonic development. Follow-up studies are needed in order to identify any functional patterns associated with the timing of DNA demethylation.

While much of the maternal and paternal genomes are demethylated after fertilization, not all loci are subject to this epigenetic erasure [67]. Imprinted differentially methylated regions (DMRs) are one category that resist demethylation in the zygote, allowing for transmission of parental imprints to the embryo, and monoallelic expression of imprinted genes [68]. In addition, some transposable and retroviral elements are also resistant to demethylation [69], implying a special need for DNA methylation maintenance at these genomic regions. Data indicate that single copy genes located proximal to retroviral elements may also be subjected to maintenance of DNA methylation after fertilization [8,70]. Importantly, while this phenomenon may have originated stochastically, it remains a possible mechanism for parental inheritance of epigenetic modifications at specific genes.

Because the male and female pronuclei are differentially demethylated, mechanisms for targeting active demethylated loci, or for protecting loci against active DNA methylation, have been sought out. A recent study revealed that DPP3A (also known as STELLA and PGC7) binds to H3K9me2 in the maternal pronucleus as well as some paternal imprinted loci, preventing 5mC oxidation [71]. In addition to DPP3A, ZFP57 has been shown to bind to methylated DNA at specific imprinted loci and recruit factors that promote heterochromatin formation. Loss of ZFP57 results in aberrant imprinting patterns, suggesting that it may protect against DNA demethylation [72]. Evidence for DPP3A and ZFP57 protecting against DNA demethylation is scarce, so further research will be necessary to distinguish mechanisms for locus-specific DNA demethylation and to determine factors upstream of these identified proteins.

Just as sperm DNA methylation patterns are erased in the early embryo, sperm-specific protamines and retained histone epigenetic modifications are also reprogrammed for embryonic development. Following fertilization, protamines from paternally inherited chromatin are quickly removed and replaced with oocyte-derived histones (Figure 2) [73]. Newly incorporated histones are then modified with H3K4, H3K9, and H3K27 monomethylation, which coincides with a pronuclear chromatin decondensation phase [74–76]. It is unclear whether histones that were retained from sperm are also cleared of epigenetic modifications at this time. However, the same developmental loci that are bivalently marked in retained histones of sperm are also bivalently marked in the early embryo [25]. This correlative evidence suggests that histone modifications from the male germ line are maintained for developmental programming, but direct evidence has yet to confirm this model.

Together, DNA methylation, histone modifications, and small RNAs all provide mechanisms for paternal epigenetic inheritance (Figures 1 and 2). Even after the genome-wide wave epigenetic reprogramming in the early zygote, it is clear that some epigenetic information is maintained. Even if a small percentage of genes are programmed with paternal epigenetic modifications, it is feasible that they could result in major gene expression changes and phenotypic consequences for the offspring. Perturbation of this epigenetic information through paternal lifestyle, aging, and environmental exposures could be inherited to the developing embryo as well.

PATERNAL AGING

It is known that paternal age is associated with increased prevalence of numerous neurological disorders including bipolar disorder, schizophrenia, autism, and impaired social behavior, among others [77–80]. Specifically, it has been shown that there is a relative risk of developing schizophrenia of approximately 1.66 in the offspring of fathers older than 50 years of age when compared to the offspring of younger fathers [81]. Studies from Denmark, Sweden, Israel, Australia, and the United States have also shown that paternal or grand paternal age is a risk factor for developing autism [82–86]. Similarly, studies have suggested that bipolar disorder is more frequently found in the offspring of older fathers when compared to offspring of younger fathers [87]. Controlled studies in mice show changes in the social and exploratory behaviors in the offspring of older fathers [80], and offspring sired by older fathers also appear to have increased risk for developing hematological and neurological cancers [88,89]. Additionally, large retrospective chart review studies from multiple investigators have demonstrated an interesting link between paternal age and birth weight, premature delivery, and various pregnancy and offspring abnormalities [90,91].

Despite these intriguing findings, the mechanisms driving these effects remain poorly defined. It is well documented that DNA methylation is altered in many somatic cell types with age and that frequently dividing cells exhibit more striking methylation changes associated with age than do cells that divide less often [92]. These pieces of evidence make sperm DNA methylation alterations with age a highly plausible mechanism to explain the increased incidence of neuropsychiatric disorders seen in the offspring of older fathers.

Very recently, studies in both humans and mice have provided some indication about the epigenetic alterations that occur with age in the sperm. In paired sperm samples collected between 9 and 19 years apart from fertile sperm donors, over 100 genomic regions have been identified that are consistently altered with age [27,93]. While globally in sperm hypermethylation was observed, at these regions, there is a strong bias toward a loss of methylation, which is a particularly interesting finding that is in opposition to what has been identified in somatic cells, namely global hypomethylation and a bias toward regional hypermethylation at sites such as promoters and CpG islands [94,95]. Intriguingly, regional DNA methylation alterations are enriched at, or very close to, genes that have been implicated in multiple neuropsychiatric disorders that have been shown to have increased incidence in the offspring of older fathers. Similar findings have been identified in mice where there are striking alterations that occur with age in the sperm epigenome at the individual CpG level [96]. Similar to human studies, these alterations are more frequently in the direction of hypomethylation rather than hypermethylation, and behavioral abnormalities were observed in offspring from aged fathers.

PATERNALLY DRIVEN ENVIRONMENTAL EFFECTS

Multiple studies have demonstrated a link between paternal exposure to various environmental agents, diet changes, or drugs and offspring abnormalities [97–99]. Specifically, it has been demonstrated that, when mated with normal females, male mice fed a low protein diet sire offspring with altered expression of many genes important in metabolism and cholesterol synthesis [97]. Similarly, metabolic alterations, specifically changes in insulin sensitivity, were also seen in the female offspring of male rats fed high fat diets [100]. Epidemiological studies utilizing data collected during, and following, massive crop failures in Sweden in the late 1800s and early 1900s provided large retrospective data sets on human populations, and investigators observed similar effects in humans as mice. These data suggest that an alteration of the quality of paternal diets may independently affect offspring health and metabolic activity [101,102].

Besides diet, other exposures have resulted in similar effects to offspring. The International Association for Research on Cancer has recently declared that paternal smoking prior to pregnancy is associated with a significantly elevated risk of leukemia in the offspring [103], suggesting that smoke-induced genetic or epigenetic changes occur in sperm that are transmitted to offspring. Similarly, preconception paternal alcohol intake (particularly in excess) is associated with increased behavioral abnormalities and cognitive delays in the offspring [104,105], both in humans and animal studies.

Together, these correlative data strongly suggest that mechanisms exist that allow the paternal genome to affect offspring phenotype. Epigenetic alteration to the sperm is a likely candidate to explain, or contribute to, the effects on the offspring identified in the above studies. The plasticity of the sperm epigenome means that chemical modifications or molecules are capable of responding to, or being altered by, a wide variety of environmental exposures. In addition, as these marks are potentially heritable, the alterations found in sperm may be directly passed to offspring, resulting in distinct consequences for offspring health.

There are many environmental agents that affect sperm functionally, genetically, or epigenetically. Environmental insults can influence everything from sperm motility and morphology to fertilizing potential. As an example, smoking and direct exposure to individual toxins in cigarette smoke has been clearly associated with various abnormalities in sperm in many studies [106–108]. Cadmium, a ubiquitous metal found in relatively large quantities in cigarette smoke, has been demonstrated to disrupt spermatogenesis by altering both Sertoli cell function as well as directly contributing to spermatozoal apoptosis [109,110]. In addition, cigarette smoking has been shown to increase sperm DNA damage and oxidative stress [111]. Adult male mice exposed to sidestream tobacco smoke display significant increases in sperm DNA mutations at expanded simple tandem repeats (ESTRs) [112], and aberrations in sperm chromatin structure [113]. In contrast, these male mice exhibit no measurable increase in somatic cell chromosome damage, indicating that germ cells may be more prone to genetic and/or epigenetic insults resulting from smoke exposure than somatic cells. Importantly, destabilization of ESTRs in the adult male germ line is well documented to be associated with transgenerational effects in the mouse genome [114–118]. In addition to cigarette smoke, recent evidence from human and animal models suggests that alcohol may affect sperm morphology [119,120], forward progression velocity, and fertilization capacity [120]. Clearly, these alterations, while dependent on degree of exposure, have the potential to impact fertility status of the exposed individual, as well as the capacity to affect offspring generated by fertilization with the affected sperm.

Of particular interest in the context of paternal epigenetic inheritance to the offspring are sperm epigenetic alterations that occur as a result of environmental exposures. Alcohol consumption has been shown to alter DNA methylation patterns subtly at various imprinted loci and imprinting control regions [121–123]. Other genes are also affected in a similar fashion, such as POMC [98] and DAT [124]. Recent findings suggest that even microRNAs in sperm are affected as a result of smoke exposure [125]. Lead, a significant component of cigarette smoke, has been shown to disrupt the packaging of sperm nuclear proteins, which may ultimately affect transgenerational inheritance via the perturbed programmed retention of histones and their associated marks [126,127]. Exposure to various forms of endocrine disrupters, namely methoxychlor and vinclozolin, significantly alters imprinted gene methylation status in the sperm though the data are somewhat controversial [128–132]. Clearly, sperm are sensitive to environmental insults. In particular, the sensitivity of the sperm epigenome to such insults makes the male gamete extremely susceptible to a high degree of alteration from the various environmental exposures that are commonly encountered throughout the reproductive lifespan of the male. As a result, it is important to thoroughly understand both the nature of the alterations (location and magnitude) that occur as the result of various exposures, as well as the heritability of these marks across multiple generations.

TRANSGENERATIONAL EPIGENETIC INHERITANCE

Epigenetic inheritance through the paternal germ line brings about the possibility of repercussions not only for a father's direct offspring but also for subsequent generations (epigenetic marks inherited over multiple generations are considered "transgenerationally inherited"). The idea that epigenetic inheritance can be mediated through the paternal germ line is not new. A number of well-known studies utilized the agouti viable yellow (Avy) locus, a retrotransposon found upstream of the agouti gene, whose expression affects mammalian coat color. Typically, this locus is turned "off" via DNA methylation enrichment, however, there are cases where these methylation marks can be removed and the agouti gene is switched "on," resulting in striking phenotypic alterations marked by a yellow coat, obesity, and increased risk of cancer and diabetes with no change to the genome [133]. Typically, this epigenetic "mutation" is a result of altered diet or environmental exposure (such as exposure to bisphenol A) in the pregnant female [134]. Intriguingly, this epigenetic alteration can be inherited, suggesting that these marks can escape embryonic epigenetic reprogramming. Phenotypes also appear to be modifiable based on environmental exposure and diet [134,135].

Of particular interest in the investigation of paternal influences on transgenerational inheritance are studies that suggest unique disease heritability through the paternal germ line via inheritance of altered regional DNA methylation profiles. This has been demonstrated as a result of many environmental exposures. One recent study showed that in utero alcohol exposure in male mice alters proopiomelanocortin (POMC) gene promoter methylation [98]. Intriguingly, this alteration is inherited only through the paternal germ line and can be seen through the F3 generation. Further studies have shown altered dopamine transport protein expression in offspring sired by ethanol-exposed fathers [124]. Similarly, studies have shown altered DNA methylation at imprinted genes and imprinting control regions in the offspring of alcohol-consuming fathers [121,122]. While not conclusive, these data support the idea that altered DNA methylation profiles in sperm, from whatever source, are capable of being inherited and affecting phenotype transgenerationally.

CONCLUSIONS

Taken together, there is strong evidence that epigenetic modifications do occur in the sperm as a result of environmental exposures. Remarkably, there are also clear indications that these alterations are, in some cases, able to escape embryonic reprogramming and affect offspring phenotype. While the ability to adjust gene expression based on environment and pass changes on to offspring may, in theory, provide some advantages, to date these changes have largely proven to result in negatively altered offspring phenotypes.

This subject matter is one that requires a great degree of further study, as it may provide a key to understanding many cases of idiopathic disease or those that lack clear genetic signatures. While we have abundant correlative evidence that implicates paternal epigenetics in offspring development, specific epigenetic alterations in sperm that are causative of phenotypic changes in offspring have not been definitively pinpointed. By better understanding the mechanisms of epigenetic inheritance, it may also become possible to modify epigenetics for offspring therapeutic options. Because the epigenome is plastic, if diet or the environment induces perturbations in sperm, then it is logical to assume that diet or environment could "correct" such alterations in the male gametes, providing us with possible intervention for what are now considered idiopathic diseases. However, it is important to note that the timing of such alterations may prove to overcome any potential rescue. Such is the case when alterations occur only in embryonic development. Exposures to toxins during embryonic development may alter the epigenome in a way that, once established, cannot be corrected in the adult organism. Certainly, these areas of study are likely to yield new information, and potentially clinical therapies, in the study of the paternal contribution to the health of offspring.

References

[1] Sinclair KD, Watkins AJ. Parental diet, pregnancy outcomes and offspring health: metabolic determinants in developing oocytes and embryos. Reprod Fertil Dev 2013;26(1):99–114.

[2] Li L, Zheng P, Dean J. Maternal control of early mouse development. Development 2010;137(6):859–70.

[3] Carrell DT. Epigenetics of the male gamete. Fertil Steril 2012;97(2):267–74.

[4] Jenkins TG, Carrell DT. The sperm epigenome and potential implications for the developing embryo. Reproduction 2012;143(6):727–34.

[5] Riggs A, Martinssen RA, Russo VEA. In: Epigenetic mechanisms of gene regulation. Cold Spring Harbor, HY: C.S.H. Press; 1996.

[6] Berger SL, et al. An operational definition of epigenetics. Genes Dev 2009;23(7):781–3.

[7] Tan M, et al. Identification of 67 histone marks and histone lysine crotonylation as a new type of histone modification. Cell 2011;146(6):1016–28.

[8] Seisenberger S, et al. Reprogramming DNA methylation in the mammalian life cycle: building and breaking epigenetic barriers. Philos Trans R Soc Lond B Biol Sci 2013;368(1609):20110330.

[9] Reik W. Stability and flexibility of epigenetic gene regulation in mammalian development. Nature 2007;447(7143):425–32.

[10] Morelli MA, Cohen PE. Not all germ cells are created equal: aspects of sexual dimorphism in mammalian meiosis. Reproduction 2005;130(6):761–81.

[11] Reik W, Dean W, Walter J. Epigenetic reprogramming in mammalian development. Science 2001;293(5532): 1089–93.

[12] Smallwood SA, Kelsey G. De novo DNA methylation: a germ cell perspective. Trends Genet 2012;28(1):33–42.

[13] Abe M, et al. Sex-specific dynamics of global chromatin changes in fetal mouse germ cells. PLoS One 2011;6(8):e23848.

[14] Feng S, Jacobsen SE, Reik W. Epigenetic reprogramming in plant and animal development. Science 2010;330(6004):622–7.

[15] Monk M, Boubelik M, Lehnert S. Temporal and regional changes in DNA methylation in the embryonic, extraembryonic and germ cell lineages during mouse embryo development. Development 1987;99(3): 371–82.

[16] Zamudio NM, Chong S, O'Bryan MK. Epigenetic regulation in male germ cells. Reproduction 2008;136(2):131–46.

[17] Nakao M. Epigenetics: interaction of DNA methylation and chromatin. Gene 2001;278(1–2):25–31.

[18] Thomson JP, et al. CpG islands influence chromatin structure via the CpG-binding protein Cfp1. Nature 2010;464(7291):1082–6.

[19] Bestor TH. Activation of mammalian DNA methyltransferase by cleavage of a Zn binding regulatory domain. EMBO J 1992;11(7):2611–7.

[20] Chédin F. The DNMT3 family of mammalian de novo DNA methyltransferases. Prog Mol Biol Transl Sci 2011;101:255–85.

[21] Bourc'his D, Bestor TH. Meiotic catastrophe and retrotransposon reactivation in male germ cells lacking Dnmt3L. Nature 2004;431(7004):96–9.

[22] Kaneda M, et al. Essential role for de novo DNA methyltransferase Dnmt3a in paternal and maternal imprinting. Nature 2004;429(6994):900–3.

[23] Molaro A, et al. Sperm methylation profiles reveal features of epigenetic inheritance and evolution in primates. Cell 2011;146(6):1029–41.

[24] Hammoud SS, et al. Chromatin and transcription transitions of mammalian adult germline stem cells and spermatogenesis. Cell Stem Cell 2014;15(2):239–53.

[25] Hammoud SS, et al. Distinctive chromatin in human sperm packages genes for embryo development. Nature 2009;460(7254):473–8.

[26] Li L, Lu X, Dean J. The maternal to zygotic transition in mammals. Mol Asp Med 2013;34(5):919–38.

[27] Jenkins TG, et al. Paternal aging and associated intraindividual alterations of global sperm 5-methylcytosine and 5-hydroxymethylcytosine levels. Fertil Steril 2013;100(4): 945–51.

[28] Kobayashi H, Kikyo N. Epigenetic regulation of open chromatin in pluripotent stem cells. Transl Res 2014;165(1):18–27.
[29] Peters AH, et al. Loss of the Suv39h histone methyltransferases impairs mammalian heterochromatin and genome stability. Cell 2001;107(3):323–37.
[30] Tachibana M, et al. Functional dynamics of H3K9 methylation during meiotic prophase progression. EMBO J 2007;26(14):3346–59.
[31] Buard J, et al. Distinct histone modifications define initiation and repair of meiotic recombination in the mouse. EMBO J 2009;28(17):2616–24.
[32] Malo AF, et al. Sperm design and sperm function. Biol Lett 2006;2(2):246–9.
[33] Aoki VW, et al. Protamine levels vary between individual sperm cells of infertile human males and correlate with viability and DNA integrity. J Androl 2006;27(6):890–8.
[34] Miller D, Brinkworth M, Iles D. Paternal DNA packaging in spermatozoa: more than the sum of its parts? DNA, histones, protamines and epigenetics. Reproduction 2010;139(2):287–301.
[35] Arpanahi A, et al. Endonuclease-sensitive regions of human spermatozoal chromatin are highly enriched in promoter and CTCF binding sequences. Genome Res 2009;19(8):1338–49.
[36] Castillo J, et al. Genomic and proteomic dissection and characterization of the human sperm chromatin. Mol Hum Reprod 2014;20(11):1041–53.
[37] Vavouri T, Lehner B. Chromatin organization in sperm may be the major functional consequence of base composition variation in the human genome. PLoS Genet 2011;7(4):e1002036.
[38] Brykczynska U, et al. Repressive and active histone methylation mark distinct promoters in human and mouse spermatozoa. Nat Struct Mol Biol 2010;17(6): 679–87.
[39] Erkek S, et al. Molecular determinants of nucleosome retention at CpG-rich sequences in mouse spermatozoa. Nat Struct Mol Biol 2013;20(7):868–75.
[40] Bernstein BE, et al. A bivalent chromatin structure marks key developmental genes in embryonic stem cells. Cell 2006;125(2):315–26.
[41] Wu S-F, Zhang H, Cairns BR. Genes for embryo development are packaged in blocks of multivalent chromatin in zebrafish sperm. Genome Res 2011;21(4):578–89.
[42] Pessot CA, et al. Presence of RNA in the sperm nucleus. Biochem Biophys Res Commun 1989;158(1):272–8.
[43] Kumar G, Patel D, Naz RK. c-MYC mRNA is present in human sperm cells. Cell Mol Biol Res 1993;39(2):111–7.
[44] Jodar M, et al. The presence, role and clinical use of spermatozoal RNAs. Hum Reprod Update 2013;19(6):604–24.
[45] Dadoune JP, et al. Identification of transcripts by macroarrays, RT-PCR and in situ hybridization in human ejaculate spermatozoa. Mol Hum Reprod 2005;11(2):133–40.
[46] Boerke A, Dieleman SJ, Gadella BM. A possible role for sperm RNA in early embryo development. Theriogenology 2007;68(Suppl. 1):S147–55.
[47] Sone Y, et al. Nuclear translocation of phospholipase C-zeta, an egg-activating factor, during early embryonic development. Biochem Biophys Res Commun 2005;330(3):690–4.
[48] Ostermeier GC, et al. Reproductive biology: delivering spermatozoan RNA to the oocyte. Nature 2004;429(6988):154.
[49] Stefani G, Slack FJ. Small non-coding RNAs in animal development. Nat Rev Mol Cell Biol 2008;9(3):219–30.
[50] Deng W, Lin H. miwi, a murine homolog of piwi, encodes a cytoplasmic protein essential for spermatogenesis. Dev Cell 2002;2(6):819–30.
[51] Kuramochi-Miyagawa S, et al. Mili, a mammalian member of piwi family gene, is essential for spermatogenesis. Development 2004;131(4):839–49.
[52] Carmell MA, et al. MIWI2 is essential for spermatogenesis and repression of transposons in the mouse male germline. Dev Cell 2007;12(4):503–14.
[53] Kumar M, et al. Novel insights into the genetic and epigenetic paternal contribution to the human embryo. Clinics (Sao Paulo) 2013;68(Suppl. 1):5–14.
[54] Aravin AA, et al. A piRNA pathway primed by individual transposons is linked to de novo DNA methylation in mice. Mol Cell 2008;31(6):785–99.
[55] Krawetz SA, et al. A survey of small RNAs in human sperm. Hum Reprod 2011;26(12):3401–12.
[56] Kawano M, et al. Novel small noncoding RNAs in mouse spermatozoa, zygotes and early embryos. PLoS One 2012;7(9):e44542.
[57] Salas-Huetos A, et al. New insights into the expression profile and function of micro-ribonucleic acid in human spermatozoa. Fertil Steril 2014;102(1):213–22.e4.
[58] Liu W-M, et al. Sperm-borne microRNA-34c is required for the first cleavage division in mouse. Proc Natl Acad Sci USA 2012;109(2):490–4.
[59] Oswald J, et al. Active demethylation of the paternal genome in the mouse zygote. Curr Biol 2000;10(8):475–8.
[60] Howell CY, et al. Genomic imprinting disrupted by a maternal effect mutation in the Dnmt1 gene. Cell 2001;104(6):829–38.
[61] Wossidlo M, et al. Dynamic link of DNA demethylation, DNA strand breaks and repair in mouse zygotes. EMBO J 2010;29(11):1877–88.
[62] Inoue A, Zhang Y. Replication-dependent loss of 5-hydroxymethylcytosine in mouse preimplantation embryos. Science 2011;334(6053):194.

[63] Wossidlo M, et al. 5-Hydroxymethylcytosine in the mammalian zygote is linked with epigenetic reprogramming. Nat Commun 2011;2:241.
[64] Hajkova P, et al. Chromatin dynamics during epigenetic reprogramming in the mouse germ line. Nature 2008;452(7189):877–81.
[65] Gu T-P, et al. The role of Tet3 DNA dioxygenase in epigenetic reprogramming by oocytes. Nature 2011;477(7366):606–10.
[66] Guo Y, et al. Characterization, isolation, and culture of mouse and human spermatogonial stem cells. J Cell Physiol 2014;229(4):407–13.
[67] Smith ZD, et al. A unique regulatory phase of DNA methylation in the early mammalian embryo. Nature 2012;484(7394):339–44.
[68] Tremblay KD, Duran KL, Bartolomei MS. A 5′ 2-kilobase-pair region of the imprinted mouse H19 gene exhibits exclusive paternal methylation throughout development. Mol Cell Biol 1997;17(8):4322–9.
[69] Lane N, et al. Resistance of IAPs to methylation reprogramming may provide a mechanism for epigenetic inheritance in the mouse. Genesis 2003;35(2):88–93.
[70] Hackett JA, Surani MA. DNA methylation dynamics during the mammalian life cycle. Philos Trans R Soc Lond B Biol Sci 2013;368(1609):20110328.
[71] Nakamura T, et al. PGC7 binds histone H3K9me2 to protect against conversion of 5mC to 5hmC in early embryos. Nature 2012;486(7403):415–9.
[72] Messerschmidt DM. Should I stay or should I go: protection and maintenance of DNA methylation at imprinted genes. Epigenetics 2012;7(9):969–75.
[73] Burton A, Torres-Padilla M-E. Epigenetic reprogramming and development: a unique heterochromatin organization in the preimplantation mouse embryo. Brief Funct Genomics 2010;9(5–6):444–54.
[74] McLay DW, Clarke HJ. Remodelling the paternal chromatin at fertilization in mammals. Reproduction 2003;125(5):625–33.
[75] Santos F, et al. Dynamic chromatin modifications characterise the first cell cycle in mouse embryos. Dev Biol 2005;280(1):225–36.
[76] Morgan HD, et al. Epigenetic reprogramming in mammals. Hum Mol Genet 2005;14 Spec No 1:R47–58.
[77] Dalman C. Advanced paternal age increases risk of bipolar disorder in offspring. Evid Based Ment Health 2009;12(2):59.
[78] Foldi CJ, et al. New perspectives on rodent models of advanced paternal age: relevance to autism. Front Behav Neurosci 2011;5:32.
[79] Naserbakht M, et al. Advanced paternal age is a risk factor for schizophrenia in Iranians. Ann Gen Psychiatry 2011;10:15.
[80] Smith RG, et al. Advancing paternal age is associated with deficits in social and exploratory behaviors in the offspring: a mouse model. PLoS One 2009;4(12):e8456.
[81] Miller B, et al. Meta-analysis of paternal age and schizophrenia risk in male versus female offspring. Schizophr Bull 2011;37(5):1039–47.
[82] Croen LA, et al. Maternal and paternal age and risk of autism spectrum disorders. Arch Pediatr Adolesc Med 2007;161(4):334–40.
[83] Frans EM, et al. Autism risk across generations: a population-based study of advancing grandpaternal and paternal age. JAMA Psychiatry 2013;70(5):516–21.
[84] Glasson EJ, et al. Perinatal factors and the development of autism: a population study. Arch Gen Psychiatry 2004;61(6):618–27.
[85] Reichenberg A, et al. Advancing paternal and maternal age are both important for autism risk. Am J Public Health 2010;100(5):772–3. Author reply 773.
[86] Reichenberg A, et al. Advancing paternal age and autism. Arch Gen Psychiatry 2006;63(9):1026–32.
[87] Frans EM, et al. Advancing paternal age and bipolar disorder. Arch Gen Psychiatry 2008;65(9):1034–40.
[88] Hemminki K, Kyyronen P, Vaittinen P. Parental age as a risk factor of childhood leukemia and brain cancer in offspring. Epidemiology 1999;10(3):271–5.
[89] Murray L, et al. Association of early life factors and acute lymphoblastic leukaemia in childhood: historical cohort study. Br J Cancer 2002;86(3):356–61.
[90] Wiener-Megnazi Z, Auslender R, Dirnfeld M. Advanced paternal age and reproductive outcome. Asian J Androl 2012;14(1):69–76.
[91] Alio AP, et al. The effect of paternal age on fetal birth outcomes. Am J Mens Health 2012;6(5):427–35.
[92] Thompson RF, et al. Tissue-specific dysregulation of DNA methylation in aging. Aging Cell 2010;9(4):506–18.
[93] Jenkins TG, et al. Age-associated sperm DNA methylation alterations: possible implications in offspring disease susceptibility. PLoS Genet 2014;10(7):e1004458.
[94] Richardson B. Impact of aging on DNA methylation. Ageing Res Rev 2003;2(3):245–61.
[95] Jung M, Pfeifer GP. Aging and DNA methylation. BMC Biol 2015;13:7.
[96] Milekic MH, et al. Age-related sperm DNA methylation changes are transmitted to offspring and associated with abnormal behavior and dysregulated gene expression. Mol Psychiatry 2014;20(8):995–1001.
[97] Carone BR, et al. Paternally induced transgenerational environmental reprogramming of metabolic gene expression in mammals. Cell 2010;143(7):1084–96.

[98] Govorko D, et al. Male germline transmits fetal alcohol adverse effect on hypothalamic proopiomelanocortin gene across generations. Biol Psychiatry 2012;72(5):378–88.
[99] Ost A, et al. Paternal diet defines offspring chromatin state and intergenerational obesity. Cell 2014;159(6):1352–64.
[100] Ng SF, et al. Chronic high-fat diet in fathers programs beta-cell dysfunction in female rat offspring. Nature 2010;467(7318):963–6.
[101] Kaati G, et al. Transgenerational response to nutrition, early life circumstances and longevity. Eur J Hum Genet 2007;15(7):784–90.
[102] Pembrey ME, et al. Sex-specific, male-line transgenerational responses in humans. Eur J Hum Genet 2006;14(2):159–66.
[103] Secretan B, et al. A review of human carcinogens–Part E: tobacco, areca nut, alcohol, coal smoke, and salted fish. Lancet Oncol 2009;10(11):1033–4.
[104] Bielawski DM, Abel EL. Acute treatment of paternal alcohol exposure produces malformations in offspring. Alcohol 1997;14(4):397–401.
[105] Gabrielli Jr WF, Mednick SA. Intellectual performance in children of alcoholics. J Nerv Ment Dis 1983;171(7):444–7.
[106] Soares SR, Melo MA. Cigarette smoking and reproductive function. Curr Opin Obstet Gynecol 2008;20(3):281–91.
[107] Sofikitis N, et al. Effects of smoking on testicular function, semen quality and sperm fertilizing capacity. J Urol 1995;154(3):1030–4.
[108] Taha EA, et al. Effect of smoking on sperm vitality, DNA integrity, seminal oxidative stress, zinc in fertile men. Urology 2012;80(4):822–5.
[109] Luca G, et al. Toxicity of cadmium on Sertoli cell functional competence: an in vitro study. J Biol Regul Homeost Agents 2013;27(3):805–16.
[110] Ji YL, et al. Crosstalk between endoplasmic reticulum stress and mitochondrial pathway mediates cadmium-induced germ cell apoptosis in testes. Toxicol Sci 2011;124(2):446–59.
[111] Esakky P, et al. Cigarette smoke condensate induces aryl hydrocarbon receptor-dependent changes in gene expression in spermatocytes. Reprod Toxicol 2012;34(4):665–76.
[112] Marchetti F, et al. Sidestream tobacco smoke is a male germ cell mutagen. Proc Natl Acad Sci USA 2011;108(31):12811–4.
[113] Polyzos A, et al. Differential sensitivity of male germ cells to mainstream and sidestream tobacco smoke in the mouse. Toxicol Appl Pharmacol 2009;237(3):298–305.
[114] Glen CD, Dubrova YE. Exposure to anticancer drugs can result in transgenerational genomic instability in mice. Proc Natl Acad Sci USA 2012;109(8):2984–8.
[115] Dubrova YE, et al. Paternal exposure to ethylnitrosourea results in transgenerational genomic instability in mice. Environ Mol Mutagen 2008;49(4):308–11.
[116] Barber RC, Dubrova YE. The offspring of irradiated parents, are they stable? Mutat Res 2006;598(1–2):50–60.
[117] Dubrova YE. Radiation-induced transgenerational instability. Oncogene 2003;22(45):7087–93.
[118] Barber R, et al. Elevated mutation rates in the germ line of first- and second-generation offspring of irradiated male mice. Proc Natl Acad Sci USA 2002;99(10):6877–82.
[119] Joo KJ, et al. The effects of smoking and alcohol intake on sperm quality: light and transmission electron microscopy findings. J Int Med Res 2012;40(6):2327–35.
[120] Anderson Jr RA, et al. Ethanol-induced male infertility: impairment of spermatozoa. J Pharmacol Exp Ther 1983;225(2):479–86.
[121] Liang F, et al. Paternal ethanol exposure and behavioral abnormities in offspring: associated alterations in imprinted gene methylation. Neuropharmacology 2014;81:126–33.
[122] Knezovich JG, Ramsay M. The effect of preconception paternal alcohol exposure on epigenetic remodeling of the h19 and rasgrf1 imprinting control regions in mouse offspring. Front Genet 2012;3:10.
[123] Stouder C, Somm E, Paoloni-Giacobino A. Prenatal exposure to ethanol: a specific effect on the H19 gene in sperm. Reprod Toxicol 2011;31(4):507–12.
[124] Kim P, et al. Chronic exposure to ethanol of male mice before mating produces attention deficit hyperactivity disorder-like phenotype along with epigenetic dysregulation of dopamine transporter expression in mouse offspring. J Neurosci Res 2014;92(5):658–70.
[125] Marczylo EL, et al. Smoking induces differential miRNA expression in human spermatozoa: a potential transgenerational epigenetic concern? Epigenetics 2012;7(5):432–9.
[126] Lazarevic K, et al. Determination of lead and arsenic in tobacco and cigarettes: an important issue of public health. Cent Eur J Public Health 2012;20(1):62–6.
[127] Wiwanitkit V. Lead and infertility. Int J Occup Environ Med 2011;2(3):182. Author reply 182–3.
[128] Paoloni-Giacobino A. Epigenetic effects of methoxychlor and vinclozolin on male gametes. Vitam Horm 2014;94:211–27.
[129] Guerrero-Bosagna C, et al. Epigenetic transgenerational inheritance of vinclozolin induced mouse adult onset disease and associated sperm epigenome biomarkers. Reprod Toxicol 2012;34(4):694–707.

[130] Guerrero-Bosagna C, et al. Epigenetic transgenerational actions of vinclozolin on promoter regions of the sperm epigenome. PLoS One 2010;5(9).
[131] Stouder C, Paoloni-Giacobino A. Transgenerational effects of the endocrine disruptor vinclozolin on the methylation pattern of imprinted genes in the mouse sperm. Reproduction 2010;139(2):373–9.
[132] Stouder C, Paoloni-Giacobino A. Specific transgenerational imprinting effects of the endocrine disruptor methoxychlor on male gametes. Reproduction 2011;141(2):207–16.
[133] Waterland RA, Jirtle RL. Transposable elements: targets for early nutritional effects on epigenetic gene regulation. Mol Cell Biol 2003;23(15):5293–300.
[134] Dolinoy DC, Huang D, Jirtle RL. Maternal nutrient supplementation counteracts bisphenol A-induced DNA hypomethylation in early development. Proc Natl Acad Sci USA 2007;104(32):13056–61.
[135] Dolinoy DC, et al. Maternal genistein alters coat color and protects *Avy* mouse offspring from obesity by modifying the fetal epigenome. Environ Health Perspect 2006;114(4):567–72.

CHAPTER

19

Linkage between In Utero Environmental Changes and Preterm Birth

Markus Velten[1], Lynette K. Rogers[2]

[1]Department of Anesthesiology and Intensive Care Medicine, Rheinische Friedrich-Wilhelms-University Bonn, Germany; [2]Center for Perinatal Research, The Research Institute at Nationwide Children's Hospital, Columbus, Ohio

OUTLINE

INTRODUCTION

Preterm birth is the leading cause of child death in the Western world and occurs in approximately 11% of pregnancies (estimates vary by country) despite preventive maternal medical checkups, prenatal care programs, and new work policies protecting pregnant women from being exposed to potential hazards. Furthermore, preterm birth is associated with factors such as adverse lifestyle and behaviors that include stress, smoking, drugs, nutrition, and living environment [1]. Direct

The Epigenome and Developmental Origins of Health and Disease
http://dx.doi.org/10.1016/B978-0-12-801383-0.00019-0

Copyright © 2016 Elsevier Inc. All rights reserved.

correlations between the maternal environment and preterm birth have thus far been difficult to establish for many reasons. First, the etiology of preterm birth is complex and even after decades of research is still poorly understood. Second, it is impossible to single out an individual factor in human studies because of the complexity of our living environment. Even with these difficulties considered, large population studies have strengthened the associations with the maternal environmental exposures, health, and nutrition and the risk for preterm birth [1–6].

Epigenetic changes can be identified as DNA methylation, changes in expression levels of microRNAs, and modifications on histone residues [7]. Current studies have identified DNA methylation as the modification most often associated with adverse environmental exposures, although others have not been extensively investigated. DNA methylation can be either dynamic or permanent [8]. Dynamic DNA methylation occurs under the regulation of normal cellular physiology to "turn genes off or on" at appropriate times. Permanent DNA methylation is pathologic and occurs when a stimulus causes activation of methyltransferase enzymes to methylate regions of the genome not under dynamic regulation or suppress demethylase or repair enzymes. Many environmental exposures have been linked to permanent DNA methylation and thus are associated with disease pathologies. While epidemiological studies have identified associations between environmental exposures and DNA methylation, there are few investigations into how these events can or will act together to cause preterm birth and promote disease susceptibility later in life.

ENVIRONMENTAL EXPOSURES

The consequences of prenatal exposure to environmental chemicals on the intrauterine environment and subsequently the developing fetus have not been fully appreciated in the context of preterm birth. Since the beginnings of the industrial revolution, manufacturing, mining, petrochemical refining, the use of combustion engines have "dumped" large quantities of chemical pollutants into the air, the ground, and drinking water. While the in utero environment plays a significant role in the health and development of the fetus, there is increasing evidence that the external environment of the mother also effects the intrauterine environment, impacts the developing fetus, and influences long-reaching epigenetic changes [7,9].

The three major classes of environmental exposures are: (1) metals, including arsenic, lead, and cadmium; (2) air pollution, including particulates, chemicals, and complex mixtures including direct and secondhand cigarette smoke; and (3) endocrine disrupters, including polychlorinated biphenyls (PCBs) and tetrachlorodibenzo-*p*-dioxin (TCDD), and bisphenol A (BPA). Exposures can occur at home both indoors and outdoors, at the workplace, or in the community and can arise from interactions with the soil, water, air, and/or food, making them impossible to avoid. In fact, some of the most concerning exposures can occur preconception (causing epigenetic changes that impact germ cells) [10] or during the first six to eight weeks of pregnancy, prior to the mother's knowledge of conception. Environmental exposures can result in fetal loss, preterm birth, small for gestational age, or birth anomalies [1,11].

CHEMICALS

Metals

Exposure to inorganic metals occurs daily through drinking water or contact with the soil. Under most circumstances, exposures to these small concentrations of metals are far below the levels known to cause health concerns even during pregnancy. They are, however, a serious matter in areas where contamination exists, such as water supplies in third-world countries and areas polluted with industrial or mining wastes. Arsenic, lead, and cadmium are the most studied metal contaminants and all present serious health risks if found in toxic quantities. Arsenic has been shown to readily cross the placenta and has been associated with low birth weight, spontaneous abortions, malformations, and stillbirths [12]. Studies have not directly linked arsenic exposure to preterm birth, however, exposure to arsenic has been strongly associated with global hypomethylation of DNA, which would result in misregulation of a multitude of genes [13].

Lead, another environmental contaminant, also readily crosses the placenta, and urinary concentrations of 10 μg/dl have been associated with a threefold risk or preterm birth and a fourfold risk for intrauterine growth retardation and babies born small for gestational age [14]. Pilsner et al. identified an inverse association between cumulative lead exposure and genomic DNA methylation [15]. While cadmium exposure is most often linked to cigarette smoking (discussed in Section Environmental Exposures), exposure from the external environment does occur by inhalation of air polluted with combustion engine exhausts. Cadmium causes direct changes in DNA methyltransferase activity and thus is linked to DNA methylation [16]. Studies by Sanders et al. have indicated that while acute exposure to cadmium induces hypomethylation much like lead exposure, chronic cadmium exposure, a much more serious human health threat, is linked to hypermethylation and gene silencing [16]. In utero exposures have been associated with low birth weight, but no studies have linked cadmium exposure to preterm birth.

Polycyclic Aromatic Hydrocarbons

Polycyclic aromatic hydrocarbon (PAH) exposures result from environmental contamination through combustion sources, however, diet is the main route of exposure [17]. Grilled, fried, smoked, or roasted meats all have measurable levels of PAHs and in combination with other sources, such as air pollution or cigarette smoke, can constitute a substantial exposure [3]. PAHs are also components of particulate matter and as such are proposed to exacerbate the toxicities associated with exposures to air pollution. Specifically, benzo(a)pyrene has been shown to cross the placenta and has been detected in cord blood, indicating a potential for a direct impact on fetal growth and gestational length [18–20]. While associations between late gestation exposure to PAHs and preterm birth have been identified, the link between PAH exposure and epigenetic changes is not well established [2]. Choi et al. have identified a strong link between PAH exposure and small for gestational age or preterm birth specifically in African American populations [21]. Mechanistic links between PAH exposure and low birth weight or preterm birth have not been identified, but researchers have speculated that the formation of DNA adducts and induction of apoptosis, the antiestrogenic effects of PAHs, or the activation of cytochrome P450 enzymes may limit oxygen and nutrients for growth [3,21]. Others have observed DNA methylation changes in cord blood that correlates to PAH exposure [22].

Endocrine Disrupters

Another class of environmental exposures is the endocrine-disrupting chemicals (EDCs). These are exogenous substances that interfere with hormone synthesis, metabolism, or action, and several have been shown to induce epigenetic changes [23]. The most extensively studied EDCs are the polychlorinated biphenyls (PCBs) and tetrachlorodibenzo-*p*-dioxin (TCDD), which are the active ingredients in several organochlorine pesticides. PCBs have been shown to interrupt androgen receptor signaling by interfering with histone demethylation [24].

PCBs including dichlorodiphenyltrichloroethane (DDT) and its metabolite dichlorodiphenyldichloroethane (DDE) have long half-lives and have been detected in human milk, serum, and tissues [25,26]. Tetrachlorodibenzo-*p*-dioxin (TCDD), another polychlorinated compound, is widely found in the environment. Multiple animal studies have indicated that TCDD is linked to decreased fertility and preterm birth, and these changes persist for multiple generations indicating a nonrepairable epigenetic modification [27,28]. Interestingly, epidemiologic evidence has not supported the association between TCDD and adverse birth outcomes, however, the numbers included in these studies were small. Larger and more extensive studies need to be conducted to eliminate the potential risk of TCDD on adverse birth outcomes [29,30].

Together the data are mixed with studies indicating no significant findings between EDCs and preterm birth, while other studies (often larger in number) have identified associations between PCB or DDE exposures and preterm births or stillbirths [30–35]. Huen et al. investigated DDT/E exposure in utero using cord blood and linked these levels to methylation of a noncoding RNA as a surrogate for global methylation [36]. If these findings are supported with future studies, exposure to EDCs could precipitate multigenerational adverse pregnancy outcomes.

Bisphenol A (BPA), a component of plastics until recently found in formula bottles, has become a topic of interest. BPA has proestrogenic properties and as such has been studied in the context of male feminization [37]. While animal and cell studies have been compelling, and have identified metabolic changes, human data have not supported these findings [38]. There are few epidemiological studies that link BPA and adverse birth outcomes or epigenetic modifications in humans [39,40]. However, most vendors now abstain from the use of bisphenol A for the production of formula bottles.

PARTICLES/AIR POLLUTION

Industrial and Combustion Engine Air Pollution

Air pollution encompasses a wide variety of environmental contributors from combustion engines, factory exhausts, farming, and cigarette smoking. In the context of these complex chemical mixtures, it becomes difficult to decipher which individual molecules are evoking toxic responses and which are working in synergy to create unique exposures. Yet, in total, air pollution has been strongly associated with adverse birth outcomes [41,42]. The chemical components of air pollution—carbon monoxide (CO), nitrogen dioxide (NO_2), and sulfur dioxide (SO_2)—are often found in diesel exhaust and factory emissions. Their influence on pregnancy outcomes has been linked to timing of gestation [43–45]. Leem et al. found that in a Korean cohort, exposures to CO, NO_2, and SO_2 during the first trimester were linked with preterm birth while exposures to CO and NO_2 during the third trimester were also linked to preterm birth [43]. Others have found that first-trimester exposure to ozone as well as NO_2 is associated with preterm birth and development of preeclampsia [44].

In addition to the chemicals found in air pollution, there is a substantial particulate component. The composition of these particulates varies but is often divided into particulate matter <10 μm in size (PM_{10}) and particulate matter <2.5 μm in size ($PM_{2.5}$). Both particle sizes have been studied independently, and both have been linked to both low birth weight and preterm birth [2]. Human epidemiological studies have identified gestational time frames where the fetus is more susceptible to particulate exposure [43,44,46,47]. Symanski et al. identified an association with exposure to $PM_{2.5}$ during the first four weeks of pregnancy and an increased risk of mildly (33–36 completed weeks of gestation), moderately (29–32 weeks of gestation), or severely (20–28 weeks of gestation) preterm birth by 16, 71, and 73%, respectively [48]. Other recent investigations have identified similar correlations between $PM_{2.5}$ exposure and preterm birth implying that the smaller particles may actually enter in the blood stream and impact the placenta in an adverse manner initiating the birth process [42,49–51]. Furthermore, most PMs contain one or more PAHs, which further compounds the complexity of the associations and likely contributes to preterm birth in either an additive or synergistic manner [2,52]. While the impact of air pollution on preterm birth is well established, the epigenetic impact of these exposures has not been extensively studied, and it is difficult to tease apart the relative contributions of the air pollution versus the influences of interventional care for a preterm infant.

Tobacco Smoke

Tobacco smoke offers another complex mixture that has been highly linked to both small for gestational age and preterm birth [53,54]. Multiple studies have identified associations between both active smoking and passive secondhand smoke exposure to DNA methylation in the placenta and in the offspring [55,56]. There is accumulating evidence that the effects of early DNA methylation may include development of asthma [57]. Maccani et al. identified the *RUNX3* gene in the placenta as a methylation target for tobacco smoking and methylation of *RUNX3* was directly correlated with gestational length [53]. Methylation of noncoding RNAs such as LINE1 and AluY68 has been observed in cord blood and may serve as biomarkers of more global methylation [58]. Subsets of genes have been analyzed for the independent effects of the components of tobacco smoke including cotinine and cadmium. DNA methylation patterns were distinct between maternal and fetal DNA but included biological pathways that control transcriptional regulation and apoptosis [13]. Few studies have focused on other forms of epigenetic modifications in relation to tobacco smoke, but Maccani et al. found changes in placental microRNA expression with specific effects on miR-146a, a regulator of inflammation, specifically TNF and IL-1 [56]. Furthermore, effects of acrolein, a component of tobacco smoke, have been shown to inhibit acetylation of histones 3 and 4 [59]. Most recently, investigations into transgenerational effects of tobacco smoke exposure have been implicated. Marczylo et al. have found that tobacco smoke alters the expression profiles of microRNAs in human spermatozoa [56].

MATERNAL HEALTH

Increasing evidence supports the observation that the intrauterine environment substantially impacts the developing fetus. An adverse intrauterine environment alters fetal development through changes in placental and fetal endocrine functions and activation of the hypothalamic-pituitary adrenal (HPA) axis. These alterations in fetal development play a major role in the so-called "fetal origin of adult disease" hypothesis [60,61]. Genetic and environmental influences impact the intrauterine environment, affecting development, growth, and survival [62]. Aberrant tissue or organ development may result in disease development later in life. Maternal nutritional status has a substantial impact on overall maternal health and subsequently the intrauterine environment, thereby affecting the developing fetus [63]. Maternal nutritional deficiencies lead to fetal deficiencies, resulting in intrauterine growth restriction (IUGR), impaired fetal development, neural defects, impaired mental status, and preterm birth [64]. The fetus is most vulnerable to maternal malnutrition when the fetus implants in early gestation. This implantation period is a time of rapid cell division and placental growth. A deficit of essential nutrients during this crucial time period has long-lasting effects for the developing fetus [63]. The fetus competes with the mother for nutrients, and fetal undernutrition can occur if the nutritional intake is not sufficient. Insufficient intake can be a result of low or imbalanced consumption as well as from increased maternal demand such as when the mother is an adolescent and still requires extra nutrients for her own growth. Preterm birth and low birth weight are significantly increased in adolescent pregnancies compared to adult pregnancies [63].

Obesity

Obesity rates have increased and have become epidemic in the Western world over the last decades. Currently overweight and obesity is considered a public health threat, jointly responsible for the development of various illnesses including cardiovascular and metabolic diseases like stroke, coronary artery disease, or diabetes. During gestation and labor, maternal weight and weight gain is an important factor for maternal health and fetal outcome [65]. Maternal obesity is increasingly a common risk factor for the development of gestational complications. In the United States, up to 40% of pregnant women are considered overweight or obese, and up to 33% are overweight or obese in Great Britain [66,67]. Both, maternal obesity and maternal underweight during pregnancy are associated with poor maternal as well as fetal outcomes. The continuum of maternal overweight and obesity has become the most common threat during pregnancy in the Western world and some developing countries. Epidemiological observations revealed that maternal overweight, defined by a BMI of ≥30 kg/m^2 during the course of pregnancy, is associated with increased rates of gestational diabetes, hypertension, cesarean delivery, shoulder dystocia, as well as premature rupture of membranes. Furthermore, fetal/neonatal outcomes like low APGAR scores, the requirement of neonatal intensive care treatment, hypoglycemia, macrosomia rates, and most importantly perinatal mortality are closely correlated with maternal obesity. In contrast, reduced birth weight is correlated with low maternal BMI rates during pregnancy [68]. The correlation between maternal obesity and maternal and fetal complication exists independent of maternal weight gain during pregnancy, indicating the importance of absolute body weight for pregnancy outcomes.

Even though various studies report a strong association between maternal obesity and maternal as well as fetal complications, the impact of maternal obesity on preterm birth has not been defined. Maternal overweight and obesity has been associated with increased, decreased, or neutral risk for preterm birth [69–71]. To determine the impact of maternal overweight and obesity on maternal health and preterm birth and low birth weight in singleton pregnancies, McDonald and coworkers performed a systematic, comprehensive, and unbiased review and meta-analyses in 2010, with a reference group of normal weight women [72]. By analyzing a total of 64 cohort and 20 case–control studies, including a total of 1,095,834 pregnant women, overweight and obesity were identified as being risk factors for preterm birth before 32 and 37 weeks of gestation. Therefore, overweight appears to be associated with overall preterm birth. In contrast to the deleterious effects of overweight on the duration of gestation, higher body weights appear to be beneficial in preventing low birth weight. This effect seems to be greater in developing countries and was not seen when studies were biased toward the Western world [72]. However, this correlation has not been evaluated in the context of the development of diseases later in life. In contrast to other causes of preterm birth, maternal overweight has the potential to be treated and thereby an option to prevent the leading cause of neonatal mortality and morbidity in childhood. Currently, consultation during pregnancy is recommended for overweight or obese women to advise them of their perinatal risk, specifically of preterm birth. Weight optimization and proper nutrition control may simply prevent many preterm births. Furthermore, overweight women should be closely monitored during pregnancy for signs of preterm labor and maternal and fetal health [73].

Diabetes

Maternal diabetes was first mentioned in 1824, when Heinrich Gottlieb Bennewitz defended his thesis "Diabetes Mellitus: A Symptom of Pregnancy" at the University of Berlin [74]. His thesis was a simple case report and literature review on the causes and treatment of diabetes at that time. Matthew Duncan published the initial study reporting severe maternal and fetal outcomes in maternal gestational diabetes in 1884. He reviewed the pregnancies of 15 diabetic mothers and reported death in 13 out of 19 fetuses. Furthermore, 9 out of these 15 diabetic mothers died of diabetes within a year after they gave birth. Later in 1957, the term *gestational diabetes mellitus* (GDM) was defined by Carrington describing carbohydrate intolerance during pregnancy that resolves after birth [75].

Estrogen, prolactin, and lactogen [76] are placental hormones that transport nutrients from the mother to the fetus, thereby protecting the fetus from undernutrition and low blood glucose levels. During the course of pregnancy, insulin resistance commonly develops. If the maternal pancreas is unable to increase insulin production accordingly, insulin resistance results with elevated blood glucose levels. This pathophysiology in GDM is coherent to type 2 diabetes. In GDM, increased maternal glucose concentrations are transferred to the fetus. Within the fetal organism, elevated blood glucose in turn stimulates the insulin production of the fetal pancreatic beta cells.

In the Western world, the incidence of GDM has continuously increased over the last decades and is the most common maternal complication during pregnancy [77]. Despite new medical strategies and improved preventive care, gestational diabetes mellitus remains a serious medical condition for the developing fetus and is associated with metabolic and physiologic complications. Studies in type 1 diabetic mothers report a correlation between unstable glucose levels during the first trimester and increased risk for neurological and cardiac defects in the offspring, while unstable diabetes during the third trimester is associated with intrauterine growth restriction. Infants born to mothers that suffered from severe type 2 diabetes are more commonly macrosomic and are subsequently obese later in life, and they are also more likely to develop cardiovascular and metabolic diseases as adults [78]. These well-documented relationships led to the preventive evaluation for gestational diabetes and therapeutic strategies for most pregnant women. The association between gestational diabetes and long-term fetal consequences is well established. Furthermore, various studies document the association between preterm birth and the development of diabetes later in life. However, studies investigating the impact of maternal diabetes on preterm birth are elusive. By analyzing registries from Norway, Eidem and coworkers identified 1307 births among women with pregestational type 1 diabetes [79]. Twenty-six of these 1307 women that were diagnosed with type 2 diabetes delivered preterm of which 42% delivered spontaneously preterm. Compared to the 7% preterm birth rate in the control population, the rate of preterm deliveries was much higher in the type 1 diabetes group. Based on these data and the emphasis that preterm delivery is the most important single factor in perinatal mortality, Eidem and coworkers speculate that an excess risk of perinatal infant death in women with diabetes could be due to preterm births [79]. Further studies are needed to verify this hypothesis. However, improving maternal nutrition and optimizing maternal metabolic conditions may be the key to preventing the detrimental effects of maternal diabetes on fetal development and preterm birth.

Nutrition

Current knowledge implies that maternal nutrition affects the developing fetus. Malnutrition substantially impacts maternal, gestational, and pregnancy outcome leading to spontaneous, noninfectious preterm birth [63]. Of particular importance seems to be the time window. A mismatch between needs and supplies can occur preconception or during early or late pregnancy with different effects for mother and fetus. Malnutrition includes (1) poor food resources, more commonly seen in developing countries, (2) overweight due to poor diet, a problem with increasing incidence in the Western world, or (3) lack of essential food components like micronutrients. Various maternal conditions like age, height, weight, and interpregnancy duration influence the nutritional status. Adolescent mothers are shorter, lighter, and have a reduced energy and micronutrients availability, much like women with short interpregnancy durations. This results in less fetal access to nutrients, impaired placental and fetal growth, and shorter gestational duration [64].

Low nutrient availability and energy is correlated with lower pregnancy duration. Rayco-Solon and coworkers reported shorter gestational duration in Gambian women who conceived between September and November, which was correlated with reduced maternal nutrients and energy due to annual food fluctuation, compared to pregnancies that started when more food and energy was available [80]. However, not only a lack of prepregnancy nutrients but excess nutrient supply may be associated with lower pregnancy duration. Even though the data for the effect of maternal overweight on preterm birth are inconclusive, Han and coworkers reported that a high maternal body mass index before pregnancy increases the risk for preterm birth [81]. Collectively, these studies indicate that maternal food consumption and weight preconception during the early stages of pregnancy impact fetal development and is associated with lower pregnancy duration and preterm birth.

In addition to the impact of quantitative nutritional supply and nutrition quality, dietary pattern also impacts the fetus and gestational duration. Evaluating associations between maternal dietary patterns before conception and preterm delivery, Grieger and coworkers reported that protein- and fruit-rich diets increased gestational duration, while diets consisting of high fat, sugar, and takeaway food are associated with shorter gestational duration and preterm birth [64]. While elevated preconceptional homocysteine levels are associated with preterm birth, the supplementation of various micronutrients seems to be preventive. Increased vitamin B 12 serum levels and preconceptional folate supplementation have been reported to prevent preterm birth. Furthermore, recommended vitamin supplementation for pregnant women reduces the risk for preterm birth.

In summary, there is evidence supporting that qualitative and quantitative maternal nutritional status impacts placental growth and the developing fetus with direct influence on pregnancy outcome and preterm birth. Of particular importance seems to be the duration of nutritional effects. While lack of nutrients both during preconception and early pregnancy is associated with preterm birth, an excessive supply in late pregnancy results in macrosomia and also preterm birth. Furthermore, there are emerging concerns regarding maternal overweight and obesity.

CONCLUSIONS

The external and internal environments both influence the pregnancy outcome as well as the health of the offspring. Exposure to environmental chemicals and nutritional variances has been reported to promote not just preterm birth but also transgenerational diseases that are associated with epigenetic modifications. Current studies have identified changes in methylation status as the primary target for external exposures on the development of disease later in life, but most of these investigations have been descriptive and observational. Future studies need to address causal relationships and how or whether these changes in methylation that are involved in disease development later in life are also involved in the preterm birth itself. Recently, newer investigations are beginning to identify changes in microRNAs as well as histone modification, but these findings are currently being validated. The long-term effects of being born prematurely are currently being discovered and include increases in metabolic dysfunction such as insulin resistance and hypertension. While it is difficult to tease apart the effects of adverse maternal health and simply being born preterm, the data would indicate that these influences may be synergistic or at least additive and are responsible for epigenetic changes that persist at least through one generation and maybe more.

Glossary Terms

BPA Bisphenol A
CO Carbon monoxide
DDE Dichlorodiphenyldichloroethane
DDT Dichlorodiphenyltrichloroethane
EDC Endocrine-disrupting chemical
GDM Gestational diabetes
IUGR Intrauterine growth restriction
NO_2 Nitrogen dioxide
PAH Polycyclic aromatic hydrocarbon
PCBs Polychlorinated biphenyls
PM Particulate matter
SO_2 Sulfur dioxide
TCDD Tetrachlorodibenzo-*p*-dioxin

References

[1] Murphy DJ. Epidemiology and environmental factors in preterm labour. Best Pract Res Clin Obstet Gynaecol 2007;21(5):773–89.
[2] Padula AM, Noth EM, Hammond SK, et al. Exposure to airborne polycyclic aromatic hydrocarbons during pregnancy and risk of preterm birth. Environ Res 2014;135C:221–6.
[3] Patelarou E, Kelly FJ. Indoor exposure and adverse birth outcomes related to fetal growth, miscarriage and prematurity—a systematic review. Int J Environ Res Public Health 2014;11(6):5904–33.
[4] Sram RJ, Binkova B, Dejmek J, Bobak M. Ambient air pollution and pregnancy outcomes: a review of the literature. Environ Health Perspect 2005;113(4):375–82.
[5] Bobak M. Outdoor air pollution, low birth weight, and prematurity. Environ Health Perspect 2000;108(2):173–6.
[6] Dadvand P, Figueras F, Basagana X, et al. Ambient air pollution and preeclampsia: a spatiotemporal analysis. Environ Health Perspect 2013;121(11–12):1365–71.
[7] Baccarelli A, Bollati V. Epigenetics and environmental chemicals. Curr Opin Pediatr 2009;21(2):243–51.
[8] Edwards TM, Myers JP. Environmental exposures and gene regulation in disease etiology. Environ Health Perspect 2007;115(9):1264–70.
[9] Patel CJ, Yang T, Hu Z, et al. Investigation of maternal environmental exposures in association with self-reported preterm birth. Reprod Toxicol 2014;45:1–7.
[10] Laubenthal J, Zlobinskaya O, Poterlowicz K, et al. Cigarette smoke-induced transgenerational alterations in genome stability in cord blood of human F1 offspring. FASEB J 2012;26(10):3946–56.
[11] Harrison E, Partelow J, Grason H. Environmental toxicants and maternal and child health: an emerging public health challenge. John Hopkins Bloomberg School of Public Health; 2009.
[12] Hopenhayn C, Ferreccio C, Browning SR, et al. Arsenic exposure from drinking water and birth weight. Epidemiology 2003;14(5):593–602.
[13] Guerrero-Preston R, Goldman LR, Brebi-Mieville P, et al. Global DNA hypomethylation is associated with in utero exposure to cotinine and perfluorinated alkyl compounds. Epigenetics 2010;5(6):539–46.
[14] Jelliffe-Pawlowski LL, Miles SQ, Courtney JG, Materna B, Charlton V. Effect of magnitude and timing of maternal pregnancy blood lead (Pb) levels on birth outcomes. J Perinatol 2006;26(3):154–62.
[15] Pilsner JR, Hu H, Ettinger A, et al. Influence of prenatal lead exposure on genomic methylation of cord blood DNA. Environ Health Perspect 2009;117(9):1466–71.
[16] Sanders AP, Smeester L, Rojas D, et al. Cadmium exposure and the epigenome: exposure-associated patterns of DNA methylation in leukocytes from mother-baby pairs. Epigenetics 2014;9(2):212–21.
[17] Bostrom CE, Gerde P, Hanberg A, et al. Cancer risk assessment, indicators, and guidelines for polycyclic aromatic hydrocarbons in the ambient air. Environ Health Perspect 2002;110(Suppl. 3):451–88.
[18] Pedersen M, Stayner L, Slama R, et al. Ambient air pollution and pregnancy-induced hypertensive disorders: a systematic review and meta-analysis. Hypertension 2014;64(3):494–500.
[19] Singh VK, Patel DK, Jyoti, Ram S, Mathur N, Siddiqui MK. Blood levels of polycyclic aromatic hydrocarbons in children and their association with oxidative stress indices: an Indian perspective. Clin Biochem 2008;41(3):152–61.
[20] Jedrychowski WA, Perera FP, Tang D, et al. The relationship between prenatal exposure to airborne polycyclic aromatic hydrocarbons (PAHs) and PAH-DNA adducts in cord blood. J Expo Sci Environ Epidemiol 2013;23(4):371–7.
[21] Choi H, Rauh V, Garfinkel R, Tu Y, Perera FP. Prenatal exposure to airborne polycyclic aromatic hydrocarbons and risk of intrauterine growth restriction. Environ Health Perspect 2008;116(5):658–65.
[22] Herbstman JB, Tang D, Zhu D, et al. Prenatal exposure to polycyclic aromatic hydrocarbons, benzo[a]pyrene-DNA adducts, and genomic DNA methylation in cord blood. Environ Health Perspect 2012;120(5):733–8.
[23] Fudvoye J, Bourguignon JP, Parent AS. Endocrine-disrupting chemicals and human growth and maturation: a focus on early critical windows of exposure. Vitamins Hormones 2014;94:1–25.

[24] Casati L, Sendra R, Poletti A, Negri-Cesi P, Celotti F. Androgen receptor activation by polychlorinated biphenyls: epigenetic effects mediated by the histone demethylase Jarid1b. Epigenetics 2013;8(10):1061–8.

[25] Khanjani N, Sim MR. Maternal contamination with PCBs and reproductive outcomes in an Australian population. J Expo Sci Environ Epidemiol 2007;17(2):191–5.

[26] Bergonzi R, De Palma G, Specchia C, et al. Persistent organochlorine compounds in fetal and maternal tissues: evaluation of their potential influence on several indicators of fetal growth and health. Sci Total Environ 2011;409(15):2888–93.

[27] Ding T, McConaha M, Boyd KL, Osteen KG, Bruner-Tran KL. Developmental dioxin exposure of either parent is associated with an increased risk of preterm birth in adult mice. Reprod Toxicol 2011;31(3):351–8.

[28] Bruner-Tran KL, Osteen KG. Developmental exposure to TCDD reduces fertility and negatively affects pregnancy outcomes across multiple generations. Reprod Toxicol 2011;31(3):344–50.

[29] Longnecker MP, Klebanoff MA, Brock JW, Guo X. Maternal levels of polychlorinated biphenyls in relation to preterm and small-for-gestational-age birth. Epidemiology 2005;16(5):641–7.

[30] Michalek JE, Rahe AJ, Boyle CA. Paternal dioxin, preterm birth, intrauterine growth retardation, and infant death. Epidemiology 1998;9(2):161–7.

[31] Lawson CC, Schnorr TM, Whelan EA, et al. Paternal occupational exposure to 2,3,7,8-tetrachlorodibenzo-p-dioxin and birth outcomes of offspring: birth weight, preterm delivery, and birth defects. Environ Health Perspect 2004;112(14):1403–8.

[32] Pathak R, Ahmed RS, Tripathi AK, et al. Maternal and cord blood levels of organochlorine pesticides: association with preterm labor. Clin Biochem 2009;42(7–8):746–9.

[33] Fenster L, Eskenazi B, Anderson M, et al. Association of in utero organochlorine pesticide exposure and fetal growth and length of gestation in an agricultural population. Environ Health Perspect 2006;114(4):597–602.

[34] Khanjani N, Sim MR. Maternal contamination with dichlorodiphenyltrichloroethane and reproductive outcomes in an Australian population. Environ Res 2006;101(3):373–9.

[35] Longnecker MP, Klebanoff MA, Zhou H, Brock JW. Association between maternal serum concentration of the DDT metabolite DDE and preterm and small-for-gestational-age babies at birth. Lancet 2001;358(9276):110–4.

[36] Huen K, Yousefi P, Bradman A, et al. Effects of age, sex, and persistent organic pollutants on DNA methylation in children. Environ Mol Mutagen 2014;55(3):209–22.

[37] Kundakovic M, Gudsnuk K, Franks B, et al. Sex-specific epigenetic disruption and behavioral changes following low-dose in utero bisphenol A exposure. Proc Natl Acad Sci USA 2013;110(24):9956–61.

[38] Singh S, Li SS. Epigenetic effects of environmental chemicals bisphenol a and phthalates. Int J Mol Sci 2012;13(8):10143–53.

[39] Cantonwine D, Meeker JD, Hu H, et al. Bisphenol a exposure in Mexico City and risk of prematurity: a pilot nested case control study. Environ Health 2010;9:62.

[40] Mileva G, Baker SL, Konkle AT, Bielajew C. Bisphenol-A: epigenetic reprogramming and effects on reproduction and behavior. Int J Environ Res Public Health 2014;11(7):7537–61.

[41] Hannam K, McNamee R, Baker P, Sibley C, Agius R. Air pollution exposure and adverse pregnancy outcomes in a large UK birth cohort: use of a novel spatio-temporal modelling technique. Scand J Work Environ Health 2014;40(5):518–30.

[42] Fleischer NL, Merialdi M, van Donkelaar A, et al. Outdoor air pollution, preterm birth, and low birth weight: analysis of the world health organization global survey on maternal and perinatal health. Environ Health Perspect 2014;122(4):425–30.

[43] Leem JH, Kaplan BM, Shim YK, et al. Exposures to air pollutants during pregnancy and preterm delivery. Environ Health Perspect 2006;114(6):905–10.

[44] Olsson D, Mogren I, Forsberg B. Air pollution exposure in early pregnancy and adverse pregnancy outcomes: a register-based cohort study. BMJ Open 2013;3(2).

[45] Maroziene L, Grazuleviciene R. Maternal exposure to low-level air pollution and pregnancy outcomes: a population-based study. Environ Health 2002;1(1):6.

[46] Sagiv SK, Mendola P, Loomis D, et al. A time-series analysis of air pollution and preterm birth in Pennsylvania, 1997–2001. Environ Health Perspect 2005;113(5):602–6.

[47] Ha EH, Lee BE, Park HS, et al. Prenatal exposure to PM10 and preterm birth between 1998 and 2000 in Seoul, Korea. J Prev Med Public Health = Yebang Uihakhoe chi 2004;37(4):300–5.

[48] Symanski E, Davila M, McHugh MK, Waller DK, Zhang X, Lai D. Maternal exposure to fine particulate pollution during narrow gestational periods and newborn health in Harris County, Texas. Maternal Child Health J 2014;18(8):2003–12.

[49] Huynh M, Woodruff TJ, Parker JD, Schoendorf KC. Relationships between air pollution and preterm birth in California. Paediatr Perinat Epidemiol 2006;20(6):454–61.

[50] Brauer M, Lencar C, Tamburic L, Koehoorn M, Demers P, Karr C. A cohort study of traffic-related air pollution impacts on birth outcomes. Environ Health Perspect 2008;116(5):680–6.

[51] Nieuwenhuijsen MJ, Dadvand P, Grellier J, Martinez D, Vrijheid M. Environmental risk factors of pregnancy

outcomes: a summary of recent meta-analyses of epidemiological studies. Environ Health 2013;12:6.
[52] Darrow LA, Klein M, Flanders WD, et al. Ambient air pollution and preterm birth: a time-series analysis. Epidemiology 2009;20(5):689–98.
[53] Maccani JZ, Koestler DC, Houseman EA, Marsit CJ, Kelsey KT. Placental DNA methylation alterations associated with maternal tobacco smoking at the RUNX3 gene are also associated with gestational age. Epigenomics 2013;5(6):619–30.
[54] Qiu J, He X, Cui H, et al. Passive smoking and preterm birth in urban China. Am J Epidemiol 2014;180(1):94–102.
[55] Lee KW, Pausova Z. Cigarette smoking and DNA methylation. Front Genet 2013;4:132.
[56] Knopik VS, Maccani MA, Francazio S, McGeary JE. The epigenetics of maternal cigarette smoking during pregnancy and effects on child development. Dev Psychopathol 2012;24(4):1377–90.
[57] Hylkema MN, Blacquiere MJ. Intrauterine effects of maternal smoking on sensitization, asthma, and chronic obstructive pulmonary disease. Proc Am Thorac Soc 2009;6(8): 660–2.
[58] Breton CV, Byun HM, Wenten M, Pan F, Yang A, Gilliland FD. Prenatal tobacco smoke exposure affects global and gene-specific DNA methylation. Am J Respir Crit Care Med 2009;180(5):462–7.
[59] Chen D, Fang L, Li H, Tang MS, Jin C. Cigarette smoke component acrolein modulates chromatin assembly by inhibiting histone acetylation. J Biol Chem 2013;288(30): 21678–87.
[60] Sullivan MC, Hawes K, Winchester SB, Miller RJ. Developmental origins theory from prematurity to adult disease. J Obstet Gynecol Neonatal Nurs 2008;37(2):158–64.
[61] Ward AM, Syddall HE, Wood PJ, Chrousos GP, Phillips DI. Fetal programming of the hypothalamic-pituitary-adrenal (HPA) axis: low birth weight and central HPA regulation. J Clin Endocrinol Metab 2004;89(3):1227–33.
[62] Gluckman PD, Cutfield W, Hofman P, Hanson MA. The fetal, neonatal, and infant environments-the long-term consequences for disease risk. Early Hum Dev 2005;81(1):51–9.
[63] Grieger JA, Grzeskowiak LE, Clifton VL. Preconception dietary patterns in human pregnancies are associated with preterm delivery. J Nutr 2014;144(7):1075–80.
[64] Bloomfield FH. How is maternal nutrition related to preterm birth? Annu Rev Nutr 2011;31:235–61.
[65] Harper LM, Tita A, Biggio JR. The Institute of Medicine guidelines for gestational weight gain after a diagnosis of gestational diabetes and pregnancy outcomes. Am J Perinatol 2015;32(3):239–46. http://dx.doi.org/10.1055/s-0034-1383846.
[66] Heslehurst N, Ells LJ, Simpson H, Batterham A, Wilkinson J, Summerbell CD. Trends in maternal obesity incidence rates, demographic predictors, and health inequalities in 36,821 women over a 15-year period. BJOG 2007;114(2): 187–94.
[67] Roman H, Goffinet F, Hulsey TF, Newman R, Robillard PY, Hulsey TC. Maternal body mass index at delivery and risk of caesarean due to dystocia in low risk pregnancies. Acta Obstet Gynecol Scand 2008;87(2):163–70.
[68] Avci ME, Sanlikan F, Celik M, Avci A, Kocaer M, Gocmen A. Effects of maternal obesity on antenatal, perinatal and neonatal outcomes. J Matern Fetal Neonatal Med 2014:1–4.
[69] Adams MM, Sarno AP, Harlass FE, Rawlings JS, Read JA. Risk factors for preterm delivery in a healthy cohort. Epidemiology 1995;6(5):525–32.
[70] Goldenberg RL, Iams JD, Mercer BM, et al. The preterm prediction study: the value of new vs standard risk factors in predicting early and all spontaneous preterm births. NICHD MFMU Network. Am J Public Health 1998;88(2):233–8.
[71] Hauger MS, Gibbons L, Vik T, Belizan JM. Prepregnancy weight status and the risk of adverse pregnancy outcome. Acta Obstet Gynecol Scand 2008;87(9):953–9.
[72] McDonald SD, Han Z, Mulla S, Beyene J. Overweight and obesity in mothers and risk of preterm birth and low birth weight infants: systematic review and meta-analyses. BMJ 2010;341:c3428.
[73] American College of O, Gynecologists. ACOG Committee opinion no. 549: obesity in pregnancy. Obstet Gynecol 2013;121(1):213–7.
[74] Bennewitz HG. De diabete mellito graviditatis symptomate. 1824.
[75] Carrington ER, Shuman CR, Reardon HS. Evaluation of the prediabetic state during pregnancy. Obstet Gynecol 1957;9(6):664–9.
[76] Duncan M. Obituary. Br Med J 1890;2(1550):655–6.
[77] Anderlova K, Krejci H, Klusackova P, et al. The alarming incidence of gestational diabetes mellitus using currently used and new international diagnostic criteria. Ceska Gynekol 2014;79(3):213–8.
[78] Hay Jr WW. Care of the infant of the diabetic mother. Curr Diabetes Rep 2012;12(1):4–15.
[79] Eidem I, Vangen S, Hanssen KF, et al. Perinatal and infant mortality in term and preterm births among women with type 1 diabetes. Diabetologia 2011;54(11):2771–8.
[80] Rayco-Solon P, Fulford AJ, Prentice AM. Maternal preconceptional weight and gestational length. Am J Obstet Gynecol 2005;192(4):1133–6.
[81] Han YS, Ha EH, Park HS, Kim YJ, Lee SS. Relationships between pregnancy outcomes, biochemical markers and pre-pregnancy body mass index. Int J Obes 2011;35(4):570–7.

CHAPTER

20

Sexual Dimorphism and DOHaD through the Lens of Epigenetics: Genetic, Ancestral, Developmental, and Environmental Origins from Previous to the Next Generation(s)

Claudine Junien, Sara Fneich, Polina Panchenko, Sarah Voisin, Anne Gabory

INRA, UMR1198 Biologie du Développement et Reproduction, Jouy-en-Josas, France

OUTLINE

The Epigenome and Developmental Origins of Health and Disease
http://dx.doi.org/10.1016/B978-0-12-801383-0.00020-7

Copyright © 2016 Elsevier Inc. All rights reserved.

INTRODUCTION

For many decades human studies and animal models have focused on describing the differences between the sexes in terms of disease occurrence, presentation, and severity and examining their relationship with different hormonal contexts. However, considering only, on the one hand, the end course, that is, the adult phenotype with distinct mechanisms of sex-dependent gene regulation in male and female, and, on the other hand, relying solely on the gonad-centric model is blatantly insufficient [1]. Indeed there is now substantial evidence that if we want to provide new cues to prevent diseases in both men and women there is an urgent need to unravel globally new aspects integrating the actual roots and mechanisms of sexual dimorphism and as far as we can go back in time and in previous generations.

Our capacity to respond to the various challenges and hazards of life, and to stress and risks of disease, during childhood and adulthood, depends on the health and human capital with which we are born [2], as supported by the concept of the developmental origins of health and disease (DOHaD) [3]. Moreover, sex-specific nongenetic and noncultural transmission of consequences of various experiences and exposures to environment to subsequent generations may also occur [4–9]. The notion that these mechanisms are able to transmit the memory of exposure to diverse environmental conditions to subsequent generations, conditioning their reactions, has excited considerable interest and has brought the long-criticized proposals of J.B. Lamarck back into the limelight (Box 1). Thus, the DOHaD covers all aspects of the life cycle, in a sex-specific manner, and three phases can be distinguished: (1) developmental plasticity, (2) long-term effects, and (3) transgenerational responses. Indeed sex-specific nongenetic and noncultural transmission of various experiences and exposures to environment consequences to subsequent generations may also occur in a sex-specific manner [4–8,10]. Sexual dimorphism in the DOHaD context is a universal transversal phenomenon affecting not just humans but all kinds of species and is little understood, except that it is widely considered to be an outcome of both genetics and epigenetics.

BOX 1 THE CARVING BEHIND THE BASE SHOWS JEAN-BAPTISTE LAMARCK AND HIS DAUGHTER, AMÉNAÏDE CORNÉLIE. IT BEARS THE INSCRIPTION: "*LA POSTÉRITÉ VOUS ADMIRERA, ELLE VOUS VENGERA, MON PÈRE*" (POSTERITY WILL ADMIRE YOU AND AVENGE YOU, FATHER)

Jean-Baptiste Pierre Antoine de Mont, Chevalier de Lamarck (1774–1829) was a French biologist/zoologist and anatomist who made a major contribution to the classification of life forms through his four laws:

> First law: Life, through its own forces, tends to increase continually the volume of any body that it possesses and to extend the dimensions of its parts to a limit that it itself defines.
> Second law: The production of a new organ in the body of an animal results from a new need that occurs and continues to be felt and a new movement that needs to be born and maintained.
> Third law: The development of the organs and their force of action is constantly consistent with the use of these organs.
> Fourth law: All that has been acquired, traced, or changed in the organization of individuals, during their lifetime, is conserved by the generation concerned and is transmitted to the new individuals produced by those that have experienced these changes.

Nongenetic transmission processes are often described as Lamarckian because they raise the possibility of inheriting characteristics acquired by previous generations. The key characteristics of Lamarckian mechanisms are (1) an environmental factor directly causing "heritable" changes; (2) the changes induced target a limited set of cell components of functional relevance; (3) the changes provide a specific adaptation to the initial challenge. However, the proof of concept for a role of epigenetic processes in Lamarckian evolution remains tenuous or fragmentary. The fourth law, which was formulated two centuries ago, may seem to go against the finding that the epigenetic marks carried by the gametes are extensively erased after fertilization, ensuring a state of totipotency that should not allow the passage of information about the experiences of parents or ancestors. However, Lamarck began from the notion that a change in environment provokes changes in the needs of the organisms living in that environment, in turn triggering changes in their behavior. These changes in behavior lead to greater or lesser use of the organ concerned, resulting in changes in the size of the organ (increases in size or disappearance) over time and generations.

There is a need to address the questions of when and where these sex-specific genetic and epigenetic processes arise, and how the latter are maintained, erased, or give rise to new ones. Not only because sex differences based on the genetic composition of a female or a male that differ by their pair of sex chromosomes (XX and XY) appear well before gonad differentiation as early as fertilization, but also because of recent data based on the new paradigm of DOHaD, these insights compel us to revise generally held notions to accommodate the prospect that biological parenting commences well before birth, even prior to conception.

The consequences of environmental factors including diet, endocrine disruptors, chemical products, or other psychoaffective, geographic, political, or socioeconomic influences can simultaneously affect at least three generations the mother and father (F0), their children (F1), and their grandchildren (F2)—through somatic and/or germ line changes in the F1 generation (Figure 1). Our experiences in utero and during the first two years of life (the 1000-days concept) are a clear determinant of our health and human capital [2]. However, the phases preceding conception, beginning with gametogenesis and distinguishing between effects on the primordial germ line cells, the gametes, are also important and must be taken into account [10] (Figure 1). Yet whether there is transmission by the mother or the father, a lack of knowledge and a great confusion persists between different types of transmission and the postulated mechanisms. Indeed, although the term *transgenerational* may be used interchangeably, one must distinguish the transgenerational effects from intergenerational or multigenerational effects [11–13] (Figure 1). Early environmental events may disturb the precisely timed processes that sculpt the embryo and fetus in a sex-dependent manner and thus influence his or her health in later life [14,15]. All environmental factors therefore can change or perturb gene expression of all genes, including sexually dimorphic genes, depending on the developmental window(s), the timing, duration, dose, and potential synergy/antagonism with other factors, the type of tissue, and most of all on whether the individual is a male or a female.

Moreover, not only the phenotypic effect of an ancestral exposure can vary across generations but nongenetic sex-specific transmission of exposure to environment to subsequent generations may also vary. We now know that what the grandparents and the parents endured or were exposed to throughout their lives, during critical windows and in specific cell types/systems, can make a difference and especially so whether we consider the grandson or the granddaughter and his/her responses to the environment and their outcomes, and how they do so whether it is transmitted through the paternal or maternal lineage [12]. Studies on sexual dimorphism for gene expression and epigenetic marks and modifiers reveal the existence of different adaptation mechanisms to these environmental factors in males and females in their somatic tissues, germ line, and reproductive systems. As highlighted particularly well in adult mouse liver studies on the potential underpinning epigenetic mechanisms, conspicuous differences in epigenetic marks and modifiers and in chromatin conformation have been revealed [16]. The challenges of unraveling the complexities associated with sex specificity are manifold due to multiple levels of interdependence between the different routes involved in transmission to the next generation(s) [12]. Thus, unraveling the origin, targets, and consequences of parent-specific transfer of information is also key to understanding the underpinning epigenetic mechanisms of sexual dimorphism.

The World Health Organization predicts a 17% increase in noncommunicable diseases (NCDs) for the next decade. Recent data demonstrate that, complementary to genetic-based approaches, the DOHaD concept now largely accounts for most of the unexplained heritability (75%) and thus represents a timely major approach for the prevention of the alarmingly

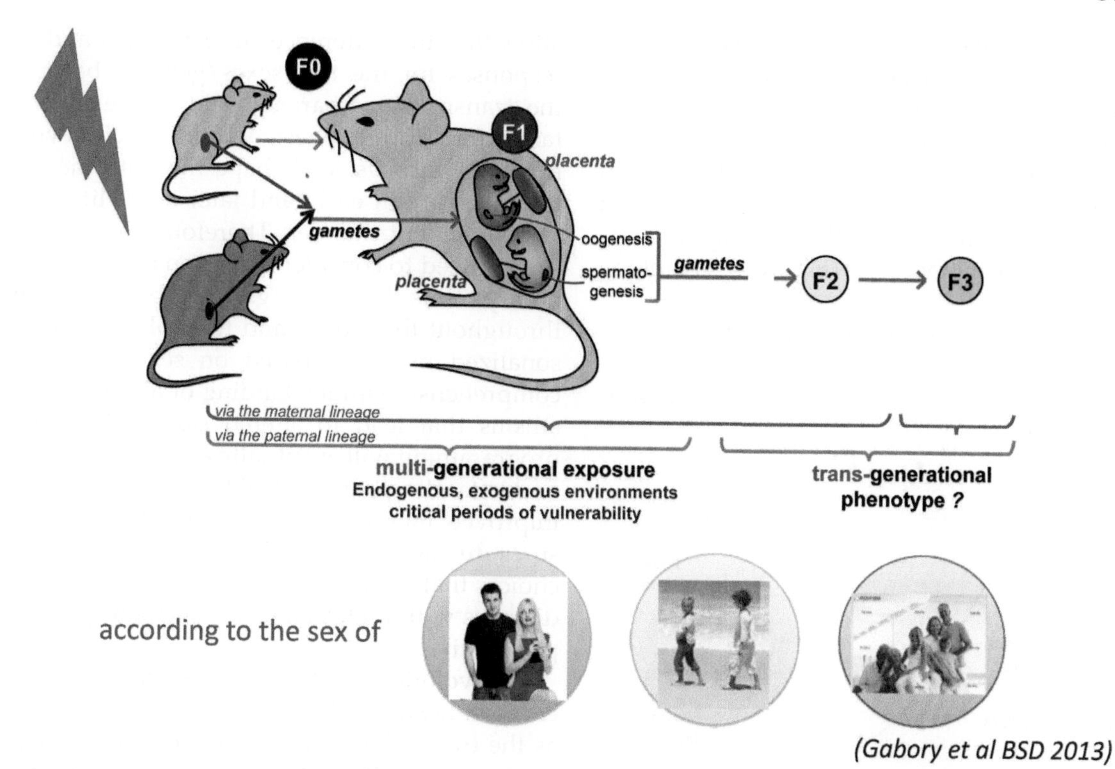

FIGURE 1 **Sex-specific transmission of the memory of exposure to environmental factors to subsequent generations.** Environmental factors, including nutrition, psychosocial stress, toxins, endocrine disruptors, tobacco, alcohol, and microbiota, affect individual (F0) epigenetic landscapes and, therefore, gene pathways and networks, in ways that differ between the sexes. For example, maternal and paternal exposures before the conception of their offspring can modify gamete quality, and information about these exposures can be transmitted to the next (F1) generation. In addition, the consequences of maternal (F0) exposure during pregnancy (stress, metabolism, diet, hormonal changes, etc.) can be transmitted from the maternal to the fetal compartment via the placenta, in a sex-specific manner, with effects on F1 tissue development. The programming of somatic tissues can lead to changes in long-term health outcomes in the first generation. Moreover, primordial germ cells, which develop and undergo reprogramming during fetal development, can also be affected by the F0 maternal environment and may transmit genetic and epigenetic information to the F2 generation. These influences are transmitted differently by the maternal and paternal lineages. In particular, multigenerational exposure in the maternal lineage can be seen in the F0, F1, and F2 generations, with a transgenerational phenotype observed in the F3, whereas, in the paternal lineage, multigenerational exposure concerns the F0 and F1 generations, and a transgenerational phenotype is seen in the F2 and F3 generations. *From Ref. [10].*

increasing incidence of NCDs [17]. While recent studies have shown the usefulness of the study of sex differences, we need to raise the awareness of scientists on the existence of sexual dimorphism. The majority, including scientists, ignores that every cell has a sex, based on the sexual pair of chromosomes, XX or XY. In a recent *Nature* paper (May, 2014) Janine A. Clayton and Francis S. Collins unveil policies to ensure that preclinical research funded by the US National Institutes of Health considers females and males [18]. What a pregnant mother consumes in her diet, her body mass and general health, her exposure to environmental chemicals, the stress she experiences, and her education and socioeconomic status have each been proposed to have subsequent health effects on her offspring. Because of the lack of plausible biological mechanisms, paternal exposures were not taken in to consideration. Recent data demonstrate, however, that even the future father may contribute to DOHaD, through environmental influences on his germ cells during the prepubertal period and on his gametes, as well as through interactions with the mothers [19] (Figure 2). Therefore, altogether the differences in exposures and in responses for the two sexes/genders both at the transmission (parents) and at the inheritance (offspring) levels, within specific vulnerable windows, will shape the outcome for the offspring in early and later adult life in a sex-dependent manner. Therefore, there is an urgent need to consider both men and women, both parents, boys and girls for better health throughout the world, and to apply truly personalized medicine based on sex/gender. A comprehensive understanding of all the mechanisms that lead to altered sex-specific fetal programming will eventually allow a shift from understanding how DOHaD leads to disease to improved methods for avoiding at-risk exposures during critical windows and to reinforce choices that favor prevention and alleviation of diseases with a DOHaD basis in a sex/gender perspective.

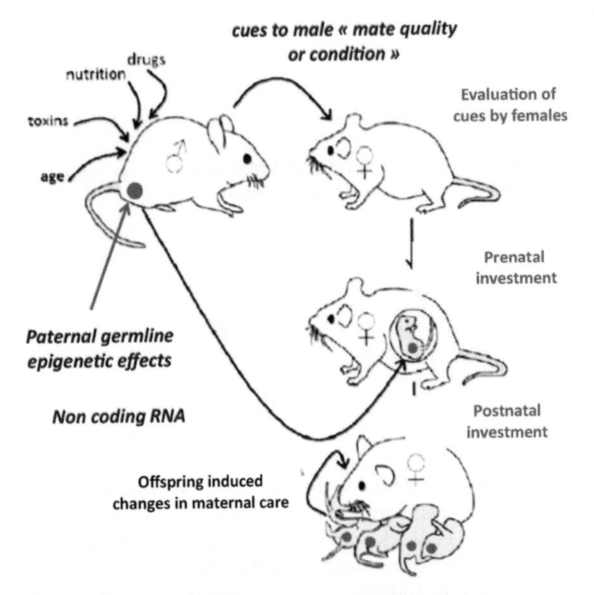

FIGURE 2 Illustration of the nongenomic pathways through which paternal effects on offspring development may occur. The experiences of males (drugs, nutrition, toxins, age, stress), particularly during early development, may lead to epigenetic alterations in the male germ line (red circle), which are then transmitted to the offspring with consequences for phenotypic variation. Alternatively, or probably in combination with these direct paternal effects, the experiences of a male before mating may lead to changes in mate quality or preference, which may be assessed by the female at the time of mating. This assessment may lead to differences in maternal prenatal and/or postnatal investment in the growth and development of the offspring generated from this mating, with consequences for phenotypic variation in the offspring. Maternal investment may also vary with paternally mediated variations in offspring phenotype, during both the prenatal and postnatal periods. Differences in maternal investment as a function of paternal experiences or offspring traits may either enhance transmission of the memory of paternal exposure or compensate for deficits in functioning due to that exposure. *From Ref. [19].*

In evolutionary terms, sex-specific differences in health and diseases can be considered as the by-products of contradictory forces that drove sex-specific selection to reach separate fitness optima in mating systems and as trade-offs between reproduction and survival [20]. The potential for sexual conflicts is predicted to increase with the magnitude of difference in mating success. However, given the differences between men and women and the recent abrupt increase in life span expectancy in most populations, it can be expected that some differences can be blurred and other may appear in the elderly [21].

The aim of this review is not to describe the abundant literature on sex differences but to go deeper into the genetic differences that lead to epigenetic differences that modulate gene expression at different stages during the life cycle. While the relevance of epigenetic processes is not universally accepted for some aspects and is still a matter of hot debate, other mechanisms have been proposed but are not within the scope of this review.

COMPLEX TRAJECTORIES DUE TO SEX SPECIFICITY TO BOTH THE TRANSMISSION AND INHERITANCE OF SUSCEPTIBILITY

Sexual Dimorphism and Nongenetic Heritability

As for the transmission to future generations, through an epigenetic mechanism, it is still not as obvious in humans, but evidence is accumulating in rodents and more in nematodes, fruit flies, and especially for plants [22–28]. So far, a large number of journals have focused on the inter- or transgenerational effects [29–32]. But this phase, though exciting, corresponds to only one of the three key phases of DOHaD.

Studies of gene expression and epigenetic marks and modifications have revealed the existence of different mechanisms of environmental adaptation in males and females, in both humans and animal models [10,12,19,22,33–35]. The effects of, and responses to programming may affect offspring of both sexes, or may affect one of the sexes more than the other [36–40]. Also, maternal exposure such as a high-fat diet (HFD) can have effects on the third-generation female body size but via the paternal lineage [41]. Furthermore, depending on the nature of the environment, and the developmental window and duration of exposure, the sex of the transmitting parent may also condition the response of the offspring to the environment. Following exposure to toxic substances, alcohol, or under- or overnutrition, during a particular developmental window or after weaning, certain phenotypic characteristics may be inherited solely from the father, solely from the mother, or from both parents equally [19,41,42].

The Overkalix cohort provides a good illustration of these differences. The risk of cardiovascular disease and diabetes in a man or a woman is dependent on the abundance or lack of food to which the grandparents, but only the paternal grandparents, were exposed before puberty [12]. The information is transmitted by the paternal grandfather to his grandsons but not to his granddaughters. Similar results have been reported for rodents, for undernutrition or the consumption of areca nuts [19,43]. The transmission of behavioral characteristics from the father to his female descendants, but not to his male descendants, has also been observed in genetically identical mice displaying phenotypic heterogeneity in terms of behavior [44].

Resilience–a Neglected Approach

The great majority of studies in animals and humans, on inter-, multi-, or transgenerational effects had their focus on deleterious effects on the offspring of deleterious behaviors or exposures of parents and grandparents. Despite long-time study of the capacity to adapt in other species, resilience were not subject to the same attention.

The "resistant" (human or animal) subjects were most of the time neglected, rejected, or not considered [45,46]. More recently surprising results suggestive of resilience resistance, adaptation mechanisms, appeared and are now receiving more attention. Thus, there are now so many examples involving epigenetic mechanisms: odor recognition by the generations that follow the exposed generation, accounting for the enigmatic "innate" fear of the predator [47,48]; social defeat in the father, causing reactions similar to those of the father, susceptibility or resilience, in the offspring [49]; resilience to trauma (PTSD): a demethylation of the DNA at a polymorphism in the DNA sequence, caused by a traumatic episode that persisted in adulthood [50–52]; physical abuse during childhood in the macaque monkey and human [53,54]. Finally, in the nematode *Caenorhabditis elegans*, several experiments demonstrate the ability to transmit a memory of odors [55], virus [56], or famine [57].

Programming of somatic tissues can lead to changes in long-term health outcomes in the first generation. Moreover, primordial germ cells, which develop and undergo reprogramming (that comprises successive phases—erasure, establishment, and maintenance—linked to zygotic erasure after fertilization) during fetal development, can also be affected by F0 maternal environment and contribute genetic and epigenetic information to the F2 generation. Maternal and paternal lineages may affect the transmission of such influences differently due to preconceptional exposure of both the mother and the father to stressful conditions [32]. In particular, multigenerational exposure on the maternal lineage can be seen in the F0, F1, and F2 generations, and transgenerational phenotype would be observed in F3, whereas on the paternal lineage multigenerational exposure concerns F0 and F1, and transgenerational phenotype in F2 and F3 generations.

Interestingly, exposure in the parent may also lead to greater resistance/resilience in the offspring. Despite the well-known deleterious effects of maternal obesity and T2D, the female offspring of obese mothers display greater resistance to the obesogenic effects of an HFD after weaning [58]. Similarly, fathers exposed to environmental factors, such as carbon tetrachloride or bile duct ligation triggering liver damage, display deleterious effects of exposure, whereas their offspring are better adapted to cope with future fibrogenic challenges to the liver or display a cocaine-resistant phenotype, respectively, in a sex-dependent manner. Resistance to liver fibrosis is observed in the son and grandson of rats treated with carbon tetrachloride [13]. The same phenomenon is observed in addiction to cocaine in rats. The male progeny of addicted fathers showed reduced cocaine reinforcement [59].

The main challenge is identifying the means by which this information linked to environmental consequences or a physiopathological change is carried and transferred from one generation to another in a sex-specific manner. The principal avenues of research have converged on certain regions of DNA (genes, repeated sequences, etc.) in which epigenetic marks may partially escape from the successive phases of reprogramming—erasure, establishment, and maintenance—linked to zygotic erasure after fertilization. It has been suggested that these regions could carry or mediate persistent changes in chromatin configuration following exposure to an environmental factor. The involvement of noncoding RNAs (short and long) is also becoming increasingly evident [60]. Also, it is clearly emerging that many new vectors, such as exosomes, prions, metabolites, pathogens, chemical substances, and the maternal microbiota, also have nonnegligible roles [22].

Transgenerational Responses to Programming: A Vicious Circle or Resilience?

The programming occurring, always under the influence of the environment during development, or inherited from parents, may be seen as the "first event." It often confers no more than a latent state, a sensitivity to a "second event," and is revealed later by an accumulation of environmental risk factors, leading to a threshold being crossed. It does not, therefore, strictly correspond to long-term "effects." Instead, it relates to all elements conditioning the "response capacity" (increases or decreases) of programmed tissues or organs conferring a predisposition to vulnerability or resilience. Throughout these processes, it is also dependent on genetic background and shows sex-specific responses.

Most phenotypic studies have been limited to explorations of the system disturbed by parental or ancestral exposure: metabolism for nutritional exposure or behavior for stress exposure, for example. However, depending on the stage of exposure, these disturbances may affect different systems or even all systems. Thus, paternal exposure to stress has been linked not only to behavioral problems but also to metabolic problems in descendants [61]. The phenotyping of descendants in animal models generally focuses on deleterious effects, thus ignoring the nonnegligible proportion of "resistant" subjects with positive adaptive responses to the exposure of their parents or grandparents [38,45]. Given our present state of ignorance, such limited analysis is unwise. Most human multigenerational cohorts have a range of measures that would allow studies to address the question of whether outcomes in descendants are limited to the F0 or also to subsequent generations.

Nevertheless, studies in *C. elegans* and *Drosophila* have revealed the existence, in some cases, of an ability to adapt, or resilience. Programming can endow networks of genes with a capacity to respond more rapidly to an environmental challenge [62]. Responses opposite to the initial effects may also be observed. For example, in the Overkalix cohort (Sweden), male malnutrition before adolescence was found to lead to a lower risk of cardiovascular death two generations later [45]. Enriched environments may also induce a favorable transgenerational response, with better performance or a better protective or compensatory response in cases of misprogramming [63]. Finally, interactions between the father and mother and between the young and their mother are relevant [19,64,65]. The responses observed in descendants may therefore be diverse and may differ from the effects of the initial impact on the parent, with anything from the vicious circle most frequently reported to an adaptation opening up new possibilities.

MECHANISMS OF UNEQUAL EXPRESSION OF X- AND Y-CHROMOSOME-LINKED GENES

Early Involvement of Sex Chromosomes in Sex Differences

Early after fertilization, the difference in chromosome makeup between mammalian males (XY) and females (XX) leads to sex-specific differences in gene expression. Even before implantation and the initiation of adrenal and gonad development, transcriptional sexual dimorphism is present in various species that has consequences for developmental competence and adult health and disease [66]. For example, in bovine blastocysts, sex determines the expression levels of one-third of all actively expressed genes [67]. Sexual dimorphism has also been observed in embryonic cells isolated from mice at E10.5. These cells responded differently to dietary stressors even before the production of fetal sex hormones [68]. This applies not only for epigenetic modifiers but also for a substantial proportion of genes encoded by both the X and Y chromosomes, in every cell. This difference in expression triggers sex-specific differences in epigenetic marks on specific sets or networks of genes and therefore for a different setting up and reading of the epigenome between men and women. These differences occur well before adrenal and gonad differentiation, in the absence of sex hormones, thus at a stage when only X- and Y-chromosome-linked genes can make the difference. They therefore shape male and female tissues/organs differently during the course of development. In all adult tissues, due to quantitative and qualitative differences in gonadal hormones and in X-linked and Y-linked genes, in various respective proportions, sex-specific differences in gene expression are observed for, on average, 30% of the genes. However, this sexual dimorphism differs between tissues, at different developmental stages, according to circadian rhythms and with aging. Little is known about the underlying mechanisms but epigenetic mechanisms are involved.

This will emphasize the need to focus on the preconception (fathers and mothers) and early life (mother and child) periods. It is therefore not surprising that from the very beginning of life and throughout life, males and females respond differently to intrinsic and extrinsic factors. One (or the other) sex is therefore more susceptible or more prone to resisting the ups and downs of life than the other sex and therefore may be more susceptible or resistant to the deleterious effects of some environmental factors or lifestyles or to developing a disease or to adapting better.

Sex Differences: Sex Hormones and/or Sex Chromosomes?

Increasing numbers of reports are challenging the traditional view regarding the influences of gonadal hormones and highlighting the potential roles for sex chromosomes (reviewed in Refs [35,69–71]). Sex-linked genes and sex hormones may work together to yield similar differences in physiology between the sexes in brain. For instance, immune responses and cytokine production, or sex-linked genes like the androgen receptor, or Y-linked genes may exhibit sex differences because they can be influenced differently by steroid hormones (reviewed in Ref. [72]). Thus, unfavorable programming, whether immediately before conception or during gestation, may result in various defects potentially translated into differences in susceptibility to disease between males and females [35,70,73–77].

Unequal Dosage and Compensation Mechanisms between Males (XY) and Females (XX)

Mammals have a very complex, tightly controlled, and developmentally regulated process of dosage compensation between males (XY) and females (XX). Two main kinds of dosage compensation exist: the first being to avoid X hyperexpression in females by equalizing the expression of the X-linked genes via inactivation of one of the two X chromosomes in females (XCI: X-chromosome inactivation), and the second leading to the balanced expression between X-linked and autosomal genes via transcriptional upregulation of the active X in both sexes, males and females. There are two forms of XCI: imprinted and random [78,79].

However, not all X-linked genes are absolutely balanced. Several X-linked genes can escape XCI. More genes escape XCI in humans than in the mouse. While it has been estimated that 15–25% of the 1400 X-linked human genes escape XCI in humans, only 3% do so in the mouse [80–82]. There are also significant differences in terms of the distribution of « escape genes » in humans and mouse, with a random distribution along the mouse X chromosome, suggesting that escape is controlled at the level of individual genes rather than chromatin domains. This suggests that men and women may demonstrate greater sex differences in X-linked gene expression than mice as a result of the large number of escape genes. In addition, the degree of escape, hence the expression levels from inactive X, can vary considerably between loci, ranging from 5% to >75% of active X levels [80]. Although there are no data on the laboratory mouse, it is interesting that in common voles, more genes were expressed on the inactive X chromosome in extraembryonic tissues than in somatic tissues [83]. Escape from XCI can vary between different tissues and/or individuals and the escape can also be developmentally regulated. In mice, silencing of some X-chromosomal regions occurs outside of the usual time window, and escape from XCI can be highly lineage specific [78,84,85].

Before implantation, X chromosome inactivation (XCI) is not accomplished and 2X are active in most blastomeres. Thus, at this stage females express a double dose of X-linked genes.

There are also additional control mechanisms to achieve balanced or unbalanced expression between the sexes. Some genes on the X chromosome are imprinted: their expression is monoallelic, depending on the parental origin of the allele. Recently, three genes have been described as imprinted and expressed from the paternal X allele: Fthl17, Rhox5, and Bex1. This monoallelic paternal expression is independent of XIC. Therefore, these genes are expressed predominantly in females [86].

Globally in bovine blastocyst these differences result in a sexually dimorphic gene expression of one-third of the genes [67]. While autosomal genes are involved, sex chromosomes are overrepresented. After random X inactivation, the genes that escape XCI and their variations of expression between tissues may play an important role in the genetic imbalance between males and females.

Male-Specific Y Chromosome Genes

Beyond its roles in testis determination and spermatogenesis, the Y chromosome is essential for male viability [87] (Figure 3). Moreover, as shown recently, with the inheritance of coronary artery disease in men associated with Y chromosome haplotypes [88], the Y Chr, with its 36 male-specific genes including SRY has unappreciated roles in phenotypic differences between the sexes in health and disease [89–91]. In addition to unequal expression of X-linked genes, the small number of expressed genes present on the Y chromosome (and therefore only expressed in males) may be involved. In humans, 29 genes are conserved in the pseudoautosomal regions of the X and Y chromosomes [92]. The nonrecombining, male-specific Y region contains about 27 protein-coding genes [93]. Some X/Y gene pairs have been retained on sex chromosomes and are referred to as paralogues [87]. In the case of X/Y pairs, in contrast to humans, for which a number of X escapees do not have a Y paralogue, all known mouse escapees do have a Y paralogue [87,94,95]. Studies in mice and rats demonstrating sex differences in placental responses to changes in the maternal environment may thus indicate a role for these escaped genes, as the placentas of female fetuses may produce small differences in the amount of the corresponding proteins compared to amounts present in male fetuses. However, there are very few studies comparing levels of mRNA and proteins for escape versus nonescape genes [69,96].

EPIGENETICS AND GENE EXPRESSION: SEX-SPECIFIC MARKS, MECHANISMS, AND DYNAMICS

Within the context of DOHaD, epigenetic marks, which respond to the environment, record the effects of the environment during development in a sex-specific manner [97]. Developmental alterations to epigenetic marks may induce long-term changes in gene

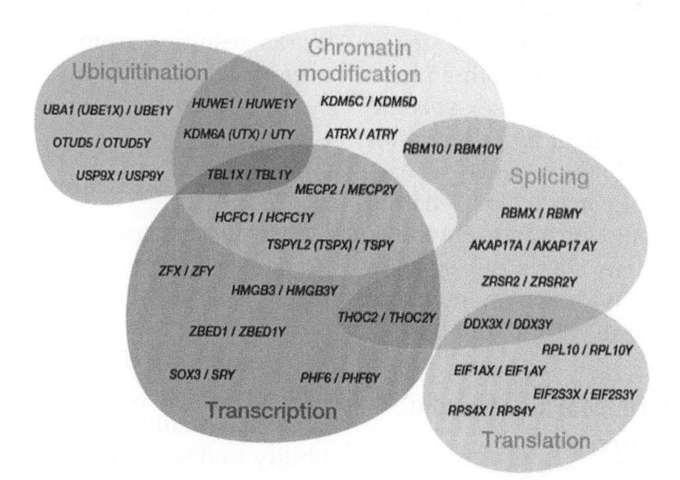

FIGURE 3 **Regulatory annotations of X–Y pair genes.** Venn diagram depicting regulatory functions predicted for selected X–Y pair genes on basis of UniProt annotations of human X homologue. Common alternatives to official gene symbols in parentheses. *From Ref. [87].*

expression, potentially leading to disease in later life [14,98]. Early after fertilization, sex-specific differences in epigenetic modifiers are observed [67,99]. These differences in gene expression may trigger sex-specific differences in epigenetic marks [100]. This occurs well before adrenal and gonad differentiation, in the absence of sex hormones, thus at a stage when only X- and Y-chromosome-linked genes can make the difference.

However, even with recent developments in this field, we still know little about the mechanisms underlying the early sex-specific epigenetic marks resulting in sex-biased gene expression of pathways and networks. Are these marks established early and permanent or are they continuously rearranged throughout life? Little is known about the timing, persistence, and continuity of gene expression required in the creation of distinct male and female phenotypes, and even less is known about how sex-specific selection pressures shift over the life cycle. In male mouse, liver developmental changes in mRNA abundance occur for all major known epigenetic modifiers [101]. In the chicken, the amount and magnitude of sex-biased expression increased as a function of age, though sex-biased gene expression was surprisingly ephemeral, with very few genes exhibiting continuous sex bias in both embryonic and adult tissues [102]. Do the sex-specific patterns observed in adult liver by D. Waxman and coworkers reflect what happened in earlier stages [16]? Epigenetic marks are dynamic throughout development, circadian rhythms, and aging. But are there stable sex-specific marks that help the individual remember its sex and environmental impacts [103,104]?

Sex-Specific Epigenetic Marks Modulate Sex-Specific Gene Expression

The study of the epigenetic marks and mechanisms underlying sex differences is in its infancy. The expression of key enzymes of the epigenetic machinery mapping to autosomes also appears to be sex dependent, even at early stages [99,105]. Levels of *Dnmt1* are similar in male and female bovine embryos, but *Dnmt3a* and *Dnmt3b* are produced in smaller amounts in female embryos [99]. Levels of DNA methylation have been reported to be lower in XX ES cell lines than in XY or XO lines, and this hypomethylation is thought to be associated with lower levels of *Dnmt3a* and *Dnmt3b* [106]. In mouse placenta, global DNA methylation is also sexually dimorphic in animals fed the CD (control diet), with lower methylation levels in the placentas of male offspring than in those of female offspring at E15.5 stage [15]. In all adult tissues examined to date, including the gonads and brain, the expression of many genes is modulated in a sex-specific manner [70,107,108]. Chromatin structure and epigenetic marks differ between male and female samples in brain [109,110]. The adult liver is the organ in which these aspects have been best characterized, with genome-wide chromatin states, DNaseI-hypersensitive sites, and sex-specific gene expression detected [16,111–114]. However, even with recent developments in this field, we still know little about the mechanisms underlying the early sex-specific expression of genes and gene networks resulting from epigenetic regulation in every tissue.

Sex-Specific Impact of Environmental Influences

Within the context of DOHaD, epigenetic marks, which respond to the environment, record the effects of the environment during development in a sex-specific manner [97]. Developmental alterations to epigenetic marks may induce long-term changes in gene expression, potentially leading to sex-specific susceptibility to disease in later life [14,98]. Efforts are now being made to determine the contribution of epigenetics to the establishment and maintenance of sex differences.

Under HFD, hypomethylation was observed only in the female placenta. Consistent with this observation, expression of the gene encoding the DNA methyltransferase cofactor *Dnmt3l* was downregulated in females only [15,76]. Clearly, further studies are needed to understand the direct effects of sex chromosomes and gonadal hormones on the regulation of genes controlling histone acetylation and methylation, coregulatory proteins, and transient and stable DNA methylation patterns.

Expression analysis has also shown that maternal HFD affects mouse placental gene expression in a sexually dimorphic manner [15]. An HFD during gestation triggers the deregulation of clusters of imprinted genes. Sexual dimorphism and sensitivity to diet were observed for nine of 20 imprinted genes, from four clusters on mouse chromosomes 6, 7, 12, and 17. An analysis of CpG methylation in the DMR of the chromosome 17 cluster revealed sex- and diet-specific differential methylation of individual CpGs in two conspicuous subregions. Bioinformatic analysis suggested that these differentially methylated CpGs might lie within recognition elements or binding sites for transcription factors or factors involved in chromatin remodeling [76]. Gregg et al. recently reported sexually dimorphic genomic imprinting in the brain, with sex-specific imprinted genes found mostly in females [115,116]. Given the importance of genomic imprinting in the brain and placenta, this provides new clues for further investigations of sexual dimorphism in the placenta.

The Special Case of X/Y Pairs of Paralogues

In the same study, transcriptomic analysis showed that both basal gene expression and response to maternal HFD were sexually dimorphic in whole placentas. The differences between the sexes in the transcriptomic response to HFD were not only quantitative but also qualitative. The biological functions and networks of genes dysregulated differed markedly in sex-specific ways, with involvement of immune cells and uptake and metabolism of amino acids in females versus the development and function of vascular system, and uptake and metabolism of glucose and fatty acids in males [15]. In this study, 11 genes displayed sexual dimorphism regardless of diet (control or HFD). Consistent with the key role of genes on the sex chromosomes, three of these genes were Y-specific, *Ddx3y*, *Eif2s3y*, and *Kdm5d* (*Jarid1d*), and were more expressed in males, and three were X-specific, *Eif2s3x*, *Kdm5c* (*Jarid1c*), and *Ogt* and were more expressed in females. Interestingly, among these six X- and Y-linked genes, there were two paralogue pairs: *Eif2s3x/y* and *Kdm5c/5d* [10,15]. Of particular interest are the X-linked genes that encode enzymes of the epigenetic machinery and transcription factors: *Kmt1a* (*Suv39h1*), *Jpx*, *Xist*, *Kdm6b* (*Jmjd3*), *Kdm5c*, *Eif2s3x*, *Kdm6a* (*Utx*), and *Ddx3x*, as well as the corresponding paralogues for the latter ones, *Kdm5d*, *Eif2s3y*, *Uty*, and *Ddx3y* that are on the Y chromosome. Sex-specific differences in expression of the histone demethylases *Utx/Uty* and *Kdm5c* have been observed in mouse brain and neurons [117,118]. Other studies have reported the male-specific expression of Y-linked genes—*Ddx3y*, *Eif2s3y*, and *Kdm5d*—in mouse hearts and human myocardium [119] (Figure 4).

In mouse brain, Reinius and coworkers recently identified four female-biased long non-coding RNAs (lncRNAs) associated with protein-coding genes that escape X inactivation, the *Ddx3x/Kdm6a* cluster, *Eif2s3x*, *2610029G223Rik*, and *Kdm5c* [91]. Given that placenta, brain, and testis could share common mechanisms involving X-linked genes [120], these lncRNAs might also be implicated in placental development or function. Moreover, these mouse escapees from X inactivation also have a paralogue on the Y chromosome. According to the authors, these lncRNAs might also escape X inactivation [91]. It would thus be interesting to investigate how

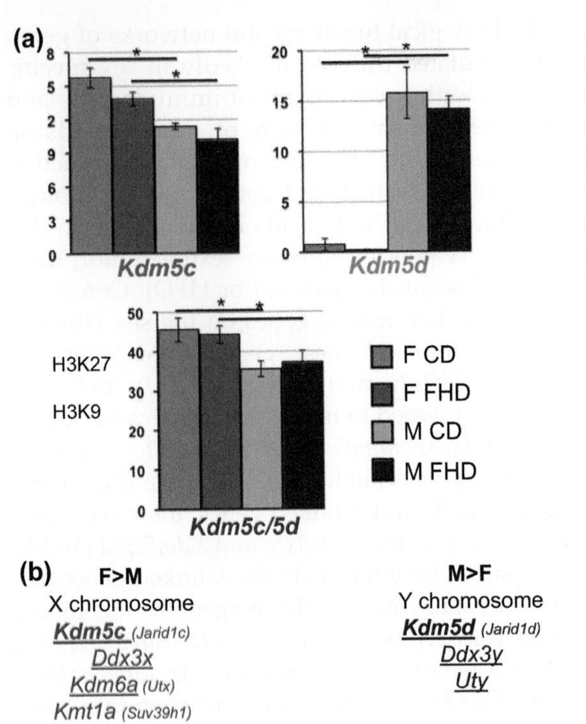

FIGURE 4 **Analysis of the mRNA expression of the *Kdm5c* and *Kdm5d* paralogues.** Three PCR primer pairs have been designed for recognizing specifically either *Kdm5c* or *Kdm5d* cDNA and for recognizing both *Kdm5c/5d* cDNA. Their expression was studied in male and female placentas in pregnant female mice fed either a control diet (CD) or a high-fat diet (HFD) from E0.5 to sacrifice at E15.5 stage: (a) in whole placentas, *Kdm5c* expression is higher in females (pink bars) than males (blue bars), and *Kdm5d* is expressed only in males, regardless of maternal diet. The *Kdm5c/5d* PCR shows that the combined expression of *Kdm5d* and *Kdm5c* expression in males is not of equivalent magnitude as the expression of *Kdm5c* from both alleles in females. *(From Ref. [10].)* (b) The expression of Y-linked with paralogues on the X chromosome: *Eif2s3x*, *Kdm5d*, *Ddx3y* sexually dimorphic genes in the heart of mice and human [119] was also sexually dimorphic in placenta layers.

these three mechanisms (escaping X inactivation, X/Y paralogues, and lncRNA) participate in sexual dimorphism.

The proteins encoded by Y-linked genes may or may not have the same functions, the same target sequences, or the same pattern of expression, according to age or tissue, as their X paralogue. *Kdm5c* and *Kdm5d* show 85% nucleotide identity and 87% amino acid identity that could account for differences in sex-, tissue-, and/or stage-specific gene expression, substrates, and targets [87]. Thus, the epigenetic enzymes produced by these two genes could mark the epigenome in a sex-specific manner, both at the quantitive and qualitative levels [121] leading to differences between the sexes in their writing, erasing, and reading their epigenome. In our study, in placenta of HFD-fed mouse mothers, the Y- and X-linked histone demethylase paralogue genes *Kdm5c* and *Kdm5d* were sexually dimorphic. In another report, in mouse brain, expression of the Y version of the gene in male mice did not compensate for the dosage imbalance between the two sexes in the expression of their X homologues escaping X inactivation. Figure 4 shows that, in placentas from mothers fed a control or HFD, the Y-linked *Kdm5d* gene expression in males is not able to compensate the expression of *Kdm5c*, its X-linked paralogue escaping XIC, in females [10,15].

In sex-biased diseases, it could be postulated that changes in the sex-specific epigenetic landscape could change the risk or resilience of one sex versus the other [122,123]. Thus, understanding the spatiotemporal [124] ontogeny, the windows of susceptibility, the dynamics, and environmental susceptibility of these marks could provide cues for the identification of the genes involved and new markers for diagnostic/prognosis [10,107,124].

THE SPECIFIC EPIGENETIC FEATURES OF EXTRAEMBRYONIC TISSUES AND PLACENTA

The DOHaD concept is consistent with the possibility that environmental influences can affect the development of sex differences early in development and in particular in the placenta, sculpting its epigenomes, and hence the epigenomes of the developing fetus [125].

The sex of the embryo affects the size of both the fetus and the placenta, and the ability of the placenta to respond to adverse stimuli [10]. But where, how, and when sex differences begin in the placenta and how they contribute to sex-specific responses of somatic tissues later in life is still poorly understood [10,75,126,127]. Female and male placentas have different routes to maximize fitness, and therefore the two sexes have different optimal transcriptomes that may affect fetal growth and later disease susceptibility or health trajectory [15]. Differences in how male and female placentas cope with stressful conditions indicate that this tissue should also be taken into account if we want to understand how it contributes to sexual dimorphism later in life [10]. The placenta may therefore be seen as an ideal system to study the sensing, by the fetus, of stresses, starvation, endocrine disruption, and obesity-prone diets or lifestyles, in a sex-specific manner [128,129].

Efforts are now being made to determine the contribution of epigenetics to the establishment and maintenance of sex differences. Most DOHaD studies have reported sex-specific transmission and/or effects, but very few have tackled the sex-specific epigenetic mechanisms involved, and especially in the placenta [10]. In a recent review, Novakovic and Saffery suggested that DNA methylation profiling highlights the unique nature of the human placental epigenome for genomic imprinting and placenta-specific gene-associated methylation. Placental cell types have a pattern of genome methylation that is significantly different from that in somatic tissues, with low methylation at some, but not all, repetitive elements (reviewed in Ref. [130]).

The Specific Features of the Placenta

The sex of the placenta and the environment have an influence on its epigenomes, and hence on the epigenomes of the developing fetus [131–133]. The epigenetic landscape required for placenta development has been described [134]. Sexually dimorphic patterns of gene expression have recently been reported for individual genes in placentas from humans and rodents, potentially accounting for differences in the sensitivity of male and female fetuses to maternal diet (reviewed in Refs [10,70]). Considering these expression studies, it is noteworthy that sex differences have been observed in the mRNA levels of housekeeping genes and of commonly used reference genes in human placenta, in a variety of mouse somatic and extraembryonic tissues, as well as in the preimplantation blastocyst and blastocyst-derived embryonic stem cells [135,136]. Although this is not surprising given the importance of sexual dimorphism in every tissue examined so far, it underlines the difficulty in choosing appropriate reference genes. Few groups have studied global sexual dimorphism in the placenta with microarrays, focusing in particular on the impact of maternal diet, asthma, or stress on placental gene expression, through systemic investigations of the relationship between diet and the expression of sexually dimorphic genes. These transcriptomic analyses showed that basal gene expression levels were sexually dimorphic in whole placentas [15,127,137]. Even fewer studies have investigated the diet-associated epigenetic changes [15,76].

The incomplete, and unstable, imprinted inactivation of the paternally inherited X-chromosome is observed in certain eutherians (e.g., rodents) at preimplantation stages of embryonic development and is retained in the extraembryonic organs that derive from the fetus. Therefore, in mice, the paternal X chromosome is inactivated in the placenta [95]. In the cells that form the tissues of the embryo proper, the paternal X chromosome is reactivated during implantation followed by a random inactivation of either the paternal or maternal X chromosome [84]. The paternal imprint in the blastocyst trophectoderm and their derivates such as placenta seems to be unique to mice, not occurring in rabbits, bovines, or humans where XCI occurs after the blastocyst stage [78].

In the mouse, detailed studies on sex chromosomal contribution to placental growth have been reported [138]. The X chromosome has been implicated in causing several malformations of the placenta. About 30% of all trophoblast-expressed genes are on the X chromosome, and alterations in many different X-linked genes could account for similar phenotypes [139,140]. Due to paternal X inactivation in trophoblast cells, mutations in these X-linked genes manifest themselves in embryonic lethality upon maternal transmission of the mutant allele in the mouse. A role for the Y chromosome in placental dysplasia has also been demonstrated [141]. It is also well established that male fetuses have a higher rate of perinatal complications attributed to placental dysfunction that may relate to the abundance of X-linked genes involved in placentogenesis [142].

Sex Differences: Sex Hormones and/or Sex Chromosomes?

Data from spotted hyena showed that the reduced expression of placental aromatase may allow the hyena placenta to convert high-circulating concentrations of androstenedione to testosterone and could explain the virilization of the fetal external genitalia in female fetuses [143]. However, current data highlight a sexually dimorphic difference in placental function that may not be conferred by classical assumptions of sex steroid regulation. Testosterone may act in a sex-specific manner in the human placenta and may be more potent in female placentas than males; however, further investigations into the role of testosterone in placental function are required [75]. Nonetheless, unequal gene expression by the sex chromosomes has an impact much earlier, beginning at conception, and may set the context for events in later life (reviewed in Refs [10,70,71,75,144]).

There are sex-specific alterations in placental gene expression in the presence of maternal asthma that may contribute to the sexually dimorphic difference in fetal growth in response to maternal asthma. Chronic maternal asthma is associated with reduced growth of the female fetus and normal growth of the male fetus [145]. Understanding the natural basic embryo sexual dimorphism for programming trajectories will help understanding of the early mechanisms of response to environmental insults [146].

DIFFERENCES BETWEEN MALE/ FEMALE GAMETOGENESIS

Even before Conception for the Future Father and the Mother

Any differentiated cell type originates after a variable number of divisions from a stem cell, thus defining sensitive windows and their boundaries, during the reproductive phase is a delicate task. Thus, the impact of the environment may have occurred both on the epigenome of a toti or pluripotent stem cell, as on the epigenome of intermediate-type cells during gametogenesis and have been preserved in a different form during the differentiation process that extends from the migration of primordial geminal cells. A recent report suggests that RNA isolated from sperm may inform the offspring of the history of early trauma in the life of a father [61]. It is therefore not surprising that preconception impacts that can go back to a few decades before conception may have an influence [12,22,34,147]. This is probably why studies of males, free of these confounding factors, although fewer in number, have been more successful [74,148–151]—especially since in the mother the effects of different types of stress, nutritional, psychoaffective, toxic, or metabolic disorders (obesity, diabetes) during pregnancy represent confounding factors that are difficult to separate from effects transmitted only through the gametes.

Differences in timing, function, gene expression, and epigenetic plasticity and dynamics between male and female gametogenesis foreshadow differences in the modes of transfer and type of information that can be transmitted to the next generation(s). Moreover, after fertilization and during the preimplantation period, the paternal and maternal genome still exhibit differences in erasure of epigenetic marks and reprogramming. The asymmetry between the paternal and maternal genome therefore suggests that exposure to certain environments may affect the germ line of the father or mother (or both), all their somatic tissues, and their reproductive systems, including the genital tract and environment

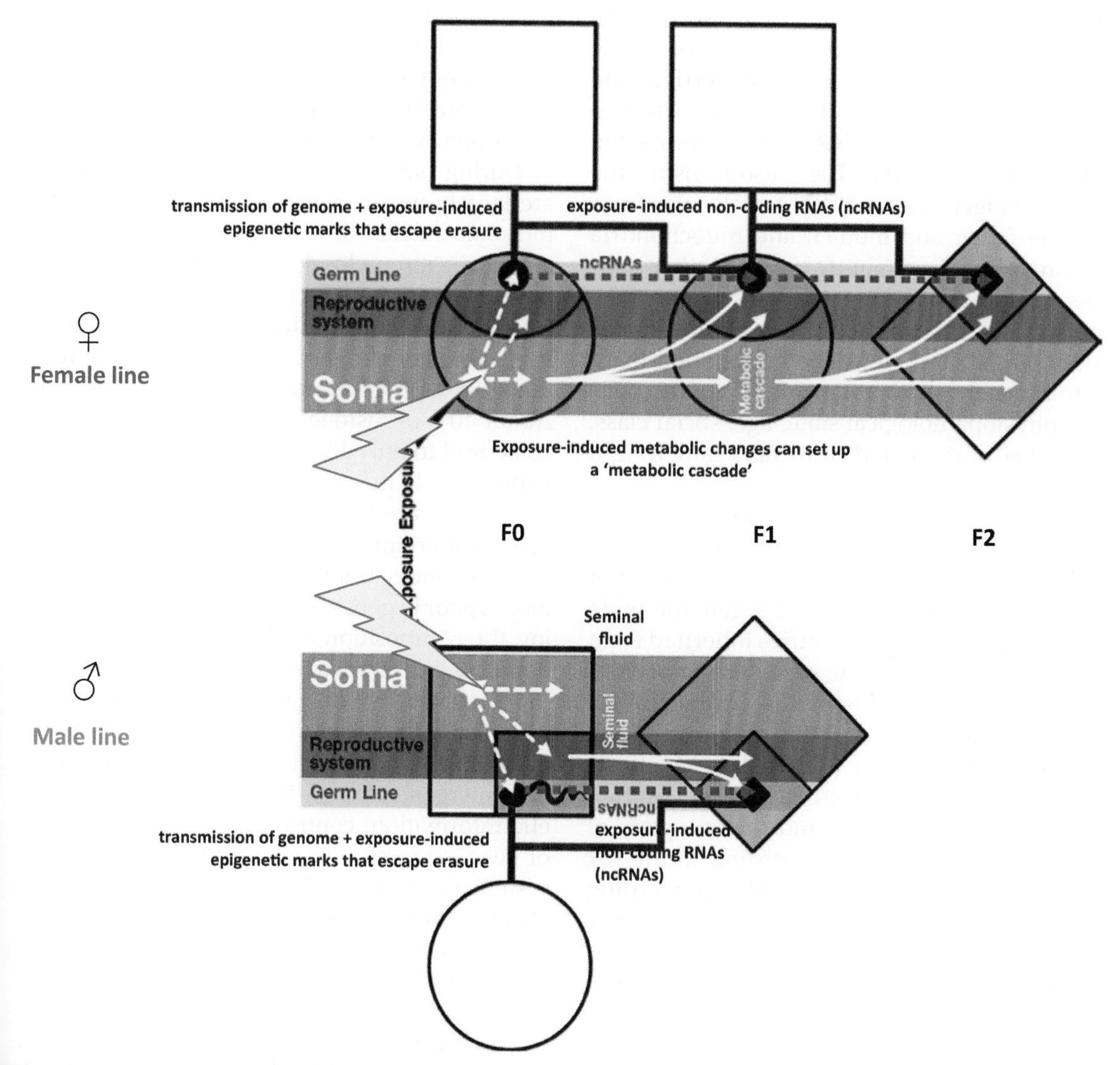

FIGURE 5 **Schematic pedigree diagram showing the main routes for the biological transmission of the effects of exposure to the subsequent generations.** Top, female line; Below, male line. The exposure can potentially affect the germ line, the reproductive system, and the somatic tissues. The traditional pedigree lines (blue) show chromosomal transmission, with the possibility of exposure-induced epigenetic marks that escape erasure and affect the development of the offspring. The germ line can potentially transmit exposure-induced noncoding RNAs (ncRNAs) that influence offspring development. Exposure-induced metabolic changes can set up a "metabolic cascade," such that changes in the reproductive tract influence early embryonic programming of the offspring or change metabolic signals across the placenta. An additional maternal route of transmission is the influence of the mother's microbiome on that of her child. *From Ref. [12].*

(Figure 5). This results in a complex dialogue between these systems that may lead to concerted transfer to subsequent generations [12,34,65,152].

The germ line and gametes display genetic (XX or XY), ontogenetic, morphological, and functional differences between the sexes. The nongenetic differences result from epigenetic asymmetry, which may persist after fertilization [153,154]. At conception, the gametes deliver the genetic heritage, DNA, which forms the genome of the embryo. They also transmit the different epigenomes and RNA molecules from both the father and mother, and mitochondria and a number of proteins from the mother only. Thus, in addition to the genetic heritage of the embryo, the parents also provide epigenetic, protein-based, and metabolic information relating to exposure to environmental factors, experience, physiopathological state, age, social class, parental education, and birth rank and weight [19,148] (Figure 5).

Given the criticality of maternal dietary behavior around the time of conception on long-term offspring health, in a sex-specific manner [155], the respective roles of the genome-wide parental epigenetic asymmetries inherited from the sperm and the oocyte represent important targets to unravel parts of the sexual dimorphism conundrum [154]. Little is known, however, about the impact and long-term effects of these asymmetric factors, their evolution (erasure or maintenance), and the ontogeny of tissue-specific expression according to the sex of the embryo. The few examples concerning smoking and obesity, in human epidemiology and in animal models, suggest that this is an important yet unexplored route for understanding sex-specific vulnerability or resilience [12,156,157].

Because the oocyte and the sperm harbor vastly different types of chromatin and carry different sets of RNA populations (Bourc'his and Voinnet, 2010 [233]; Gill, Erkek, and Peters [188]), the potential for additional parental asymmetry in the early embryo is tremendous [154,158]. A survey of parental epigenetic asymmetries in mammals, organized in a developmental time frame, from their establishment in gametes, to their selective consolidation or erasure immediately after fertilization, to their further restriction during embryonic development was recently provided [154]. The differences in hormonal, temporal, structural, and functional control of germ cell development shape gametic epigenomes in a highly sexually dimorphic manner (Figure 6).

During spermiogenesis canonical histones are largely replaced by protamines, small basic molecules that allow the formation of DNA structures more closely compacted and important for normal sperm function [159]. However, recent studies suggest that all histones organized in nucleosomes are not replaced. Thus the resistance to reprogramming is more widespread. About 10% of histones are retained in humans, compared to only 1% in mice. The idea that prevails today is that nucleosomes not replaced by protamines are located primarily at the level of genes important for development [160] at regulatory sequences, but were also found at repeated blocks poor in genes [161]. A set of criteria including the composition of the DNA sequence, the non-CpG island methylation of CpG-rich at the promoter, the dynamic replacement of canonical histones with histone variants and H3K27me3 mark suggests a structure may be involved in the intergenerational transmission of epigenetic information, contributing to the regulation of gene expression in offspring [162]. The same systematic study on human and mouse sperm shows a conservation in the evolution of the basic principles of the retention of nucleosomes that might be a case for their role in epigenetic inheritance between generations [162].

Moreover, Lesch et al. have demonstrated that a set of genes critical for the early and later development is maintained in an ambiguous state epigenetically (poised) from embryonic stages until the end of the meiosis characterized by the presence of divalent and lacking expression marks as in ES cells. These data suggest that this condition (poised) for these development genes is a

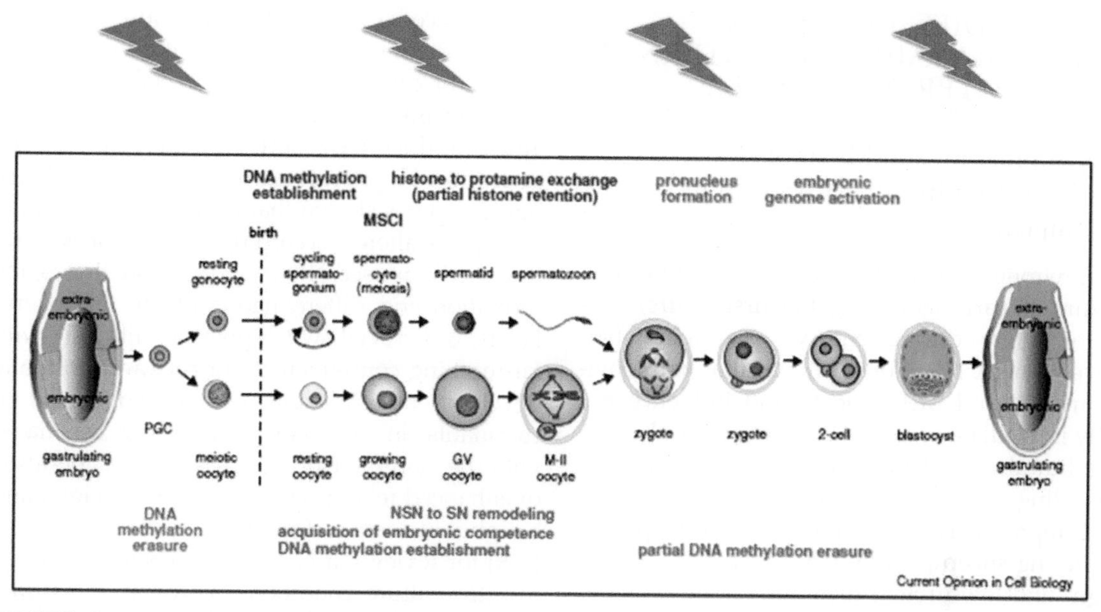

FIGURE 6 **Life cycle of mammalian gametogenesis and embryogenesis.** Primordial germ cells (PGCs) arise from proximal epiblast cells. They undergo extensive erasure of DNA methylation and chromatin changes during migration to and entry into the gonad. Directed by the somatic gonadal environment, germ cells are destined for a male or female fate. Male germ cells, initially called gonocytes, have arrested cell cycles and they begin to establish male-specific DNA methylation patterns. During the subsequent meiotic prophase, the X and Y chromosomes undergo meiotic sex chromosome inactivation (MSCI) characterized by major chromatin remodeling events. Following meiotic divisions, haploid spermatids undergo extensive nuclear and morphological changes, including an almost genome-wide replacement of histones with protamines. However, nucleosomes are retained on regulatory sequences, providing a potential means of epigenetic inheritance. Female germ cells enter meiotic prophase in the embryo and complete their meiotic divisions upon hormonal induction in the adult ovary and fertilization by sperm. During the growing phase, oocytes establish DNA methylation at genes and imprinting control regions, undergo chromatin remodeling, and acquire competence for the direction of embryogenesis. Upon fertilization, the parental genomes form two pronuclei that are epigenetically different, reflecting the history of parental germ line-specific chromatin remodeling events. The paternal and maternal genomes undergo active and passive erasure of DNA methylation. The asymmetry of the chromatin states of paternal and maternal chromosomes may potentially regulate the activation and repression of de novo gene expression in preimplantation embryos, thereby directing embryogenesis. A latent epigenetic state, characterized by the presence of H3K4me3 and H3K27me3 bivalent marks in the promoters of genes involved in development, not expressed at these stages, is a fundamental property of the nucleus of mammalian germ line cells, enabling differentiated gametes to initiate a totipotency program immediately after fertilization [89]. *From Ref. [188].*

fundamental property of the nucleus of the germ line of mammals, giving the gametes differentiated the ability to "unleash" without waiting a totipotence program just after fertilization [89]. These data were demonstrated for the female to the initiation stage of meiosis, and for the male to the stage of meiotic germ cells and post meiotic in humans and mouse [151,160,163]. However, probably for technical reasons potentially linked to the composition of histone variants, the various studies performed did not identify the same types of sequence. The nucleosomes identified were located principally at genes critical for early or late development [160], and at regulatory sequences, but some were also found at repeated sequences containing few genes [164]. Nonetheless, these sequences are potential candidates for epigenetic heredity.

DIFFERENCES IN REPROGRAMMING OF MATERNAL/PATERNAL GENOME

Environmental and Metabolic Programming from Fertilization to Preimplantation

Environmental and metabolic programming occurs as early as during the first hours/days in ontogeny during embryo preimplantation development. It has been shown that both environmental and metabolic factors influence not only immediate events of blastocyst morphogenesis but also the fetal and postnatal phenotype, including behavior, cardiovascular function, and reproductive function in several species, including sheep [73,165–176]. In mice, maternal high- and low-protein diets reduce the number of inner cell mass (ICM) cells, lower the mitochondrial membrane potential, and elevate reactive oxygen species levels in blastocysts [177]. Kwong and coworkers [173] showed that feeding a low-protein diet to pregnant rats during the preimplantation period (0–4.25 days after mating) was sufficient to disturb subsequent embryo development. Blastocysts showed significantly reduced cell numbers, first within the ICM and later within both blastocyst cell lineages (ICM/embryoblast and trophectoderm/trophoblast). These changes, induced by a slower rate of cellular proliferation and not by increased apoptosis, were subsequently followed by sex-dependent long-term effects such as excess growth and hypertension in adulthood [73,165,166,173]. Similarly, protein deprivation during in vitro culture [165] led to altered growth and hypertension in offspring and/or altered expression of metabolic regulatory enzymes.

In mice, Watkins and coworkers [165] showed that embryo culture from the two-cell to the blastocyst stage in medium supplemented or not with a protein source induced reduced trophectoderm and ICM cell numbers compared with in vivo–produced embryos. Embryo culture also led to an enhanced systolic blood pressure in adults, together with increased activity of enzymatic regulators of cardiovascular and metabolic physiology, serum angiotensin converting enzyme, and the gluconeogenetic key enzyme, hepatic phosphoenolpyruvate carboxykinase. In rabbits, our group has shown that gene expression in the embryo is altered according to the embryo culture medium used [178]. Gastrulation, that is, the formation and differentiation of the three germ layers, is a crucial switch point in ontogeny with far-reaching consequences for pre- and postnatal development of all cells, tissues, and organs. In ruminants, in vitro culture has been associated with the "large offspring syndrome," a condition of enhanced fetal growth, large birth weight, and a higher incidence of perinatal mortality (see Ref. [179] for review). It has recently been shown that in vitro culture conditions, as found in assisted reproduction technology for humans, may affect global patterns of DNA methylation and gene expression. Katari et al. highlighted the association between in vitro conception and changes in DNA methylation, potentially affecting the long-term pattern of expression of genes involved in chronic metabolic disorders, such as obesity and T2D [180]. Identifying the specific features and functions of the epigenetic buildup at these stages and determining the mechanistic pathways by which environmental factors may affect them in the long term will be a major milestone in the domain of DOHaD investigation [181,182].

The Various Phases of Reprogramming and the Twists of Mark Erasure

Two principal phases of reprogramming (comprising erasure—reestablishment—and maintenance of marks) have been studied; the first occurs in the zygote, just after fertilization, and the others occur in the germ line, when the primordial germ cells migrate toward the genital ridges before sexual differentiation [183] (Figure 7). It was commonly agreed that, apart from the imprinted genes, methylation of genes

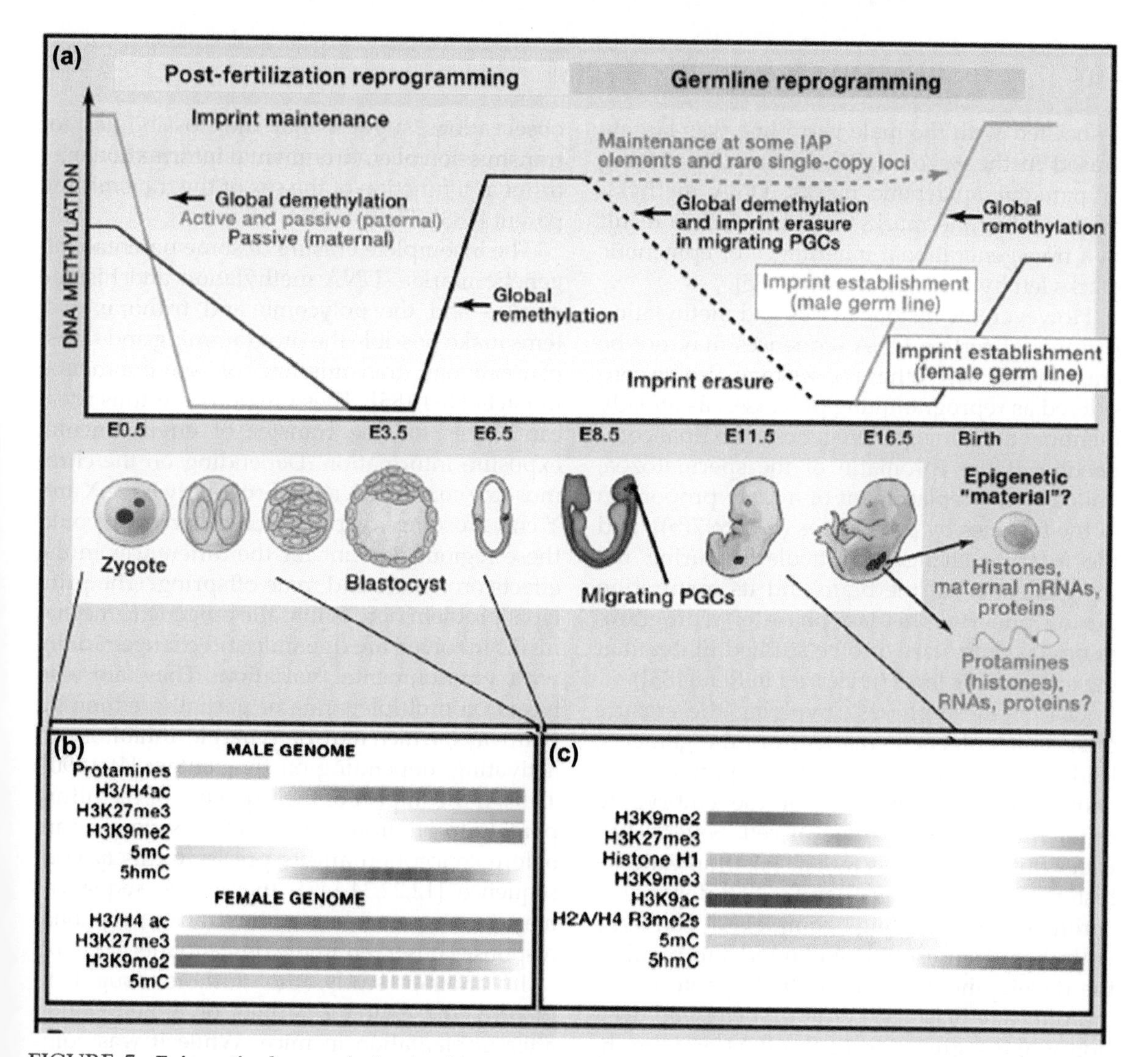

FIGURE 7 **Epigenetic changes during in vivo reprogramming.** (a) DNA methylation dynamics during developmental reprogramming. After fertilization, the paternal genome (blue line) is rapidly demethylated by active mechanisms, whereas the maternal genome (red line) is passively demethylated. Differentially methylated regions (DMRs) associated with imprinted genes are protected from this erasure (dashed green line). De novo methylation occurs after implantation (black line), but primordial germ cells (PGCs) are not specified until the epiblast stage (shading at the top of the figure). This methylation must be reset in PGCs. The figure shows the methylation dynamics, from E6.5, of the cells forming the germ line only. Most sequences are demethylated by E9.5 in PGCs. Some sequences are subject to late demethylation and are not reprogrammed until after PGC migration. These sequences include, but are not limited to, the imprinted DMRs. Intracisternal A particles (IAPs) are resistant to demethylation during both the postfertilization and PGC waves of reprogramming. Variably erased CG islands (VECs) can resist erasure during PGC reprogramming, but their methylation status during postfertilization reprogramming is unclear. Following sex determination, the germ cells undergo de novo methylation, but the dynamics are sex specific. Methylation is completed in the prospermatogonia before birth, whereas methylation in the oocytes is established during postnatal growth. In adulthood, the gametes are appropriately methylated to form a new zygote and to restart the cycle of methylation dynamics. We show below the developmental windows investigated in three key studies, with the specific time points analyzed indicated. blast., blastocyst; d5, day 5 oocytes; GV, germinal vesicle oocytes, MII, metaphase I oocytes. *(From Ref. [183].)* (b) Epigenetic changes during in vivo reprogramming. Schematic diagram of the global DNA and histone modifications leading to transcriptional activation of the embryonic genome between the late zygote (paternal genome only) and the two-cell stage. Gamete genomes undergo different epigenetic programs after fertilization, with the paternal genome mostly subject to epigenetic remodeling at the zygote stage and the maternal genome gradually losing repressive modifications during subsequent cleavages. (c) Global epigenetic changes during germ line development from PGC specification (E6.5) to mitotic/meiotic arrest at E13.5. Two major reprogramming phases can be distinguished during PGC migration toward the genital ridges (E7.5–E10.5) and upon their arrival in the gonads (E10.5–E12.5). *(From Ref. [232].)*

associated with the male germ line was largely erased in the zygote [184]. Incomplete erasing of parental epigenetic marks, DNA methylation and histone marks, could therefore result in a transgenerational inheritance of epigenetic marks left by the environment [185].

However, the histone marks and methylation patterns of certain DNA sequences may not be erased [185]. Two other phases may also be considered as reprogramming processes: as already mentioned during gametogenesis the final compaction of the chromatin of the spermatozoa, linked to the replacement of a large proportion of the histones by protamines (Figure 7(b)), and the massive changes, particularly during the reorganization of the brain and its maturation during puberty. This last phase of reprogramming has only started to be studied in detail at the epigenomic level (reviewed in Ref. [186]).

One of these phases, involving the erasure of specific epigenetic marks from the gametes, leads to the acquisition of a totipotent epigenome, allowing the cells of the embryo to differentiate into any type of cell. Several key transcription factors as well as posttranslational histone modifications and ATP-dependent chromatin remodeling complexes converge with these transcriptional networks to regulate the development of a totipotent zygote into a multilineage blastocyst. The first cell-fate decision is the segregation of the ICM and trophectoderm cell lineages [187](Figure 7(b)). Some sequences, such as those of genes subject to parental imprinting, escape this process. Following another phase of reprogramming linked to the germ line, the remethylation of DNA occurring after sex determination facilitates the acquisition of a very specific expression program, including imprinted genes, for gamete differentiation (Figure 7(c)). Given the epigenetic asymmetry of the gametes of the father and the mother, sensitive unerased marks may differ between chromosomes of paternal and maternal origin in the zygote. The mechanisms involved have yet to be determined, but these observations suggest that the possibilities for transmission of environmental information may differ as a function of the sex of the transmitting parent [188] (Figure 6).

The incomplete erasure of some parental epigenetic marks—DNA methylation and histone marks—and the polycomb and trithorax systems make possible the programming and transgenerational transmission of environmental impacts [184,185]. These regions are thus ideal candidates for the transfer of environmental exposure information. Depending on the chromosome concerned, and particularly if the X and Y chromosomes display such differences, could these regions account for the differences in the effects on female and male offspring? The principal problem here is that the epigenetic mechanisms involved are dynamic and change rapidly with environmental variation. They are also based on multiple strata of partially redundant pathways, which may be synergic, inhibitory, or activating, depending on the context [189,190]. Thus, if the impact of the exposure of the future parents to environmental factors several years before conception affects precisely this type of sequence [12,22,34,147], then these sequences may be responsible for the transgenerational responses observed in the descendants.

In another study, the authors sought to identify the genomic targets of remethylation after implantation in mice. While it was commonly believed that only the imprinted genes escaped the process of postfertilization demethylation, they demonstrated that other genes (not imprinted) could carry on the parental gamete methylation pattern in their promoter. Escaping the erasure of post-fertilization methylation is more prevalent than generally accepted in the mouse genome [191]. Certain genes, particularly genes expressed in the germ line but also genes of somatic tissues, are resistant to the overall demethylation after fertilization and inherit the parental gametes methylation at their promoter. So far, through the genes tested, these data indicate that the transmission of DNA methylation after fertilization comes from the oocyte [191]. In humans, very recent studies show that as in mice the number of methylated sites is much more important than in the sperm into the oocyte [192,193]. An extended but site-specific active and passive demethylation occurs in the male pronucleus and in the female pronucleus isolated at stages PN3-PN4 [192,193].

For repeated sequences, the best known example, in mice, is that of IAP sequences (intracisternal A particle retrotransposon), involved in the transmission of methylated states that interfere with the expression of several loci including the two loci, viable yellow agouti, A^{vy}, and Axin-Fused, AxinFu. These sequences are resistant to postfertilization demethylation [194]. In humans, a single locus shows a similar phenomenon [28].

It is now well established that environmental exposures during any window during the life of a female or a male can induce different phenotypes in offspring. All these studies show that we are beginning to know the genes and sequences that could explain differences in transmission between a male and a female. Yet there is very little research focusing on the effects of the environment on these processes to understand how the memory of these events can be transmitted, what are the supports mediating this information? The preservation of certain mechanisms between human and mouse opens up interesting avenues of investigation.

Mysterious Intermediaries Passing on the Message from Generation to Generation

At fertilization, in both humans and mice, there are many more methylated sites in the spermatozoa than in the oocyte [193]. There is extensive, but site-specific demethylation in the male and female pronuclei after fertilization. This process involves both active and passive mechanisms, depending on the parental origin of the chromosome [193]. It is widely accepted that only imprinted genes escape this process of demethylation. However, one recent study showed that other genes are also resistant [195]. In this mouse model of undernutrition in the grandmother (F0), the spermatozoa of the father (F1) display a disturbance of the methylome in differentially methylated regions (DMRs), with effects on the metabolism of his descendants (F2) [195]. Interestingly, 43% of the DMRs hypomethylated in the F1 were also hypomethylated in the F2 generation and therefore had the potential to affect the development of this subsequent generation. Many of the genes affected are expressed in the germ line, but some are also expressed in somatic tissues. However, although this differential methylation was lost from the F2 generation by the end of gestation, major differences persisted in the expression of genes involved in metabolism located close to these DMRs. It therefore seems unlikely that these changes in expression are directly controlled by DNA methylation [195]. A similar process has been reported for the repercussions in the second generation of the effects of diet-induced maternal obesity [196].

These examples show that epigenetic profiles deregulated early in development are capable of passing the torch to other entities, thereby inducing other, as yet unidentified changes that might affect chromatin architecture, networks of transcription factors, or the differentiation or structure of tissues. In the model of resistance to cocaine addiction, the same modification (histone acetylation) to the same gene (*Bdnf*) was observed in the spermatozoa of the father and in the prefrontal cortex of his resistant male progeny [59]. As histone acetylation is a mark associated with expression, this observation cannot be seen as proof that this is the mechanism responsible for information transfer.

The two examples cited above only appear to be contradictory; they do not in any way exclude the possible involvement of an epigenetic process. The pertinent epigenetic marks involved have probably either not been studied or have not been studied at the appropriate stage.

Given the dialogue known to occur between marks, we would expect more than one type of mark to be involved, together with other, non-epigenetic processes. Are these associations the cause or a consequence of the dynamics of these marks? The key question to be addressed here remains that of the true causal link between epigenetic marks and the observed phenotypes.

Noncoding RNAs

During fertilization, the spermatozoid not only provides the paternal haploid genome but it also releases 24,000 noncoding RNAs (ncRNAs: siRNA, piRNA, and miRNA...) into the oocyte. Sperm RNA has been shown to transmit acquired characters in rodents. In particular, the use of sperm from maltreated animals has been shown to reproduce metabolic or behavioral changes in the descendants similar to those observed in the father [25,30,61,197–201].

One recent report suggested that RNA isolated from sperm might provide the progeny with information about the history of precocious trauma (through maternal stress) in the life of the father, with the effects and responses persisting until the third generation [61]. However, once again, the absence of presumed causal epigenetic alteration suggests that the initial mark may be transposed to other marks or epigenetic complexes in a relay. The epigenetic modifications present in the sperm cells following exposure to maternal stress may thus be converted into other marks, which may or may not be epigenetic in nature, for subsequent transmission [202,203].

The involvement of ncRNA in transgenerational effects and responses was recently demonstrated in an invertebrate species lacking DNA methylation, *C. elegans* [31]. Exposure to viral particles led to the appearance of ncRNAs derived from the virus, which inhibited the expression of the viral genome, by RNA interference mechanisms, over several generations, thereby conferring a transmissible "immunity" [56]. A lack of food during the larval stage also leads to the appearance of microRNAs (miRNAs) targeting transcripts for proteins involved in nutrition and leading to an increase in the longevity of the third generation. These miRNAs cope with all eventualities, as some also target genes that are normally switched off but may be induced in response to stress [57].

Finally, seminal fluid can play different roles and in particular can warn the mother of previous exposure of the father (infection) and affect anxiety, body composition, and metabolic phenotype of the offspring [65]. The oviductal fluid surrounding the embryo varies according to maternal nutrition, metabolic status, and inflammatory parameters, providing a microcosm that reflects the outside world [34].

A comprehensive understanding of all the mechanisms that compromise sex-specific fetal programming will eventually allow a shift from understanding how DOHaD leads to disease to improved methods for avoiding such factors during sex-specific critical windows and to reinforce choices that favor prevention and alleviation of diseases with a DOHaD basis in a sex/gender perspective. Beyond epigenetics stricto sensu, it is the whole set of systems biology that can help to explore the mechanisms involved in the nongenetic transfer of information, memory exhibitions: metabolites, seminal liquid, prions, etc. These concepts compel us to revise the generally accepted ideas to make room for the idea that biological parenthood begins well before birth, even before conception [34].

Skip in Generation

As for genetic diseases, there may be a skip of generation in transmission. A recent study showed that undernutrition in utero disrupts the male methylome and is associated with effects on the intergenerational metabolism associated with DMRs that escape reprogramming [195]. Indeed, according to a Radford et al. study, 43% of DMRs persist and thus the potential to affect the development of the next generation. Yet although the differential methylation is lost in the F2 late in pregnancy, significant tissue-specific differences in the expression of genes involved in metabolism and in the vicinity of these DMRs persist. It is therefore unlikely that these expression changes are directly under the control of DNA methylation. More likely, the effects of epigenetic profiles deregulated earlier during development must be capable of triggering secondary alterations in chromatin architecture, in the regulatory networks of transcription, differentiation, and tissue structure [195].

A similar process was observed in the impact of the effects of maternal obesity induced by diet, on the second generation [196]. In another model of maternal stress [61], there is also a persistence of the effects on the F3. But again, the absence of the alleged causal epigenetic alteration, noncoding RNAs, observed at the previous generation, suggests a transposition of the initial marking to other epigenetic marks or other substrate. It is thus possible that epigenetic changes that occurred in the sperm cells, following exposure to maternal stress, were transferred to other supports, epigenetic or not, such as DNA methylation or MPT histone for maintenance and subsequent transmission [202,203]. Similarly, in humans, access to abundant or restricted food for paternal grandfathers (F0) during childhood result in variations in the risk of diabetes or longevity in the grandson (F2) without apparent effects for the father (F1) [5].

Persistence?

When, how, who can transmit some "acquired characteristics" or "souvenirs" and for how many generations? The number of generations studied depends on the model and generation time. Humans studies generally do not exceed two generations and therefore correspond to intergenerational, not transgenerational effects [5,12,22,204–206].

In contrast, in rodents, there is one model of glucose intolerance by overnutrition over more than 12 generations, but most studies proceed to the F2 and, much more rarely, to the F3 or F4 [6,7,41,207–213]. However, for animals with much shorter generation time as the nematode *C. elegans* [27,56], or *Drosophila*, persistence is observed on several dozen generations. Nonetheless, there is no question of permanence, indeed chromatin was shown to preserve the ability to return to its initial state [214].

WHAT LEVELS OF EVIDENCE: THE LIMITING AND CONFOUNDERS FACTORS

There is still a lack of a unifying model concerning the involvement of epigenetics in inter/transgenerational effects. This easily explains the skepticism or controversy with respect to the precise roles of epigenetics. The most commonly cited example of epigenetic transmission concerns the effect of maternal diet during pregnancy on the coat color in mice Agouti AVY (Viable yellow) [215,216], but is not strictly speaking Lamarckian. Indeed, a change in coat color acquired during the life of the mother is somehow encoded in the germ cells and passed on to offspring. Moreover, the probable duration of these effects is incompatible with the timescale of evolution. The first explanation for this apparent ability of the genome of the offspring to be programmed in response to the maternal environment before birth is that it should be beneficial to the offspring already born with a metabolic response pattern most appropriate for the environment in which it is called to live. However, the most logical explanation is that since the genome should be reprogrammed each generation to optimize its response to the environment, this could rule out a multigenerational effect operating on timescales of evolution [217].

However, in another case—in utero undernutrition for the father—while changes in DNA methylation or DMR were found for a large number of genes in the sperm, these DMR were not found in the offspring. Although in agreement with the observed phenotype, disruption of the expression of metabolic genes persisted. These two examples are contradictory only in appearance and in any case do not exclude the role of epigenetic process involving other marks or other superimposed nonepigenetic mechanisms.

Simply relevant epigenetic marks were probably not studied and at the appropriate stages. Given the number and the dialogues between marks, it would be surprising only one type of mark is involved. This issue was well explored by Kaufman and Rando, who in a recent review on the topic, make a critical analysis of some examples and indicate when a process other than epigenetic may be incriminated or when the facts are not as well established as one might think [218]. There are several ways to check the involvement of an epigenetic mechanism, easily achievable in an animal model, but much more difficult in humans. Ideally, before concluding to a nongenetic transgenerational effect, we should also proceed to (1) massive sequencing to assess the new mutation rate transmitted by maternal or paternal gametes or occurring at a very early stage. It is about 100 mutations per generation, but anyway some areas of the genome (DNA repeats) remain inaccessible with current sequencing methods or are difficult to highlight (duplications, deletions) to verify that no new mutation is causing the observed phenomenon and its persistence over several generations. (2) Massive sequencing to study variations in copy number of repeated elements (CNV) in the genome that could account for the phenotype observed in the offspring, and could be transmitted indefinitely as a new mutation. However, because of their repeated feature, it is not easy from a technical point of view. (3) An embryo transfer, which can separate the early effects related to the influence of the mother; (4) Cross-adoption, the pups being exchanged at birth. (5) In vitro fertilization to separate what is transmitted through semen, the effects due to the male behavior and its interactions with the female, which may lead to a change in the resources allocated by the mother to her young, as has been observed in rats [219] or for the Finch [220]. If the proof of concept does exist, there is a long way to go to have epigenetic markers validated [221], let alone reliable customized nutritional interventions in humans [222]. The main challenge for the individual and especially for the government is "how to avoid these transmissions?" This challenge remains open, but is not yet, apparently, on the agenda!

CONCLUSIONS

The influences of environmental factors on epigenetic processes have revolutionized our view of developmental plasticity and of transgenerational transmission of information or environmental impacts, but several key questions remain unanswered: How do differences linked to the sex of the parents impose sexual dimorphism on the progeny and, even, on subsequent generations? What is the true nature of the impact of environmental factors? What is the nature of the targets of these factors (marks and/or conformation)? What is the nature of the targets to which the information is transferred? How does the stored information persist over generations? What are the windows of sensitivity or insensitivity to these factors? [10,12,22,28,30].

Several critical issues remain to be addressed for unraveling the sexually dimorphic nature of programming in utero. We still know little about the mechanisms underlying the early sex specific expression of genes and gene networks resulting from epigenetic regulation in every tissue, even cell types. Elucidation of the biological basis of differences in male and female development will improve our understanding of the respective contributions of hormones

X- and Y-linked genes, autosomal genes, and their possible synergistic or antagonistic interactions [72,114,223]. Moreover, sexual dimorphism is not based just on different gene expression but also in levels and turnover of proteins, on metabolites, therefore requiring a systems biology approach that has not yet been developed [224]. An understanding of these factors and of the sex-specific genetic and epigenetic architecture of human disease might also reveal the existence of sex-specific protective mechanisms that could be exploited in novel treatments [225].

Nevertheless, the issue that remains to be solved is that of the true causal link between epigenetic marks undoubtedly "associated" with exposure and the phenotype observed and relevant marks observed in the offspring. Is this association, the real cause or a mere consequence of the dynamics of marks [226]? Thus, there is sometimes a same modification (histone acetylation) on the same gene in the father's sperm—exposed to cocaine—and in the prefrontal cortex, indeed a relevant tissue, for offspring who resist cocaine [59]. If the proof of concept does exist, there is a long way to go to have epigenetic markers validated [221] let alone reliable customized nutritional interventions in humans [222].

The genes and sequences escaping reprogramming and the mechanisms involved are beginning to be identified and are good candidates for involvement in transgenerational effects. Studies of the effects of the environment would make it possible to determine whether these sequences carry a memory of these effects or whether other sequences can gain the same capacity to resist mark erasure. By contrast, the processes, epigenetic or otherwise, by which the information is propagated are unknown, as are those underlying the differences in transmission from the father and the mother. Above all, very few studies have focused on the effects of the environment on these processes, to determine how the memory of events can be transmitted and to reveal the nature of the successive intermediary supports. Most reprogramming studies have been carried out in mice [183,227]. The conservation of certain mechanisms between species opens up interesting possibilities.

Thus, sexual dimorphism in gene expression exists in all tissues at any stage and can be perturbed by most environmental factors at a given stage. It makes sense to anticipate that comparing one sex (protected) with the other (susceptible) will provide new strategies to discover epigenetic mechanisms, genes, and pathways involved in disease etiology. In addition, reversal and recovery of sexual dimorphism in gene expression, which can be observed when comparing males and female patients, represent new avenues of investigations for diagnostic, prognostic, and treatment purposes [228,229]. Thus, sex reversal can unveil sequences/mechanisms involved in diseases. Moreover, explorations of changes in sexual dimorphism will be useful to end the underrepresentation of women/females for the development of different treatment strategies more precisely adapted to males and females.

The need to identify relevant sex-dependent epigenetic marks, to monitor their regulation during development and in response to various environmental stress, and to use them as potential markers of disease risk thus remains a major challenge. Given the tissue and stage specificity of sexual dimorphism, animal experiments are instrumental to unravel the role of epigenetic inheritance and noncoding RNAs, in terms of what lessons can be learned for designing and interpreting human studies. Given the sex differences in disease occurrence, severity, and resilience, and in response to different environmental factors, it is important to convince the scientific community working on experimenanimal models to take sex, generation into account and to look at both parents to study both sexes, to take into account resilience, skip in generation, and to study different traits, not just the initial effect, which altogether is far from being the case.

The striking sexual dimorphism for programming trajectories necessitates a considerable revision of current dietary intervention protocols. The identification of sex-specific explanations of the responses and adaptation of males and females to dietary quality, quantity, and other environmental factors should help physicians and patients anticipate the major challenges likely to occur during the patient's lifetime. In that context, placental analyses could be used to identify children at risk of adverse programming. If some characteristics are actually transmitted, the main challenge for the individual and especially for the government is "how to avoid these transmissions?" This challenge remains open, but is not yet, apparently, on the agenda!

Owing to the flexibility of epigenetic processes, the DOHaD and their underlying epigenetic mechanisms offer a new possibility to envisage a comprehensive and evidence-based plan of nutritional, behavioral, and socioeconomic recommendations to apply new cost-effective preventive actions against NCDs, in a sex-specific manner [32,230,231].

References

[1] Arnold AP. The end of gonad-centric sex determination in mammals. Trends Genet 2011;28(2):55–61.

[2] Heckman JJ. The developmental origins of health. Health Econ 2012;21(1):24–9.

[3] Barker DJP, Osmond C. Infant mortality, childhood nutrition and ischaemic heart disease in England and Wales. Lancet 1986;327(8489):1077–81.

[4] Crews D, Gillette R, Scarpino SV, Manikkam M, Savenkova MI, Skinner MK. Epigenetic transgenerational inheritance of altered stress responses. Proc Natl Acad Sci USA 2012;109(23):9143–8.

[5] Pembrey ME. Male-line transgenerational responses in humans. Hum Fertil (Camb) 2010;13(4):268–71.

[6] Nilsson E, Larsen G, Manikkam M, Guerrero-Bosagna C, Savenkova MI, Skinner MK. Environmentally induced epigenetic transgenerational inheritance of ovarian disease. PLoS One 2012;7(5):e36129.

[7] Govorko D, Bekdash RA, Zhang C, Sarkar DK. Male germline transmits fetal alcohol adverse effect on hypothalamic proopiomelanocortin gene across generations. Biol Psychiatry 2012;72(5):378–88.

[8] Pembrey ME, Bygren LO, Kaati G, Edvinsson S, Northstone K, Sjostrom M, et al. Sex-specific, male-line transgenerational responses in humans. Eur J Hum Genet 2006;14(2):159–66.

[9] Koonin EV, Wolf YI. Is evolution Darwinian or/and Lamarckian? Biol Direct 2009;4:42.

[10] Gabory A, Roseboom TJ, Moore T, Moore LG, Junien C. Placental contribution to the origins of sexual dimorphism in health and diseases: sex chromosomes and epigenetics. Biol Sex Differ March 21, 2013;4(1):5.

[11] McCarrey JR. Distinctions between transgenerational and non-transgenerational epimutations. Mol Cell Endocrinol 2014;398(1-2):13–23.

[12] Pembrey M, Saffery R, Bygren LO. Human transgenerational responses to early-life experience: potential impact on development, health and biomedical research. J Med Genet 2014;51(9):563–72.

[13] Zeybel M, Hardy T, Wong YK, Mathers JC, Fox CR, Gackowska A, et al. Multigenerational epigenetic adaptation of the hepatic wound-healing response. Nat Med 2012;18(9):1369–77.

[14] Attig L, Gabory A, Junien C. Early nutrition and epigenetic programming: chasing shadows. Curr Opin Clin Nutr Metab Care 2010;13(3):284–93.

[15] Gabory A, Ferry L, Fajardy I, Jouneau L, Gothie JD, Vige A, et al. Maternal diets trigger sex-specific divergent trajectories of gene expression and epigenetic systems in mouse placenta. PLoS One 2012;7(11):e47986.

[16] Sugathan A, Waxman DJ. Genome-wide analysis of chromatin states reveals distinct mechanisms of sex-dependent gene regulation in male and female mouse liver. Mol Cell Biol 2013;33(18):3594–610.

[17] Teh AL, Pan H, Chen L, Ong ML, Dogra S, Wong J, et al. The effect of genotype and in utero environment on interindividual variation in neonate DNA methylomes. Genome Res 2014;24(7):1064–74.

[18] Clayton JA, Collins FS. Policy: NIH to balance sex in cell and animal studies. Nature 2014;509(7500):282–3.

[19] Curley JP, Mashoodh R, Champagne FA. Epigenetics and the origins of paternal effects. Horm Behav 2011;59(3):306–14.

[20] Morrow EH. The evolution of sex differences in disease. Biol Sex Differ 2015;6:5.

[21] Maklakov AA, Lummaa V. Evolution of sex differences in lifespan and aging: causes and constraints. Bioessays 2013;35(8):717–24.

[22] Grossniklaus U, Kelly WG, Ferguson-Smith AC, Pembrey M, Lindquist S. Transgenerational epigenetic inheritance: how important is it? Nat Rev Genet 2013;14(3):228–35.

[23] Sela M, Kloog Y, Rechavi O. Non-coding RNAs as the bridge between epigenetic mechanisms, lineages and domains of life. J Physiol 2014;592(Pt 11):2369–73.

[24] Cuzin F, Rassoulzadegan M. Non-Mendelian epigenetic heredity: gametic RNAs as epigenetic regulators and transgenerational signals. Essays Biochem 2010;48(1):101–6.

[25] Rassoulzadegan M, Grandjean V, Gounon P, Vincent S, Gillot I, Cuzin F. RNA-mediated non-mendelian inheritance of an epigenetic change in the mouse. Nature 2006;441(7092):469–74.

[26] Somer RA, Thummel CS. Epigenetic inheritance of metabolic state. Curr Opin Genet Dev 2014;27C:43–7.

[27] Rechavi O, Houri-Ze'evi L, Anava S, Goh WS, Kerk SY, Hannon GJ, et al. Starvation-induced transgenerational inheritance of small RNAs in *C. elegans*. Cell 2014;158(2):277–87.

[28] Heard E, Martienssen RA. Transgenerational epigenetic inheritance: myths and mechanisms. Cell 2014;157(1):95–109.

[29] Meaney MJ, Ferguson-Smith AC. Epigenetic regulation of the neural transcriptome: the meaning of the marks. Nat Neurosci 2010;13(11):1313–8.

[30] Daxinger L, Whitelaw E. Understanding transgenerational epigenetic inheritance via the gametes in mammals. Nat Rev Genet 2012;13(3):153–62.

[31] Lim JP, Brunet A. Bridging the transgenerational gap with epigenetic memory. Trends Genet 2013;29(3):176–86.

[32] Ferguson-Smith AC, Patti ME. You are what your dad ate. Cell Metab 2011;13(2):115–7.

[33] Junien C, Gabory A, Attig L. Sexual dimorphism in the XXI(st) century. Med Sci (Paris) 2012;28(2):185–92.

[34] Lane M, Robker RL, Robertson SA. Parenting from before conception. Science 2014;345(6198):756–60.

[35] Dunn GA, Morgan CP, Bale TL. Sex-specificity in transgenerational epigenetic programming. Horm Behav 2010;59(3):290–5.

[36] Drake AJ, Walker BR. The intergenerational effects of fetal programming: non-genomic mechanisms for the inheritance of low birth weight and cardiovascular risk. J Endocrinol 2004;180(1):1–16.

[37] Anderson LM, Riffle L, Wilson R, Travlos GS, Lubomirski MS, Alvord WG. Preconceptional fasting of fathers alters serum glucose in offspring of mice. Nutrition 2006;22(3):327–31.

[38] Attig L, Vige A, Gabory A, Karimi M, Beauger A, Gross MS, et al. Dietary alleviation of maternal obesity and diabetes: increased resistance to diet-induced obesity transcriptional and epigenetic signatures. PLoS One 2013;8(6):e66816.

[39] Delahaye F, Wijetunga NA, Heo HJ, Tozour JN, Zhao YM, Greally JM, et al. Sexual dimorphism in epigenomic responses of stem cells to extreme fetal growth. Nat Commun 2014;5:5187.

[40] Aiken CE, Ozanne SE. Sex differences in developmental programming models. Reproduction 2013;145(1): R1–13.

[41] Dunn GA, Bale TL. Maternal high-fat diet effects on third-generation female body size via the paternal lineage. Endocrinology 2011;152(6):2228–36.

[42] Anway MD, Skinner MK. Epigenetic programming of the germ line: effects of endocrine disruptors on the development of transgenerational disease. Reprod Biomed Online 2008;16(1):23–5.

[43] Martinez D, Pentinat T, Ribo S, Daviaud C, Bloks VW, Cebria J, et al. In utero undernutrition in male mice programs liver lipid metabolism in the second-generation offspring involving altered Lxra DNA methylation. Cell Metab 2014;19(6):941–51.

[44] Alter MD, Gilani AI, Champagne FA, Curley JP, Turner JB, Hen R. Paternal transmission of complex phenotypes in inbred mice. Biol Psychiatry 2009;66(11):1061–6.

[45] Kaati G, Bygren LO, Edvinsson S. Cardiovascular and diabetes mortality determined by nutrition during parents' and grandparents' slow growth period. Eur J Hum Genet 2002;10(11):682–8.

[46] Wu G, Feder A, Cohen H, Kim JJ, Calderon S, Charney DS, et al. Understanding resilience. Front Behav Neurosci 2013;7:10.

[47] Dias BG, Ressler KJ. Parental olfactory experience influences behavior and neural structure in subsequent generations. Nat Neurosci 2014;17(1):89–96.

[48] Szyf M. Lamarck revisited: epigenetic inheritance of ancestral odor fear conditioning. Nat Neurosci 2014;17(1):2–4.

[49] Franklin TB, Linder N, Russig H, Thony B, Mansuy IM. Influence of early stress on social abilities and serotonergic functions across generations in mice. PLoS One 2011;6(7):e21842.

[50] Jordan B. Is resilience an epigenetic phenomenon? Med Sci (Paris) 2013;29(3):325–8.

[51] Klengel T, Mehta D, Anacker C, Rex-Haffner M, Pruessner JC, Pariante CM, et al. Allele-specific FKBP5 DNA demethylation mediates gene-childhood trauma interactions. Nat Neurosci 2013;16(1):33–41.

[52] Scofield MD, Kalivas PW. Forgiving the sins of the fathers. Nat Neurosci 2013;16(1):4–5.

[53] Provencal N, Suderman MJ, Guillemin C, Vitaro F, Cote SM, Hallett M, et al. Association of childhood chronic physical aggression with a DNA methylation signature in adult human T cells. PLoS One 2014;9(4):e89839.

[54] Guillemin C, Provencal N, Suderman M, Cote SM, Vitaro F, Hallett M, et al. DNA methylation signature of childhood chronic physical aggression in T cells of both men and women. PLoS One 2014;9(1):e86822.

[55] Remy JJ. Stable inheritance of an acquired behavior in *Caenorhabditis elegans*. Curr Biol 2010;20(20):R877–8.

[56] Rechavi O, Minevich G, Hobert O. Transgenerational inheritance of an acquired small RNA-based antiviral response in *C. elegans*. Cell 2011;147(6):1248–56.

[57] Rechavi O. Guest list or black list: heritable small RNAs as immunogenic memories. Trends Cell Biol 2014;24(4):212–20.
[58] Attig L, Jais JP, Vigé A, Beauger A, Gross M-S, Gallou-Kabani C, et al. Locus-specific epigenetic changes associated with peripheral leptin-resistance in increased resistance to a high-fat diet in mice born to obese mothers fed a control diet during gestation. J DOHaD 2013;4(Suppl 1):S1–S2.
[59] Vassoler FM, White SL, Schmidt HD, Sadri-Vakili G, Pierce RC. Epigenetic inheritance of a cocaine-resistance phenotype. Nat Neurosci 2013;16(1):42–7.
[60] Yan W. Potential roles of noncoding RNAs in environmental epigenetic transgenerational inheritance. Mol Cell Endocrinol 2014;398(1–2):24–30.
[61] Gapp K, Jawaid A, Sarkies P, Bohacek J, Pelczar P, Prados J, et al. Implication of sperm RNAs in transgenerational inheritance of the effects of early trauma in mice. Nat Neurosci 2014;17(5):667–9.
[62] Junien C, et al. Le nouveau paradigme de l'Origine développementale de la santé et des maladies (DOHaD), Epigénétique, Environnement: preuves et chaînons manquants. Med Sci 2015, in press.
[63] Arai JA, Feig LA. Long-lasting and transgenerational effects of an environmental enrichment on memory formation. Brain Res Bull 2011;85(1–2):30–5.
[64] Junien C. L'empreinte parentale: de la guerre des sexes à la solidarité entre générations. Médecine/Sciences 2000;3:336–44.
[65] Bromfield JJ, Schjenken JE, Chin PY, Care AS, Jasper MJ, Robertson SA. Maternal tract factors contribute to paternal seminal fluid impact on metabolic phenotype in offspring. Proc Natl Acad Sci USA 2014;111(6):2200–5.
[66] Bermejo-Alvarez P, Rizos D, Lonergan P, Gutierrez-Adan A. Transcriptional sexual dimorphism during preimplantation embryo development and its consequences for developmental competence and adult health and disease. Reproduction 2011;141(5):563–70.
[67] Bermejo-Alvarez P, Rizos D, Rath D, Lonergan P, Gutierrez-Adan A. Sex determines the expression level of one third of the actively expressed genes in bovine blastocysts. Proc Natl Acad Sci USA 2010;107(8):3394–9.
[68] Penaloza C, Estevez B, Orlanski S, Sikorska M, Walker R, Smith C, et al. Sex of the cell dictates its response: differential gene expression and sensitivity to cell death inducing stress in male and female cells. FASEB J 2009;23(6):1869–79.
[69] Howerton CL, Bale TL. Prenatal programing: at the intersection of maternal stress and immune activation. Horm Behav 2012;62(3):237–42.
[70] Gabory A, Attig L, Junien C. Sexual dimorphism in environmental epigenetic programming. Mol Cell Endocrinol May 25, 2009;304(1–2):8–18.
[71] Davies W, Wilkinson LS. It is not all hormones: alternative explanations for sexual differentiation of the brain. Brain Res 2006;1126(1):36–45.
[72] Xu J, Disteche CM. Sex differences in brain expression of X- and Y-linked genes. Brain Res 2006;1126(1):50–5.
[73] Kwong WY, Miller DJ, Wilkins AP, Dear MS, Wright JN, Osmond C, et al. Maternal low protein diet restricted to the preimplantation period induces a gender-specific change on hepatic gene expression in rat fetuses. Mol Reprod Dev 2007;74(1):48–56.
[74] Ng SF, Lin RC, Laybutt DR, Barres R, Owens JA, Morris MJ. Chronic high-fat diet in fathers programs beta-cell dysfunction in female rat offspring. Nature 2010;467(7318):963–6.
[75] Clifton VL. Review: sex and the human placenta: mediating differential strategies of fetal growth and survival. Placenta 2010;31(Suppl):S33–9.
[76] Gallou-Kabani C, Gabory A, Tost J, Karimi M, Mayeur S, Lesage J, et al. Sex- and diet-specific changes of imprinted gene expression and DNA methylation in mouse placenta under a high-fat diet. PLoS One 2010;5(12):e14398.
[77] Eriksson JG, Kajantie E, Osmond C, Thornburg K, Barker DJ. Boys live dangerously in the womb. Am J Hum Biol 2009;22(3):330–5.
[78] Okamoto I, Patrat C, Thepot D, Peynot N, Fauque P, Daniel N, et al. Eutherian mammals use diverse strategies to initiate X-chromosome inactivation during development. Nature 2011;472(7343):370–4.
[79] Berletch JB, Yang F, Xu J, Carrel L, Disteche CM. Genes that escape from X inactivation. Hum Genet 2011;130(2):237–45.
[80] Carrel L, Willard HF. X-inactivation profile reveals extensive variability in X-linked gene expression in females. Nature 2005;434(7031):400–4.
[81] Prothero KE, Stahl JM, Carrel L. Dosage compensation and gene expression on the mammalian X chromosome: one plus one does not always equal two. Chromosome Res 2009;17(5):637–48.
[82] Yang F, Babak T, Shendure J, Disteche CM. Global survey of escape from X inactivation by RNA-sequencing in mouse. Genome Res 2010;20(5):614–22.
[83] Dementyeva EV, Shevchenko AI, Anopriyenko OV, Mazurok NA, Elisaphenko EA, Nesterova TB, et al. Difference between random and imprinted X inactivation in common voles. Chromosoma 2010;119(5):541–52.
[84] Chow J, Heard E. X inactivation and the complexities of silencing a sex chromosome. Curr Opin Cell Biol 2009;21(3):359–66.
[85] Patrat C, Okamoto I, Diabangouaya P, Vialon V, Le Baccon P, Chow J, et al. Dynamic changes in paternal X-chromosome activity during imprinted X-chromosome inactivation in mice. Proc Natl Acad Sci USA 2009;106(13):5198–203.

[86] Kobayashi S, Fujihara Y, Mise N, Kaseda K, Abe K, Ishino F, et al. The X-linked imprinted gene family Fthl17 shows predominantly female expression following the two-cell stage in mouse embryos. Nucleic Acids Res 2010;38(11):3672–81.
[87] Bellott DW, Hughes JF, Skaletsky H, Brown LG, Pyntikova T, Cho TJ, et al. Mammalian Y chromosomes retain widely expressed dosage-sensitive regulators. Nature 2014;508(7497):494–9.
[88] Charchar FJ, Bloomer LD, Barnes TA, Cowley MJ, Nelson CP, Wang Y, et al. Inheritance of coronary artery disease in men: an analysis of the role of the Y chromosome. Lancet 2012;379(9819):915–22.
[89] Lesch BJ, Dokshin GA, Young RA, McCarrey JR, Page DC. A set of genes critical to development is epigenetically poised in mouse germ cells from fetal stages through completion of meiosis. Proc Natl Acad Sci USA 2013;110(40):16061–6.
[90] Reinius B, Jazin E. Prenatal sex differences in the human brain. Mol Psychiatry 2009;14(11):987, 8–9.
[91] Reinius B, Shi C, Hengshuo L, Sandhu KS, Radomska KJ, Rosen GD, et al. Female-biased expression of long non-coding RNAs in domains that escape X-inactivation in mouse. BMC Genomics 2010;11:614.
[92] Ross MT, Grafham DV, Coffey AJ, Scherer S, McLay K, Muzny D, et al. The DNA sequence of the human X chromosome. Nature 2005;434(7031):325–37.
[93] Skaletsky H, Kuroda-Kawaguchi T, Minx PJ, Cordum HS, Hillier L, Brown LG, et al. The male-specific region of the human Y chromosome is a mosaic of discrete sequence classes. Nature 2003;423(6942):825–37.
[94] Disteche CM, Filippova GN, Tsuchiya KD. Escape from X inactivation. Cytogenet Genome Res 2002;99(1–4):36–43.
[95] Berletch JB, Yang F, Disteche CM. Escape from X inactivation in mice and humans. Genome Biol 2010;11(6):213.
[96] Xu J, Watkins R, Arnold AP. Sexually dimorphic expression of the X-linked gene Eif2s3x mRNA but not protein in mouse brain. Gene Expr Patterns 2006;6(2):146–55.
[97] Heijmans BT, Tobi EW, Lumey LH, Slagboom PE. The epigenome: archive of the prenatal environment. Epigenetics 2009;4(8):526–31.
[98] Wu Q, Suzuki M. Parental obesity and overweight affect the body-fat accumulation in the offspring: the possible effect of a high-fat diet through epigenetic inheritance. Obes Rev 2006;7(2):201–8.
[99] Bermejo-Alvarez P, Rizos D, Rath D, Lonergan P, Gutierrez-Adan A. Epigenetic differences between male and female bovine blastocysts produced in vitro. Physiol Genomics 2008;32(2):264–72.
[100] Page-Lariviere F, Sirard MA. Spatiotemporal expression of DNA demethylation enzymes and histone demethylases in bovine embryos. Cell Reprogram 2014;16(1):40–53.
[101] Lu H, Cui JY, Gunewardena S, Yoo B, Zhong XB, Klaassen CD. Hepatic ontogeny and tissue distribution of mRNAs of epigenetic modifiers in mice using RNA-sequencing. Epigenetics 2012;7(8):914–29.
[102] Mank JE, Nam K, Brunstrom B, Ellegren H. Ontogenetic complexity of sexual dimorphism and sex-specific selection. Mol Biol Evol 2010;27(7):1570–8.
[103] Xu H, Wang F, Liu Y, Yu Y, Gelernter J, Zhang H. Sex-biased methylome and transcriptome in human prefrontal cortex. Hum Mol Genet 2014;23(5):1260–70.
[104] Ammerpohl O, Bens S, Appari M, Werner R, Korn B, Drop SL, et al. Androgen receptor function links human sexual dimorphism to DNA methylation. PLoS One 2013;8(9):e73288.
[105] Bermejo-Alvarez P, Lonergan P, Rath D, Gutierrez-Adan A, Rizos D. Developmental kinetics and gene expression in male and female bovine embryos produced in vitro with sex-sorted spermatozoa. Reprod Fertil Dev 2010;22(2):426–36.
[106] Zvetkova I, Apedaile A, Ramsahoye B, Mermoud JE, Crompton LA, John R, et al. Global hypomethylation of the genome in XX embryonic stem cells. Nat Genet 2005;37(11):1274–9.
[107] Kang HJ, Kawasawa YI, Cheng F, Zhu Y, Xu X, Li M, et al. Spatio-temporal transcriptome of the human brain. Nature 2012;478(7370):483–9.
[108] Yang X, Schadt EE, Wang S, Wang H, Arnold AP, Ingram-Drake L, et al. Tissue-specific expression and regulation of sexually dimorphic genes in mice. Genome Res 2006;16(8):995–1004.
[109] Qureshi IA, Mehler MF. Genetic and epigenetic underpinnings of sex differences in the brain and in neurological and psychiatric disease susceptibility. Prog Brain Res 2010;186:77–95.
[110] McCarthy MM, Auger AP, Bale TL, De Vries GJ, Dunn GA, Forger NG, et al. The epigenetics of sex differences in the brain. J Neurosci 2009;29(41):12815–23.
[111] Waxman DJ, Holloway MG. Centennial perspective: sex differences in the expression of hepatic drug metabolizing enzymes. Mol Pharmacol 2009;76:215–228.
[112] Ling G, Sugathan A, Mazor T, Fraenkel E, Waxman DJ. Unbiased, genome-wide in vivo mapping of transcriptional regulatory elements reveals sex differences in chromatin structure associated with sex-specific liver gene expression. Mol Cell Biol 2010;30(23):5531–44.
[113] van Nas A, Guhathakurta D, Wang SS, Yehya N, Horvath S, Zhang B, et al. Elucidating the role of gonadal hormones in sexually dimorphic gene coexpression networks. Endocrinology 2009;150(3):1235–49.
[114] Wauthier V, Sugathan A, Meyer RD, Dombkowski AA, Waxman DJ. Intrinsic sex differences in the early growth hormone responsiveness of sex-specific genes in mouse liver. Mol Endocrinol 2010;24(3):667–78.

[115] Gregg C, Zhang J, Weissbourd B, Luo S, Schroth GP, Haig D, et al. High-resolution analysis of parent-of-origin allelic expression in the mouse brain. Science 2010;329(5992):643–8.

[116] Gregg C, Zhang J, Butler JE, Haig D, Dulac C. Sex-specific parent-of-origin allelic expression in the mouse brain. Science 2010;329(5992):682–5.

[117] Xu J, Deng X, Disteche CM. Sex-specific expression of the X-linked histone demethylase gene Jarid1c in brain. PLoS One 2008;3(7):e2553.

[118] Xu J, Deng X, Watkins R, Disteche CM. Sex-specific differences in expression of histone demethylases Utx and Uty in mouse brain and neurons. J Neurosci 2008;28(17):4521–7.

[119] Isensee J, Witt H, Pregla R, Hetzer R, Regitz-Zagrosek V, Noppinger PR. Sexually dimorphic gene expression in the heart of mice and men. J Mol Med (Berl) 2008;86(1):61–74.

[120] Wilda M, Bachner D, Zechner U, Kehrer-Sawatzki H, Vogel W, Hameister H. Do the constraints of human speciation cause expression of the same set of genes in brain, testis, and placenta? Cytogenet Cell Genet 2000;91(1–4):300–2.

[121] Xu J, Burgoyne PS, Arnold AP. Sex differences in sex chromosome gene expression in mouse brain. Hum Mol Genet 2002;11(12):1409–19.

[122] Kelsey G. Epigenetics and the brain: transcriptome sequencing reveals new depths to genomic imprinting. Bioessays 2011;33(5):362–7.

[123] Yuan Y, Chen YP, Boyd-Kirkup J, Khaitovich P, Somel M. Accelerated aging-related transcriptome changes in the female prefrontal cortex. Aging Cell 2012;11(5):894–901.

[124] Lister R, Mukamel EA, Nery JR, Urich M, Puddifoot CA, Johnson ND, et al. Global epigenomic reconfiguration during mammalian brain development. Science 2013;341(6146):1237905.

[125] Moritz KM, Cuffe JS, Wilson LB, Dickinson H, Wlodek ME, Simmons DG, et al. Review: sex specific programming: a critical role for the renal renin-angiotensin system. Placenta 2010;31(Suppl.):S40–6.

[126] Clifton VL, Hodyl NA, Murphy VE, Giles WB, Baxter RC, Smith R. Effect of maternal asthma, inhaled glucocorticoids and cigarette use during pregnancy on the newborn insulin-like growth factor axis. Growth Horm IGF Res 2010;20(1):39–48.

[127] Mao J, Zhang X, Sieli PT, Falduto MT, Torres KE, Rosenfeld CS. Contrasting effects of different maternal diets on sexually dimorphic gene expression in the murine placenta. Proc Natl Acad Sci USA 2010;107(12):5557–62.

[128] Goel N, Bale TL. Examining the intersection of sex and stress in modelling neuropsychiatric disorders. J Neuroendocrinol 2009;21(4):415–20.

[129] Barouki R, Gluckman PD, Grandjean P, Hanson M, Heindel JJ. Developmental origins of non-communicable disease: implications for research and public health. Environ Health 2012;11:42.

[130] Novakovic B, Saffery R. The ever growing complexity of placental epigenetics - role in adverse pregnancy outcomes and fetal programming. Placenta 2012; 33(12):959–70.

[131] Nielsen CH, Larsen A, Nielsen AL. DNA methylation alterations in response to prenatal exposure of maternal cigarette smoking: a persistent epigenetic impact on health from maternal lifestyle?. Arch Toxicol 2014.

[132] Zhao Y, Shi HJ, Xie CM, Chen J, Laue H, Zhang YH. Prenatal phthalate exposure, infant growth, and global DNA methylation of human placenta. Environ Mol Mutagen 2015;56(3):286–92.

[133] Marsit CJ. Influence of environmental exposure on human epigenetic regulation. J Exp Biol 2015;218(Pt 1):71–9.

[134] Hemberger M. Epigenetic landscape required for placental development. Cell Mol Life Sci 2007;64(18): 2422–36.

[135] Cleal JK, Day PL, Hanson MA, Lewis RM. Sex differences in the mRNA levels of housekeeping genes in human placenta. Placenta 2010;31(6):556–7.

[136] Lucas ES, Watkins AJ, Cox AL, Marfy-Smith SJ, Smyth N, Fleming TP. Tissue-specific selection of reference genes is required for expression studies in the mouse model of maternal protein undernutrition. Theriogenology 2011;76(3):558–69.

[137] Sood R, Zehnder JL, Druzin ML, Brown PO. Gene expression patterns in human placenta. Proc Natl Acad Sci USA 2006;103(14):5478–83.

[138] Ishikawa H, Rattigan A, Fundele R, Burgoyne PS. Effects of sex chromosome dosage on placental size in mice. Biol Reprod 2003;69(2):483–8.

[139] Hemberger M. The role of the X chromosome in mammalian extra embryonic development. Cytogenet Genome Res 2002;99(1–4):210–7.

[140] Zechner U, Hemberger M, Constancia M, Orth A, Dragatsis I, Luttges A, et al. Proliferation and growth factor expression in abnormally enlarged placentas of mouse interspecific hybrids. Dev Dyn 2002;224(2): 125–34.

[141] Hemberger M, Kurz H, Orth A, Otto S, Luttges A, Elliott R, et al. Genetic and developmental analysis of X-inactivation in interspecific hybrid mice suggests a role for the Y chromosome in placental dysplasia. Genetics 2001;157(1):341–8.

[142] Walker MG, Fitzgerald B, Keating S, Ray JG, Windrim R, Kingdom JC. Sex-specific basis of severe placental dysfunction leading to extreme preterm delivery. Placenta 2012;33(7):568–71.

[143] Yalcinkaya TM, Siiteri PK, Vigne JL, Licht P, Pavgi S, Frank LG, et al. A mechanism for virilization of female spotted hyenas in utero. Science 1993;260(5116):1929–31.
[144] Al-Khan A, Aye IL, Barsoum I, Borbely A, Cebral E, Cerchi G, et al. IFPA Meeting 2010 Workshops Report II: placental pathology; trophoblast invasion; fetal sex; parasites and the placenta; decidua and embryonic or fetal loss; trophoblast differentiation and syncytialisation. Placenta 2011;32(Suppl. 2):S90–9.
[145] Osei-Kumah A, Smith R, Jurisica I, Caniggia I, Clifton VL. Sex-specific differences in placental global gene expression in pregnancies complicated by asthma. Placenta 2011;32(8):570–8.
[146] Laguna-Barraza R, Bermejo-Alvarez P, Ramos-Ibeas P, de Frutos C, Lopez-Cardona AP, Calle A, et al. Sex-specific embryonic origin of postnatal phenotypic variability. Reprod Fertil Dev 2012;25(1):38–47.
[147] Brydges NM, Jin R, Seckl J, Holmes MC, Drake AJ, Hall J. Juvenile stress enhances anxiety and alters corticosteroid receptor expression in adulthood. Brain Behav 2014;4(1):4–13.
[148] Rando OJ. Daddy issues: paternal effects on phenotype. Cell 2012;151(4):702–8.
[149] Lecomte V, Youngson NA, Maloney CA, Morris MJ. Parental programming: how can we improve study design to discern the molecular mechanisms? Bioessays 2013;35(9):787–93.
[150] Carone BR, Fauquier L, Habib N, Shea JM, Hart CE, Li R, et al. Paternally induced transgenerational environmental reprogramming of metabolic gene expression in mammals. Cell 2010;143(7):1084–96.
[151] Puri D, Dhawan J, Mishra RK. The paternal hidden agenda: epigenetic inheritance through sperm chromatin. Epigenetics 2010;5(5):386–91.
[152] Alminana C, Caballero I, Heath PR, Maleki-Dizaji S, Parrilla I, Cuello C, et al. The battle of the sexes starts in the oviduct: modulation of oviductal transcriptome by X and Y-bearing spermatozoa. BMC Genomics 2014;15:293.
[153] Hackett JA, Surani MA, Beyond DNA. programming and inheritance of parental methylomes. Cell 2013;153(4):737–9.
[154] Duffie R, Bourc'his D. Parental epigenetic asymmetry in mammals. Curr Top Dev Biol 2013;104:293–328.
[155] Watkins AJ, Lucas ES, Wilkins A, Cagampang FR, Fleming TP. Maternal periconceptional and gestational low protein diet affects mouse offspring growth, cardiovascular and adipose phenotype at 1 year of age. PLoS One 2011;6(12):e28745.
[156] Northstone K, Golding J, Davey Smith G, Miller LL, Pembrey M. Prepubertal start of father's smoking and increased body fat in his sons: further characterisation of paternal transgenerational responses. Eur J Hum Genet 2014;22(12):1382–6.
[157] Miller LL, Pembrey M, Davey Smith G, Northstone K, Golding J. Is the growth of the fetus of a non-smoking mother influenced by the smoking of either grandmother while pregnant? PLoS One 2014;9(2):e86781.
[158] Puschendorf M, Terranova R, Boutsma E, Mao X, Isono K, Brykczynska U, et al. PRC1 and Suv39h specify parental asymmetry at constitutive heterochromatin in early mouse embryos. Nat Genet 2008;40(4):411–20.
[159] Johnson GD, Lalancette C, Linnemann AK, Leduc F, Boissonneault G, Krawetz SA. The sperm nucleus: chromatin, RNA, and the nuclear matrix. Reproduction 2011;141(1):21–36.
[160] Hammoud SS, Nix DA, Zhang H, Purwar J, Carrell DT, Cairns BR. Distinctive chromatin in human sperm packages genes for embryo development. Nature 2009;460(7254):473–8.
[161] Carone BR, Hung JH, Hainer SJ, Chou MT, Carone DM, Weng Z, et al. High-resolution mapping of chromatin packaging in mouse embryonic stem cells and sperm. Dev Cell 2014;30(1):11–22.
[162] Erkek S, Hisano M, Liang CY, Gill M, Murr R, Dieker J, et al. Molecular determinants of nucleosome retention at CpG-rich sequences in mouse spermatozoa. Nat Struct Mol Biol 2013;20(7):868–75.
[163] Brykczynska U, Hisano M, Erkek S, Ramos L, Oakeley EJ, Roloff TC, et al. Repressive and active histone methylation mark distinct promoters in human and mouse spermatozoa. Nat Struct Mol Biol 2010;17(6):679–87.
[164] Saitou M, Kurimoto K. Paternal nucleosomes: are they retained in developmental promoters or gene deserts? Dev Cell 2014;30(1):6–8.
[165] Watkins AJ, Platt D, Papenbrock T, Wilkins A, Eckert JJ, Kwong WY, et al. Mouse embryo culture induces changes in postnatal phenotype including raised systolic blood pressure. Proc Natl Acad Sci USA 2007;104(13):5449–54.
[166] Watkins AJ, Wilkins A, Cunningham C, Perry VH, Seet MJ, Osmond C, et al. Low protein diet fed exclusively during mouse oocyte maturation leads to behavioural and cardiovascular abnormalities in offspring. J Physiol 2008;586(8):2231–44.
[167] Thompson JG, Gardner DK, Pugh PA, McMillan WH, Tervit HR. Lamb birth weight is affected by culture system utilized during in vitro pre-elongation development of ovine embryos. Biol Reprod 1995;53(6):1385–91.
[168] Sinclair KD, Allegrucci C, Singh R, Gardner DS, Sebastian S, Bispham J, et al. DNA methylation, insulin resistance, and blood pressure in offspring determined by maternal periconceptional B vitamin and methionine status. Proc Natl Acad Sci USA 2007;104(49):19351–6.
[169] Ecker DJ, Stein P, Xu Z, Williams CJ, Kopf GS, Bilker WB, et al. Long-term effects of culture of preimplantation mouse embryos on behavior. Proc Natl Acad Sci USA 2004;101(6):1595–600.

[170] Fernandez-Gonzalez R, Moreira P, Bilbao A, Jimenez A, Perez-Crespo M, Ramirez MA, et al. Long-term effect of in vitro culture of mouse embryos with serum on mRNA expression of imprinting genes, development, and behavior. Proc Natl Acad Sci USA 2004;101(16):5880–5.

[171] Fleming TP, Kwong WY, Porter R, Ursell E, Fesenko I, Wilkins A, et al. The embryo and its future. Biol Reprod 2004;71(4):1046–54.

[172] Sinclair KD, Singh R. Modelling the developmental origins of health and disease in the early embryo. Theriogenology 2007;67(1):43–53.

[173] Kwong WY, Wild AE, Roberts P, Willis AC, Fleming TP. Maternal undernutrition during the preimplantation period of rat development causes blastocyst abnormalities and programming of postnatal hypertension. Development 2000;127(19):4195–202.

[174] Ashworth CJ, Toma LM, Hunter MG. Nutritional effects on oocyte and embryo development in mammals: implications for reproductive efficiency and environmental sustainability. Philos Trans R Soc Lond B Biol Sci 2009;364(1534):3351–61.

[175] Edwards D. Non-linear normalization and background correction in one-channel cDNA microarray studies. Bioinformatics 2003;19(7):825–33.

[176] Symonds ME, Pearce S, Bispham J, Gardner DS, Stephenson T. Timing of nutrient restriction and programming of fetal adipose tissue development. Proc Nutr Soc 2004;63(3):397–403.

[177] Mitchell M, Schulz SL, Armstrong DT, Lane M. Metabolic and mitochondrial dysfunction in early mouse embryos following maternal dietary protein intervention. Biol Reprod 2009;80(4):622–30.

[178] Duranthon V, Evsikov A, Khan D, Bui LC, Leandri R, Rodrigues A, et al. Differential regulation of LTR retrotransposons during the transition from totipotency to pluripotency in mammalian embryos. Retrovirology 2009;6.

[179] McEvoy JM, Doherty AM, Finnerty M, Sheridan JJ, McGuire L, Blair IS, et al. The relationship between hide cleanliness and bacterial numbers on beef carcasses at a commercial abattoir. Lett Appl Microbiol 2000;30(5):390–5.

[180] Katari S, Turan N, Bibikova M, Erinle O, Chalian R, Foster M, et al. DNA methylation and gene expression differences in children conceived in vitro or in vivo. Hum Mol Genet 2009;18(20):3769–78.

[181] Corry GN, Tanasijevic B, Barry ER, Krueger W, Rasmussen TP. Epigenetic regulatory mechanisms during preimplantation development. Birth Defects Res C Embryo Today 2009;87(4):297–313.

[182] Aranda P, Agirre X, Ballestar E, Andreu EJ, Roman-Gomez J, Prieto I, et al. Epigenetic signatures associated with different levels of differentiation potential in human stem cells. PLoS One 2009;4(11):e7809.

[183] Cowley M, Oakey RJ. Resetting for the next generation. Mol Cell 2012;48(6):819–21.

[184] Hajkova P, Erhardt S, Lane N, Haaf T, El-Maarri O, Reik W, et al. Epigenetic reprogramming in mouse primordial germ cells. Mech Dev 2002;117(1–2):15–23.

[185] Holland ML, Rakyan VK. Transgenerational inheritance of non-genetically determined phenotypes. Biochem Soc Trans 2013;41(3):769–76.

[186] Morrison KE, Rodgers AB, Morgan CP, Bale TL. Epigenetic mechanisms in pubertal brain maturation. Neuroscience 2014;264:17–24.

[187] Paul S, Knott JG. Epigenetic control of cell fate in mouse blastocysts: the role of covalent histone modifications and chromatin remodeling. Mol Reprod Dev 2014;81(2):171–82.

[188] Gill ME, Erkek S, Peters AH. Parental epigenetic control of embryogenesis: a balance between inheritance and reprogramming? Curr Opin Cell Biol 2012;24(3):387–96.

[189] Riising EM, Comet I, Leblanc B, Wu X, Johansen JV, Helin K. Gene silencing triggers polycomb repressive complex 2 recruitment to CpG islands genome wide. Mol Cell 2014;55(3):347–60.

[190] Festenstein R, Chan JP. Context is everything: activators can also repress. Nat Struct Mol Biol 2012;19(10):973–5.

[191] Borgel J, Guibert S, Li Y, Chiba H, Schubeler D, Sasaki H, et al. Targets and dynamics of promoter DNA methylation during early mouse development. Nat Genet 2010;42(12):1093–100.

[192] Guo F, Li X, Liang D, Li T, Zhu P, Guo H, et al. Active and passive demethylation of male and female pronuclear DNA in the mammalian zygote. Cell Stem Cell 2014;15(4):447–58.

[193] Smith ZD, Chan MM, Humm KC, Karnik R, Mekhoubad S, Regev A, et al. DNA methylation dynamics of the human preimplantation embryo. Nature 2014;511(7511):611–5.

[194] Lane N, Dean W, Erhardt S, Hajkova P, Surani A, Walter J, et al. Resistance of IAPs to methylation reprogramming may provide a mechanism for epigenetic inheritance in the mouse. Genesis 2003;35(2):88–93.

[195] Radford EJ, Ito M, Shi H, Corish JA, Yamazawa K, Isganaitis E, et al. In utero effects. In utero undernourishment perturbs the adult sperm methylome and intergenerational metabolism. Science 2014;345(6198):1255903.

[196] King V, Dakin RS, Liu L, Hadoke PW, Walker BR, Seckl JR, et al. Maternal obesity has little effect on the immediate offspring but impacts on the next generation. Endocrinology 2013;154(7):2514–24.

[197] Saab BJ, Mansuy IM. Neuroepigenetics of memory formation and impairment: the role of microRNAs. Neuropharmacology 2014;80C:61–9.

[198] Liu WM, Pang RT, Chiu PC, Wong BP, Lao K, Lee KF, et al. Sperm-borne microRNA-34c is required for the first cleavage division in mouse. Proc Natl Acad Sci USA 2012;109(2):490–4.
[199] Wagner KD, Wagner N, Ghanbarian H, Grandjean V, Gounon P, Cuzin F, et al. RNA induction and inheritance of epigenetic cardiac hypertrophy in the mouse. Dev Cell 2008;14(6):962–9.
[200] Abramowitz LK, Bartolomei MS. Genomic imprinting: recognition and marking of imprinted loci. Curr Opin Genet Dev 2012;22(2):72–8.
[201] Rodgers AB, Morgan CP, Bronson SL, Revello S, Bale TL. Paternal stress exposure alters sperm microRNA content and reprograms offspring HPA stress axis regulation. J Neurosci 2013;33(21):9003–12.
[202] Drake AJ, Seckl JR. Transmission of programming effects across generations. Pediatr Endocrinol Rev 2011;9(2):566–78.
[203] Sharma A. Bioinformatic analysis revealing association of exosomal mRNAs and proteins in epigenetic inheritance. J Theor Biol 2014;357:143–9.
[204] Matthews SG, Phillips DI. Transgenerational inheritance of stress pathology. Exp Neurol 2012;233(1):95–101.
[205] Newbold RR, Padilla-Banks E, Jefferson WN. Adverse effects of the model environmental estrogen diethylstilbestrol are transmitted to subsequent generations. Endocrinology 2006;147(Suppl. 6):S11–7.
[206] Kalfa N, Paris F, Soyer-Gobillard MO, Daures JP, Sultan C. Prevalence of hypospadias in grandsons of women exposed to diethylstilbestrol during pregnancy: a multigenerational national cohort study. Fertil Steril 2011;95(8):2574–7.
[207] Skinner MK, Haque CG, Nilsson E, Bhandari R, McCarrey JR. Environmentally induced transgenerational epigenetic reprogramming of primordial germ cells and the subsequent germ line. PLoS One 2013;8(7):e66318.
[208] Burdge GC, Hoile SP, Uller T, Thomas NA, Gluckman PD, Hanson MA, et al. Progressive, transgenerational changes in offspring phenotype and epigenotype following nutritional transition. PLoS One 2011;6(11):e28282.
[209] Nilsson EE, Anway MD, Stanfield J, Skinner MK. Transgenerational epigenetic effects of the endocrine disruptor vinclozolin on pregnancies and female adult onset disease. Reproduction 2008;135(5):713–21.
[210] Waterland RA, Travisano M, Tahiliani KG. Diet-induced hypermethylation at agouti viable yellow is not inherited transgenerationally through the female. Faseb J 2007;21(12):3380–5.
[211] Chang HS, Anway MD, Rekow SS, Skinner MK. Transgenerational epigenetic imprinting of the male germ-line by endocrine disruptor exposure during gonadal sex determination. Endocrinology 2006;147(12):5524–41.
[212] Benyshek DC, Johnston CS, Martin JF. Glucose metabolism is altered in the adequately-nourished grand-offspring (F3 generation) of rats malnourished during gestation and perinatal life. Diabetologia 2006;49(5):1117–9.
[213] Boucher BJ, Ewen SW, Stowers JM. Betel nut (Areca catechu) consumption and the induction of glucose intolerance in adult CD1 mice and in their F1 and F2 offspring [see comments]. Diabetologia 1994;37(1):49–55.
[214] Seong KH, Li D, Shimizu H, Nakamura R, Ishii S. Inheritance of stress-induced, ATF-2-dependent epigenetic change. Cell 2011;145(7):1049–61.
[215] Jirtle RL, Skinner MK. Environmental epigenomics and disease susceptibility. Nat Rev Genet 2007;8(4):253–62.
[216] Daxinger L, Whitelaw E. Transgenerational epigenetic inheritance: more questions than answers. Genome Res 2010;20(12):1623–8.
[217] Rando OJ, Verstrepen KJ. Timescales of genetic and epigenetic inheritance. Cell 2007;128(4):655–68.
[218] Kaufman PD, Rando OJ. Chromatin as a potential carrier of heritable information. Curr Opin Cell Biol 2010;22(3):284–90.
[219] Dietz DM, Nestler EJ. From father to offspring: paternal transmission of depressive-like behaviors. Neuropsychopharmacology 2012;37(1):311–2.
[220] Pryke SR, Griffith SC. Genetic incompatibility drives sex allocation and maternal investment in a polymorphic finch. Science 2009;323(5921):1605–7.
[221] Lock M. The lure of the epigenome. Lancet 2013;381(9881):1896–7.
[222] Burdge GC, Lillycrop KA. Bridging the gap between epigenetics research and nutritional public health interventions. Genome Med 2010;2(11):80.
[223] Waddell J, McCarthy MM. Sexual differentiation of the brain and ADHD: what is a sex difference in prevalence telling us? Curr Top Behav Neurosci 2012;9:341–60.
[224] Arnold AP, van Nas A, Lusis AJ. Systems biology asks new questions about sex differences. Trends Endocrinol Metab 2009;20(10):471–6.
[225] Arnold AP, Lusis AJ. Understanding the sexome: measuring and reporting sex differences in gene systems. Endocrinology 2012;153(6):2551–5.
[226] Martin DI, Cropley JE, Suter CM. Epigenetics in disease: leader or follower? Epigenetics 2011;6(7):843–8.
[227] Reik W, Kelsey G. Epigenetics: cellular memory erased in human embryos. Nature 2014;511(7511):540–1.
[228] Wu Q, Laloë D, Jaffrezic F, Attig L, Vigé A, Beauger A, et al. Sexual dimorphism of hepatic epigenetic marks and machinery in offspring of obese and diabetic mothers fed a control diet (CD) during periconceptional/gestation/lactation period. J DOHaD 2013;4(Suppl 1):S1–S2.

[229] Jessen HM, Auger AP. Sex differences in epigenetic mechanisms may underlie risk and resilience for mental health disorders. Epigenetics 2012;6(7):857–61.

[230] Doyle O, Harmon CP, Heckman JJ, Tremblay RE. Investing in early human development: timing and economic efficiency. Econ Hum Biol 2009;7(1):1–6.

[231] Hanson MA, Gluckman PD, Ma RC, Matzen P, Biesma RG. Early life opportunities for prevention of diabetes in low and middle income countries. BMC Public Health 2012;12(1):1025.

[232] Cantone I, Fisher AG. Epigenetic programming and reprogramming during development. Nat Struct Mol Biol 2013;20(3):282–9.

[233] Bourc'his D, Voinnet O. A small-RNA perspective on gametogenesis, fertilization, and early zygotic development. Science 2010;330(6004):617–22.

CHAPTER

21

Transgenerational Epigenetic Inheritance: Past Exposures, Future Diseases

Carlos Guerrero-Bosagna

Avian Behavioral Genomics and Physiology Group, IFM Biology, Linköping University, Linköping, Sweden

OUTLINE

INTRODUCTION

The phenomena of transgenerational epigenetic inheritance involve environmental exposures that alter epigenetic patterns in organisms, which correlate with the emergence of altered phenotypes in the later generations. In order for this process to occur, however, it is required that germline epigenetic patterns be altered by environmental exposures and that these alterations then be passed on to next generations through reproduction. Germline, environmentally induced

The Epigenome and Developmental Origins of Health and Disease
http://dx.doi.org/10.1016/B978-0-12-801383-0.00021-9

Copyright © 2016 Elsevier Inc. All rights reserved.

epigenetic alterations will then influence the epigenome of derived somatic cells in the progeny and ultimately their phenotype [1]. Transgenerational epigenetic inheritance has recently gained attention owing to the implications that exposure to environmental toxicants or other stressors can have in lineages of organisms or human populations [2,3]. In animal models, exposure to environmental toxicants such as fungicides, pesticides, or plastic compounds have been shown to produce abnormal reproductive or metabolic phenotypes that are transgenerationally transmitted. These phenotypes observed in animal models include noncommunicable diseases of increasing incidence in human populations, such as obesity, polycystic ovary syndrome (PCOS), pregnancy defects, or fertility impairments [4–6]. Because of this, the study of transgenerational epigenetic inheritance will have a fundamental role in the near future when addressing the causes of noncommunicable disease, as well as the consequences for future generations of current human exposures to environmental toxicants [7]. Moreover, this increased incidence of diseases will be correlated with long-lasting epigenomic alterations in human populations in both gametic and somatic cells.

In order to properly understand the process of transgenerational epigenetic inheritance, it is important to make the distinction between mitotic and meiotic epigenetic inheritance. Mitotic epigenetic inheritance takes place at the cellular level and involves the maintenance in the daughter cells of the epigenetic patterns (e.g., DNA methylation) observed in mother cells. Meiotic epigenetic inheritance, in contrast, involves the maintenance of the epigenetic patterns observed in germ cells of parents in the germ cells of the offspring. Disruptions in mitotic epigenetic inheritance can lead to disease in the exposed individual, but since it is a somatic process it will not affect future generations. Disturbances in meiotic epigenetic inheritance, however, can have consequences that will transcend generations owing to the involvement of gametes. Therefore, disturbances in the germline are fundamental for the process of transgenerational epigenetic inheritance to occur. Only environmentally induced alterations in the germline, whether genetic or epigenetic, will be intrinsically transmitted to next generations and then associate with the formation of altered phenotypes in the progeny (Figure 1). These germline epigenetic alterations, however, can be generated by direct actions from environmental exposures (i.e., endocrine disruptors) that cross the uterine and/or the placental barriers, or indirect actions that alter the physiological responses of somatic cells that are in close contact with the germ cells (i.e., Sertoli cells).

This distinction between somatic and germ cell epigenetic effects is important when interpreting the results that emerge from transgenerational experiments. In the strict sense, a transgenerational effect is visible only in the first generation in which the germline was not developmentally exposed to the environmental stimuli being studied. Since the exposure as an embryo can generate both somatic and germline epigenetic effects, it is not rare to find that F1 effects would differ from F3 effects, which sometimes are even opposite. This is due to the fact that the F1 generation phenotype will express maternally mediated effects that generate somatic epigenetic changes, while the F3 generation phenotype will express effects derived from the germline epigenetic alterations that are ancestrally induced and transgenerationally transmitted. This chapter describes the advances to date that show or suggest transgenerational epigenetic inheritance in this sense, i.e., in which epigenetic effects are observed in the germline after environmental exposures and are transmitted to future generations, influencing phenotype formation and mitotic epigenetic processes in the progeny.

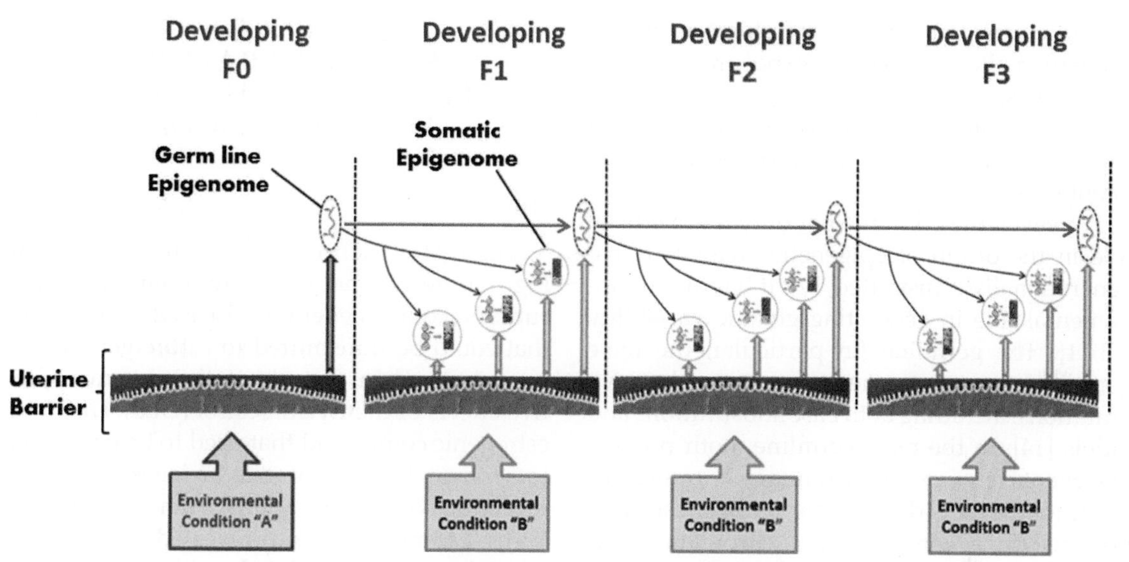

FIGURE 1 Schematic representation of an environmentally induced transgenerational epigenetic process, when altered environmental conditions affect the F1 offspring inside an F0 mother. Environmental stimuli crossing the uterine barrier will affect the developing germline and somatic cells in the embryo. Epigenetic patterns generated are concordant with a commonly experienced environmental condition (A, green arrows). However, a new environmental condition (B, purple arrows) will generate altered epigenomes in both the germline and somatic cells. The altered *somatic* epigenome will be observed only in the F1 generation, which is reflected in transcriptomic and phenotypic changes. The altered *germline* epigenome, in turn, is maintained in subsequent generations (red arrows) and will affect the epigenome and consequently the transcriptome of derived somatic cells (blue arrows). Therefore, the phenotypes observed in F1 are due to disturbances in somatic epigenomes and will not last for several generations (unless this alteration has a secondary effect on the germline). Germline epigenomic alterations, however, will last for several generations and influence the somatic epigenomes, transcriptomes, and phenotypes of the descendants. In summary, somatic epigenetic and phenotypic effects observed in the F1 (somatic) generation might differ from the effects observed in subsequent generations, which are derived from germline epigenomic alterations. In some instances, the effects do not appear in the developmentally exposed individuals at all, but can be seen in their offspring. The specific molecular mechanisms for this are still unknown and deserve future investigation.

A CRUCIAL ROLE FOR THE GERMLINE IN EPIGENETIC AND GENETIC INHERITANCE

The germline development has a crucial role in the phenomena of transgenerational epigenetic inheritance, because germline epigenetic alterations induced by environmental exposures occurring early during development can be transmitted to next generations [8,9]. Epigenetic disturbances are generated early in the development. For example, DNA methylation and expression patterns can be disrupted in primordial gem cells when a maternal exposure to vinclozolin occurs between the onset of gonadal sex determination and after testis cord formation [10]. Altered DNA methylation and expression patterns in primordial germ cells are maintained in subsequent generations [10,11] and will influence the epigenome of the mature sperm. These epigenetic germline alterations are then transgenerationally transmitted and will affect gene expression and epigenetic patterns in somatic tissues [4,12]. This has been shown for somatic tissues that are in close contact with the male or the female germ cells, such as Sertoli [12] and granulosa [4] cells, which present altered epigenome and transcriptome in the F3 generation after an ancestral exposure to vinclozolin. These altered somatic epigenetic marks and gene expression are involved in generating altered phenotypes that are transgenerationally maintained [8,9].

However, the germline not only has a crucial role in the origin of epigenetic changes transgenerationally transmitted, but it has also a fundamental role in generating genetic variability [13,14]. The germline (in particular, the male germline) is very prone to generate de novo mutations, including both base substitutions and indels [14]. In the male germline, both premeiotic events (e.g., oxidative damage or numerous replication events during spermatogenesis) and postmeiotic events (massive chromatin rearrangements or error-prone nonhomologous end-joining repair processes) can lead to increased mutation rates [14]. Interestingly, the normally increased frequency of transitions in CpG sites regarding nonCpG seems to be even higher in the germline [15]. Another example of the role of the male germline in generating genetic variability is the origin of translocations associated with diseases. It has been shown that translocations responsible for the Emmanuel syndrome (non-Roberstsonian translocations) originate exclusively in the germline (in palindromic AT-rich repeats) and are paternally transmitted to the offspring [16,17]. Furthermore, evidence shows that germline DNA mutations or instability can be induced by environmental factors [18] or by early developmental epigenomic changes [19]. Overall, it is becoming increasingly evident that environmental disturbances in the germline can have profound consequences not only in epigenetic marks but also in generating genetic instability in the germline. Moreover, environmental exposures that affect early developmental stages of the germline could induce both genetic and epigenetic variability that generates unpredictable phenotypic consequences in the descendants originating from that developmentally perturbed germline.

EVIDENCE FOR TRANSGENERATIONAL EPIGENETIC INHERITANCE IN ANIMAL MODELS

Initial Studies

The first evidence in animal models indicating that gestational exposure to endocrine disruptors could generate diseased phenotypes that could be transmitted to future generations was obtained in a study testing for reproductive effects of diethylstilbestrol (DES). DES is an estrogenic compound that used to be prescribed to women to prevent miscarriages and other reproductive complications from the 1930s to 1970s. Later observations revealed that this DES administration associated with reproductive defects in the offspring of the women treated (reviewed in Giusti et al. [20]). In the year 2000, while studying reproductive effects of a gestational exposure to DES in rodents, Newbold and collaborators [21] found that administration of DES to pregnant rats during early postimplantation development generated reproductive abnormalities in the following generations. These abnormalities in the F1 and F2 generation male offspring included increased susceptibility for proliferative lesions in the *rete testis* that are usually of rare incidence, as well as increased tumor formation in other reproductive tract tissues [21]. These results led the authors to suggest that epigenetic alterations could be involved in the generation of these reproductive abnormalities and their transmission to future generations.

In the year 2005, while reproductive effects of the fungicide vinclozolin were being studied, there came the first description of an environmentally induced transgenerational inheritance process confirmed to be related to epigenetic changes in the germline [22]. Vinclozolin is a fungicide with antiandrogenic effects (widely used in crops around the world) [23] that had been shown to generate increased apoptosis in spermatogenic cells in the offspring of mothers treated with intraperitoneal injections of the compound during stages of their gestation [24]. Follow-up experiments demonstrated that this increased apoptosis in spermatogenic cells was also observed in the next three generations in the absence of any further exposure [22,25,26]. The transgenerational transmission of this phenotype was attributed to disturbances in the DNA methylation reprogramming that occur during the migration of primordial germ cells (at initial stages of sex determination) [10], which ultimately affected the epigenome of the mature sperm. The epigenomic changes in the germline were detected three generations after this initial maternal exposure to vinclozolin [22,27,28].

Exposure to Other Environmental Toxicants and Drugs

Other abnormalities were then shown to be generated in the male offspring after exposures to a variety of environmental toxicants, which were subsequently transmitted through transgenerational epigenetic inheritance. Transgenerationally transmitted kidney disease was induced after ancestral exposures to vinclozolin [25], dioxin [29], DDT [5], and methoxychlor [30]. Transgenerationally transmitted prostate disease was observed after an ancestral exposure to vinclozolin [25]. Transgenerationally transmitted pubertal abnormalities were induced after ancestral exposures to dioxin [29]. Transgenerationally transmitted testis abnormalities were generated after ancestral exposures to a mixture of pesticides (permethrin and DEET) [31] and DDT [5].

Female reproductive abnormalities were also observed to be induced by developmental exposure to environmental toxicants. Transgenerationally transmitted ovary and kidney abnormalities, as well as polycystic ovary disease, were observed after ancestral exposure to DDT [5] or to methoxychlor [30]; transgenerationally transmitted pubertal abnormalities, primordial follicle loss, and polycystic ovary disease were observed after ancestral exposure to a mixture of plastic compounds (bisphenol A (BPA) and phthalates) [32], to a mixture of pesticides (permethrin and DEET) [31], or to dioxin [29]; transgenerationally transmitted primordial follicle loss and polycystic ovary disease were observed after an ancestral exposure to the hydrocarbon jet fuel 8 [33]. In addition, transgenerationally transmitted obesity was observed in both females and males after ancestral exposure to DDT [5], to methoxychlor [30], or to a mixture of plastic compounds (BPA and phthalates) [32]. Ancestral exposure to all these compounds induced epimutations (i.e., DNA methylation changes) in the male germline in the F3 generation after the initial exposure. The fact that all the compounds tested generated a different set of germline epigenetic alterations [5,6,30,34] opens the conceptual possibility that sperm epimutations can be indicators of both ancestral exposures and susceptibility to diseases [5–7].

Experiments showing transgenerational epigenetic inheritance in animal models have been performed independently by different laboratories around the world. For example, vinclozolin-induced transgenerational effects in the sperm epigenome have also been reported in imprinted genes [35]. The same group also reported that an early developmental exposure to methoxychlor altered sperm DNA methylation in imprinted genes, but this effect persisted only for two generations, disappearing in the third generation after the exposure [36]. Another group showed that an early developmental exposure to di-(2-thylhexyl) phthalate induced transgenerationally transmitted disruption of testicular germ cell associations, reduced sperm count, and decreased sperm motility [37]. An early developmental exposure to the endocrine disruptor DDE, which is a degradation product of DDT, generated increased testicular germ line apoptosis from the F1 to the F3 generations [38]. Additionally, increased rate of infertility and incidence of small testes were observed in the F3 but not in the F1 or F2 generations [38]. In the sperm, decreased expression of IGF2 and a concomitant increase in the expression of H19 was observed from the F1 to the F3, together with altered methylation patterns in the IGF2/H19 DMR [38].

Gestational exposure to the chemotherapeutic drug doxorubicin generated transgenerational effects seen even in the F6 generation females [39]. These included despair-like behaviors, delivery complications, decreased primordial follicle pool, reduced reproductive capacity, and increased rates of neonatal and maternal pregnancy-related deaths, as well as incidence of physical and chromosomal abnormalities [39].

Dietary Exposures and Exercise

In addition to environmental toxicants (generally endocrine disruptors), dietary exposures have also been shown to promote transgenerational epigenetic inheritance of disease states. For example, a gestational dietary exposure to either ethinyl estradiol or high fat have shown to increase susceptibility to mammary tumors in the next generations. For the ethinyl estradiol group, this increased susceptibility was accompanied by transgenerational increases in the occurrence of terminal end buds and in the activity of Dnmt1 in mammary glands [40]. Overfeeding during lactation has been shown to generate many features of metabolic syndrome in mice, including obesity, insulin resistance, and glucose intolerance [41]. Interestingly, some of the phenotypes that emerged were transgenerationally transmitted, such as fasting hyperglycemia and glucose intolerance [41]. Maternal high-fat diet has also been shown to generate increased body weight in the offspring that is, however, only transmitted transgenerationally through the paternal lineage [42]. Interestingly, the F3 male offspring exhibited improved glucose tolerance relative to controls [42]. A study evaluating effects of a sustained exposure across generations found that increases in dietary energy intake that are maintained across generations correlate with progressive alterations in the glucose and fatty acid metabolism [43].

On the other hand, undernutrition has also been shown to promote transgenerational effects. Maternal nutrient restriction has been shown to produce cardiovascular dysfunction in the offspring, which was transmitted for at least three generations, in the absence of any further nutritional restriction [44]. Dietary protein restriction during pregnancy was shown to generate transgenerational transcriptomic effects in the liver of the offspring, which persisted for at least three generations, also in the absence of any further dietary restriction [45]. Recently, it has been shown that undernutrition of pregnant dams generates offspring with low birth weights and metabolic defects. Interestingly, the metabolic phenotype was also observed in the F2 generation, and differentially methylated regions were identified genome-wide in the sperm of F1 males [46].

Maternal consumption of alcohol has been shown to generate transgenerational effects in the expression and DNA methylation of the *POMC* (proopiomelanocortin) gene in the arcuate nucleus of rats [47]. *POMC*-derived peptides have a lower function in patients with a family history of alcoholism and are involved in stress control and in metabolic and immune functions [47]. Moreover, DNA methylation is transgenerationally altered in the male germline but not in the female germline, for three generations [47].

Exercise has also been shown to induce transgenerational effects. A study in mice evaluating transgenerational consequences of premating exercise in the 8-week-old F1 and F2 offspring found that transcriptomic alterations in the gastrocnemius muscle in males were more prominent in the F2 than in the F1 generations [48]. Although the study evaluated the offspring at a very young age, it certainly opens up the possibility that maternal exercise might have important effects that could be observed at older age in future generations. This is an area in which further investigation is needed.

Behavioral Effects

In rats, mate preference has also been reported to be transgenerationally affected by ancestral exposures to endocrine disruptors. In an experiment evaluating mate preference in rats from a lineage ancestrally exposed to vinclozolin, it was found that F3 generation females from either the control or vinclozolin lineage would strongly prefer F3 males from the control lineage over the vinclozolin lineage [49]. This is extremely interesting, since it indicates that females can somehow perceive if an animal descends from a linage in which ancestors were exposed to vinclozolin. The mechanism through which females can perform this sort of discrimination is unknown. More recently, a study showed that both the F1 mice developmentally exposed to BPA and the unexposed F3 descendants displayed higher levels of investigation than controls [50]. F3 BPA lineage mice were also more active than controls and the males, in particular, spent more time investigating urine odor than BPA lineage females and control animals [50]. Recently, it was found that paternal fear conditioning of adult mice with an odorant derived in an increased sensitivity to the odor, which was transmitted to the next two generations. Moreover, this was accompanied by changes in the anatomy of the olfactory bulb and altered germline DNA methylation in the *Olfr151* gene [51].

Fish and Invertebrates

Transgenerational effects have also been reported in other organisms, such as in *Drosophila*, *Artemia*, zebrafish, and *Caenorhabditis elegans*. Exposing fly embryos to heat shock generated somatic and germline chromatin changes in the individuals exposed and somatic chromatin changes that were observed a few generations later. The number of generations to which the somatic effects persisted and the intensity of the effects depended on how many generations were previously exposed [52]. In *Artemia*, exposure to heat shock after hatching and before the reproductive period generated increased survival that lasted for at least three generations. Interestingly, a massive change in global DNA methylation was observed only in the F1 generation, indicating that the heat stress strongly influenced DNA methylation reprogramming [53]. Even though specific DNA methylation changes were not analyzed, it is expected that they would be produced and maintained in subsequent generations. Evidence for this is the level of expression of Hsp70, which was transgenerationally altered [53]. In zebrafish, exposure of juveniles to dioxin during the gonadal differentiation period was shown to generate many changes that lasted for at least two generations. These include higher female-to-male ratio in the population, incidence of otherwise rare scoliosis-like skeleton abnormalities, and reduction in fertility parameters [54,55]. In *C. elegans*, exposure to starvation induced changes in the small RNA expression profile that were observed even three generations later, together with a transgenerational increase in lifespan [56].

EVIDENCE FOR TRANSGENERATIONAL EPIGENETIC INHERITANCE IN HUMANS

In addition to the transgenerational epigenetic effects reported in laboratory animals, recent studies in human populations suggest that past exposures could have an important role in current disease trends [57], which would have occurred through transgenerational epigenetic inheritance [7]. For obvious reasons, transgenerational experiments in humans cannot be performed in a controlled manner as in animal studies. However, a few historical cases of massive exposures of humans to environmental contamination or adverse conditions have been investigated in recent years. These studies have provided valuable insights on the consequences of human exposures for future generations.

As in animal models, one of the first pieces of evidence for transgenerational effects in humans also appeared due to exposures to DES. As previously mentioned, DES started to be prescribed for women in the late 1930s for many reasons. These include prevention of miscarriages, treatment of gonorrheal or atrophic vaginitis, reduction of menopausal symptoms, and also suppression of postpartum lactation (to prevent breast engorgement). It is estimated that between 5 million and 10 million women were exposed to DES during the approximately 40 years of its being used in the United States, until its approval to be prescribed was removed in 1971 [20], after a critical study showed increased incidence of cervicovaginal clear cell cancers in daughters exposed in utero to DES [58]. Other studies followed on the effects of DES. In women directly administered DES, increased risk for breast cancer was observed [20]. This and many other reproductive effects were later seen in daughters of women treated with DES. These included increased risk of malformations and dysfunctions of reproductive organs, poor pregnancy outcome, and immune disorders [20,59]. Not only were the female offspring affected, but also the incidence of hypospadias in the sons of these exposed women was correlated with the gestational exposure to DES [60–62].

In recent years, a series of studies based on historical health records of the Northern Swedish population of Överkalix [57] have produced valuable evidence for the transgenerational phenomena in humans. These studies are based on the particularity that during the nineteenth century food availability in Northern Sweden populations was highly dependent on their crops, due to the lack of connectivity with other regions for food supplies [63]. Therefore, in the first half of the nineteenth century, high variability in food intake from year to year was common and was dependent on the success of crops, which was recorded. Using this information, investigations were performed on the exposure of humans (grandparents) to drastic changes in the availability of food and the consequences in the risk for cardiovascular mortality in the descendants (grandsons) [63]. Interestingly, an increased risk in cardiovascular mortality was observed only in granddaughters whose paternal grandmother was exposed [63]. A previous study from the same group showed that early paternal smoking (before age 11) was associated with increased body mass index in sons but not daughters, that paternal grandfather food supply was related to overall mortality risk of grandsons, and that the paternal grandmother food supply was related to the granddaughter mortality risk [64]. Interestingly, these effects were observed when the exposure of grandparents was during their slow growth period (mid-childhood) or as a fetus or infant, but not when the exposure was during puberty [64]. Altogether, these studies demonstrate that exposure of humans to environmental conditions can have transgenerational consequences, that sensitive periods to these exposures exists during life, and that gender specificity exists in the transmission of these effects. Moreover, this evidence suggests that germline epigenetic processes are involved in the transmission of these environmentally induced effects.

Another historic event in which food restriction was involved was the Dutch Hunger Winter (1944–1945) during the Nazi occupation of parts of the Netherlands in the Second World War. Using well-preserved health-care registries from that time, a study determined that siblings that were in utero during the famine time had reduced blood DNA methylation in the differentially methylated region 2 of the IGF2 gene when compared to their unexposed same-sex siblings [65]. Other studies based on this historical event have revealed that individuals exposed to famine during gestation had higher risk of metabolic disorders (i.e., diabetes and obesity) [66–69], increased blood pressure in response to stress [70], and differences in blood cell DNA methylation in a variety of genes implicated in growth and metabolism [71].

ENVIRONMENTAL EXPOSURES AFFECTING THE MALE GERMLINE: ADULT VERSUS DEVELOPMENTAL EXPOSURES

In addition to early developmental exposures to environmental toxicants, exposures after birth have also been evaluated for their transgenerational effects. Perinatal exposure to BPA was found to impair spermatogenesis and fertility [72] and to induce transgenerational alterations in the expression of steroid receptors and their co-regulators in testis [73]. Also, exposure of six-week-old male mice to benzo(a)pyrene (oral administration) generated impairments in several fertility parameters that were observed up until the F2 generation [74]. The fertility impairments observed included testicular malformations, decreased sperm count, and reduced number of seminiferous tubes with elongated spermatids [74].

Paternal adult exposures can also generate epigenetic modifications in the germline that could be transgenerationally transmitted. Factors such as stress in parents were recently shown in both humans and animal models to generate physiological and somatic epigenetic effects in the offspring [75–77]. Some stress effects have even been reported in the following generations after paternal exposure. It has been shown that paternal stress in male mice affects the hypothalamic–pituitary–adrenal (HPA) axis and micro-RNA expression in the offspring [78]. Another study shows that stress during early pregnancy produces brain dysmasculinization and altered gene and micro-RNA expression in the brain of the F2 male offspring [79]. Also, fear conditioning of adult male mice with an odorant has consequences in the olfactory sensitivity to the conditioning smell and in the neural anatomy in the next two generations [51]. These effects are possibly transmitted through epigenetic alterations in the germline, as it has been described by some of these studies [51,78].

Paternal diet has also been shown to generate offspring effects that would have to be transmitted through the germline. For example, paternal chronic consumption of high-fat diet generates pancreatic beta-cell dysfunction in their female offspring and impaired insulin secretion and glucose tolerance that increased as they aged [80]. A recent study using flies controlled for genetic variability for several generations found that both very high and very low consumption of sugar by parents lead to obesity in the offspring and also to somatic epigenetic changes in the chromatin structure [81]. Interestingly, chromatin alterations were also observed in the paternal sperm, again suggesting that the paternal diet can interfere with germline epigenetic marks that will influence future generations [81]. Epigenetic changes in DNA methylation were also analyzed in the germline of mouse fathers in response to consumption of a low-protein diet [82]. In this case, no major changes in DNA methylation were found, although epigenetic and transcriptomic changes were observed in the liver of the offspring [82].

This evidence on adult exposures that result in transgenerational effects represents new theoretical challenges [83], given that it means that disturbance of major DNA methylation resetting can generate not only transgenerational effects, but also interferences in postnatal spermatogenesis. Although the mechanism through which paternal exposures would affect epigenetic marks in the sperm is not known, it is known that important epigenetic changes take place during adult spermatogenesis, such as in DNA methylation, histones, and chromatin structure [84,85]. Therefore, just as early developmental periods of epigenetic reprogramming are windows of sensitivity to environmental exposures, the epigenetic changes taking place during spermatogenesis may also represent a process particularly sensitive to environmental exposures. This will certainly be an exciting research field in the near future.

CONCLUSIONS AND PERSPECTIVES

The field of developmental origins of health and disease studies altered developmental processes that lead to variable phenotypes [86,87]. These altered phenotypes are induced by environmental exposures that affect developmental processes that involve epigenetic mechanisms. Increasing evidence emerging in animal models shows that disruption of developmental processes affects the adult phenotype not only in one generation, but also in future generations. For example, in animal models diseases such as polycystic ovary, impairments in reproductive function, or obesity can be both environmentally induced and transmitted to next generations. Parallels exist between the phenotypes observed in these animal studies and the noncommunicable diseases of increasing incidence in human populations. Evidence is accumulating toward an important role of developmental origins of health and disease in the etiology of many common human conditions. Nowadays, it is becoming increasingly evident that the etiologies of most noncommunicable diseases are known to be related to environmental exposures during embryonic development or infancy rather than being explained by specific genetic variations [1]. The results obtained from recent investigations on historical events of massive exposure of humans to environmental contamination or to adverse conditions confirm that the environment experienced by the parents does affect the propensity for diseases in their descendants. Diseases such as metabolic abnormalities [88] or female reproductive tract tumors [59] are examples of these effects reported in humans. When diseases present an important transgenerational component after being induced by environmental exposures, then germline epigenetic alterations can be expected. In this sense, epimutations could be used for human health as biomarkers of ancestral exposures and as biomarkers of future propensity to diseases.

Investigations in human health will give increasing importance to transgenerational epigenetic inheritance in the near future. This is because of the repercussions of environmental exposures for future generations and also because of the panorama of global environmental contamination due to industrialization. The new knowledge on environmentally induced transgenerational epigenetic inheritance will also pose new challenges for regulations on the production and use of environmental toxicants. In this sense, regulating agencies should invest in developing screening tools that consider both developmental disturbances and transgenerational effects in order to help develop future regulations and recommendations, with the aim of minimizing exposures of human populations to harmful compounds. The banning of DES and DDT in the United States, and of BPA in other countries, are good initiatives, but represent too little considering the amount of harmful compounds to which human populations are currently exposed and the potential for harmful consequences for future generations.

Acknowledgments

The author greatly appreciates critical revision of the manuscript by Dr Per Jensen and Dr Anita Öst. The author is grateful for funding support by the European Research Council Advanced Research Grant Genewell 322206, held by Dr Per Jensen.

References

[1] Guerrero-Bosagna C, Skinner MK. Environmentally induced epigenetic transgenerational inheritance of phenotype and disease. Mol Cell Endocrinol 2012; 354:3–8.

[2] Burton T, Metcalfe NB. Can environmental conditions experienced in early life influence future generations? Proc R Soc B 2014;281:20140311.

[3] Guerrero-Bosagna C, Skinner MK. Environmentally induced epigenetic transgenerational inheritance of male infertility. Curr Opin Genet Dev 2014;26C: 79–88.

[4] Nilsson E, et al. Environmentally induced epigenetic transgenerational inheritance of ovarian disease. PLoS One 2012;7:e36129.

[5] Skinner MK, et al. Ancestral dichlorodiphenyltrichloroethane (DDT) exposure promotes epigenetic transgenerational inheritance of obesity. BMC Med 2013;11.

[6] Manikkam M, Guerrero-Bosagna C, Tracey R, Haque MM, Skinner MK. Transgenerational actions of environmental compounds on reproductive disease and epigenetic biomarkers of ancestral exposures. PLoS One 2012;7:e31901.

[7] Guerrero-Bosagna C, Jensen P. Globalization, climate change and transgenerational epigenetic inheritance: will our descendants be at risk? Clin Epigen 2015;7(1):8.

[8] Skinner MK. Environmental epigenetic transgenerational inheritance and somatic epigenetic mitotic stability. Epigenetics 2011;6:838–42.

[9] Skinner MK, Manikkam M, Guerrero-Bosagna C. Epigenetic transgenerational actions of environmental factors in disease etiology. Trends Endocrinol Metab 2010;21:214–22.

[10] Skinner MK, et al. Environmentally induced transgenerational epigenetic reprogramming of primordial germ cells and the subsequent germ line. PLoS One 2013;8:e66318.

[11] Anway MD, Rekow SS, Skinner MK. Transgenerational epigenetic programming of the embryonic testis transcriptome. Genomics 2008;91:30–40.

[12] Guerrero-Bosagna C, Savenkova M, Haque MM, Nilsson E, Skinner MK. Environmentally induced epigenetic transgenerational inheritance of altered Sertoli cell transcriptome and epigenome: molecular etiology of male infertility. PLoS One 2013;8:e59922.

[13] Crow JF. The origins, patterns and implications of human spontaneous mutation. Nat Rev Genet 2000;1:40–7.

[14] Gregoire MC, et al. Male-driven de novo mutations in haploid germ cells. Mol Hum Reprod 2013;19:495–9.

[15] Kong A, et al. Rate of de novo mutations and the importance of father's age to disease risk. Nature 2012;488:471–5.

[16] Kurahashi H, Emanuel BS. Unexpectedly high rate of de novo constitutional t(11;22) translocations in sperm from normal males. Nat Genet 2001;29:139–40.

[17] Ohye T, et al. Paternal origin of the de novo constitutional t(11;22)(q23;q11). Eur J Hum Genet 2010;18:783–7.

[18] Olsen AK, et al. Environmental exposure of the mouse germ line: DNA adducts in spermatozoa and formation of de novo mutations during spermatogenesis. PLoS One 2010;5:e11349.

[19] Molaro A, et al. Sperm methylation profiles reveal features of epigenetic inheritance and evolution in primates. Cell 2011;146:1029–41.

[20] Giusti RM, Iwamoto K, Hatch EE. Diethylstilbestrol revisited: a review of the long-term health effects. Ann Intern Med 1995;122:778–88.

[21] Newbold RR, et al. Proliferative lesions and reproductive tract tumors in male descendants of mice exposed developmentally to diethylstilbestrol. Carcinogenesis 2000;21:1355–63.

[22] Anway MD, Cupp AS, Uzumcu M, Skinner MK. Epigenetic transgenerational actions of endocrine disruptors and male fertility. Science 2005;308:1466–9.

[23] Wong C, Kelce WR, Sar M, Wilson EM. Androgen receptor antagonist versus agonist activities of the fungicide vinclozolin relative to hydroxyflutamide. J Biol Chem 1995;270:19998–20003.

[24] Uzumcu M, Suzuki H, Skinner MK. Effect of the antiandrogenic endocrine disruptor vinclozolin on embryonic testis cord formation and postnatal testis development and function. Reprod Toxicol 2004;18:765–74.

[25] Anway MD, Leathers C, Skinner MK. Endocrine disruptor vinclozolin induced epigenetic transgenerational adult-onset disease. Endocrinology 2006;147:5515–23.

[26] Anway MD, Memon MA, Uzumcu M, Skinner MK. Transgenerational effect of the endocrine disruptor vinclozolin on male spermatogenesis. J Androl 2006;27:868–79.

[27] Guerrero-Bosagna C, et al. Epigenetic transgenerational inheritance of vinclozolin induced mouse adult onset disease and associated sperm epigenome biomarkers. Reprod Toxicol 2012;34:694–707.

[28] Guerrero-Bosagna C, Settles M, Lucker B, Skinner MK. Epigenetic transgenerational actions of vinclozolin on promoter regions of the sperm epigenome. PLoS One 2010;5.

[29] Manikkam M, Tracey R, Guerrero-Bosagna C, Skinner MK. Dioxin (TCDD) induces epigenetic transgenerational inheritance of adult onset disease and sperm epimutations. PLoS One 2012;7:e46249.

[30] Manikkam M, Haque MM, Guerrero-Bosagna C, Nilsson EE, Skinner MK. Pesticide methoxychlor promotes the epigenetic transgenerational inheritance of adult-onset disease through the female germline. PLoS One 2014;9:e102091.

[31] Manikkam M, Tracey R, Guerrero-Bosagna C, Skinner MK. Pesticide and insect repellent mixture (Permethrin and DEET) induces epigenetic transgenerational inheritance of disease and sperm epimutations. Reprod Toxicol 2012;34:708–19.

[32] Manikkam M, Tracey R, Guerrero-Bosagna C, Skinner MK. Plastics derived endocrine disruptors (BPA, DEHP and DBP) induce epigenetic transgenerational inheritance of adult-onset disease and sperm epimutations. PLoS One 2012;8:e55387.

[33] Tracey R, Manikkam M, Guerrero-Bosagna C, Skinner MK. Hydrocarbon (Jet fuel JP-8) induces epigenetic transgenerational inheritance of adult-onset disease and sperm epimutations. Reprod Toxicol 2012;36:104–16.

[34] Guerrero-Bosagna C, Weeks S, Skinner MK. Identification of genomic features in environmentally induced epigenetic transgenerational inherited sperm epimutations. PLoS One 2014;9:e100194.

[35] Stouder C, Paoloni-Giacobino A. Transgenerational effects of the endocrine disruptor vinclozolin on the methylation pattern of imprinted genes in the mouse sperm. Reproduction 2010;139:373–9.

[36] Stouder C, Paoloni-Giacobino A. Specific transgenerational imprinting effects of the endocrine disruptor methoxychlor on male gametes. Reproduction 2011;141:207–16.

[37] Doyle TJ, Bowman JL, Windell VL, McLean DJ, Kim KH. Transgenerational effects of di-(2-ethylhexyl) phthalate on testicular germ cell associations and spermatogonial stem cells in mice. Biol Reprod 2013;88:112.

[38] Song Y, et al. Transgenerational impaired male fertility with an Igf2 epigenetic defect in the rat are induced by the endocrine disruptor p,p′-DDE. Hum Reprod 2014;29:2512–21.

[39] Kujjo LL, et al. Chemotherapy-induced late transgenerational effects in mice. PLoS One 2011;6:e17877.

[40] de Assis S, et al. High-fat or ethinyl-oestradiol intake during pregnancy increases mammary cancer risk in several generations of offspring. Nat Commun 2012;3:1053.

[41] Pentinat T, Ramon-Krauel M, Cebria J, Diaz R, Jimenez-Chillaron JC. Transgenerational inheritance of glucose intolerance in a mouse model of neonatal overnutrition. Endocrinology 2010;151:5617–23.

[42] Dunn GA, Bale TL. Maternal high-fat diet effects on third-generation female body size via the paternal lineage. Endocrinology 2011;152:2228–36.

[43] Burdge GC, et al. Progressive, transgenerational changes in offspring phenotype and epigenotype following nutritional transition. PLoS One 2011;6:e28282.

[44] Ponzio BF, Carvalho MH, Fortes ZB, do Carmo Franco M. Implications of maternal nutrient restriction in transgenerational programming of hypertension and endothelial dysfunction across F1–F3 offspring. Life Sci 2012;90:571–7.

[45] Hoile SP, Lillycrop KA, Thomas NA, Hanson MA, Burdge GC. Dietary protein restriction during F0 pregnancy in rats induces transgenerational changes in the hepatic transcriptome in female offspring. PLoS One 2011;6:e21668.

[46] Radford EJ, et al. In utero effects. In utero undernourishment perturbs the adult sperm methylome and intergenerational metabolism. Science 2014;345:1255903.

[47] Govorko D, Bekdash RA, Zhang C, Sarkar DK. Male germline transmits fetal alcohol adverse effect on hypothalamic proopiomelanocortin gene across generations. Biol Psychiatry 2012;72:378–88.

[48] Guth LM, et al. Sex-specific effects of exercise ancestry on metabolic, morphological and gene expression phenotypes in multiple generations of mouse offspring. Exp Physiol 2013;98:1469–84.

[49] Crews D, et al. Transgenerational epigenetic imprints on mate preference. Proc Natl Acad Sci USA 2007;104: 5942–6.

[50] Wolstenholme JT, Goldsby JA, Rissman EF. Transgenerational effects of prenatal bisphenol A on social recognition. Hormones Behav 2013;64:833–9.

[51] Dias BG, Ressler KJ. Parental olfactory experience influences behavior and neural structure in subsequent generations. Nat Neurosci 2014;17:89–96.

[52] Seong KH, Li D, Shimizu H, Nakamura R, Ishii S. Inheritance of stress-induced, ATF-2-dependent epigenetic change. Cell 2011;145:1049–61.

[53] Norouzitallab P, et al. Environmental heat stress induces epigenetic transgenerational inheritance of robustness in parthenogenetic Artemia model. FASEB J 2014;28:3552–63.

[54] Baker TR, King-Heiden TC, Peterson RE, Heideman W. Dioxin induction of transgenerational inheritance of disease in zebrafish. Mol Cell Endocrinol 2014;398: 36–41.

[55] Baker TR, Peterson RE, Heideman W. Using zebrafish as a model system for studying the transgenerational effects of dioxin. Toxicol Sci 2014;138:403–11.

[56] Rechavi O, et al. Starvation-induced transgenerational inheritance of small RNAs in *C. elegans*. Cell 2014;158:277–87.

[57] Pembrey M, Saffery R, Bygren LO, Network in Epigenetic Epidemiology. Human transgenerational responses to early-life experience: potential impact on development, health and biomedical research. J Med Genet 2014;51:563–72.

[58] Herbst AL, Ulfelder H, Poskanzer DC. Adenocarcinoma of the vagina. Association of maternal stilbestrol therapy with tumor appearance in young women. N Engl J Med 1971;284:878–81.

[59] Newbold RR. Lessons learned from perinatal exposure to diethylstilbestrol. Toxicol Appl Pharmacol 2004;199:142–50.

[60] Kalfa N, Paris F, Soyer-Gobillard MO, Daures JP, Sultan C. Prevalence of hypospadias in grandsons of women exposed to diethylstilbestrol during pregnancy: a multigenerational national cohort study. Fertil Steril 2011;95:2574–7.
[61] Klip H, et al. Hypospadias in sons of women exposed to diethylstilbestrol in utero: a cohort study. Lancet 2002;359:1102–7.
[62] Pons JC, Papiernik E, Billon A, Hessabi M, Duyme M. Hypospadias in sons of women exposed to diethylstilbestrol in utero. Prenat Diagn 2005;25:418–9.
[63] Bygren LO, et al. Change in paternal grandmothers' early food supply influenced cardiovascular mortality of the female grandchildren. BMC Genet 2014;15:12.
[64] Pembrey ME, et al. Sex-specific, male-line transgenerational responses in humans. Eur J Hum Genet 2006;14:159–66.
[65] Heijmans BT, et al. Persistent epigenetic differences associated with prenatal exposure to famine in humans. Proc Natl Acad Sci USA 2008;105:17046–9.
[66] de Rooij SR, et al. Impaired insulin secretion after prenatal exposure to the Dutch famine. Diabetes Care 2006;29:1897–901.
[67] de Rooij SR, et al. Glucose tolerance at age 58 and the decline of glucose tolerance in comparison with age 50 in people prenatally exposed to the Dutch famine. Diabetologia 2006;49:637–43.
[68] Lumey LH, Stein AD, Kahn HS, Romijn JA. Lipid profiles in middle-aged men and women after famine exposure during gestation: the Dutch Hunger Winter Families Study. Am J Clin Nutr 2009;89:1737–43.
[69] Veenendaal MV, et al. Transgenerational effects of prenatal exposure to the 1944–45 Dutch famine. BJOG 2013;120:548–53.
[70] Painter RC, et al. Blood pressure response to psychological stressors in adults after prenatal exposure to the Dutch famine. J Hypertens 2006;24:1771–8.
[71] Tobi EW, et al. DNA methylation differences after exposure to prenatal famine are common and timing- and sex-specific. Hum Mol Genet 2009;18:4046–53.
[72] Salian S, Doshi T, Vanage G. Perinatal exposure of rats to Bisphenol A affects fertility of male offspring–an overview. Reprod Toxicol 2011;31:359–62.
[73] Salian S, Doshi T, Vanage G. Impairment in protein expression profile of testicular steroid receptor coregulators in male rat offspring perinatally exposed to Bisphenol A. Life Sci 2009;85:11–8.
[74] Mohamed el, SA, et al. The transgenerational impact of benzo(a)pyrene on murine male fertility. Hum Reprod 2010;25:2427–33.
[75] Goerlich VC, Natt D, Elfwing M, Macdonald B, Jensen P. Transgenerational effects of early experience on behavioral, hormonal and gene expression responses to acute stress in the precocial chicken. Horm Behav 2012;61: 711–8.
[76] Champagne FA. Epigenetic mechanisms and the transgenerational effects of maternal care. Front Neuroendocrinol 2008;29:386–97.
[77] Matthews SG, Phillips DI. Minireview: transgenerational inheritance of the stress response: a new frontier in stress research. Endocrinology 2010;151:7–13.
[78] Rodgers AB, Morgan CP, Bronson SL, Revello S, Bale TL. Paternal stress exposure alters sperm microRNA content and reprograms offspring HPA stress axis regulation. J Neurosci 2013;33:9003–12.
[79] Morgan CP, Bale TL. Early prenatal stress epigenetically programs dysmasculinization in second-generation offspring via the paternal lineage. J Neurosci 2011;31: 11748–55.
[80] Ng SF, et al. Chronic high-fat diet in fathers programs beta-cell dysfunction in female rat offspring. Nature 2010;467:963–6.
[81] Ost A, et al. Paternal diet defines offspring chromatin state and intergenerational obesity. Cell 2014;159:1352–64.
[82] Carone BR, et al. Paternally induced transgenerational environmental reprogramming of metabolic gene expression in mammals. Cell 2010;143:1084–96.
[83] Sharma A. Transgenerational epigenetic inheritance: focus on soma to germline information transfer. Prog Biophys Mol Biol 2013;113:439–46.
[84] Rajender S, Avery K, Agarwal A. Epigenetics, spermatogenesis and male infertility. Mutat Res 2011;727:62–71.
[85] Vlachogiannis G, et al. The Dnmt3L ADD domain controls cytosine methylation establishment during spermatogenesis. Cell Rep 2015;10:944–56.
[86] Godfrey KM, Lillycrop KA, Burdge GC, Gluckman PD, Hanson MA. Epigenetic mechanisms and the mismatch concept of the developmental origins of health and disease. Pediatr Res 2007;61:5R–10R.
[87] Hanson MA, Gluckman PD. Developmental origins of health and disease: new insights. Basic Clin Pharmacol Toxicol 2008;102:90–3.
[88] Somer RA, Thummel CS. Epigenetic inheritance of metabolic state. Curr Opin Genet Dev 2014;27:43–7.

CHAPTER

22

The Placenta and Developmental Origins of Health and Disease

Jane K. Cleal, Rohan M. Lewis

Institute of Developmental Sciences, University of Southampton, Southampton, UK

OUTLINE

INTRODUCTION

The placenta is the interface between the mother and the fetus and its functions create the *in utero* environment in which the fetus develops. The maternal environment can modulate the structure and function of the placenta throughout gestation and thus may affect the development of the fetus and, subsequently, the long-term effects of fetal development on

The Epigenome and Developmental Origins of Health and Disease
http://dx.doi.org/10.1016/B978-0-12-801383-0.00022-0

Copyright © 2016 Elsevier Inc. All rights reserved.

lifelong health. Placental adaptations may act to buffer the effects of adverse maternal environments on the fetus (e.g., under- or over-nutrition, toxic exposures). However, placental adaptations may not be sufficient to protect the fetus and in some cases may have an adverse effect on fetal development. Over time, placental adaptations to changes in the maternal environment have selected for those that were advantageous for survival in terms of evolutionary fitness. The placenta may not, however, be well adapted to cope with new dangers, as demonstrated by the tragedy of thalidomide exposure [1].

The placenta selectively mediates the transfer of nutrients to the fetus and removes toxins and waste products from the fetal circulation [2]. These processes are regulated by factors such as hormones, nutrient levels, blood flow, maternal body composition, and epigenetic mechanisms. Appropriate placental function is vital for optimal fetal development, and failure of the placenta to protect or nourish has significant consequences for the baby. This has long-term implications, as a reduction or acceleration in fetal growth due to an adverse intrauterine environment can influence susceptibility to chronic diseases in adulthood [3]. Placental function, as mediator between maternal environment and fetal growth, has been proposed to be a major determinant of fetal development and the programming of lifelong health [4].

The aim of this chapter is to outline the placental responses to changes to the maternal environment and subsequent influence on fetal programming. Epidemiologic studies provided the initial link between the placenta and the developmental origins hypothesis. In order to understand how the placenta mediates fetal programming, placental function and regulation will first be described. We will then discuss the evidence for placental involvement in programming offspring health, together with the maternal and environmental factors that may influence placental function; the role of epigenetic mechanisms in contributing to these events will also be addressed. Finally, the implications for the identification of individuals at future risk of disease due to poor placental function and the potential opportunities for interventions will be discussed.

EPIDEMIOLOGY AND THE PLACENTA

From the epidemiologic evidence we know that an individual's risk of developing chronic disease in adult life is in part determined by how they grew *in utero*. Babies who do not reach their growth potential *in utero*, or who grow too big, are at increased risk of disease. As the placenta is a central determinant of fetal growth, its structure and function could have an important influence on fetal programming.

Early studies reported associations between cardiovascular disease (CVD) in adulthood and either increased or decreased placental weight [5]. However, subsequent larger studies suggest an association between CVD in adulthood and a higher placental-to-birth-weight ratio (a disproportionately large placenta) rather than placental weight per se [6,7]. Adverse perinatal outcome [8], higher systolic blood pressure in childhood [6,9], and increased rates of coronary heart disease in adulthood [7] were also associated with a higher placental-to-birth-weight ratio. In these pregnancies, the fetus has not grown to its optimum size despite having a relatively large placenta, suggesting that maternal or other environmental factors are decreasing placental efficiency. In addition to weight, placental shape has been reported to be associated with postnatal health issues such as CVD [10,11] and cancer [12]. The shape of the placenta can influence fetal growth in a sex-specific manner [13] and may also reflect the ability of the endometrium to support placental growth. It could therefore be the babies who grow poorly relative to their genetic growth potential, rather than those that are just small, who are at risk of later disease, indicating a graded effect across the normal birth weight range. Fetal overgrowth also represents a risk factor for developing obesity and diabetes later in life and can occur in pregnancies complicated by gestational diabetes mellitus (GDM) [14]. Maternal diabetes can increase placental weight [15], which could increase placental nutrient transport and therefore fetal growth. These changes may contribute to the increased fetal growth and fat deposition seen in offspring of women with GDM [16].

The placenta is important for fetal growth, yet few large studies show associations between placental weight and postnatal phenotype. Placental weight may therefore not be the best marker of the placenta's ability to support the fetus. Placental function and placental efficiency (transport capacity per gram of placental tissue) are more likely to be the features vital for regulating fetal growth and influencing how the fetus copes with changes to the maternal environment. Both these factors are subject to maternal and fetal regulation, responding to maternal and fetal signals in order to balance maternal supply with fetal demand. Measuring placental function directly in humans is difficult; however, placental markers that reflect placental function more directly than placental weight could be identified. Such "placental phenotypes" will be important for identifying those individuals exposed to environmental changes during gestation. For example, placental amino acid transfer capacity is regulated by both maternal and fetal signals; therefore, it may act as a marker of placental adaption to an altered maternal environment [17–19].

PLACENTAL FUNCTION

In order to understand the role of the placenta in fetal programming, it is important to understand its role in regulating fetal growth. The placental barrier protects the fetus from substances in maternal blood, controls the flow of maternal nutrients to the fetus, and removes toxins and waste products from the fetal circulation. As an endocrine organ, the placenta regulates maternal and fetal physiology. In performing these essential functions, the placenta determines the health of the fetus, both *in utero* and in later life. This section will outline the functions of the placenta and how it facilitates the transport of substances required by the fetus while preventing transport of those that are not.

The Placental Barrier

The placenta is a fetal tissue embedded in the wall of the uterus and perfused with maternal blood from the uterine spiral arteries. The structural components of the placenta that mediate transport function are the placental villi; these form tree-like structures that float in a pool of maternal blood and absorb the nutrients. The placental villi form a physical barrier between the maternal and fetal circulations, preventing any mixing of the two circulations. The villi consist of a core of fetal blood vessels surrounded by a layer of syncytiotrophoblast (Figure 1). The syncytiotrophoblast is formed by fusing cytotrophoblast cells (the cell type found in the outer cell mass of the embryo) into a single multinucleated syncytium. This syncytium forms the primary placental barrier to placental transfer and is a transporting epithelium with two polarized membranes. The maternal-facing microvillus membrane (MVM) is in direct contact with maternal blood and increases the surface area available for nutrient exchange, whereas the basal plasma membrane (BM) faces the fetal capillaries. Once across the BM of the syncytiotrophoblast, nutrients can diffuse through the discontinuous layer of cytotrophoblast, connective tissue, and fetal capillary endothelial junctions into fetal blood [20] (Figure 2).

The syncytiotrophoblast is the primary structure that determines which substances cross the placenta (e.g., nutrients and oxygen) and which substances do not (e.g., maternal hormones and certain toxins). This placental barrier protects

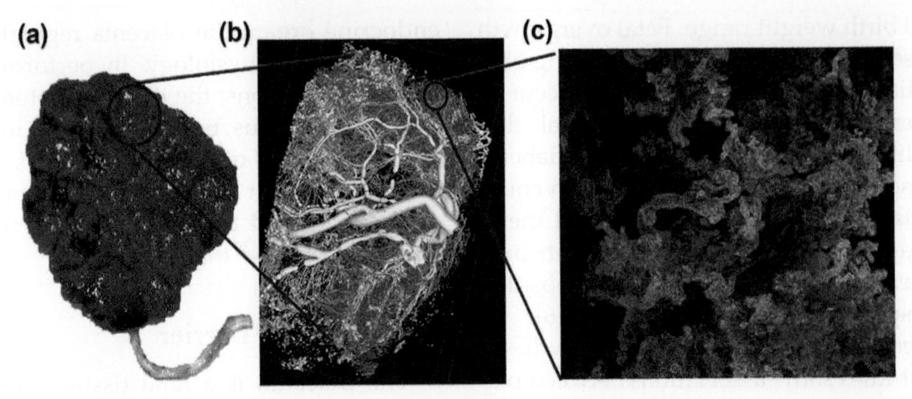

FIGURE 1 **The structural organization of the human placenta.** (a) The maternal-facing side of a human placenta. (b) A micro–computed tomography image of a placental cotyledon. (c) Placental villi with the syncytiotrophoblast highlighted in red and the fetal capillaries in green.

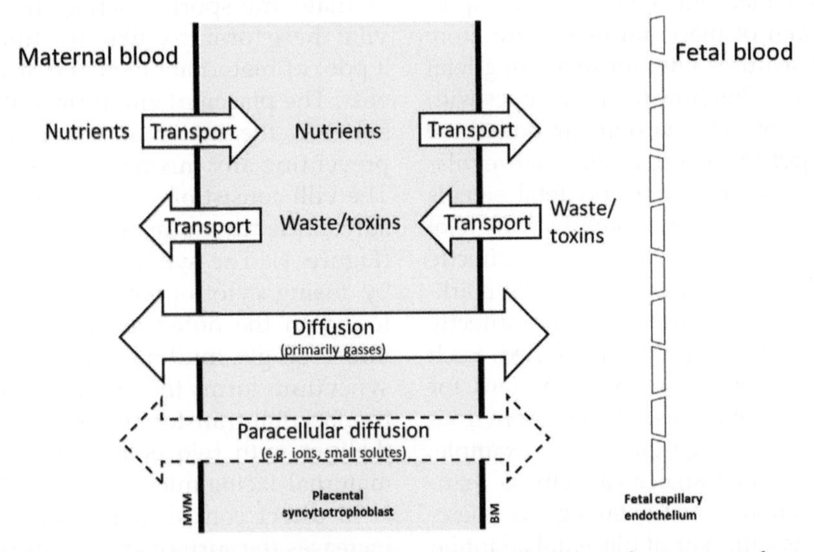

FIGURE 2 **Placental transfer across the syncytiotrophoblast.** Most solutes must be transported across both maternal-facing microvillus (MVM) and fetal-facing basal membranes (BM) of the syncytiotrophoblast. Gases and some lipophilic substances (e.g., anesthetics) can cross by diffusion. Ions and small molecules may also be transferred by paracellular route; the anatomical nature of this is uncertain.

the fetus from many unwanted substances and has specific mechanisms to mediate nutrient transport to the fetus and the removal of fetal waste products to the mother. Small lipophilic molecules, such as oxygen and carbon dioxide, can diffuse across the placenta relatively easily, whereas transfer of cells and molecules that cannot diffuse through lipid membranes is blocked.

Placental Drug Transfer

Drug administration during pregnancy is often necessary to treat the mother for conditions such as diabetes or protect the fetus from conditions such as human immunodeficiency virus (HIV). Pregnant women can also be exposed to environmental toxins such as cigarette smoke.

These potentially harmful drugs can diffuse across the placenta to reach the fetus and in some cases they can be transported by placental transport proteins in the MVM and BM. The ATP binding cassette (ABC) efflux transporters P-glycoprotein (ABCB1) and breast cancer resistance protein (ABCG2) are expressed in human placenta and transport drugs from the placenta back into the maternal circulation, thus protecting the fetus [21,22]. Other members of the multidrug resistance-associated protein (MRP) group are expressed in placenta, such as MRP2 (ABCC2), which appears to protect the fetus from drug exposure [23]. Solute carrier transporters such as organic cation transporters, organic anion transporters, carnitine transporters, and organic anion-transporting polypeptides are expressed in placenta and can mediate drug transfer across the placenta [24]. The placenta also contains phase I and II drug-metabolizing enzymes, including the cytochrome P450 (CYP) superfamily, which play a role in placental drug detoxification and therefore influence how much actually reaches the fetus [25]. Placental drug transporters and metabolizing enzymes are regulated by many factors including diseases during pregnancy, such as GDM [26].

Placental Nutrient Transfer

The placenta transports macro- and micronutrients from mother to fetus, in order to sustain appropriate fetal development. All nutrients require specific transport proteins to be transported across both the MVM and BM of the placental syncytiotrophoblast at levels adequate for the needs of the growing fetus (Figure 2). The expression of these transporters in the placenta mediates nutrient transfer and influences fetal growth. For instance, amino acid transporter expression in human placenta has been identified as a key mediator of fetal growth [27], with impaired placental to fetal amino acid transfer being associated with intrauterine growth restriction [28]. In the human placenta, decreased System A amino acid transporter activity in the MVM is associated with intrauterine growth restriction (IUGR) [29–31]. Animal studies suggest that decreased placental amino acid transport precedes and therefore contributes directly to fetal growth restriction, thus matching the fetal growth to the environment to enhance fetal (and possibly maternal) survival [32]. Studies also indicate that placental nutrient transport can be upregulated in fetal overgrowth. Placental lipases associated with the MVM make fatty acids available for transport and are upregulated with maternal obesity and diabetes [33]. Fatty acid binding proteins within the syncytiotrophoblast are also increased in diabetic pregnancies [34], and both these and lipases are associated with increased fetal growth. It is also thought that in maternal diabetes the placental transport of key nutrients such as glucose, fat, and amino acids is increased, resulting in fetal overgrowth, which is associated with adulthood disease risk [35].

Placental nutrient metabolism is also an important determinant of nutrient transfer to the fetus. The placenta consumes nutrients to meet its own metabolic requirements and can alter the composition of nutrients available for transfer to the fetus. For instance, the human placenta is known to synthesize glutamine, which can be transported to both the maternal and fetal circulations [36]. In maternal diabetes, impaired uptake of long-chain polyunsaturated fatty acids is associated with the placental expression of long-chain acyl CoA synthetase, which may contribute to the effects of diabetes on the offspring [37].

In summary, the placenta has important roles in providing the fetus with nutrients, removing waste products, and protecting the fetus from toxins. Inappropriate nutrient supply or inefficient waste product or toxin removal could affect fetal development and induce long-term effects on adult physiology. Placental function is regulated by signals from both the mother and the fetus, possibly involving hormonal signaling and the placenta sensing maternal plasma nutrient levels. The placenta's specialized structure and functions adapt to changes in the maternal environment in order to maintain fetal growth and protect the fetus from harmful *in utero* exposures.

ENVIRONMENTAL INFLUENCES ON PLACENTAL FUNCTION

The placenta is exposed to numerous maternal and fetal influences, which it perceives via signals from both the maternal and fetal circulations. Hormonal, nutrient, hemodynamic, and epigenetic influences can affect placental development and function, as well as factors such as exercise, drug or alcohol consumption, and smoking during pregnancy [38–40]. The placenta is thought to buffer these harmful changes in the maternal environment in order to protect the fetus and sustain fetal growth in a sex-specific manner [41].

Maternal and Fetal Signals

The maternal environment can influence placental function so as to limit fetal growth to match the mother's ability to sustain the pregnancy, termed "maternal constraint." In humans, birth weight is more closely related to mother's rather than father's birth weight, despite their equivalent genetic contribution, providing evidence for maternal constraint [42]. This is demonstrated by studies that transferred embryos from large horse breeds into mares from smaller breeds, resulting in offspring smaller than their genetic potential [43]. These studies suggest that maternal signals could be controlling the fetal size, or alternatively the size of the uterus is restricting space for fetal growth. It is unclear as to what the maternal signals are, but they could be hormonal, such as glucocorticoids, or metabolic signals, such as the levels of specific nutrients [17]. Maternal hormones such as leptin and IGF2 can increase placental nutrient uptake [44], and efflux of toxins from the placenta can also be hormonally regulated. For example, the multidrug resistance gene ABCB1, which transports drugs out of the fetal circulation, is regulated by progesterone and estrogen [45]. Placental function is thought to be up- or downregulated in response to maternal plasma nutrient levels. Nutrient availability can regulate placental nutrient transporters such as amino acid transporters [32] through intracellular signaling systems such as mTOR [46]. Interestingly, it has been shown that mTOR regulates placental transport of leucine and that the activity of the mTOR signaling pathway in placenta is downregulated in IUGR [47].

It is difficult to study fetal signals to the placenta, however, they are thought to regulate placental function in order to control fetal growth. Indeed, a placental-specific *Igf2* knockout in mice indicates that the fetus can regulate placental function to sustain its own growth [48]. This gene knockout resulted in reduced placental size, with these placentas able to transport more nutrients per gram of placenta, possibly in response to fetal signals. Although the final result was growth restriction, fetal growth was initially sustained owing to increased placental capacity [48]; this parallels observations in human studies [49].

Maternal Body Composition

Maternal body composition could be a determinant of placental function possibly reflecting the mother's nutrient reserves and capacity to support the pregnancy, although the mechanisms remain unclear [50,51]. Epidemiologic studies indicate that maternal body composition is related to adult health in the offspring and may therefore affect placental function and fetal development. Low weight gain during pregnancy and maternal thinness have been associated with reduced neonatal bone mass and raised blood pressure in the offspring [52–55], whereas maternal adiposity during pregnancy, particularly in shorter women, has been linked with cardiovascular and metabolic disorders in the offspring in later life [56–58]. Mothers with a higher prepregnancy body mass index (BMI) can also have children with increased neonatal fat mass, which has further health implications [59,60].

The associations between maternal body composition and fetal growth may involve changes in placental function, with the placenta acting as a mediator between maternal supply and fetal demand for nutrients. Maternal obesity can be associated with larger babies, possibly mediated by the corresponding increase in placental amino acid transporters [61]. Conversely, maternal obesity has been associated with smaller babies and a corresponding reduction in the transport of the amino acid taurine [62]. Placental fatty acid uptake and transporter expression has been shown to be reduced in the placentas of male but not female babies following maternal obesity, suggesting sex-specific placental adaptations [63]. An increased maternal BMI at the start of pregnancy has also been shown to result in impaired placental vascular function [64] and reduced placental efficiency [65]. Maternal thinness is thought to induce impaired materno-placental nutrient supply that alters blood flow distribution in the fetus, with long-term consequences for cardiovascular and metabolic function [66]. Maternal body composition before pregnancy, specifically a larger muscle mass, is associated with increased placental amino acid transport activity in the term placenta [50]. Maternal prepregnancy body composition may represents the metabolic environment to which the placenta is exposed during development. This could alter placental function, possibly through the action of nutrient-sensing pathways such as the mTOR pathway [67].

Maternal Diet

Alterations in the macronutrient content of the maternal diet (carbohydrate, fat, and protein) in animal models have long-term effects on the offspring [68,69], and human studies suggest similar effects [68,69]. In addition, the balance of nutrients in the maternal diet has been associated with physiological changes such as hypertension [70,71], insulin deficiency, and impaired glucose tolerance in the offspring in human cohorts [72]. Undernutrition, overnutrition, and dietary balance of nutrients can alter placental weight, structure, and function, indicating a role for the placenta in the programming effects.

Placental growth can be either decreased or increased by maternal undernutrition, depending on the severity and timing of the insult [73,74]. Maternal nutrient restriction throughout pregnancy or periods of fasting, such as during Ramadan, can influence placental growth [75,76]. In rats, maternal dietary protein restriction has been shown to affect placental structure [32,77] and reduce amino acid transporter expression [32]. Placental transport capacity decreases before fetal growth restriction, indicating that the placenta has causal effects on fetal development [32]. The placenta can therefore alter its function in response to the maternal environment in order to regulate fetal growth. While fetal growth restriction may have adverse long-term consequences, the reduced growth may be a necessary adaptation to ensure fetal survival, as it reduces the demands on the undernourished mother so as to maintain the pregnancy.

If the placenta senses increased nutrient levels it could upregulate glucose and key amino acid transporters, as seen with hyperglycemia during pregnancy [78]. In the pregnant rat, a transient hyperglycemia in early gestation resulting from glucose injections has been shown to increase both placental size and fetal growth, even though placental amino acid transport was reduced [79]. The balance of nutrients in the maternal diet is important for both placental function and fetal development. A high-carbohydrate diet in women during early pregnancy leads to lower placental and birth weights, especially when combined with low protein intake in late pregnancy [80]. Increased placental weight was also associated with famine exposure in early pregnancy and high food intake in mid–late pregnancy in the Dutch famine study [81].

Micronutrient shortage affects placental function and ability to support fetal development. Maternal iron restriction, for example, alters placental structure in rats [77], and placental transfer of micronutrients such as zinc, iron, copper, calcium, and vitamins A, D, and E may play a role in fetal programming [82]. Vitamin D insufficiency is common in women of childbearing age and can result in reduced fetal growth, changes in the shape of the fetal femur, and reduced childhood bone mineral density [83,84]. Placental transport of vitamin D and the effects of maternal vitamin D on placental function are likely to be important for fetal development. Indeed, human FGR is associated with reduced placental expression of the vitamin D receptor, which mediates the metabolism and transcriptional effects of vitamin D [85]. Vitamin D regulates the expression of the calcium transporter *PMCA3*, whose expression in placenta is positively associated with offspring bone mass development [86,87]. Many of the components of the vitamin D pathway in human placenta are subject to epigenetic modifications [88]. These changes could influence placental vitamin D transport and the actions of vitamin D within the placenta, which could have subsequent effects on fetal growth.

Maternal undernutrition has been reported to have different effects on placentas from male and female fetuses in terms of placental size and the relationships between placental shape and hypertension in the offspring [13,76]. These sex differences reflect different evolutionary strategies in males and females. Males appear to maintain placental function and keep growing, risking more severe growth restriction or death, whereas female fetuses adapt growth rate to optimize survival in a poor postnatal environment [41]. Overnutrition can also induce sex-specific effects on the placenta. For example, in rats increased maternal caloric intake decreased placental weight only in placentas of female fetuses [89]. This suggests that different interactions between the maternal environment and placental function underlie differences in the fetal programming of lifelong health in males and females.

Uteroplacental and Umbilical Blood Supply

Placental function requires sufficient blood flow through both the maternal uteroplacental and fetal umbilical circulations. A sufficient uteroplacental blood supply during pregnancy is achieved by increased blood flow to the uterus and remodeling of the spiral arteries. Failure to fulfill these or other maternal adaptations, such as increased cardiac output, reduced vascular resistance, and altered hematocrit levels, can result in reduced fetal growth and cardiovascular adaptations in the offspring [90,91]. In mice, reduced uteroplacental blood supply results in altered placental structure, vasculature, and amino acid transport that may underlie the effects on fetal growth [92,93]. Women that have undergone uterine artery embolization can maintain a pregnancy, suggesting that compensatory remodeling can occur in uterine vasculature [94]. Although the volume of blood flow to the intervillous space is important, the efficiency with which it mixes around the villi may also be important. Indeed, modeling of blood flow through the spiral arteries suggests that failure to remodel in preeclampsia may not impair blood flow but instead may alter the speed at which blood enters the intervillous space and the efficiency of mixing [90]. The effects of impaired umbilical blood flow are most evident in the case of FGR with absent or reversed end diastolic blood flow [95]. In these pregnancies, impaired development of the placental villi creates a high-resistance circulation and the fetal heart is unable to create enough pressure to maintain blood flow throughout the cardiac cycle [96].

Maternal Exercise

Moderate exercise is recommended during pregnancy, as it is thought to reduce the risk of pregnancy complications including back pain and mood disorders and to regulate excessive pregnancy weight gain, which can have adverse effects on the mother and the offspring [97,98]. During maternal exercise, there are transient reductions in uterine blood flow to the placenta, which reduce nutrient availability to the placenta and fetus [99]. However, the overall improvement in maternal cardiovascular function resulting from the exercise may increase placental perfusion and be beneficial to the fetus. Moderate exercise does not appear to adversely affect the fetus, as it is not associated with any major decrease in fetal or placental weight. Indeed, there is some evidence that moderate exercise may be protective against the occurrence of small- and large-for-gestational-age babies [100] and may reduce offspring adiposity at birth and in later childhood [101]. Exercise-induced changes to maternal body composition, blood flow to the placenta, or metabolism may affect placental function, although currently there is little evidence. One controlled study suggests that exercise in early but not late pregnancy can increase placental size, with effects on fetal growth leading to a bigger baby [102].

Maternal Stress and Glucocorticoids

Glucocorticoids such as cortisol are released during maternal illness or stress, and their effects on placental function and placental transfer have implications for fetal maturation and lifelong health [103]. Fetal exposure to maternal glucocorticoids is influenced by placental levels of 11-beta-hydroxysteroid dehydrogenase type 1 (11β-HSD1) and type 2 (11β-HSD2) [104]. 11β-HSD2, which converts cortisol into inactive cortisone, decreases during placental infection, allowing more cortisol to cross to the fetal circulation [105]. In rodents, increased placental transfer of glucocorticoids leads to hypertension in the offspring [106], and in humans decreased placental expression of both enzymes results in reduced birth weight [107,108]. Consistent with this, the methylation status of placental 11β-HSD2 and the glucocorticoid receptor promoter is inversely related to birth weight [109,110]. In mice, glucocorticoid exposure also results in altered placental amino acid transporter expression and reduced glucose transport and transporter expression, which may underlie the effects of glucocorticoids on fetal growth [106,111].

Placental function may be affected by steroid medications used during pregnancy. Betamethasone is used to treat babies that are likely to be premature, as this matures the fetal lungs in preparation for birth. Betamethasone induces alterations in placental 11β-HSD2 activity [112], but is not deactivated by 11β-HSD2 and therefore can cross the placenta and influence the fetus. Inhaled steroids used for asthma treatment have also been linked with altered placental structure and 11β-HSD2 activity [113,114]. The long-term effects of maternal steroid treatment on the offspring are not known, but reduced glucose tolerance was observed in children of mothers treated during pregnancy with betamethasone in one follow-up study [115]. There may be sex differences in the placental and fetal response to maternal stress and glucocorticoids. Indeed, in humans, the placentas of female fetuses are more sensitive to environmental changes in glucocorticoid levels [112], possibly due to sex differences in placental glucocorticoid receptor expression [116].

Environmental Toxins

The placenta cannot always protect the fetus from exposure to environmental toxins and the placenta itself may be affected by the exposure, thus influencing its function. Smoking is one such example, whereby nicotine alters placental nutrient transport capacity and reduces uterine blood flow and placental nutrient supply, resulting in smaller babies [117,118]. The placenta can protect the fetus by transferring some harmful substances from the fetal circulation back to the mother using transporters such as the ATP transporters. Unfortunately, some life-threatening substances may be transported directly to the fetus by other transport proteins in the placental barrier.

Alcohol can cross the placenta and may directly affect fetal development or result in fetal alcohol syndrome, with growth retardation, facial-cranial and organ anomalies [119], and effects on central nervous system development [120]. Maternal alcohol consumption has been associated with decreased placental size, nutrient transport capacity, and endocrine function, which has implications for fetal development [121].

Smoking during pregnancy causes infant morbidity and mortality and reduced birth weight and may influence later disease risk [122]. Maternal smoking also affects placental structure, transporter, and enzyme activity [118], and these effects not only may make the baby small but may also have longer-term effects on its health, particularly in relation to skeletal development. Maternal smoking can decrease bone mass, but not fat mass, in childhood [59,123]; and if this effect persists, it could affect peak bone mass attainment. It could be that cadmium present in tobacco smoke is competing with calcium for transporter binding, which therefore affects fetal bone development [124]. Interestingly, maternal smoking alters the methylation of placental and embryonic genes, which suggests a mechanism by which the effects of smoking are transmitted into later life [125].

Fetal development can be seriously affected by recreational drug use during pregnancy, with the consequences depending on the drug, the dose, and the gestational age of the fetus. Many recreational drugs cross the placenta and so can act on the fetus. For example, the maternal use of opiates or cocaine results in changes to the placental vasculature that could underlie the associated changes in fetal growth [126]. Delta-9-tetrahydrocannabinol in cannabis crosses the placenta [127], and as cannabinoid receptors are expressed in the placenta, this drug may directly affect placental function [128,129]. The effects of cannabis include fetal growth restriction [130], placental abruption, preterm birth, stillbirths, and spontaneous miscarriages [127,131]. Placental transfer of drugs can be of benefit to the fetus in some cases; for example, maternal-to-fetal transfer of antiretroviral drugs helps protect the fetus from vertical transmission of HIV [132].

In summary, the maternal environment may influence placental function, with consequences for fetal development. Understanding how the maternal environment affects placental function is particularly important as this could lead to interventions to improve placental function, with long-term benefits for the baby. The mechanisms involved in regulating placental function are not well understood. They may, however, involve epigenetic changes affecting placental development or nutrient transport–related genes mediated by maternal factors during pregnancy [133].

EPIGENETICS AND THE PLACENTA

Changes to the intrauterine environment may result in epigenetic changes to the placenta or fetus, with the subsequent effects persisting throughout the lifespan. The placental epigenome is important for the regulation of placental development and function and any alterations may therefore affect subsequent fetal development [134]. Maternal factors thought to influence placental epigenetics include nutrition [135], obesity [136], depression and anxiety [137], and smoking [138] during pregnancy. The placenta contains many unique imprinted genes that have been linked to changes in fetal growth [139]. Imprinted genes have specific epigenetic marks that could make them more susceptible to environmental programming. The epigenetic profile of the placenta may contribute to the "placental phenotype" and help identify those individuals exposed to environmental changes during gestation.

The Placental Epigenome

The placenta is subject to epigenetic regulation in terms of normal placental development and parental imprinting and as a consequence of the maternal environment. Epigenetic regulation includes DNA methylation within control regions of genes and histone modifications that affect the packaging of DNA within the cell. The effects of micro-RNAs (miRNA) on gene transcription are also recognized as a mechanism of epigenetic gene regulation.

There is extensive demethylation and remethylation of the genome during gametogenesis and in the preimplantation embryo, meaning these may be critical windows for the laying down of epigenetic modifications [140]. When the DNA methylation levels are reestablished following blastocyst hatching, the trophectoderm-derived cells remain hypomethylated compared to inner cell mass-derived cells [141]. However, this early methylation remodeling does not affect the DNA methylation patterns of the imprinted genes [142]. The trophectoderm mediates implantation of the embryo and develops into the placenta; this trophoblast migration and invasion involves genes that can be regulated by DNA methylation [143]. The early environment in the fallopian tubes and uterus could therefore influence epigenetic modifications in the trophectoderm, which may affect placental development and function. Altered placental development may, in turn, affect how the fetus develops and programme its risk of adult disease.

Compared to other tissues, human placenta has low global DNA methylation: 2.5–3% compared to 4–5% 5-methylcytosine content [144]. The DNA methylation profile is also different in the placenta, with the hypomethylation mainly at repetitive DNA elements such as long interspersed elements (*LINE-1*) [144,145]. The reasons for this hypomethylation are not fully understood, but could be to allow the production of placental-specific transcripts from alternative promoters [145]. There is some epigenetic variation across the placental structure that occurs early in development (e.g., *LINE-1* methylation) and after villus tree formation (e.g., methylation at the *H19/IGF2* imprinting control region) [146]. The different cell types that make up the placenta could have different epigenetic profiles according to their specific function and developmental lineage; however, studies suggest that placental methylation generally reflects that of the cytotrophoblast cells [147]. Throughout gestation, there are also changes in methylation levels within the placenta, which may underlie the observed changes in gene expression across pregnancy [148]. There is generally an increase in both promoter-specific and global methylation toward term, which includes many immune-related genes reflecting adaptation to pregnancy [149,150]. Also, key nutrient transporters such as the glucose transporters (*GLUT3*) are regulated by methylation across gestation, suggesting effects on placental function [151]. Changes in methylation levels across gestation and increased variation in methylation toward term may contribute to the programming effects of adverse exposures at specific time points during pregnancy [149].

Environmental Influences on Placental Epigenetics

The unique epigenetic profile of the placenta can be modified by the maternal factors previously associated with effects on the offspring [133]. Global methylation in the human placenta is reduced by maternal conditions that influence fetal development, such as GDM and preeclampsia, and increased by maternal obesity [152,153]. Other studies have shown differentially methylated genes in placentas exposed to GDM, with the genes affected being those involved in metabolic disease pathways [154]. Indeed, impaired maternal glucose tolerance results in altered DNA methylation profiles in the placental IGF system, potentially affecting the fetal growth [155]. Increased maternal glucose levels have also been associated with reduced methylation in specific placental genes involved in metabolism, suggesting that both maternal adiposity and nutrition are important factors regulating placental epigenetics [156]. The effects of specific alterations to the maternal diet on placental epigenetics have been studied in a small number of human cohorts; for example, increased choline intake during pregnancy is associated with higher placental gene methylation [157]. In rodent models, maternal calorie restriction during pregnancy is associated with global hypomethylation in the placenta [158], whereas a high-fat diet fed to pregnant mice resulted in sex-specific differential methylation in the placenta [159]. Calorie restriction in mice, resulting in fetal growth restriction, has been shown to be accompanied by reduced placental glucose transporter expression mediated by increased DNA methylation [160].

Other modifiable maternal factors that could affect the placental epigenome include the consumption of alcohol or smoking during pregnancy. Maternal cigarette smoking can differentially alter DNA methylation at repetitive elements (*LINE-1*) and specific genes within the placenta, while maternal alcohol intake is associated with increased placental *LINE-1* DNA methylation [161]. One gene affected by maternal smoking is *cytochrome P450 1A1* (*CYP1A1*), which is important for detoxifying the harmful compounds in tobacco smoke. DNA methylation within the promoter region of *CYP1A1* is reduced in the placentas of women who smoke, which corresponds with increased gene expression [138]. This could represent a placental protective response to the increased exposure to toxins.

Placental miRNAs, considered to act as translational silencers [162], have been implicated in placental function and pathogenesis. The chromosome 19 miRNA cluster shows placental-specific expression and is involved in the regulation of trophoblast migration [163]. There is altered expression of specific miRNAs in placentas from women with preeclampsia [164]; for example, there is dysregulation of the miR-106a~363 cluster [165] and alterations in the expression of other miRNAs associated with trophoblast migration [166]. GDM also affects placental miRNA expression, with the placental-specific miR-518 upregulated alongside downregulation of its target protein PPARα [167]. Smoking during pregnancy can affect the expression of miRNAs in human placenta, with the developmentally important microRNAs miR-16, miR-21, and miR-146a downregulated in the placentas of smokers [168]. Interestingly, there are placental-specific miRNAs that show potential as biomarkers, as these can be measured in the maternal plasma [169]. These studies highlight the importance of the placental epigenome in terms of fetal development.

Placental Epigenetic Influences on the Fetus

The epigenetic profile of the placenta could reflect the *in utero* environment and corresponding changes in fetal growth and development. Placental DNA methylation may also explain the associations between placental size or function and fetal growth. In humans, decreased or increased methylation of specific genes has been demonstrated in placentas from both small- and large-for-gestational-age babies [109,170]. The reduction in placental size in the low-protein rat model is also accompanied by changes in the methylation of placental genes associated with IUGR [171]. These may be placental adaptive mechanisms to protect the fetus from an adverse environment. Increased methylation at repetitive elements of the genome is associated with increased birth weight in normal pregnancies [161] and with the placental overgrowth in partial hydatidiform moles [145]. Whereas increased methylation in the promoter of genes important for placental development and protection from maternal hormones is associated with lower birth weight [172,173]. Owing to the difficulties in collecting placental samples, many studies use umbilical cord samples instead. Using umbilical cord as a representative fetal tissue, studies have shown that methylation levels in the retinoid X receptor-α and endothelial nitric oxide synthase promoter can predict childhood fat mass [174]. Several candidate genes have been identified supporting the idea that placental methylation could be used as a marker of fetal and postnatal development and the subsequent risk of disease in adulthood. Studies are now also investigating the effects of other placental epigenetic modifications on fetal outcome and placental miRNAs have been associated with changes in fetal growth [175].

Placental Imprinting

Genomic imprinting involves specific mammalian genes that are monoallelically expressed in a parent-of-origin–specific manner owing to epigenetic marks laid down in the germ cell [176]. Imprinted genes are expressed in the placenta and are essential for normal placental development, with many imprinted genes being important for fetal growth [177]. For example, paternally inherited genes such as *IGF2* have epigenetic modifications that enhance nutrient transport and fetoplacental growth, while maternally inherited genes such as *IGF2R* have epigenetic modifications that limit fetal growth [178,179]. Such imprinting in the placenta may have evolved owing to conflict between the parental genomes for allocation of maternal resources to fetal growth. Studying these genes will help us understand the central role of the placenta in controlling maternal–fetal nutrient partitioning and fetal growth [180]. This mechanism may act to balance resource allocation between the current pregnancy, the mother, and her future pregnancies as an evolutionary strategy for the species. However, investing less in a single pregnancy may have lifelong health consequences for the individual offspring. Altered expression of imprinted genes leads to developmental abnormalities in the placenta and fetus [181]. Imprinted genes in the placenta are responsive to changes in the maternal environment and may help adapt placental phenotype to match the prevailing nutritional conditions. Indeed, increased placental expression of imprinted genes (*Igf2, Dlk1, Snrpn, Grb10,* and *H19*) is seen with excessive consumption of sugar and fat during pregnancy in mice, alongside reduced placental weight and altered fetal growth [182]. A maternal low-protein diet in mice increased the placental expression of imprinted genes involved in fetoplacental growth, such as *Igf2*, with the placental-specific *Igf2P0* transcript influencing the placental adaptation to undernutrition [183,184]. No changes to DNA methylation were observed in these studies, but methylation on the paternal allele at the *Igf2/H19* imprinting control region has been observed following changes to the maternal environment. This region has reduced methylation following alcohol exposure during gestation in mice, which was associated with placental growth restriction [185]. Placental imprinted genes may act as nutrient sensors that epigenetically regulate the placental phenotype and mediate fetal acquisition of nutrients. This will have long-term effects on the risk of developing adulthood disease.

Placental Influences on Fetal Epigenetics

Changes to the maternal environment are transmitted to the fetus by the placenta and can alter the epigenetic status within the genome of the offspring. For example, animal studies demonstrate that maternal diet [186] or uterine blood flow [187] during pregnancy can regulate DNA methylation and subsequent expression of specific genes, which influences the phenotype of the offspring.

Placental nutrient transport function will determine the nutrient availability to the fetus, which may then programme epigenetic changes in the fetus. More specifically, the placenta may influence fetal epigenetic programming through its control of the supply of folate to the fetus. Placental folate transfer is important owing to its role as a methyl donor in DNA methylation and epigenetic programming [186]. A lack of folate in the maternal diet, or impaired placental folate transport, may adversely affect normal epigenetic regulation in the fetus. This has been elegantly demonstrated in agouti mice, where maternal dietary folate concentration alters the coat color in the offspring through epigenetic modifications [188,189]. Changes to specific genes in rat offspring (GR and PPARα) following a maternal low-protein diet are also prevented by supplementing the maternal diet with folic acid [186].

THE "PLACENTAL PHENOTYPE," INTERVENTIONS, AND TREATMENTS

The association between impaired growth *in utero* and chronic disease in later life means that in assessing an individual's risk of later ill health, it may be important to know how they developed before birth. This would allow the provision of more personalized health advice to those at increased risk. If there were evidence that particular individuals were at greater risk, this may encourage those individuals to alter their lifestyles in more healthy ways from an early age.

One way to determine how the baby grew *in utero* could be to study the placenta, as birth weight is a poor proxy for fetal nutrition. It has been proposed that there are "placental phenotypes" (identified from a combination of nutrient transporter activity/expression, blood flow, and morphology at birth) that distinguish between fetal growth restriction or overgrowth *in utero*, as opposed to genetically small or large babies [190]. If placental biomarkers could be identified and validated, they could be routinely measured to indicate reduced or accelerated fetal growth during gestation. These biomarkers may provide better information on the risk of ill health in later life, as it is likely to be those with altered fetal growth who will suffer the effects of programming.

Interventions

As our understanding of placental function improves, it may become possible to intervene to improve placental function and therefore fetal growth. Identifying placental phenotypes may also allow targeting of *in utero* exposures that are important for fetal growth. Animal studies demonstrate that interventions can prevent the adverse effects of a poor intrauterine environment. For example, giving rat dams folate during pregnancy prevents some of the adverse long-term effects of a low-protein maternal diet [186]. Translating the findings of animal studies into humans has obvious challenges in terms of demonstration of safety.

CONCLUDING REMARKS AND FUTURE DIRECTIONS

The ability of the placenta to support fetal growth has long-term health consequences for the offspring. If placental function is inadequate, or if placental efficiency is down- or upregulated, the development of the fetus may be compromised, leading to increased risk of postnatal disease. Placental regulation of fetal development in response to maternal environmental changes may be mediated by short-term humoral and hormonal changes or longer-term alterations in placental epigenetic status or structure. As the

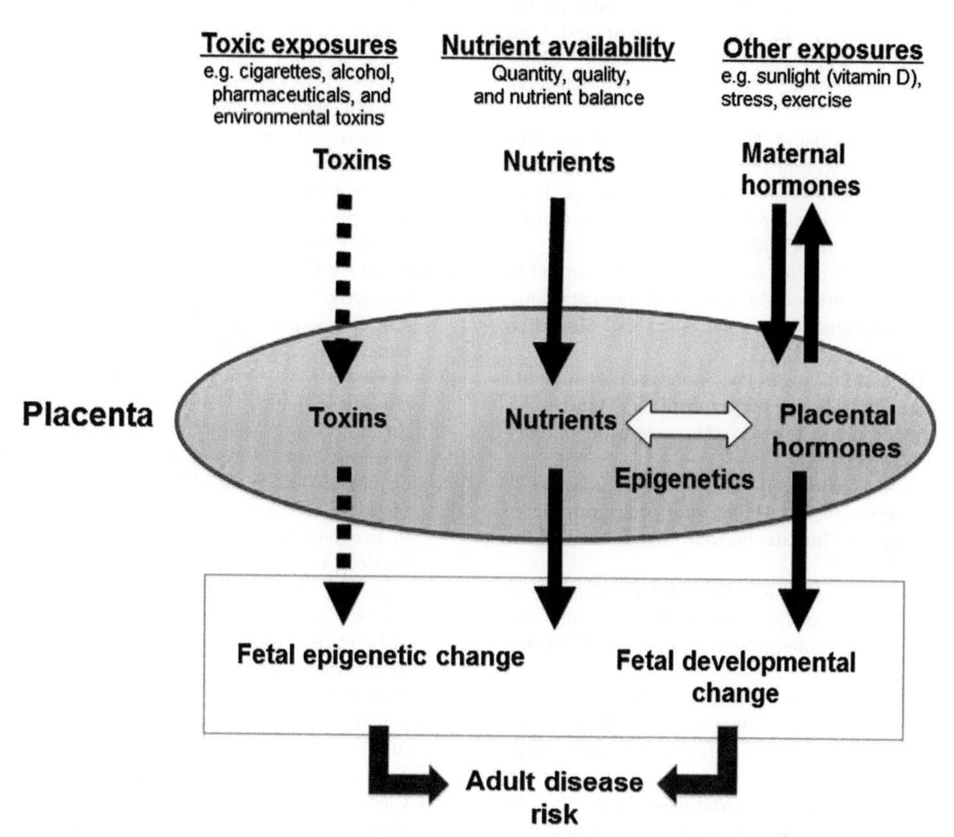

FIGURE 3 **The interactions between the maternal environment, placenta, fetal development, and disease risk.**

placenta is fetal in origin, it may provide epigenetic biomarkers that represent the changes made by the fetus in response to the changes in the maternal environment (Figure 3).

We now have a greater appreciation of the active role played by the placenta in determining fetal nutrient supply. Future research needs to be targeted on understanding the maternal determinants of placental function. Understanding the causes of placental dysfunction and the subsequent fetal programming effects may allow the development of better-informed dietary and lifestyle guidelines for women before and during pregnancy.

References

[1] Dally A. Thalidomide: was the tragedy preventable? Lancet April 18, 1998;351(9110):1197–9.
[2] Jansson T, Myatt L, Powell TL. The role of trophoblast nutrient and ion transporters in the development of pregnancy complications and adult disease. Curr Vasc Pharmacol October 2009;7(4):521–33.
[3] Hanson MA, Gluckman PD. Early developmental conditioning of later health and disease: physiology or pathophysiology? Physiol Rev October 2014;94(4):1027–76.
[4] Lewis RM, Cleal JK, Hanson MA. Review: placenta, evolution and lifelong health. Placenta February 2012;(33 Suppl):S28–32.
[5] Godfrey KM. The role of the placenta in fetal programming-a review. Placenta April 2002;(23 Suppl A):S20–7.

[6] Hemachandra AH, Klebanoff MA, Duggan AK, Hardy JB, Furth SL. The association between intrauterine growth restriction in the full-term infant and high blood pressure at age 7 years: results from the Collaborative Perinatal Project. Int J Epidemiol August 2006;35(4):871–7.

[7] Risnes KR, Romundstad PR, Nilsen TI, Eskild A, Vatten LJ. Placental weight relative to birth weight and long-term cardiovascular mortality: findings from a cohort of 31,307 men and women. Am J Epidemiol September 1, 2009;170(5):622–31.

[8] Shehata F, Levin I, Shrim A, Ata B, Weisz B, Gamzu R, et al. Placenta/birthweight ratio and perinatal outcome: a retrospective cohort analysis. BJOG May 2011;118(6):741–7.

[9] Wen X, Triche EW, Hogan JW, Shenassa ED, Buka SL. Association between placental morphology and childhood systolic blood pressure. Hypertension January 2011;57(1):48–55.

[10] Winder NR, Krishnaveni GV, Hill JC, Karat CL, Fall CH, Veena SR, et al. Placental programming of blood pressure in Indian children. Acta Paediatr May 2011;100(5):653–60.

[11] Barker DJ, Thornburg KL, Osmond C, Kajantie E, Eriksson JG. The surface area of the placenta and hypertension in the offspring in later life. Int J Dev Biol 2010;54(2–3):525–30.

[12] Barker DJ, Osmond C, Thornburg KL, Kajantie E, Eriksson JG. The shape of the placental surface at birth and colorectal cancer in later life. Am J Hum Biol July 2013;25(4):566–8.

[13] Alwasel SH, Harrath AH, Aldahmash WM, Abotalib Z, Nyengaard JR, Osmond C, et al. Sex differences in regional specialisation across the placental surface. Placenta June 2014;35(6):365–9.

[14] Lewis RM, Demmelmair H, Gaillard R, Godfrey KM, Hauguel-de MS, Huppertz B, et al. The placental exposome: placental determinants of fetal adiposity and postnatal body composition. Ann Nutr Metab 2013;63(3):208–15.

[15] Lao TT, Lee CP, Wong WM. Placental weight to birthweight ratio is increased in mild gestational glucose intolerance. Placenta March 1997;18(2–3):227–30.

[16] de Santis MS, Taricco E, Radaelli T, Spada E, Rigano S, Ferrazzi E, et al. Growth of fetal lean mass and fetal fat mass in gestational diabetes. Ultrasound Obstet Gynecol September 2010;36(3):328–37.

[17] Roos S, Powell TL, Jansson T. Placental mTOR links maternal nutrient availability to fetal growth. Biochem Soc Trans February 2009;37(Pt 1):295–8.

[18] Ericsson A, Hamark B, Jansson N, Johansson BR, Powell TL, Jansson T. Hormonal regulation of glucose and system A amino acid transport in first trimester placental villous fragments. Am J Physiol Regul Integr Comp Physiol March 2005;288(3):R656–62.

[19] Constancia M, Angiolini E, Sandovici I, Smith P, Smith R, Kelsey G, et al. Adaptation of nutrient supply to fetal demand in the mouse involves interaction between the Igf2 gene and placental transporter systems. Proc Natl Acad Sci USA December 27, 2005;102(52):19219–24.

[20] Leach L, Firth JA. Fine structure of the paracellular junctions of terminal villous capillaries in the perfused human placenta. Cell Tissue Res June 1992;268(3):447–52.

[21] Ceckova-Novotna M, Pavek P, Staud F. P-glycoprotein in the placenta: expression, localization, regulation and function. Reprod Toxicol October 2006;22(3):400–10.

[22] Hahnova-Cygalova L, Ceckova M, Staud F. Fetoprotective activity of breast cancer resistance protein (BCRP, ABCG2): expression and function throughout pregnancy. Drug Metab Rev February 2011;43(1):53–68.

[23] St-Pierre MV, Serrano MA, Macias RI, Dubs U, Hoechli M, Lauper U, et al. Expression of members of the multidrug resistance protein family in human term placenta. Am J Physiol Regul Integr Comp Physiol October 2000;279(4):R1495–503.

[24] Staud F, Cerveny L, Ceckova M. Pharmacotherapy in pregnancy; effect of ABC and SLC transporters on drug transport across the placenta and fetal drug exposure. J Drug Target November 2012;20(9):736–63.

[25] Pavek P, Smutny T. Nuclear receptors in regulation of biotransformation enzymes and drug transporters in the placental barrier. Drug Metab Rev February 2014;46(1):19–32.

[26] Anger GJ, Cressman AM, Piquette-Miller M. Expression of ABC efflux transporters in placenta from women with insulin-managed diabetes. PLoS One 2012;7(4):e35027.

[27] Cleal JK, Glazier JD, Ntani G, Crozier SR, Day PE, Harvey NC, et al. Facilitated transporters mediate net efflux of amino acids to the fetus across the basal membrane of the placental syncytiotrophoblast. J Physiol February 15, 2011;589(Pt 4):987–97.

[28] Cetin I. Placental transport of amino acids in normal and growth-restricted pregnancies. Eur J Obstet Gynecol Reprod Biol September 22, 2003;110(Suppl. 1):S50–4.

[29] Mahendran D, Donnai P, Glazier JD, D'Souza SW, Boyd RD, Sibley CP. Amino acid (system A) transporter activity in microvillous membrane vesicles from the placentas of appropriate and small for gestational age babies. Pediatr Res November 1993;34(5):661–5.

[30] Glazier JD, Cetin I, Perugino G, Ronzoni S, Grey AM, Mahendran D, et al. Association between the activity of the system A amino acid transporter in the microvillous plasma membrane of the human placenta and severity of fetal compromise in intrauterine growth restriction. Pediatr Res October 1997;42(4):514–9.

[31] Jansson T, Ylven K, Wennergren M, Powell TL. Glucose transport and system A activity in syncytiotrophoblast microvillous and basal plasma membranes in intrauterine growth restriction. Placenta May 2002;23(5):392–9.

[32] Jansson N, Pettersson J, Haafiz A, Ericsson A, Palmberg I, Tranberg M, et al. Down-regulation of placental transport of amino acids precedes the development of intrauterine growth restriction in rats fed a low protein diet. J Physiol November 1, 2006;576(Pt 3):935–46.
[33] Gauster M, Hiden U, van PM, Frank S, Wadsack C, Hauguel-de MS, et al. Dysregulation of placental endothelial lipase in obese women with gestational diabetes mellitus. Diabetes October 2011;60(10):2457–64.
[34] Magnusson AL, Waterman IJ, Wennergren M, Jansson T, Powell TL. Triglyceride hydrolase activities and expression of fatty acid binding proteins in the human placenta in pregnancies complicated by intrauterine growth restriction and diabetes. J Clin Endocrinol Metab September 2004;89(9):4607–14.
[35] Pettitt DJ, Nelson RG, Saad MF, Bennett PH, Knowler WC. Diabetes and obesity in the offspring of Pima Indian women with diabetes during pregnancy. Diabetes Care January 1993;16(1):310–4.
[36] Day PE, Cleal JK, Lofthouse EM, Goss V, Koster G, Postle A, et al. Partitioning of glutamine synthesised by the isolated perfused human placenta between the maternal and fetal circulations. Placenta December 2013;34(12):1223–31.
[37] Araujo JR, Correia-Branco A, Ramalho C, Keating E, Martel F. Gestational diabetes mellitus decreases placental uptake of long-chain polyunsaturated fatty acids: involvement of long-chain acyl-CoA synthetase. J Nutr Biochem October 2013;24(10):1741–50.
[38] Weissgerber TL, Davies GA, Roberts JM. Modification of angiogenic factors by regular and acute exercise during pregnancy. J Appl Physiol (1985) May 2010;108(5):1217–23.
[39] Malek A, Obrist C, Wenzinger S, von MU. The impact of cocaine and heroin on the placental transfer of methadone. Reprod Biol Endocrinol 2009;7:61.
[40] Wang N, Tikellis G, Sun C, Pezic A, Wang L, Wells JC, et al. The effect of maternal prenatal smoking and alcohol consumption on the placenta-to-birth weight ratio. Placenta July 2014;35(7):437–41.
[41] Clifton VL. Review: sex and the human placenta: mediating differential strategies of fetal growth and survival. Placenta March 2010;(31 Suppl):S33–9.
[42] Ounsted M, Scott A, Ounsted C. Transmission through the female line of a mechanism constraining human fetal growth. Int J Epidemiol April 2008;37(2):245–50.
[43] Allen WR, Wilsher S, Turnbull C, Stewart F, Ousey J, Rossdale PD, et al. Influence of maternal size on placental, fetal and postnatal growth in the horse. I. Development in utero. Reproduction March 2002;123(3):445–53.
[44] Fowden AL, Sferruzzi-Perri AN, Coan PM, Constancia M, Burton GJ. Placental efficiency and adaptation: endocrine regulation. J Physiol July 15, 2009;587(Pt 14):3459–72.
[45] Petropoulos S, Kalabis GM, Gibb W, Matthews SG. Functional changes of mouse placental multidrug resistance phosphoglycoprotein (ABCB1) with advancing gestation and regulation by progesterone. Reprod Sci May 2007;14(4):321–8.
[46] Roos S, Kanai Y, Prasad PD, Powell TL, Jansson T. Regulation of placental amino acid transporter activity by mammalian target of rapamycin. Am J Physiol Cell Physiol January 2009;296(1):C142–50.
[47] Roos S, Palmberg I, Saljo K, Powell TL, Jansson T. Expression of placental mammalian target of rapamycin (mTOR) is altered in relation to fetal growth and mTOR regulates leucine transport. Placenta September 2005;26(8–9):A9.
[48] Angiolini E, Coan PM, Sandovici I, Iwajomo OH, Peck G, Burton GJ, et al. Developmental adaptations to increased fetal nutrient demand in mouse genetic models of Igf2-mediated overgrowth. FASEB J May 2011;25(5):1737–45.
[49] Godfrey KM, Matthews N, Glazier J, Jackson A, Wilman C, Sibley CP. Neutral amino acid uptake by the microvillous plasma membrane of the human placenta is inversely related to fetal size at birth in normal pregnancy. J Clin Endocrinol Metab September 1998;83(9):3320–6.
[50] Lewis RM, Greenwood SL, Cleal JK, Crozier SR, Verrall L, Inskip HM, et al. Maternal muscle mass may influence system A activity in human placenta. Placenta March 4, 2010;31(5):418–22.
[51] Winder NR, Krishnaveni GV, Veena SR, Hill JC, Karat CL, Thornburg KL, et al. Mother's lifetime nutrition and the size, shape and efficiency of the placenta. Placenta November 2011;32(11):806–10.
[52] Clark PM, Atton C, Law CM, Shiell A, Godfrey K, Barker DJ. Weight gain in pregnancy, triceps skinfold thickness, and blood pressure in offspring. Obstet Gynecol 1998;91(1):103–7.
[53] Godfrey KM, Forrester T, Barker DJ, Jackson AA, Landman JP, Hall JS, et al. Maternal nutritional status in pregnancy and blood pressure in childhood. Br J Obstet Gynaecol May 1994;101(5):398–403.
[54] Adair LS, Kuzawa CW, Borja J. Maternal energy stores and diet composition during pregnancy program adolescent blood pressure. Circulation August 28, 2001;104(9):1034–9.
[55] Margetts BM, Rowland MG, Foord FA, Cruddas AM, Cole TJ, Barker DJ. The relation of maternal weight to the blood pressures of Gambian children. Int J Epidemiol December 1991;20(4):938–43.
[56] Fall CH, Stein CE, Kumaran K, Cox V, Osmond C, Barker DJ, et al. Size at birth, maternal weight, and type 2 diabetes in South India. Diabet Med 1998;15(3):220–7.
[57] Forsen T, Eriksson JG, Tuomilehto J, Teramo K, Osmond C, Barker DJ. Mother's weight in pregnancy and coronary heart disease in a cohort of Finnish men: follow up study. BMJ October 4, 1997;315(7112):837–40.

[58] Forsen T, Eriksson J, Tuomilehto J, Reunanen A, Osmond C, Barker D. The fetal and childhood growth of persons who develop type 2 diabetes. Ann Intern Med August 1, 2000;133(3):176–82.

[59] Godfrey K, Walker-Bone K, Robinson S, Taylor P, Shore S, Wheeler T, et al. Neonatal bone mass: influence of parental birthweight, maternal smoking, body composition, and activity during pregnancy. J Bone Miner Res September 2001;16(9):1694–703.

[60] Gale CR, Javaid MK, Robinson SM, Law CM, Godfrey KM, Cooper C. Maternal size in pregnancy and body composition in children. J Clin Endocrinol Metab October 2007;92(10):3904–11.

[61] Jansson N, Rosario FJ, Gaccioli F, Lager S, Jones HN, Roos S, et al. Activation of placental mTOR signaling and amino acid transporters in obese women giving birth to large babies. J Clin Endocrinol Metab January 2013;98(1):105–13.

[62] Ditchfield AM, Desforges M, Mills TA, Glazier JD, Wareing M, Mynett K, et al. Maternal obesity is associated with a reduction in placental taurine transporter activity. Int J Obes (Lond) 2015;39(4):557–64.

[63] Brass E, Hanson E, O'Tierney-Ginn PF. Placental oleic acid uptake is lower in male offspring of obese women. Placenta June 2013;34(6):503–9.

[64] Hayward CE, Higgins L, Cowley EJ, Greenwood SL, Mills TA, Sibley CP, et al. Chorionic plate arterial function is altered in maternal obesity. Placenta March 2013;34(3):281–7.

[65] Wallace JM, Horgan GW, Bhattacharya S. Placental weight and efficiency in relation to maternal body mass index and the risk of pregnancy complications in women delivering singleton babies. Placenta August 2012;33(8):611–8.

[66] Haugen G, Hanson M, Kiserud T, Crozier S, Inskip H, Godfrey KM. Fetal liver-sparing cardiovascular adaptations linked to mother's slimness and diet. Circ Res January 7, 2005;96(1):12–4.

[67] Wen HY, Abbasi S, Kellems RE, Xia Y. mTOR: a placental growth signaling sensor. Placenta April 2005;(26 Suppl A):S63–9.

[68] Shiell AW, Campbell-Brown M, Haselden S, Robinson S, Godfrey KM, Barker DJ. High-meat, low-carbohydrate diet in pregnancy: relation to adult blood pressure in the offspring. Hypertension December 1, 2001;38(6):1282–8.

[69] McMillen IC, Robinson JS. Developmental origins of the metabolic syndrome: prediction, plasticity, and programming. Physiol Rev April 2005;85(2):571–633.

[70] Roseboom TJ, van der Meulen JH, van Montfrans GA, Ravelli AC, Osmond C, Barker DJ, et al. Maternal nutrition during gestation and blood pressure in later life. J Hypertens January 2001;19(1):29–34.

[71] Campbell DM, Hall MH, Barker DJ, Cross J, Shiell AW, Godfrey KM. Diet in pregnancy and the offspring's blood pressure 40 years later. Br J Obstet Gynaecol 1996;103(3):273–80.

[72] Shiell AW, Campbell DM, Hall MH, Barker DJ. Diet in late pregnancy and glucose-insulin metabolism of the offspring 40 years later. BJOG July 2000;107(7):890–5.

[73] Robinson JS, Hartwich KM, Walker SK, Erwich JJ, Owens JA. Early influences on embryonic and placental growth. Acta Paediatr Suppl November 1997;423:159–63.

[74] McCrabb GJ, Egan AR, Hosking BJ. Maternal undernutrition during mid-pregnancy in sheep. Placental size and its relationship to calcium transfer during late pregnancy. Br J Nutr March 1991;65(2):157–68.

[75] Alwasel SH, Abotalib Z, Aljarallah JS, Osmond C, Alkharaz SM, Alhazza IM, et al. Changes in placental size during Ramadan. Placenta July 2010;31(7):607–10.

[76] van Abeelen AF, de Rooij SR, Osmond C, Painter RC, Veenendaal MV, Bossuyt PM, et al. The sex-specific effects of famine on the association between placental size and later hypertension. Placenta September 2011;32(9):694–8.

[77] Lewis RM, Doherty CB, James LA, Burton GJ, Hales CN. Effects of maternal iron restriction on placental vascularization in the rat. Placenta July 2001;22(6):534–9.

[78] Jansson T, Ekstrand Y, Bjorn C, Wennergren M, Powell TL. Alterations in the activity of placental amino acid transporters in pregnancies complicated by diabetes. Diabetes July 2002;51(7):2214–9.

[79] Ericsson A, Saljo K, Sjostrand E, Jansson N, Prasad PD, Powell TL, et al. Brief hyperglycaemia in the early pregnant rat increases fetal weight at term by stimulating placental growth and affecting placental nutrient transport. J Physiol June 15, 2007;581(Pt 3):1323–32.

[80] Godfrey K, Robinson S, Barker DJ, Osmond C, Cox V. Maternal nutrition in early and late pregnancy in relation to placental and fetal growth. BMJ 1996;312(7028):410–4.

[81] Lumey LH. Compensatory placental growth after restricted maternal nutrition in early pregnancy. Placenta January 1998;19(1):105–11.

[82] Ashworth CJ, Antipatis C. Micronutrient programming of development throughout gestation. Reproduction October 2001;122(4):527–35.

[83] Harvey NC, Javaid MK, Poole JR, Taylor P, Robinson SM, Inskip HM, et al. Paternal skeletal size predicts intrauterine bone mineral accrual. J Clin Endocrinol Metab May 2008;93(5):1676–81.

[84] Javaid MK, Crozier SR, Harvey NC, Dennison EM, Boucher BJ, Arden NK, et al. Maternal vitamin D status during pregnancy and childhood bone mass at age nine years: a longitudinal study. Lancet 2006;367(9504):36–3.

[85] Nguyen TP, Yong HE, Chollangi T, Borg AJ, Brennecke SP, Murthi P. Placental vitamin D receptor expression is decreased in human idiopathic fetal growth restriction. J Mol Med (Berl) 2015;93(7):795–805.

[86] Lewis RM, Cleal JK, Ntani G, Crozier SR, Mahon PA, Robinson SM, et al. Relationship between placental expression of the imprinted PHLDA2 gene, intrauterine skeletal growth and childhood bone mass. Bone January 2012;50(1):337–42.

[87] Martin R, Harvey NC, Crozier SR, Poole JR, Javaid MK, Dennison EM, et al. Placental calcium transporter (PMCA3) gene expression predicts intrauterine bone mineral accrual. Bone May 2007;40(5):1203–8.

[88] Anderson CM, Ralph JL, Johnson L, Scheett A, Wright ML, Taylor JY, et al. First trimester vitamin D status and placental epigenomics in preeclampsia among Northern Plains primiparas. Life Sci 2015;15(129):10–5.

[89] Vickers MH, Clayton ZE, Yap C, Sloboda DM. Maternal fructose intake during pregnancy and lactation alters placental growth and leads to sex-specific changes in fetal and neonatal endocrine function. Endocrinology April 2011;152(4):1378–87.

[90] Burton GJ, Woods AW, Jauniaux E, Kingdom JC. Rheological and physiological consequences of conversion of the maternal spiral arteries for uteroplacental blood flow during human pregnancy. Placenta June 2009;30(6):473–82.

[91] Gaillard R, Steegers EA, Tiemeier H, Hofman A, Jaddoe VW. Placental vascular dysfunction, fetal and childhood growth, and cardiovascular development: the generation R study. Circulation November 12, 2013;128(20):2202–10.

[92] Habli M, Jones H, Aronow B, Omar K, Crombleholme TM. Recapitulation of characteristics of human placental vascular insufficiency in a novel mouse model. Placenta December 2013;34(12):1150–8.

[93] Kusinski LC, Stanley JL, Dilworth MR, Hirt CJ, Andersson IJ, Renshall LJ, et al. eNOS knockout mouse as a model of fetal growth restriction with an impaired uterine artery function and placental transport phenotype. Am J Physiol Regul Integr Comp Physiol July 1, 2012;303(1):R86–93.

[94] Berkane N, Moutafoff-Borie C. Impact of previous uterine artery embolization on fertility. Curr Opin Obstet Gynecol June 2010;22(3):242–7.

[95] Forouzan I. Absence of end-diastolic flow velocity in the umbilical artery: a review. Obstet Gynecol Surv March 1995;50(3):219–27.

[96] Mayhew TM, Charnock-Jones DS, Kaufmann P. Aspects of human fetoplacental vasculogenesis and angiogenesis. III. Changes in complicated pregnancies. Placenta February 2004;25(2–3):127–39.

[97] Bhardwaj A, Nagandla K. Musculoskeletal symptoms and orthopaedic complications in pregnancy: pathophysiology, diagnostic approaches and modern management. Postgrad Med J August 2014;90(1066):450–60.

[98] Harris ST, Liu J, Wilcox S, Moran R, Gallagher A. Exercise during pregnancy and its association with gestational weight gain. Matern Child Health J March 2015;19(3):528–37.

[99] Clapp III JF, Stepanchak W, Tomaselli J, Kortan M, Faneslow S. Portal vein blood flow-effects of pregnancy, gravity, and exercise. Am J Obstet Gynecol July 2000;183(1):167–72.

[100] Juhl M, Olsen J, Andersen PK, Nohr EA, Andersen AM. Physical exercise during pregnancy and fetal growth measures: a study within the Danish National Birth Cohort. Am J Obstet Gynecol January 2010;202(1):63–8.

[101] Clapp III JF. Morphometric and neurodevelopmental outcome at age five years of the offspring of women who continued to exercise regularly throughout pregnancy. J Pediatr December 1996;129(6):856–63.

[102] Clapp III JF, Kim H, Burciu B, Schmidt S, Petry K, Lopez B. Continuing regular exercise during pregnancy: effect of exercise volume on fetoplacental growth. Am J Obstet Gynecol January 2002;186(1):142–7.

[103] Cottrell EC, Seckl JR. Prenatal stress, glucocorticoids and the programming of adult disease. Front Behav Neurosci 2009;3:19.

[104] Chapman K, Holmes M, Seckl J. 11beta-hydroxysteroid dehydrogenases: intracellular gate-keepers of tissue glucocorticoid action. Physiol Rev July 2013;93(3):1139–206.

[105] Johnstone JF, Bocking AD, Unlugedik E, Challis JR. The effects of chorioamnionitis and betamethasone on 11beta hydroxysteroid dehydrogenase types 1 and 2 and the glucocorticoid receptor in preterm human placenta. J Soc Gynecol Investig May 2005;12(4):238–45.

[106] Wyrwoll CS, Seckl JR, Holmes MC. Altered placental function of 11beta-hydroxysteroid dehydrogenase 2 knockout mice. Endocrinology March 2009;150(3):1287–93.

[107] Muramatsu-Kato K, Itoh H, Kobayashi-Kohmura Y, Murakami H, Uchida T, Suzuki K, et al. Comparison between placental gene expression of 11beta-hydroxysteroid dehydrogenases and infantile growth at 10 months of age. J Obstet Gynaecol Res February 2014;40(2):465–72.

[108] Dy J, Guan H, Sampath-Kumar R, Richardson BS, Yang K. Placental 11beta-hydroxysteroid dehydrogenase type 2 is reduced in pregnancies complicated with idiopathic intrauterine growth restriction: evidence that this is associated with an attenuated ratio of cortisone to cortisol in the umbilical artery. Placenta February 2008;29(2):193–200.

[109] Banister CE, Koestler DC, Maccani MA, Padbury JF, Houseman EA, Marsit CJ. Infant growth restriction is associated with distinct patterns of DNA methylation in human placentas. Epigenetics July 2011;6(7):920–7.
[110] Marsit CJ, Maccani MA, Padbury JF, Lester BM. Placental 11-beta hydroxysteroid dehydrogenase methylation is associated with newborn growth and a measure of neurobehavioral outcome. PLoS One 2012;7(3):e33794.
[111] Belkacemi L, Jelks A, Chen CH, Ross MG, Desai M. Altered placental development in undernourished rats: role of maternal glucocorticoids. Reprod Biol Endocrinol 2011;9:105.
[112] Stark MJ, Wright IM, Clifton VL. Sex-specific alterations in placental 11beta-hydroxysteroid dehydrogenase 2 activity and early postnatal clinical course following antenatal betamethasone. Am J Physiol Regul Integr Comp Physiol August 2009;297(2):R510–4.
[113] Clifton VL, Rennie N, Murphy VE. Effect of inhaled glucocorticoid treatment on placental 11beta-hydroxysteroid dehydrogenase type 2 activity and neonatal birthweight in pregnancies complicated by asthma. Aust N Z J Obstet Gynaecol April 2006;46(2):136–40.
[114] Mayhew TM, Jenkins H, Todd B, Clifton VL. Maternal asthma and placental morphometry: effects of severity, treatment and fetal sex. Placenta April 2008;29(4):366–73.
[115] Dalziel SR, Walker NK, Parag V, Mantell C, Rea HH, Rodgers A, et al. Cardiovascular risk factors after antenatal exposure to betamethasone: 30-year follow-up of a randomised controlled trial. Lancet May 28, 2005;365(9474):1856–62.
[116] Saif Z, Hodyl NA, Hobbs E, Tuck AR, Butler MS, Osei-Kumah A, et al. The human placenta expresses multiple glucocorticoid receptor isoforms that are altered by fetal sex, growth restriction and maternal asthma. Placenta April 2014;35(4):260–8.
[117] Lambers DS, Clark KE. The maternal and fetal physiologic effects of nicotine. Semin Perinatol April 1996;20(2):115–26.
[118] Jauniaux E, Burton GJ. Morphological and biological effects of maternal exposure to tobacco smoke on the feto-placental unit. Early Hum Dev November 2007;83(11):699–706.
[119] Kelly JJ, Davis PG, Henschke PN. The drug epidemic: effects on newborn infants and health resource consumption at a tertiary perinatal centre. J Paediatr Child Health June 2000;36(3):262–4.
[120] Floyd RL, O'Connor MJ, Sokol RJ, Bertrand J, Cordero JF. Recognition and prevention of fetal alcohol syndrome. Obstet Gynecol November 2005;106(5 Pt 1):1059–64.
[121] Burd L, Roberts D, Olson M, Odendaal H. Ethanol and the placenta: a review. J Matern Fetal Neonatal Med May 2007;20(5):361–75.
[122] Mund M, Louwen F, Klingelhoefer D, Gerber A. Smoking and pregnancy—a review on the first major environmental risk factor of the unborn. Int J Environ Res Public Health December 2013;10(12):6485–99.
[123] Javaid MK, Godfrey KM, Taylor P, Shore SR, Breier B, Arden NK, et al. Umbilical venous IGF-1 concentration, neonatal bone mass, and body composition. J Bone Miner Res January 2004;19(1):56–63.
[124] Lin FJ, Fitzpatrick JW, Iannotti CA, Martin DS, Mariani BD, Tuan RS. Effects of cadmium on trophoblast calcium transport. Placenta May 1997;18(4):341–56.
[125] Markunas CA, Xu Z, Harlid S, Wade PA, Lie RT, Taylor JA, et al. Identification of DNA methylation changes in newborns related to maternal smoking during pregnancy. Environ Health Perspect 2014;122(10):1147–53.
[126] Ortigosa S, Friguls B, Joya X, Martinez S, Marinoso ML, Alameda F, et al. Feto-placental morphological effects of prenatal exposure to drugs of abuse. Reprod Toxicol August 2012;34(1):73–9.
[127] Hatch EE, Bracken MB. Effect of marijuana use in pregnancy on fetal growth. Am J Epidemiol December 1986;124(6):986–93.
[128] Helliwell RJ, Chamley LW, Blake-Palmer K, Mitchell MD, Wu J, Kearn CS, et al. Characterization of the endocannabinoid system in early human pregnancy. J Clin Endocrinol Metab October 2004;89(10):5168–74.
[129] Park B, Gibbons HM, Mitchell MD, Glass M. Identification of the CB1 cannabinoid receptor and fatty acid amide hydrolase (FAAH) in the human placenta. Placenta November 2003;24(10):990–5.
[130] Zuckerman B, Frank DA, Hingson R, Amaro H, Levenson SM, Kayne H, et al. Effects of maternal marijuana and cocaine use on fetal growth. N Engl J Med March 23, 1989;320(12):762–8.
[131] Felder CC, Glass M. Cannabinoid receptors and their endogenous agonists. Annu Rev Pharmacol Toxicol 1998;38:179–200.
[132] McCormack SA, Best BM. Protecting the fetus against HIV infection: a systematic review of placental transfer of antiretrovirals. Clin Pharmacokinet November 2014;53(11):989–1004.
[133] Reynolds RM, Jacobsen GH, Drake AJ. What is the evidence in humans that DNA methylation changes link events in utero and later life disease? Clin Endocrinol (Oxf) June 2013;78(6):814–22.
[134] Novakovic B, Saffery R. The ever growing complexity of placental epigenetics—role in adverse pregnancy outcomes and fetal programming. Placenta December 2012;33(12):959–70.
[135] Heijmans BT, Tobi EW, Stein AD, Putter H, Blauw GJ, Susser ES, et al. Persistent epigenetic differences associated with prenatal exposure to famine in humans. Proc Natl Acad Sci USA November 4, 2008;105(44):17046–9.

[136] Hoyo C, Fortner K, Murtha AP, Schildkraut JM, Soubry A, Demark-Wahnefried W, et al. Association of cord blood methylation fractions at imprinted insulin-like growth factor 2 (IGF2), plasma IGF2, and birth weight. Cancer Causes Control April 2012;23(4):635–45.

[137] Oberlander TF, Weinberg J, Papsdorf M, Grunau R, Misri S, Devlin AM. Prenatal exposure to maternal depression, neonatal methylation of human glucocorticoid receptor gene (NR3C1) and infant cortisol stress responses. Epigenetics March 2008;3(2):97–106.

[138] Suter M, Abramovici A, Showalter L, Hu M, Shope CD, Varner M, et al. In utero tobacco exposure epigenetically modifies placental CYP1A1 expression. Metabolism October 2010;59(10):1481–90.

[139] St-Pierre J, Hivert MF, Perron P, Poirier P, Guay SP, Brisson D, et al. IGF2 DNA methylation is a modulator of newborn's fetal growth and development. Epigenetics October 2012;7(10):1125–32.

[140] Reik W, Dean W, Walter J. Epigenetic reprogramming in mammalian development. Science August 10, 2001;293(5532):1089–93.

[141] Santos F, Hendrich B, Reik W, Dean W. Dynamic reprogramming of DNA methylation in the early mouse embryo. Dev Biol January 1, 2002;241(1):172–82.

[142] Reik W, Dean W. DNA methylation and mammalian epigenetics. Electrophoresis August 2001;22(14): 2838–43.

[143] Rahnama F, Shafiei F, Gluckman PD, Mitchell MD, Lobie PE. Epigenetic regulation of human trophoblastic cell migration and invasion. Endocrinology November 2006;147(11):5275–83.

[144] Gama-Sosa MA, Wang RY, Kuo KC, Gehrke CW, Ehrlich M. The 5-methylcytosine content of highly repeated sequences in human DNA. Nucleic Acids Res May 25, 1983;11(10):3087–95.

[145] Perrin D, Ballestar E, Fraga MF, Frappart L, Esteller M, Guerin JF, et al. Specific hypermethylation of LINE-1 elements during abnormal overgrowth and differentiation of human placenta. Oncogene April 12, 2007;26(17):2518–24.

[146] Penaherrera MS, Jiang R, Avila L, Yuen RK, Brown CJ, Robinson WP. Patterns of placental development evaluated by X chromosome inactivation profiling provide a basis to evaluate the origin of epigenetic variation. Hum Reprod June 2012;27(6):1745–53.

[147] Grigoriu A, Ferreira JC, Choufani S, Baczyk D, Kingdom J, Weksberg R. Cell specific patterns of methylation in the human placenta. Epigenetics March 2011;6(3):368–79.

[148] Mikheev AM, Nabekura T, Kaddoumi A, Bammler TK, Govindarajan R, Hebert MF, et al. Profiling gene expression in human placentae of different gestational ages: an OPRU Network and UW SCOR Study. Reprod Sci November 2008;15(9):866–77.

[149] Novakovic B, Yuen RK, Gordon L, Penaherrera MS, Sharkey A, Moffett A, et al. Evidence for widespread changes in promoter methylation profile in human placenta in response to increasing gestational age and environmental/stochastic factors. BMC Genomics 2011;12:529.

[150] Chavan-Gautam P, Sundrani D, Pisal H, Nimbargi V, Mehendale S, Joshi S. Gestation-dependent changes in human placental global DNA methylation levels. Mol Reprod Dev March 2011;78(3):150.

[151] Novakovic B, Gordon L, Robinson WP, Desoye G, Saffery R. Glucose as a fetal nutrient: dynamic regulation of several glucose transporter genes by DNA methylation in the human placenta across gestation. J Nutr Biochem January 2013;24(1):282–8.

[152] Nomura Y, Lambertini L, Rialdi A, Lee M, Mystal EY, Grabie M, et al. Global methylation in the placenta and umbilical cord blood from pregnancies with maternal gestational diabetes, preeclampsia, and obesity. Reprod Sci January 2014;21(1):131–7.

[153] El HN, Pliushch G, Schneider E, Dittrich M, Muller T, Korenkov M, et al. Metabolic programming of MEST DNA methylation by intrauterine exposure to gestational diabetes mellitus. Diabetes April 2013;62(4):1320–8.

[154] Ruchat SM, Houde AA, Voisin G, St-Pierre J, Perron P, Baillargeon JP, et al. Gestational diabetes mellitus epigenetically affects genes predominantly involved in metabolic diseases. Epigenetics September 2013;8(9):935–43.

[155] Desgagne V, Hivert MF, St-Pierre J, Guay SP, Baillargeon JP, Perron P, et al. Epigenetic dysregulation of the IGF system in placenta of newborns exposed to maternal impaired glucose tolerance. Epigenomics April 2014;6(2):193–207.

[156] Bouchard L, Hivert MF, Guay SP, St-Pierre J, Perron P, Brisson D. Placental adiponectin gene DNA methylation levels are associated with mothers' blood glucose concentration. Diabetes May 2012;61(5):1272–80.

[157] Jiang X, Yan J, West AA, Perry CA, Malysheva OV, Devapatla S, et al. Maternal choline intake alters the epigenetic state of fetal cortisol-regulating genes in humans. FASEB J August 2012;26(8):3563–74.

[158] Chen PY, Ganguly A, Rubbi L, Orozco LD, Morselli M, Ashraf D, et al. Intrauterine calorie restriction affects placental DNA methylation and gene expression. Physiol Genomics July 15, 2013;45(14):565–76.

[159] Gallou-Kabani C, Gabory A, Tost J, Karimi M, Mayeur S, Lesage J, et al. Sex- and diet-specific changes of imprinted gene expression and DNA methylation in mouse placenta under a high-fat diet. PLoS One 2010;5(12):e14398.

[160] Ganguly A, Chen Y, Shin BC, Devaskar SU. Prenatal caloric restriction enhances DNA methylation and MeCP2 recruitment with reduced murine placental glucose transporter isoform 3 expression. J Nutr Biochem February 2014;25(2):259–66.

[161] Wilhelm-Benartzi CS, Houseman EA, Maccani MA, Poage GM, Koestler DC, Langevin SM, et al. In utero exposures, infant growth, and DNA methylation of repetitive elements and developmentally related genes in human placenta. Environ Health Perspect February 2012;120(2):296–302.
[162] Bartel DP. MicroRNAs: genomics, biogenesis, mechanism, and function. Cell January 23, 2004;116(2):281–97.
[163] Xie L, Mouillet JF, Chu T, Parks WT, Sadovsky E, Knofler M, et al. C19MC microRNAs regulate the migration of human trophoblasts. Endocrinology December 2014;155(12):4975–85.
[164] Zhu XM, Han T, Sargent IL, Yin GW, Yao YQ. Differential expression profile of microRNAs in human placentas from preeclamptic pregnancies vs normal pregnancies. Am J Obstet Gynecol June 2009;200(6):661–7.
[165] Zhang C, Li Q, Ren N, Li C, Wang X, Xie M, et al. Placental miR-106a approximately 363 cluster is dysregulated in preeclamptic placenta. Placenta February 2015;36(2):250–2.
[166] Li X, Li C, Dong X, Gou W. MicroRNA-155 inhibits migration of trophoblast cells and contributes to the pathogenesis of severe preeclampsia by regulating endothelial nitric oxide synthase. Mol Med Rep July 2014;10(1):550–4.
[167] Zhao C, Zhang T, Shi Z, Ding H, Ling X. MicroRNA-518d regulates PPARalpha protein expression in the placentas of females with gestational diabetes mellitus. Mol Med Rep June 2014;9(6):2085–90.
[168] Maccani MA, Avissar-Whiting M, Banister CE, McGonnigal B, Padbury JF, Marsit CJ. Maternal cigarette smoking during pregnancy is associated with downregulation of miR-16, miR-21, and miR-146a in the placenta. Epigenetics October 1, 2010;5(7):583–9.
[169] Kotlabova K, Doucha J, Hromadnikova I. Placental-specific microRNA in maternal circulation—identification of appropriate pregnancy-associated microRNAs with diagnostic potential. J Reprod Immunol May 2011;89(2):185–91.
[170] Filiberto AC, Maccani MA, Koestler D, Wilhelm-Benartzi C, Avissar-Whiting M, Banister CE, et al. Birthweight is associated with DNA promoter methylation of the glucocorticoid receptor in human placenta. Epigenetics May 2011;6(5):566–72.
[171] Reamon-Buettner SM, Buschmann J, Lewin G. Identifying placental epigenetic alterations in an intrauterine growth restriction (IUGR) rat model induced by gestational protein deficiency. Reprod Toxicol June 2014;45:117–24.
[172] Alikhani-Koopaei R, Fouladkou F, Frey FJ, Frey BM. Epigenetic regulation of 11 beta-hydroxysteroid dehydrogenase type 2 expression. J Clin Invest October 2004;114(8):1146–57.
[173] Ferreira JC, Choufani S, Grafodatskaya D, Butcher DT, Zhao C, Chitayat D, et al. WNT2 promoter methylation in human placenta is associated with low birthweight percentile in the neonate. Epigenetics April 2011;6(4):440–9.
[174] Godfrey KM, Sheppard A, Gluckman PD, Lillycrop KA, Burdge GC, McLean C, et al. Epigenetic gene promoter methylation at birth is associated with child's later adiposity. Diabetes May 2011;60(5):1528–34.
[175] Wang D, Na Q, Song WW, Song GY. Altered expression of miR-518b and miR-519a in the placenta is associated with low fetal birth weight. Am J Perinatol October 2014;31(9):729–34.
[176] Reik W, Walter J. Genomic imprinting: parental influence on the genome. Nat Rev Genet January 2001;2(1):21–32.
[177] Renfree MB, Hore TA, Shaw G, Graves JA, Pask AJ. Evolution of genomic imprinting: insights from marsupials and monotremes. Annu Rev Genomics Hum Genet 2009;10:241–62.
[178] Coan PM, Burton GJ, Ferguson-Smith AC. Imprinted genes in the placenta—a review. Placenta April 2005;(26 Suppl A):S10–20.
[179] Moore T, Mills W. Evolutionary theories of imprinting–enough already! Adv Exp Med Biol 2008;626:116–22.
[180] Fowden AL, Coan PM, Angiolini E, Burton GJ, Constancia M. Imprinted genes and the epigenetic regulation of placental phenotype. Prog Biophys Mol Biol July 2011;106(1):281–8.
[181] Lim AL, Ferguson-Smith AC. Genomic imprinting effects in a compromised in utero environment: implications for a healthy pregnancy. Semin Cell Dev Biol April 2010;21(2):201–8.
[182] Sferruzzi-Perri AN, Vaughan OR, Haro M, Cooper WN, Musial B, Charalambous M, et al. An obesogenic diet during mouse pregnancy modifies maternal nutrient partitioning and the fetal growth trajectory. FASEB J October 2013;27(10):3928–37.
[183] Coan PM, Vaughan OR, Sekita Y, Finn SL, Burton GJ, Constancia M, et al. Adaptations in placental phenotype support fetal growth during undernutrition of pregnant mice. J Physiol February 1, 2010;588(Pt 3):527–38.
[184] Sferruzzi-Perri AN, Vaughan OR, Coan PM, Suciu MC, Darbyshire R, Constancia M, et al. Placental-specific Igf2 deficiency alters developmental adaptations to undernutrition in mice. Endocrinology August 2011;152(8):3202–12.
[185] Haycock PC, Ramsay M. Exposure of mouse embryos to ethanol during preimplantation development: effect on DNA methylation in the h19 imprinting control region. Biol Reprod October 2009;81(4):618–27.
[186] Lillycrop KA, Phillips ES, Jackson AA, Hanson MA, Burdge GC. Dietary protein restriction of pregnant rats induces and folic acid supplementation prevents epigenetic modification of hepatic gene expression in the offspring. J Nutr June 2005;135(6):1382–6.

[187] Pham TD, MacLennan NK, Chiu CT, Laksana GS, Hsu JL, Lane RH. Uteroplacental insufficiency increases apoptosis and alters p53 gene methylation in the full-term IUGR rat kidney. Am J Physiol Regul Integr Comp Physiol November 2003;285(5):R962–70.

[188] Wolff GL, Kodell RL, Moore SR, Cooney CA. Maternal epigenetics and methyl supplements affect agouti gene expression in Avy/a mice. FASEB J August 1998;12(11):949–57.

[189] Dolinoy DC. The agouti mouse model: an epigenetic biosensor for nutritional and environmental alterations on the fetal epigenome. Nutr Rev August 2008;66(Suppl. 1):S7–11.

[190] Sibley CP, Turner MA, Cetin I, Ayuk P, Boyd CA, D'Souza SW, et al. Placental phenotypes of intrauterine growth. Pediatr Res November 2005;58(5):827–32.

CHAPTER

23

The Moral and Legal Relevance of DOHaD Effects for Pregnant Mothers

Michele Loi[1], *Marianna Nobile*[2]

[1]Centro de Estudos Humanísticos, Universidade do Minho, Campus de Gualtar, Braga, Portugal;
[2]Universita' degli Studi di Milano-Bicocca, Dipartimento dei Sistemi Giuridici, Milano, Italy

OUTLINE

INTRODUCTION: DEVELOPMENTAL ORIGINS OF HEALTH AND DISEASE RESEARCH AND LEGISLATIVE INTERVENTIONS

The concept of developmental origins of health and disease (DOHaD) has been used to suggest that "the action of a stimulus or insult during a specific critical period in utero or early postnatal development leads to 'programmed' alterations in tissue structure and function predisposing the individual to later disease" [1]. It seems that negative health outcomes can result from adverse conditions during prenatal development, such as prenatal exposure to

The Epigenome and Developmental Origins of Health and Disease
http://dx.doi.org/10.1016/B978-0-12-801383-0.00023-2

Copyright © 2016 Elsevier Inc. All rights reserved.

fungicides, pesticides, and other harmful chemical substances; air pollution; parental under- and overnutrition; smoke; stress and psychological trauma; neurobehavioral disorders; obesity; and an unbalanced diet *in the pregnant mother*.[1] While epidemiological evidence of transgenerational paternal transmission of DOHaD effects exists (moreover, recent discoveries in epigenetics have begun to unravel molecular mechanisms for the transmission of environmental influences through the germline), here we will focus on maternal influences during pregnancy.

Such discoveries have several implications for individual morality, law, and policy. For example, the American Congress of Obstetricians and Gynecologists and the American Society of Reproductive Medicine issued a statement in 2012 entitled *Toxic Environmental Agents* [2]. The committee opinion states that "the scientific evidence over the last 15 years shows that exposure to toxic environmental agents before conception and during pregnancy can have significant and long-lasting effects on reproductive health" and that "reducing exposure to toxic environmental agents is a critical area of intervention for health care professionals."

Does the aim of protecting the fetus (or of future generations) from harm entail positive duties by women to protect the health of the fetus inside them? Should these duties be regarded as matters of individual morality, or should they be enforced by the state through criminal law? These are the main questions in this chapter.

Different statutes (child neglect, child abuse, delivery of drugs to a minor, child endangerment, etc.) have been applied to circumstances related to fetal injuries, even if the law originally aimed at ruling injuries caused to children after their birth. More recently, cases of conviction of the woman during pregnancy in order to protect fetal health have been recorded.[2] Civil liability for injuries to the fetus and the criminalization of maternal behavior (such as drug or alcohol abuse, the noncompliance of medical prescriptions, careless conduct, etc.) extend the power of the state to control women during pregnancy. As we will argue, these legal developments are problematic. Moreover, we will argue that DOHaD effects should not be conflated with prenatal injuries because of the probabilistic nature of the "harm" imposed to future generations in the context of DOHaD.

Society should not approach these political questions exclusively through the legal and moral categories of the "fetal rights" debate. More is at stake in DOHaD than the conflict between fetal rights and women's right to autonomy or control over their bodies. As we will argue, society must also ask to what extent morality and the legal system should take into consideration and protect the interests and rights of future persons.

The investigation in this paper takes into account two different levels of inquiry. We analyze juridical arguments in favor of and against ascribing legal personality to the fetus. We also explore moral arguments for duties owed to future people that require mothers to exercise caution when their behavior affects fetal health. We do not imply that these moral arguments ought to find expression in the law and explore arguments claiming that "the tendency to transform a *moral* duty of non-maleficence into a *legal* duty to limit autonomy" [3] is highly problematic.

For the sake of brevity, we will leave aside the broader legal, political, and philosophical questions concerning disease prevention during pregnancy. While criminal statutes have emphasized the role of individual responsibility, other authors, who regard DOHaD from a public health perspective, have emphasized communal and political responsibilities. Some authors have argued, on the basis of theories of distributive justice, that society as a whole is in charge of promoting and protecting the health of mothers, on account of its effect on the health of future generations [4–6]. DOHaD can be used to justify public health policies, as well as broad social changes influencing the social determinants of health. These are requirements to improve the social circumstances in which women live, as opposed to emphasizing individual responsibilities.

In what follows, we begin by highlighting the problematic nature of fetal rights, which have led in some contexts to the criminalization of women. In the following two sections, we briefly consider the issue of fetal rights as it emerges in the legal system with particular attention to the US context. In subsequent sections, we review moral arguments supporting a moral obligation to protect the health of the fetus independent of the intrinsic moral worth of fetal life. These moral duties are owed to the future person whose health would be affected by adverse developmental conditions. Finally we move back from morality to law, analyzing arguments for and against the legal enforcement of such duties.

THE LEGAL RECOGNITION OF FETAL RIGHTS

The debate concerning fetal rights is especially prominent in Anglo-American countries, in particular in the United States. The expansion of fetal rights, including not only the right to live but also the right of the fetus to be born healthy and to receive medical treatments, requires a reflection on the need to identify limits in the protection of such rights.[3]

In the North American case law, the maternal–fetal relationship shapes different aspects of the daily life of a woman during pregnancy:

> The relationship between a pregnant woman and her fetus is unlike any other in law, medicine, or ethics. Within the same body, there exist one person and one potential person with both similar and separate interests and, for the fetus, developing rights [3].

Only upon birth is a child considered to acquire legal standing, and, consequently, to hold rights as a separate person. But, when the fetus is becoming a person, does it have interests independent of its mother? And if these interests are conflicting, can the fetus claim their protection against its mother? Does the state have the duty to intervene on behalf of the fetus? The ascription of rights to fetuses gives rise to the so-called maternal–fetal conflict, i.e., a conflict between the right of the fetus to be born healthy and the autonomy of the woman in making the choices related to medical treatments and her lifestyle during pregnancy.

The existence of fetal rights is usually discussed in the context of abortion legislation, stating that the legislator, prohibiting abortion after the fetus is considered viable, demonstrates its interest in promoting prenatal life. Furthermore, it is argued that we can infer the power of the state to impose coercive medical treatments aimed to defend fetal life from the prohibition of abortion in the final phase of pregnancy [7].

The basis for shifting the discussion about abortion to the issue of fetal rights was laid by *Roe v Wade*. In this constitutional decision, in fact, we can find not only the premises for the guarantee of the woman's right to seek a legal, medically supervised abortion during the first trimester of pregnancy, but also the groundwork for contest of interest between a pregnant woman and the fetus she carries. But notice that the Court limited the woman's right to terminate her pregnancy in declaring that it "is not absolute, and may to some extent be limited by the state's legitimate interests in safeguarding the woman's health, in maintaining proper medical standards, and in protecting potential human life."[4] The Court stated that the interest in potential human life has to be measured against the woman's right to privacy, and that the fetus at a certain point in development acquires the potentiality for personhood and, therefore, the right to protection. This point is identified with the moment of fetal viability. This ambiguity allows the arguments that, interpreting the viable fetus as a legal person, restrict the woman's right in the name of the potential child. Furthermore, a few years after the recognition of the rights of a viable fetus, some courts began expanding the doctrine by stating that the fetus has rights even before viability.[5]

The legislator usually expresses its interest in preserving the health of fetuses; still, in the US legal system abortion has been defended by the Supreme Court in name of privacy and of equality. The privacy argument in the US constitutional tradition is connected with the Fourteenth Amendment's guarantee of due process. Since *Roe v Wade*, this procedural right has been interpreted by the Supreme Court as entailing women's right to privacy, i.e., "to decide for themselves ethical and personal issues arising from marriage and procreation," which includes the right to decide whether or not to carry a pregnancy to term. The equality argument, grounded in the equal-protection clause of the same amendment, is based on the recognition that antiabortion laws oppress women, invalidating their ability to act as autonomous agent in their private life [8]. Furthermore, the constitutional interpretation of this clause "combats the abuses to which women have been subjected in the name of sexual freedom by identifying reproductive difference as a difference that mandates protective interference" [8].

A different perspective grounds the existence of fetal rights on recognizing the woman's free choice to be pregnant as the basis for her civil and criminal responsibility. From this point of view, a woman's decision to carry a pregnancy to term is viewed as a voluntary abrogation of her civil rights for the duration of pregnancy. [6] A pregnant woman thus assumes the same duty of care toward the fetus as she would toward any stranger: a duty to refrain from causing harm. If the woman foregoes her right to abort, she voluntarily enters a special relationship with the fetus and is held accountable for a stricter duty of care. Thus, in the discussion of criminal prosecutions of pregnant women it has been argued that case law demonstrates that the state's interest in protecting the fetus may overcome a woman's right to autonomy and to privacy, to advance a policy of fetal protection and parental responsibility [9].

While the United States Supreme Court has not recognized the fetus as a "person" within the meaning of the Fourteenth Amendment, the protection of fetal rights has evolved significantly outside constitutional law in common law. [7] Fetal right doctrine often grants implicit legal status to the unborn, what is sometimes referred to as "contingent legal personhood" [10]. [8] Courts have also recognized the fetal right to be medically treated, independently from the will of the mother. This entails the recognition of the corresponding mother's duty to allow every medical treatment, i.e., to undergo any clinical intervention necessary to safeguard health and integrity of the fetus. For example, in *Jefferson v Griffin Spalding County Hospital Authority*, for the first time the Supreme Court of Georgia affirmed a lower court order requiring a pregnant woman to submit to a cesarean section and other medical procedures necessary to save her unborn child's life. The court found that the state's interest in protecting the viable fetus outweighed the pregnant mother's right to religious practice, the right to refuse medical treatment, and parental autonomy. [9]

Moreover, courts bring wrongful death actions against a third party, considering the fetus that has died in utero to be a "person" under wrongful death statutes. Even if the "born alive" rule[10] governed criminal law until the mid-nineteenth century, recently fetal death has been compared to an in utero homicide [11]. For instance, if a woman is involved in a traffic accident that causes fatal injuries to the fetus, the person responsible for the accident may be charged for causing the death of the fetus. In some cases, the criminal sanctions imposed for the death of the fetus are identical to those imposed for the murder of a person.[11] The first court to bring a wrongful death action against a third party was the Massachusetts Supreme Judicial Court in *Commonwealth v Cass*. In that case, a driver struck a pregnant woman, causing the death of the fetus, and was charged with vehicular homicide.[12] Recognizing wrongful death actions for the death of fetuses serves in most cases to compensate parents for the loss of their expected child and to protect the interest of the woman to carry her pregnancy to term.[13]

THE CRIMINALIZATION OF THE "BAD MOTHER"

The recognition of fetal rights led to charging the mother when she is responsible for causing the death of the fetus, identifying the fetus rather than the woman as the locus of the right, as in *State v McKnight*, where a woman, after giving birth to a stillborn child, was indicted for homicide by child abuse and sentenced to 20 years for causing the death of the viable fetus through the assumption of cocaine.[14] According to Dawn Johnsen, "the law no longer recognizes the fetus only in those cases where it is necessary to protect the interest of the subsequently born child and her or his parents. Rather, the law has conferred rights upon the fetus *qua* fetus" [12]. The introduction of new criminal cases of fetus homicide represents the most drastic expression of the legislator's will to recognize fetal rights in conflict with maternal rights.

Since the mid-twentieth century, women have been prosecuted in two-thirds of the United States for fetal abuse, a crime that does not exist in any state's criminal statutes.[15] District attorneys used a variety of existing criminal statutes, grounding fetal abuse on the inaccurate analogy between child and fetus, implementing laws on delivering drugs to a minor, child neglect, child abuse, and child endangerment. According to Bell, the most commonly used statute is "delivery or distribution of an unlawful substance to a minor" [13], which was the basis for the first criminal conviction of a mother for fetal injury. In 1989, under Florida's general drug statute, Jennifer Clarise Johnson was convicted for delivering illegal drugs to her fetus. Ms Johnson, who gave birth to a son in 1987 and a daughter in 1989, admitted that she had used cocaine during both pregnancies and specifically stated that she had used cocaine on the day before her son's birth and on the morning of her daughter's birth. Florida charged Johnson with two counts on the theory that "delivery" occurred when the cocaine passed through the umbilical cord to the child during the 60–90 seconds between delivery and the cutting of the cord. Johnson was sentenced to 1 year in a drug treatment program, 14 years probation, and 200 hours of community service.[16]

In 1993, the Iowa Legislature passed a bill primarily targeted at children born to drug-using pregnant women. It required physicians to report positive drug tests of newborns to authorities.[17] In 1996, the South Carolina Supreme Court upheld the conviction of Cornelia Whitner for criminal child neglect for taking cocaine during her pregnancy that might harm a viable fetus. The court explained that "the word 'child' as used in [the state's child abuse] statute includes viable fetus."[18]

Furthermore, some judges authorized the custody and the civil commitment of the mother during pregnancy, according to the child protection statutes, giving sanctions in case the fetus suffered damage and resorting to detention in order to control pregnancy and protect the fetus. In *New Jersey v Ikerd*, the court overturned a sentence handed down to Simmone Ikerd, who at the time of her sentencing was undergoing methadone treatment for a heroin addiction. She was also in the first trimester of her pregnancy. Even though she had no extensive criminal record, prosecutors asked for a prison sentence to protect the fetus. Middlesex County Superior Court Judge Phillip Paley said he did not want to send Ikerd to prison but felt compelled to do so to ensure she remained drug-free during her pregnancy. He sentenced her to a three-year prison term, 18 months of which had to be served before parole eligibility.[19]

Criminalization of the mother is the direct consequence of the recognition of fetal rights. Given the complete physical dependence of the fetus on the body of the woman, virtually every act of the pregnant woman has some effect on the fetus. Furthermore, the current possibilities of controlling the fetus from the first stages of pregnancy increase the responsibility of the mother for injuries originating from diseases that have not been treated. In this perspective, a woman can be held criminally liable for fetal injuries resulting from maternal negligence or from any behavior that has a potentially adverse effect on the fetus (such as using nonprescription and illegal drugs, smoking, drinking alcohol, exposing herself to workplace hazards, etc.). The risk is that, by raising the status of a fetus so that it is considered equal to that of a "human being who has been born and is alive,"[20] the woman's fundamental rights are diminished and she is treated as simply an incubator for potential life [14].

So the duty to protect fetal health needs to be questioned. No constitutional precedent establishes such a duty, although in some cases, as *Jefferson v Griffin Spalding County Hospital Authority*, the court recognized the duty to allow a viable fetus to be born alive, deriving it from a compelling state interest. Even if this case indicates a trend toward imposing some duty to protect fetal health, it does not indicate how far this duty should be extended [15].

The constitutional rights of due process, privacy, and equal protection seem to be violated by rulings designed to protect the fetus. First of all, the use of child abuse statutes to prosecute a pregnant woman for fetal abuse violates the woman's right to due process because on the one hand, as noted above, it considers the fetus a child, assuming it is a person; on the other hand, the extension of these statutes to the fetus violates the guarantee that a criminal law must clearly and explicitly indicate the conduct to be prohibited. Preventive incarceration is particularly controversial because the woman is, in essence, being punished for a crime that she has not committed yet.

Furthermore, moral obligations should not be confused with legal duties, as the result would be a wrongful notion of state responsibility to promote fetal health by intervening in maternal conduct. A woman's decision to carry her pregnancy to term does not mean that she will put the interest of the fetus *ahead* of her own and it does not give rise to a state obligation to control maternal behavior for the unborn well-being.

Women's rights advocates think that the recent expansion of fetal rights infringes on the fundamental women's right to privacy. In fact, if a woman's reproductive privacy right extends to her decision about contraception and abortion, that right should also include privacy around her health during pregnancy as well as pregnancy itself. Mandatory reporting laws, allowing a physician to conduct drug tests on women and requiring him or her to report the positive test to governmental authorities, contravene the privacy right [16]. Moreover, drug addiction is not per se a criminal behavior, but a disease that is generally considered confidential under the doctor–patient privilege. In addition, the state's attempts to impose the "right behavior" during pregnancy deprive women of their right to control their lives, which arguably means depriving them of their legal personhood (albeit for a limited period of time).

Finally, the right to equal protection is violated, as pregnant low-income and no-white women are more likely subject to forced medical treatment, as well as punishment for substance abuse, because they are often treated at public hospitals and tested for drug use.[21]

In his dissenting opinion in *Whitner v State*, Justice Moore noted the failure to properly examine the state's abortion statute, which is the only law that specifically regulates the conduct of a mother toward the fetus. According to him, this failure undermined the goal of "equal treatment of viable fetuses and children." In fact, "a pregnant woman now faces up to 10 years in prison for ingesting drugs during pregnancy but can have an illegal abortion and receive only a two-years sentence for killing her viable fetus."[22]

Conviction for criminal conduct requires mens rea, or criminal intent, which is very difficult to establish in these cases. Thus, if we exclude a criminal intent, we must deal with unintentional responsibility for damages occurring to the fetus. So here we are dealing with charges of negligence, which are extremely problematic in this context. It has been noticed that the liability to punishment for involuntary harm on the fetus entails an excessive responsibility for the mother, because it is inevitably linked to noteworthy restrictions to her liberty. Furthermore, the legal system should guarantee the minimum and fundamental notions concerning prenatal care, providing the woman with adequate medical and legal advice.[23]

CARING FOR FETUSES VERSUS CARING FOR FUTURE PERSONS

In the previous section the focus has been on the potential conflict between the interests of the mother and the fetus, and the way this conflict has been coded within the law. It may seem that a mother's moral duty to protect fetal health is incompatible with a moral framework recognizing the permissibility of abortion. If so, the recognition of fetal rights can only be seen as part of a political agenda opposing abortion. According to some pro-choice positions on abortion, behavior during pregnancy should always be seen as a matter of personal choice [17].

But this ignores an important difference between abortion, on the one hand, and non-lethal fetal injury, or the developmental programming of diseases, on the other. When the mother's behavior does not terminate the life of a fetus, her behavior during pregnancy may not affect the fetus, and yet may affect the life prospects of a future person. For those who maintain that a fetus and the future person have a different moral status, this is a morally relevant difference [17]. (This is denied by some views on abortion, such as Judith Jarvis Thomson's [18], which maintains that the mother has a right to abort a fetus even if it is a person. If that position is correct, the mother has an unrestricted right to put her own interests ahead of those of the fetus *even when* the baby will live and eventually become a full moral agent, and be negatively affected by her choices.)[24]

One should also distinguish fetal injury and developmental programming ("DOHaD"). When fetal injury occurs, the fetus and later person are harmed, sometimes seriously; by contrast, DOHaD effects do not harm the fetus and only increase the relative *probability* of disease or premature death in the adult offspring. For the limited purpose of the present argument, this difference can be safely ignored; the probabilistic nature of DOHaD-related harm will be discussed in a later section of this chapter.

Suppose that the fetus is not a person and *therefore* the mother is allowed to abort it. Should the mother still be allowed to behave in a way that could harm a fetus, or cause DOHaD effects, if she intended to keep the child? The fetal programming of diseases links fetal development and adult health. If the mother does not abort the fetus, it will later develop into a full person with full moral rights. This person could be harmed, or have imposed a higher risk of harm, as a result of the behavior of her mother during pregnancy. Fetal health could be worthy of protection because the interests of future children and adults matter as much as the interests of pregnant mothers.

It is extremely important not to confuse the abortion case with the DOHaD (or fetal injury) cases. Suppose that an activity the mother enjoys causes third–trimester fetal pain or discomfort, which would have no effect after birth. Those concerned only with the rights and interests of the future person would find nothing wrong with the woman causing such temporary harm for her own comfort. If, on the contrary, the same activity had caused pleasure in the fetus, but also programmed the organism for a higher likelihood of postnatal disease, those concerned only with the rights and interests of the future persons may find that behavior morally objectionable. According to the view that the fetus is not a person, fetal injury and DOHaD can be more serious grounds for concern than abortion itself, even if death is much worse than fetal injury, or the developmental programming of disease, *for the fetus*.

This argument invokes the future rights and interests of future people. This idea will not be altogether unfamiliar to most readers. In the context of environmental ethics, it is often claimed that present generations have duties to preserve a healthy environment and to avoid consuming more than a fair share of natural resources, for the sake of future generations. It would be immoral to plant now a time bomb that will explode in 130 years, even if none of the people affected presently exists. If such moral duties make sense, future-oriented reasons to care about fetuses also make sense. If we are concerned with preventing future diseases and future diseases are also influenced by fetal development, we ought to be concerned with the health of present fetuses for the sake of promoting the future health of (born) individuals.

Summing up:

1. It seems intuitive and it is widely acknowledged that we all have moral duties toward future born persons.[25]
2. Pregnant mothers, as well as other people, have duties to future persons whose future health is causally related to events in current fetuses.
3. The interest in autonomy or happiness of present mothers may conflict with the interest in health of future people. It is true that most of the time interest of mothers, fetuses, and future people are aligned, for what tends to benefit the health of the pregnant mother also tends to benefit the health of the fetus and, through DOHaD effects, the health of a future adult person. The conflict can exist because a mother may desire to avoid social control and engage in risky behavior, even though she is aware of the fact that this puts her own health, the health of the fetus, and the future health of a future person at risk. This is the essence of the DOHaD mother–fetus conflict.

BENEFICENCE, NONMALEFICENCE, AND FUTURE PEOPLE

Future-oriented reasons for caring about fetuses require a way of conceiving maternal–fetal conflicts that is not very familiar. Thus, we shall introduce here a thought experiment, as philosophers often do to test our moral intuitions.

The Magic Doll Analogy

> Imagine a world similar to ours in all respects, except that persons in this world reproduce in a very different way. Reproduction is asexual and takes place by parthenogenesis from the female ovus, after a ritual in which the male gives a doll (a simple piece of cloth) to the female and the female keeps the doll with her for nine months. Moreover, this doll is a magic doll: whatever damages the doll in those nine months will also damage the future offspring in a similar way. For example, if someone hits or breaks the leg of the magic doll, the leg of her child will be hit or broken when he or she is an adult. If the magic doll is hit by the rain, the adult offspring will also be, etc. If the mother destroys the doll, the mother loses her ovum and parthenogenesis does not begin.

There is one important similarity between the doll world and the actual world. In the doll world, individuals are affected by what happens to magic dolls, even if, clearly, (a) the magic doll is not a person and (b) the new person developing by parthenogenesis from the mother's ovum is *a distinct entity* from the magic doll kept by her mother during the nine months preceding parthenogenesis. In the actual world, we can all agree that future persons are affected by what happens to fetuses, irrespective of our metaphysical and moral disagreements concerning whether (a) fetuses are persons and (b) fetuses are the same persons as the persons into whom they develop. Arguably, the empirical facts described by DOHaD have moral implications, since they affect the health of future persons, irrespective of the truth of metaphysical and moral opinions about the moral and metaphysical status of fetuses. So, if it is reasonable to claim that mothers have duties to treat their magic dolls in a certain way (owed to future persons), it should also be reasonable to claim that mothers have duties to pay attention to events that may affect how the fetuses inside them develop (owed to future persons). These duties do not depend on assigning any intrinsic moral status to fetuses (or magic dolls).

The main difference between the two cases is that the relation between the states of the doll and the states of the future person are deterministic, while the programming of future health and disease is not (we shall explore the implications of this in the next section).

According to common-sense morality, people have a moral duty to choose, at least among alternatives equivalent in all other respects, what would benefit others. This is the duty of *beneficence*. If the mother left the doll where it might be damaged, a stranger who came upon it would have a duty of returning it to her, if at a negligible cost to herself. In this case (easy rescue), the duty of beneficence applies to the stranger. But if the mother had exposed the doll to risk in the first place, she would have owed it to the future person to retrieve it, even at considerable inconvenience to herself. This would be a duty of nonmaleficence, corresponding to a negative right of the future person to be inflicted no harm. The mother is causally and morally responsible for leaving the doll in a dangerous place, so she has a serious moral duty to rescue it, even when the rescue is costly for her. This duty would not be owed to the doll, but to the future person that would be inflicted harm if the doll were damaged.

BOX 1 THE DUTY OF BENEFICENCE

A mother has a moral duty to promote the good of her (future) offspring, at least if she can do that at a negliwgible cost to herself (for instance, by taking folic acid supplements during pregnancy).

BOX 2 THE DUTY OF NONMALEFICENCE

Arguably, mothers have a duty to avoid harming their future children by exposing them to increased health risk, by virtue of their choices during pregnancy, which may program the fetus for adult diseases.

Although the idea of the rights of future persons is to some extent controversial, recent advances in moral and in political philosophy have helped to clarify it. Consider the following example by Derek Parfit:

> Suppose that I [for no good reason; senselessly] leave some broken glass in the undergrowth of a wood. A hundred years later this glass [foreseeably] wounds a child. My act harms this child. If I had safely buried the glass, this child would have walked through the wood unharmed [19].

It is clearly morally wrong for me to leave some glass in the wood. It also seems entirely plausible to explain this intuition by appealing to the negative right to be inflicted no harm of the future person. For it also seems plausible to say, a 100 years after my negligent act, that my leaving some glass (an event in the past) caused the (present) right (of being inflicted no harm) of an actual child to be violated.

The moral duty of maleficence has wide-ranging implications. Women have duties to future people to be careful about the circumstances of fetal development and to exercise caution when dealing with any potential factor that could affect development in a way that affects disease risk.

It might seem paradoxical that a woman should be free to abort and yet be considered morally blameworthy for facts concerning fetal development, on account of duties owed to future persons. This only makes sense if one supposes, as many philosophers do (at least, concerning early fetuses), that fetuses cannot be harmed at all, or that the interests that they have are so weak that, in any conflict with the interests of mothers (who are persons), they are always defeated. By contrast, in the DOHaD cases, the interests of women are set against interests of other persons, more specifically their future offspring. So the moral duty of exercising caution emerges.

On the other hand, the limitation on a woman's freedom necessary to avoid *any behavior* that might increase the risk of harm to future people might be serious. This is an extremely problematic implication of this account, in both the real world and the imaginary world of magic dolls. In this fanciful example, women can avoid all risk of harm to persons only if they avoid *every* action that increases the risk of damaging the magic doll, for example hiding themselves in the safest place where doll-carrying women can live. This seems unjust, as it places too heavy a burden on women. In the actual world, women risk being asked similar sacrifices if they are required to subject themselves to frequent health checks and, as required by prevention goals, to strict control over their diet and other aspects of their lifestyle.

It is therefore reasonable to consider the future person's negative right to be inflicted no harm as merely prima facie. A prima facie right or duty is one that can be overridden by more urgent, or important, rights and duties. For example, suppose that I have borrowed a knife from you and promised to give it back. I have a prima facie duty to return it to you when you ask me to. But suppose that you need your knife to stab your wife. In that case my duty is overridden by a more urgent duty to save an innocent life. Analogously, the prima facie right of the future person to be inflicted no risk of harm can be overridden by more important rights of the mother. It seems unreasonable to require women to do *everything* that is in their power, as individuals, to avoid any increase in disease risk, no matter how severe the limitation of women's freedom that this would entail.

It might be objected that mothers (parents more generally) have special duties with respect to their children: morality can demand of parents a greater sacrifice for the sake of the well-being of their own child than it can demand of strangers. This might be true, but it still fails to establish an absolute duty of the mother not to harm the offspring that does not take into consideration the burden to the mother, which is due to increased precautions necessary to avoid any increase in the risk of harm.

The question may be conceptualized as one of weighing advantages and disadvantages of stricter or more relaxed attitudes toward women's behavior during pregnancy. Social attitudes that emphasize the importance of fetal health for the future well-being of the child may harm the status of women, since they seem to place women's function of creating new life and caring for it above any other aspect of their person, and restrict their freedom [20]. If societal attitudes are too protective, pregnant women may face demands to submit themselves to extensive testing and bodily invasion, and other forms of medicalization [21], in order to prevent harm to a wanted child. More permissive attitudes can protect women from this potential source of physical and psychological discomfort.

DOHaD EFFECTS ARE NOT DETERMINISTIC

Unlike the imagined relationship between magic dolls and future persons, causation in the DOHaD paradigm is not deterministic. The mother who suffers from caloric restrictions during pregnancy will not necessarily have a child who will suffer from cardiovascular disease or die prematurely. What the DOHaD paradigm in nutrition shows is only an increased risk, in comparison to mothers whose nutrition was adequate. For most conditions considered in the DOHaD literature, such as cardiovascular diseases, diabetes, premature death, and the like, the fetal environment merely adds additional risk factors to the overall risk profile of a person. Whether a person will develop cardiovascular problems depends also on his or her lifestyle, socioeconomic position, and genetic endowments. With so many variables amenable to be affected by morality, law, and policy, it may be argued that the influence of maternal lifestyle should be ignored.

We do not believe that DOHaD factors should be ignored, but rather that they should be given a proper weight in deliberation. The fact that the harm involved is probabilistic does not protect individuals from being considered responsible, at least, for that portion of increased DOHaD risk to which they contribute through their behavior. Parfit's example of leaving a piece of broken glass in the woods can be used to show this, since it can be reformulated to involve probable, rather than certain, harm. Moreover, the relationship between tobacco smoking and lung cancer is also probabilistic. If probabilistic harm did not count at all, tobacco companies could not be held responsible for the harm accrued by heavy smokers. The problem is to assess what the proper weight of DOHaD factors should be.

The issue of risk in the DOHaD paradigm must be considered when weighting future benefits for future children with present harm for pregnant mothers. A restriction of freedom and autonomy during pregnancy may appear like a very minor inconvenience if compared with conditions such as diabetes or cardiovascular disease affecting the future child. But maternal behavior only affects *the probability* of this future harm and, according to present DOHaD evidence, to a limited degree. In many cases, those who are worst off in the risk distribution are only slightly more likely to suffer from complex diseases. By contrast, the freedom and autonomy of women is immediately limited, even in the absence of legal coercion, by framing any level of risk imposed to the fetus as a result of maternal behavior during pregnancy, no matter how minor, as indicating serious maternal irresponsibility [20]. When we weigh a small added risk of a serious disease with a less severe but certain limitation of women's freedom, it is often not very clear which of the two should take priority. Some people may feel that preventing the certain, but minor harm to the mother ought to take priority over avoiding the small risk of the greater harm to the child.

DOHaD AND THE NONIDENTITY PROBLEM

Mothers have a prima facie duty to choose lifestyles that promote the well-being of future born persons via their beneficial effects on fetal health. This duty has common features with the (controversial) duty of procreative beneficence defended by Julian Savulescu, according to which:

> Couple (or single reproducers) should select the child, of the possible children they could have, who is expected to have the best life, or at least as good a life as the others, based on the relevant, available information [22].

In the genetic testing case discussed by Savulescu (e.g., selecting away the embryo with a predisposition to asthma), benefits and harms are merely probabilistic. The DOHaD case is similar to the genetic case, since the relationship between fetal circumstances and diseases is probabilistic in a way similar to the relationship between genetic factors and diseases. According to Savulescu's argument, there is a duty of beneficence in the genetic case, in spite of the probabilistic nature of genetic risk. Choosing the embryo with a predisposition to asthma, as opposed to an embryo with no known predispositions, is like playing wheel of fortune and choosing a wheel in which you have worse chances of a good outcome [22].

BOX 3 THE NONIDENTITY PROBLEM

Most reproductive choices, in particular choices to delay or anticipate conception, do not simply influence *how well off* the future child will be, but also *who will exist*. When human choices change the identity of future persons, it is more difficult to criticize them from the moral point of view.

The duty discussed by Savulescu falls, however, within the scope of the nonidentity problem. This is a term of art for a philosophical problem making our duties toward future people harder to justify.

Consider a fourteen-year-old orphan girl who wants to get pregnant and keep the child. She is advised to postpone conception until she is more mature, for her own good and for the good of her child. But how can the postponement of conception be good for the child she would otherwise have had? If the mother postpones conception for four years, a different child will be born. It is biologically not possible for the *same* child who would be conceived now to be conceived four years later. If conception is delayed, a different combination of egg and sperm will take place and generate a distinct person. This is typically labeled a "different person" case.

The case of procreative beneficence discussed by Savulescu is a different person case, since different embryos correspond to different people. Suppose that the embryo with a predisposition to asthma is selected. This grows into a person, called Mary, who actually develops asthma. Mary is not harmed by her mother's choice to bring her into existence, since the alternative for her would not have been to exist without harm but not to exist at all. Some people (including many philosophers) think we only have duties toward people who can be harmed (or benefited) by our actions. If that is correct, mothers do not have a moral duty to postpone conception when they are considered too young to have children and parents do not have duties to select the best genetic endowments for their children. For, if they do otherwise, they cause the existence of a person who is not harmed by their decision to cause her existence. (This can be plausibly denied if the parent chooses an embryo with a genetic defect so severe that it leads to a life not worth living [23].)

What about DOHaD? Some DOHaD cases are the same person cases. Suppose that a mother must choose whether to take folic acid supplements, which have a protective effect against several harmful DOHaD effects [24,25]. Taking supplements does not cause the existence of a (numerically) different person; by taking them the mother potentially *benefits* a numerically distinct individual, who will develop from the fetus inside her.

It is not difficult, however, to conceive of cases in which DOHaD choices have identity-affecting consequences. Consider an obese woman planning to have a child as soon as she can. She is told that according to a recent study, the offspring of obese mothers is 1.35 times more likely to die (of any cause) and 1.29 times more likely to be hospitalized for a cardiovascular event compared to the offspring of mothers with normal body mass index, after adjusting for known confounders [26]. Since it is difficult to lose so much weight during pregnancy, the mother understands that, if she wants to reduce the risk of adverse health outcomes in her offspring, she must try to lose weight *before* attempting to have a child. As a result of delayed conception, a different child will be born. In this context, duties of beneficence and nonmaleficence (as defined here) do not apply, so different moral principles should be invoked, such as Savulescu's (controversial) principle of *procreative* beneficence or the egalitarian principles invoked by Del Savio and his colleagues [6].

DO WOMEN HAVE A DUTY TO ABORT IN ORDER TO AVOID CAUSING DEVELOPMENTAL ORIGINS OF HEALTH AND DISEASE HARM TO FUTURE PERSONS?

If abortion is morally permissible, while causing harm to future persons is not, it might seem that abortion provides an acceptable remedy when a woman, due to some imprudent act, has caused prenatal injury, or an increase of DOHaD risk in the offspring. Certainly no one could seriously recommend abortion as a desirable strategy to avoid having a child with a somewhat increased disease risk due to DOHaD effects. In the literature, at least one author has argued that society has the right to force pregnant drug addicts to abort that fetus [27,28]. But notice that the author's argument presupposes serious teratogenic effects, produced by significant and protracted substance abuse, such as those that would lead to fetal alcohol syndrome. Teratogenic effects are different from developmental programming effects. The former involve a departure from the physiological path of development and lead with certainty (or very high probability) to serious pathologies. But notice that potentially teratogenic substances, such as alcohol, may not lead to serious disease if used in small quantities. For instance, there is no conclusive evidence of the harmful effects on offspring health of moderate alcohol use and even occasional binge drinking [29].

Here we are dealing with DOHaD cases where maternal behavior does not involve a serious injury to fetal health and is only probabilistically associated with later diseases. The issue, still, deserves discussion because it highlights some theoretical difficulties for moral frameworks that treat abortion as permissible, while considering the imposition of health risks to the future offspring as a possible moral wrong.

Here the analogy between the DOHaD case and the "magic doll" case breaks apart. The magic doll is not an entity capable of being harmed or benefited by the choices of the mother. The mother can destroy the doll at any time before the end of the nine months, even if the stipulated implication (in this fanciful world) is that she will fail to generate a new person. Clearly, no one is harmed or wronged if the woman acts in this way. Failing to cause the existence of a child with a good life is not morally analogous to killing a child.

In the fetal case, by contrast, the fetus and the person that develops from it are the same individual. If the mother does not abort, a new person will exist and the good things in her life will offset the harm in it, assuming one is dealing here with a life that is worth living [30]. Unlike the magic doll case, where the doll and the future person are two distinct entities, in this case the fetus and the future person with rights *are* the same entity (an entity acquiring different natural and moral properties as it develops). The abortion of the fetus does not merely fail to produce a life worth living; rather, it prevents *someone* (who already exists as a fetus) to develop into a person with a life that would be good *for her*. Thus, according to McMahan [30], one should argue *both* that causing fetal injury might *not* be permissible in a context in which abortion is permitted *and* that abortion does *not* provide a morally acceptable remedy to prenatal injury (although explaining how the two claims can be held together is difficult and raises further paradoxes).

But there is a further complication. McMahan also defends the view that we do not begin to exist at conception, but at some later point in fetal development. The human organism that exists from conception to that point is not identical with the person who may later develop and therefore would have no interest in the continued life of that person. According to this theory, a fetus in the first trimester is like the magic doll of our analogy. Aborting it does not deprive *it* of a life worth living, but merely prevents the coming into existence of a new individual with a life worth living. As McMahan points out, in this case the mother would have an impersonal moral reason to abort the early fetus that was injured, in order to conceive again, and produce a life with better expectations of well-being [30]. Plausibly, however, this reason is always outweighed by the psychological burden of abortion to the mother. We are not recommending that mothers should abort early fetuses as a way to avoid having children with increased disease risk or endorsing laws providing incentives for mothers to abort.

IS THE CRIMINALIZATION OF MOTHERS JUSTIFIABLE?

Here we shall consider the question of the legal criminalization of mothers. By legal criminalization we mean coercing women to adopt healthier behavior for the sake of fetal health or the health of future persons through the threat of legal sanctions. Our view is that the impact of fetal abuse prosecutions is particularly alarming, because they do not work as a deterrent. [26] Even if the need for greater efforts to protect prenatal life is well grounded morally, the value of such protection does not entail that legal punishment is justifiable.

In the legal literature, fetal abuse prosecutions are mostly related to drug abuse by pregnant women, alleged to have caused damage to the fetus. Some emphasize the undesirable consequences of a criminal prosecution in this case, due to the context in which penal law operates: the lack of prevention, of proven effectiveness in protecting prenatal life, the fact that the values protected by those laws are not widely shared, the broader costs of sanctions for families and society as a whole, the loss of trust between doctor and patient, the ineffectiveness of services dedicated to assistance and prenatal care, and the discrimination of certain categories of women [31]. The absence of direct causality from fetal circumstances to later health outcomes has been used as the main objection against the criminal prosecution of women through negligence in drug exposure cases [32].

It has also been argued that the criminalization of women would be self-defeating. Criminal sanctions may give women an incentive to abort in order to avoid being prosecuted. Arguably, it might in some cases be in the interest of a woman who acted against the law to avoid the burdens associated with criminal prosecution for fetal injury by aborting the fetus, at least where abortion is legal [16].

How should this argument be evaluated? Whether sanctions are self-defeating depends on what the purpose of the law is. If the law's ultimate rationale were to protect fetal life for its own sake, then clearly a sanction providing an incentive to abort would be self-defeating. But suppose that the law's rationale is not to protect fetal life, but to protect the health of future persons. Earlier, we described duties toward future people that could ground a *moral* duty of the mothers to avoid acting negligently and to protect the health of the fetus that will develop into a person with full moral and legal rights. Should this moral duty be backed by the coercive sanctions of the state?

One argument against turning the moral duty to future generations into a law is that punishment provides an incentive for mothers to abort. As discussed in the preceding section, McMahan argues that it would not be ethical for mothers to abort as a means of escaping the moral responsibility for causing harm to a future person. But he also argues that abortion should be permitted [30]. In this perspective, a law providing an incentive to abort would be problematic. Moreover, abortion has a significant psychological and physical cost to mothers; the infliction of such pain to women in relation to DOHaD may appear out of proportion with the intended health benefit to future generations.

Other consequentialist arguments against criminalizing women are perhaps more persuasive. Here the debate on the criminalization of mothers for drug abuse is especially instructive. Preventive incarceration is likely to have self-defeating consequences, as women, without knowing how to access programs and services that could help them to overcome their drug addiction, will likely return to their old habits upon release from prison [16]. According to Logan, convicted pregnant women and their fetuses "face conditions hazardous to fetal health, including overcrowding, poor nutrition and exposure to contagious disease" [33]. Additionally, in-prison prenatal care is generally little or is not provided at all [13]. Moreover, criminal sanctions can deter pregnant women who use drugs from seeking prenatal care.[27] Similar predictions concern the effects of a possible criminalization of women on account of DOHaD effects. Notice that consequentialist arguments rely on premises that need to be established empirically. More research may be needed into the public health implications of incarceration, or weaker legal sanctions, to ground a definitive view on the matter.

CONCLUSIONS

Discoveries related to DOHaD will extend liability for all potentially harmful maternal behavior, such as drinking alcohol, smoking cigarettes, having an unhealthy diet, being exposed to chemical and dangerous substances, and doing heavy work, i.e., the whole range of diseases explored in the scientific chapters of this book. Criminalizing women for such a wide a range of legal circumstances concerning fetal health might keep pregnant woman away from health centers. Consequently, from the legal perspective, it seems that criminal punishment, instead of acting as a deterrent, could negatively impact the health of future people in the long run. Morally speaking, it could be argued that women have a duty to exercise caution during pregnancy, and that behavior leading to increased disease risk in the offspring should be avoided if possible. This duty is only prima facie, because it can be defeated by considerations pertaining to the woman's interests in autonomy. We conclude that the debate on the *moral* implications of DOHaD discoveries should be encouraged without associating it with controversial attempts to control pregnancy through the law.

Endnotes:

1. For scientific evidence concerning these epidemiologic facts, see esp. Chapters 2, 6, 7, 8, 12, and 13.
2. In Italy and in other EU states similar precedents are not evidenced, but it is predictable that the progress of medicine and the increased attention to prenatal life will emphasize situations of maternal–fetal conflict.
3. See Weinberg S.R., "A Maternal Duty to Protect Fetal Health", *Indiana Law J* 58, no. 3 (1983), Article 4.
4. See Roe v Wade, 93 Supreme Court Reporter 705 (U.S. Supreme Court 1973), 148.
5. For example, in Gloria C. v William C. (476 N.Y.S.2d 991 (Fam. Ct. 1984)) the court found that a "fetus is a person for the purpose of issuing protective order." The court further stated that the decision "in no way conflicts with [the woman's] privacy right to freely decide what to do with her pregnancy."
6. See In Re AC, 573 A.2d 1235, 1990 D.C. App., when the court stated that "the right of a woman to an abortion is different and distinct from her obligation to the fetus once she has decided not to timely terminate her pregnancy." It also maintained that, although a pregnant woman may have an interest in bodily integrity, there is an overriding state interest in potential fetal life, which creates a legal duty to subvert a woman's right to bodily integrity when it clashes with interests in the viable-fetus.
7. In 1973, the United States Supreme Court established a woman's constitutional right to have an abortion in Roe v Wade. Although the Court did not resolve the difficult question of when life begins, it stated that "the word 'person', as used in the Fourteenth Amendment does not include the unborn." The Court did not accept the notion that an unborn child has rights entitled to constitutional protection. See Roe v Wade, 410 U.S. 113, 154 (1973).
8. The legal history of fetal personhood is not a recent issue, but has a long and complex history in the United States that is deeply analyzed in Dobow S, Ourselves Unborn: A History of the Fetus in Modern America (Oxford University Press; 2010).

9. See Jefferson v Griffin Spalding County Hospital Authority, 274, S.E. 2d 457 (1981).
10. The common law "born alive" rule only gave legal standing to born human life; the killing of a fetus was not homicide because the fetus did not have an existence separate from its mother. See C. L. Leventhal, Comment, The Crimes Against the Unborn Child: Reorganizing Potential Human Life in Pennsylvania Criminal Law, 103 DICK. L Rev. 173, 176 (1998).
11. See, e.g., Cal Penal Code § 187 (West Supp. 1986), "murder is the unlawful killing of a human being, or a fetus, with malice aforethought."
12. See Commonwealth v Cass, 467 N.E.2d 1324, 1329 (Mass. 1984). Similarly, in Hughes v State the court considered the fetus a human being who may be the victim of a homicide under a first-degree manslaughter statute. The court also stated that "if a person were to commit violence against a pregnant woman and destroy the fetus within her, we would not want the death of the fetus to go unpunished." See Hughes v State, 868 P.2d at 733 (Okla. Crim. App. 1994).
13. See, e.g., Volk v Baldazo, 103 Idaho 570, 574, 651 O.2d 11, 15 (1982), "It is clear, therefore, that [the wrongful death statute] confers upon parents a cause of action for the wrongful death of a 'child' and thus protects the rights and interests of the parents, and not those of the decedent child".
14. See State v McKnight, 576 S.E.2d 168 (S.C. 2003). In 2008, however, the South Carolina Supreme Court unanimously ruled that Regina McKnight did not have a fair trial.
15. For a wide examination of the cases in which a woman's pregnancy was a necessary factor leading to attempted and actual deprivations of her liberty, see Paltrow L.M. and Flavin J, "Arrests and forced interventions on pregnant women in the United States, 1973-2005: Implications for women's legal status and public health," Jorunal of Health Politics, Policy and Law 39, no. 6 (2013): 299-343.
16. See Johnson v State, 602 so. 2nd 1288, 1290 (Fla. 1992).
17. Iowa Code § 232.77(2) (1999); see also Statehouse Notes, Des Moines Reg. Apr. 29, at M1, describing the Senate approval of a bill to help children "who are found to have alcohol or other drugs in their systems".
18. See Whitner v State, 328 S.C. 1, 492 S.E. 2d 777 (1997). The extension of the child neglect regulation to cases of maternal–fetal conflict is the result of a simplification that put on the same ground two unequaled hypotheses. In fact, the relationship between parent–child and mother–fetus is hugely different.
19. See New Jersey v Ikerd, 850 A. 2d 516 (N.J. Sup. Ct. App. V.Div. 2004).
20. Section 1.07 of the Texas Penal Code previously defined an "individual" as "a human being who has been born and is alive." See Act of June 19, 1993, 73rd Leg., R.S., ch. 900, § 1.01, 1993 Tex. Gen. Laws 3586, 3589 (current version at Tex. Penal Code Ann. § 1,07 (Vernon Supp. 2004-05)).
21. Although this chapter focuses on legal issues, these are tied to women's real lives. Social scientists have done important work on showing the racial and class implications of fetal rights cases. See Roberts D. Killing the Black Body. Knopf Doubleday Publishing Group; 1998. See also Morgan L.M. and Michaels M.W. Fetal Subjects, Feminist Positions. Philadelphia: University of Pennsylvania Press; 1999. These works focuses on the importance of women's reproductive agency and on the increasing significance of fetal subjects in public discourse and private experience.
22. See Whitner v State, 492 S.E. 2d 787 (1997), Moore J. dissenting opinion.
23. For an anthropological study of the deployment of risk discourse regarding fetal endangerment through maternal diet, exercise, lifestyle choices, and personal habits, investigated using popular advice manuals directed at pregnant women, in opposition to the practice of "self-regulation" that occurs in advanced liberal rule, see Ruhl L, "Liberal governance and prenatal care: risk and regulation in pregnancy," *Economy and Society*. (February 1999): 95–117.
24. Thomson has influentially argued that a mother has a right to abort even if the fetus is a person. She has a right to decide what happens in and to her body, because the person inside her does not have the right to continued use of her body. Thomson's justification of the permissibility of abortion is incompatible with any duty of the mother to avoid prenatal injuries or DOHaD effects. Suppose that the fetus is a person. If the mother may permissibly kill a person (the fetus) to protect her own interests, she is also permitted to harm that person by injuring him, as long as he is in her body. Suppose that a mother's selfish interest is better served by working in a dangerous and unsafe environment during the third trimester of her pregnancy. The mother has three choices: A) to look for a less lucrative job while keeping the child; B) to keep the dangerous job, disregarding the risk of harming the future person; C) to abort the fetus. Based on Thomson's rationale for defending abortion, B must be permissible if C is, since the harm procured by the mother to the other person by B is less than that procured by C. Based on Thomson's premises, fetal injury is impermissible only if abortion is impermissible. In what follows we shall suppose, on the contrary, that the mother does not have a right to kill persons.
25. This claim needs to be qualified by adding "at least if the identity of future persons does not change as a result of our actions." This will be clarified in a later section.
26. "There is no evidence that these latter [criminal] sanctions prevent in utero drug exposure or help drug-exposed children after birth. Without strong evidence [that criminal involvement has any benefit] such intervention is unjustifiable." See Committee on Substance Abuse, Drug-Exposed Infants, 86 Pediatrics 639, 640 (1990) at 641.
27. For example, South Carolina has charged eighteen women with either child neglect or distributing drugs

to minors. Since these criminal sanctions have begun, South Carolina has had a dramatic increase in the number of births at home, in taxis, and in bathrooms, because the women have been too fearful to go to hospital to get proper care and assistance during pregnancy and delivery. See Pregnant and Newly Delivered Women Jailed on Drug Charges, Reproductive Rights Update, Feb. 1, 1990, at 6 (A.C.L.U. publication). See also Poland M.L. et al, Punishing pregnant drug users: enhancing the flight from care, Drug and Alcohol Dependence, 31(1993), 199–203 at 202. "The key finding of this survey is that our sample of low-income mothers in Detroit strongly believed that punitive legislation would further alienate pregnant substance-using women from needed health care. [...] This study suggests that punitive laws may have an opposite effect".

References

[1] Drake AJ, Liu L. Intergenerational transmission of programmed effects: public health consequences. Trends Endocrinol Metab April 2010;21(4):206–13.

[2] ACOG Committee Opinion No 575. Exposure to toxic environmental agents. Fertil Steril 2013;100(4):931–4.

[3] Post LF. Bioethical consideration of maternal-fetal issues. Fordham Urban Law J 1997;24(4):757–75.

[4] Loi M, Savio LD, Stupka E. Social epigenetics and equality of opportunity. Publ Health Ethics July 1, 2013;6(2):142–53.

[5] Kollar E, Loi M. Prenatal equality of opportunity. J Appl Philos 2015;32(1):35–49.

[6] Del Savio L, Loi M, Stupka E. Epigenetics and future generations. Bioethics 2015. http://dx.doi.org/10.1111/bioe.12150. http://onlinelibrary.wiley.com/doi/10.1111/bioe.12150/abstract.

[7] Lenow JL. The fetus as a patient: emerging rights as a person? Am J Law Med 1983;9(1):1–29.

[8] Poovey M. The abortion question and the death of man. In: Butler J, Scott JW, editors. Feminists theorize the political. London: Routledge; 2013. p. 239–56.

[9] Balisy SS. Maternal substance abuse: the need to provide legal protection for the fetus. South Calif Law Rev 1987;60(4):1209–38.

[10] Robertson JA. The right to procreate and in utero fetal therapy. J Leg Med 1982;3(3):333–66.

[11] Schroedel JR, Fiber P, Snyder BD. Women's rights and fetal personhood in criminal law. Duke J Gend Law Policy 2000;7:89.

[12] Johnsen DE. The creation of fetal rights: conflicts with women's constitutional rights to liberty, privacy, and equal protection. Yale Law J 1986;95(3).

[13] Bell HF. In utero endangerment and public health: prosecution vs. treatment. Tulsa Law J 2001;36:649–76.

[14] Haynes M. Inner turmoil: redefining the individual and the conflict of rights between woman and fetus created by the prenatal Protection act. Tex Wesley Law Rev 2004;131. [Fall].

[15] Weinberg SR. A maternal duty to protect fetal health? Indiana Law J 1983;58(3):531–46.

[16] Millis MD. Fetal abuse prosecutions: the triumph of reaction over reason. DePaul Law Rev 1998;47:989–1040.

[17] Steinbock B. Mother-fetus conflict. [Internet]. In: Kuhse H, Singer PA, editors. A companion to bioethics [Internet]. 2nd ed. London: Wiley & Blackwell; 2009. [cited 2015 Apr 15]. p. 149–60. Available from:. http://onlinelibrary.wiley.com/doi/10.1002/9781444307818.ch15/summary.

[18] Thomson JJ. A defense of abortion. Philos Public Aff October 1, 1971;1(1):47–66.

[19] Parfit D. Reasons and persons. Oxford: Clarendon Press; 1984.

[20] Richardson SS, Daniels CR, Gillman MW, Golden J, Kukla R, Kuzawa C, et al. Society: don't blame the mothers. Nature August 13, 2014;512(7513):131–2.

[21] Kukla R, Wayne K. Pregnancy, birth, and medicine. [Internet] In: Zalta EN, editor. The Stanford Encyclopedia of philosophy. Spring; 2011. [cited 2015 Apr 7]. Available from: http://plato.stanford.edu/archives/spr2011/entries/ethics-pregnancy/.

[22] Savulescu J. Procreative beneficence: why we should select the best children. Bioethics 2001;15(5–6):413–26.

[23] Boonin D. The non-identity problem and the ethics of future people. Oxford: Oxford University Press; 2014. 321 p.

[24] Timmermans S, Jaddoe VWV, Hofman A, Steegers-Theunissen RPM, Steegers EAP. Periconception folic acid supplementation, fetal growth and the risks of low birth weight and preterm birth: the Generation R Study. Br J Nutr September 2009;102(05):777–85.

[25] Lillycrop KA, Burdge GC. Epigenetic changes in early life and future risk of obesity. Int J Obes 2011;35(1):72–83.

[26] Reynolds RM, Allan KM, Raja EA, Bhattacharya S, McNeill G, Hannaford PC, et al. Maternal obesity during pregnancy and premature mortality from cardiovascular event in adult offspring: follow-up of 1 323 275 person years. aug13 1 BMJ August 13, 2013;347:f4539.

[27] Schedler G. Does society have the right to force pregnant drug addicts to abort their fetuses? Soc Theory Pract October 1, 1991;17(3):369–84.

[28] Schedler G. Forcing pregnant drug addicts to abort: rights-based and utilitarian justifications. Soc Theory Pract October 1, 1992;18(3):347–58.

[29] Kesmodel U, Bertrand J, Støvring H, Skarpness B, Denny C, Mortensen E, et al. The effect of different alcohol drinking patterns in early to mid pregnancy on the child's intelligence, attention, and executive function. BJOG 2012;119(10):1180–90.

[30] McMahan J. Paradoxes of abortion and prenatal injury. Ethics 2006;116(4):625–55.
[31] Thompson E. The criminalization of maternal conduct during pregnancy: a decisionmaking model for lawyers. 3 Indiana Law J 1989;64(2):357–74.
[32] Smith T, Maccani M, Knopik V. Maternal smoking during pregnancy and offspring health outcomes: the role of epigenetic research in informing legal policy and practice. Hastings Law J 2013;64(6):1619–48.
[33] Logan E. The wrong race, committing crime, doing drugs, and maladjusted for motherhood: the nation's fury over "crack babies.". San Franc Soc Justice 1999;26(1): 115–38.

Bibliography

Annas GJ. Pregnant women as fetal containers. Hastings Cent Rep 1986;16:13–4.
Chavkin W, Breitbart V, Elman D, Wise PH. National survey of the states: policies and practices regarding drug-using pregnant women. Am J Public Health 1998;88:117–9.
Comittee on Substance Abuse. Drug-exposed infants. Pediatrics 1990;86:639.
Dailard C, Nash E. State RESPONSES to substance abuse among pregnant women. Guttmacher Rep Public Policy 2000;3:1–4.
Feinman C. The criminalization of a woman's body. New York: Harrington Park Press; 1992.
Gomez LE. Misconceiving mothers. legislators, prosecutors, and the politics of prenatal drug exposure. Philadelphia: Temple University Press; 1997.
Janssen ND. Fetal rights and the prosecution of women for using drugs during pregnancy. Drake Law Rev 2000;48:741–67.
Parks KT. Protecting the fetus: the criminalization of prenatal drug Use. Wm Mary J Women L 1998;5:245–71.
Poland M L, Dombrowski MP, Ager JW, Sokol RJ. Punishing pregnant drug users: enhancing the flight from care. Drug Alcohol Depend 1993;31:199–203.
Rippey SE. Criminalizing substance abuse during pregnancy. N Eng J Crim Civ Confinement 1991;17:69–106.
Suppé R. Pregnancy on trial: the Alabama supreme court's erroneous application of Alabama chemical endangerment law in ex parte ankrom. Health Law Policy Brief 2014;7:49–75.
Vassoler FM, Byrnes EM, Pierce RC. The impact of exposure to addictive drugs on future generations: physiological and behavioral effects. Neuropharmacology 2014;76:269–75.
Weinberg SR. A maternal duty to protect fetal health. Indiana Law J 1983;58:531–46.

CHAPTER

24

Introduction to *Moms in Motion* (MIM)

Cheryl S. Rosenfeld

Department of Bond Life Sciences Center, Department of Biomedical Sciences, Genetics Area Program, Thompson Center for Autism and Neurobehavioral Disorders, University of Missouri, Columbia, MO, USA

OUTLINE

It is clear from the prior chapters that the lifestyles of the mother and father can lead to dramatic developmental origin of health and disease (DOHaD) effects on their offspring. Given the studies to date, the next logical step is to encourage good habits in child-bearing age women. However, this is easier said than done. It is difficult to break bad habits, even when ample evidence exists to indicate that they are bad for our own health, let alone the health of our children, and several generations to follow.

The challenges become even more daunting when dealing with low-income, less-educated women. Yet, the children of these women are the most vulnerable and something has to be done to break the vicious cycle of metabolic and other diseases that will continue to plague these populations for generations to come. Even for those seeking physician's advice, health care options are limited. Many low-income areas are also considered nowadays as "food deserts," geographical areas lacking supermarkets selling affordable and nutritious food. Without easy access to such grocery stores, many low-income individuals resort to eating unhealthy and processed food; this likely only exacerbates underlying metabolic disorders. Further, many such regions lack bicycle and walking trails and low-cost gym access. Even when such amenities are available, women living in poverty often are reluctant to go outside or engage in physical activity in neighborhoods generally plagued by violent crimes. The physical inactivity even further perpetuates the risk of metabolic disorders in these women.

The Epigenome and Developmental Origins of Health and Disease
http://dx.doi.org/10.1016/B978-0-12-801383-0.00024-4

Copyright © 2016 Elsevier Inc. All rights reserved.

With all of these factors in mind, Dr Mei-Wei Chang, the author of the next chapter, discusses the heroic attempt to curb obesity and metabolic disorders in a low-income population of women residing in Michigan. This ongoing study is being conducted in partnership with the Special Supplemental Nutrition Program for Women, Infants, and Children. The community-based approach has been given the appropriate name of *Mothers in Motion* (*MIM*). The overarching goal of this program is to help young, overweight and obese mothers prevent weight gain by promoting healthy eating, physical activity, and stress management. Once mothers establish a healthy lifestyle, their children are more likely to eat healthier and be more physically active. When the women become pregnant, the prevailing notion is that they will have better maternal and birth outcomes. The study includes a combined strategy where the women are provided DVDs on topics ranging from nutrition to physical activity, and stress management, as well as peer support group teleconferences. Although the work is currently ongoing, Dr Chang discusses the historical context of the *MIM* program, the pilot *MIM* program that helped to form the basis of the current study, and the challenges and remedies that occurred in the pilot and ongoing *MIM* study. As mentioned, the study has posed several challenges due to sociodemographic limitations of the participants, such as disconnected phones, moving from place to place, and lack of computer access. Also, the author provides strategies to overcome potential limitations, e.g., emphasizing the free home-based program and encouraging facilitators to demonstrate compassion and respect to the women involved in the study. Dr Chang and her group are to be commended for working to find solutions to troubleshoot the hurdles and help these vulnerable women. Many women who would be otherwise hard to reach have remained in the study by the investigator's employing personal touches, including cards on special occasions, and generally showing an interest in their general well-being.

The findings from the *MIM* cohort studies will likely provide broader insight into community-based approach tactics that might be employed to stem the dramatic rise in obesity and metabolic disorders in expectant mothers and the population as a whole. At least 35% of the US population is currently obese, with costs for treating ailments relating to this condition in 2008 estimated at $147billion [1]. More than one-third of children are obese with a predisposition to type 2 diabetes, and most of them will remain so as adults. The World Health Organization has estimated that close to 350million people worldwide already have type 2 diabetes and that this number is increasing annually, particularly in the developing countries, as populations gain greater access to a so-called Western-style diet and become more sedentary [2]. A 2014 report by the Institute of Medicine Roundtable on Obesity and Solutions suggested that obesity has been curbed in some US states, likely due to increased physical activity and improved nutrition [3]. Yet, this summary warned that obesity rates continue to rise in other states, and importantly, the incidence of morbid obesity has escalated dramatically. Further reports indicate that children are less physically active now than they were in past decades, which may be due to a convenient lifestyle with automated transportation, reduced accessibility to parks and other areas to play, and increasing the amount of time engaged in sedentary activities [4–6].

While the investigators in the current *MIM* study have focused on healthy eating, physical activity, and stress management, the approach could serve as a foundation to educate the women about other factors, such as environmental chemicals, smoking, and drinking, that could lead to harmful DOHaD effects and what they can do in their daily lives to minimize such risks to their unborn offspring. A branch study (Fathers in Motion, FIM) may be envisioned to educate future dads about how their daily habits affect their unborn offspring, as it is clear that the paternal environment can lead to epigenetic and phenotypic changes in the father's sons and daughters [7–10]. The *MIM* study provides an important framework for such future work in this area and will provide key results as to whether we can induce positive lifestyle changes in obese pregnant women and thereby promote long-term health of their children and descendants.

ABBREVIATIONS

DOHaD Developmental origin of health and disease
FIM *Fathers in motion*
MIM *Moms in motion*
WIC Women, infants, and children

References

[1] http://www.cdc.gov/obesity/data/adult.html.
[2] Scully T. Diabetes in numbers. Nature 2012;485(7398):S2–3.
[3] Roundtable on Obesity S, et al. The current state of obesity solutions in the United States: Workshop summary. National Academies Press (US); 2014. Copyright 2014 by the National Academy of Sciences. All rights reserved. Washington (DC).
[4] Brownson RC, Boehmer TK, Luke DA. Declining rates of physical activity in the United States: what are the contributors? Annu Rev Public Health 2005; 26:421–43.
[5] Gray CE, et al. Are we driving our kids to unhealthy habits? Results of the active healthy kids Canada 2013 report card on physical activity for children and youth. Int J Environ Res Public Health 2014;11(6): 6009–20.
[6] Ziviani J, et al. A place to play: socioeconomic and spatial factors in children's physical activity. Aust Occup Ther J 2008;55(1):2–11.
[7] Bromfield JJ, et al. Maternal tract factors contribute to paternal seminal fluid impact on metabolic phenotype in offspring. Proc Natl Acad Sci USA 2014; 111(6):2200–5.
[8] Crean AJ, Dwyer JM, Marshall DJ. Adaptive paternal effects? Experimental evidence that the paternal environment affects offspring performance. Ecology 2013;94(11):2575–82.
[9] Rodgers AB, et al. Paternal stress exposure alters sperm microRNA content and reprograms offspring HPA stress axis regulation. J Neurosci 2013;33(21): 9003–12.
[10] Rando OJ. Daddy issues: paternal effects on phenotype. Cell 2012;151(4):702–8.

CHAPTER

25

Reversing Harmful Developmental Origins of Health and Disease Effects

Mei-Wei Chang[1], *Susan Nitzke*[2], *Roger Brown*[3], *M Jean Brancheau Egan*[4], *Kobra Eghtedary*[4], *Cheryl S. Rosenfeld*[5]

[1]Michigan State University College of Nursing, East Lansing, MI, USA; [2]Department of Nutritional Sciences, University of Wisconsin-Madison, Madison, WI, USA; [3]School of Nursing, University of Wisconsin-Madison, Madison, WI, USA; [4]Michigan Department of Community Health, WIC Division, Lansing, MI, USA; [5]Department of Bond Life Sciences Center, Department of Biomedical Sciences, Genetics Area Program, Thompson Center for Autism and Neurobehavioral Disorders, University of Missouri, Columbia, MO, USA

OUTLINE

[T]he Epigenome and Developmental Origins of Health and Disease
[ht]tp://dx.doi.org/10.1016/B978-0-12-801383-0.00025-6

Copyright © 2016 Elsevier Inc. All rights reserved.

INTRODUCTION

Over 54% of low-income women of childbearing age in the United States (US) are overweight or obese [1], placing them at risk for several chronic diseases including type 2 diabetes and cardiovascular disease (CVD) [2–5]. The postpartum period is a critical time when weight retention and weight gain can lead to long-term increases in adiposity [6,7], exacerbating health problems related to overweight and obesity. While women of all income levels retain at least 2.2–4.4 lbs after 6–18 months postpartum or beyond [8,9], low-income overweight and obese women tend to retain 13.2 lbs [10,11]. Also, low-income overweight and obese women are at risk of major weight gain (>25 lbs) 10 years after delivery [11,12].

Unfortunately, current weight loss programs remain relatively ineffective for treating obesity, as most participants ultimately regain their lost weight [13,14]. Also, few programs that have targeted the unique needs of low-income mothers have been tested [15]. Thus, preventing weight gain in women with a high risk of major weight gain is a high priority [4,16]. Excessive weight gain in early adulthood (ages 20–30) increases CVD risk factors [17], whereas preventing weight gain is associated with either improving or not significantly changing the risk of developing CVD [4,18], and is generally easier to achieve than weight loss [4]. Also, preventing an increase of even 1 body mass index (BMI) unit (~6 lbs) can prevent one's risk for developing type 2 diabetes by about 13% [2]. Moreover, preventing excessive weight gain has societal benefits by reducing the cost burden of diabetes treatment and losses in employee productivity [19].

The Diabetes Prevention Program (DPP) demonstrated that healthy lifestyle (healthy eating and physical activity) promotes an average weight loss of 7% among adults with prediabetes [20]. However, the DPP's intervention model cannot be easily replicated in most community-based programs for low-income audiences because it includes 16 intensive counseling sessions [21]. Similarly, model programs of this nature often lack cultural competence and underutilize information technology [22].

We developed *Mothers in Motion (MIM)* with the potential for implementation in community-based settings, e.g., the Special Supplemental Nutrition Program for Women, Infants, and Children (WIC). WIC is a federally funded program that provides nutrition consultation and other services to nearly 9.7 million low-income pregnant, postpartum, and breast-feeding women and children less than five years old in the US [23].

To address cultural misunderstandings, *MIM* DVDs were designed to include deep and surface structure components of low-income African American and white culture. Deep structure reflects how cultural, social, psychological, historical, and environmental factors influence health behaviors differently across racial populations [24,25]. The deep structure requires understanding how the target audience perceives the cause, course, and treatment of illnesses (e.g., overweight and obesity) and how their perceptions influence specific healthy lifestyle behaviors (e.g., not being physically active because of time constraints) [26]. The surface structure includes incorporating an appropriate channel of intervention and using suitable media elements such as people, language, foods, clothing, music, and locations that are familiar to and preferred by the target audience.

To overcome underutilization of information technology, we use DVDs and peer support group teleconferences to deliver the intervention. Previous researchers use an interactive CD-ROM to deliver a single session of nutrition education to WIC mothers in WIC offices [27–30] and found significant improvements in nutrition attitudes [27,28] and self-efficacy [29], but not dietary behavior change [29]. Subsequently, a study of diabetic patients using a combination of interactive CD-ROM and individual telephone consultation found significantly reduced dietary fat intake, increased fruit and vegetable intake, and increased physical activity [31]. These studies suggest that frequent contact and a combination of these modes might be needed. Since repeated contact with interactive CD-ROMs is not feasible in most WIC clinics [29], an alternative is to provide DVDs that participants can view in their homes, given the fact that most American households including low-income households have a television (99%) [32]. Our pilot *MIM* showed that more than 92% of WIC mothers had a DVD player at home in 2007 [33]. Most low-income people list TV as a major source of health information [34]. The DVD, which does not require conventional prose literacy skills [35], is an effective means to deliver motivational and educational messages to low-income populations and is cost effective [36].

Interventions utilizing peer support group teleconferences (PSGTs) have resulted in significant gains in the ability to cope, support satisfaction, perceived support and information [37–39], and decreased feelings of isolation and loneliness [37,40]. The teleconference environment is comfortable and convenient, provides privacy, is readily accessible without transportation, and can increase participation [41,42]. Also, group intervention can be effective because it provides empathy, social support, and healthy competition [13]. Thus, the combination of DVDs and PSGTs has great potential for promoting healthy lifestyle and meeting the needs of low-income mothers, as shown in our pilot test of *MIM* [33].

In collaboration with the State of Michigan WIC and several local WIC agencies, we conducted a pilot (feasibility) test [35] and then a full-scale study to test the effectiveness of the *MIM* intervention. *MIM* is a community-based intervention aimed to prevent weight gain among low-income overweight and obese mothers 18–39 years old by promoting stress management, healthy eating, and physical activity. We included stress management based on our prior work with the target audience (described later) and suggestions from local WIC agencies.

Our academic research team has a long history of collaborating with State of Michigan WIC and several local WIC agencies. Instead of presenting the principles of community-based participatory research, which can be found in numerous textbooks and research articles, this chapter will summarize the perspectives and experiences of WIC community partners who have worked closely with academic researchers on the *MIM* projects. We will also share how we worked with other stakeholders at the community level.

Community Partners' Perspectives Working with Academic Researchers

WIC Perspectives

Representatives from State of Michigan WIC have maintained collaborative partnerships with academic researchers to develop and test *MIM* materials and methodology over a 12-year period. Although WIC is a public health service–oriented program, WIC is interested in collaborating with researchers to enhance service delivery and improve health outcomes of the clients that they serve. The value of academic partnerships with WIC is enhanced when researchers share findings via timely reports and presentations. WIC appreciates the opportunities for the intellectual stimulation, sharing "field-based" experiences to help research projects and creation of potential resources for local WIC agencies to better serve their clients. Also, WIC values the opportunity to reciprocate by sharing their statistics with researchers in a mutual effort to improve service delivery and health outcomes of clients nationwide.

Prior history of successful collaboration with WIC is the most important factor when WIC makes decisions on collaboration with academic researchers. Success is enhanced when researchers who collect data from WIC share their preliminary findings, ask WIC to assist in interpreting results, and consider WIC suggestions prior to submission of abstracts and articles for publication or formal presentations. When these critical steps are missed, misinterpretations may occur and WIC service capabilities may be curtailed. In some instances, researchers have appeared to be uncaring or condescending toward WIC clients, possibly due to the researchers' incomplete knowledge of WIC policies, procedures, and populations. Those instances have had negative effects on WIC's public image and may have affected WIC's ability to secure funding. Just as researchers need to obtain funding to sustain their labs, each WIC entity at the state, district, or tribal level needs to secure funding to maximize services for their clients.

WIC is interested in adopting evidence-based interventions that can enhance WIC service delivery and/or improve health outcomes of WIC clients. However, WIC must always consider the cost of training, maintaining, and sustaining the intervention when making such decisions. The following section summarizes specific actions and strategies that foster successful academic/community partnerships, as suggested by state WIC partners in the *MIM* research team.

Suggested Strategies for Working with Community Partners Such as WIC

Prior to conducting any research project, gaining an understanding of state WIC policy, procedures, goals, and health indicators is an important first step toward successful collaboration. For potential WIC partners, such data are available via USDA, the national WIC association, and state WIC websites. For example, academic professionals and students should be familiar with WIC's most important health indicators, e.g., percent of clients gaining access to WIC services during the first trimester, and rates of pregnancy/infant conditions such as low birth weights, childhood obesity, low hemoglobin levels, and excessive gestational weight gain.

Maintaining open, honest, timely, and effective communication is one of the most important requirements for working with community partners such as WIC at all levels. To ensure continuity of a project, researchers must know whom to involve in important decisions. In other words, the staff members that you know personally may not be authorized to make commitments on behalf of an agency or organization.

Being respectful to WIC personnel and client needs and time constraints is crucial. WIC appreciates it when researchers provide well-organized materials, a clear agenda and purpose of the meeting, and plan everything in advance to allow WIC sufficient time to make a decision or provide feedback. Having research personnel handle subject recruitment and data collection is desired because WIC staff must concentrate on providing maximum service to clients. Be aware that Institutional Review Board (IRB) approval may be required by your state or local WIC agencies in addition to your academic institution. Moreover, being open and willing to collect additional data for WIC is greatly appreciated.

Following through on agreements and commitments is important. Researchers need to consult with WIC collaborators when they experience challenges and need to make adjustments to their protocols. This is imperative since some changes and/or adjustments may affect WIC's critical content or delivery methods. Being respectful and having a positive attitude are essential when working with WIC. For example, ask for feedback or suggestions from WIC (not necessarily at state level) and be willing to make improvements as needs arise. This helps everyone to ensure safety/privacy of and positive experiences for WIC clients and address the logistical needs of WIC as it strives to fulfill its service mission.

Factors Affecting WIC Involvement

Many factors facilitate State of Michigan WIC's involvement in research projects. For example, intervention designs need to be culturally sensitive to WIC clients and need to be easily adopted by WIC with minimum cost. If staff training is required, it needs to be easily incorporated into WIC's existing training system. The State of Michigan WIC has been actively involved in *MIM* because it has these positive features.

WIC's interest in involvement with collaborative research projects is diminished when studies require intensive efforts from the state and local WIC personnel, require large space allocations, and/or call for significant commitments of facilities or WIC administrators' time. Also, some studies create negative experiences for WIC clients. When this happens, WIC clients may develop negative perceptions about WIC, thus discouraging WIC clients from returning to their follow-up appointments or receiving further services and food assistance.

The above principles are exemplified by the following challenges that were encountered by state and local WIC agencies working on the *MIM* project. In the early stages, WIC policy and procedures were hindered when materials were submitted to WIC's IRB for approval without proper organization and language clarity. Also, the *MIM* team experienced some challenges regarding appropriate supervision of peer recruiters who were working at collaborating WIC agencies. Overall, the *MIM* research team has generally been open to suggestions and WIC concerns to resolve issues.

Researchers' Reflections on Successfully Working with WIC

Establishing trust is the most important element of working with community agencies such as WIC. Trust is strengthened by showing respect to everyone working at WIC. To maintain trust and mutual satisfaction, we take care to follow through with our commitment to WIC in a timely manner. We keep all partners well informed about the project progress via monthly updates that include testimonial narratives from *MIM* study participants.

Our key research leader (principal investigator, PI) stays directly involved and demonstrates respect to WIC staff and clients consistently. Being a good listener requires being patient and open minded. We listen to suggestions and concerns about our projects from WIC staff at all levels (from stakeholders and WIC administrators to receptionists). When concerns become apparent, we ask for advice from collaborating WIC agencies to solve problems and improve procedures. Since we have been willing to listen and make improvements as suggested, WIC has helped us anticipate and avoid difficult situations. In summary, respecting WIC's policies and procedures and demonstrating personal interest in working with WIC at all levels helps maintain a healthy working relationship among all key collaborators in our *MIM* research.

We have learned some team-building approaches that may be helpful to other researchers. First of all, one should not assume that administrative approval will assure enthusiastic cooperation from all personnel involved in the project. Our PI met with the WIC staff to explain the study purpose and emphasize the fact that our plan placed no additional burden on WIC staff time. As part of normal counseling, we did ask WIC staff to refer potential participants to our recruiters by informing their clients that *MIM* might be potentially helpful to them. We continually acknowledge the importance of WIC's contributions, which help our team achieve the common goal between WIC and the academic partners—improved health outcomes of WIC clients.

We show our appreciation for WIC inputs on the design and operation of the collaborative projects by inviting key WIC collaborators to be coauthors on publications [43–45], presentations, and/or poster presentations at national conferences. We also collaborate with them to share findings at WIC conferences. We always acknowledge the health department, WIC administrators and staff in our publications and presentations about our collaborative projects, and WIC is given an opportunity to review drafts of manuscripts and abstracts prior to submission.

MIM is funded by the National Institute of Diabetes and Digestive and Kidney Diseases (NIDDK), through March 2016. Results and conclusions cannot yet be made as we write this chapter, but we can share design of the *MIM* and modifications that helped strengthen our project's recruitment, adherence to intervention, and retention.

Primary Aim and Hypotheses

Our aim is to test the effectiveness of *MIM* intervention. We hypothesize that the intervention group will have lower weight gain (difference between the baseline and three months postintervention) than the comparison group by an average of at least 2.8 lbs, due to a combination of prevention of weight gain in the intervention group and a slight weight gain in the comparison group. Our secondary hypothesis is that at three months postintervention, the intervention group will have greater improvements than comparison group in dietary fat, fruit, and vegetable intake, physical activity, stress, affect, self-efficacy, emotional coping response, social support, and autonomous motivation.

Prior Studies

We conducted two studies to help development of *MIM*: focus group discussion with overweight and obese WIC mothers [43] and pilot *MIM* [33]. We used the results of focus group discussions with the target audience to inform the design of the pilot *MIM*, funded by NIDDK, 2006–2009. We then used results and lessons learned from the pilot *MIM* to design and refine the current *MIM*.

Focus Group Discussions with Overweight and Obese WIC Mothers

We conducted eight focus group discussions to identify barriers and motivators for prevention of weight gain, recruitment, and retention strategies [43]. We also identified target audience preferences and needs regarding intervention contents and delivery.

METHOD

Participants (N = 80) were nonpregnant overweight or obese WIC mothers aged 18 and 35 years who self-identified as African American and non-Hispanic white. These women were recruited from six collaborating local WIC agencies in Michigan in 2004. Depending on the size of the WIC agency, some may have multiple clinics. Each focus group lasted two hours and was audio recorded. The audio recordings were transcribed and codes were established to identify common themes.

RESULTS

Emotional stress was reported to be a key factor preventing women from eating healthier and being physically active. Participants indicated that young mothers often used large portions of high-fat and high-calorie foods to cope with stress. Also, when women experienced stress, they were less likely to engage in physical activity. Motivators to prevent weight gain were health concerns, personal appearance, inability to fit in clothes, and difficulty in playing with their children [43]. When asked about what components in DVDs would be helpful, they suggested the topics of diet, physical activity, and stress management. Suggested topics included food label reading, grocery shopping tips, meal preparation, enjoyable home exercise/activities with children, stress management skills, and ways to avoid emotional eating. Examples in the DVDs must be simple and concrete and be easy to apply in their daily lives. The people in the DVDs should be WIC mothers and their children wearing casual clothing. These mothers should have a similar body size and life experience as participants do. These mothers must share their positive and negative experiences and the process of their behavioral and weight changes. The DVD should include a variety of settings, e.g., home, grocery store, and neighborhood. Each module should be brief (10–15 min) and entertaining to watch. The mothers also suggested a way of answering questions at the end of each module to monitor future participants' adherence to watching DVDs.

Focus group participants indicated that a PSGT would motivate them to participate actively, modify lifestyle behavior, and prevent weight gain. When asked how to motivate women to participate in the PSGT, they indicated that the group members must be mothers that they could relate to, e.g., similar in body size and age. We also asked how to recruit and retain participants; women suggested that we should tell the potential participants exactly what our program is about. Monetary incentives were suggested, especially cash, grocery coupons, and gift certificates for diapers or clothing. Providing a weekly calendar was also suggested. Moreover, telephone and postcard reminders were recommended. We incorporated the study findings when designing the pilot *MIM*.

Pilot Mothers in Motion

The pilot *MIM* (P-*MIM*) aimed to evaluate the feasibility of conducting the project in WIC settings. We focused on identifying successful strategies for recruitment, adherence to intervention, and retention. We also evaluated the acceptability of the project by the study participants and WIC staff. Moreover, we analyzed preliminary indicators of anticipated outcomes, such as dietary intake, physical activity, stress, body weight, and blood glucose (via figure poke) comparing intervention participants to the control group [33].

METHOD

Collaborating with State of Michigan WIC and three local WIC agencies, we piloted *MIM*. We enhanced recruitment by emphasizing confidentiality and enlisting positive support from WIC personnel at the collaborating WICs. Recruiters who were members of our research staff were trained to be culturally sensitive, speak clearly, and listen carefully. They explained the study's purpose, requirements, confidentially, flexible scheduling, benefits of participation (e.g., a free program to help manage stress), and cash/gift incentives. They also emphasized that the pilot *MIM* was a collaborative effort between WIC and Michigan State University. The recruiters invited potential participants to enroll at WIC clinics while women waited for their WIC appointment. Women who were interested in participating filled out the demographic and screening surveys. The recruiters measured their height and weight with light clothing and without shoes at a private room at WIC clinics. Eligible women who consented to participate in the study were invited to participate. Participants were also required to be nonpregnant African American or white who had measured BMIs between 25.0 and 39.9, were free of type 1 or 2 diabetes, spoke and understood English, could walk more than one block without resting, and were WIC mothers between 18 and 35 years old.

DVD development for the pilot study was similar to the *MIM*'s DVD development (described later). The DVD featured four overweight and obese WIC mothers, hereafter referred to as featured mothers. The PSGT moderators who were WIC dietitians attended a one-day training class and followed a specialized instruction manual to lead the discussions. After completion of baseline phone interviews, we randomly assigned participants to an intervention (n=64) or control (n=65) following an algorithm to balance race (African American and white) and BMI categories (25.0–29.9, 30.0–34.9, and 35.0–39.9). Both groups received usual WIC care. The intervention group received usual WIC care and a 10-week intervention that called for them to watch five chapters in a DVD (10–15 min/chapter) at home and dial in to five PSGTs (30 min/session) at any convenient location every other week. After watching each chapter, the intervention participants were asked to answer three quizzes on a worksheet and returned to the study office using a prestamped envelope. Thus, we were able to monitor adherence to watching the chapter. During each PSGT, an assistant moderator dialed in to keep track of time and record who dialed in [33].

We collected data at three time points: baseline, two months (six months from baseline), and eight months (one year from baseline) post the 10-week intervention. Trained telephone interviewers collected survey data (~60 min). WIC staff measured body weight and obtained blood glucose (via finger stick) when participants returned to WIC clinics for their children's appointment or coupon pickup. The primary outcomes were body weight and blood glucose via finger stick. The mediators included dietary fat, fruit, and vegetable intake, physical activity, stress, and affect/feelings [33].

At the completion of the study, we sought suggestions to enhance/evaluate enrollment, adherence, retention, and acceptability by conducting four focus group discussions with the intervention participants (n=12) who had watched the DVD and called in to the PSGTs. We also conducted individual telephone interviews with three PSGT moderators to evaluate the PSGT implementation in WIC settings. Moreover, we conducted five focus groups with 25 WIC administrators and staff from State of Michigan WIC and three collaborating WIC agencies to evaluate the feasibility of implementing DVD and PSGT in WIC settings and recruitment.

RESULTS

RECRUITMENT STRATEGIES BASED ON RECRUITERS' LOGS By showing respect, establishing rapport, and providing incentives, we were able to recruit young, low-income overweight, and obese mothers that are often considered "hard to reach." Babysitting was critical because WIC personnel and mothers perceived it as a sign of caring. In fact, some WIC mothers who had refused to participate during the initial invitation decided to participate after they saw the positive interaction between the recruiters and their children. We also found that talking to each woman individually was more successful than recruiting two or more participants at the same time.

ADHERENCE TO INTERVENTION We mailed all intervention materials in one package to the participants' homes via certified mail (signature required) at the start. Of the 60 participants who received the intervention package, 65.5% provided evidence of watching the DVD chapters and applying learned skills and knowledge from the DVD to daily life as assessed using returned weekly worksheets (mean=3.2 worksheets, SD=1.6). The main reason for not watching the DVD was misplacing the intervention package. For the PSGT, 48% called in to participate in one or more PSGTs (mean=2.17, SD=1.33); the main reasons for not calling were time conflicts and forgetfulness [33].

RETENTION Of 129 women, nine reported becoming pregnant, and these women were excluded from the analyses. Of 118 participants,

70 (59.3%, six-month follow-up) and 48 (40.7%, one-year follow-up) completed telephone interviews and 58 (49.2%, six-month follow-up) and 39 (33.0%, one-year follow-up) returned to the WIC clinics for body weight measurement. The main reasons for dropout were loss to follow-up (54/118, 45.8%; mainly due to disconnected phones) and self-reported loss of interest (16/118, 13.6%) [46].

PRIMARY OUTCOMES AND MEDIATORS We did not find significant differences in primary outcomes and mediators between the intervention and control groups at six-month and one-year follow-up. However, our pilot data showed apparent trends in body weight changes and differences in blood glucose levels that were consistent with the study's hypotheses.

FOCUS GROUP DISCUSSION WITH THE INTERVENTION PARTICIPANTS The *P-MIM* intervention participants said that the DVD was about having a happy and healthy family so that the key message to recruit their peers should be "happy and healthy family" rather than "prevention of weight gain." Lack of understanding study requirements and incentives affected retention rate. Mothers stated that they could not remember the study requirements and incentives after signing a consent form because of all the distractions at WIC clinics. They suggested providing a flyer with key study requirements and incentives for participants to take home and stating reminders at the end of each telephone interview. Every mother said that she watched all five chapters but not everyone returned all worksheets because of misplacing them. They made seven suggestions to improve the DVD:

1. Limiting characters in each DVD chapter to two or three featured mothers
2. Showing featured mothers going through their daily routine and how their routine differed before and after making healthy lifestyle changes and how they involved family members in making positive changes
3. Expanding the interactive information, steps for making changes and problem solving skills, and certain topics in the DVD (e.g., adding information on self-control during holidays)
4. Providing more concrete messages and examples of short- and long-term goals
5. Emphasizing key strategies from the DVD for healthy lifestyle changes rather than general information for the quizzes
6. Adding recipes
7. Rearranging sequence, i.e., presenting goal setting right after interactive information.

PSGT MODERATORS The moderators enjoyed leading the PSGTs but found it difficult to lead the PSGTs at prearranged times. Also, the PSGT scripts needed flexibility.

FOCUS GROUPS WITH WIC ADMINISTRATORS AND STAFF To increase retention rate, WIC personnel suggested having newly recruited study participants returned to their WIC clinics where they had been recruited to pick up the first study package as a condition of enrollment. Potential participants who returned to this first step were more likely to follow through and actively participate throughout the study. They perceived high feasibility of DVD implementation in WIC settings but were less encouraging for the PSGTs due to severe time constraints of WIC staff. They suggested collaborating with Michigan State University Extension (MSUE) to lead the PSGTs. We incorporated findings from the P-MIM to design and refine the current *MIM* program.

METHOD

Overview of the Ongoing *Mothers In Motion* Program

The ongoing *MIM* study delivers theory-based, culturally sensitive intervention messages focusing on healthy eating, physical activity, and stress management via a combination of DVDs (featuring African American and non-Hispanic white overweight or obese WIC mothers) and PSGTs. The PSGT moderators are trained in motivational interviewing (MI) and group facilitation skills. Community and peer advisory groups and target audience representatives have reviewed rough-cut DVDs to ensure that they are culturally sensitive to the target audience and to provide feedback for revisions. The full team (WIC and the academic research partners) continue to collaborate on implementation, evaluation, and dissemination of the study findings. Randomization occurs at the individual participant level. The intervention group receives a 16-week intervention on a weekly or every other week basis. Participants watch 10 DVD chapters at home (10–15 min/per chapter) and join 10 PSGT sessions by phone (30 min/session). The comparison group receives reading materials about healthy eating, physical activity, and stress management. Both groups receive usual WIC care.

Community and Peer Advisory Groups

The community advisory group includes six WIC administrators from the collaborating agencies and the State of Michigan WIC. The peer advisory group includes six African American and white overweight or obese WIC mothers. *MIM* utilizes these two groups to assist in planning and evaluating intervention and to review rough-cut DVDs. The community advisors also help with *MIM* evaluation and implementation.

Figure 1 presents a study design of *MIM*. The study (recruitment, intervention, and data collection) is being conducted with the approval of two IRBs at Michigan State University and the Michigan Department of Community Health. We apply community-based participatory approach to design the intervention and study methods [47].

The Conceptual Model

MIM applies the concepts of social cognitive and self-determination theories to systematically address key personal and environmental factors [48] to promote healthy lifestyle behaviors with the ultimate goal of preventing weight gain. Concepts from the social cognitive theories are self-efficacy (one's confidence to perform a specific behavior), emotional coping response (strategies used to manage stress), and social support. The autonomous motivation concept from the self-determination theory is defined as a person's values and beliefs to guide behavioral change [49]. Personal factors are self-efficacy, emotional coping response, and autonomous motivation. The environmental factor emphasized in *MIM* is social support.

Intervention

Culturally Sensitive DVDs

TRAINING OF WIC (FEATURED) MOTHERS IN THE DVDs

We designed the *MIM* DVDs to present videos that are compelling, entertaining, engaging, and innovative. The video portrays the experience of four overweight and obese mothers (three African Americans and one white) who met the study criteria. They were counseled and encouraged for 16 weeks (~3–4 h of counseling per week) to make healthy lifestyle changes in three key areas of daily life: stress management, healthy eating, and physical activity. The PI applied MI techniques to counsel these featured mothers. Each session included (1) setting small, realistic, and measurable goals, (2) asking open-ended questions to help featured mothers think about their current lifestyle and identify root causes of their difficulties, and (3) developing problem-solving skills. All counseling sessions and filming took place at the featured mothers' homes, playgrounds, neighborhoods, and local grocery stores. We filmed the 4 featured mothers, their significant others, and young children

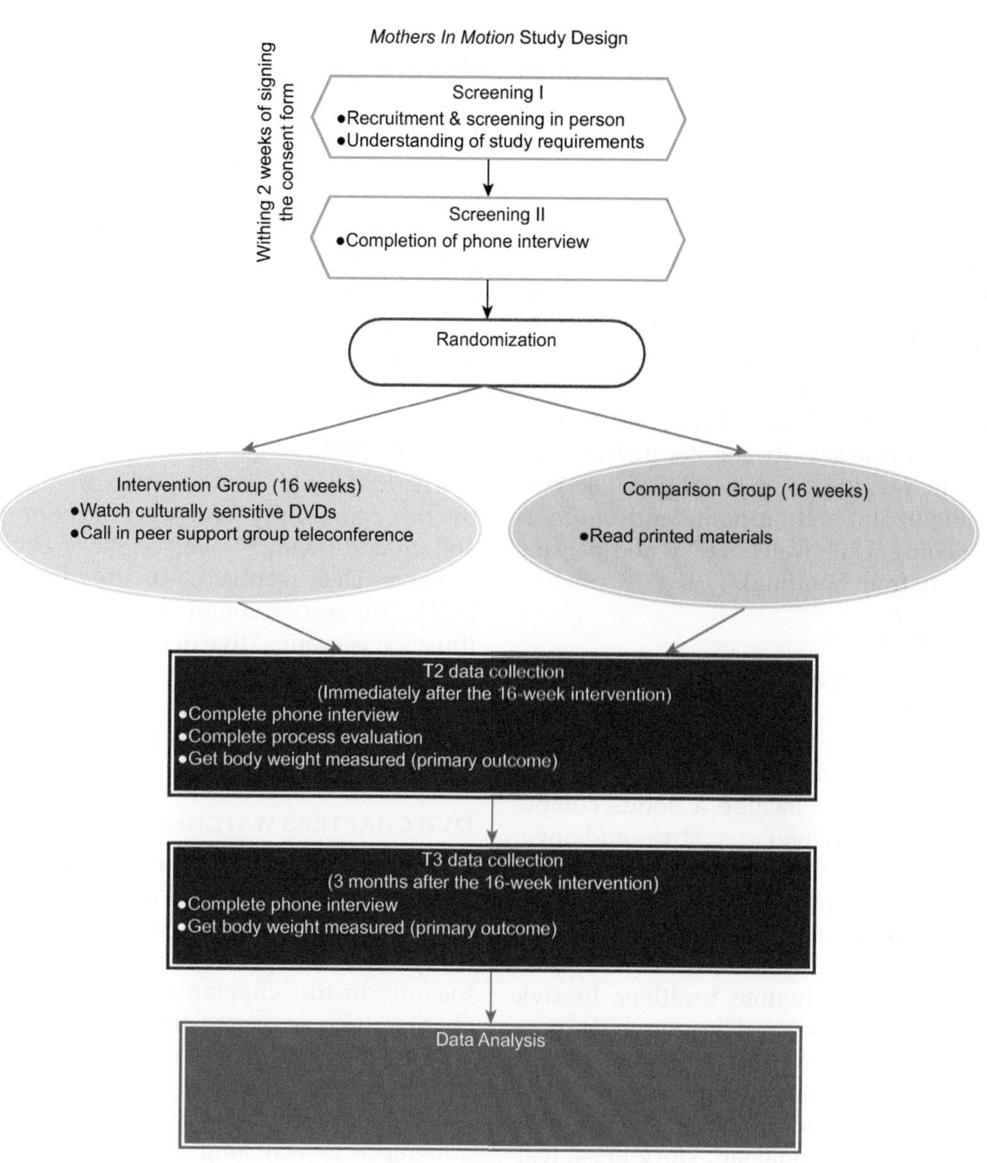

FIGURE 1 Study design of *MIM*.

over 1 year to document changes before, during, 3 and 6 months after the 16-week counseling. Each filming included two parts: interview narratives and B-roll (action). The interview questions were developed based on the principle of MI and B-rolls were based on the content of interviews.

THE DVD CHAPTERS

The final DVD set has 11 intervention chapters. Each topic is designed to be independent with no overlapping content. **Introductory chapter** (~6 min) presents the study purpose, a preview of the *MIM* DVDs, and goal setting that emphasizes importance of and confidence in making positive lifestyle changes, values, and small/gradual change. **Ten intervention chapters** address stress management (n=4), healthy eating (n=5), and lifestyle physical activity (n=1). The stress management chapters include better ways to handle everyday responsibilities and hassles (e.g., laundry and dishes), time-saving tips (e.g., setting priority), ways to handle negative feelings (e.g., emotional eating), and parenting tips. The healthy eating chapters cover effective ways to reduce junk food intake, read food labels, plan meals, shop for groceries, and cook healthier meals/snacks with children. The physical activity chapter includes fun and realistic ways to get more physical activity (low to moderate) both at home and outdoors (e.g., marching while watching TV and parking farther away from building).

DVD COMPONENTS

Each video chapter is about 20min. Each chapter has three components with a similar structure: interactive information (~2min), culturally sensitive vignettes (~17min), and action plan (~40s). We also include a bonus component (1–2min per chapter, e.g., stress and anger logs, recipes, and credible websites for further information). **Interactive information**: The interactive information is designed to correct misperceptions, increase self-awareness of current lifestyle, and promote healthier lifestyle choices. We use "pause" and "continue" buttons to allow viewers to stop and think about their current habits/behaviors. **Culturally sensitive vignettes** are designed to engage the attention of the audience with poignant story lines, real-world examples, and testimonials that model healthy lifestyle choices. The featured mothers demonstrate ways to identify root causes of their difficulties, positive and negative experiences in making changes, and how their lifestyle choices affect their family members. We include testimonials from the featured mothers, their significant others, and young children (three to nine years). We also present real stories before, during, and after making changes and the benefits therein (e.g., impact on young children) over a one-year period. Moreover, we show strategies to overcome procrastination and challenges of making change. **Action plan**: Each section concludes with an action plan that is presented in a multiple choice format and is used to monitor participants' adherence to watching each DVD chapter. The action plan focuses on key strategies related to self-efficacy from each chapter and helps viewers make action plans for healthy lifestyle changes.

DVD REVIEW

The first rough cuts of DVDs were reviewed by the community and peer advisor groups and an additional 50 overweight or obese WIC mothers. Their feedback was used to revise the DVDs. The second rough cuts were reviewed by three focus groups (five or six African American and white overweight or obese WIC mothers per group). Their feedback was used to finalize the DVDs, and the featured mothers approved the DVD content.

DVD CHAPTERS WATCHED AT HOME

During the 16-week intervention, participants watched one DVD chapter weekly (weeks 1–4) or every other week (weeks 5–16) at home, except the first week when they also watched the introductory chapter. After watching a DVD chapter, they answered three questions (action plan) and mailed the action plan worksheet to the study office using a prestamped envelope. We used the returned worksheet to monitor adherence to watching the intervention chapters. The participants also used a weekly worksheet to set one or two personal goal(s) and to make plans for behavioral changes and self-monitor progress for seven days.

Peer Support Group Teleconference

TRAINING OF MODERATORS

The moderators were peer educators and dietitians who work in the community. Michigan

Motivational Interviewing uses an interpersonal, egalitarian, empathetic, and client-centered approach to help individuals overcome ambivalence about behavior change, solve their own barriers, and explore potential untapped sources of motivation [50]. The training comprised three components. First, MI moderators received a two-day training via a mix of didactic and experiential/practice activities. We inculcated skills and practice with real-time constructive feedback. Examples of didactic elements include introduction to MI spirit and core MI skills: asking open questions, affirming, reflective listening, summarizing [50]. The moderators also received training on group facilitation skills. For the practice activity, we used the authentic WIC scenarios provided by the State of Michigan WIC and local WIC agencies and unscripted testimonies from the *MIM* DVD filming. Moderators received a certificate of completion at the end of the two-day training.

Before leading the actual PSGT, each moderator took part in four supervised sessions by conducting practice PSGTs with overweight or obese WIC mothers. At the end of each practice call, we reviewed recordings of the practice PSGTs and debriefed with the PSGT participants to assess their satisfaction with each call. We shared the debriefing summary with the moderators and identified strengths as well as specific areas for improvement.

Each month, the MI trainer, a master prepared psychologist, randomly monitored selects sessions (25%) for each moderator and used a tracking sheet created by *MIM* to monitor treatment fidelity. Each moderator listened and scored her audiotapes on a monthly basis. The MI trainer and moderators both scored these recording independently using the fidelity tracking sheet. The MI trainer made specific notes for strengths and weaknesses. We then sent the fidelity sheet scored by the MI trainer to the moderator who compared the score, strengths, and weakness comments to their own evaluation. When a moderator's performance was below expectation, the MI trainer developed an individualized plan for further integrating MI to the moderator's style of conducting a PSGT. The MI trainer was also available for moderators for questions related to MI and group facilitation skills. Moderators attended booster trainings via phone (1.5–2h in months 3 and 9) or met in person (4–5h in months 6 and 12) to receive additional information and enhance skills.

PSGTs DIAL IN AT ANY CONVENIENT LOCATION

Based on the pilot *MIM*, [33] we created groups comprising 10 women who remain in the same cohort for the 16-week intervention. The participants dialed into the PSGT weekly or every other week to discuss contents in the designated DVD chapter watched. To increase participation, they were provided two different times (one in mid-morning or early afternoon and one in the evening) to call into a PSGT. Each PSGT session lasted about 30min. The moderators were online as participants called in. Their role was to establish a safe, nonconfrontational, and supportive climate where the participants feel comfortable discussing the stress in their lives and expressing both positive and negative aspects of their current behavior, and the pros and cons for change. They also helped participants explore their ambivalence about change and motivated them to address their own barriers. The moderators followed a semistructured "roadmap protocol," e.g., introduction, rapport building, assessment of current behaviors and progress, and providing feedback with permission from participants. Assistant moderators recorded the session, took notes, kept time, and used a log to keep track of who called in to monitor PSGT adherence.

Comparison Group

The comparison group receives printed materials from standard reliable sources such as MyPlate.gov for stress management, healthy eating, and physical activity. They also receive a food and home safety DVD (~10 min). We provide printed materials to the comparison group to avoid confusion and make both groups' information similar in scope. The comparison participants receive the *MIM* intervention materials including DVDs when they complete the final data collection.

Setting, Participants, and Recruitment

Recruitment started in September 2012. We anticipate closing enrollment in December 2014. Participants are recruited from five collaborating local WIC agencies (comprising nine WIC clinics) in Michigan. **Inclusion criteria**: Participants must be nonpregnant African Americans or white, understand and speak English, be 18–39 years old, be at least six weeks postpartum, be willing to provide valid contact information for three backup people, be willing to update their contact information monthly, be willing to participate in *MIM* for nine months, be willing to make three additional trips to the WIC clinics where they are recruited for randomization and body weight measurement, have a working DVD player or computer at home, and have ready access to a working phone. Also, participants must have a measured BMI between 25.0 and 39.9 kg/m^2. Height is measured to the nearest 0.1 cm using a wall-mounted stadiometer, with participants wearing no shoes. Weight is measured using an electronic digital scale (described later). **Exclusion criteria**: We exclude women who plan to become pregnant or move to a location outside of study sites during the trial, have self-reported type 1 or 2 diabetes, or are unable to walk more than one block without resting or having shortness of breath.

Recruiter Training and Recruitment Strategies

Peer recruiters are WIC mothers and receive a five-day training with focus on hands-on experience and adherence to study protocol. Also, they are trained to be culturally sensitive and relate to the potential participants. We encourage recruiters to use a gentle and caring voice, demonstrate understanding, empathy, and excitement about the study while maintaining eye contact, and offer assistance with small tasks [46]. Women coming to the collaborating WIC clinics during the data collection dates are personally invited by the recruiters to participate while waiting for appointments. The recruiters explain the study purpose, confidentiality and study requirements, and explain *MIM*'s flexibility and no-cost availability. The potential participants who are not interested in participating in the program are asked to fill out the demographic survey. Thus, we will be able to evaluate the representativeness of the study sample. In order to determine eligibility, interested participants fill out demographic and screening surveys, and recruiters measure their height and weight.

Sequential Screening

To minimize dropouts and increase participants' understanding of the study requirements and incentives, we are implementing a sequential screening based on lessons learned from the pilot *MIM*. [33] **Screening I**: We conduct cognitive testing with more than 20 WIC mothers to develop an easy-to-read flyer that provides a study overview. Eligible participants read the flyer, then are interviewed (~2 min) by the recruiter to assess their interest and commitment and to determine if the study goals and expectations are a good fit. The recruiters ask eligible participants to sign a consent form only if they can verbalize full understanding of the study requirements. **Screening II**: Eligible participants are enrolled after they complete a baseline phone interview and return to WIC clinics where they are recruited in person to pick up the study package within two weeks after signing the consent form.

Recruitment Challenges and Modification

Initially (September to October 2012), we encountered four recruitment challenges.

Challenge #1: We could not reach many participants to schedule the baseline (T1) interview because they rejected calls that came from outside their area code. Once this situation was identified, recruiters started reminding participants of the area code of the study office number so that we could reach the majority to schedule the T1 interview. **Challenge #2**: The study recruitment flyer was ineffective because it presented information about healthy eating and physical activity that was too general and had limited appeal for our potential participants. Also, in our recruitment efforts, the emphasis on healthy eating and physical activity were not effective motivators for potential participants. We addressed this challenge by modifying the flyer to reflect an emphasis on stress management and to offer a free home-based program. In subsequent recruitment efforts, we emphasized stress management and improved eating habits and physical activity as ways to achieve happier and healthier families.

Challenge #3: The recruiters emphasized the study incentives, e.g., a large box of diapers. Many participants called the study office to request a specific diaper size and indicated that their participation depended upon diapers or specific incentives. Also, some participants dropped immediately after receiving the diapers. We reduced this problem by eliminating diapers from the incentives and asking recruiters to give less emphasis to incentives as a reason to enroll. **Challenge #4**: Several peer recruiters had to be terminated because of excessive tardiness and absences or inability to follow the recruitment protocol. Our Data Safety Monitoring Board suggested that we expand our hiring practices without limiting the pool to WIC mothers only. We also developed more effective strategies to monitor recruiters and identify potential problems in a more timely manner.

After implementing new recruitment strategies for nearly one year, two challenges continue to impede our overall progress. **Challenge #1**: Some of the non-WIC mothers that we hired as recruiters were unwilling to talk cordially to the potential participants and some left for more lucrative positions. Thus, we continue to hire WIC mothers. Still, we have had limited success with recruiters at one inner-city location. We learned that the mothers that we recruited lived in extremely dangerous neighborhoods. That local WIC agency has been closed to recruitment, and data from this agency will be eliminated from the data set for this study.

To minimize recruitment impediments, we hire peer recruiters with at least one year working experience with the same employer in the past recent years. We also increased hourly pay rate of recruiters. In addition, the primary investigator (PI) applies MI to enhance communication with the recruiters. The PI leads teleconference with the recruiters on a weekly basis. During the teleconference, the PI reviews study protocol and recruiters share their success and brainstorm challenges of recruitment and retention with each other.

Challenge #2: Personnel changes and building remodeling projects hindered procedures at one local WIC agency. We worked with State of Michigan WIC to replace that agency and continue recruitment. **Challenge #3**: We observed that some participants consent to participate in the study to please the recruiters who are nice and talk about the program with excitement. Therefore, all recruiters have been trained to encourage participation based on the program's advantages for potential enrollees and to deemphasize the recruiter's desire for a successful enrollment.

Randomization

Within two weeks of signing the consent form, participants who complete baseline measures (described later) return to the WIC clinics to be randomized. The recruiters use a laptop computer equipped with WIFI to perform randomization. Then, participants receive either an intervention or comparison study package and study incentives (e.g., cooking utensil or laundry detergent).

Retention Strategies

We ask study participants to give us updated pregnancy status and contact information monthly, even if there are no changes. Every month, we mail a specially designed greeting card and project update. We also mail birthday, sympathy, and Mother's day cards. Our recruiters call the study participants on a monthly basis to greet them and remind them of the date for their next data collection. We also work closely with the collaborating WIC agencies to identify WIC appointments for our study participants who would have otherwise been lost because of lapses in contact information. In a few cases where WIC return visits are not feasible, recruiters have done home visits to measure body weights.

Measurements

Data are collected via phone or in person (body weight measured at WIC clinics) at three time points: baseline (T1), immediately (T2), and three months (T3) after the 16-week intervention.

Primary Outcome

Body weight is measured to the nearest 0.2 lbs on an electronic digital scale (Seca 869, Germany) with the participants wearing light clothing and no shoes. **Mediators**: We measure dietary fat, fruit, and vegetable intake; physical activity; perceived stress; and affect (emotion/mood). To measure dietary fat, fruit, and vegetable intake, we use a Rapid Food Screener (24 items: 17 items for dietary fat, seven items for fruit and vegetable intake), which has adequate validity and reliability [51]. Each month, the first seven women who complete the T1 phone interview also complete Block's short-form Food Frequency Questionnaire [52]. Physical activity is measured using Pregnancy Infection Nutrition 7-day physical activity recall (asking frequency, duration, and intensity) with established validity and reliability [53]. Perceived stress is measured using a nine-item Perceived Stress Scale [54]. Affect is measured using the Positive Affect and Negative Affect Scale (18 items) [55].

Additional Mediators

Mediators include self-efficacy, emotional coping response, social support, and autonomous motivation. Self-efficacy is measured by asking participants to rate their confidence (1 = not confident at all, 4 = very confident) in eating healthy foods (8 items), becoming more physically active (10 items), and managing their stress (9 items) [56,57]. Emotional coping response is measured by asking participants about strategies used to cope with emotional eating (6 items), physical activity (4 items), and stress management (7 items) [56,58]. Social support is measured by asking participants to report social support from their family members, friends, or other people to eating healthy foods (6 items), engaging in physical activity (4 items), and managing stress (6 items) [56]. Autonomous motivation for healthy eating (17 items), physical activity (17 items), and stress management (17 items) are measured using a Treatment Self-Regulation Questionnaire [59]. **Process evaluation**: We perform process evaluation (T2 only, e.g., program satisfaction, and reasons not watching DVDs or dialing in the PSGT) to help explain intervention impact.

Sample Size and Statistical Methods

The sample size calculation was based on a two-sided test of significance at $\alpha < 0.05$, 80% power, a difference in mean change in body weight between groups of 2.8 pounds and standard deviation of weight change = 10.7 pounds [33], and an allocation rate of 2 (intervention):1 (comparison). Since we anticipated 30% attrition and 15% pregnancy [60] occurrence during the study, we will oversample up to 525 women to reach a final sample size of 290. We will use the Mplus modeling package to impute data (that are missing either at completely random or at random) by implementing the full information maximum likelihood algorithm. We will perform intention-to-treat analysis. Also, we will use appropriate statistical tests (e.g., t-test and analysis of covariance model for continuous variables and chi-squared test for categorical variables) to confirm the success of randomization. To assess two hypotheses at three months postintervention, a general linear model will be used covarying mean centered baseline measures, including a treatment condition dummy variable as well as covariates.

DISCUSSION

WIC has a strong interest in obesity prevention. Thus, this study's findings should be of interest to a broad audience of policy makers and could lead to a variety of dissemination initiatives. Based on the results of this study, the academic research team and WIC partners will conduct a series of activities aimed at sharing the findings and collaboratively interpreting and translating the findings into their usual practice with a set of policy recommendations. Anticipating effectiveness, the academic research team will work with WIC to refine the *MIM* components. This may include refining elements and revising content of the *MIM* DVDs and the PSGT training manuals, and other materials as needed before the materials are presented to WIC for broader implementation. Knowing that recruiting mothers into this type of program is a challenge, the research team will work with community and peer advisory groups to create a culturally sensitive recruitment DVD and flyers/pamphlets and postcards/subscription system to encourage African American and white overweight or obese mothers to enroll in *MIM*. To meet the ultimate translational goals, we will initially target WIC at state/district level. The research team will also consider targeting more clinically oriented programs that serve this audience (e.g., pediatric and OB/GYN offices) and to other settings (e.g., community health centers, churches, and Head Start). The research team will offer trainings via interactive Internet technology (webinars) to representatives from WIC and others at the state/district level. The webinars will be videotaped for those who cannot participate. The team uses a "train the trainer" model to organize the training conference, thus representatives would complete training of the PSGT protocol and implementation of the *MIM* DVDs. They would then be charged to train other key users at their state/district. Additionally, a set of recommendations for effective dissemination and implementation of the *MIM* will be distributed via print, Internet, and other channels.

The research team will work with State of Michigan WIC to link the *MIM* data with WIC's existing data set to examine pregnancy outcomes. For example, we will explore if there is a higher proportion of *MIM* participants who become pregnant and meet the Institute of Medicine's pregnancy weight gain guidelines than the non-*MIM* participants. We are also interested in assessing the pattern of weight gain among children whose mothers participate in *MIM* compared to children's mothers not enrolled in the *MIM*.

Limitations

There are two limitations to *MIM*. For maximum external validity, overweight or obese WIC mothers should be randomly selected from the target population. In this study we enroll self-selected participants, resulting in possible selection bias that may limit external validity. However, this mimics how WIC would likely engage clients to participate in prevention of weight gain in real-world settings, which could enhance external validity of our findings. We will continue to enroll participants with characteristics of high attrition [61,62]. The anticipated high attrition rate is attributable to loss to follow up due to high household mobility, instability in the living situations, and frequent phone disconnections or number changes. Strategies to minimize attrition are described above.

Conclusions

The academic research team has worked closely with State of Michigan WIC and local WIC agencies to develop, design, and implement *MIM*. *MIM* is designed to be easily adopted, implemented, maintained, and sustained in community settings serving low-income populations. If successful, *MIM* could be implemented in WIC, clinical practice (e.g., OB/GYN office), and other settings (e.g., community health centers) nationwide. Our methodology can be amenable to future implementation by researchers and community stakeholders who promote healthy lifestyle behaviors for low-income women. If we are able to document *MIM*'s effectiveness, it will be a valuable resource for public health and community programs. Also, it may have concomitant benefits on childhood obesity because promotion of mothers' healthy lifestyle can help their young children establishing healthy eating behaviors and physical activity, which are major factors in preventing childhood obesity [63–65].

Future Studies

Future studies may consider collecting dietary intake and physical activity data and body weight from study participants' young children and following these children to their adulthood. This may allow researchers to examine how external environment changes impact on these children and their offspring. Also, health behavior researchers may collaborate with researchers in epigenetics to determine whether/how an intervention like *MIM* might have beneficial intergenerational effects.

ABBREVIATIONS

BMI Body mass index
CVD Cardiovascular disease
DPP Diabetes Prevention Program
IRB Institutional Review Board
MI Motivational interviewing
MIM *Mothers in Motion*
NIDDK National Institute of Diabetes and Digestive and Kidney Diseases
PI Principal investigator
P-MIM Pilot *Mothers in Motion*
PSGT Peer support group teleconference
WIC Special Supplemental Nutrition Program for Women, Infants, and Children

References

[1] Hinkle SN, Sharma AJ, Kim SY, et al. Prepregnancy obesity trends among low-income women, United States, 1999-2008. Matern Child Health J 2012;16(7):1339–48.
[2] Burke JP, Williams K, Narayan KM, Leibson C, Haffner SM, Stern MP. A population perspective on diabetes prevention: whom should we target for preventing weight gain? Diabetes Care 2003;26(7):1999–2004.
[3] Kannel WB, D'Agostino RB, Cobb JL. Effect of weight on cardiovascular disease. Am J Clin Nutr 1996;63(3 Suppl):419S–22S.
[4] Lloyd-Jones DM, Liu K, Colangelo LA, et al. Consistently stable or decreased body mass index in young adulthood and longitudinal changes in metabolic syndrome components: the Coronary Artery Risk Development in Young Adults Study. Circulation 2007;115(8):1004–11.
[5] NIH. Clinical guidelines on the identification, evaluation, and treatment of overweight and obesity in adults. National Heart, Lung, and Blood Institute in cooperation with the National Institute Diabetes and Digestive and Kidney Diseases; 1998.
[6] Gore SA, Brown DM, West DS. The role of postpartum weight retention in obesity among women: a review of the evidence. Ann Behav Med 2003;26(2):149–59.
[7] Kim SY, Dietz PM, England L, Morrow B, Callaghan WM. Trends in pre-pregnancy obesity in nine states, 1993-2003. Obesity 2007;15(4):986–93.
[8] Keppel KG, Taffel SM. Pregnancy-related weight gain and retention: Implications of the 1990 Institute of Medicine guidelines. Am J Public Health 1993;83(8):1100–3.
[9] Williamson DF, Kahn HS, Remington PL, Anda RF. The 10-year incidence of overweight and major weight gain in US adults. Arch Intern Med 1990;150(3):665–72.

[10] Rosenberg L, Palmer JR, Wise LA, Horton NJ, Kumanyika SK, Adams-Campbell LL. A prospective study of the effect of childbearing on weight gain in African-American women. Obes Res 2003;11(12):1526–35.

[11] Wolfe WS, Sobal J, Olson CM, Frongillo Jr EA, Williamson DF. Parity-associated weight gain and its modification by sociodemographic and behavioral factors: a prospective analysis in US women. Int J Obes Relat Metab Disord 1997;21(9):802–10.

[12] Williamson DF, Madans J, Pamuk E, Flegal KM, Kendrick JS, Serdula MK. A prospective study of childbearing and 10-year weight gain in US white women 25 to 45 years of age. Int J Obes Relat Metab Disord 1994;18(8):561–9.

[13] Wadden TA, Butryn ML, Byrne KJ. Efficacy of lifestyle modification for long-term weight control. Obes Res 2004;(12 Suppl):151S–62S.

[14] Jeffery RW, Drewnowski A, Epstein LH, et al. Long-term maintenance of weight loss: current status. Health Psychol 2000;19(1 Suppl):5–16.

[15] Guelinckx I, Devlieger R, Beckers K, Vansant G. Maternal obesity: pregnancy complications, gestational weight gain and nutrition. Obes Rev 2008;9(2):140–50.

[16] Kumanyika S, Jeffery RW, Morabia A, Ritenbaugh C, Antipatis VJ. Obesity prevention: the case for action. Int J Obes Relat Metab Disord 2002;26:425–36.

[17] Norman JE, Bild D, Lewis CE, Liu K, West DS. The impact of weight change on cardiovascular disease risk factors in young black and white adults: the CARDIA study. Int J Obes Relat Metab Disord 2003;27(3):369–76.

[18] Truesdale KP, Stevens J, Lewis CE, Schreiner PJ, Loria CM, Cai J. Changes in risk factors for cardiovascular disease by baseline weight status in young adults who maintain or gain weight over 15 years: the CARDIA study. Int J Obes (Lond) 2006;30(9):1397–407.

[19] Hillier TA, Pedula KL. Characteristics of an adult population with newly diagnosed type 2 diabetes: the relation of obesity and age of onset. Diabetes Care 2001;24(9):1522–7.

[20] Knowler WC, Barrett-Connor E, Fowler SE, et al. Reduction in the incidence of type 2 diabetes with lifestyle intervention or metformin. N Engl J Med 2002;346(6):393–403.

[21] National Institutes of Health. Translational research of the prevention and control of diabetes and obesity (R18). 2006. http://grants.nih.gov/grants/guide/pa-files/PAR-06-532.html. [accessed 28.08.06].

[22] Garfield SA, Malozowski S, Chin MH, et al. Considerations for diabetes translational research in real-world settings. Diabetes Care 2003;26(9):2670–4.

[23] Women, Infants, and Children (WIC) participant and program characteristics. 2012. http://www.fns.usda.gov/sites/default/files/WICPC2012.pdf. [accessed 26.03.14].

[24] Resnicow K, Baranowski T, Ahluwalia JS, Braithwaite RL. Cultural sensitivity in public health: defined and demystified. Ethn Dis 1999;9(1):10–21.

[25] Resnicow K, Braithwaite RL, Dilorio C, Glanz K. Applying theory to culturally diverse and unique populations. In: Glanz K, Lewis FM, Rimer BK, editors. Health behavior and health education: theory, research, and practice. 3rd ed. San Francisco: Jossey-Bass Publishers; 2002. p. 485–509.

[26] Campbell MK, Hudson MA, Resnicow K, Blakeney N, Paxton A, Baskin M. Church-based health promotion interventions: evidence and lessons learned. Annu Rev Public Health 2007;28:213–34.

[27] Jantz C, Anderson J, Gould SM. Using computer-based assessments to evaluate interactive multimedia nutrition education among low-income predominantly Hispanic participants. J Nutr Educ Behav 2002;34(5):252–60.

[28] Gould SM, Anderson J. Using interactive multimedia nutrition education to reach low-income persons: an effectiveness evaluation. J Nutr Educ 2000;32(4):204–13.

[29] Campbell MK, Carbone E, Honess-Morreale L, Heisler-Mackinnon J, Demissie S, Farrell D. Randomized trial of a tailored nutrition education CD-ROM program for women receiving food assistance. J Nutr Educ Behav 2004;36(2):58–66.

[30] Block G, Miller M, Harnack L, Kayman S, Mandel S, Cristofar S. An interactive CD-ROM for nutrition screening and counseling. Am J Public Health 2000;90(5):781–5.

[31] Estabrooks PA, Nelson CC, Xu S, et al. The frequency and behavioral outcomes of goal choices in the self-management of diabetes. Diabetes Educ 2005;31(3):391–400.

[32] Energy information administration: Official energy statistics from the U.S. Government. Residential energy consumption survey: Home energy uses and costs. 2005. http://www.eia.doe.gov/emeu/recs/ [accessed 22.08.08].

[33] Chang M, Nitzke S, Brown R. Design and outcomes of a *Mothers in Motion* behavioral intervention pilot study. J Nutr Educ Behav 2010;42(3 Suppl):S11–21.

[34] Schwitzer G. Ten troublesome trends in TV health news. BMJ 2004;329(7478):1352.

[35] Ward JA, Anderson DM, Pundik CG, Redrick A, Kaufman R. Cancer Information Service utilization by selected U.S. ethnic groups. J Natl Cancer Inst Monogr 1993;14:147–56.

[36] Gould SM, Anderson J. Economic analysis of bilingual interactive multimedia nutrition education. J Nutr Educ Behav 2002;34(5):273–8.

[37] Stewart MJ, Hart G, Mann K, Jackson S, Langille L, Reidy M. Telephone support group intervention for persons with hemophilia and HIV/AIDS and family caregivers. Int J Nurs Stud 2001;38:209–25.

[38] Goodman C, Pynoos J. Telephone networks connect caregiving families of Alzheimer's victims. Gerontologist 1988;28:602–5.
[39] Nokes MM, Chew L, Altman C. Using a telephone support group for HIV-positive persons aged 50+ to increase social support and health-related knowledge. AIDS Patient Care STDs 2003;17(7):345–51.
[40] Curran VR, Church JG. A study of rural women's satisfaction with a breast cancer self-help network. J Telemed Telecare 1999;5(1):47–54.
[41] Rosenfield SN, Stevenson JS. Perception of daily stress and oral coping behaviors in normal, overweight, and recovering alcoholic women. Res Nurs Health 1988;11(3):165–74.
[42] Rosenfield M. Electronic technology for social work education and practice: the application of telephone technology to counseling. J Technol Hum Serv 2002;20:173–81.
[43] Chang M, Nitzke S, Guilford E, Adair C, Hazard D. Motivators and barriers to healthful eating and physical activity among low-income overweight and obese mothers. J Am Diet Assoc 2008;108(6):1023–8.
[44] Chang MW, Nitzke S, Buist D, Cain D, Horning S, Eghtedary K. I am pregnant and want to do better but I can't: focus groups with low-income overweight and obese pregnant women. Matern Child Health J 2015; 19(5):1060–70.
[45] Chang MW, Brown R, Nitzke S, Smith B, Eghtedary K. Stress, sleep, depression and dietary intakes among low-income overweight and obese pregnant women. Matern Child Health J 2015;19(5):1047–59.
[46] Chang M, Brown R, Nitzke S. Participant recruitment and retention in a pilot program to prevent weight gain in low-income overweight and obese mothers. BMC Public Health 2009;9:424.
[47] Methods in community-based participatory research for health. San Francisco (CA): Jossey-Bass; 2005.
[48] Baranowski T, Perry CL, Parcel GS. How individuals, environments, and health behavior interact: social cognitive theory. In: Glanz K, Lewis FM, Rimer BK, editors. Health behavior and health education: theory, research, and practice. 3rd ed. San Francisco: Jossey-Bass Publishers; 2002. p. 246–79.
[49] Markland D, Ryan RM, Tobin VJ, Rollnick S. Motivational interviewing and self-determination theory. J Soc Clin Psychol 2005;24(6):811–31.
[50] Miller WR, Rollnick S. Motivational interviewing: helping people change. 3rd ed. New York (NY): Guilford Press; 2013.
[51] Block G, Gillespie C, Rosenbaum EH, Jenson C. A rapid food screener to assess fat and fruit and vegetable intake. Am J Prev Med 2000;18(4):284–8.
[52] Boucher B, Cotterchio M, Kreiger N, Nadalin V, Block T, Block G. Validity and reliability of the Block98 food-frequency questionnaire in a sample of Canadian women. Public Health Nutr 2006;9(1):84–93.
[53] Evenson KR, Wen F. Measuring physical activity among pregnant women using a structured one-week recall questionnaire: evidence for validity and reliability. Int J Behav Nutr Phys Act 2010;7:21.
[54] Cohen S, Kamarck T, Mermelstein R. A global measure of perceived stress. J Health Soc Behav 1983;24(4):385–96.
[55] Watson D, Clark LA, Tellegen A. Development and validation of brief measures of positive and negative affect: the PANAS scales. J Pers Soc Psychol 1988;54(6):1063–70.
[56] Chang M, Brown R, Nitzke S. Scale development: factors affecting diet, exercise, and stress management (FADESM). BMC Public Health 2008;8(76).
[57] Chang M, Nitzke S, Brown R, Baumann L, Oakley L. Development and validation of a self-efficacy measure for fat intake behaviors in low-income women. J Nutr Educ Behav 2003;35(6):302–7.
[58] Chang M, Brown R, Nitzke S, Baumann L. Development of an instrument to assess predisposing, enabling, and reinforcing constructs associated with fat intake behaviors of low-income mothers. J Nutr Educ Behav 2004;36(1):27–34.
[59] Pelletier LG, Tuson KM, Haddad NK. Client Motivation for Therapy Scale: a measure of intrinsic motivation, extrinsic motivation, and amotivation for therapy. J Pers Assess 1997;68(2):414–35.
[60] Michigan Department of Community Health. Links of WIC Resources for Providers of WIC Services, http://www.michigan.gov/documents/mdch/2006.state.pnss.tables_215430_7.pdf [accessed 27.06.08].
[61] Lohse B, Stotts JL, Bagdonis J. Income sub stratification within a low income sample denotes dropout and completion patterns in nutrition education intervention for young adults. FASEB J 2006;20(5):A1312.
[62] Damron D, Langenberg P, Anliker J, Ballesteros M, Feldman R, Havas S. Factors associated with attendance in a voluntary nutrition education program. Am J Health Promot 1999;13(5):268–75.
[63] Klohe-Lehman DM, Freeland-Graves J, Clarke KK, et al. Low-income, overweight and obese mothers as agents of change to improve food choices, fat habits, and physical activity in their 1-to-3-year-old children. J Am Coll Nutr 2007;26(3):196–208.
[64] Wardle J, Guthrie C, Sanderson S, Birch L, Plomin R. Food and activity preferences in children of lean and obese parents. Int J Obes Relat Metab Disord 2001;25(7):971–7.
[65] Burke V, Beilin LJ, Dunbar D. Family lifestyle and parental body mass index as predictors of body mass index in Australian children: a longitudinal study. Int J Obes Relat Metab Disord 2001;25(2):147–57.

CHAPTER

26

Informational Resources for Developmental Origins of Health and Disease Research

Cheryl S. Rosenfeld

Department of Bond Life Sciences Center, Department of Biomedical Sciences, Genetics Area Program, Thompson Center for Autism and Neurobehavioral Disorders, University of Missouri, Columbia, MO, USA

OUTLINE

The Epigenome and Developmental Origins of Health and Disease
http://dx.doi.org/10.1016/B978-0-12-801383-0.00026-8

Copyright © 2016 Elsevier Inc. All rights reserved.

KEY CONCEPTS

- Developmental origins of health and disease (DOHaD) research findings have mushroomed since the early years of the twenty-first century.
- Those interested in DOHaD need to be cognizant of the various resources to keep abreast of advancements in this field.
- Current DOHaD-related resources, including scientific societies where DOHaD is the central theme or those where it is a sub-part but still integral to the mission of the organization, scientific meetings where unpublished and published data in the DOHaD field are presented.
- Scientific journals in which DOHaD may be the primary or related focus, and social media resources on DOHaD that are springing up, are likely to become even more plentiful in coming years as a mainstay to disseminate current findings.
- The reliability of the resource needs to be considered, along with whether findings are anecdotal or based on rigorous testing in an animal model or large human datasets, with the resulting manuscript subjected to peer review.
- This chapter is a current snapshot of existing external resources to learn more about DOHaD and to stay informed on developments in this area that may impact a wide range of noncommunicable diseases in humans and animals.

INTRODUCTION

The previous chapters reveal that the field of developmental origins of health and disease (DOHaD) is rapidly expanding and branching off in various directions. Almost all species and systems can be impacted by this phenomenon. Therefore, it is nearly impossible for a single book to discuss all aspects of this phenomenon. Further, with the astronomical pace of the research, the current findings will likely become outdated in the coming years. Therefore, the overarching goals of this book are to provide the reader a critical review of the current knowledge in several DOHaD areas and available resources to learn more and keep informed of advancements in this field, which has far-reaching human- and animal-health implications.

The organizational resources presented in this chapter are broken into the following categories: (1) scientific societies where DOHaD is the central theme or those where it is a sub-part but still integral to the mission of the organization, (2) scientific meetings where unpublished and published data in the DOHaD field are presented, (3) scientific journals in which DOHaD may be the primary or related focus, and (4) social media resources (Facebook, Twitter, and YouTube videos). As with any resource, the reader needs to consider the reliability of the information and whether findings are anecdotal or based on rigorous testing in an animal model or large human datasets, with the resulting manuscript subjected to peer review. To aid the reader in judging the merits of the information presented, the last section of this chapter includes recommendations on judging this aspect. Given the creativity of the human mind and our quest for rapid information, it is difficult to envision future resources where DOHaD studies will be featured, but hopefully the chapter will provide readers other resources to learn more about this intriguing topic and gauge the reliability of new findings.

DOHaD-RELATED SCIENTIFIC ORGANIZATIONS

The flagship organization for DOHaD is the International Society for Developmental Origins of Health and Disease (DOHaD Society, http://www.dohadsoc.org/). As detailed in the organization's Website, the primary goals of this society are as follow.

- "To promote co-ordination of a research strategy in different countries, for the scientific exploration of early human development in relation to chronic disease in later life
- To promote the development and application of public health strategies to prevent chronic disease
- To advocate for the need for funds from governmental and non-governmental sources for research in the developmental origins of health and disease
- To champion training opportunities for scientists and clinicians
- To foster regular meetings to discuss research findings and potential intervention
- To promote the interchange of ideas, staff and expertise between laboratories across the world
- To make representations to government, NGOs and other relevant agencies concerning the health implications of DOHaD."

The society also produces the *Journal of Developmental Origins of Health and Disease* and sponsors an every-other-year meeting, which is held in a variety of countries. The journal and meeting are discussed in more detail below. There are no restrictions on memberships, and reduced rates are offered for trainees and for those in developing countries, where noncommunicable diseases (NCD) due to DOHaD effects are also of significant concern. Membership benefits include subscription to the associated journal; periodic newsletters where cutting-edge research and initiatives in the area are discussed; summaries of the workshops, meetings, and conferences; reduced rates at the World Congress meetings; opportunity to be part of an interdisciplinary scientific group; and member access to the restricted portion of the Website, which includes recent publications in the area, contact details for other members, and the ability to view and participate in ongoing discussion forums. The organization's site also allows members to post notifications for position openings at various levels and links to affiliated societies, such as the Chinese DOHaD Society, Japan Society for DOHaD (http://square.umin.ac.jp/Jp-DOHaD/index.html), and SF-DOHaD. The organization is overseen by council members from various countries.

While not the central focus, the DOHaD topic relates to other scientific organizations. We will consider four examples: the Endocrine Society, Society for the Study of Reproduction, Society of Toxicology, and American Society of Andrology. There are assuredly many more examples, and readers of other basic and clinical science organizations are encouraged to consider methods promote this topic.

As discussed in Chapter 5, endogenous hormones, especially testosterone and estrogen, are essential in programming various systems, including the reproductive, cardiovascular, and central nervous systems. Developmental exposure to endocrine-disrupting chemicals can thus lead to detrimental DOHaD effects. Through various resources, the Endocrine Society (http://www.endocrine.org/) communicates the latest discoveries in this area. Members of this society, where continuing medical education credits are offered, include clinicians and basic scientists. Various levels of membership are offered, including full for basic scientists, clinical scientists, and physicians-in-practice; early career for these same three categories; in-training associate for students, residents, and fellows; and associate for those industry, research, or healthcare professionals not holding a doctoral degree. The Endocrine Society supports several journals (*Endocrine Reviews*, *Endocrinology*, *Molecular Endocrinology*, *Hormones and Cancer*, *The Journal of Clinical Endocrinology & Metabolism (JCEM)*, and *Translational Endocrinology & Metabolism (TEAM)*), all of which publish DOHaD-related manuscripts. The organization hosts one annual meeting, which rotates throughout the United

States. In 2014, the Endocrine Society sponsored the Prenatal Programming and Toxicity (PPTox) meeting on Environmental Stressors in Disease and Implications for Human Health, held in Boston, Massachusetts, October 26–29 (http://www.endocrine.org/meetings/pptox-iv). Further details on this every-other-year meeting are discussed below.

The mission of the Society for the Study of Reproduction (SSR; https://www.ssr.org/) *"is to advance scientific knowledge by promoting outstanding research and training in reproductive sciences and to protect and preserve human and animal reproductive health."* The reproductive state of both parents can contribute to DOHaD effects in their offspring. Therefore, this topic is of great interest to SSR. The Society offers several levels of membership: regular, associate, trainee, sustaining associate, emeritus, honorary, and affiliate. One annual SSR meeting is held in various North American cities and US territories; in 2015 the meeting took place in San Juan, Puerto Rico. SSR publications include *Biology of Reproduction* and a newsletter distributed in February, June, and October.

The Society of Toxicology (SOT; http://toxicology.org/) focuses on how chemical, physical, or biological agents impact humans, animals, and the environment. Such exposures *in utero* or during the antenatal period can lead to detrimental DOHaD effects. Therefore, this Society and its members are actively involved in this area. Membership levels include full, associate, postdoctoral, and graduate student. Benefits of joining this organization include reduced registration costs for the annual SOT meeting, which rotates to various US cities; free electronic access to *The Toxicologist* and *Communiqué*, and reduced rates for the printed version of the former and other related journals; printed copy of the annual meeting program and membership directory; access to the members-only section on the SOT site; opportunity to participate in one of 18 regional chapters; career resource and developmental services for trainees; and mentor match, which pairs mentees with SOT mentors who can provide valuable career and general advice. The organization publishes the monthly journal *Toxicological Sciences* (ToxSci), the *Communiqué* Newsletter, and an annual SOT meeting program (The Toxicologist).

The primary interest of the American Society of Andrology (ASA; http://andrologysociety.org/) is interested in male reproduction in all species. It is becoming increasingly apparent that the paternal environment can result in DOHaD effects in his offspring. This might be due to direct effects on the male germ cells or male secretions, such as through alteration of the seminal fluid. The ASA organization offers various levels of membership, including active, associate, trainee, life, and emeritus. Membership includes a subscription to the ASA-produced journal, *Andrology*; reduced registration fees at the annual meeting, held in various US cities; Testis Workshop; and Andrology Lab Workshop. Additional benefits are an online members-only section, which includes continuing education, a bulletin board, and a searchable membership directory. The Andrology Press oversees the monthly journal and has published the Andrology Handbook, 2nd Edition.

DOHaD-RELATED SCIENTIFIC MEETINGS

The International Society for Developmental Origins of Health and Disease held its last meeting in Singapore in 2013 (http://www.dohad2013.org/), and the next one is scheduled to take place in Cape Town, South Africa in 2015 (http://www.dohad2015.org/). The upcoming meeting will also include a full day of satellite workshops (http://www.dohad2015.org/images/DOHaD20201520SATELLITE20workshops.pdf), where various DOHaD-related topics will be covered. By holding the 2015 meeting in South Africa, the meeting organizers seek to increase the potential interaction and collaborations between scientists, clinicians, policy leaders, and those in nongovernmental organizations (NGOs) from Western and developing African countries working in the DOHaD field. In many of these African nations, malnutrition of both parents can exert long-term effects on their children. With NGO support, many of these African nations are employing chemicals that are now banned in the United States and other Western countries, such as the insecticide dichlorodiphenyltrichloroethane (DDT), to curb malarial transmission. Such chemicals can, however, lead to harmful DOHaD and even transgenerational effects. Therefore, it is important that NGOs, clinicians, and scientists working in these areas be educated on the current findings. The last DOHaD conference included 1200 attendees who spanned the globe. Similar attendance is predicted at the upcoming meeting, where 350 speakers will be featured.

The Developmental Origins of Health and Disease Society of Australia and New Zealand established an annual meeting, with the 2015 meeting held in Melbourne on April 17–19. The focus of this meeting was "A healthy start for the human race." The program included a wide range of speakers in this field (https://custom.cvent.com/FE8ADE3646EB4896BCEA8239F12DC577/files/2f572d36f1ea4ec08fe56d76a0b0021f.pdf).

In March of 2014, there was a two-day meeting on "The Power of Programming: International Conference on Developmental Origins of Adiposity and Long-Term Health" in Munich, Germany. Nutrition was the focus of this conference, which was in conjunction with the Early Nutrition Project and the Early Nutrition Academy. There are, however, no current plans to hold another conference. The combined conference attracted about 350 individuals.

In 2012, there was a Keystone Symposia was convened on the topic of Nutrition, Epigenetics, and Human Disease (http://www.keystonesymposia.org/index.cfm?e=web.Meeting.Program&meetingid=1212). Several of the talks discussed how maternal and postnatal nutritional disruptions may result in harmful DOHaD effects. The approximate attendance was 400 individuals.

Another past meeting where DOHaD was prominently featured was the two-day conference in 2012 held by the New York Academy of Sciences on "Fetal Programming and Environmental Exposures: Implications for Prenatal Care and Pre-Term Birth" (http://www.nyas.org/Events/Detail.aspx?cid=0080f200-94a0-412f-b4f9-d5a19971f489). It is not clear, however, if there is any intention to repeat this conference. Attendance for this conference was 130 individuals, who represented scientists from various universities across North America, the Food and Drug Administration, and the National Institutes of Health (NIH), as well as clinicians.

The Prenatal Programming and Toxicity (PPTox) meetings are convened on an every-other-year basis in various international cities. The one in 2012 took place in Paris, France, and the most recent meeting (October 2014) occurred in Boston, Massachusetts, and focused on environmental stressors in disease. The 2014 meeting was sponsored by the Endocrine Society (http://www.endocrine.org/meetings/pptox-iv). While the majority of symposia and poster sessions detailed how developmental exposure to environmental chemicals may lead to DOHaD effects, the most recent meeting also included talks and posters on maternal and paternal diets and maternal stress alone and in combination with exposure to environmental chemicals (or as it may be considered, the "two-hit hypothesis"). Attendance at the past few meetings has been approximately 150 individuals, with participants including trainees, public health specialists, basic scientists, and clinicians. A meeting consensus statement, general conclusions, and review articles based on select talks will be published as an Endocrinology Supplement in the summer of 2015. The 2016 PPTox meeting is slated to be in Japan.

There is currently not a Gordon Conference devoted solely to DOHaD, although a proposal for such a topic has been submitted. If approved, the inaugural Gordon Conference DOHaD meeting would be in the winter of 2017. There are other related Gordon Conferences, including those on "Epigenetics," "Environmental Endocrine Disruptors," "Mammalian Reproduction," "Cellular & Molecular Mechanisms of Toxicity," "Hormone-Dependent Cancers," "Drinking Water Disinfection By-products," "Mycotoxins & Phycotoxins," and "Oceans & Human Health." A complete list of Gordon Conferences for the current and upcoming years is available at https://www.grc.org/home.aspx. While select speakers might discuss DOHaD at some of these Gordon Conferences, it is not the central theme of the meeting. Another limitation of these current conferences is that they are confined in scope, in the sense that they fail to bring those with diverse interests in the DOHaD topic together to discover commonalities across disciplines and systems and do not include a full schedule of experts on this topic.

Other meetings where a handful of talks and posters on DOHaD and related topics are generally presented include the Endocrine Society, SSR, SOT, ASA, Society for Neuroscience, Society for Behavioral Neuroendocrinology, American Society for Nutrition, and the larger Experimental Biology, to list a few examples. Links to these other meetings are included below. Readers who are members of these and other scientific organizations are encouraged to extend speaker invitations to scientists and clinicians working in this field. It is also important that there are meetings where there is cross-disciplinary exchange and sharing of data and ideas for those working in diverse DOHaD disciplines.

Such interdisciplinary meetings will serve several purposes. First, they raise awareness of how preconceptional and antenatal factors can perturb offspring phenotypes. Second, such meetings will allow for an exploration of common underpinning mechanisms in a range of systems. Such talks may also provide a bridge between how various maternal and paternal extrinsic factors lead to DOHaD effects, thereby informing attendees of the prevailing thoughts on potential overlapping mechanisms, prevention, and remediation strategies. Finally, such an approach would identify important knowledge gaps in the DOHaD field.

DOHaD-RELATED JOURNALS, DATABASES, AND OTHER INFORMATIVE RESOURCES

Consistent with the surge of interest in this area, the number of DOHaD-related articles has dramatically risen since the early years of the twenty-first century (Figure 1). The number of research and review articles and journals devoted to this topic is predicted to escalate dramatically in the coming years. Example journals where DOHaD-related studies are prominently featured include *PLoS One*, *Journal of Developmental Origins of Health and Disease*, *American Journal of Obstetrics and Gynecology*, *Environmental Health Perspectives*, *Journal of Maternal Fetal and Neonatal Medicine*, and other journals listed in Table 1. To locate other articles in this area, searches on this topic can be performed through PubMed, Google Scholar, Scopus, Biological Abstracts, Current Contents, and Zoological Record. Search terms that may identify other cutting-edge articles in this area are listed in Table 2. These broad search strategy approaches allow retrieval of research and review articles in a variety of journals and simultaneous identification of human epidemiologic and animal-model studies. Both sets of studies are needed to advance the field. The human studies allow assessments to be performed on the strength of the correlation between various factors and later DOHaD effects, but initial causation- and mechanistic-based approaches are generally first investigated in rodents or other laboratory animal models. The World Health Organization (WHO) and Endocrine Society have published their own DOHaD-related reports (listed below).

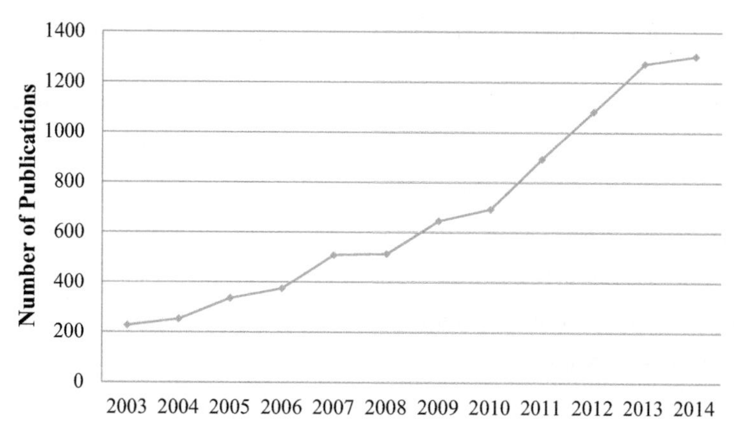

FIGURE 1 Scopus search of DOHaD-related articles from 2003 to 2014. The graph reveals that the number of articles in this area has dramatically risen in that time period.

SOCIAL MEDIA RESOURCES REPORTING DEVELOPMENTAL ORIGINS OF HEALTH AND DISEASE FINDINGS

Social media is a useful tool for rapid communication of scientific findings, including in the DOHaD area, but it can also be a purveyor of sensationalized and false claims. On the positive side, such sites provide a template to foster national and international discussions on this topic. The primary concern, though, is reliability of the information (discussed in more detail below). Even so, readers are encouraged to consult credible social media sites to learn more about the ongoing work in this area. US government Facebook and Twitter pages that address issues related to DOHaD include: the Eunice Kennedy Shriver National Institute of Child Health and Human Development (https://www.facebook.com/nichdgov?fref=nf and https://twitter.com/search?q=NIChD&src=typd), the National Institute of Environmental Health Sciences (https://www.facebook.com/NIH.NIEHS and https://twitter.com/NIEHS), and the primary National Institutes of Health page (https://www.facebook.com/nih.gov?fref=nf and https://twitter.com/search?q=NIH&src=typd). The Australian government's National Health and Medical Research Council (https://twitter.com/nhmrc) includes funding and research information in this area. Other valid government, NGO, and university/scientific institute Facebook and Twitter sites may be identified with the terms listed in Table 2.

Many organizations also now have Facebook and Twitter pages associated with their scientific meetings, e.g., the 9th World Congress of Developmental Origins of Health and Disease (https://www.facebook.com/dohad2015.org and https://twitter.com/DOHaD2015). Such meeting sites are useful, as participants may comment and publicly discuss the content of allowable presentations, additional relevant resources may be linked to such sites or listed in other recommended Twitter accounts. Individuals who cannot attend the meeting may become aware of some of the unpublished data being presented. For instance, the Australia and New Zealand branch of the Developmental Origins of Health and Disease established its own Twitter account (https://twitter.com/ANZDOHaD) for the 2014 and 2015 conferences.

With the growing interest in this area, many Universities and private Institutes are also sponsoring DOHaD-related conferences and establishing associated Facebook and Twitter pages for such events. For instance, as part of the Life Sciences and Society Program (LSSP), the University of Missouri recently had a symposium titled "The

TABLE 1 Top Journals for DOHaD-related Articles and Number of Articles Published in Each from 2003 through April 2, 2015

Journal	Number of Articles
1. PLos One	210
2. Journal of Developmental Origins of Health and Disease	209
3. American Journal of Obstetrics and Gynecology	92
4. Environmental Health Perspectives	80
5. Journal of Maternal Fetal and Neonatal Medicine	73
6. British Journal of Nutrition	69
7. Placenta	68
8. American Journal of Physiology Regulatory Integrative and Comparative Physiology	66
9. Biology of Reproduction	65
10. Endocrinology	63
11. American Journal of Clinical Nutrition	60
12. Toxicological Sciences	60
13. Journal of Nutrition	59
14. BMC Pregnancy and Childbirth	59
15. Early Human Development	52
16. Neurotoxicology and Teratology	51
17. Paediatric and Perinatal Epidemiology	50
18. Journal of Animal Science	50
19. International Journal of Obesity	49
20. Reproductive Toxicology	48
21. Pediatrics	44
22. Proceedings of the National Academy of Sciences of the United States of America	44
23. Obstetrics and Gynecology	44
24. Acta Obstetricia Et Gynecologica Scandinavica	42
25. Pediatric Research	41
26. BJOG: An International Journal of Obstetrics and Gynaecology	39
27. Physiology and Behavior	37
28. Toxicology and Applied Pharmacology	37
29. Obesity	36

TABLE 1 Top Journals for DOHaD-related Articles and Number of Articles Published in Each from 2003 through April 2, 2015—cont'd

Journal	Number of Articles
30. Neurotoxicology	36
31. American Journal of Physiology Endocrinology and Metabolism	35
32. Psychoneuroendocrinology	35
33. Hypertension	35
34. Hormones and Behavior	34
35. European Journal of Clinical Nutrition	33
36. Maternal and Child Health Journal	31
37. Journal of Physiology	31
38. FASEB Journal	31
39. European Journal of Obstetrics Gynecology and Reproductive Biology	30
40. Journal of Clinical Endocrinology and Metabolism	30
41. Brain Research	29
42. Reproduction	28
43. Birth Defects Research Part A Clinical and Molecular Teratology	27
44. Journal of Nutritional Biochemistry	27
45. Acta Paediatrica International Journal of Paediatrics	26
46. Proceedings of the Nutrition Society	26
47. Journal of Obstetrics and Gynaecology	25
48. Physiological Genomics	25
49. Nutrition Reviews	25
50. Toxicology	25
51. International Journal of Developmental Neuroscience	24
52. Diabetes Care	24
53. Human Reproduction	24
54. American Journal of Perinatology	24
55. Neuroscience and Biobehavioral Reviews	23
56. Developmental Psychobiology	23
57. Epigenetics	23
58. American Journal of Epidemiology	23

Continued

TABLE 1 Top Journals for DOHaD-related Articles and Number of Articles Published in Each from 2003 through April 2, 2015—cont'd

Journal	Number of Articles
59. Archives of Gynecology and Obstetrics	23
60. Neuroscience	23
61. British Journal of Midwifery	22
62. Behavioural Brain Research	22
63. Australian and New Zealand Journal of Obstetrics and Gynaecology	22
64. Advances in Experimental Medicine and Biology	22
65. Journal of Pediatrics	21
66. Journal of Perinatal Medicine	21
67. Journal of Midwifery and Women's Health	21
68. Obstetrical and Gynecological Survey	21
69. Archives of Women's Mental Health	21
70. Journal of Endocrinology	21
71. Seminars in Fetal and Neonatal Medicine	21
72. Prenatal Diagnosis	20
73. Diabetologia	20
74. Biological Psychiatry	20
75. Reproductive Sciences	20
76. Clinical Science	20
77. Fertility and Sterility	20
78. Journal of Perinatology	20
79. Molecular and Cellular Endocrinology	20
80. International Journal of Epidemiology	20
81. Infant Mental Health Journal	19
82. Annals of Nutrition and Metabolism	19
83. Stress	19
84. Maternal and Child Nutrition	19
85. Seminars in Reproductive Medicine	19
86. European Journal of Nutrition	18
87. BMC Genomics	18
88. American Journal of Human Biology	18
89. Stem Cells	18

TABLE 1 Top Journals for DOHaD-related Articles and Number of Articles Published in Each from 2003 through April 2, 2015—cont'd

Journal	Number of Articles
90. Journal of Child Psychology and Psychiatry and Allied Disciplines	18
91. Reproduction Fertility and Development	18
92. International Journal of Gynecology and Obstetrics	17
93. Aquatic Toxicology	16
94. Transgenic Research	16
95. Pediatric Allergy and Immunology	16
96. Alcoholism Clinical and Experimental Research	16
97. Diabetes	16
98. Public Health Nutrition	16
99. Journal of Hypertension	16
100. Annals of the New York Academy of Sciences	15
101. Nutrition Research	15
102. Journal of Autism and Developmental Disorders	15
103. Clinical Obstetrics and Gynecology	15
104. Expert Review of Obstetrics and Gynecology	15
105. Midwifery	15
106. American Journal of Physical Anthropology	14
107. Journal of Allergy and Clinical Immunology	14
108. Reproductive Biomedicine Online	14
109. BMJ Online	14
110. Journal De Gynecologie Obstetrique Et Biologie De La Reproduction	13
111. Indian Journal of Pediatrics	13
112. Infant Behavior and Development	13
113. Journal of Reproductive and Infant Psychology	13
114. Nature Reviews Endocrinology	13
115. Brain Behavior and Immunity	13
116. Current Opinion in Obstetrics and Gynecology	13
117. Animal	13
118. BMC Medicine	13
119. Social Science and Medicine	13

Continued

TABLE 1 Top Journals for DOHaD-related Articles and Number of Articles Published in Each from 2003 through April 2, 2015—cont'd

Journal	Number of Articles
120. Journal of Neuroendocrinology	13
121. Metabolism Clinical and Experimental	13
122. Obesity Reviews	13
123. Best Practice and Research Clinical Obstetrics and Gynaecology	13
124. Current Diabetes Reports	13
125. Toxicology Letters	13
126. Nutrition	13
127. Nature	12
128. Clinical and Experimental Allergy	12
129. Life Sciences	12
130. American Journal of Respiratory and Critical Care Medicine	12
131. Reproductive Biology and Endocrinology	12
132. Poultry Science	12
133. Journal of Clinical Investigation	12
134. Stem Cells and Development	12
135. Epidemiology	12
136. Journal of Neuroscience	12
137. Neuroscience Letters	12
138. Methods in Molecular Biology	12
139. Nutrients	12
140. American Journal of Physiology Heart and Circulatory Physiology	11
141. American Journal of Physiology Renal Physiology	11
142. Molecular Human Reproduction	11
143. Archives De Pediatrie	11
144. Journal of Women's Health	11
145. Molecular Reproduction and Development	11
146. Journal of Pediatric Gastroenterology and Nutrition	11
147. Genesis	11
148. Journal of Paediatrics and Child Health	11
149. Seminars in Perinatology	11

TABLE 1 Top Journals for DOHaD-related Articles and Number of Articles Published in Each from 2003 through April 2, 2015—cont'd

Journal	Number of Articles
150. Prostaglandins Leukotrienes and Essential Fatty Acids	11
151. Obstetrics Gynaecology and Reproductive Medicine	11
152. European Journal of Epidemiology	10
153. Journal of Intellectual Disability Research	10
154. Archives of Toxicology	10
155. American Journal of Reproductive Immunology	10
156. Epigenomics	10
157. Gynakologe	10
158. Developmental Neuroscience	10
159. JOGNN Journal of Obstetric Gynecologic and Neonatal Nursing	10

Epigenetics Revolution: Nature, Nurture and What Lies Ahead" (https://www.facebook.com/lsspmu?fref=nf and https://twitter.com/DecodeSci), where speakers discussed the impact of altered maternal and paternal environments (diet, stress, and environmental chemicals) on offspring DOHaD effects. Facebook and Twitter pages for other related scientific organizations, publications, and news sites in this area are listed below.

YouTube videos provide valuable information and insight on this field from speakers considered the leading authorities in this area and popular-press book authors who are able to communicate complicated scientific DOHaD discoveries to non-scientists. Example of such videos include the now-deceased father of DOHaD, Sir David Barker, on "Developmental Origins of Health and Disease" (https://www.youtube.com/watch?v=795u8zl82GI), Dr Mark Hanson "The Birth and Future Health of DOHaD" (https://www.youtube.com/watch?v=ihl82QKq2ME); the several videos on this topic published by the NBIC49 Animal Function and Environmental Adaptation course at the University of Linköping, Sweden, e.g., "Developmental Origins of Health and Disease. 1. Epidemiological Evidence" (https://www.youtube.com/watch?v=OvGd1XwV5ZQ), "The Fetal Origins of Success" by Douglas Almond at TEDxMiamiUniversity (https://www.youtube.com/watch?v=qqlgH8OAJiI), and Annie Murphy Paul (scientific writer and author of Origins): "What We Learn Before We're Born TED Talks" (https://www.youtube.com/watch?v=stngBN4hp14). Additional YouTube presentations may be identified by using the search terms listed in Table 2.

JUDGING THE RELIABILITY OF DEVELOPMENTAL ORIGINS OF HEALTH AND DISEASE INFORMATIONAL RESOURCES

While social media and other non-scientific sites might provide useful and up-to-date information, exaggerated statements and misinformation may also be routed through such venues, especially in this revolutionary and ever-expanding field. Therefore, metrics are needed to gauge the credibility of the resource. First and foremost, the conclusions drawn need to be based on robust animal model or human epidemiologic studies. In the case of

TABLE 2 Example Search Terms to Identify Scientific Journal Articles and Information on Social Media Sites Relating to DOHaD Research Findings

	Search Term
1.	Developmental origins of health and disease
2.	DOHaD
3.	Fetal programming
4.	FOAD or Fetal origins of adult disease
5.	Barker hypothesis
6.	Maternal diet
7.	Paternal diet
8.	Maternal stress
9.	Paternal stress
10.	Environmental chemicals and fetal exposure
11.	Fetal exposure and endocrine disruption
12.	Maternal obesity
13.	Paternal obesity
14.	Germline transmission
15.	Placenta
16.	Maternal metabolic disorder
17.	*In utero* environment
18.	*In utero* environmental factors
19.	Perinatal period and disease risk
20.	Antenatal period
21.	Fetal sex differences
22.	Periconceptional period and offspring development
23.	Developmental exposure
24.	Fetal epigenome
25.	Fetal transcriptome

human epidemiologic studies, where fetuses and neonates can be exposed to a wide range of factors, it is nearly impossible to establish direct causation between a single maternal or paternal factor and DOHaD effects. Instead, statistical analyses can only be performed to establish the strength of the association or correlation between these factors. In general, Facebook and Twitter sites representing government health and scientific organizations may be considered more reliable than sites of private individuals. Government health organizations, such as NIH and its branches, often report on publications from individuals funded through the organizations or members of the scientific societies. Studies mentioned on such sites will have undergone peer review prior to being published. In the case of YouTube or social media sites, the credentials of the individuals should be considered. Are the individuals highly esteemed scientists or physicians with years of experience working in a DOHaD-related field? Another consideration is whether the sites filter comments to avoid erroneous and misleading statements. At some scientific meetings, attendees are discouraged from "tweeting" or commenting on unpublished scientific data. Such sessions are generally demarcated with a "No tweet" or "No Twitter" symbol. The other concern with social media sites is that often they may be re-reporting a prior media article. As with DNA replication, fidelity of the story depends upon stringent proofreading, and this is becoming increasingly rare in the rapid communication era. Instead, mutations are now commonly introduced with each reiteration of the original story. It is thus best to identify the original source and, preferably, the published scientific study to verify the conclusions and actual strength of the data for oneself.

CONCLUSIONS

The DOHaD concept might account for many NCD in animals and humans. This book has aimed to provide readers a thorough review of our current understanding of how various extrinsic and intrinsic factors may insidiously or beneficially affect almost all body systems. However, even by the time this book is published, there will assuredly be additional knowledge and understanding of this complex and multidisciplinary field. Therefore, readers are encouraged to join relevant scientific societies and consult other resources described herein to keep abreast of current findings. Because of its ease of use and widespread reach, social media sites (Facebook, Twitter, and YouTube) are increasingly being employed as adjunct sources for scientific information, including in DOHaD-based fields. As with any resource, it is always best to consult the original source and consider the credentials of the individual or organization communicating the current findings. While such sites may be exploited to propagate extravagant claims, they can also be a useful tool to communicate current data to fellow scientists, clinicians, students and the public. It will be intriguing to follow the new discoveries in this area and what new avenues are developed to speedily transmit research updates on a global scale. The greater swiftness with which clinicians in varying countries, including developing nations, can be educated on current DOHaD results might equate to rapid procedural and policy adjustments and greater ability to diagnose and implement prevention/remediation strategies for NCD with DOHaD origins.

Glossary

ASA American Society of Andrology
ASN American Society for Nutrition
CME Continuing Medical Education
DDT dichlorodiphenyltrichloroethane
DOHaD Developmental Origins of Health and Disease
EB Experimental Biology
EDCs Endocrine-disrupting chemicals
ENA Early Nutrition Academy
LSSP Life Sciences and Society Program
NCD Noncommunicable diseases
NGO Nongovernment organization
NICHD National Institute of Child Health and Human Development
NIEHS National Institute of Environmental Health Sciences
NIH National Institutes of Health
PNAS Proceedings of the National Academy of Sciences
PPTox Prenatal Programming and Toxicity
SBN Society for Behavioral Neuroendocrinology
SFN Society for Neuroscience
SOT Society of Toxicology
SSR Society for the Study of Reproduction

Acknowledgments

The author is grateful to Ms Kate Anderson's assistance in identifying DOHaD informational resources and relevant search terms for this scientific area.

References

Example DOHaD-Related Organizations

[1] International Society for Developmental Origins of Health and Disease, http://www.dohadsoc.org/.
[2] Japan Society for DOHaD, http://square.umin.ac.jp/Jp-DOHaD/index.html.
[3] Endocrine Society, http://www.endocrine.org/.
[4] Society for the Study of Reproduction (SSR), https://www.ssr.org/.
[5] Society of Toxicology (SOT), http://toxicology.org/.
[6] American Society of Andrology (ASA), http://andrologysociety.org/.

Example DOHaD-Related Scientific Meetings

[7] The International Society for Developmental Origins of Health and Disease, in Singapore in 2013 (http://www.dohad2013.org/), in Cape Town, South Africa in 2015 (http://www.dohad2015.org/) and Satellite Workshops for 2015 (http://www.dohad2015.org/images/DOHaD20201520SATELLITE20workshops.pdf).
[8] Developmental Origins of Health and Disease Society of Australia and New Zealand, https://custom.cvent.com/FE8ADE3646EB4896BCEA8239F12DC577/files/2f572d36f1ea4ec08fe56d76a0b0021f.pdf.
[9] Keystone meeting on Nutrition, Epigenetics and Human Disease, http://www.keystonesymposia.org/index.cfm?e=web.Meeting.Program&meetingid=1212.
[10] New York Academy of Sciences on "Fetal Programming and Environmental Exposures: Implications for Prenatal Care and Pre-Term Birth", http://www.nyas.org/Events/Detail.aspx?cid=0080f200-94a0-412f-b4f9-d5a19971f489.
[11] Prenatal Programming and Toxicity (PPTox) 2014 meeting, http://www.endocrine.org/meetings/pptox-iv.
[12] Gordon Conferences related to DOHaD, https://www.grc.org/home.aspx.
[13] Endocrine Society Annual Meeting, http://www.endocrine.org/endo-2016.
[14] Society for the Study of Reproduction Annual Meeting, http://www.ssr.org/15Meeting.

[15] Society of Toxicology Annual Meeting, http://www.toxicology.org/AI/MEET/am2015/supporters2015.asp.
[16] American Society of Andrology Annual Meeting, http://andrologysociety.org/Home.aspx.
[17] Society for Neuroscience (SFN) Annual Meeting, http://www.sfn.org/annual-meeting/neuroscience-2015.
[18] Society for Behavioral Neuroendocrinology (SBN) Annual Meetin, http://www.sbn.org/Meetings/Annual-Meeting.aspx.
[19] American Society for Nutrition (ASN) Annual Meeting at Experimental Biology, http://www.nutrition.org/meetings/asn-scientific-sessions-at-eb-2016/.
[20] Experimental Biology Annual Meeting, http://experimentalbiology.org/2015/Home.aspx.

Journals and Databases to Find DOHaD Articles

[21] The Journal of Developmental Origins of Health and Disease, http://journals.cambridge.org/action/displayJournal?jid=DOH.
[22] Endocrine Society publications, http://press.endocrine.org/bestofbasicresearch/2014.
[23] Biology of Reproduction, http://www.biolreprod.org/.
[24] Toxicological Sciences (ToxSci), http://toxsci.oxfordjournals.org/.
[25] The Toxicologist (SOT annual meeting program, Supplement to Toxicological Sciences), http://www.toxicology.org/AI/PUB/Tox/2015Tox.pdf.
[26] Andrology,http://onlinelibrary.wiley.com/journal/10.1111/(ISSN)2047-2927.
[27] Endocrine Disruptors, http://www.tandfonline.com/loi/kend20#/guidelines.
[28] Epigenetics, http://www.tandfonline.com/loi/kepi20#.VRDNsuoo7X4.
[29] PLoS One, http://www.plosone.org/.
[30] Environmental Health Perspectives, http://ehp.niehs.nih.gov/.
[31] PubMed, http://www.ncbi.nlm.nih.gov/pubmed.
[32] Google Scholar, http://scholar.google.com/.
[33] Scopus, http://www.scopus.com/.
[34] Biological Abstracts, http://www.ebscohost.com/academic/biological-abstracts.
[35] Current Contents Connect, http://thomsonreuters.com/en/products-services/scholarly-scientific-research/scholarly-search-and-discovery/current-contents-connect.html.
[36] Zoological Record, http://thomsonreuters.com/en/products-services/scholarly-scientific-research/scholarly-search-and-discovery/zoological-record.html.
[37] World Health Organization (WHO) DOHaD informational resources, http://www.who.int/quantifying_ehimpacts/publications/preventingdisease.pdf?ua=1http://www.unep.org/chemicalsandwaste/Portals/9/EDC/SOS%202012/EDC%20report%20cover%20final%20low%20res%20070213.pdf.
[38] Endocrine Society, Introduction to Endocrine Disrupting Chemicals (EDCs): A Guide for Public Interest Organizations and Policy-Makers (covers DOHaD-related topics), http://www.endocrine.org/~/media/endosociety/Files/Advocacy%20and%20Outreach/Important%20Documents/Introduction%20to%20Endocrine%20Disrupting%20Chemicals.pdf.

Social Media Resources for DOHaD Topics

Facebook Pages

[39] Eunice Kennedy Shriver National Institute of Child Health and Human Development, https://www.facebook.com/nichdgov?fref=nf.
[40] National Institute of Environmental Health Sciences, https://www.facebook.com/NIH.NIEHS.
[41] National Institutes of Health, https://www.facebook.com/nih.gov?fref=nf.
[42] The 9th World Congress of Developmental Origins of Health and Disease, https://www.facebook.com/dohad2015.org).
[43] University of Missouri Life Sciences and Society Program (LSSP) Symposium, "The Epigenetics Revolution: Nature, Nurture and What Lies Ahead", https://www.facebook.com/lsspmu?fref=nf.
[44] Endocrine Society, https://www.facebook.com/EndocrineSociety?fref=nf.
[45] Society for the Study of Reproduction (SSR), https://www.facebook.com/SSRepro?fref=nf.
[46] Society for Toxicology (SOT),https://www.facebook.com/pages/Society-of-Toxicology/112502275429041?fref=ts.
[47] American Society of Andrology (ASA), https://www.facebook.com/AndrologyASA?fref=ts.
[48] Society for Neuroscience (SFN), https://www.facebook.com/societyforneuroscience?fref=ts.
[49] Society for Behavioral Neuroendocrinology (SBN), https://www.facebook.com/groups/183497986274/.
[50] World Health Organization (WHO), https://www.facebook.com/WHO?fref=ts.
[51] Gordon Research Conferences (GRC), https://www.facebook.com/GordonResearchConferences?fref=ts.
[52] PLoS, https://www.facebook.com/PLoS.org.
[53] PNAS, https://www.facebook.com/pages/PNAS/18262365099.
[54] EurekAlert, https://www.facebook.com/EurekAlert?fref=ts.
[55] Futurity, https://www.facebook.com/futuritynews?fref=ts.

Twitter Pages
[56] Eunice Kennedy Shriver National Institute of Child Health and Human Development, https://twitter.com/search?q=NIChD&src=typd.
[57] National Institute of Environmental Health Sciences, https://twitter.com/NIEHS.
[58] National Institutes of Health, https://twitter.com/search?q=NIH&src=typd.
[59] 9th World Congress of Developmental Origins of Health and Disease, https://twitter.com/DOHaD2015.
[60] Australia and New Zealand branch of the Developmental Origins of Health and Disease 2014 Conference, https://twitter.com/ANZDOHaD.
[61] University of Missouri Life Sciences and Society Program (LSSP) Symposium, "The Epigenetics Revolution: Nature, Nurture and What Lies Ahead", https://twitter.com/DecodeSci.
[62] Endocrine Society, https://twitter.com/TheEndoSociety.
[63] Society for the Study of Reproduction (SSR), https://twitter.com/SSRepro.
[64] Society of Toxicology (SOT), https://twitter.com/SOToxicology.
[65] American Society of Andrology (ASA), https://twitter.com/AndrologyASA.
[66] Society for Neuroscience (SFN), https://twitter.com/search?q=society%20for%20neuroscience&src=tyah.
[67] World Health Organization (WHO), https://twitter.com/WHO.
[68] Gordon Research Conferences (GRC), https://twitter.com/search?q=Gordon%20Research%20Conferences&src=typd.
[69] PLoS, https://twitter.com/PLOS.
[70] PNAS News, https://twitter.com/PNASNews.
[71] EurekAlert, https://twitter.com/EurekAlertAAAS.
[72] Futurity News, https://twitter.com/FuturityNews.

Example DOHaD YouTube Videos

[73] Sir David Barker on Developmental Origins of Health and disease, https://www.youtube.com/watch?v=795u8zl82GI.
[74] Dr Mark Hanson on the birth and future health of DOHaD, https://www.youtube.com/watch?v=ihl82QKq2ME.
[75] Several videos on this topic published by the NBIC49 Animal Function and Environmental Adaptation course at the University of Linköping, Sweden, e.g., "Developmental Origins of Health and Disease. 1. Epidemiological Evidence", https://www.youtube.com/watch?v=OvGd1XwV5ZQ.
[76] The Fetal Origins of Success: Douglas Almond at TEDxMiamiUniversity, https://www.youtube.com/watch?v=qqlgH8OAJiI.
[77] Annie Murphy Paul (scientific writer and author of Origins): What we learn before we're born (TED Talks), https://www.youtube.com/watch?v=stngBN4hp14.

Index

Note: Page numbers followed by "f", "t" and "b" indicate figures, tables and boxes respectively.

E

F

O

P

R

T

U

Printed and bound by CPI Group (UK) Ltd, Croydon, CR0 4YY
08/05/2025
01864999-0002